# 9th International Unimas Stem Engineering Conference (ENCON 2016)



## Innovative Solutions for Engineering and Technology Challenges

MATEC Web of Conferences Volume 87 (2016)

Sarawak, Malaysia
26 - 28 October 2016

**Editors:**

**A. Hasan**
**A.A. Khan**
**Md. A. Mannan**
**C.N. Hipolito**
**N. Mohamed Sutan**
**Al-K. Hj. Othman**
**M.R. Kabit**
**N. Abdul Wahab**

ISBN: 978-1-5108-3363-0

**Printed from e-media with permission by:**

Curran Associates, Inc.
57 Morehouse Lane
Red Hook, NY 12571

proceedings.com

**Some format issues inherent in the e-media version may also appear in this print version.**



Printed by Curran Associates, Inc. (2017)

For additional information, please contact EDP Sciences – Web of Conferences
at the address below.

EDP Sciences – Web of Conferences
17, Avenue du Hoggar
Parc d'Activité de Courtabœuf
BP 112
F-91944 Les Ulis Cedex A
France

Phone: +33 (0) 1 69 18 75 75
Fax: +33 (0) 1 69 28 84 91

contact-edps@webofconferences.org

**Additional copies of this publication are available from:**

Curran Associates, Inc.
57 Morehouse Lane
Red Hook, NY 12571 USA
Phone: 845-758-0400
Fax: 845-758-2633
Email: curran@proceedings.com
Web: www.proceedings.com

# TABLE OF CONTENTS

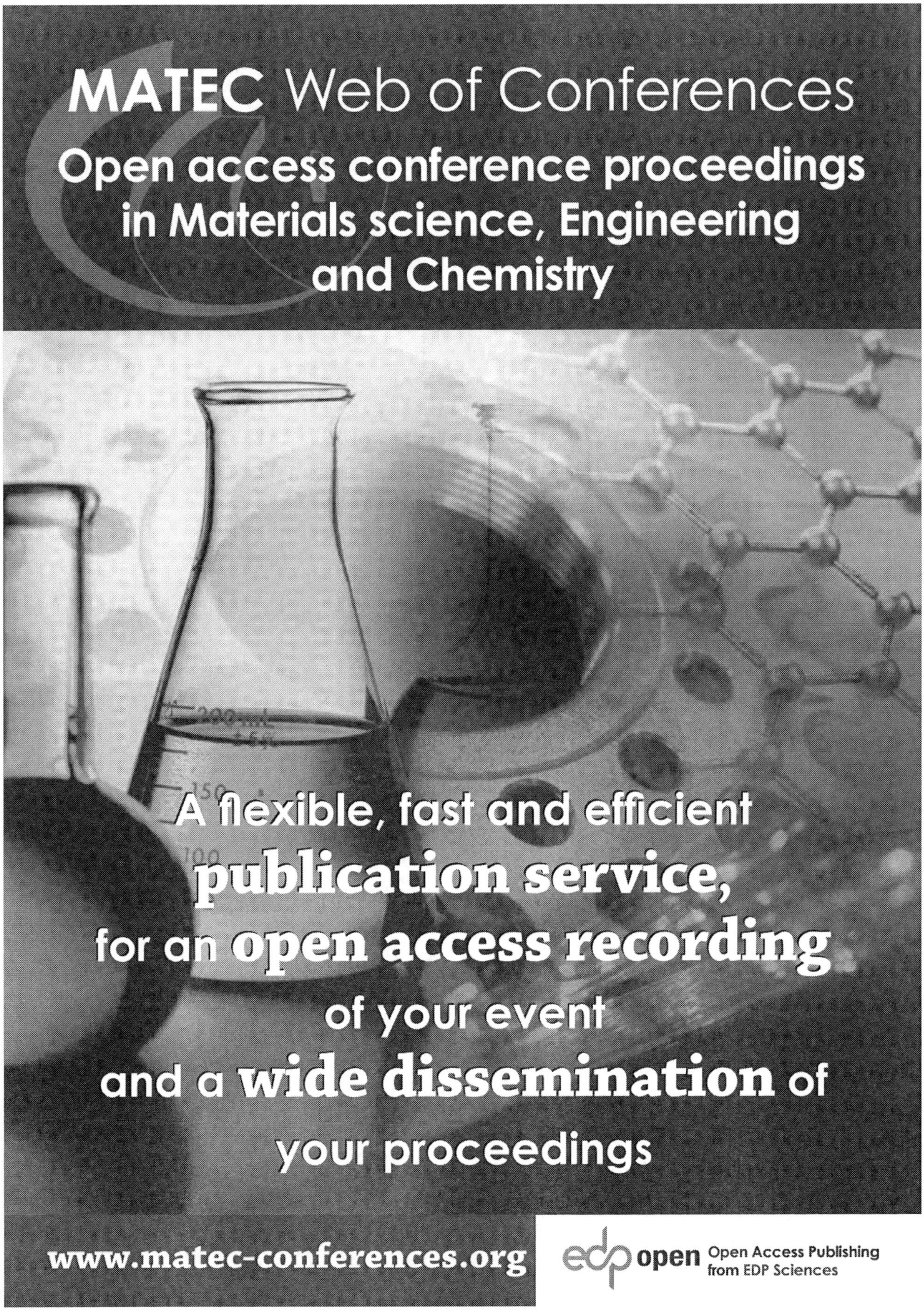
MATEC Web of Conferences
Open access conference proceedings in Materials science, Engineering and Chemistry
A flexible, fast and efficient publication service, for an open access recording of your event and a wide dissemination of your proceedings
www.matec-conferences.org
edp open
Open Access Publishing from EDP Sciences

edp sciences

MATEC Web of Conferences

# Statement of Peer review

In submitting conference proceedings to MATEC Web of Conferences, I certify to the Publisher that I adhere to the **Policy on Publishing Integrity** of the journal in order to safeguard good scientific practice in publishing.

1. All articles have been subjected to peer review administered by the proceedings editors.
2. Reviews have been conducted by expert referees, who have been requested to provide unbiased and constructive comments aimed, whenever possible, at improving the work.
3. Proceedings editors have taken all reasonable steps to ensure the quality of the materials they publish and their decision to accept or reject a paper for publication has been based only on the merits of the work and the relevance to the journal.

Title, date and place of the conference

THE 9th INTERNATIONAL UNIMAS STEM ENGINEERING CONFERENCE (ENCON) 2016
Theme: "Innovative Solutions for Engineering and Technology Challenges"
26-28 October 2016
Riverside Majestic Hotel, Kuching, Sarawak, MALAYSIA

Proceedings editor(s):

Chief Editor : Dr Alsidqi Hasan

Editors : Prof Amir Azam Khan
Prof Md Abdul Mannan
Dr Cirilo Nolasco Hipolito
Dr Norsuzailina bt. Mohamed Sutan
Dr Al-Khalid bin Hj. Othman
Dr Mohamad Raduan Kabit
Noraziah bt. Abdul Wahab

Date and editor's signature

16/11/2016

Chief Editor : Dr Alsidqi Hasan

DOI: 10.1051/matecconf/20178700001

# PREFACE

The 9th International Unimas STEM Engineering Conference (Encon) 2016 was held at Riverside Majestic Hotel in the beautiful city of Kuching, Sarawak, Malaysia during 26-28 October 2016. The conference was organized by The Faculty of Engineering Universiti Malaysia Sarawak, Malaysia Ministry of Higher Education, Jabatan Pendidikan Sarawak, and Tabung Ekonomi Gagasan Anak Bumiputera Sarawak (TEGAS). With the theme 'Innovative Solutions for Engineering and Technology Challenges', the conference has successfully gathered presenters and participants from across the region and from different backgrounds including researchers, academics, engineers, industrial professionals as well as graduate students to share their experience and research findings. This would enhance networking opportunities among the participants in the field of STEM, engineering and technology.

Papers presented during the conference are peer-reviewed and selected among papers that were submitted to the organizers prior to the conference. Chief editor together with associate editors conducted normal refereeing procedure by assigning two to three competent reviewers to review each paper in their respective area of expertise. The papers in this proceeding are classified into three categories: Civil Engineering, Mechanical Engineering, Chemical Engineering and STEM Education. Editors would like to thank all reviewers for their voluntary contribution to review the papers.

Dr Alsidqi Hasan
Universiti Malaysia Sarawak

# Review in formulating the standard form of contract for Industrialized Building System (IBS) construction approach in Malaysia

Mohd Ashraf Mohd Fateh[1a], Mohammad Fadhil Mohammad[2] and Ani Saifuza Abd Shukor[2]

[1]*Faculty Engineering and Technology Infrastructure, Infrastructure University Kuala Lumpur (IUKL) 40300 Kajang, Selangor, Malaysia*
[2]*Faculty Architecture Planning and Surveying, Universiti Teknologi Mara (UiTM) 40450 Shah Alam, Selangor, Malaysia*

**Abstract.** Industrialised Building System (IBS) is the common term used to represent the prefabrication concept in Malaysia. Based on previous research being done locally and internationally, it is certain that using IBS offers huge benefits in terms of cost, time, productivity and quality. The Malaysian construction industry has been urged to use innovative construction techniques. It is being highlighted under the present Construction Industry Transformation Programme 2016 - 2020 (CITP). These initiatives illustrate the Malaysia government's commitment in addressing the IBS agenda. Although some efforts have been taken to enhance the IBS adoption in Malaysia, various contract issues have emerged to become major concern amongst the IBS players that have led to delays, disputes, payment and cash flow issues in their projects. Formulating a dedicated standard form of contract for IBS construction approach is necessary to address the concerns. Therefore, the aim of the main research is to formulate the standard form of contract for IBS projects that will assist authorities and IBS industry players in mitigating the issues and challenges from the contractual perspectives. However, this paper will only present the literature reviews and findings from a preliminary survey conducted to establish pertinent contractual issues and challenges on the subject matter.

## 1 INTRODUCTION

The Malaysian construction industry has played a pivotal role in both its consumption of a wide range of goods and services, as well as through its provision of core infrastructure that has initiated growth and development throughout the country. Over the past five (5) years, Malaysia has experienced significant economic growth; having grown at a steady rate of 6% per annum as reported by the (Construction Industry Development Board (CIDB), 2015a). As reported by the (Department of Statistics, 2013) the local construction industry is an integral component of the Malaysian economy. It accounts for 3.7% of the total Gross Domestic Product (GDP). Moreover, the construction industry offers large-scale positive spill over effects onto both the manufacturing and services sectors. Based on the report by the (CIDB), 2015c), the construction industry accounts for 15% of the total output of Malaysia's manufacturing sector. Beyond this, the construction industry plays a critical role in initiating and sustaining economic progress through the provision of infrastructure. The construction industry has been a real catalyst for the economic development enjoyed in recent years, as infrastructure development has been at the core of both the Economic Transformation Program (ETP) and the Tenth Malaysia Plan. By this figures, it is clear that the demand from the construction sector does a lot to propel the economy forward. In 2006, CIDB launched the Construction Industry Master Plan (CIMP) as a strategic roadmap to chart the path for the Malaysian construction industry towards becoming world-class. The success of CIMP underlined the necessity of having a blueprint for the construction industry. Through these achievements, CIDB is pleased to present Construction Industry Transformation Programme 2016 – 2020 (CITP). CITP'S vision is to transform the industry to become highly productive, sustainable and globally competitive, while raising the professional bar in the industry. 18 initiatives were introduced to support the CITP. One of the key initiatives in raising the productivity of the construction workforce is to drive higher technology adoption both in terms of the technology used in advanced construction methods (IBS). The Malaysian government is also committed not only in addressing the IBS agenda but also meeting its target and adopting innovations in the near future as reported by (Kamaruddin et al. 2013). (Shukor, Mohammad, & Mahbub, 2011) also reported that the necessity of the Malaysian government to embark on rigorous initiatives and adopt the Industrialised Building System (IBS) as an innovative approach is a must to ensure Malaysia is globally competent, competitive and remain relevant. This suggestion is in line with P3 CITP's initiatives. Although the efforts mentioned above have been developed to push IBS adoption, the adoption rate is still low as reported by (CIDB), 2010). In order to

[a] mohd.ashraf.fateh@gmail.com

accelerate the adoption of IBS, it needs more support from the stakeholders and construction players in the industry. The reason is that this new concept of construction uses a lot of innovative technology and the knowledge of construction is very crucial. From this scenario, it provokes this research to be conducted in line with the national construction industry agenda on P3 initiative, an acceleration adoption of modern construction method (IBS). This research will focus on the procurement and contract of implementation of IBS in the Malaysian construction industry. This research will give an insight of information and guidance to all the related stakeholders such as CIDB, Public Works Department (PWD), expert panels, academicians and related IBS players. The challenges in boosting the adoption of IBS projects would be looking at development of a standard form of contract for IBS projects, which may have a value-added impact on the success of IBS project implementation and delivery.

This paper is structured into two parts. Firstly, the review covers a broad range of literature, providing a concise overview on the general approach of IBS in the Malaysian construction industry, low adoption of IBS in Malaysia, lack of integration among IBS players, lack of standards for IBS projects and no standard form of contract for IBS projects. The second part discusses the findings and conclusions derived from evidence from the literature review and the preliminary study.

## 2 LITERATURE REVIEW

### 2.1 Industrialised Building System (IBS) and the Malaysia Construction Industry

In general, Industrialised Building System (IBS) is the term used to represent the prefabrication concept in Malaysia. The term was invented to shift from the typical paradigm of prefabricated systems. IBS is defined as a construction technique in which components are manufactured in a controlled environment (on or off site), transported, positioned and assembled into a structure with minimal additional site work. As reported by (Kamar et.al, 2009) some successful IBS implementations in the world are Sekisui Home (Japan), Living Solution (United Kingdom), Open House (Sweden) and Wenswonen (Netherlands).

It is the opposite of a conventional construction which is a common practice in Malaysia whereby most of the building components (beam, column, wall and etc) are cast in situ using timber framework while steel reinforcement is fabricated off-site. This method is labour intensive involving formwork fabrication, steel bending and concreting. It requires many wet trades on site such as skilled carpenters, plasterers and brick workers. The process can be hampered by quality issues, unfavourable site conditions, skilled labour shortages and bad weather conditions as reported by (CIDB), 2010). Based on previous research being done by local and international reseachers by (Construction Research Institute of Malaysia (CREAM), 2011), (Sarja, 1998), (Gibb, 2001), (The UK government Department of Trade and Industry (DTI), 2004), (Nawi, Lee, Kamar, & Hamid, 2011), ( Kamar et.al, 2009), (CIDB), 2015a), (Mohammad, 2013), (Shukor et al., 2011), (Musa, Mohammad, Yusof, & Mahbub, 2015), (Halil, 2015) and (Kamaruddin et al. 2013) is certain that using IBS offers huge benefits in terms of cost and time certainty, attaining better construction quality and productivity, reducing risk related to occupational safety and health, alleviating issues on skilled workers and dependency on manual foreign labour and achieving the ultimate goal of reducing overall cost of construction. As reported by (CREAM), 2011) the Malaysian Construction Industry has been urged to use innovative construction technique and to shift from traditional practice to IBS construction. (CIDB), 2015a) reported that one key factor in raising the productivity of the construction workforce is to drive higher technology adoption both in terms of the technology used in advanced construction methods, e.g. IBS. In parallel with the national and construction industry agenda, CIDB is pleased to present the CITP. In this master plan, acceleration adoption of modern construction methods (especially using IBS) is one of the initiatives, initiatives P3. Therefore, this research is initiated to propel the acceleration adoption of modern constructions methods (IBS) parallel with the CITP.

### 2.2 Low adoption in IBS

Although IBS has been introduced for over 40 years, with well-documented benefits and strong support from the government, the pace of implementation and usage of IBS is still slow and below the government target as reported by (CIDB), 2015a) and (Nawi et al., 2011). Significant efforts have been put in place to push IBS adoption. IBS has been a strong agenda propounded by the YAB Prime Minister of Malaysia as evidenced by the government's commitment to deploy IBS for 34 schools in Sabah and Sarawak with further consideration to expand its usage. However, these efforts have yet to yield sufficient impact and the adoption rate is still low as reported by (CIDB), 2015b). The IBS roadmap 2011 – 2015 by (CIDB), 2010) targeted an adoption rate of 70% for public sector projects and 50% for private sector projects; therefore the weighted average for IBS adoption should be at approximately 55%. A survey conducted by CIDB in 2012 shows that Klang Valley by (CIDB), 2015a) –is seen to have to have the highest IBS adoption rate – has achieved a 46% IBS adoption rate, which is below target. In other regions in Malaysia, the gap is wider. In reality, IBS adoption is still relatively low and is still at its the growing stages as stated by (Halil, 2015) and (Shukor et al., 2011)

### 2.3 Lack of integration among IBS players

Further investigation by previous researchers (Kamar et.al, 2009), (Chung, 2006), (Haron et.al, 2005) and (Nawi et al., 2011) identified that one of the main barriers of IBS implementation in the Malaysian construction industry is related to poor integration and communication among stakeholders involved during the design and contract stage (Planning & Regulation elements). As reported by (Hussein, 2007) the lack of integration among relevant players in the design and contract stage has resulted in a need for redesign and additional costs to be incurred if IBS is adopted. (Kamar et.al, 2009) reported that the pre-casters and contracting firm in Malaysia are only involved after the tender stage of the construction value chain. Whereby there is no integration during the design and contract stage. (CREAM), 2011) also stated that the integration of all parties in a project is the element in IBS that cannot be separated. The IBS players can improve better with proper planning, policies and management. Furthermore, (CREAM), 2011) stated, that the integration of various parties and supply chain to cooperate closely is also important. This requires very careful definition and management of interfaces between contractors and suppliers, and good communication channels. Furthermore, (Nawi et al., 2011) reported that improving procurement systems and supply chains is one of the key factors to achieving IBS success in the Malaysian construction industry. The report by (Halil, 2015) stated that one of the key reasons that the adoption of IBS in the Malaysian construction industry, has not been successfully implemented was due to lack of integration between the design and the construction stage. (Shukor et al., 2011) also stated that the fragmented scenario in the construction industry leaves the IBS supply chain players with noticeable difficulties in terms of competitiveness and efficiency. (CIDB), 2006), (Shukor et al., 2011) and (Mohammad et.al, 2014) also reported that the industry players were facing a difficult task to establish integration and cooperation between parties involved.

One of the critical success factors for IBS as reported by (Yusof et.al, 2014) is the management of supply chain and logistic activities, where a proper arrangement, in a manner that allows all parties to bind together and have the full control of the process. Therefore, integration is a very important aspect in construction. Without it, the project might fail. Due to this scenario, a development of a standard form of contract to act as a formal contract arrangement is a must to ensure all parties bind together, this it will promote and encourage integration among all.

## 2.4 Lack of standards for IBS projects

A research conducted by (Kamar et.al, 2009) stated that poor knowledge among the approving authorities has resulting misunderstanding and misinterpretation of IBS. It is also mentioned by (Chung & Kadir, 2007) that most of the local authorities in Malaysia are unlikely to change local building regulations to suit IBS, as the amendment will consume a lot of time and cost. Even though the use of IBS assures valuable advantages in context of the local construction scenario, but new innovative management and procurement in IBS has still not been fully realised by the industry players and approving authorities as reported by (Shukor et al., 2011). Furthermore, (Shukor et al., 2011) also highlighted that poor knowledge and unfamiliarity with IBS concepts and processes as one of the factors in slow adoption of IBS in Malaysia. Therefore, the adoptions of IBS stay low as all the approving authorities are reluctant to migrate to use IBS because there are no a proper guidelines in term of contractual matters (standard form of contract). A standard form of contract is a very critical matter it will help all the approving authorities to decide and manage when handling an IBS project thus it might give them better understanding and ease the process of IBS adoption in Malaysia.

## 2.5 No standard form of contract for IBS projects

As reported by (CREAM), 2011) stated that contractors, manufacturers and suppliers agreed that factor that can contribute to the success of IBS in the Malaysia construction industry will be managing and coordinating IBS properly, not only from the construction perspective but also through the procurement and contract routine. For example, re-thinking the construction mechanism i.e. payment terms and deliveries as suggested by (CREAM), 2011). (Musa et al., 2015) also agreed that the payment and procurement mechanism needs to be reviewed to tailor to IBS activities which are reliable and for safer payment and procurement. The IBS players require safer and more reliable payment mechanism and contracts as reported by (CREAM), 2011). Change in construction methods and processes from conventional to IBS will affect the change in mode of payment and any related clauses in the contract. At this time, there is no security of payment designed for IBS projects and contractors need to deposit to manufacturers, which results in cash flows issues as reported by the (CREAM), 2011),(CIDB, 2015a) and (Shukor et al., 2011). (CIDB), 2015b) is proposing change in the public procurement approach. The existing procurement structures constraint contractors by requiring them to have high upfront financial capital to be able to procure IBS components from suppliers. (CIDB, 2015a) also reported the slow uptake of IBS is due to constraints or loopholes in regulations, specifications, requirements and standards that suits IBS environment and processes. Furthermore, a report by (Yusof et al., 2014) stated that processes and management and skill sets behind IBS differs from the traditional construction method. As reported by (Yusof et al., 2014), IBS requires partnership and close relationship with the suppliers and contractors from the early stages of construction. Most of the previous research by (CIDB, 2006, 2010; CREAM, 2011; Haron et.al, 2005; Hussein, 2007; Kamar et.al, 2009; Kamar et.al, 2009; Nawi et al., 2011), mainly touches on the IBS roadmap in Malaysia, the comparison between conventional and IBS construction, IBS challenges and benchmark reviews on the United Kingdom and Malaysia construction industry in IBS. On the other perspectives, previous research by (Halil, 2015; Kamaruddin et al., 2013; Mohammad et al., 2014; Mohammad, 2013; Musa et al., 2015; Shukor et al., 2011; Yusof et al., 2014) touches on challenges in the integrations of supply chain in IBS projects, economic perspectives on mechanisation and automation in IBS, the way forward for IBS in Malaysia, strategic partnering in IBS projects and key success factors in IBS projects. While, this research mainly focuses on contract and procurement aspect of IBS adoption. Due to this issue that have been brought up, all the industry players does not feel safe and secure as there are many contractual arrangement that is still vague. There are no proper guidelines to refer if things do not go as planned. Therefore, a dedicated standard form of contract that suits the IBS processes is necessary. By having it, all the clauses will be tailored to the IBS processes. For example clauses of deliverables, payment and materials on site will be covered. All the problems stated can be mitigated and addressed by developing a standard form of contract tailored to IBS projects. As to date, there is no dedicated standard form of contract for IBS projects in Malaysia. (CIDB, 2015a) reported that low adoption of IBS is due to a board definition of IBS and lack of standards. (CIDB, 2015a) also reported that there is room to potentially to establish a standard form of contract that reflects in a IBS environment and its processes. (Mohammad et al., 2014) also agreed that the procurement system plays an important role in the adoption of IBS. Furthermore, (Yusof et al., 2014) stated the introduction of different project procurement systems is needed to ensure more efficient and speedier project delivery systems and better project performance as one of the critical success factors in the adoption of IBS. A different procurement system will bring changes not only to the process and procedure of project delivery but also the aspects of management and organization. Findings from (Yusof et al., 2014) stated the establishment of integration among construction supply chains must take place to overcome the problems encountered.

## 3 METHODOLGY

Literature review and preliminary study with consenting respondents were used in data collection. By conducting the preliminary study to a variety of IBS players namely clients, contractors and manufacturers the seriousness of this issue and the significance of the research to the industry is genuine and valid were identified. However, since this research is still on going as this paper was being written, therefore the data presented in this paper is only the findings from the preliminary study.

Literature review was the first phase of the research with the secondary data derived from relevant books, journals articles, thesis and dissertations, conference proceedings and reports. The second phase involved the collection of primary data, wherein the information was collected through preliminary study. Table 1 illustrate the background of the preliminary study:

**Table 1**.Preliminary study background

| Description | Remarks |
|---|---|
| Target respondents | • Clients<br>• Contractors<br>• Manufacturers |
| Sampling frame | 10 respondents from CIDB's records |
| Duration | 2 month |
| Methodology | • Structured interview<br>• Discussion |
| Software of analysis | Microsoft excel 2010 |

## 4 FINDINGS AND DISCUSSION

60% (6) of the respondents were clients, followed by 20% (2) were contractors and the last 20% (2) were manufacturers/distributors. Figure 1 summarizes on the respondent's sector. Most of the respondents that participated in the survey have vast experience in the construction industry and IBS projects. According to figure 2, 50% (5) of the respondents have more than 10 years of experience in the construction, 30% (3) with 6 to 10 years of experience in the construction industry followed by 20% (2) of respondents who have only 1 to 5 years of experience in the construction. A majority of the respondents have involved with more than 10 IBS projects. Figure 3 summarizes the respondent's involvement in IBS projects, which illustrates that 50% (5) respondents have been involved in more than 10 IBS projects as the highest percentage. The data collected represents that respondents are experienced both in the construction industry and involvement in IBS projects. Therefore, the data collected is considered to be reliable and good for the preliminary study.

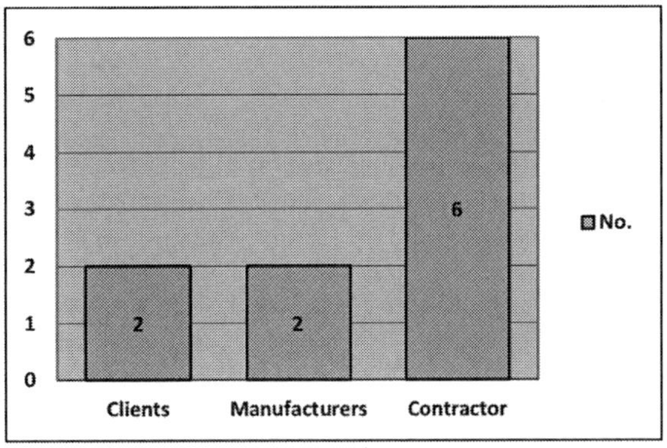


**Figure 1**.Respondent's sector

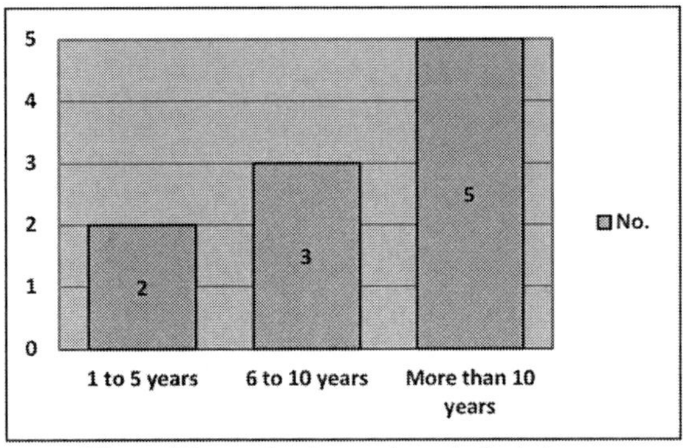


**Figure 2**.Respondent's experience in the industry

DOI: 10.1051/matecconf/20178701001

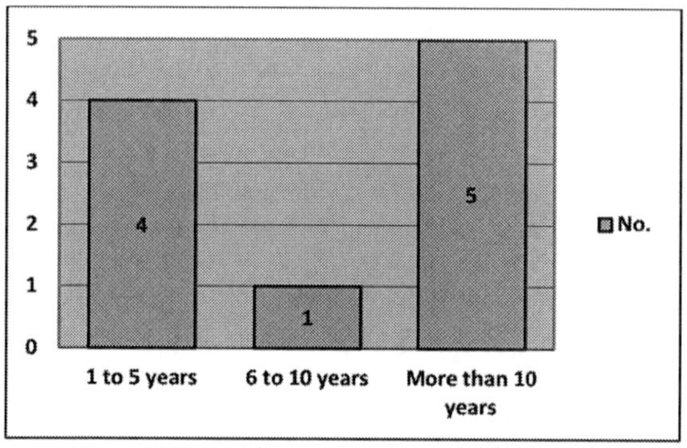


**Figure 3**.Respondent's involvement in IBS projects

Even though IBS has been in Malaysia for quite sometime, all the players are still facing some issues and challenges when adopting IBS, thus it contributes to the low adoption of IBS in Malaysia. 100% (10) of the respondents admit that they have experienced any issues and challenges in adopting with IBS construction. Majority of the respondents admit that there is no standard form of contract which specifically tailors to the IBS construction approach, and activities is one of the major challenges that they are facing when implementing IBS construction approach. 90% (9) of the respondents agree that there is no standard form of contract which tailored to IBS construction approach and activities. Table 2 summarizes the percentage of issues and challenges respondents have experienced in adopting the IBS construction approach. All respondents 100% (10) agree it is a necessity to formulate a standard form of contract which specifically tailors to IBS construction approach.

**Table 2**. Issues and challenges the respondents experienced when adopting IBS construction

| Issues and challenges | Rank | No. | (%) |
|---|---|---|---|
| No standard form of contract to tailor IBS construction | 1 | 9 | 90 |
| Lack of knowledge from industry players | 2 | 7 | 70 |
| Lack of integration among IBS players | 3 | 6 | 60 |
| Lack of proper contract arrangement | 3 | 6 | 60 |
| Expensive | 5 | 4 | 40 |
| Lack of standard for IBS projects | 5 | 4 | 40 |
| Lack of machinery and supplier | 7 | 3 | 30 |

Open ended question were asked to the respondents to comment and suggest on the current issues and challenges of IBS construction approach in Malaysia. The findings are summarize in table 3.

**Table 3**. Issues and challenges the respondents experienced when adopting IBS construction

| Theme | Comments / suggestion |
|---|---|
| Contractual Clause | 1. Payment security and offsite (start from factory manufacturing to delivering to site)<br>2. The specification for IBS is quite ambiguously spelled out in any contract thus leads to confusion and argument between the contractual parties. There should be clauses in IBS contracts so that implementation in the industry would be smooth.<br>3. Develop a standard form on contract for IBS project to suit the decoupling contract concept.<br>4. The appointed consultants should be well versed with the contractual aspects with regards to IBS construction to avoid unnecessary delay when issuing instruction to contractors. A special or dedicated clause needs to be introduced for IBS project. |
| Procurement arrangement | 1. Decoupling contract between client to IBS manufacturer and contractors, may results in:<br>- The reduction of price of component -> direct between clients to manufacturer<br>- Effective payment system<br>- Avoid project delay during construction<br>2. Other elements, for example: integrated procurement & BIM are equally important in the success of IBS delivery.<br>3. Most Malaysian contractors have little knowledge and experience in handling IBS system construction. The work is considered specialized and needs attention as to the process of construction from pre-planning up to the defect liability and maintenance period. Like the trades in construction such as piling, IBS should be introduced and trained as a specialist subcontractor. Likewise in the form of contract for IBS scope of work.<br>4. Dedicated procurement arrangements need to be introduced to cater for IBS project. It cannot rely solely on traditional/ conventional arrangement. |
| Integration & supply chain | 1. It is important to understand that key success to delivery of IBS projects is the early integration of IBS during the planning & design stage. |

## 5 CONCLUSION

From the preliminary study conducted, it shows that the industry is still facing issues and challenges when adopting IBS construction. The study also finds that formulating a standard form of contract for IBS construction approach is necessary to ensure smooth implementation as the activities are different from a normal construction approach. Previous research by (Shukor et al., 2011) and (Mohammad, 2013) have stated that, when there is a changes of technology or industrialisation in the industry, new and appropriate project procurement is needed. This is to ensure it can cater efficiently towards innovative activities. As every project is unique and dynamic in term of processes, resource allocation, risk exposure and responsibilities between all parties therefore the necessity of having a standard form of contract for IBS. For example, for a traditional design and build project, both approaches are different and each approach has their own standard form of contract which tailors to specific activities and construction processes. Therefore, it is crucial to conduct a research that is expected to give a guideline to the industry by developing a standard form of contract for IBS projects in Malaysia. The output of this research will give a good insight to the industry and help to accelerate the adoption of IBS construction in Malaysia as a whole.

## REFERENCES

1. Chung, L. P. (2006). *Implementation Strategy for Industrialized Building System*. Johor.
2. Construction Industry Development Board (CIDB). (2006). Construction Industry Master Plan 2006-2015 (CIMP 2006-2015).
3. Construction Industry Development Board (CIDB). (2010). IBS Road Map, (72).
4. Construction Industry Development Board (CIDB). (2015a). Construction Industry Master Plan 2 (CIMP 2 2016-2020), (January 2015).
5. Construction Industry Development Board (CIDB). (2015b). *Construction Industry Transformation Programme 2016 - 2020*. Kuala Lumpur.
6. Construction Industry Development Board (CIDB). (2015c). *Input-Output Table*.
7. Construction Research Institute of Malaysia (CREAM). (2011). Proceedings of 1st IBS Roundtable Workshop. *Vasa*, 19. Retrieved from http://medcontent.metapress.com/index/A65RM03P4874243N.pdf
8. Department of Statistics. (2013). *Ecomonic Census of Contruction*.
9. Gibb, A. G. F. (2001). *Pre-assembly in Construction: A review of recent and current industry and research initiatives on pre-assembly in construction In Construction Research & Innovation Strategy Panel CRISP Consultancy Commission 00/19*. London.
10. Halil, F. M. (2015). SUSTAINABLE VALUE TFIROUGH THE IMPLEMENTATION OF STRATEGIC PARTNERING IN DTDUSTRIALISED BUILDING SYSTEM ( IBS ). In *International Symposium in Developing Economies: Commonalities Among Diversities* (pp. 188 – 200).
11. Haron, N.A. Hassim, S. Kadir, M. R.A., Jaafar, M. S. (2005). Building Costing Comparison Between Conventional and Framework System. *Journal of Technology*, (43(B)).
12. Hussein, T. S. J. (2007). Industrialised Building System: The Challenges and Way Forward. *Construction Industry Research Achievements International Conference*, (Ciraic), 1–12.
13. Kamar, K. A. M., Alshawi, M. and Hamid, Z. A. (2009). BARRIERS TO INDUSTRIALIZED BUILDING SYSTEM ( IBS ): THE CASE OF MALAYSIA. In *9th International Postgraduate Research Conference (IPGRC)*. United Kingdom.
14. Kamaruddin, S., Mohammad, M., Mahbub, R., & Yunus, R. (2013). Perception Towards Cost Implication of Mechanisation and Automation Approcah in IBS Projects in Malaysia. *29th Annual ARCOM Conference*, (September 2013), 213–222.
15. Kamaruddin, S. S., Mohammad, M. F., & Mahbub, R. (2013). Enhancing the Quality of Life by Adopting IBS: An Economic Perspective on Mechanisation and Automation. *Procedia - Social and Behavioral Sciences*, *101*, 71–80. doi:10.1016/j.sbspro.2013.07.180
16. Kamarul Anuar Mohamad Kamar, Mustafa Alshawi, Zuhairi Abd. Hamid, Mohd Nasrun Mohd Nawi, Ahmad Tarmidzi Haron, M. R. A. (2009). Industrialised Building System (IBS): A Review of Experiences in Uk and Malaysian Construction Industry. *2nd Construction Industry Research Achievement International Conference (CIRAIC)*, 1–12.
17. Mohammad, M. F. (2013). Construction Environment: Adopting IBS Construction Approach Towards Achieving Sustainable Development. *Procedia - Social and Behavioral Sciences*, *85*, 8–15. doi:10.1016/j.sbspro.2013.08.332
18. Mohammad, M. F., Saifuza, A., Shukor, A., Mahbub, R., & Halil, F. M. (2014). Challenges in the Integration of Supply Chains in IBS Project Environment in Malaysia. *Procedia - Social and Behavioral Sciences*, *153*, 44–54. doi:10.1016/j.sbspro.2014.10.039
19. Musa, M. F., Mohammad, M. F., Yusof, M. R., & Mahbub, R. (2015). The Way Forward for Industrialised Building System (IBS) in Malaysia, (MAY). doi:10.1007/978-981-287-290-6
20. Nawi, M. N. M., Lee, A., Kamar, K. A. M., & Hamid, Z. A. (2011). A critical literature review on the concept of team integration in industrialised building System (IBS) project. *Malaysian Construction Research Journal*, *9*(2), 1–17.
21. Sarja, A. (1998). Open & Industrialised Building, CIB (the International Council for Building Research Studies and Documentation).
22. Shukor, A. S. A., Mohammad, M. F., & Mahbub, R. (2011). Supply Chain Integration Challenges in Project Procurement. *Management and Innovation*

DOI: 10.1051/matecconf/20178701001

*for a Sustainable Built Environment*, (June).
23. The UK government Department of Trade and Industry (DTI). (2004). *Modern methods of construction in Germany – playing the off-site rule In Global Watch Mission Report.* United Kingdom.
24. Yusof, M. R., Nawawi, A. H., Mohammad, M. F., & Musa, M. F. (2014). Key Success Factors in IBS Project Management Key Success Factors in IBS Project Management, (APRIL).

MATEC Web of Conferences 87, 01002 (2017) DOI: 10.1051/matecconf/20178701002
*ENCON 2016*

# Engineering Behavior of Concrete with Recycled Aggregate

Afizah Ayob[1, a], Mohd. Ekhwan Razali[1], Salina Alias[2], Abdul Ghapar Ahmad[1], Dayang Siti Hazimmah Ali[3]

[1] *Department of Civil Engineering, School of Environmental Engineering, Universiti Malaysia Perlis, 02600 Arau, Perlis, Malaysia*
[2] *Faculty of Civil Engineering, Kampus Permatang Pauh, UiTM Pulau Pinang, 03500 Permatang Pauh, Pulau Pinang, Malaysia*
[3] *School of Engineering and Technology, University College of Technology Sarawak, 868 Persiaran Brooke, 96000 Sibu, Sarawak, Malaysia*

**Abstract.** Concrete is extensively used as construction materials in Malaysia. Concrete contributes suitable feature for construction industry for instance durability, adequate compressive strength, fire resistance, availability and is economic as compared to other construction materials. Depletion of natural resources and disposal of construction and demolition waste remarkably claim environmental threat. In this paper, the engineering behavior, durability, and concrete microstructure of recycled concrete aggregates (RCA) on short-term concrete properties were investigated. The studied concrete at design mix proportion of 1:0.55:2.14:2.61 (weight of cement :coarse aggregates :sand :water) used to obtain medium-high compressive strength with 20%, 50%, and 100% of RCA. Results show that for the same water/cement ratio, RCA replacement up to 50% still achieved the targeted compressive strength of 25 MPa at 28 curing days. Addition, at similar RCA replacement, the highest carbonation depth value was found at 1.03 mm which could be attributed to the pozzolanic reaction, thus led to lower carbonation resistance. Scanning electron microscopy microstructure shows that the RCA surface was porous and covered with loose particles. Moreover, the interfacial transition zone was composed of numerous small pores, micro cracks, and fissures that surround the mortar matrix. On the basis of the obtained results, recommendable mineral admixtures of RCA are necessary to enhance the quality of concrete construction.

## 1 Introduction

A large amount of construction and demolition waste (CDW) is being generated by the construction industry in Malaysia due to demands of small and major infrastructure projects, that includes commercial building, and housing development planning. Nonusable and inert waste is the result of enormous construction and demolition activities. It consists abundantly of concrete, bricks and block, aggregate, soil and sand, stones, pebbles, graded organic particles and wood, among others. Providing land fill sites for CDW is not the main solution for this problem. Technology and science must play a main role in fostering the successful implementation of waste management. Factors, such as environmental sustainability, economic condition, land fill space depletion, and quality of construction basic raw material, are essential for recycling/reusing CDW [1]. Most CDW in Malaysia was formerly assumed nonusable and illegally discarded. To date, these wastes are totally utilize as recycled/reused concrete aggregates and are benefited for various applications in construction [2, 3].

Recycled concrete aggregate (RCA) is an environmentally friendly material that employ CDW, especially demolished concrete waste as alternative aggregates to take place of natural aggregates. This RCA is produced partly or totally by processes of cleaning, pulverizing, classifying and proportional mixing. However, the application is complex because the properties of RCA are different from natural aggregates. In addition, the quality of RCA varies among different sources. The physical properties of RCA mainly determine on water absorption, adhered mortar quality, amount of adhered mortar, contaminant, chemical constituent and the interfacial transition zone (ITZ) between the natural aggregates and original cement mortar. Adhered mortar is a porous material, and its porosity rely on the originally selected ratio of water to cement (w/c) [3, 4]. The high porosity of RCA is responsible for its low resistance toward mechanical and chemical actions. The quantity of adhered mortar significantly increases with decreasing aggregate size. RCA exhibits lower bulk density and greater water absorption than natural aggregates because of the adhered mortar amount. Meanwhile, ITZ is the weakest section in concrete due to its less porosity and crack resistance than either hardened cement paste or aggregate [3]. Crushing procedure usually affects the amount of adhered mortar and ITZ in the concrete behavior with recycled aggregate.

RCA application remains unfavorable, particularly in countries where natural aggregates are still abundant. RCA replacement in concrete is inversely associated to its workability. Several works have been conducted to investigate the essential properties of RCA, such as compressive strength, flexural strength, tensile strength, elastic modulus, and durability. The actual concrete strength differs from compressive strength of the original concrete used as recycled aggregates. The reduction in compressive strength is not prominent, when the RCA replacement is relatively up to 30% [5]. The compressive strength of a recycled aggregate concrete becomes unsatisfactory when high RCA percentage is used to substitute natural aggregates. However, a suitable concrete mix design enables the RCA to achieve target strength in wide range of application in construction. Kwan et al. (2012) [2] achieved target strength even when 80% of the total coarse aggregate content was replaced by RCA. The same authors have reported that, even though the porous materials of adhered mortar could significantly increase the permeability, the curing period would control the intrinsic permeability of RCA for the long term durability of the concrete structure.

The concept of reusing and utilizing RCA from CDW has been used since the World War II in Europe as unbound sub-base or main load-bearing layer material of the pavement [5]. Natural aggregates cannot be reproduced in a short period of time, and aggregates basically hold 60% to 75% of the concrete volume; thus no doubt RCA is suggested for construction applications. The use of RCA not only solves the absence of natural resources but also saves land disposal for CDW. Therefore, this study aims to report the engineering behavior of RCA in a concrete construction mixture. Engineering properties, durability, and concrete microstructure of hardened concrete used as replacement of natural aggregates with RCA were discussed.

## 2 Materials and Methods

### 2.1 Materials

Ordinary Portland cement (OPC) with a specific gravity of 3.24 was employed as the main binder for the experiment. Table 1 shows the typical chemical constituent of OPC.

Natural coarse aggregates were crushed granite with sizes of 4.75 to 20 mm were obtained from local quarries. The aggregates were washed with water to remove any unwanted particles, such as clay, dirt, and dust, and were then air dried. Fine aggregates (with less than 5.0 mm size) used were uncrushed quartzite natural river sand. RCA was obtained from a demolished building (Fig. 1) and was manually crushed to achieve the required grain size distribution of 10–20 mm.

**Table 1.** Major chemical constituents of OPC

| Constituent of OPC | % |
|---|---|
| Lime (CaO)_ | 68.85 |
| Silica ($SiO_2$) | 15.00 |
| Alumina ($Al_2O_3$) | 4.30 |
| Sulfur trioxide ($SO_3$) | 4.00 |
| Iron oxide ($Fe_2O_3$) | 3.96 |
| Potassium oxide ($K_2O$) | 1.08 |
| Magnesia (MgO) | 1.30 |
| Barium oxide (BaO) | 0.34 |
| Titanium dioxide ($TiO_2$) | 0.34 |
| Others | 0.73 |

**Figure 1.** (a) RCA samples from CDW

Grain sizes of aggregates were reported to influence density and water adsorption. Aggregate size distribution was determined in accordance with BS 882 (1992) [6], and the result of sieve analysis shown in Table 2. Given the high water absorption capacity of RCA, aggregates were maintained in a saturated surface dry condition before the start of the mixing process.

**Table 2.** Aggregate size distribution

| Type of Aggregates | Grain size (mm) |
|---|---|
| Fine | 0-5 |
| Natural coarse | 4.75 to 20 |
| RCA | 4-8 |
| | 8-20 |

### 2.2 Concrete Mixes

Four types of mixtures were prepared by replacing the coarse aggregate with 20%, 50% and 100% of RCA. The percentage of replacement was determined based on the total weight of the coarse aggregate content. Normal specimen was denoted as 0% of RCA replacement. The replacement of natural coarse aggregate with RCA is expected to increase remarkably the w/c ratio [7]. Thus, the design mix proportion was set up as shown in Table 3 [8] for design targeted compressive strength of 25 MPa at 28 curing days. All specimens were casted in 150 mm × 150 mm × 150 mm molds under laboratory condition for 28 h, demolded, and then completely submerged in water for curing until testing.

**Table 3. The design mix proportion**

| Constituents | Proportion ($kg/m^3$) |
|---|---|
| Cement | 418 |
| Natural coarse aggregates /RCA | 987 |
| Fine aggregate | 745 |
| Water | 2.3 |

The water absorption and compressive strength tests were performed according to BS EN 12390-3 (2009) [9] and BS 1881-122 (2011) [10], respectively, by using three cubes to obtain the average value of the specimens.

Carbonation test was conducted to investigate the concrete durability [11]. Carbonation depth was measured after spraying a solution of phenolphthalein on a freshly broken concrete surface (Fig. 2 (a)). The carbonated region with decreased alkalinity remained colorless, whereas the uncarbonated region with high alkalinity turned its color to pink. Examples of six spots reading location are shown in Fig. 2 (b).

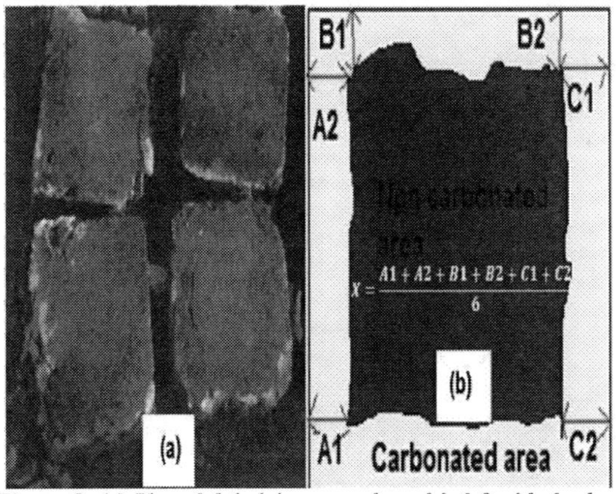


**Figure 2.** (a) Phenolphthalein sprayed on dried freshly broken concrete surface; (b) readings location spot of carbonation depth (*X*)

The concrete microstructure properties on the RCA surface were observed after 28 curing days under a scanning electron microscope (SEM; JEOL JSM 6460 LA).

## 3 Results and discussion

### 3.1. Water absorption

Fig. 3 shows the influence of RCA content on water absorption rate. The water absorption rates of concrete containing the RCA were higher than that of the normal concrete. In this study, the water absorption rate of natural course aggregate recorded at 2.6%. Water absorption rate of concrete shows and increasing profile in relation to higher RCA content (Fig. 3).

The highest water absorption value was obtained by 100% RCA specimen, which is 0.6 times higher than that of normal. It was due to the higher absorption capacity of the RCA itself resulting from cement mortar remains attaching to its surface, and the substitution level of RCA, which then has developed higher osmosis pressure within the concrete matrix, that imposes a high risk towards the durability of concrete.

At the same time, the RCA is a porous material that correlates to high permeability properties for the concrete and greater probability of interconnection between the micro structure systems of the specimen. Additionally, the pressure-induced flow would contribute the easiest way for the transmission of molecules as the internal flow paths are not blocked.

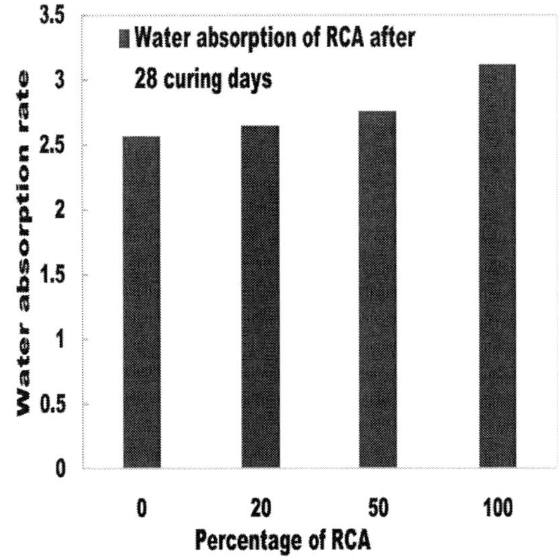


**Figure 3**. Behavior of the water absorption rate related to RCA content after 28 curing days

These reveal the highest water absorption at 100% RCA sample that occurs through the capillary effects by the high osmosis pressure when the dry samples were submerged into the water.

### 3.2 Compressive strength

Fig. 4 shows the influence of RCA content on the average results of concrete compressive strength after 7 and 28 curing days. The figure shows the similar profile of strength development as the normal concrete, where the compressive strength increased with the number of curing days for 20% and 50% of RCA replacement. This result was due to the increased amount of hydration products produced during prolonged curing period. The RCA content in concrete greatly affected the compressive strength of the specimen. The normal concrete (0% RCA) achieved the highest strength, followed by 20%, 50%, and 100% of RCA replacement (Fig. 4).

The results indicate a decreasing trend in compressive strength toward high RCA content for 7 and 28 curing days. This finding was consistent to that of Kwan et al. (2012) [2] who reported that RCA influences concrete design and durability. This trend was not observed by Lotfy and Al-Fayez, (2015) [13] when they used high-quality RCA produced on a commercial scale. Their RCA performed equally to virgin aggregates up to 30% coarse replacement and 20% granular replacement.

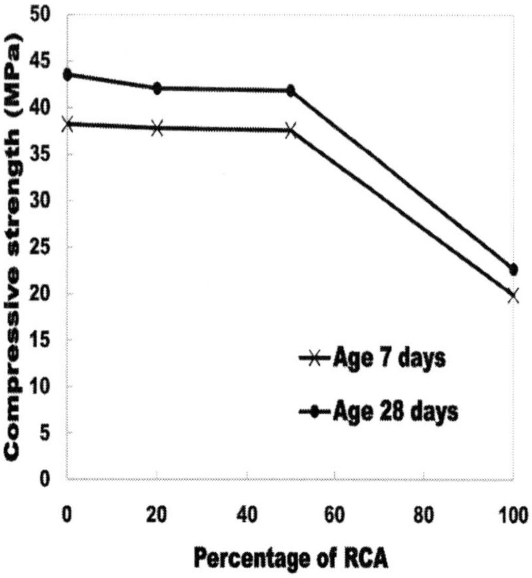


**Figure 4**. Behavior of the concrete compressive strength related to RCA content for age 7 days and 28 days

The inverse trend between the RCA content and compressive strength was probably due to the poor quality of the adhered mortar, which underwent crushing process, that weakened the ITZ in the concrete. The presence of adhered residual cement mortar on the RCA surface was in accordance to porous, rough [2] and weak interfacial bond strengths between the aggregate and the presence of micro cracks and fissures, which could decrease the bonding with the surrounding matrix [5]. Therefore, by giving ultimate load to the concrete, the weak sections would have greater probability to be interconnected between each other as the RCA content increases. The new ITZ between the RA and cement paste waste was highly efficient. This ITZ exhibited a lower aggregate w/c ratio than the adhered mortar on the RCA surface. This ITZ is thus denser than the old cement paste, which was accordingly weaker [14].

In addition, the RCA has high hydrated and non-hydrated levels that can increase its saturation level and then later affect the concrete strength. At higher saturation levels, the mechanical bonding between the cement paste and RCA became weaker, which led to low compressive strength, as observed in 100% of RCA replacement (Fig. 4). Thereby, replacing the coarse aggregate with up to 50% RCA is the optimum level for the mix proportion because of the absence of significant reduction in the compressive strength.

### 3.3 Carbonation depth

The results of carbonation depths in specimens with varied RCA contents throughout exposure time are shown in Fig. 5. A high influence of RCA content was observed in all studied admixtures. However, for different curing days, the incorporation of RCA had no significant effect on carbonation depth.

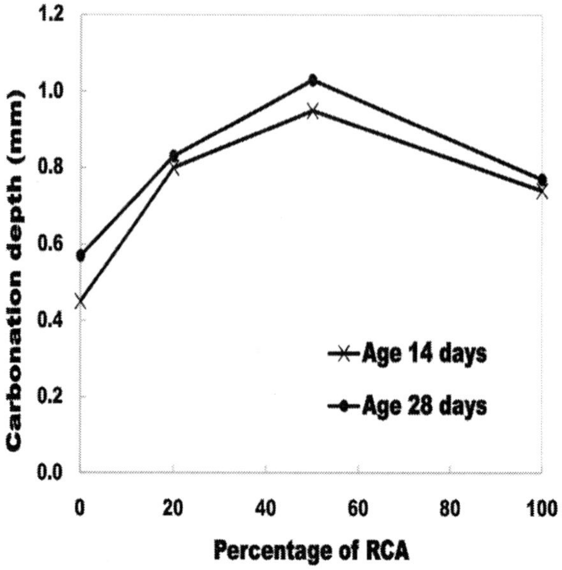


**Figure 5.** Behavior of carbonation depth related to the RCA content for age 14 days and 28 days

The highest difference of carbonation depth value was found in 50% RCA replacement, which depicted almost similar behavior at 14 and 28 curing days. The measured carbonation depths for normal samples were 0.45 and 0.57 mm, and the depths increased 0.95 and 1.03 mm with 50% RCA replacement, respectively, at 14 and 28 curing days. The negative affect of 50% RCA in carbonation depth could be attributed to the pozzolanic reaction. The pozzolanic material in the concrete specimen consumes calcium hydroxide (CH) and hydrated calcium silicate (CS). Calcium hydroxide and hydrated calcium silicate are the main hydration products of the adhered mortar on RCA. $CO_2$ can penetrate into the pores of the adhered mortar and react with the reaction products as follows:

$$Ca(OH)_2 + CO_2 \longrightarrow CaCO_3 + H_2O \quad (1)$$

$$C\text{-}S\text{-}H + CO_2 \longrightarrow CaCO_3 + SiO_2.nH_2O \quad (2)$$

Carbonation of CH and C-H-S (calcium silicate hydrates) provides 11.0% and 23.1% of solid volume, respectively [15], therefore, carbonation could decrease the porosity of the adhered mortar and consequently reduce the alkalinity of concrete [16, 17]. At the same time, C-H-S formed during hydration could contribute a bridging effect within the structure that wraps the aggregate. Subsitutely, the dissolved silicates could seep through the wrap and precipitate as an outer layer, thus generating a dense micro structures and decreasing the number of pores. In addition, carbonation enhancement of RCA, which is based on the chemical reaction between $CO_2$ cement minerals and hydration product, promoted the adhesion of mortars, as well the original and new ITZs around the RCA surface.

Carbonation depth was affected by water adsorption, crushing value of RCA [4] alkalinity of binding medium, and aggregate size. In this case, the aggregate size for 50% RCA replacement showed higher probability that the concrete mixes that composed of fine aggregates exhibited higher carbonation depths than those of large RCA. The greater porosity of fine RCA was due to the greater amount of adhered mortar. The carbonation development of 100% RCA replacement decreased to 0.74 and 0.77 mm over time for 14 and 28 curing days, respectively (Fig. 5).

### 3.4 RCA microstructure

Fig. 6 shows the microstructure of RCA samples after 28 curing days. The angular conventional concrete shape has cleaner surface aggregate, and is less porous and free from loose particles, such as dust (Fig. 6(a)). The microstructures of the concrete with 20% RCA content and the conventional concrete were not significantly different (Fig. 6(b) (c)).

Fig.6. SEM images of concrete specimen after 28 curing days (a) 0% RCA content; (b) 20% RCA content; (c) 50% RCA content; (d) 100% RCA content.

Meanwhile, the angular shape and rough surface of concrete with 50% and 100% of RCA replacement appeared to be more complicated, as shown in Fig. 6(d). The surface was porous and covered with loose crumbly particles, either dust or sources from adhered mortar attached to the coarse RCA after crushing (Fig. 6(b), (c)). The porosity and amount of loose particles covered on the surface depend on the RCA content used in the original concrete.

The ITZ thickness of RCA may range from 10 μm to 50 μm around the aggregate; this thickness is linked with the degree of hydration and content of adhered mortar (Figs. 6 (c), (d)). The ITZ consists of less unhydrated particle and greater concentration of $Ca(OH)_2$ and ettringite. The microstructure of the specimen also shows that the ITZ of RCA was granular in nature and composed of small pores, micro cracks and fissures, which significantly lowered the bond strength in the area surrounding the mortar matrix

## 4 Conclusions

The engineering and durability performance of RCA was predominantly inferior to conventional concrete, but RCA has been recently gaining wide attention as a construction material. In this study, an experimental program was conducted to investigate the use of RCA as partial replacement of coarse aggregate in concrete production. The replacement level of natural coarse aggregate with RCA reduced the compressive strength of the concrete. However, RCA replacement up to 50% still achieved medium-high target strength. The water absorption value was directly proportional to the level of RCA replacement. The depth of carbonation increased with increasing percentage of RCA in concrete, which shows that the resistance to carbonation is reduced. Given that carbonation depth is closely related to concrete porosity, incorporating increasingly porous RCA evidently created significantly higher permeable concrete and hence greater carbonation depths. The SEM microstructure of RCA concrete was porous, and its surface was coated with loose particles. Moreover, the ITZ of RCA was composed of numerous tiny pores, micro cracks, and fissures, which relatively weakened the surrounding hardened mortar matrix. Enhancing the quality of concrete by modifying the weak ITZ of RCA and the bulk concrete matrix with mineral admixture, such as fly ash, nano-silica, and meta kaolin, is necessary for future RCA applications. These mineral admixtures behave as micro filler, filling the ITZ between the aggregate surface and the matrix. Additionally, a lack of reliable information on corrosion behavior on RCA has to be established

## References

1. N. Jain, M. Garg, A. K. Minocha, J. Waste Manage, 281048, (2015).
2. W. H. Kwan, M. Ramli, K. J. Kam, M. Z. Sulieman, Construc. Build. Mater. **26** (1), 565-573 (2012).
3. S. Ismail, M. Ramli, Construc. Build. Mater. **44**, 464-476 (2013).
4. J. Zhang, C. Shi, Y. Li, X. Pan, C. Poon, Z. Xie, J. Mater. Civ. Eng. **27**(11), 04015029 (2015).
5. M. Behera, S.K. Bhattacharyya, A.K. Minocha, R. Deoliya, S. Maiti, Construc. Build. Mater. **68**, 501-516 (2014).
6. BS 882, British Standards Institution, London, (1992).
7. L. Evangelista, J. de. Brito, Cement Concrete Comp. 29, 397-401 (2007).
8. BS 1881-125, British Standards Institution, London, (1986).

DOI: 10.1051/matecconf/20178701002

9. BS EN 12390-3, British Standards Institution, London, (2009).
10. BS 1881-122, British Standards Institution, London, (2011).
11. BS EN 14630, British Standards Institution, London, (2006).
12. W. Li, J. Xiao, C. Shi, C. P. Poon, Adv. Struct. Eng. **18**(6), 919-938, (2015).
13. Lotfy, M. Al-Fayez, Cement Concrete Comp. **61**, 36-43, (2015).
14. M. Etxeberria, E. Vázquez, A. Marí, M. Barra, Cem. Concr. Res. **35**(5), 735-742, (2007).
15. C. Shi, Y. Li, J. Zhang, W. Li, L. Chong, Z. Xie, J. Clean. Prod. **112**(1), 1-7, (2015).
16. S. P. Arredondo-Rea, R. Corral-Higuera, J. M. Gomez-Soberon, J. H. Castorena-Gonzalez, V. Orozco-Carmona, J. L. Almaral-Sanchez, Int. J. Electrochem. Sc. **7**(2), 1602-1610, (2012).
17. R. V. Silva, J. de Brito, R. K. Dhir, Cement Concrete Comp. **62**, 201-217, (2015).

MATEC Web of Conferences **87**, 01003 (2017) DOI: 10.1051/matecconf/20178701003
*ENCON 2016*

# Study on Concrete Containing Recycled Aggregates Immersed in Epoxy Resin

Suraya Hani Adnan[1], Alshrif Abdslam Alsonosi[2], Suzana Wati Adnan[3]

[1]*Faculty of Engineering Technology[1],Universiti Tun Hussein Onn Malaysia, Malaysia*
[2]*Faculty of Civil and Environmental Engineering, Universiti Tun Hussein Onn Malaysia, Malaysia*
[3]*Civil Engineering Department, Polytechnic Kuching Sarawak, Malaysia*

**Abstract.** In recent decades, engineers have sought a more sustainable method to dispose of concrete construction and demolition waste. One solution is to crush this waste concrete into a usable gradation for new concrete mixes. This not only reduces the amount of waste entering landfills but also alleviates the burden on existing sources of quality natural concrete aggregates. There are too many kinds of waste but here constructions waste will be the priority target that should be solved .It could be managed by several ways such as recycling and reusing the concrete components, and the best choice of these components is the aggregate, because of the ease process of recycle it. In addition, recycled aggregates and normal aggregates were immersed in epoxy resin and put in concrete mixtures with 0%, 5%, 10% and 20% which affected the concrete mixtures properties. The strength of the concrete for both normal and recycled aggregates has increased after immersed the aggregates in epoxy resin. The percentage of water absorption and the coefficient of water permeability decreased with the increasing of the normal and the recycled aggregates immersed in epoxy resin. Generally the tests which have been conducted to the concrete mixtures have a significant results after using the epoxy resin with both normal and recycled aggregates.

## 1 Introduction

Recycled Aggregate (RA) can be defined as an aggregate that has been produced from used material as coarse or fine aggregate. RA can be produced from different sources like broken precast elements and cubes after testing, and demolished concrete buildings (Rao *et al.*, 2007). Olorunsogo & Padayachee (2002) revealed that RA consisting materials like dust, mortar, brick, stone and mortar. In RA, the lowest percentage of proportion is dust which accounted 1.90% and the highest proportion is stone and mortar which is 84.60% (Olorunsogo & Padayachee, 2002). RA can be produced from large amount of waste material from construction industry which this waste material can lead into environmental problem. Thus, Ravindrarajah (1987), Adnan *et al.* (2007) and Abdel-Hay (2015) expressed their opinions that using RA would reduce the use of natural aggregate and landfill for disposal.

Nowadays the whole world focus on reducing the constructions and demolition waste that will affect our environment and the sustainability due to the principle of sustainable construction must prudent the natural resources and encourage this approach in all government activities (Cakir and Sofyanh, 2015) .

"Construction waste" means any substance, matter or thing which is generated as a result of construction work and abandoned. It is a mixture of surplus materials arising from site clearance, excavation, construction, refurbishment, renovation, demolition and road works.

Over 90% of construction waste are inert and are known as public fill. Public fill includes debris, rubble, earth and concrete which is suitable for land reclamation and site formation. If they just separate certain materials such as concrete and asphalt these materials can be used again or recycled for constructions usages. Approximately 110 million tonnes of construction, demolition and excavation waste (C&DW) are generated in UK yearly, accounting for over 60% of the country's total waste. Increasing interest in RA application is because the increasing cost of landfill, anxiety of decreasing in natural resources and increasing in aggregate requirement for construction work (Khatib, 2005).

One of the main wastes produced by constructions is unused concrete like beams and columns which is produced by demolition process and also concrete cube produced from the concrete tests. Hence, for sustainable purpose all those unused concrete could be recycle or reuse so that the consumption of natural resources could be reduce. Concrete could be useful by reusing its aggregates instead of dispose .There are too many reasons to reuse the recycled aggregate such as conservation of our environment and sustainability and economic reasons .Recycling is the process of reusing materials for further use to develop new product. It includes the breaking of concrete to produce smaller size of fragments .Hence, this study is conducted to determine the properties of recycled aggregate concrete which using the aggregate that has been immersed in epoxy resin.

Epoxy resins are generally recognized as workhorse products among the category of thermosetting polymers due to their outstanding mechanical properties and good handling characteristic. The use of epoxy resins in industry extends back over fifty years since their introduction commercially and they find an extremely wide range of applications as coatings, adhesives and matrix resins in different fields G like electronics, defense, aerospace industry etc. The main advantages of epoxy resins are their good mechanical properties, minimum shrinkage after cure and suitable weather,

DOI: 10.1051/matecconf/20178701003

chemical and electrical resistace (Gilat *et al*, 2007 and Li *et al.*, 2015).

# 2 Materials and Methods

## 2.1 Materials

The materials that have used in this study are cement, recycled aggregates, normal aggregates, resin to immerse the recycled aggregates and the normal aggregates on it, water and the sand .The source for recycled aggregates is the waste cubes from the Laboratory of Concrete in Universiti Tun Hussein Onn Malaysia (UTHM).

## 2.2 Epoxy Resin

Epoxy resins have a wide variation in properties. Consequently, it is possible to formulate many different compound with each compound having slightly different characteristics. As in example of the range of properties that an epoxy resin may have, it is possible for an epoxy to be as brittle as glass or as mobile as water. Hence, for this study Epoxy Resin has been used as a water resistance by immersing both the normal and the recycled aggregates on it. Figure 1 illustrated the normal and recycled aggregates immersed in various percentage of resin (0, 5, 10 and 20).

**Figure 1**. Normal and recycled aggregates immersed in resin

## 2.3 Aggregate

Fine aggregates and coarse aggregates have been used in this study for producing the concrete specimens. According to Neville (2012) the size of fine aggregates were the aggregates which passed through the sieve 4.75 mm. Two types of coarse aggregate that have been used in this study are Normal Aggregates (NA) and the Recycled Aggregates (RA). Recycled Aggregates (RA) is as shown in Figure 2. The normal and the recycled aggregates were place in the room temperature for 24 hours after immersed in the resin in order to dry well and take the same shape with the aggregate. The size for coarse aggregates that have used in this study is between (5 mm – 20 mm).

### 2.3.1 Aggregate Test

In this studyt there are several tests that have been conducted for the aggregates before concrete were cast such as Aggregates Impact Value (AIV) Test and Aggregates Crushing Value (ACV) Test. The aim for these tests is to assess the physical properties of the aggregates that have used in this study which are normal aggregates, recycled aggregates, normal aggregates immersed in resin and recycled aggregates immersed in resin.

**Figure 2**. Recycled Aggregates (RA)

## 2.4 Concrete Mix Design

For this study, the concrete mixes were designed according to Department of Environment (DOE) method. The concrete mixes were designed to achieve the target strength 25 MPa at 28 days. The concrete cubes with size 100 mm x 100 mm x 100 mm were prepared. 72 samples of concrete cubes were prepared. Table 1 presents the concrete mix which has been used for this study.

**Table 1.** Mix Proportions for Concrete Mixes

| Resin Percentages (%) | Cement (kg/ m$^3$) | Water (kg/ m$^3$) | Fine aggregates (kg/ m$^3$) | Coarse Aggregates (kg/ m$^3$) |
|---|---|---|---|---|
| | | | | NA/RA* |
| 0 | 360 | 180 | 681 | 1159 |
| 5 | 360 | 180 | 681 | 1101 |
| 10 | 360 | 180 | 681 | 1043 |
| 20 | 360 | 180 | 681 | 927 |

* NA = Normal Aggregate
* RA = Recycled Aggregate

## 2.5 Compressive Strength Test

This test was conducted according to BS EN 12390-3:2002.

### 2.6 Water Absorption Test

This test was conducted according to BS 1881: Part 122 at 28 days age of concrete.

## 3 RESULTS AND DISCUSSIONS

### 3.1 Aggregate Impact Value

Table 2 shows the results of aggregate impact value test. This test was conducted in highway laboratory at UTHM for normal aggregates, recycled aggregates, normal aggregates immersed in epoxy resin and recycled aggregates immersed in epoxy resin.

It is revealed that AIV obtained for RA is higher compared to NA therefore indicating higher strength of NA compared to RA and the results for present study is in agreement with other researchers (Ridzuan *et al.*, 2001 and Limbachiya *et al.*, 2000). According to BS 882:1992, the maximum value of AIV of aggregate for application of heavy-duty floors is 25%, 30% for wearing surfaces and 45% for other concretes used. Thus, the higher value of AIV for RA indicates that this aggregate is suitable to be used for other concretes purposes. The mortar attached to RA could most probably influence the amount of broken particles size smaller than 2.36 mm size after the blasting. Thus, this leads into higher value of AIV for RA than that of NA.

**Table 2.** Aggregate impact value results

| Sample | AIV (%) |
|---|---|
| NA | 10.9 |
| RA | 20.8 |
| NA immersed in epoxy resin | 8.1 |
| RA immersed in epoxy resin | 13.9 |

### 3.2 Aggregate Crushing Value

Table 3 shows the results of aggregate crushing value test which were conducted in highway laboratory in UTHM for the normal aggregates, recycled aggregates, normal aggregates immersed in epoxy resin and recycled aggregates immersed in epoxy resin.

From this table, it revealed that the RA was the weakest with the ACV is 31.1% while the best aggregate was the NA immersed in epoxy resin with ACV is 15.6%. These results showed that the NA strength is higher than that of the RA. In addition, the epoxy resin gave an extra strength to the NA which enhance the strength aggregate. These results illustrated that the NA and the RA immersed in epoxy resin is better compared to the NA and RA without immersed in epoxy resin. According to Neville & Brooks (2012), the value of ACV influences the compressive strength of concrete. Higher value of ACV, the lower the compressive strength will be and vice verse.

**Table 3.** Aggregate crushing value results

| Sample | ACV (%) |
|---|---|
| NA | 16.7 |
| RA | 31.1 |
| NA immersed in epoxy resin | 15.9 |
| RA immersed in epoxy resin | 26.4 |

### 3.3 Compressive Strength

Figure 3 shows the results for the concrete mixtures with different percentages of NA immersed in epoxy resin. The Figure 3 illustrated the relationship between the compressive strength of the concrete mixtures at 7 and 28 days and the increase of the percentage of NA immersed in epoxy resin. The strength of the concrete mixtures started to increase from the mixture which contain (5%) of normal aggregates immersed in epoxy resin for both ages. After that, it continues to increase obviously for both ages by increasing the percentage of (NA) immersed in epoxy resin. The highest concrete mixture strength was recorded for the mixture with (20%) NA immersed in epoxy resin and it was 39 MPa.

These results indicate that the NA immersed in epoxy resin were stronger compared to the NA without immersed in epoxy resin. This result was expected due to the ACV result. It shows that epoxy resin improved the compressive strength of concrete.

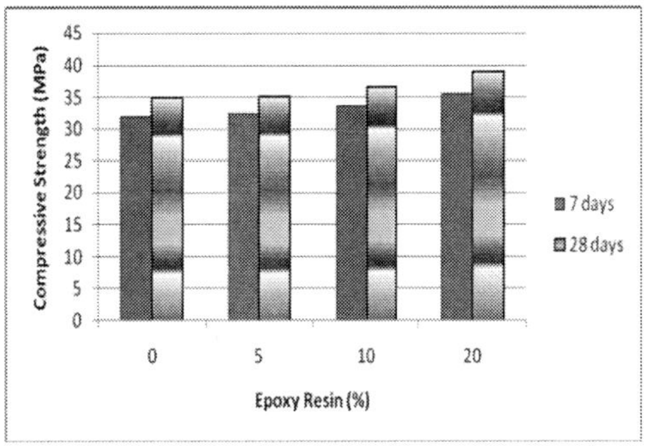


**Figure 3.** Compressive Strength Result For Normal Aggregate Concrete

DOI: 10.1051/matecconf/20178701003

On the other hand, Figure 4 shows the relationship between the compressive strength of the RA concrete immersed with various percentages of epoxy resin for 7 and 28 days. From this figure, it can be seen that the strength of the concrete mixture which contain of RA immersed in epoxy resin slightly increased for both ages. It shows that the strength increases as percentage of epoxy resin increases.
It can be stated that epoxy resin contribute to the better bonding behavior and extra strength towards compressive. The highest concrete mixture strength was recorded for the mixture with (20%) RA immersed in epoxy resin and it was 37 MPa.

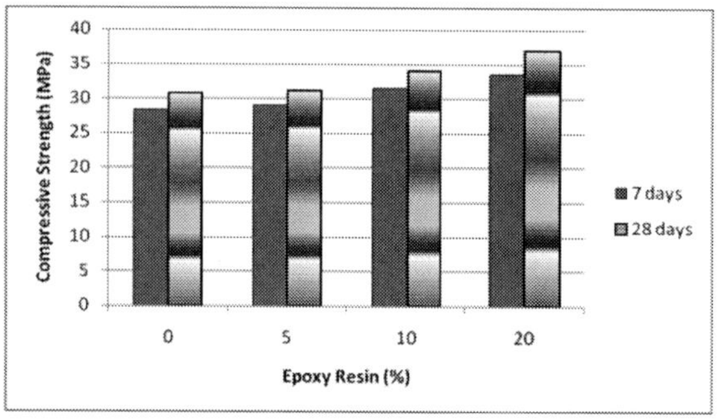


**Figure 4**. Compressive Strength Result For Recycled Aggregate Concrete

### 3.4 Water Absorption

Table 4 shows water absorption results for all concrete samples. For concrete mixtures used RA showed that the absorption decreased by increasing the percentage of epoxy resin. The recycled aggregates concrete have higher water absorption compared to normal aggregate concrete. This may be due to the mortar that is attached to RA which led into greater porosity and lead into higher water absorption characteristic of RA (Chen *et al.*, 2003). On the other hand, Cakur and Sofyanh (2015) revealed that the high absorption characteristic of RA that developed the higher osmosis pressure. This higher osmosis pressure have a tendency to absorb more water from the surrounding of the samples.
In this study, RA immersed in epoxy resin have lower in water absorption compared to RA without immersed in epoxy resin. As the epoxy resin increases, the water absorption of RA concrete is decreases. However, the concrete with NA without immersed in epoxy resin obtained lower in water absorption compared to NA immersed in epoxy resin. Although the NA without immersed in epoxy resin attain lower in water absorption compared to NA immersed in epoxy resin, as the epoxy resin content is increases the water absoprtion of concrete is decreases.

**Table 4.** Water absorption results for all concrete mixes

| Percentage of epoxy resin | Aggregate type | Absorption (%) |
|---|---|---|
| 0% | NA | 3.39 |
| | RA | 5.58 |
| 5% | NA | 4.40 |
| | RA | 5.18 |
| 10% | NA | 4.38 |
| | RA | 4.62 |
| 20% | NA | 3.92 |
| | RA | 4.15 |

## 4 CONCLUSION

Based on the results of this study, the following conclusions are made:

1) Aggregate impact value for recycled aggregate is higher compared to normal aggregate. Aggregate immersed in epoxy resin obtained lower in impact value than that of aggregate without immersed in epoxy resin.

2) Aggregate crushing value for recycled aggregate is higher than that of normal aggregate. Aggregate immersed in epoxy resin attained lower in crushing value compared to aggregate without immersed in epoxy resin.

3) The compressive strength value increased for all concrete mixtures with the increase of the normal and recycled aggregates immersed in epoxy resin percentage in the concrete mixture.

4) The water absorption of all the concrete mixtures have decreased with the increase of the normal and recycled aggregates immersed in epoxy resin percentage in the concrete mixture.

## References

1. S.H. Adnan, Y.L., Lee, I.A. Rahman, H. Mohd Saman, M.W. Soejose, Compressive strength of recycled aggregate concrete with various percentage of recycled aggregate. Proceeding of National Seminar on Civil Engineering Research (SEPKA 2007), Universiti Teknologi Malaysia, (2007)
2. J.M. Khatib, Properties of concrete incorporating fine recycled aggregate. Cement and Concrete Research, 35 (4), 763-769, (2005)

3. A.M.Neville, Properties of Concrete, Prentice Hall, (2012)
4. A. Gilat, R. Goldberg and G. Roberts, Strain Rate Sensitivity of Epoxy Resin in Tensile and Shear Loading. J. Aerosp. Eng., 10.1061/(ASCE)0893-1321(2007)20:2(75), 75-89, (2007)
5. A. Rao, K.N. Jha, & S. Misra, Use of aggregates from recycled construction and demolition waste in concrete. Resources, Conservation and Recycling, 50(1), 71-81, (2007)
6. F.T. Olorunsogo, & N. Padayachee, Performance of Recycled Aggregate Concrete Monitored By Durability Indexes. Cement and Concrete Research, 32 (2), 179-185, (2002)
7. A.S. Abdel-Hay, Properties of Recycled Concrete Aggregate Under Different Curing Conditions, Housing and Building National Research Center, http://dx.doi.org/10.1016/j.hbrcj.2015.07.001, (2015)
8. O. Cakir, O.O.Sofyanh, Influence of Silica Fume On Mechanical and Physical Properties of Recycled Aggregate Concrete, Housing and Building National Research Center, 11, 157-166, (2015)
9. Y. Li, X. Liu, J. Yuan, M. Wu, Toughness Improvement of Epoxy Resin Mortar By Incorporation of Ground Calcium Carbonate,Construction and Building Materials, 100, 122-128, (2015)
10. A.R.M. Ridzuan, M.F. Nuruddin, A.B.M. Diah, K.B. Kamarulzaman, (2001). Early Compressive Strength and Drying Shrinkage of Recycled Aggregate Concrete. Proceeding of Seventh International Conference On Concrete Engineering and Technology. Selangor Malaysia, pp. 51-58, (2001)
11. M.C. Limbachiya, T. Leelawat, & R.K. Dhir. Use of recycled concrete aggregate in high-strength concrete. Materials and Structures, 33, pp. 574-580, (2000).

# 1D Compressibility of DMS Treated With Cement-GGBS Blend

Suaathi Kaliannan[1a], Chee-Ming Chan[2], Azeanita Suratkon[2]

[1]*Master's Student, University of Tun Hussein Onn Malaysia, 86400, Parit Raja, Johor, Malaysia*
[2]*Senior Lecturer, University of Tun Hussein Onn Malaysia, 86400, Parit Raja, Johor, Malaysia*

**Abstract.** Great quantities of dredged marine soils (DMS) have been produced from the maintenance of channels, anchorages and for harbour development. DMS have the potential to pose ecological and human health risks and it is also considered as a geowaste. Malaysia is moving towards the sustainability approach and one of the key factors to achieve it is to reduce waste. Thus, this geowaste should be generated as a new resource to substitute soil for civil works such as for land reclamation and backfilling. This shows the improved settlement of consolidation in treated DMS. DMS is referred to as a cohesive soil which includes clayey silt, sandy clay, silty clay and organic clay. This type of soil has low strength and high compressibility. The objectives were achieved through literature review analysis and also laboratory test which was one dimensional oedometer test. On the other hand, treated DMS with more ground granulated blast furnace slag (GGBS) gives a lower settlement compared to specimen with higher percentage of cement in a treated soil. Thus this shows that cement content can be reduced in soil solidification when GGBS is added. The optimum binder ratio found was 3:7 where 3 is cement and 7 is GGBS.

## 1 Introduction

Dredging can be described as underwater excavation of soils. It is essential to maintain existing coastal areas, ports and water channels. The need of increase in waterway depths might be due to the increased demand for transporting by water. Besides that, dredging process is also used in flood control measures to maintain or improve the river or channels flow capacities. The soil dislodged from the waterways, whether from the sea, river or port is known as dredged soil. Mostly, such dredged materials consist of sands, silts, clay and other material from underwater [1-4]. The dredged soils are very similar to soft clay on land, with low loadbearing capacity, high compressibility and low permeability. These properties make the soil unsuitable for construction activities. Besides, continuous disposal of the material, whether offshore on inland, does not promote sustainable practice, considering that the material could be reused, and that the dumping procedures almost always destroys one or other environmental concerns [5]. Thus, if the dredged material is to be reused as a geo- material, the naturally weak and soft soil needs to be treated. A possible solution is the solidification technique, where binders are admixed with the wet soil to dry and stiffen it chemically. A journal reports the use of improved soil (a mixture of dredged soil and converter slag) for land reclamation in Japan [6]. This is an example of evidence that dredged materials can be treated to improve its suitability, either by improving its environmental properties or providing economic benefits [7].

**Figure 1.** Clamshell Dredger

## 2 Sample Collections

The soil samples were collected from Kuala Perlis, Perlis. The soils were dredged from the sea by using clamshell dredger. The soils were dredged at 6 – 7 m depth from the sea level. The dredged soil was temporarily stored in a barge. The soils were scooped out from the barge and placed into the sampling buckets and then transported from Kuala Perlis to laboratory. The soil samples were stored at UTHM laboratory. The soils were stored indoors to avoid sunlight and heat.

[a] Corresponding author: suaathikaliannan@gmail.com

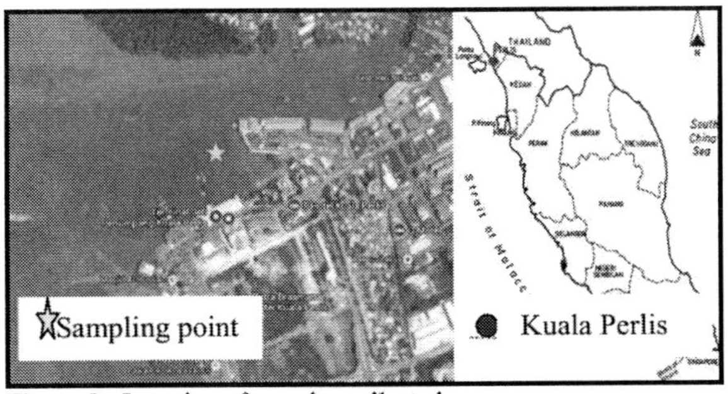

**Figure 2.** Location of samples collected

High temperature of surrounding area can destroy the organic matter, causing loss of mass, contributes to water loss and result in the inaccuracy of moisture content measurement. Then the soils were transferred from sampling bucket to 5 covered storage containers.

# 3 DMS Classification

Physical and chemical characterisations are important for describing the properties of DMS. Basic characteristics of the soil were obtained using the classification test referring to British Standard. Table 1 shows the physical and chemical characteristics of DMS sample. DMS consists of 11 % silt, 38 % sand and 51 % of clay as seen in Figure 3. According to Unified Soil Classification System (USCS), DMS falls in High Plasticity clay (CH) category.

**Table 1.** Soil Classification Result

| Parameters | Values |
|---|---|
| Moisture Content | 147 % |
| Specific Gravity | 2.66 |
| Liquid Limit | 70 % |
| Plastic Limit | 33.33 % |
| Plasticity Index | 36.67 % |
| Loss on Ignition | 10.6 |
| pH | 7.28 |
| Soil Classification | CH |

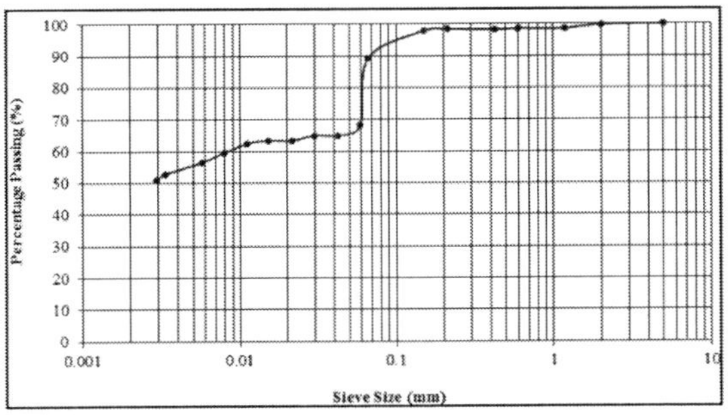

**Figure 3.** Particle size distribution of DMS

# 4 Types of Solidifying Agent

The binders are cement and ground granulated blast furnace slag (GGBS). Cement was acquired from UTHM's laboratory while GGBS was provided by YTL Corporation from the Pasir Gudang plant. All the materials were oven dried for 24 hours at 105°C and kept in an air-tight container.

## 4.1 Cement

Cement solidification has been widely used to improve the engineering properties of the clayey soils [8-10]. Cement is a hydraulic type solidifying agent. According to [11], there are two major chemical reactions which are induced by the addition of cement to clay and govern the soil cement solidification process which are the primary hydration reaction of the cement and water, and the secondary pozzolanic reactions between the limes released by the cement and the clay minerals. These cement particles bind the adjacent cement grains together during hardening and form a hardened skeleton matrix, which covers virgin soil particles. Many studies have focused on using cement-treated dredged soil for ground improvement. [8,12-14] studied the characteristics of soft clay treated cement and found that treating soft clay with a little quantity of cement can effectively reduce settlement. Cement acts as an additive and improves the quality of soil for the purpose of increasing strength and durability.

## 4.2 Ground Granulated Blast Furnace Slag

GGBS is a by-product from the blast-furnaces used to make iron. These operate at a temperature of about 1,500 degrees centigrade and are mixed with a carefully controlled mixture of iron-ore and limestone. The iron ore is reduced to iron and the remaining materials form a slag that floats on top of the iron. This slag is consistently tapped off as a molten liquid and if it is to be used for the manufacture of GGBS it has to be rapidly quenched in large volumes of water. The quenching optimises the cementitious properties and produces granules similar to coarse sand. [15]. This 'granulated' slag is then dried and ground to a fine powder. The major use of GGBS is in ready-mixed concrete. In the production of ready-mixed concrete, GGBS replaces a certain portion of the normal cement content, generally about 50% to 70%. The higher the proportion, the better is the durability. The disadvantage of the higher replacement level is that early-age strength development is somewhat slower GGBS is not only used in concrete and other applications include the in-situ solidification of soil. In South Africa, ggbs activated lime, is commonly used in binder for soil solidification [16] and there is 40 years of experience of its use.

# 5 Curing Period

The improved engineering properties of cement-treated soils are mainly attributed to the cement hardening effect. It is due to hydration, where the moisture of the soil reacts with cement for example curing period. Due to pozzolanic reaction with time, it is generally accepted that the longer the curing period, the better the strength development. The curing time also has an effect on the specimens in consolidated-drained triaxial compression test, with a longer curing time leading to a lower strain to peak strength and higher peak strength for the specimens[10]. The 7-day strength increases almost proportionately with the cement content throughout the range of cement content investigated in the test of unconfined compressive strength of treated clays [17]. As reported by [18], cement-treated Singaporean marine clay showed significant curing effect in the physical properties, unconfined compressive strength and compressibility characteristics, such as the decrease of compression index (Cc) with longer curing period.

## 6 Details of Test Specimens

Table 2 shows the proportions of binders added to the DMS. All the amounts were calculated using dry weight of DMS. Percentage of DMS was maintained at 100% for each test. The curing period for each specimen is 7 days.

**Table 2.** Test specimen for binders + DMS

| Mix Propo -rtion (C:G) | Cement content (g) | | | GGBS content (g) | | | Specimen |
|---|---|---|---|---|---|---|---|
| | 10 % | 20 % | 30 % | 10 % | 20 % | 30 % | |
| 0 : 0 | - | - | - | - | - | - | 0C0G_# |
| 0 : 100 | 0 | 0 | 0 | 20.4 | 40.8 | 61.2 | 0C10G_# |
| 15 : 85 | 3.06 | 6.13 | 9.20 | 17.4 | 34.7 | 52.0 | 1.5C8.5G # |
| 30 : 70 | 6.12 | 12.3 | 18.4 | 14.3 | 28.6 | 42.9 | 3C7G_# |
| 70 : 30 | 14.3 | 28.6 | 42.9 | 6.12 | 12.3 | 18.4 | 7C3G_# |
| 100 : 0 | 20.4 | 40.8 | 61.2 | 0 | 0 | 0 | 10C0G_# |

## 7 One -Dimensional Consolidation Test

The consolidation test, also called as one dimensional oedometer test, is used to estimate the amount of settlement under a structured load. According to BS1377-5:1990, Test 3, this method covers the determination of the magnitude and the rate of the consolidation of a saturated specimen of soil in the form of disc confined laterally, subjected to vertical axial pressure, and allowed to drain freely from the top and bottom surfaces. In this test, the soil specimen is loaded axially in increments of applied stress. Each stress increment is held constant until the primary consolidation has ceased. During this process water drains out of the specimen, resulting in a decrease in height which is measured at suitable intervals. These measurements are used for the determination of the relationship between compression (or strain) or voids ratio and effective stress, and for the calculation of parameters which describe the amount of compression and the rate at which it takes place. The test procedure is conducted according to the BS 1377-5:1990, clause 3.0.

### 7.1 Preparations of Specimens

The DMS were mixed with the required amount of binders at their natural water contents. DMS and binders were thoroughly mixed in a conventional food mixer according to the mix ratios. In order to avoid spillage, the food mixer was initially run at a low speed for 1 minute. After that, the mixing process was stopped to scrape off the materials stucked on the mixing paddle and sides of the bowl. Then, the mixing process was resumed for 4 minutes, where a homogeneous clay-additive paste was formed by then. The total mixing process took approximately 5 minutes, as recommended by [19]. The mixture was next compacted into the consolidation ring of 75 mm diameter and 20 mm height. The total amount of mixture needed to be compacted into the ring was approximately 160 g. It is to standardize the densities of oedometer specimens to ensure a fair assessment of the test results. The mixture was divided into 2 lumps of 80g each. Each lump was tamped 40 times by hand with a tamping rod to form a layer in the ring. After compaction of the 2 layers, the end of specimen was trimmed flat by using straight edged spatulas. The test specimen was in the form of a disc of proportions specified for the consolidation ring in which it is to be tested. The specimens with binders were left to cure for 7 days.

## 8 Results and Discussions

### 8.1 Analysis of $t_{50}$ and $t_{90}$

Theoretically $t_{50}$ and $t_{90}$ can be identified in all plots of settlement curves but practically $t_{50}$ and $t_{90}$ could not be determined in every plot of settlement curves. Table 3(a,b,c) shows the data of $t_{50}$ from all the samples and note that some samples could not be derived. If the specimen has sustained a certain load previously, and a smaller or same load is being applied again, the settlement curve will not have the immediate settlement, primary and secondary consolidation stages. The higher the percentage of binders, the higher the pressure needed to be applied. It is observed in time settlement plot that when a specimen is solidified, it takes a larger load to show a visible settlement process. For example in a specimen which has 10 % binders, an evident time plot curve can be seen when 25 kPa of load is applied unlike specimen with 20 % binders, the clear settlement process begins to appear at the load of 50 kPa. As for the specimen with 30 % binders, the clear curve can be seen only after 200 or 400 kPa.

**Table 3 Compilation of data for all specimens**

| a) $t_{50}$ (10 % binders) | | | | | | |
|---|---|---|---|---|---|---|
| **Load (kPa)** | **DMS** | **0C 10G** | **1.5C 8.5G** | **3C 7G** | **7C 3G** | **10C 0G** |
| 12.5 | 32 | 25 | 3.5 | 7 | 11 | 1.8 |
| 25 | 20 | 38 | 10 | 17 | 11 | 7.5 |
| 50 | 20 | 25 | 7.5 | 16 | 42 | 7 |
| 100 | 15 | 18 | 5.5 | 8 | 42 | 4.5 |
| 200 | 14 | 18 | 9 | 9.5 | 45 | 5.2 |
| 400 | 12 | 7.5 | 3.5 | 4.5 | 33 | 4.3 |
| 800 | 8.5 | 6.5 | 3.8 | 3.5 | 17 | 4.8 |

| b) $t_{50}$ (20 % binders) | | | | | | |
|---|---|---|---|---|---|---|
| **Load (kPa)** | **DMS** | **0C 10G** | **1.5C 8.5G** | **3C 7G** | **7C 3G** | **10C 0G** |
| 12.5 | 32 | 25 | - | 49 | - | - |
| 25 | 20 | 38 | - | 1.4 | - | - |
| 50 | 20 | 25 | 2.5 | 2 | 2.1 | 3.4 |
| 100 | 15 | 18 | 4 | 3.8 | 5.6 | 2.7 |
| 200 | 14 | 18 | 6.5 | 2 | 3.5 | 4 |
| 400 | 12 | 7.5 | 4 | 7.5 | 7 | 8 |
| 800 | 8.5 | 6.5 | 4 | 10 | 5.4 | 4 |

| c) $t_{50}$ (30 % binders) | | | | | | |
|---|---|---|---|---|---|---|
| **Load (kPa)** | **DMS** | **0C 10G** | **1.5C 8.5G** | **3C 7G** | **7C 3G** | **10C 0G** |
| 12.5 | 32 | 50 | - | - | - | - |
| 25 | 20 | 60 | - | - | - | - |
| 50 | 20 | 80 | - | - | - | 1.5 |
| 100 | 15 | 25 | 8 | 1.1 | - | 4.5 |
| 200 | 14 | 16 | 5.5 | 1.7 | 2 | 6 |
| 400 | 12 | - | 10 | 1.4 | 1.8 | 15 |
| 800 | 8.5 | 4 | 16 | 1.7 | 2.1 | 90 |

## 8.2 Comparison of Compression Curves

Referring to Figure 4, 5 and 6 the percentage of binders added was 10 %, 20 % and 30 % respectively. Binders refer to cement (C) and ground granulated blast furnace slag (G) In all the figures, data of 3C7G, 7C3G, 10C0G, 0C10G, 1.5C8.5G and 0C0G were compared. The ratios are 3:7, 7:3, 10:0, 0:10, 1.5:8.5 and 0:0 In Figure 4, specimen 3C7G, 7C3G and 1.5C8.5G does not show much difference whereas 10C0G remained as the lowest settlement value. This may be due to the percentage of binders (10%) which is quite minimal, thus the obvious reaction could not be noticed. The reaction most probably occurred because of the cement hydration process and also the percentage of binder was too small to activate the reaction in the specimen. On the other hand, Figure 5 and 6 have more or less the same pattern but different value of settlement. Specimen 3C7G in both graphs has a lower settlement compared to the other specimens with value of 14 % and 9 % each. Specimen 3C7G of both figure 5 and 6 has greater yield strength despite of lower cement content in it.

Initially specimen 3C7G recorded more settlement as time goes it can be observed that plot of 7C3G had a drastic drop which is nearly twice of the value of 3C7G specimen in both figure 7 and 8. Besides, specimen 0C10G which does not contain any cement mixture in all 3 figures shows a very poor resistance towards the load applied. This is because GGBS itself is not as reactive as when cement is added to it. Overall settlement reduction is shown by specimen 3C7G_20%.

The compression index (Cc) of this specimen is 0.015 where it is the lowest if compared among the other specimens in binder percentage of 20. Even though specimen 3C7G_30% has a lower Cc value which is 0.009 but the settlement reduction of specimen 3C7G_20% is sufficient thus raw materials and cost can be cut down. In general both specimens exhibited a decrease in the compression index in their respective binder percentage but not in a large scale. This can be observed in figure 7.

When soil undergoes compressive force, the development of hydration is destroying the structure and furthermore water is being dissipated, thus consolidation effects overwhelmed the. development of structure. This overall shows that a lower cement content mixed with a higher ratio of GGBS helps in lowering the settlement value. The United Kingdom it's quite common to replace between 40 and 70% of cement[20]. Cement has an apparent activation energy of 34 kJ/mol while this figure increases to 60 kJ/mol when a binder consisting of 70% GGBS and 30% Portland Cement was used [21].

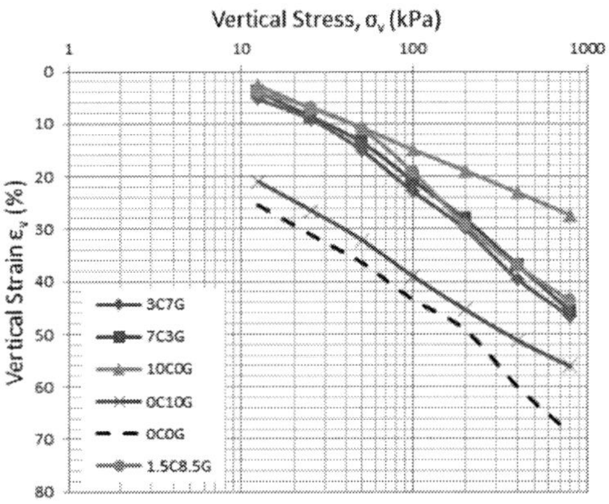


**Figure 4.** Compression curve of 10 % binders

 DOI: 10.1051/matecconf/20178701004

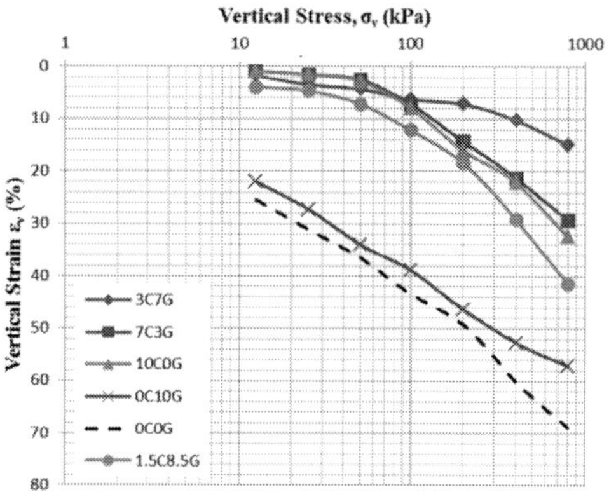


**Figure 5.** Compression curve of 20 % binders

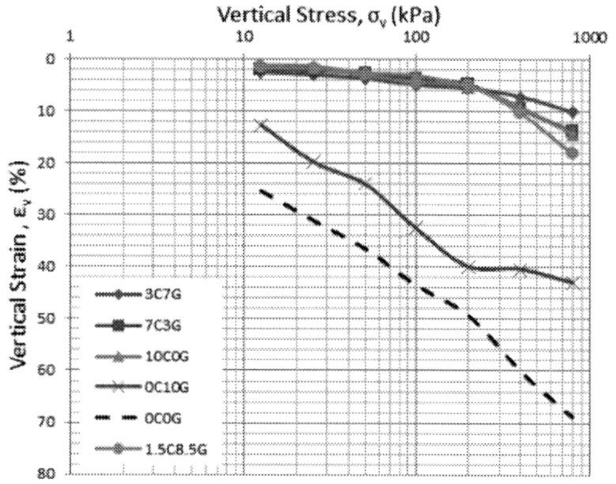


**Figure 6.** Compression curve of 30 % binders

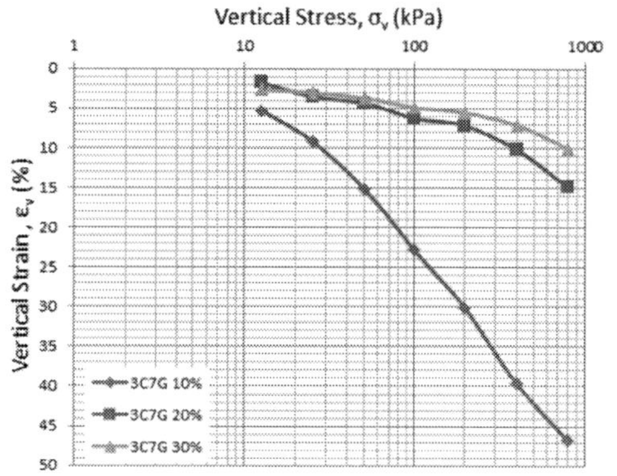


**Figure 7.** Comparison of Compression curve of 3C7G of 10%, 20% and 30% of binders

## 8.3 Cement Ratio Effects

Referring to Figure 8, x-axis represents the cement dosage in a binder which affects the choice of an optimum binder whereas y-axis shows the final settlement of each specimen according to percentage of the binders. Binders are combination of cement and GGBS. Figure 8 shows the ratio for 10%, 20% and 30% of total binder. So, notice that in Figure 8 the plot for 10% specimen is unclear as the percentage of binder is insufficient for the binder to react towards the soil whereas for plot of 20% and 30% binder, it is seen that there is an optimum ratio of cement portion which is 3. Apparently, cement ratio with 3 of 20% binder shows the optimum and a visible difference in the entire cement portion as seen in Figure 8. From this plot can be concluded that any cement portion less than 3 or more than 3 is not good. Thus the optimum mixture of binder is determined from these plots which are 20 % binder and cement portion of 3. Material usage can be saved for onsite application and it will be also cost-effective and less pollution.

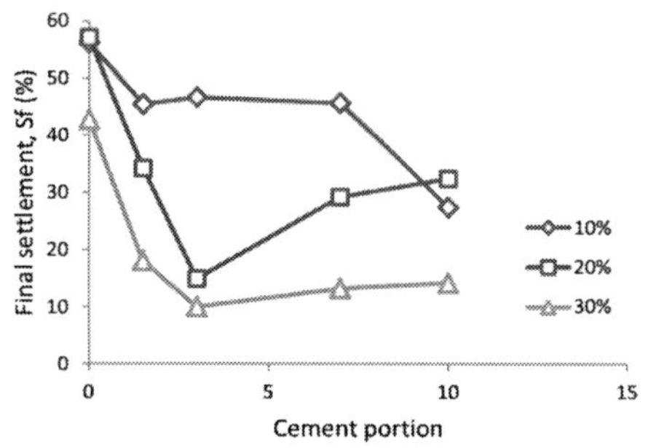


**Figure 8.** Comparison of Cement dosage in 10%, 20 % and 30 % binders

## 8.4 Coefficient of volume compressibility ($m_v$) of DMS + Binders

The coefficient volume of compressibility ($m_v$) is influenced by the chemical and physical properties of soil materials such as clay mineral composition, particle size, pore water chemistry, organic matter etc. Based on Figures 9-11, the analysis of various values of vertical effective stress, the coefficient of volume compressibility ($m_v$) values, for all solidified DMS specimens have been determined.

For DMS specimen, the value of $m_v$ decreases from 203 $cm^2/kN$ to 30 $cm^2/kN$ as the pressure increases from 12.5 kPa to 800 kPa, which shows that volume of compressibility of soil decreases with the increase of effective stress. Specimens with more percentage of binders are less compressible. It can be observed that 0C10G specimens of all percentage of binders has more or less the same pattern of plots with the DMS specimen which shows that GGBS alone is not giving a stiffening effect to the soil. The $m_v$ for DMS specimen is 203 $cm^2/kN$ and it is in the range of >15 $cm^2/kN$ classified as very high compressibility clay referring to Bella (2007). The $m_v$ value varies between 203 $cm^2/kN$ to 1 $cm^2/kN$ for various percentages of binders. The specimen that has lowest values of $m_v$ in 20 % and 30 % of binder is 3C7G which shows that the specimen is stiffer compared to other specimens. A significant trend of decrease in the value of $m_v$ with the increase of stress was observed but the decreasing rate is smaller as the percentage of binders increased. This can be attributed to the addition of cement to the DMS because the stiffer the soil the less compressible it will be. Volume compressibility of soil decreases with cement stabilization. Besides, $m_v$ decreases with increase in consolidation pressure.

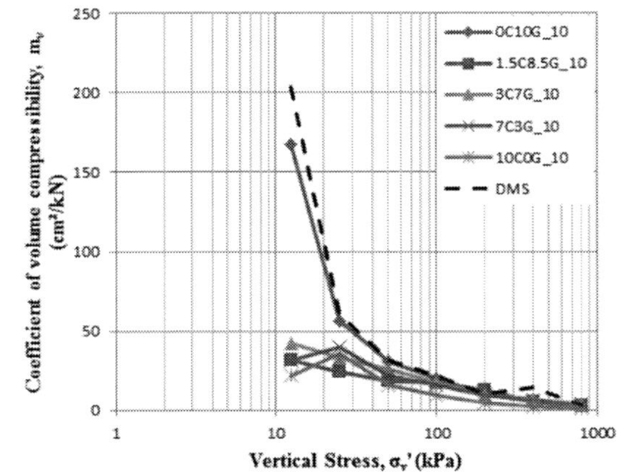


**Figure 9.** Coefficient of volume compressibility in DMS + Binder (10 %)

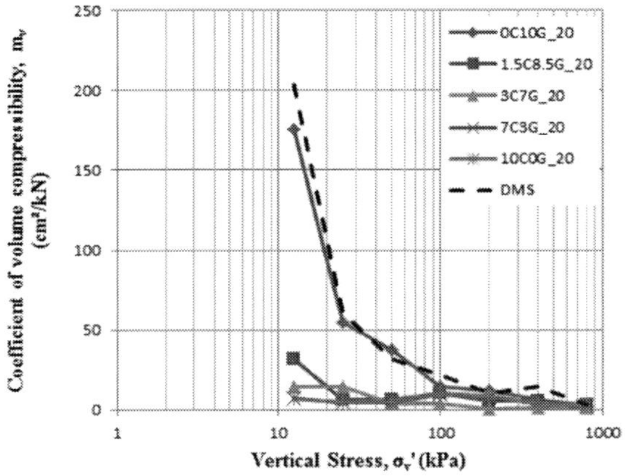


**Figure 10.** Coefficient of volume compressibility in DMS + Binder (20 %)

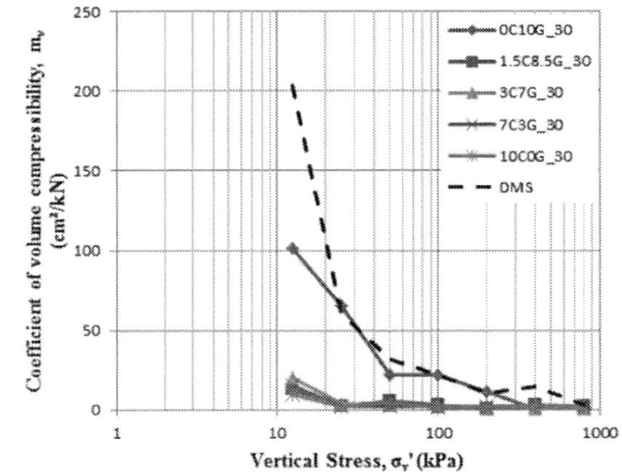


**Figure 11.** Coefficient of volume compressibility in DMS + Binder (30 %)

## 9 Conclusion

Solidification is a method which aims to improve the engineering properties of the soil. Without solidification, DMS has a very low stiffness, low loadbearing capacity, high compressibility and low permeability. The mixing of cement and GGBS helped to improve the characteristics of the DMS. The optimum binder content for the best solidification result was identified which is 3C7G_20% specimen. As seen GGBS gives an effective replacement to cement but it cannot replace cement completely. But even though it replaces partially it gives very good results and a greener approach in construction and sustainable development.

## Acknowledgement

Author thank authorities of Universiti Tun Hussein Onn Malaysia for providing the laboratory facilities and also to the field and laboratory staff. This work was supported by Office for Research, Innovation, Commercialization and Consultancy Management (ORICC), UTHM.

## References

1. Vinothkumar, R., & Arumairaj, P. D. Green Stabilization of Coimbatore Clay, 2(11), 234–240.Christine R. 1999. The Eye of the Beholder–Designing for Colour-blind Users. *British Telecommunications Engineering*. 17: 291–295. (2013).
2. Wang, D., Abriak, N. E., & Zentar, R. Strength and deformation properties of Dunkirk marine sediments solidified with cement, lime and fly ash. *Engineering Geology*, *166*, 90–99. doi:10.1016/j.enggeo.2013.09.007.(2013).
3. Azhar, A., Chan, C., & Abdkarim, A. T. Solidification Potential of Fine-Grained Dredged Marine Soils : Water-Binder Ratio Effects, *4*, 48–53. doi:10.5923/c.jce.201402.07. (2014).
4. Jong, S., & Chan, C. The Fundamental Compressibility Characteristics of Solidified Dredged Marine Soil, 597–603(2013).
5. Butt, T. E., Lockley, E., & Oduyemi, K. O. K.. Risk assessment of landfill disposal sites--State of the art. *Waste Management (New York, N.Y.)*, *28*(6), 952–64. doi:10.1016/j.wasman.2007.05.012. (2008).
6. Matsumoto.A, Tanaka.y, Nakagawa.M, Hiroki.k and Yamagoshi.Y Land reclamation using calcium oxide (cao) improved soil in japan. An information update from the Terra et Aqua,134. (2014).
7. IADC "Trailing Suction Hopper Dredgers", Factsabout - An information update from the IADC – Number 1 - 2014, International Association of Dredging Companies (IADC). (2014).
8. Feng, T. W., Lee, J.Y. & Lee, Y. J. Consolidation Behaviour of A Soft Mud Treated With Small

Cement Content. *Journal of Engineering Geology (59):* Elsevier Ltd., pp. 327-335. (2001).

9. Lorenzo, G. A. & Bergado, D. T. Fundamental Characteristics of Cement-Admixed Clay in Deep Mixing. *Journal of Materials in Civil Engineering, Vol. 18, No. 2*: ASCE, pp. 161-174. (2006).
10. Xiao, H., & Lee, F. Curing time effect on behavior of cement treated marine clay. *Int. J. Eng. Phys. Sci*, *2*(7), 427–434. (2009).
11. Bergado, D.T., Anderson, L.R., Miura, N. & Balasubramaniam, A.S. *Soft Ground Improvement In Lowland And Other Environments*. New York: ASCE Press. (1996).
12. Watabe, Y., Tsuchida, T., Furuno, T., and Yuasa, H. "Mechanical characteristics of a cement treated dredged soil utilized for waste rec- lamation land fill." Coastal geotechnical engineering in practice,A. Nakasa and T. Tsuchida, eds., Balkema, Rotterdam, Netherlands, 739–745(2000).
13. Miura, N., Horpibulsuk, S. & Nagaraj, T.S.. Engineering behavior of ce- ment stabilised clay at high water content. Soils and Foundations, Vol. 41, No. 5, pp. 33-45. (2001).
14. Kim CK, Cho WB, Lee SL, Choi WJ A study on the Consolidation Characteristic of Cohesive Soil by Plastic Index. J. Korean Geotech. Society., 24(8): 99-109. (in Korean) (2008).
15. Ha. W., & Velu Saraswathy,. Studies on the corrosion resistance of reinforced steel in concrete with ground granulated blast-furnace slag-An overview. *Journal of Hazardous Materials*, *138*(2), 226–233. doi:10.1016/j.jhazmat.2006.07.022.(2006).
16. Sherwood, P.T.,. Effect of sulphates on cement and lime treated soils, HRB Bull. No. 353 1982 pp. 98–107. (1982)
17. Chew, S. H., Kamruzzaman, a. H. M., & Lee, F. H. Physicochemical and Engineering Behavior of Cement Treated Clays. *Journal of Geotechnical and Geoenvironmental Engineering*, *130*(7), 696–706. doi:10.1061/(ASCE)1090-0241(2004)130:7(696) .(2004).
18. Kamaruzzaman, A. H. M., Chew, S. H. &Lee, F. H. Engineering behaviour of cement treated Singapore marine clay. Department of Civil Engineering, National University of Singapore, Paper 1190. (1998).
19. EuroSoilStab. *Development of Design and Construction Methods to Stabilize Soft Organic Soils (Design Guide Soft Soil Stabilization).* Brussels: CT97-0351. (2002).
20. Higgins, D. Soil stabilisation with ground granulated blastfurnace slag. *UK Cementitious Slag Makers Association (CSMA)*, (September), 1–15. Retrieved from http://www.ukcsma.co.uk/files/csma report_on_soil_ stabilisation.pdf. (2005).
21. Barnett, S. J., Soutsos, M. N., Millard, S. G., & Bungey, J. HStrength development of mortars containing ground granulated blast-furnace slag: Effect of curing temperature and determination of apparent activation energies. *Cement and Concrete Research*, *36*(3), 434–440. doi:10.1016/j.cemconres.2005.11.002. (2005).

# Suitability of Coconut Shell Concrete for Precast Cool Wall Panel-A Review

Shahiron Shahidan[1,a], Alif Syazani Leman[2], Mohamad Syamir Senin[3], Nurul Izzati Raihan Ramzi Hannan[4]

[1,2,3,4]*University Tun Hussein Onn Malaysia 86400 Batu Pahat, Johor, Malaysia*

**Abstract.** The cost for producing concrete has been increasing steadily. Therefore, the demand for using agricultural waste as a replacement or admixture inside concrete is also increasing gradually. Many researchers are doing research on agricultural waste such as palm oil, corn cobs, rice husks and coconut shells. Agricultural waste is increasingly being utilized in cement, concrete and other construction materials and provides numerous indirect benefits such as reduction in land fill cost, saving energy and protecting the environment from possible pollution. The aim is to produce concrete with improved properties at a lower cost and to maintain environmental sustainability. In this review paper, the suitability of coconut shell for concrete cool wall panels will be the main focus. Coconut shells can be used as aggregates in concrete. The characteristic properties of coconut shell concrete such as workability, bulk density, compressive strength, flexural tensile, water absorption and thermal performance were reviewed in this paper. This paper attempts to answer whether coconut shell is suitable to be used in concrete to produce a concrete cool wall panel in order to reduce heat transfer inside a building.

## 1 Introduction

The most critical issue in environment protection and natural resource conservation is waste management [1][2]. Changes in environment and an increase in population are the main causes of the many processes of deterioration which have altered the ecosystem of our planet, including the generation of municipal solid waste (MFS) [3][4]. Therefore, there is a need to reuse waste to create a greener and healthier place on earth. The usage of agricultural waste will be emphasised in this research. Being renewable, low-cost, lightweight, having high specific strength and stiffness have made agricultural waste ideal for use as construction materials [5]. Coconut shell, oil palm shell, oil palm clinker, corncob ash, and rice husk ash are all agricultural by-products. Although some of these materials can be used as animal feed or fuel in biomass power plants or boilers of various industrial sectors to produce steam, a lot of these materials are still disposed off into landfills or burnt. This leads to serious environmental problems [6][7].

Coconut is grown in more than 90 countries in an area of 14.231 million hectares with a total production in terms of copra equivalent of 11.04 million (mt). Indonesia (25.63%), Philippines (23.91%), and India (19.20%) are the major coconut-producing countries in the world [8]. Due to the increased dumping of agricultural waste, coconut was used in concrete in this research as a filler in a precast wall.

According to Kardooni et al. [9], as the most visited country in Southeast Asia, the climate in Malaysia shows that the temperature increased by 0.18℃ per decade for over 40 years. From the data received, it is crucial to ensure that the people of Malaysia live in a cool house as the temperature has risen. It happens because of convective heat transfer from internal room surface which has a major effect on the thermal comfort, air movement and heating and cooling loads for the room [10]. Recent studies have shown the convective heat transfer varies from surface to surface in buildings, as well as with time, in response to local air flow patterns. The calculation of convection coefficients must correspond to local flow conditions in a dynamic fashion in order to accurately capture this effect [11][12].

Therefore, in order to achieve a comfortable environment, this research will emphasize on utilizing coconut shell fillers to form composite materials in constructing a wall panel that is strong and can be used to replace conventional walls. Doing so will reduce the cost of construction and make the room cooler. It will also help to reduce agricultural waste and make the world more sustainable.

## 2 Coconut Shell Concrete

Coconut shell is one of the solid disposal wastes from agricultural activities. The use of coconut shells as one of the composite materials in the production of concrete was driven by the problem caused by the disposal of solid waste. According to Gunasekaran et. al [13], coconut shells represent more than 60% of domestic waste volume. Coconut shells present serious disposal problems

[a] Corresponding author: shahironshahidan@gmail.com

DOI: 10.1051/matecconf/20178701005

for local environments. However, these wastes can be used as potential material or replacement material in the construction industry.

Previous research by Osei et. al. [14] stated that coconut shells are by-products of coconut oil production. Coconut shells are used in the production of activated carbon due to its hardness and high carbon content. Besides, Kambli et al. [15] mentioned that coconut shells are potential candidates for the development of new composite material in concrete mix design because of their high strength and modulus properties.

In addition, the use of coconut shells in concrete production will give more benefits compared to conventional materials. If coconut shells are used for structural applications, it would not only be advantageous towards the environment, but also towards low-income families, especially in the surrounding areas of coconut plantations [16]. Figure 1.0 shows an example of coconut shells.

Figure 1.0 : Example Of Coconut Shells[12]

## 3 Properties Of Concrete

To utilize agricultural waste such as coconut shells as an admixture or a replacement in concrete mixtures.

### 3.1 Fresh Concrete

The tests were all conducted under the standard procedure of ASTM that include workability and bulk density.

#### *3.1.1 Workability*

Yerramala et al. [17] in his study examined the slump value between control concrete and coconut shell concrete that have a range between 20 to 26 mm. However, the slump value gradually decreased with the increase of coconut shell content inside the concrete. Apart from that, Gunasekaran et al. [13] concluded that coconut shell concrete has excellent workability compared to conventional concrete because of the smooth surface on one side of the shells as well as the size of coconut shells used. This is also because of the low water-cement ratio. Besides that, Rahman et al. [18] also found that the workability of coconut shell concrete is much better compared to control concrete. The reason is that the coconut shells bind better with concrete. D. Osei [14] in his research provided that the workability of true slump obtained with coconut shell added to concrete. It seems that all researches show that the workability of coconut shell concrete is far away better than normal concrete. True workability means that it is easier to handle the concrete during the mixing process.

#### *3.1.2 Bulk Density*

Pooja et al. [19] found that concrete added with coconut shells has a lower density. The 28-day air dry density was less than 2000 kg/m3 and it was within the range of structural lightweight concrete. The research is proven as M. Kaur et al. [20] concluded that coconut shells suffice as a lightweight aggregate in concrete. The replacement of coarse aggregate with coconut shells has made the concrete lightweight because the coconut shell density itself is lower than the density of coarse aggregates. Ahmeed et al. [21] found that at 30% of replacement, the density of coconut shell concrete is 2240 kg/m3. It also stated that the density of the concrete containing coconut shells decreases as the percentage of replacement increases. Apart from that Shafigh et al. [22] stated that the replacement of coarse aggregates with coconut shells can produce a lightweight concrete since it shows a value that is less than normal concrete. Besides that, a study conducted by Ganiron et al. [23] shows that a value of 1213.59 kg/m3 was obtained when coconut shells were used to replace coarse aggregate. The value that obtained by the researcher was for construction purposes such as wall panels and partitions.

### 3.2 Hardened Concrete

Hardened concrete was also tested under the standard procedure of ASTM that includes compressive strength, flexural tensile and moisture content.

#### *3.2.1 Compressive Strength*

According to Gunasekaran et al. [24] the compressive strength of the concrete containing coconut shell as a coarse aggregate replacement depends on its matrix and the particle tensile strength of the aggregate. The statement was supported when Kumar et al. [25] found that the maximum compressive strength was attained at 0% of replacement and the minimum strength was obtained at 10% replacement. The result shows that the strength decreases as the percentage of replacement increases due to the bond between them. Apart from that, Olan et al. [26] also studied the substitution of coarse aggregate with coconut shell at the gradation of 0%, 25%, 50%, 75% and 100%. The result shows the same pattern as the findings of the previous researcher that is the compressive strength also decreases as the replacement percentage increases. Besides that, B. Rajeevan et. al [27] reviewed the strength of concrete replaced with 20%,.

30%, 40%, 50% and 100% also showed a decrease pattern. Meanwhile, D. Ahlawat et. al. [28] have found that as the content of coconut shell increases, the surface area also increases and required more cement for proper bonding. It shows that since the cement content is constant, there is no extra bonding and thus the strength is reduced. However, there are certain percentages of replacement that can increase concrete strength. From the review, the range of replacement must be under 10% so that the concrete containing coconut shells has a higher strength.

#### *3.2.2 Flexural Strength*

Based on Jaya et al. [29], it was found that the flexural tensile strength value obtained was 6.52 MPa. This is due to the size of coconut shell used (12.5 – 10 mm) which was tougher and harder than small sized particles. The bigger size used (10 mm and above) will result in lower abrasion and higher impact value as compared to smaller (0 – 5 mm) and medium size (5 – 10mm). According to R. Abinesh et al. [30] coconut shell concrete can be used as the load factor against failure for reinforcement ratios for flexural strength is up to 3.14 %. Therefore, it will simultaneously solve environmental problems. In another study carried out by Shraddha et al. [31], flexural strength decreases as the content of coconut shell inside the concrete increases. The result was consistent with the ACI-1985 predicted results. Apart from that, Vishwas et al. [32] showed that the flexural behaviour of coconut shell concrete is similar to all other lightweight concrete. All concrete showed typical failure in flexural strength as vertical cracks appeared during pure bending tests.

#### *3.2.3 Water Absorption*

The water absorption capacity is a measure of the porosity of an aggregate. According to Olan et al. [26] water absorption for coconut shell concrete was found to be at 6.17 %. These low values show that coconut shells absorb very little amount of water during the mixing process. Besides that, Anju et. al [33] found that 6.42 % of water absorption rate was obtained for concrete grade M25 with the addition of coconut shell in it. The size used was the same as the aggregate that was replaced. In contrast, Kakade et al. [34] found that the water absorption of coconut shell concrete was found to be at 24 % . To prevent the coconut shell from absorbing more water, the coconut shells were soaked for 24 hours in water and were in a surface saturated dry condition during mixing to prevent absorption of mixing water. Besides Kakade et al. [34], Shaikh et al. [35] also found that the water absorption of coconut shell concrete was at 24 %. This is due to the higher porosity in its shell structure.

### 3.3 Thermal Performance

For a normal concrete, it is basically can resist at 300℃ and will start to rapidly lose it strength if being subjected to the hot temperature for a long period of time [36]. Therefore, changes needed to be done so that the concrete can resist at a higher temperature. According to Ganiron et al. [23], coconut shell concrete can resist high temperature conditions. It was subjected to compressive strength to check whether the strength still stayed the same and the result shows that it was still the same at a load capacity of 50.47 kN. Apart from hot temperature, Ganiron et al. [23] placed the sample in cold weather to check if coconut shells can resist freezing. It showed that the concrete has a load capacity of 96.69 kN. Besides that, Muthusamy et al. [37] stated that coconut shell concrete will reduce in strength after being exposed to high temperatures ranging from 100℃ to 800℃. This is due to the expulsion of free water from micro pores and the dehydration of concrete due to high temperature. Another research done by O.M et al. [38] stated that as the percentage of admixture increases, the thermal conductivity and the thermal diffusivity increases. However, the specific heat capacity and the thermal effusivity will decrease. It shows that the admixture or replacement has a serious effect on the cooling load of a building space. Besides that, Zuki et al. [39] had founded that the size of specimen effected the temperature. The larger the size posses greater mass of concrete thus lowering the overall temperature.

## 4 Discussion

Concrete is considered as the major cause of natural material deployment [40-44]. Therefore, many researchers have done studies on agricultural waste to overcome the problem. One of the most famous agricultural waste used is the coconut shell. Based on all the studies that has been reviewed above, many conclusions can be made which include:

### 4.1 Workability

The workability of the coconut shell concrete was superb in comparison to normal concrete. The workability here refers to the true slump that was obtained during the slump test. The workability will decrease as the percentage of replacement or admixture increases. This means that an optimum value of mixing must be obtained in order to obtain better results. The contradiction between Yerramala and Gunasekaran statement is due to the type of coconut shell used. In some countries the properties might differ because of the soil type that affect the result

### 4.2 Bulk Density

The bulk density of concrete that is added with coconut shell will have a value below 2400 kg/m$^3$. It shows that the concrete will fall under a lightweight concrete class. It is good for construction because lightweight concrete gives more advantages and is easier to handle.

DOI: 10.1051/matecconf/20178701005

### 4.3 Compressive Strength

The strength of coconut shell concrete will reduce when the percentage of coconut shell increases. The size that was used for testing the concrete was 20 mm. However, an observation was made where the optimum percentage to be used as an admixture or replacement was below 10 %. The range used will increase the compressive strength and thus makes coconut shells suitable to be used in concrete.

### 4.4 Flexural Strength

The flexural strength of normal concrete and coconut shell concrete is similar. Coconut shells of a bigger size yield better results since there is lower abrasion and higher impact value. However, a conflict in previous studies was observed when other researchers found that the addition of coconut shell to concrete will reduce its flexural strength. Therefore, further research needs to be conducted to evaluate the flexural strength of coconut shell concrete.

### 4.5 Water Absorption

The water absorption of coconut shell concrete shows a lower value that means that the coconut shell absorbs a small amount of water during the mixing process. Other researchers found that the coconut shell absorbs about 24 % of water. This problem, however, can be countered by soaking the coconut shell in water for about 24 hours. This treatment allows the coconut shell to absorb only a small amount of water during mixing. The high contrast between the researchers shows that it has a different method used by both researchers. The coconut shell that is not treated will absorb more water compared to the threated coconut shell. The method used played an important role during the mixing process.

### 4.6 Thermal Performance

The thermal performance of coconut shell concrete yields a better result than normal concrete. It can withstand a heat of about 800℃ and only reduces a small amount of compressive strength. It also has some issues in thermal conductivity, thermal diffusivity, specific heat capacity and thermal effusivity. It can be controlled by adding a suitable amount of coconut shells to concrete.

## 5 Conclusions

Based on the review above, the coconut shell is indeed suitable to be used as an admixture or as a replacement for making a concrete cool wall panel. Further research needs to be conducted to find the optimum value and size of particles that need to be used as there is a gap in the literature.

## Acknowledgements

The authors would like to thank Universiti Tun Hussein Onn Malaysia for Grant GPPS U449 and Grant Vot U523.

## References


1 M. A. Mir, P. T. Ghazvinei, N. M. N. Sulaiman, N. E. A. Basri, S. Saheri, N. Z. Mahmood, A. Jahan, R. A. Begum, and N. Aghamohammadi, "Application of TOPSIS and VIKOR improved versions in a multi criteria decision analysis to develop an optimized municipal solid waste management model," J. Environ. Manage, vol. **166**, pp. 109–115, (2016).

2 M. Abdul Rahim, N. M. Ibrahim, Z. Idris, Z. M. Ghazaly, S. Shahidan, N. L. Rahim, L. A. Sofri, and N. F. Isa, "Properties of Concrete with Different Percentange of the Rice Husk Ash (RHA) as Partial Cement Replacement," Mater. Sci. Forum, vol. **803**, pp. 288–293, (2014).

3 M. Osman, M. Nasir, and M. A. Mujeebu, "Assessment of municipal solid waste generation and recyclable materials potential in Kuala Lumpur, Malaysia," Waste Manag., vol. **29**, no. 7, pp. 2209–2213, (2009).

4 S. Shahidan, R. Pullin, K. M. Holford, M. B. N, and N. Nor, "Quantitative Evaluation of the Relationship between Tensile Crack and Shear Movement in Concrete beams," Adv. Mater. Res., vol. **626**, pp. 355–359, (2013).

5 Y. Zhao, C. Xu, C. Xing, X. Shi, L. M. Matuana, H. Zhou, and X. Ma, "Fabrication and characteristics of cellulose nanofibril films from coconut palm petiole prepared by different mechanical processing," Ind. Crops Prod., vol. **65**, pp. 96–101, (2015).

6 A. Rashad, "Cementitious materials and agricultural wastes as natural fi ne aggregate replacement in conventional mortar and concrete," J. Build. Eng., vol. **5**, pp. 119–141, (2016)

7 S. A. Kudus, N. M. Bunnori, S. R. Basri, S. Shahidan, M. N. M. Jamil, and N. M. Noor, "An Overview Current Application of Artificial Neural Network in Concrete," Adv. Mater. Res., vol. **626**, pp. 372–375, 2012.

8 D. Verma and P. C. Gope, The use of coir/coconut fibers as reinforcements in composites. (2015).

9 R. Kardooni, S. Binti, and F. Binti, "Renewable energy technology acceptance in Peninsular Malaysia," Energy Policy, vol. **88**, pp. 1–10, (2016).

10 H. B. Awbi, "Calculation of convective heat transfer coefficients of room surfaces for natural

convection," Energy Build., vol. **28**, no. 2, pp. 219–227, (1998).

11 a. G. Muoz, S. B. Saidman, and J. B. Bessone, "The adaptive simulation of convective heat transfer at internal building surfaces," Build. Environ, vol. **37**, no. 8–9, pp. 791–806, (2002).

12 S. Shahidan, N. M. Bunnori, N. Md Nor, and S. R. Basri, "Damage severity evaluation on reinforced concrete beam by means of acoustic emission signal and intensity analysis," in 2011 IEEE Symposium on Industrial Electronics and Applications, (2011), pp. 337–341.

13 K. Gunasekaran, R. Annadurai, and P. S. Kumar, "Long term study on compressive and bond strength of coconut shell aggregate concrete," Constr. Build. Mater, vol. **28**, no. 1, pp. 208–215, (2012).

14 D. Y. Osei, C. Engineering, and C. Coast, "Experimental assessment on coconut shells as aggregate in concrete," vol. **2**, no. 5, pp. 7–11, (2013).

15 P. S. Kambli and S. R. Mathapati, "Application of Coconut Shell as Coarse Aggregate in Concrete : A Technical Review," vol. **4**, no. 3, pp. 498–501, (2014).

16 K. Gunasekaran, "Study on reinforced lightweight coconut shell concrete beam behavior under shear," Mater. Des., vol. **50**, pp. 293–301, Sep. (2013).

17 A. Yerramala, "Properties of Concrete with Coconut Shells as Aggregate Replacement," vol. **1**, no. 6, pp.21–31, (2012).

18 M. M. Rahman and K. Kadirgama, "Investigate the Combination of Coconut Shell and Grained," Test, no. October, pp. 49–58, (2009).

19 S. P. G. Akshay S. Shelke, Kalyani R. Ninghot, Pooja P. Kunjekar, "Coconut Shell as Partial Replacement of Coarse Aggregate in Concrete," vol. **5**, no. 3, pp. 211–214, (2014).

20 M. Kaur and M. Kaur, "A review on utilization of coconut shell as coarse aggregates in mass concrete," Int. J. Appl. Eng. Res., vol. **7**, no. 11 SUPPL., pp. 2063–2065, (2012).

21 I. Ahmed and N. Munigal, "Study on Strength Characteristics of Concrete with Partial Replacement of Cement with Flyash and Coarse Aggregate with Coconut Shell," vol. **1**, no. 12, pp. 417–425, (2015).

22 P. Shafigh, H. Bin Mahmud, M. Z. Jumaat, and M. Zargar, "Agricultural wastes as aggregate in concrete mixtures - A review," Constr. Build. Mater., vol. **53**, pp. 110–117, (2014).

23 T. U. Ganiron, "Sustainable management of waste coconut shells as aggregates in concrete mixture," J. Eng. Sci. Technol. Rev., vol. **6**, no. 5, pp. 7–14, (2013).

24 K. Gunasekaran, P. S. Kumar, and M. Lakshmipathy, "Mechanical and bond properties of coconut shell concrete," Constr. Build. Mater., vol. **25**, no. 1, pp. 92–98, (2011).

25 N. Kumar, "Strength Properties of Coconut Shell as Coarse Aggregate In Concrete," (2012).

26 E. A. Olanipekun, K. O. Olusola, and O. Ata, "A comparative study of concrete properties using coconut shell and palm kernel shell as coarse aggregates," Build. Environ., vol. **41**, no. 3, pp. 297–301, (2006).

27 B. Rajeevan and K. M. Shamjith, "A Study on the Utilization of Coconut Shell as Coarse Aggregate in Concrete," Int. J. Eng. Res. Technol., vol. **4**, no. 07, pp. 77–80, (2015).

28 D. Ahlawat and L. G. Kalurkar, "Coconut Shell as Partial Replacement of Coarse Aggregate in Concrete," IOSR J. Mech. Civ. Eng., vol. **2014**, pp. 61–64, (2014).

29 A. Jaya Prithika and S. K. Sekar, "Mechanical and fracture characteristics of Eco-friendly concrete produced using coconut shell, ground granulated blast furnace slag and manufactured sand," Constr. Build. Mater., vol. **103**, pp. 1–7, (2016).

30 A. R and G. K, "Study On Flexural Behavior of Steel-Coconut Shell Concrete - Steel Sandwich Beam using Quarry Dust," vol. **1**, no. 11, pp. 433–439, (2015).

31 D. Shraddha, F. Hitali, D. Pradeep, and S. Varpe, "Sustainable Concrete by Partially Replacing Coarse Aggregate Using Coconut Shell," J. Today's Ideas - Tomorrow's Technol., vol. **2**, no. 1, pp. 41–54, (2014).

32 S. kumar B. G. Vishwas P. Kukarni, "Comparative Study on Coconut Shell Aggregate with Conventional Concrete," Int. J. Eng. Innov. Technol., vol. **2**, no. 12, pp. 67–70, (2013).

33 Anju Mary Ealias, Rajeena A P, Sivadutt S, Asst. Prof. Life John, and and Asst. Prof. Anju Paul, "Improvement of Strength of Concrete with Partial Replacement Of Course Aggregate With Coconut Shell and Coir Fibres," IOSR J. Mech. Civ. Eng. , vol. **11**, no. 3, pp. 16–24, (2014).

34 S. A. Kakade and A. W. Dhawale, "Light Weight Aggregate Concrete By Using Coconut Shell," Int. J. Tech. Res. Appl., vol. **3**, no. 3, pp. 127–129, (2015).

35 A. S. Shaikh, S. B. Thorat, R. V Unde, P. S. Shirse, A. S. Shaikh, C. Engineering, and A. Polytechnic, "Advance Concrete-Aggregate replaced by Coconut Shell," (2015).

36 S. Shahidan, S. Salwa, M. Zuki, C. K. Keong, and J. Jayaprakash, "Repaired of Fire-Damaged Concrete-Filled Double Skin Steel Tubular (CFDST) Columns With Fiber Reinforced Polymer (FRP)," ARPN J. Eng. Appl. Sci., vol. **11**, no. 6, pp. 3718–3725, (2016).

37 S. Muthusamy and P. Kolandasamy, "Samozbijajui lagani beton na visokim temperaturama," Gradjevinar, vol. **67**, no. 4, pp. 329–338, (2015).

38 K. O.M, O. E.C, and S. O.J, "1St International Conference on Heat Transfer, Fluid Mechanics and Thermodynamics," Exp. Therm. Fluid Sci., vol. **23**, no. 3–4, pp. I–V, (2000).

39 N. Md Nor, N. Muhamad Bunnori, A. Ibrahim, S. Shahidan, and S. N. M. Saliah, "An investigation on acoustic wave velocity of reinforced concrete beam in-plane source," in Proceedings - 2011 IEEE 7th

International Colloquium on Signal Processing and Its Applications, CSPA 2011, 2011, pp. 19–22
40 S. Shahidan, H. B. Koh, A. M. S. Alansi, and L. Y. Loon, "Strength Development and Water Permeability of Engineered Biomass Aggregate pervious Concrete," vol. 7, pp. 2–7.
41 S. Shahidan, I. Isham, and N. Jamaluddin, "A Review on Waste Minimization by Adopting in Self Compacting Concrete," MATEC Web Conf., vol. **47**, pp. 1–7, (2016).
42 N. Salleh, A. R. Mohd Sam, and J. Mohd Yatim, "Flexural Behavior of GFRP RC Beam Strengthened with Carbon Fiber Reinforced Polymer (CFRP) Plate: Cracking Behavior," Appl. Mech. Mater., vol. 752–753, pp. 610–616, 2015.
43 M. H. Wan Ibrahim, N. Jamaludin, J. M. Irwan, P. J. Ramadhansyah, and a. Suraya Hani, "Compressive and Flexural Strength of Foamed Concrete Containing Polyolefin Fibers," Adv. Mater. Res., vol. 911, no. October, pp. 489–493, 2014.
44 R. P. Jaya, B. H. Abu Bakar, M. A. M. Johari, M. H. W. Ibrahim, M. R. Hainin, and D. S. Jayanti, "Strength and microstructure analysis of concrete containing rice husk ash under seawater attack by wetting and drying cycles," Adv. Cem. Res., vol. 26, no. MAY, pp. 145–154, 2014.

# TRMM Satellite Algorithm Estimates to Represent the Spatial Distribution of Rainstorms

Marina Patrick [1,a], Yau Seng Mah [1], Frederik Josep Putuhena [2], Yin Chai Wang [1] and Onni Suhaiza Selaman [1]

[1] *Universiti Malaysia Sarawak (UNIMAS), Jalan Datuk Mohammad Musa, 94300 Kota Samarahan, Sarawak, Malaysia*
[2] *Universitas Pancasila, Jl. Srengseng Sawah, Jagakarsa, Jakarta Selatan 12640, Indonesia*

**Abstract.** On-site measurements from rain gauge provide important information for the design, construction, and operation of water resources engineering projects, groundwater potentials, and the water supply and irrigation systems. A dense gauging network is needed to accurately characterize the variation of rainfall over a region, unfitting for conditions with limited networks, such as in Sarawak, Malaysia. Hence, satellite-based algorithm estimates are introduced as an innovative solution to these challenges. With accessibility to dataset retrievals from public domain websites, it has become a useful source to measure rainfall for a wider coverage area at finer temporal resolution. This paper aims to investigate the rainfall estimates prepared by Tropical Rainfall Measuring Mission (TRMM) to explain whether it is suitable to represent the distribution of extreme rainfall in Sungai Sarawak Basin. Based on the findings, more uniform correlations for the investigated storms can be observed for low to medium altitude (>40 MASL). It is found for the investigated events of Jan 05-11, 2009: the normalized root mean square error (NRMSE = 36.7 %); and good correlation (CC = 0.9). These findings suggest that satellite algorithm estimations from TRMM are suitable to represent the spatial distribution of extreme rainfall.

## 1 Introduction

The understanding of spatial distribution for rainfall, especially during extreme conditions is important for water resources planning, river basin management, hydrological and ecological applications, assessment of groundwater potential, and the design of water supply and irrigation systems [1,2].

In-situ measurements from rain gauge have been implied as the ground truth observations where it yields relatively reliable point records of precipitation [3]. However, the precipitation depth collected by the rain gauge is affected by local turbulence, variations in vertical velocities, and rapid short-term fluctuations in the moisture that flow into the area above the rain gauge [4]. There is also the possibility of sampling errors when measuring these instabilities, particularly when extrapolating the probable maximum yield. In addition, a dense network of rain gauges is also required to accurately characterize the variation of rainfall pattern over a region. Unfortunately, this is not an ideal condition for areas with a limited network or sparse distribution of gauging and meteorological stations in the tropics, such as in Sarawak, Malaysia. This has resulted to engineering and technological challenges in understanding the spatial distribution of rainfall especially during storm events.

Hence, remotely sensed satellite-based precipitation (SBP) estimates are introduced as an innovative solution to these challenges. With the free accessibility to retrieve this dataset from public domain websites, it has become a useful data source to measure rainfall for a wider coverage area at finer temporal resolution.

Among various SBP products is the Tropical Rainfall Measuring Mission (TRMM). The TRMM system is the pioneer of satellite-based program mission specifically to observe precipitation in the tropics. Operating at a low-altitude orbit of 402 km and declination of 35° to the equator, the satellite orbits the globe 16-times over the tropical regions within the 24-hour duration. TRMM Multi-Satellite Precipitation Analysis (TMPA) provides a sequential scheme of combined rainfall estimates from multiple satellites, which is also calibrated with on-ground rain gauge observations. TMPA has 0.25° x 0.25° scale of spatial resolution and temporal resolution of 3-hourly scales over the tropical region between 50° N–50° S [5].

Several studies have evaluated the different types of SBP products. For example, three SBP products: TMPA 3B42 real time (RT), TMPA 3B42 v7 and Climate Prediction Center Morphing Technique (CMORPH) were assessed to analyze the seasonal rainfall variability in Pakistan [6]. The TMPA 3B42 v7, its predecessor v6, and the North-Western South America Retrospective Simulation (OA-NOSA30) in the Pacific-Andean region of Ecuador and Peru were evaluated in order to compensate for data scarcity [7], whereby the

[a] Corresponding author: 15010110@siswa.unimas.my

performance of SBP products: TMPA 3B42 V6, V7 and RT, the NOAA/Climate Prediction Center Morphing technique (CMORPH), Hydroestimator (HYDRO) and the Combined Scheme algorithm (CoSch) were evaluated by [8]. The analysis was conducted in southern South America. In Malaysia, the performance of SBP products: TMPA 3B42 RT and v7, Global Precipitation Climatology Project (GPCP-1DD), Precipitation Estimation from Remotely Sensed Information using Artificial Neural Networks (PERSIANN-CDR), CMORPH and Asian Precipitation Highly Resolved Observational Data Integration towards Evaluation of the Water Resources (APHRODITE) were assessed by [9]. They maintained that TMPA 3B42 v7 has better accuracy to estimate rain with lower bias results. They also concluded that TMPA 3B42 products showed the closest estimations during flood events induced by heavy precipitation.

However, despite datasets calibration between ground observations and SBP products, discrepancies still do exist. Sampling frequency, the satellite's spatial and temporal resolution and width of swath, non-uniform field or sensor's view and the uncertainty to quantify the precipitation retrieval algorithm may attribute to these errors [10]. Furthermore, studies also found that modulating orographic and relief variations in higher ground elevations or mountainous regions could also affect SBP performance. For example, TRMM precipitation radar (PR) were evaluated over the mountainous regions in the southern of continental United States [11]. It was found large biases and low correlation in the satellite precipitation rates.

Thus, there is a need to evaluate the remote sensing satellite observations with ground surface of varying elevation in the tropics to verify its accuracy. Moreover, there has yet been any studies evaluating the suitability of the TRMM satellite algorithm estimates to represent the distribution of rainstorms especially during flood events in Sarawak. This paper's aim is to investigate the mentioned ground surface elevation on rainfall estimates prepared by TMPA 3B42 v7 on its suitability to represent the spatial distribution of extreme rainfalls in Sungai Sarawak Basin (SSB).

## 2 Study area

SSB is located in the south-western part of Sarawak state with an approximate coverage of 2,459 km$^2$ comprising of 2 principal tributaries, namely Sungai Sarawak Kiri and Sungai Sarawak Kanan. The two rivers confluence at Batu Kitang while the main stem flows through the capital, Kuching City. Sungai Sarawak spreads over 120 km in length separating the capital into northern and southern regions, before finally flowing into the South China Sea [12]. The site is situated in the tropics between the latitude of 1°0' N and 2°0' N and longitude of 109°30' E and 111°0' E. Digital elevation model (DEM) of SSB are extracted from NASA's Shuttle Radar Topography Mission (SRTM) and the positions of the available rain gauge stations and the study area is depicted in Figure 1.

All year round, SSB experiences wet and humid climate, characterized by high annual rainfall, humidity, and temperature. There are two monsoon seasons experienced by the state, the North-East Monsoon (NEM), comes with heavy rainfall from November to March. Meanwhile, the lesser rainfall seasons, South-West Monsoon (SWM) occurs from May to September and two transition periods occur in April and October. Average rainfall varies between 3000 to 5500 mm annually and temperature fluctuates within the range of 20° C to 36° C [13-15].

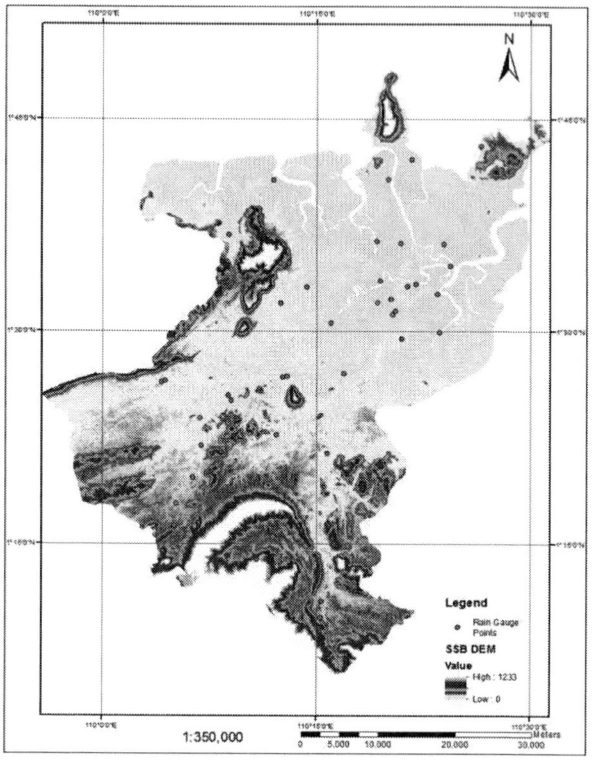


**Figure 1.** Locations of rain gauges over the DEM of SSB. The blue dot is representing the rain gauge point.

## 3 Methods

Rainfall Spatial Modelling is prepared using a Geographic Information System (GIS), whereby Table 1 summarizes the GIS thematic data layers, the data format or representation, and their sources applied in the research.

**Table 1.** GIS thematic data layers

| Data Layer | Data Representation | Data Sources |
|---|---|---|
| Digital elevation model (DEM) | 90m x 90m resolution raster | Shuttle Radar Topography Mission (SRTM) |
| Base map | JPEG Image file | Topographic map of 1:240,000 scale collated from the |

**Table 1.** (continued)

| | | |
|---|---|---|
| | | Department of Irrigation and Drainage (DID) |
| Precipitation Accumulation Maps | NetCDF | TRMM Multi-satellite Precipitation Analysis (TMPA 3B42 v7) |
| Rain Gauge Stations | Vector-point | Department of Irrigation and Drainage (DID) |

### 3.1. Ground surface observations

According to Sarawak Hydrological Station Inventory, there are 43 gauging stations available in SSB [16]. The current overall network density for the catchment area is about 57 $km^2$ per gauge, which is relatively dense within the guideline by the World Meteorological Organization (WMO). For tropical zones, the suggested minimum density of rain gauge stations are (600~900) $km^2$ and (100~250) $km^2$ for flat areas and mountainous areas, respectively [17]. Nonetheless, the gauge density in SSB is below the ranges of the typical urban rain gauge catchments, which is one gauge per (10~20) $km^2$ [18].

Hourly rainfall data collected from DID was screened through for outliers and any data that shows inconsistency was removed. Consequently, due to insufficient records, only 25 rain gauge stations were considered. Table 2 shows the details of the rain gauge stations such as the station names and the geographical coordinates of their respective locations.

**Table 2.** Details of selected rain gauge stations

| No. | Station Name | Station ID | Lat / ° | Lon / ° | El / MASL[a] |
|---|---|---|---|---|---|
| 1 | Batu Kitang | 1402047 | 1.45 | 110.28 | 60 |
| 2 | Kampung Git | 1302078 | 1.36 | 110.26 | 25 |
| 3 | Kuching Airport | 1403001 | 1.49 | 110.35 | 25 |
| 4 | Bau | 1401005 | 1.42 | 110.15 | 55 |
| 5 | Buan Bidi | 1301002 | 1.40 | 110.11 | 11 |
| 6 | Krokong | 1301074 | 1.37 | 110.11 | 58 |
| 7 | Kampung Monggak | 1301001 | 1.33 | 110.10 | 50 |
| 8 | Kampung Opar | 1400001 | 1.44 | 110.07 | 30 |
| 9 | Siniawan Water Works | 1402001 | 1.45 | 110.21 | 40 |
| 10 | Buntal DID | 1703001 | 1.70 | 110.36 | 2 |
| 11 | Kuching Saberkas | 1503004 | 1.54 | 110.34 | 80 |
| 12 | Kuching Third Mile | 1503083 | 1.52 | 110.34 | 25 |
| 13 | Siol JPS | 1603002 | 1.60 | 110.35 | 10 |
| 14 | Telok Assam | 1704013 | 1.72 | 110.44 | 5 |
| 15 | Ulu Maong | 1503008 | 1.52 | 110.34 | 5 |
| 16 | Semariang Fisheries | 1603001 | 1.61 | 110.32 | 2 |
| 17 | Rampangi | 1603058 | 1.68 | 110.33 | 10 |
| 18 | Bako Causeway | 1503001 | 1.60 | 110.40 | 13 |
| 19 | Sungai China | 1601003 | 1.61 | 110.19 | 120 |
| 20 | Matang | 1502026 | 1.58 | 110.21 | 120 |
| 21 | Sungai Rayu | 1601001 | 1.61 | 110.15 | 5 |
| 22 | Kampung Sagah | 1502003 | 1.53 | 110.21 | 39 |
| 23 | Sebubut | 1502001 | 1.58 | 110.22 | 30 |
| 24 | Padawan | 1102019 | 1.16 | 110.26 | 45 |
| 25 | Semban | 1201076 | 1.23 | 110.18 | 460 |

MASL[a] is unit of elevation in meters above sea level

Flood disasters have been recurring in Sarawak, especially in Kuching City where significant floods were reported to occur in Kuching during February 2003, January 2004, and January 2009 [19]. Expectedly, all these events had occurred during the NEM season. According to the Annual Flood Report prepared by DID, continuous heavy rain influenced by the NEM season coincided with high tide that increased the Sungai Sarawak water level and thus causing massive floods. In addition, it was reported in [20] as to how the two episodes of extreme rainfall events in early 2009 (8 – 11 January & 29 January) had caused severe flooding that had a statewide effect. It was known as the extreme year for Sarawak.

Therefore, this study is prompted to take the mentioned extreme rain events for the year 2003, 2004 and 2009 after further deliberation to investigate the distribution of rainstorms by means of spatial analysis.

### 3.2 Satellite-based precipitation

TMPA 3B42 v7 rainfall dataset retrievals are obtained from the public domain website, (http://giovanni.sci.gsfc.nasa.gov/giovanni/).

The TRMM multi-satellite orbits the globe with a 35° inclination to the equator. Therefore, the data is rearranged based on the Universal Traverse Mercator (UTM) geographical coordinate system with the datum of World Geodetic System of 1984 (WGS84). For mapping purposes, the reference coordinate system is projected to the Rectified Skew Orthomorphic (RSO) map projection system. The projected coordinate system is referred as the Timbalai 1948 RSO Borneo Meters with False Easting of 2,000,000 m and False Northing of 5,000,000 m in the transformation parameters. The RSO map projection system provides an optimum solution to reduce distortion whilst remaining conformal to suit the coordinate's computation and cadastral mapping [21].

The mission time for TRMM is measured based on Coordinated Universal Time (UTC) [22]. Therefore, it needs to be adjusted to Malaysian local time: UTC + 0800 hour [23]. The acquired data at similar instants in the local time were stratified by the local standard time at a 3-hourly time interval.

Accumulation maps of TMPA 3B42 v7 have total rainfalls computed over time for a given grid cell of continuous data variables with few or no gaps. Since the gaps are treated as the same values as 0, this may result in low bias in data with gaps [22]. Rainfall accumulation maps are generated from the website in Network Common Data (NetCDF) format. NetCDF is an interface for array-oriented data access that presents data access libraries freely for various programming languages such as C, C++, Fortran, and Java [24]. The gridded data files containing both the geographical coordinates and the accumulated rainfall depth for each grid cells are imported into the GIS platform. Each grid cell has a fine spatial resolution of 0.25° x 0.25° scale at approximate distance of 15' x 15' ≈ 27.73 km x 27.73 km. The TMPA grid cell maps have 20 cells being considered on the SSB as labelled in Figure 2. By using ArcGIS, the retrieved rainfall accumulation maps are converted as point maps to obtain rainfall accumulation depth for each grid cell.

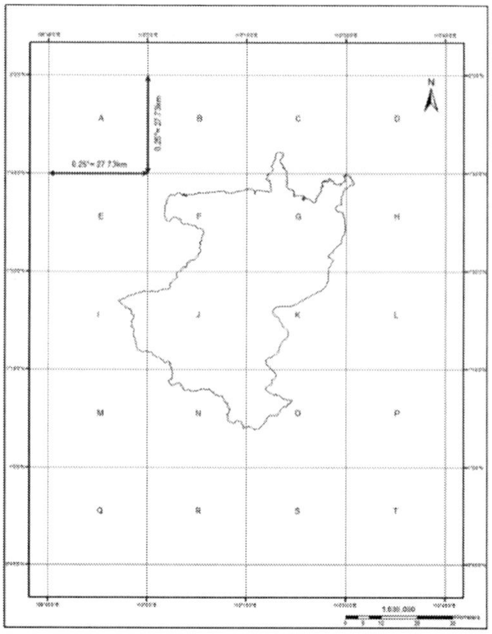


**Figure 2.** TMPA grid cells map layered on SSB

## 4 Results and discussion

### 4.1 Assessment of TMPA 3B42 v7

#### *4.1.1 Validation of TMPA 3B42 v7 with rain gauge*

Percentage Errors (PE) are differences between satellite algorithm estimates and rain gauge observations for individual stations. The output values can be either positive or negative values, where negative PE values would indicate underestimation of rainfall depth while positive PE values would indicate otherwise. The relative bias (RB) describes the systematic bias of SBP. Negative percentage of RB indicates underestimation by TMPA 3B42 v7 while, positive percentage of RB means overestimation by TMPA 3B42 v7. The root mean square error (RMSE) is used to measure the magnitude of the average error. Due to the RMSE values being inappropriate to define mean error and easily misinterpreted [25], therefore the normalized root mean square error (NRMSE), or also known as the relative RMSE, is introduced. The statistic index is used to measure the reliability of the satellite algorithm precipitation estimates. The recommended value of NRMSE is (<50 %) [26]. The Pearson's Correlation Coefficient (CC) is used to measure the agreement between precipitation estimations from satellite and observations from rain gauge. Ideally the values of CC range from (-1< CC <+1). The CC value of +1 specifies perfect positive fit while values of -1 indicate a perfect negative fit. If there is no linear correlation or weak linear correlation then the CC is near 0.

For validation of the TMPA 3B42 v7, rainfall observations from the ground surface (rain gauge) that are located within each grid cell are compared with the rainfall estimation from satellite (TMPA 3B42 v7). The accuracy of the TRMM satellite algorithm estimates are determined by the NRMSE and the CC statistic indexes are used in this study. When the estimates are deem reliable, then it is considered suitable to represent the spatial distribution of rainstorms.

The following continuous statistical methods by [9] are used to evaluate the quality of the TMPA 3B42 v7 3-hourly precipitation:

$$PE = \frac{(P_{obs} - P_{est})}{P_{obs}} \times 100 \quad (1)$$

$$RB = \frac{\sum_{i=1}^{N}(P_{est} - P_{obs})}{\sum_{i=1}^{N} P_{obs}} \times 100 \quad (2)$$

$$RMSE = \sqrt{\frac{\sum_{i=1}^{N}(P_{est} - P_{obs})^2}{N}} \quad (3)$$

$$NRMSE = \frac{RMSE}{\left(\frac{\sum_{i=1}^{N} P_{obs}}{N}\right)} \times 100 \quad (4)$$

$$CC = \frac{Cov(P_{obs}, P_{est})}{\sigma(P_{obs}) \cdot \sigma(P_{est})} \quad (5)$$

where, $Cov$ = Covariance; $\sigma(P_{obs})$ = Standard deviation of rain accumulation from ground surface observations (rain gauge); $\sigma(P_{est})$ = Standard deviation of rain accumulation from satellite estimations (TMPA 3B42 v7); $P_{obs}$ = Precipitation accumulation from ground surface observations (rain gauge); and $P_{est}$ = Precipitation accumulation from satellite estimations (TMPA 3B42 v7).

### *4.1.2 Performance assessment using continuous statistics*

Presented in Table 3, the percentage error assessment can be observed for TMPA 3B42 v7 during extreme rainfall events.

The PE between ground surface and satellite observations for storm period: a) Feb 01-07, 2003 ranges from (−49.38 to +42.12) %, b) Jan 23-29, 2004 ranges from (−2.75 to +50.11) %, c) Jan 05-11,2009 ranges from (−66.71 to +52.09) %, and d) Jan 24-30, 2009 ranges from (−46.83 to +19.49) % (Table 3).

**Table 3.** Percentage error assessments for storm events in each gauging station

| | | **Storm Period** | | | |
|---|---|---|---|---|---|
| | | **Feb 01-07, 2003** | **Jan 23-29, 2004** | **Jan 05-11, 2009** | **Jan 24-30, 2009** |
| **No.** | **Station Name** | **PE / %** | **PE / %** | **PE / %** | **PE / %** |
| 1 | Batu Kitang | 34.31 | 43.25 | 25.45 | 15.79 |
| 2 | Kampung Git | 42.12 | 42.54 | -2.07* | 1.88 |
| 3 | Kuching Airport | 31.14 | 32.38 | 15.04 | 8.84 |
| 4 | Bau | 13.13 | 9.67 | 45.34 | -28.53* |
| 5 | Buan Bidi | NA | NA | 39.85 | -28.28* |
| 6 | Krokong | 3.60 | -0.80* | 38.8 | -19.08* |
| 7 | Kampung Monggak | -7.17* | 5.83 | 39.33 | -32.37* |
| 8 | Kampung Opar | 17.05 | 34.00 | 52.09 | -37.57* |
| 9 | Siniawan Water Works | 19.17 | 14.60 | 35.46 | 11.99 |
| 10 | Buntal DID | -24.63* | 24.73 | 32.24 | -11.12* |
| 11 | Kuching Saberkas | -10.80* | 22.60 | 35.68 | -4.15* |
| 12 | Kuching Third Mile | 6.86 | 33.55 | -14.87* | -27.79* |
| 13 | Siol JPS | NA | 30.72 | 27.33 | -34.99* |
| 14 | Telok Assam | 9.06 | -2.75* | 30.70 | -16.06* |
| 15 | Ulu Maong | 4.20 | 30.15 | 13.99 | -25.79* |
| 16 | Semariang Fisheries | -39.06* | 30.09 | -40.70* | 14.60 |
| 17 | Rampangi | -49.38* | 26.04 | 26.89 | -16.98* |
| 18 | Bako Causeway | -4.94* | 30.47 | 20.74 | -18.87* |
| 19 | Sungai China | -10.18* | 20.59 | -66.71* | -29.66* |
| 20 | Matang | 5.86 | 19.23 | 35.94 | 1.86 |
| 21 | Sungai Rayu | -35.94* | 9.66 | -1.56* | -46.83* |
| 22 | Kampung Sagah | NA | NA | 38.42 | -30.31* |
| 23 | Sebubut | -7.93* | 9.94 | 22.79 | -17.28* |
| 24 | Padawan | 6.52 | 50.11 | -5.47* | -33.15* |
| 25 | Semban | 4.85 | 29.11 | 30.99 | 19.49 |

**Table 3.** (continued)

*% value is overestimation by TMPA 3B42 v7

Referring to Table 4, the RB indicates systematic underestimations conducted by the TMPA 3B42 v7 is observed during storm periods: Feb 01-07, 2003, Jan 23-29, 2004 and Jan 05-11, 2009, meanwhile systematic overestimation by TMPA 3B42 v7 is observed during storm period: Jan 24-30, 2009. The magnitude of average error and the relative RMSE are the highest during the storm period of Jan 05-11, 2009 (RMSE = 215.4 mm, NRMSE = 37.2 %). However, the RMSE does not affect the statistical correlation and all storm events have a positive fit of linear correlation whereby the storm period from Jan 23-29, 2004 has the nearest to perfect linear correlation (CC = 0.8).

**Table 4.** Statistical assessments for storm events

| | **Storm Period** | | | |
|---|---|---|---|---|
| | **Feb 01-07, 2003** | **Jan 23-29, 2004** | **Jan 05-11, 2009** | **Jan 24-30, 2009** |
| **RB / %** | -2.7 | -23.8 | -26.1 | 13.6 |
| **RMSE / mm** | 126.2 | 176.5 | 215.4 | 75.4 |
| **NRMSE / %** | 22.1 | 27.9 | 37.2 | 22.5 |
| **CC** | 0.3 | 0.8 | 0.6 | 0.7 |

To assess the performance of the satellite estimates during heavy precipitation for different elevations, the datasets are classified into 2 groups: 1) for lower altitude of (<40 MASL) elevation (Table 5); and 2) low to medium altitude of (>40 MASL) elevation (Table 6).

For lower altitudes, the RB indicates systematic overestimations by TMPA 3B42 v7 is observed during storm periods: Feb 01-07, 2003 and Jan 24-30, 2009 but systematic underestimations for the rest of the storm periods (Table 5). The RMSE and the NRMSE are the highest during the storm period of Jan 05-11, 2009, and followed by Jan 23-29, 2004, Feb 01-07, 2003 and Jan 24-30, 2009. All storm events have a positive fit of linear correlation, except for the event during Feb 01-07, 2003 which has the weakest relation (CC = -0.3), and during the storm period from Jan 23-29, 2004 has the nearest to perfect linear correlation (CC = 0.8).

For low to medium altitude, the RB is within (±33.4 %) and indicates a majority of systematic underestimations by TMPA 3B42 v7 for the rest of the storm periods, except for storm period: Jan 24-30, 2009 (Table 6). The magnitude of average error and the relative RMSE are the highest during the storm period of Jan 05-11, 2009 (RMSE = 236.5 mm, NRMSE = 36.7 %) and has the best fit of correlation (CC = 0.9). Notably, a higher CC value also leads to a larger RMSE value in this case. However, more uniform and positive correlation for all storm events can be observed for low to medium altitude (>40 MASL). Furthermore, all random errors of the TMPA 3B42 v7 estimations are (<50 %) of the measured rain gauge observations. Therefore, the satellite algorithm estimates are deemed reliable. These findings also suggest that satellite algorithm estimations from TMPA 3B42 v7 are suitable for low to medium ground elevations (>40 MASL) during extreme rainfalls. This proves that TMPA 3B42 v7 estimations have better agreement with gauging observations for higher ground levels than lower levels, which matches with other's study findings [27-29].

**Table 5.** Statistical assessments for elevation <40 MASL for storm events

| | Storm Period | | | |
|---|---|---|---|---|
| | **Feb 01-07, 2003** | **Jan 23-29, 2004** | **Jan 05-11, 2009** | **Jan 24-30, 2009** |
| **RB / %** | 0.4 | -26.7 | -19.2 | 17.5 |
| **RMSE / mm** | 150.2 | 184.7 | 197.2 | 84.9 |
| **NRMSE / %** | 25.3 | 30.7 | 37.4 | 24.0 |
| **CC** | -0.3 | 0.8 | 0.6 | 0.6 |

**Table 6.** Statistical assessments for elevation >40 MASL for storm events

| | Storm Period | | | |
|---|---|---|---|---|
| | **Feb 01-07, 2003** | **Jan 23-29, 2004** | **Jan 05-11, 2009** | **Jan 24-30, 2009** |
| **RB / %** | -6.7 | -20.5 | -33.4 | 7.8 |
| **RMSE / mm** | 89.2 | 165.1 | 236.5 | 61.3 |
| **NRMSE / %** | 16.4 | 24.6 | 36.7 | 19.7 |
| **CC** | 0.7 | 0.8 | 0.9 | 0.7 |

Generally, overall results indicate the tendency for underestimations by TMPA 3B42 v7 when compared to rain gauge observations during seasonal heavy precipitation, regardless of varying elevations. The differences between ground surface observations and satellite estimations are anticipated because the TMPA 3B42 v7 satellite grid are accumulated average SBP product with a spatial resolution of 27.73 km by 27.73 km distance, but compared to the ground observations, which is a measurement from single point rain gauge.

### 4.2 Spatial distribution of rainstorms

Spatial distribution of rain accumulation from ground surface observations (rain gauge) and satellite algorithm estimations (TMPA 3B42 v7) for the four rainstorm events are presented in Appendix A. Spatial interpolation by means of the Kriging estimation approach is used to represent the rainfall distribution during rainstorm events. Based on comparison of the rainfall maps, the locations of the storm eye for each storm period are quite similar. These results reinstate that the TRMM satellite algorithm plot has represented the spatial distribution of extreme rainfalls in SSB effectively.

Notably, for storm periods of Jan 05-11, 2009, the variability of the spatial distribution differs greatly between TMPA 3B42 v7 and rain gauge. The maximum rain accumulation during this period is recorded at 455.4 mm (TMPA 3B42 v7) and 950.5 mm (rain gauge). A large difference of TMPA 3B42 v7 and rain gauge estimation (-495.1 mm) indicates a significant gap between the two.

## 5 Conclusion

This paper compares the rainfall distribution from ground surface observations (rain gauge) and satellite algorithm estimations (TRMM). Precipitation accumulation for the 6-day duration during rainstorm events is retrieved from TMPA 3B42 v7 dataset. An investigation is then conducted to determine the suitability of TRMM satellite algorithm estimates to represent the distribution of extreme rainfalls in SSB.

Based on the investigation, the satellite algorithm datasets from TMPA 3B42 v7 are reliable and suitable for extreme rainfall estimations. Moreover, the TRMM satellite algorithm plots have also represented the spatial distribution of extreme rainfalls in SSB effectively. Nevertheless, precaution must be considered when using the data, where the results indicating the tendency of underestimations by TMPA 3B42 v7.

This research has utilized research method sourcing alternative data from public domain satellite (TRMM) that are accessible to public. Furthermore, the alternative method also provides the platform for innovative solutions to engineering and technology challenges that contribute to understanding the variation pattern of rain during extreme conditions. The findings are relevant to regions with limited gauging stations, but satellite algorithm estimates (TRMM) do not necessarily substitute the observations from rain gauge and is only useful for additional information or as an alternative data source. Hence, the accuracy and robustness of the TRMM satellite product needs to be further explored in hydrologic modelling which can determine theoretically ideal conditions.

## Acknowledgement

This study was supported by the research fund, *Dana Pelajar PhD* (DPP) (Grant number F02/DPP64/1445/2016/11) of Universiti Malaysia Sarawak (UNIMAS). The TMPA 3B42 v7 precipitation datasets are provided by the NASA/Goddard Space Flight Center and PPS, which develop and compute the TMPA 3B42 v7 dataset as a contribution to TRMM, and archived at the NASA GES DISC. The base map and rain gauge datasets are acquired from the DID.

DOI: 10.1051/matecconf/20178701006

## References

1. A. Mair & A. Fares. Comparison of rainfall interpolation methods in a mountainous region of a tropical island. J. Hydrol. Eng., **16**(4), 371–383. (2011)
2. F. Othman, A. Akbari, & A. A. Samah. Spatial rainfall analysis for an urbanized tropical river basin. Int. J. Phys. Sci., **6**(20), 4861–4868. (2011)
3. T. V. Omotosho, J. S. Mandeep, M. Abdullah, & A. T. Adediji. Distribution of one-minute rain rate in Malaysia derived from TRMM satellite data. In Ann. Geophys. **31** (11), 2013–2022. Copernicus GmbH. (2013)
4. C. J. Wiesner. *Hydrometeorology*. London, United Kingdom: Chapman and Hall Ltd. (1970)
5. G. J. Huffman, R. F. Adler, D. T. Bolvin, & E. J. Nelkin. Satellite Rainfall Applications for Surface Hydrology. In M. Gebremichael & F. Hossain (Eds.), (pp. 3–22). Dordrecht: Springer Netherlands. (2010).
6. S. I., Khan, Y., Hong, J. J. Gourley, M. U. K. Khattak, B. Yong, & H. J. Vergara. Evaluation of three high-resolution satellite precipitation estimates: Potential for monsoon monitoring over Pakistan. Adv. Space Res., **54**(4), 670–684. (2014)
7. A. Ochoa, L. Pineda, P. Crespo, & P. Willems. Evaluation of TRMM 3B42 precipitation estimates and WRF retrospective precipitation simulation over the Pacific-Andean region of Ecuador and Peru. Hydrol. Earth Syst. Sci., **18**(8), 3179–3193. (2014)
8. P. Salio, M. P. Hobouchian, Y. García Skabar, & D. Vila. Evaluation of high-resolution satellite precipitation estimates over southern South America using a dense rain gauge network. Atmos. Res., **163**, 146–161. (2015)
9. M. L. Tan, A. L., Ibrahim, D. Zheng, A. P. Cracknell, V. Chaplot, 2015. Evaluation of Six High-Resolution Satellite and Ground-Based Precipitation Products over Malaysia. Remote Sens., **7**(2), 1504–1528. (2015)
10. F. A. Semire, R. Mohd-Mokhtar, W. Ismail, N. Mohamad, & J. S. S. Mandeep. Ground validation of space-borne satellite rainfall products in Malaysia. Adv. Space Res., **50**(9), 1241–1249. (2012)
11. S. Chen, P. E. Kirstetter, Y. Hong, J. J. Gourley, Y. D. Tian, Y. C. Qi, Q. Cao, J. Zhang, K. Howard, J.J. Hu, & X.W. Xue. Evaluation of spatial errors of precipitation rates and types from TRMM spaceborne radar over the southern CONUS. J. Hydrometeorol., **14**(6), 1884–1896. (2013)
12. K. K., Kuok, Z. Z., Liew, & P. Chiu. Flood map development by coupling satellite maps and three-dimensional drafting software: Case study of the Sarawak River Basin. Water SA, **39**(1), 175–182. (2013)
13. Y. H. Lim, & L. M. Lye. Regional flood estimation for ungauged basins in Sarawak, Malaysia. Hydrol. Sci. J., **48**(1), 79–94. (2003)
14. S. L. Muhammad, W. A. W. Z. Abidin, W. Y. Chai, A. Baharun, & T. Masri. Development of Wind Mapping Based on Artificial Neural Network (ANN) for Energy Exploration in Sarawak. Int. J. Renew. Energy Res. (IJRER), **4**(3), 618–627. (2014)
15. SPU (State Planning Unit Sarawak). All about Sarawak. [Online]. http://www.spu.sarawak.gov.my/geography.html (2011) [Accessed on 19 April 2016]
16. DID (Department of Irrigation and Drainage), *Sarawak Hydrological Year Book 2014 Volume 41*. Kuching, 497p. (2016)
17. WMO (World Meteorological Organization), *Guide to Hydrometeorological Practices*, WMO 168. TP.82. *World Meteorological Organization*. Geneva, Switzerland, 323 pp. (1965)
18. B. E. Vieux. *Distributed Hydrologic Modeling Using GIS*. 2nd Edition, Vol. **48**. Springer Netherlands. (2004)
19. D. Y. S. Mah, S. H. Lai, R. Aldrino, & F. J. Putuhena. Investigative modelling of the flood bypass channel in Kuching, Sarawak, by assessing its impact on the inundations of Kuching-Batu Kawa-Bau Expressway. Struct. Infrastruct. E., **8**(7), 705–714. (2010)
20. A. Dindang, A. Taat, E. B. Phuah, A. B. M. Alwi, A. A. Mandai, S. F. Adam, F.S. Othman, D. N. Awang Bima, D. Lah. *Statistical and trend analysis of rainfall data in Kuching, Sarawak from 1968-2010*. Malaysian Meteorological Department (MMD), Ministry of Science, Technology and Innovation (MOSTI). (2013)
21. JUPEM (Jabatan Ukur dan Pemetaan Malaysia), *Garis Panduan Mengenai Sistem Rujukan Koordinat di dalam Penggunaan Global Navigation Satellite System (GNSS) Bagi Tujuan Ukur dan Pemetaan*. JUPEM, Malaysia. (2009)
22. GSFC (Goddard Space Flight Center). *Tropical Rainfall Measuring Mission Office*. [Online]. http://trmm-fc.gsfc.nasa.gov/trmm_gv/information/brochure/brochure.html. (2016). [Accessed on 20 January 2016]
23. H. Varikoden, B. Preethi, A. A. Samah, & C. A. Babu, Seasonal variation of rainfall characteristics in different intensity classes over Peninsular Malaysia. J. Hydrol., **404**(1-2), 99–108. (2011)
24. R. Rew, G. Davis, S. Emmerson, H. Davies, E. Hartnett, D. Heimbigner, & W. Fisher. Introduction and Overview. [Online]. From: http://www.unidata.ucar.edu/software/netcdf/docs/index.html (2016). [Accessed on 18 May 2016]
25. T. Chai, & R. R. Draxler, Root mean square error (RMSE) or mean absolute error (MAE)? -Arguments against avoiding RMSE in the literature. Geosci. Model Dev., **7**(3), 1247–1250. (2014).
26. S. H. Franchito, V. B. Rao, A. C. Vasques, C. M. E. Santo, & J. C. Conforte. Validation of TRMM precipitation radar monthly rainfall estimates over Brazil. J. Geophys. Res. – Atmos., **114**(2). (2009).
27. F. A. Hirpa, M. Gebremichael, & T. Hopson. Evaluation of high-resolution satellite precipitation products over very complex terrain in Ethiopia. J. Appl. Meteorol. Climatol., **49**(5), 1044–1051. (2010)
28. N. Y. Krakauer, S. M. Pradhanang, T. Lakhankar, & A. K. Jha. Evaluating satellite products for

precipitation estimation in mountain regions: A case study for Nepal. Remote Sens., **5**(8), 4107–4123. (2013)

29. S. Rana, J. McGregor, & J. Renwick. Precipitation Seasonality over the Indian Subcontinent: An Evaluation of Gauge, Reanalyses, and Satellite Retrievals. J. Hydrometeorol., **16**(2), 631–651. (2015)

DOI: 10.1051/matecconf/20178701006

## Appendix

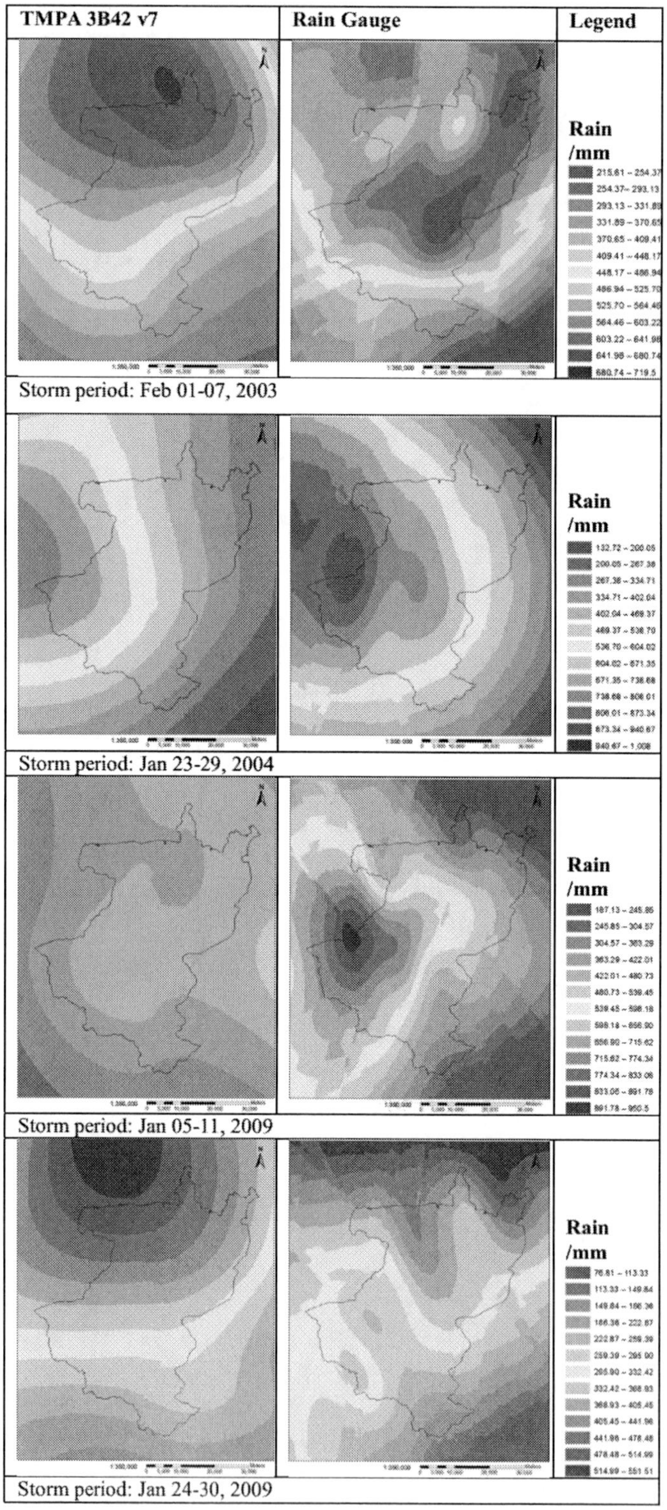


**Appendix A.** Spatial distribution of rainstorms over SSB

# A Comprehensive Review on the Effectiveness of Existing Noise Barriers commonly used in the Railway Industry

Shahiron Shahidan[1,a], Nurul Izzati Raihan Ramzi Hannan[1], Mohamad Zulkhairi Md Maarof[2], Alif Syazani Leman[1] and Mohamad Syamir Senin[1]

[1]*Faculty of Civil and Environmental Engineering, Universiti Tun Hussein Onn Malaysia, 86400 Batu Pahat, Johor, Malaysia.*
[2]*Faculty of Engineering Technology, Universiti Tun Hussein Onn Malaysia, 86400 Batu Pahat, Johor, Malaysia.*

**Abstract.** Nowadays, advanced development and sophisticated new technology have led to various types of environmental pollution such as water, air, land, thermal pollution and so on. Recently, however, noise pollution is becoming one of the major threats to the world especially in urban areas where it adversely affects the quality of life of the public. In Malaysia, the Department of Environment has identified that the average transportation noise levels in major cities in peninsular Malaysia are 71.6 dB (A) and 70.4 dB (A) during the day and night respectively. The noise is usually emitted by airplanes, trains, vehicles, motorcycles, trucks and etc. Even though rail transport requires less energy and emits less hazardous substances, it has contributed to noise pollution issues and several health hazards among urban inhabitants such as deafness, nervous breakdowns, mental disorder, heart troubles, high blood pressure, headaches, dizziness, inefficiency and insomnia. Therefore, many studies attempt to reduce noise pollution by applying noise barriers at noise polluted areas via various approaches. This paper aims to explore the effectiveness of noise barriers using noise absorption performance due to several factors such as type of absorbent materials, material thickness, density, porosity and design. This research has found that the thicker the specimen and the denser the absorbent material, the better the sound absorption performance. Besides that, barrier design also plays a major role in determining its effectiveness, where the effectiveness of noise barriers should be high and long enough to break the line-of-sight between the sound source and the receiver. There are several methods that can be used to measure the effectiveness of noise barriers such as the Adrienne Method (in-situ measurement method) and impedance tube method (laboratory measurement method) to measure the acoustic absorption. Nevertheless, the impedance tube measurement method provides the most precise results with the least measurement uncertainty as it only required small samples of the material.

## 1 Introduction

In a densely populated city with many high-rise buildings, noise pollution is one of the environmental problems that has been addressed with high priority [1]. A rapidly increasing population around the world accompanied by increasing consumption and a developing economy continue to generate noise pollution, where almost 80% of city traffic noises come from vehicles such as trains, cars, buses and motorcycles which are the major sources of noise in a modern environment. According to H. B. Huang et. al [2], there are various sources which contribute to vehicle noise which including axle-gear, tires, wind and engine. Meanwhile, Pultznerov and Izvolt [3] has identified railways as one of the most important modes of transportation for developed countries such as Korea, Japan and etc. Nowadays, rail transport has contributed to the biggest environmental noise issue even though it requires less energy and emits less hazardous substances [4]. Based on Comision & Energy [5], railways noise originates from the rolling noise, noise from traction, auxiliary system and aerodynamic noise. Noise is primarily defined as "disagreeable or undesired sound" or other disturbance. Bruneau [6] has indicated that noise and sound are constituted in the same phenomenon of atmospheric pressure fluctuations about the mean atmospheric pressure, where the differentiation is greatly subjective.

According to Hanidza et al. [7], the average level of sound pressure at night of 40 dB and 55 dB as an interim target should be the target to be achieved in order to prevent nocturnal noise deleterious health consequences which is recommended by the WHO regional office for Europe. In Malaysia, the Department of Environment has identified that the average transportation noise levels in major cities in peninsular Malaysia are 71.6 dB (A) and 70.4 dB (A) during the day and night respectively. However, a noise level above 55 dB (A) is considered as noise pollution [8]. If noise above this level lasts for an extended period of time, the efficiency and well-being of a person will be reduced [9].

Noise pollution issues have reduced the quality of life of human beings. Bistrup et al.[10] found that the

[a] Corresponding author: shahironshahidan@gmail.com

exposure of noise to humans depends on the emission of sound, how the sound is received by the human body and the setting for the emission and perception of sound. Noise in the range of 65 to 75 dB (A) causes stress to the body, which can lead to arterial hypertension (high blood pressure), cardiovascular disease and myocardial infarction (heart attack) [11-13]. Besides that, the highest level of noise will also reduce gastric secretion and cause stomach ulcers [14].

Popović et al. [15] have listed several health problems caused by noise such as uneasiness, irritability, tendency towards depression, insomnia, digestive problems, cardio-vascular diseases and deafness. Even though noise can be classified as a slow and subtle killer, it could also be so severe that it may lead to permanent loss of memory or a psychiatric disorder [16-18]. Therefore, many studies on noise pollution using different approaches are actively being conducted at the moment [19][20].

In order to reduce noise pollution caused by noise from rail transport and to protect the health and the quality of life of people living close to railroads, noise barriers could be used as it is considered as one of the best noise reduction devices. Maffei and Luigi [21] also stated that noise barriers are the main solution for noise mitigation and very often provides a good insulation. All related devices that act on airborne sound propagation (road/rail covers, claddings and added devices) are collectively called Noise Reducing Devices (NRDs) [22].

Noise barrier is a structure that is built between a noise source and a receiver which may be living or non-living. However, there are several types of materials that could be used for noise barriers such as wood, concrete, masonry, metal, gabion and transparent materials [23], where the materials used for barriers must be of an impervious material with a minimum surface density of 4lb/sq. ft to ensure that the noise barrier is effective for reducing noise levels. The effectiveness of certain noise barriers is determined by its dimensions and material used, where the noise barrier should have a transmission loss of at least 10 dB (A) greater than desired noise reduction[24]. Though the noise barrier could mitigate or reduce the noise level, it cannot block the noise from the source to the receiver completely. However, the evaluation of effectiveness of noise barriers could be determined by using the insertion loss method which is defined as differences between the measured sound pressure levels behind existing barriers and without barriers [25]. However, little emphasis has been placed on the evaluation of the effectiveness of noise barriers. This, this paper aims to determine the effectiveness of different types of existing noise barriers in the industry in terms of materials used, parameters and shapes identified by using the insertion loss method.

## 2 Methodology

According to P. Guidorzi and M. Garai [26], several methods can be used to determine the effectiveness of noise barriers. The Adrienne method is one of the methods that can be used to measure in-situ sound reflection, diffraction and airborne sound insulation. N. D Jambrosic [27] has indicated that the Adreienne method (in-situ measurement method) has been developed to measure the acoustic absorption or reflection coefficients in-situ, where this method uses one single-driver loudspeaker and one microphone in a fixed position from the loudspeaker. However, this method also has two measurements systems; in-situ measurement absorption coefficient and the second method which is basically similar to the first method, but it is used with the multiplication of impulse responses with the ratio between the estimated time of the arrival of the direct incoming sound impulse and the estimated time of the reflected incoming sound impulse (used in some commercial acoustic measurement application) such as Easera software [27] as shown in Figure 1 and Figure 2. Both systems introduce significant harmonic distortion but the Adrienne method produces less harmonic distortion than second method.

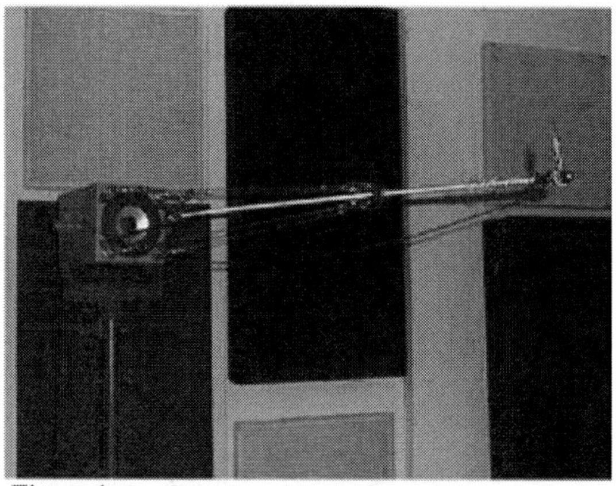

**Figure 1:** Configuration of the Adrienne method for in-situ measurement of the absorption coefficient.

**Figure 2:** Configuration of the second method

Sound sources emitted the sound wave in both methods that travel past the microphone position to the device under the test and are then reflected on it [27]. The surface undergoing the test can be an absorbing material such as noise barrier material to determine its absorption coefficient. A hard surface with a absorption coefficient of less than 0.05 is selected as a reference to be compared with the tested material. Figure 3 shows the measurement set up. The source and the microphone are 125 cm apart and the microphone is placed 25 cm away from the measured surface.

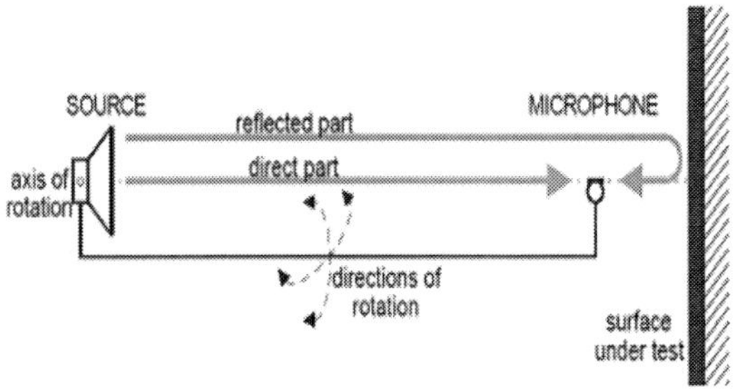


**Figure 3:** Measurement set up

The signal subtraction technique is used to separate the direct component and the reflected component by extracting the reflected component from the overall impulse response after the direct component is removed by subtracting the identical signal. For the Adrienne method, the reflected part of the impulse is being multiplied with time due to the longer time taken compared to the direct part. Meanwhile, for the second method, the correction factor is obtained by dividing the arrival time of the reflected component with the arrival time of the direct component [27]. Figure 4 shows the direct comparison of measurement between the two methods which is between the Adrienne method and the Easera software. It is obvious that the absorption coefficient is higher when the sound travels at a greater angle due to the thicker area of absorption material.

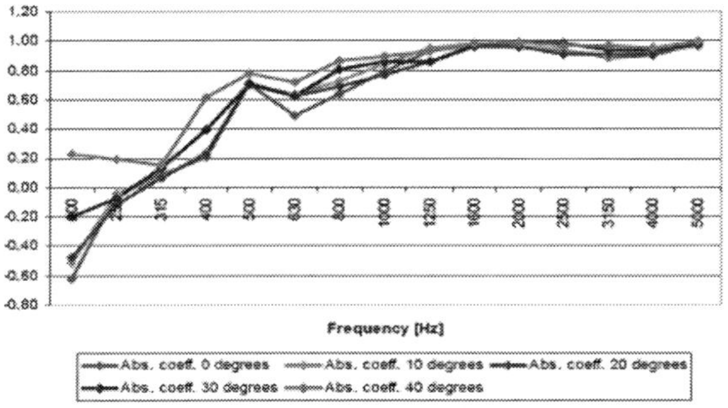


**Figure 4:** Absorption coefficient measured for various angles of incidence

Both methods are compared and they show a good agreement if the measured absorption coefficient is very high, but it shows disagreement for the lower values [27]. Besides that, laboratory measurement method also can been conducted in order to determine accurate measurement of the normal incident acoustic impedance and requires just a small sample of absorbing material to be tested, which used a standing wave tube or called as the impedance tube method (as shown in Figure 5). Using this method, acoustic waves produced by a loud speaker travel down a pipe and are reflected from the test sample. The standing wave pattern in the pipe is formed by the phase interference between the waves which are incident upon and reflected from the test sample material. The incident and reflected waves are considered to have the same amplitude if 100% of the incident wave is reflected. The incident and reflected wave have different amplitudes when some of the incident wave energy is absorbed by the test material. During the test, the sound source is moved towards the specimen and the first sound pressure level minimum is detected and the corresponding voltage is read off the measurement amplifier.

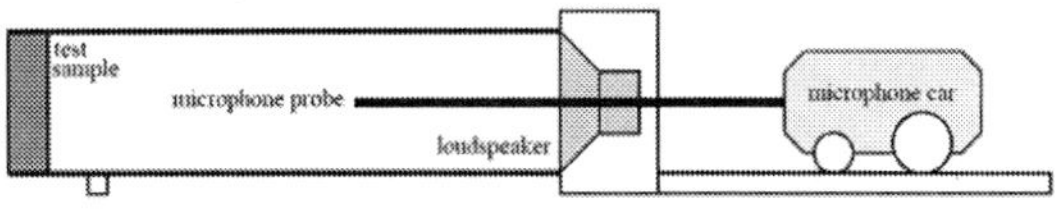


**Figure 5:** Standing wave impedance tube

According to C. Heed [28], the incidence sound absorption coefficient can be determined based on the ratio of the maximum and the minimum voltage corresponding to the measurements of the maximum to minimum sound pressure levels used to calculate the incidence sound. Therefore, it can be concluded that all methods listed are accurate measurements that have been used to measure the effectiveness of noise barriers. Nonetheless, the in-situ measurements of the absorption coefficient like the first and second method are not easy to carry out due to unfavourable measurement conditions. In addition, the main limitation of all systems based on the pass-by method and Adrienne method is the geometry of the measuring space [27], even though it is more economical in terms of measurement equipment and easy to be implemented. So, it can be concluded that the impedance tube measurement method provides the most precise results with the least measurement uncertainty where it only requires small samples of the materials to be measured either by using the standing wave ratio for discrete frequencies or by using transfer functions between two fixed points inside the tube where the pressure is measured using equalized microphones.

## 3 Noise Barrier Material and Design

### 3.1. Sound absorption material

For the last few decades, sound absorption materials have been used for reducing noise and reducing echoes in enclosed spaces [29]. Usually, absorptive materials such as porous materials are always used to reduce the reflection of sound by controlling airborne sounds. Besides that, there are several types of material which have the potential to absorb sound from several sources such as porous absorptive materials, fibrous materials and lightweight materials. Generally, porous materials are classified as fibrous medium or porous foam which is produced using various types of materials. On the other hand, fibrous media consists of high acoustics absorption and fire proof properties for several materials such as glass or rock wool [30-33]. According to T. Morimoto [34], more than 90% of air on porous absorbing materials leads to the dissipation of sound propagation through the small pores on the porous material. The performance of conventional absorbers depends on the pores and fibre structure. Conventional sound absorption materials can be expensive. However, nowadays many researchers are interested to take a similar approach in the development of sustainable sound absorbers by using waste material which has created environmental pollution.

Due to the abundance of waste materials, many researchers have taken innovative steps by exploring the potential of a variety of material as sound barriers. A. Amanda et al. [35] have investigated the potential of sugarcane bagasse for sound absorption application. Here, the comparison was done between three types of panels with and without sugarcane bagasse. The potential of coir as agriculture waste has also been investigated [36]. It was found that coir with the addition of recycled rubber is capable of absorbing sound with an average of 0.6 (sound absorption coefficient) which is caused by filler content. S. Ersoy and H. Küçük [37] have also investigated the sound absorption performance by using tea-leaf fibre and found the comparison between the different configuration of tea-leaf fibre with and without backing provided by a single layer of woven textile cloth. Meanwhile, H. K. Kim and H. K. Lee [38] have investigated the influence of lightweight aggregates on the acoustics properties of porous concrete due to different layers. The result has shown that the sound absorption coefficient for porous concrete has achieved the maximum value which is approximately 1.00. The minimum absorption coefficient of the 'double-layered porous concrete' structure is shown to be more than 0.60 with a frequency of 400 Hz or above, considering the tolerant error.

Therefore, acoustical material plays an important role in acoustics engineering in reducing the noise levels in various field. Usually, sound absorptive material is used to overcome unwanted effects of sound reflection and is also helpful in reducing the reverberation noise level [39]. Hence, the selection of sound absorbing materials was done based on several factors that will be discussed in this paper.

### 3.2 Factors influencing acoustics performance

Many researchers have conducted studies on the acoustics performance of noise barriers by using various types of sound absorptive materials. They have found several factors that will influence the absorption performance of noise barriers such as the types of materials, thickness, density, placement or position of the sound absorptive medium (noise barriers), compression, porosity and etc [29]. However, in this paper only six physical factors that have the greatest potential to influence the acoustics performance of noise barriers which are choice of materials, size of materials, porosity, thickness, density and design will be discussed.

### 3.3 Choice of materials

Noise barriers can be divided into reflective types, absorptive types, earth landscape and mixed types. Table 1 shows the details of each type mentioned.

**TABLE 1:** Types of noise barriers [24]

| Types | Descriptions |
|---|---|
| Reflective type | Transparent and non-transparent |
| Absorptive type | Sound absorbent materials and possible finishes of absorptive panels |
| Earth landscape | Nature landscaped mound and retaining structures |
| Mixed type | A combination of all types |

Despite the categorization in Table 1, the noise barriers are often divided into two groups which are either reflective or absorptive. The Environmental Impact Assessment (EIA) and Noise Impact Assessment (NIA) already determined the types of noise barriers such as reflective noise barriers, absorptive noise barriers or a combination of both.

### 3.4 Size of materials

The diameter of materials is one of the most important factors that need to be considered as it will influence the absorption coefficient during the test. According to Kouzumi et al.[40], the increase in fiber diameter tend to increase the coefficient of sound absorption as the movement of sound waves in thinner fibres occur more easily compared to thicker fibre. Meanwhile, P. R. Tahir et al, [36] used 5mm to 10 mm of recycled rubber particles as additional material and 10 to 20 mm of coir fibre in the mixture of composite boards with different percentages of fillers. It was found that the increase in the percentage of coconut coir up to 20% is suitable to be made as sound absorbing material due to the good properties of coconut coir which are cellular, light and porous. A. Putra et al. [41] conducted a study on sugarcane wasted fibres as a sustainable acoustic absorber with a diameter of 5 to 10 mm in length and it was found that sugarcane absorbers that possess ½ inch in thickness affects the acoustical performance which is better than commercial sound absorptive material. Moreover, the size of materials will also influence the acoustic absorption. H. K. Kim and H. K. Lee [38] discovered that 4 to 19 mm of aggregates have resulted in low absorption coefficients. This was due to a decrease in airflow resistance by means of friction of viscosity through the vibration of the air.

### 3.5 Thickness

Stein and Reynolds [42] concluded that the thickness of absorbent material is one of the important parameters that will affect the acoustic properties. A study on the acoustical performance of oyster shells has been conducted by E. Setyowati and G. Hardiman [43], with various dimensions of the samples which were tested to determine the absorption coefficient or sound transmission loss at 500 Hz to 1000 Hz. From the studies, it was found that the mixture with a thickness of 1 cm containing oyster shells and white cement has the largest potential to absorb sound. Meanwhile, A. Putra et al. [41] also investigated the effect of the thickness of an innovative absorbent made from tea leaf fibre and it was found that the sound absorption coefficient of tea leaf fibre with a thickness of 10mm is similar when compared with traditional absorbers such as polyester. The increase in sound absorption was due to the increase in sample thickness. Thicker samples will absorb more sound due to a longer travel distance by the impinged wave which causes it to lose more energy [44]. Besides that, H. K. Kim and H. K. Lee [45] discovered that the arrangement of material such as aggregates as absorbent material will affect the thickness of air gaps in panels. A decrease in frequency occurred due to the increasing thickness of the specimen. However, according to E. M. Samsudin et al. [29], the amount of absorption is not always proportional to the thickness of the absorbent material. It also depends on the type of materials and method of installation.

### 3.6 Density

The best sound absorption behavior depends on the density of the material which it is an important factor that needs to be considered in a research. According to A. Putra et al. [46], increasing amounts of sound absorber material added into a mixture will increase its density and subsequently improve the absorption at a higher frequency. Besides that, the density of absorptive material has shown an effect on flow resistivity. By adding the amount of absorption material into a mixture, the pores will become closer to each other which ultimately reduce the absorption capability. Meanwhile, L. Peng et al. [30] found that when the density of composite decreases, the acoustic wave reflection will be reduced due to the presence of larger voids (pores) inside the material.

Therefore, the acoustic waves in a less dense material will propagate more easily compared to denser material. In addition, the frictional and viscous resistance among the air particles and fibers inside the composite will also decrease due to the internal surface area [47-49]. Thus, most denser materials absorb more sound energy compared to less dense materials. However, in some cases, materials with higher density will absorb less sound energy due to non-fibrous characteristics of the materials. Materials that are more compact and dense are low in porosity which significantly affects its sound absorption performance

### 3.7 Porosity

Porosity is also an important factor that should be considered in this research. E. M. Samsudin et al. [29] has defined porosity as the ratio of the volume of voids to the total volume of the samples. By knowing the density of absorbent material, several methods can be used to identified the porosity of a particular absorbent material; dynamic method, static method and also by simple calculation [50]. Z. Hong et al. [51] have studied porosity by using several types of polymer perforated panels with different pore size and found that a panel containing rubber particles with a thickness of 3mm and pore size of 5 mm has a porosity of 4.75%. From the research, it is obvious that the porosity eventually affects the sound absorption performance of samples. The rubber particles have improved the sound absorption ability just like traditional porous materials such as PU foam or glass wool. According to S. Mahzan et al. [52], the absorbency of 60% recycled rubber will increase the porosity value up to 98.62% while a specimen with 100% recycled rubber has shown a reduction in porosity which is 86.59%. Therefore, it can be said that a higher content of absorbent material in a sample will fill up more voids and decrease the porosity of samples. In terms of absorption capability, when the porosity value is more than 70%, the value of absorption will usually remain constant. H. S. Seddeq [53] suggests that enough pores on the surface of material will allow sound waves to penetrate the porous material for maximum energy dissipation by friction.

### 3.8 Noise barrier design

H. Berger et al. [2] also stated that there are various factors that need to be considered in the production of noise such as traffic noise, construction noise, underwater noise as well as environmental and structural factors. The existing noise pollution mitigation methods could be improved by improving materials, parameters and shapes of existing noise barriers. Therefore, the designs of barriers also play a major role in determining its effectiveness [23][24]. Many designs have been proposed including classic wall type barrier and gabions [8]. Covering a T-shape barrier with a layer of mineral wool type absorbent material (flow resistivity r = 20,000 Nsm and thickness T = 0.1 m) could improve the performance of a single rigid barrier by 2.5 dB which is better than the performance of purely absorptive barriers [25]. The barrier must be high enough and long enough to break the line-of-sight between the sound source and the receiver [19]. Modification of the top edges of the noise barriers has been shown to be capable of increasing noise attenuation over that of a simple barrier of the same height [28]. However, the height of a noise barrier does not necessarily have to be too high but it can be as high as 3m or 4 m [20][23]. Nevertheless, noise barriers which are too long and too high cause other problems such as loss of sunlight and visual impact [26]. Therefore, the low height of noise barriers is mostly suitable for trains as the noise produced mainly originates from the rail tracks and bogie areas, which are close to the ground.

## 4 Conclusions

Effective noise barriers are usually able to reduce noise levels by 5 to 10 dB(A) and heavily depend on its dimensions as well as the location of the noise sources and the noise receivers [54-56] The ability of environmental noise barriers to reduce A-weighted noise levels depends on its design, materials, density of absorbent material, porosity and thickness [23]. A reduction as much as 5 dB(A) in noise level produced can be achieved if the noise barrier surface density exceeds 20kg/m2 and possesses a height tall enough to break the line of sight from the road to the receiver. An additional 1.5 dB (A) reduction can be achieved for each additional meter of height[23]. Therefore, some important remarks on this literature review for this research are:

1. The materials used to build certain noise barriers play a major role in reducing noise levels existing in a real situation. Different types of sound absorbent materials would result in different absorption coefficients.
2. The decrease in thickness and size of the absorbent material will affect the increase in the sound absorption coefficient.
3. Most of the findings show that denser materials will absorb more sound energy compared to less dense materials. However, when the density of the absorbent material is higher with low porosity, it will usually absorb less sound energy. Therefore, it can be concluded that density and porosity are related to each other.
4. Porosity influences the sound absorption performance of samples. Porosity decreases due to a high content of absorbent material which fills up more voids in a particular sample. However, some research has indicated that the absorption coefficient will remain constant when the porosity value is more than 70%.
5. Barrier design also play a major role where effective noise barriers should be high and long enough to break the line-of-sight between the sound source and the receiver
6. Most of the previous studies believe that the choice of materials, size of materials, porosity, thickness, density and design will influence the acoustic performance of noise barriers.
7. Various methods can be used to measure the effectiveness of noise barriers such as the Adrienne Method (in-situ measurement method) and the impedance tube method (laboratory measurement method). These methods can be used to measure acoustic absorption. However, the Adrienne method is not easy to carry out due to unfavourable measurement conditions.
8. Therefore, the impedance tube measurement method provides the most precise results with the least measurement uncertainty where it only requires small samples of the material.

To sum up, very little research focuses on the application of noise barriers for the railway industry even though it is one of the biggest contributors to noise pollution. Therefore, this warrants further research on this matter.

## Acknowledgement

The authors would like to thank Universiti Tun Hussein Onn Malaysia and fellowship for the UTHM Scholarship and GPPS (U448).

## References

1. K. M. Li, M. K. Law, and M. P. Kwok, "Absorbent parallel noise barriers in urban environments," *J. Sound Vib.*, vol. **315**, no. 1–2, pp. 239–257,(2008).
2. H. B. Huang, R. X. Li, X. R. Huang, M. L. Yang, and W. P. Ding, "Sound quality evaluation of vehicle suspension shock absorber rattling noise based on the Wigner-Ville distribution," Appl. Acoust., vol. **100**, pp. 18–25, (2015).
3. Pultznerová. A, and Ižvolt. L, "Structural Modifications, Elements and Equipment For Railway Noise Reduction", Procedia Engineering, (2014).
4. I. Tsukernikov, I. Shubin, N. Ivanov, and T. Nevenchannaya, "Features of railway noise rationing and assessment in housing estate territory in Russia," Procedia Eng., vol. **117**, no. 1, pp. 362–367, (2015).
5. E. COmision and D. G. Energy, "Assessment Study on Rail Noise Abatement Measures," Evaluation, pp. 1–129, (2007).
6. M. Bruneau, "Fundamentals of Acoustics," Fundam. Acoust., (2010).
7. Hanidza, T. I. T., Jan, A. A. M., Abdullah, R., & Ariff, M., A Preliminary Study of Noise Exposure among Grass Cutting Workers in Malaysia. Social and Behavioral Sciences, (2013)
8. N. Ozkurt, D. Sari, A. Akdag, M. Kutukoglu, and A. Gurarslan, "Modeling of noise pollution and estimated human exposure around Istanbul Atatürk Airport in Turkey," *Sci. Total Environ.*, vol. **482–483**, no. 1, pp. 486–492, (2014).
9. M. S. Hammer, T. K. Swinburn, and R. L. Neitzel, "Environmental noise pollution in the United States : Developing an effective public health response," *Environ. Health Perspect.*, vol. **122**, no. 2, pp. 115–119, (2014).
10. Bistrup, Marie Louise et al. "Health Effects Of Noise On Children And Perception Of The Risk Of Noise" (2001).
11. W. Babisch, B. Beule, M. Schust, N. Kersten, and H. Ising, "Traffic Noise and Risk of Myocardial Infarction," *Epidemiology*, vol. **16**, no. 1, pp. 33–40, (2005).
12. E. A. King, E. P. Bourdeau, X. Y. K. Zheng, and F. Pilla, "A combined assessment of air and noise pollution on the High Line , New York City," *Transp. Res. Part D*, vol. **42**, pp. 91–103, (2016).
13. N. M. Noor, H. Hamada, Y. Sagawa, and D. Yamamoto, "Effect of crumb rubber on concrete strength and chloride ion penetration resistance," J. Teknol., vol. **77**, no. 32, pp. 171–178, (2015).
14. M. Gouveia, M. Borges, J. Costa, and A. V. Carneiro, "Burden of disease from hypercholesterolemia in Portugal.," *Rev. Port. Cardiol.*, vol. **23,** no. 2, pp. 255–270, (2004).
15. Popović, Zdenka, Luka Lazarević, Ljiljana Brajovic, and Milica Vilotijević, "The Importance of Rail Inspections In The Urban Area -Aspect of Head

Checking Rail Defects", Procedia Engineering. (2015).
16. Savale, Pa."Effect of Noise Pollution On Human Being: Its Prevention and Control", Journal of Environmental Research and Development (2014).
17. S. A. Kudus, N. M. Bunnori, S. R. Basri, S. Shahidan, M. N. M. Jamil, and N. M. Noor, "An Overview Current Application of Artificial Neural Network in Concrete," Adv. Mater. Res., vol. **626**, pp. 372–375, (2012).
18. S. Shahidan, N. M. Bunnori, N. Md Nor, and S. R. Basri, "Damage severity evaluation on reinforced concrete beam by means of acoustic emission signal and intensity analysis," in 2011 IEEE Symposium on Industrial Electronics and Applications, pp. 337–341, (2011).
19. K. Mehraby, "Numerical and Analytical Investigation in Radiated Noise by a Shock-Absorber," *Int. J. Eng.*, vol. **26**, no. 12 (C), pp. 1525–1534, (2012).
20. M. Attarchi, M. Mazloumi, S. K. Sadrnezhaad, A. Jafari, and M. Asadi, "Formation and rupture of carbonate film: an electrochemical noise approach," *Anti-Corros. Methods Mater.*, vol. **56**, no. 2, pp. 103–109, (2009).
21. Maffei and Luigi, "Influence of The Design Of Railway Noise Barriers On Soundscape Perception" (2012).
22. Clairbois, Jean-Pierre and Massimo Garai. "The European Standards For Roads And Railways Noise Barriers : State of The Art 2015" (2015).
23. B. K. and C. English, "Environmental Noise Barriers. A Guide to their Acoustic and Visual Design, second edition," *J. Sound Vib.*, vol. **329**, no. 10, pp. 1982–1984, (2010).
24. E. P. D. EPD and H. D. HD, "Guidelines on Design of Noise Barriers," no. January, p. 4, (2003).
25. E. Van Haaren and P. H. Van Tol, "Validation of Ray Acoustics Applied for the Modelling of Noise Barriers," *J. Sound Vib.*, vol. **231**, no. 3, pp. 681–688, (2000).
26. P. Guidorzi and M.Garai,"Advancements in Sound Reflection and Airborne Sound Insulation Measurement on Noise Barriers." (2013).
27. K. Jambrosic, "In-Situ Measurement of the Absorption Coefficient." 1–8.
28. C. Heed , "Sound Absorption and Acoustic Surface Impedance." (1940).
29. E. M. Samsudin, L. H. Ismail, and A. A. Kadir, "A Review on Physical Factors Influencing Absorption Performance of fibrous Sound Absorption Material From Natural Fibers," vol. **11**, no. 6, pp. 3703–3711, (2016).
30. C. Wang and J. Torng, "Experimental study of the absorption characteristics of some porous fibrous materials," Appl. Acoust., vol. **62**, pp. 447–459, (2001).
31. R. P. Jaya, B. H. Abu Bakar, M. A. M. Johari, M. H. W. Ibrahim, M. R. Hainin, and D. S. Jayanti, "Strength and microstructure analysis of concrete containing rice husk ash under seawater attack by wetting and drying cycles," Adv. Cem. Res., vol. **26**, no. May, pp. 145–154, (2014).
32. S. R. Abdullah, W. R. Wan Zainal Abidin, and S. Shahidan, "Strength of Concrete Containing Rubber Particle As Partial Cement Replacement," MATEC Web Conf., vol. **47**, pp. 2–5, (2016).
33. M. Abdul Rahim, N. M. Ibrahim, Z. Idris, Z. M. Ghazaly, S. Shahidan, N. L. Rahim, L. A. Sofri, and N. F. Isa, "Properties of Concrete with Different Percentange of the Rice Husk Ash (RHA) as Partial Cement Replacement," Mater. Sci. Forum, vol. **803**, pp. 288–293, (2014).
34. T. Morimoto, "Sound absorbing materials," J. Acoust. Soc. Am., vol. **94**, no. 5, p. 3037, (1993).
35. A. Amanda, J. B. Flóreza, and F. A. Moria, "Acoustic Characterization of Sugarcane Bagasse Particleboard Panels ( Saccharum officinarum L )," vol. **18**, no. 4, pp. 821–827, (2015).
36. P. R. Tahir, A. M. A. Zaidi, N. Arsat, R. H. Mohamad Khairul Hadi Che Mohamad Khalib, and Z. F. Z. Abidin, "Potential Usage Of Agriculture Waste (Coconut Coir) For Sound Absorbing Material In Automotive Industry," (1991).
37. S. Ersoy and H. Küçük, "Investigation of industrial tea-leaf-fibre waste material for its sound absorption properties," Appl. Acoust., vol. **70**, no. 1, pp. 215–220, (2009).
38. H. K. Kim and H. K. Lee, "Influence of cement flow and aggregate type on the mechanical and acoustic characteristics of porous concrete," Appl. Acoust., vol. **71**, no. 7, pp. 607–615, (2010).
39. N. Md Nor, N. Muhamad Bunnori, A. Ibrahim, S. Shahidan, and S. N. M. Saliah, "An investigation on acoustic wave velocity of reinforced concrete beam in-plane source," in Proceedings - 2011 IEEE 7th International Colloquium on Signal Processing and Its Applications, CSPA 2011, pp. 19–22 (2011).
40. Koizumi, T., N. Tsujiuchi and A Adachi, "The Development of Sound Absorbing Materials Using Natural Bamboo Fibers, High Performance" WIT Press (2002).
41. A. Putra, Y. Abdullah, H. Efendy, W. M. Farid, M. R. Ayob, and M. S. Py, "Utilizing sugarcane wasted fibers as a sustainable acoustic absorber," *Procedia Eng.*, vol. **53**, pp. 632–638, (2013).
42. Stein, B. and Reynolds, J. S. Mechanical and Electrical Equipment for Building. New York: John Wiley and Sons (2000).
43. E. Setyowati and G. Hardiman, "The Acoustical Performances of Oyster Shell Waste Based Green Concrete Materials," J. Eng. Technol., vol. **3**, no. 3, (2015).
44. AL-Rahman, L.A. et al. Acoustic properties of innovative material from date palm fibre. American Journal of Applied Sciences, 9(9), pp. 1390-1395, (2012).

45. H. K. Kim and H. K. Lee, "Acoustic absorption modeling of porous concrete considering the gradation and shape of aggregates and void ratio," *J. Sound Vib.*, vol. **329**, no. 7, pp. 866–879, (2010).
46. A. Putra, Y. Abdullah, H. Efendy, W. M. F. W. Mohamad, and N. L. Salleh, "Biomass from paddy waste fibers as sustainable acoustic material," *Adv. Acoust. Vib.*, (2013).
47. L. Peng, B. Song, J. Wang, and D. Wang, "Mechanic and acoustic properties of the sound-absorbing material made from natural fiber and polyester," *Adv. Mater. Sci. Eng.*, (2015).
48. S. Shahidan, R. Pullin, K. M. Holford, M. B. N, and N. Nor, "Quantitative Evaluation of the Relationship between Tensile Crack and Shear Movement in Concrete beams," Adv. Mater. Res., vol. **626**, pp. 355–359, (2013).
49. S. Shahidan, H. B. Koh, A. M. S. Alansi, and L. Y. Loon, "Strength Development and Water Permeability of Engineered Biomass Aggregate Pervious Concrete," MATEC Web Conf., vol. **47**, pp. 2–7, (2016).
50. Wassilieff, C. Sound absorption of wood-based materials. Applied Acoustics, 48(4), pp. 339-356, (1996).
51. Z. Hong, L. Bo, H. Guangsu, and H. Jia, "A novel composite sound absorber with recycled rubber particles," J. Sound Vib., vol. **304**, no. 1–2, pp. 400–406, (2007).
52. S. Mahzan, a M. A. Zaidi, N. Arsat, M. N. M. Hatta, M. I. Ghazali, and S. R. Mohideen, "Study on Sound Absorption Properties of Coconut Coir Fibre Reinforced Composite with Added Recycled Rubber," Integr. Eng., no. Mechanical, Materials and Manufacturing, pp. 1–6, (2009).
53. H. S. Seddeq, "Factors Influencing Acoustic Performance of Sound Absorptive Materials," vol. **3**, no. 4, pp. 4610–4617, (2009).
54. DuPont, "Acoustic : Different kinds of materials for traffic noise barriers," *Urban Acoust.*, (2009).
55. N. Cooperative, "Guidelines for Selection and Approval of Noise Barrier Products Final Report," no. July, (2008).
56. J. Clairbois, foort de Roo, M. Garai, M. Conter, J. Defrance, crina A. Oltean, and C. Durso, "Guidebook to Noise Reducing Devices optimisation," no. December 2012, pp. 1–56, (2013).

# Conceptual Model for Systematic Construction Waste Management

Mohd Hilmi Izwan Abd Rahim[1,a] and Narimah Kasim[2]

[1]*Department of Construction Management, Faculty of Technology Management and Business, Universiti Tun Hussein Onn Malaysia, 86400 Parit Raja, Batu pahat, Johor Malaysia*
[2]*Department of Construction Management, Faculty of Technology Management and Business, Universiti Tun Hussein Onn Malaysia, 86400 Parit Raja, Batu Pahat, Johor Malaysia*

**Abstract.** Development of the construction industry generated construction waste which can contribute towards environmental issues. Weaknesses of compliance in construction waste management especially in construction site have also contributed to the big issues of waste generated in landfills and illegal dumping area. This gives sign that construction projects are needed a systematic construction waste management. To date, a comprehensive criteria of construction waste management, particularly for construction project in developing countries such as Malaysia is still not systematically defined. Therefore, construction waste management needs further investigation. This paper aims to explain a concept model for systematic construction waste management in the Malaysian construction industry. The methodology of this paper is through literature and previous research focusing on current practices of construction waste management, the barriers of construction waste management implementation and the additional factor for implementing systematic construction waste management in order to develop a conceptual model. Findings from this study will help to find out the elements required to implement systematic construction waste management in Malaysian construction industry. This action seems significant to the participants in prioritizing construction waste management more systematically. The implementation also will increase the awareness and better understanding of the construction players of the importance of construction waste management in construction projects.

## 1 Introduction

The construction industry has contributed to the negative effect of environment. The demand of the construction development especially in developing countries, has generated construction waste [1][2][3][4]. Generally, waste is a substance or item which is discarded or is expected for transfer or is required to be discarded by the procurements of laws [5]. Additionally in another opinion, waste also can be defined as unwanted or discard material produced by households, commercial, institutional and industrial activities [6][7].

Construction waste is a part of waste which can contribute to the environmental issues. Malaysia Solid Waste and Public Cleansing Management Act 2007 (Act 672) stated that the usual method for waste management in Malaysia are through directly disposal into dump or landfills [8][3]. Overall process of construction waste management in the Malaysian construction industry started from waste production in construction site, then all the waste combined and directly disposed into landfill. There is no segregation process in construction site and specialize construction waste landfill in Malaysia [9].

Previous researches had also highlighted, construction waste management practices in Malaysia are improper, which is lack of awareness among the construction parties and increasing number of illegal dumping generated from the construction waste [3][4][10]. It shows the critical construction waste management issues which is increasing year by year. Presently, per capita generation in Peninsular Malaysia is approximately 1.2 kg of municipal waste (MSW) which amounts to 19,000 tons of waste per day and the figure is set to rise to 30, 000 tons daily by 2020, in tandem with the development of country [3][11][9]. The wastes are including construction waste and all waste dumps into same landfill. From the stated data, it shows Malaysian construction industry needs systematic of waste management in order to reduce the number of waste towards the year 2020.

Additionally, the lack of construction waste management also can contributes negative impact to environmental issues, social and economy [3][4][12]. Unsystematic construction waste management also can contribute to the health and safety issue of construction workers [13]. The unsystematic of construction waste is because of the lack of awareness among the construction parties to manage the construction waste [9][11]. From the statements, it shows Malaysia needs systematic waste management to reduce the number of waste towards the year 2020. Table 1 shows average total of construction waste generated.

[a] Corresponding author: hilmi.uthm@gmail.com

DOI: 10.1051/matecconf/20178701008

**Table 1.** Average Total of Construction Waste Generated [9]

| Year | Total (Tonne/Daily) |
|---|---|
| 2014 | 7.2 |
| 2015 | 28.6 |

From the data given by SW Corp Malaysia, the critically increasing number of construction waste was generated from year 2014 to 2015. It shows that there are no reduction in numbers of produced waste from year to year. Several studies have been conducted in various states in Malaysia ranging from year 2004 until 2015 by SW Corp Malaysia [9], showed there are no reduction of construction waste but it increased year by year. Hence, construction waste management is a critical issue have to be emphasized in order to reduce the generated waste in Malaysia. Therefore, this paper is to develop a conceptual systematic model of construction waste management in order to reduce the construction waste management issues.

## 2 Literature review

This section discusses the literature of current practices and trends of construction waste in Malaysia, barriers for implementing construction waste management, and influencing factors for a systematic model on construction waste management.

### 2.1 Malaysian construction waste: Trends and current practices

Rapid development in the construction industry as a result from the expansion of the lifestyle standards, demands of construction, as well as the increasing number of population has positively contributed to increasing number of waste generation [3][14]. Construction wastes are generated by building debris, rubble, earth, concrete, steel, timber and mixed site clearance materials, arising from various construction activities [15]. Construction wastes can be hazardous, for example, asbestos which is generated during demolition of existing structures [16]. Hence, a proper and well defined policy and technologies used in managing wastes produced from the construction activities is required to reduce the adverse effect that can arose on environment, social, and economic aspects. Figure 1 shows the illegal dumping generated from construction project in Malaysia.

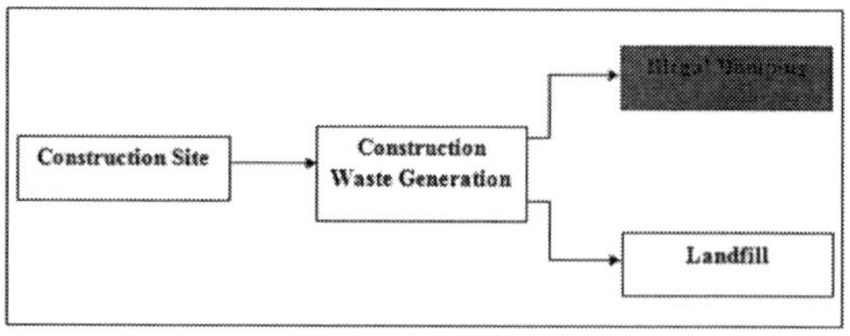


**Figure 1.** Generation of illegal dumping in construction

Figure 1 shows the current practices of Malaysian construction waste flow which all the waste directly dump into landfill and also the construction waste dump into illegal dump. Hence, the issues of illegal dumping from the construction which are not dumping to the specific landfill [8][3][9]). Furthermore, the information given by SWCorp [9] stated that every year, illegal dumping site increases which is illustrated in Figure 2. In Malaysia, most of the construction site takes the alternative ways to throw their waste to the illegal dumping site. In that case, the attitudes of the person's will contribute to the environmental pollution. Previous research has stated that landfill is the ultimate waste disposal method that can deal with many types of materials [3][17].

Due to this reason, it can be seen in Table 2, the percentage of waste disposed through landfill is increasing proportionally till the targeted year of 2020. Most of landfills in Malaysia are of small scale operations with varying design. In the light to it, most of the landfills are poorly maintained. Land filling also contributes the highest potentials to global warming [18][19]. Among all the methods used, other disposal sites achieve the highest percentage of waste followed by sanitary landfill whilst the least incineration. Other disposal sites are the open dumpsites that are illegal or have accumulated a large quantity of garbage [20].

### 2.2 Barriers for implementing construction waste management

Malaysia also faces waste problems due to the increasing number construction development [3] [14] Up to the present time, there are many factors contributed to construction waste which were identified from the literature [21-36] as follow:

- Poor site management
- Lack of experience
- Inadequate planning and scheduling
- Mistakes and errors in design
- Incompetent subcontractors
- Rework
- Frequent design changes
- Labour productivity
- Inadequate monitoring and control
- Inaccurate quantity take-off
- Shortage of site workers
- Lack of coordination between parties
- Slow information flow between parties
- Shortage of technical personal (Skilled labor)
- Changes in material specification and type
- Equipment availability and failure

- Effect of weather

There are 18 identified factors from the literature. These factors have shown many problems in the construction industry that generated the construction waste. As can be seen from the factors, most of the factors are due of the improper planning, which can be concluded as improper construction waste management. All of these factors contributed to the increasing number of construction waste in the Malaysian construction industry. Therefore, it is important to overcome the factors in order to increase systematic construction waste management in construction project. Additionally the main barriers in generating of construction waste has been grouped into seven main categories based on the literature review [25-32][36]. There are:

- Design – Changes in Design
- Procurement – Ordering errors
- Management – Poor Planning
- Material – Material storage
- Environment – Weather Effects
- Worker – Worker's mistakes
- Site Area – Leftover materials on site

## 2.3 Factors influencing construction waste management

Review of the literature has identified the factors influencing systematic construction waste management. The factors encompass design changes, consideration of reducing construction waste in design, investment on construction waste management, regulation, guidelines, space to perform waste management, adoption of technologies, waste reduction cost and waste management culture within an organization [3][9][14].

### *2.3.1 Design changes*

Design changes are one of the influencing factors of systematic construction waste management [28][25][14]. The design change can be identified as an error of construction materials which have been purchased based on the original design. After the design changes has been taken, the material cannot be resold or returned to the supplier or vendor and this situation will contribute to the generation of waste [14].

### *2.3.2 Consideration of construction waste reduction in design*

From the previous research [36], it has been reported that approximately 33% of waste in construction site are related to project design. It shows the big percent contributed to the construction waste is from the early stage of construction. To reduce the generation of waste in construction site, well prepared design stage will directly reduce the possibility of waste generation in construction projects. In order to minimize and proper waste reduction strategies, practices are taken into the design stage such as design for standard size and add the prefabricated or adopting modular [37][38][14].

### *2.3.3 Investment in construction waste management*

This factor can help to promote the construction waste management practices among the construction players, stakeholders, and worker in construction site [14]. Additionally, this investment will make workers more responsible to implement the method to control the waste, sorting technique and segregation. At the same time, this factor will motivate the practitioners in the construction industry of the importance to minimize the construction waste.

### *2.3.4 Regulation and guidelines*

SWCorp Malaysia, has mentioned in their report the ever increasing number of generated construction waste year by year [9]. In Malaysia, there are no focus or special regulation or guidelines to be followed by the construction industry. In a developed country such as Britain, Australia, Japan, and others country, they have specific guidelines and regulation for construction waste management [3].

Let's have a look at Japan which had adopted an integrated waste and material management approach that promotes dematerialisation and resource efficiency. The government's 'sound material cycle society' initiative launched in 2000 brought with it a number of new regulatory codes including specific laws targeting construction materials [10].

In 2002 Japan introduced the construction waste recycling law, which has resulted in high rates of recycling. For example, 99% of concrete had been recycled in 2006. The law enforces the recycling of a broad range of construction and demolition materials. Demolition contractors are required to separate and recycle specific construction wastes such as concrete, asphalt, and timber [10]

### *2.3.5 Site space for performing waste management*

In developing country, one of the methods for managing the construction waste in site, is to provide space for waste management. Site space refers to the space for sorting and segregation, handling and manage the waste. Sorting at the construction site is widely useful as an effective way in achieving a higher rate of waste reuse and recycling [14][15].

Developed country such as Hong Kong found that given site space as an important factor than the influence of systematic construction waste management [39] Previous study also prove that on site waste sorting could increase the rate of reuse and recycling and directly reduce the logistic cost and dumping cost [39][40][9].

### *2.3.6 Adoption of technology*

Use of technologies could help to reduce, reuse or recycle construction waste [14]. The implementation of technologies such as innovation in construction project will contribute to the generation of construction waste. Previous studies proved that the employment of technologies such as prefabrication, IBS, innovation of formwork and others technologies can minimize the

DOI: 10.1051/matecconf/20178701008

errors and construction waste being produced [38][41][14][9].

### *2.3.7 Waste management culture*

Waste management culture is related to the human behaviour such as awareness of construction parties to get involved in construction waste management [14]. It has been found that the awareness of waste management among practitioners and construction parties can save the resources and protect the environment [42][36][14]. Study has been conducted by [43] for revealing the understanding of both managers and site workers about waste reduction.

Furthermore, in 2006, study done by [36] find out a way for assessing architects' view on the origins of design waste that can minimize the waste design practices in the UK. Finding from this study shows that the awareness and attitudes of the practitioners and construction parties in waste management are big factors that contributes to the reduction of waste management and systematic waste management.

## 3 Methodology

This section describes the methodology adopted throughout the research project. The purpose of this paper is to explain a conceptual model for systematic construction waste management in Malaysia construction industry. Figure 2 shows the methodology for this paper.

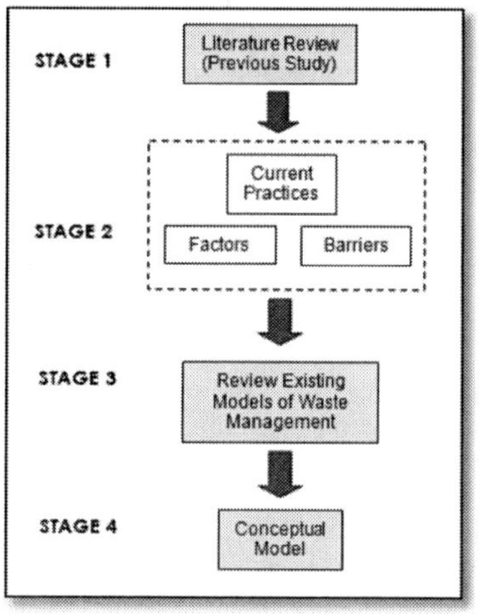


**Figure 2.** Methodology

The methodology was designed to pursue the aim of this paper as shows in Figure 2, which is to explain a conceptual model for systematic construction waste management in the Malaysia construction industry. Firstly, literature search and previous study was made to review the current practices of construction waste management in Malaysia, the barriers of implementing construction waste management and factors potentially influence the implementation of systematic construction waste management in the Malaysian construction industry.

Next, in stage 2, the elements from the literature review and previous study were analysed to find out the current practices of construction waste management, the barriers of implementing construction waste management and the potential factor in implementing of systematic construction waste management. These three elements are needed to develop a conceptual model in this paper. Stage 3 involved the review process of existing model or related model of construction waste management to find out the similarity between the model and literature. Finally, stage 4 is the process of developing conceptual model. The conceptual model is helpful in improving existing approach of construction waste management in the Malaysia construction industry.

## 4 Conceptual model of systematic construction waste management

The conceptual model is a process, which provides convenient label and categories that helps in explaining and developing detailed description and a coherent analysis of a research. Figure 3 shows the conceptual model for this paper.

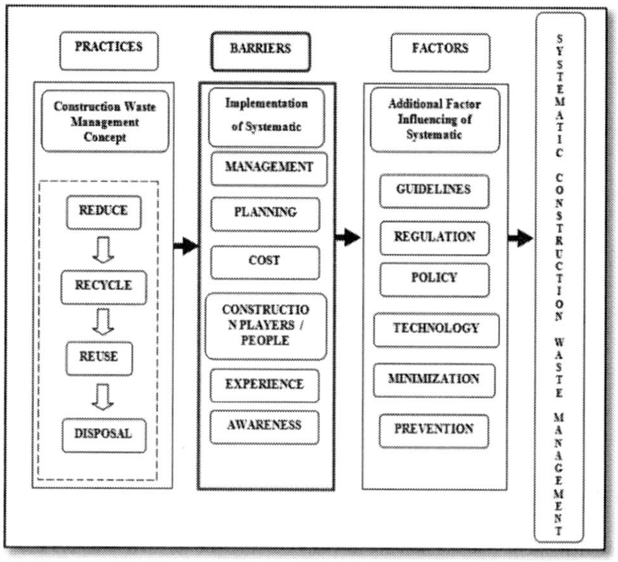


**Figure 3.** Conceptual model of systematic construction waste management

This paper proposes a conceptual model of systematic construction waste management in the Malaysian construction industry. The conceptual model is developed based on literature review of waste management, construction waste management, construction industry and gap found from the literature. This conceptual model as illustrated in Figure 3, is to investigate the current practices of waste management, barriers and factor influencing systematic construction waste.

From the literature review, it has been mentioned that most developed country such as the United Stated of America (USA), Australia, Singapore and others had implemented four basic concepts of waste management [14][32-33][36]. The four basic concepts of waste management are, reduce, recycle, reuse, and disposal. The four basic concepts as mentioned, come in stages which the big portion for implementation is reduce, followed by recycle, then reuse and a small portion for disposal method. But for developing country, it still not clear which practices had been applied. [9], reported that there are no reduction in number of construction waste generation in Malaysia.

Then from the literature, several barriers can be found that hinders the implementation of construction waste management. The barriers are, poor management in construction, improper planning of construction, construction player or people who are involved in construction projects, high of cost, low of experience and lack of awareness among construction parties on the importance of construction waste management. Previous research [20] and [14] mentioned that the lack of awareness among the construction players will contribute to the failure of implementing waste management. Malaysia as a developing country now focuses on construction waste management to reduce the environmental issues. These six major barriers can be dealt with the several factors influencing the implementation of systematic construction waste management.

The factors that influence the implementation of systematic construction waste management are design, guidelines, regulations, investment, technology, reduction and prevention [21-36]. All of these factors will help to improve the current issues on construction waste management and it will produce systematic construction waste management.

## 5 Conclusion and recommendation

It is well recognised that construction waste can be reduced and well managed by a systematic model of construction waste management as proposed in this paper. The purpose of this study is to explain a conceptual model for systematic construction waste management in the Malaysia construction industry. Findings from the literature have found several barriers that influence implementation of construction waste management in the Malaysian construction industry. From the barrier issue, factors influencing implementation of construction waste management was determined.

The most significant factors potentially influencing implementation of systematic construction waste management is design, guidelines, regulation, investment, technology, reduction and prevention. Thus, this on-going research helps to identify the factors that influence systematic construction waste management. This paper proved that the factors having a strong positive impact to improve the construction waste management in Malaysia construction industry. Additionally these findings will give a better solution to the construction players and increase the awareness among them to implement systematic construction waste management in future construction projects.

## Acknowledgement

The author would like to thank Ministry of Education (MOE), Graduate Students Office Universiti Tun Hussein Onn Malaysia, Department of Construction Management, Faculty of Technology Management and Business, UTHM, for supporting this research. This paper was partly sponsored by the Centre for Graduate Studies UTHM.

## References

[1] Nasaruddin, F.H., Ramli, N.H.M., and Ravana, S.D. 2008, E-Construction waste exchange in malaysia: preliminary study, Proceedings of Information Technology International Symposium 2008, Kuala Lumpur, Malaysia

[2] Siti, N. M., and Noor, Z. M. 2008, Approach in construction industry: A study on prefabrication method as a tool for waste minimization, Proceeding of International Conference on Environmental Research and Technology (ICERT), Penang, Malaysia

[3] Nagapan S., I. A. Rahman, A. Asmi, A. H. Memon, and I. Latif,. (2012). Issues on construction waste: The need for sustainable waste management," 2012 IEEE Colloq. Humanit. Sci. Eng., no. Chuser, pp. 325–330, Dec. 2012.

[4] Seow Ta Wee. (2016). Senario Pengurusan Sisa Pepejal di Malaysia. Penerbit Universiti Tun Hussein Onn Malaysia.

[5] Begum, R. A., Satari, S. K. and Pereira, J. J., (2010). Waste Generation and Recycling: Comparison of Conventional and Industrialized Building Systems. American Journal of Environmental Sciences, 6(4), 383-388.

[6] Ferguson, J., Kermode, N., Nash, C. L., Sketch, W.A.J., and HUxford, R.P. 1995. Minimizing Construction Waste: A Practical Guide. Institute of Civil Engineers, London.

[7] Winkler, G. 2010. Recycling Construction and Demolition Waste. Publisher; McGraw-Hill.

[8] The Ingenieur. 2011. Reduce, reuse and recycle, The Institution of Engineers Malaysia (IEM),

[9] SWCorp Malaysia. 2015. Progress Report 2015. SWCorp Malaysia.

[10] Ismam, J. N. and Ismail, Z. 2014. Sustainable Construction Waste Strategic Implementation Model. WSEAS Transactions on Environment and Development. ISSN:2224-3496. Vol. 10.

[11] Sreenivasan J., Marthandran G., and Malarzhi C. A. N. 2012. Domestic Solid Waste Minimization in Malaysia-A Behavioural Study. Faculty of Management Multimedia University.Konstantakopoulou, E., Rodriguez-Carmona M., and Barbur J. L. 2012. Processing of Color Signals in Female Carriers of Color Vision Deficiency. Journal of Vision. 12(2): 1–11.

[12] Udawarta, N., Zuo. J., Chiveralls, K., and Zillante, G. 2015. Improving waste management in construction projects: an Australian study. Elsievier 101 (2015) 73-83.

[13] Seow, T. W. & Mohamad, A. H. 2007. Pengurusan Sisa Pembinaan Di Tapak Binaan. Retrieved on March 20, 2012.

[14] Yuan, H. 2013. Key indicators for assessing the effectiveness of waste management in construction projects. Ecological Indicators 24(2013) 476-484.

[15] Shen, L. Y., Hao, J. L., Tam, V. W. Y., and Hou, H. 2007. A checklist for assessing sustainable performance of construction project. J. Civil Eng. Manage. 13(4), 273-281. Augmented Reality. Massachusetts: A K Peters.

[16] Hassan, M.N. 2002. Solid waste management in Malaysia: can we charter future strategies? In: Proceedings of International Conference Environmental Management: Ten Years after Rio. Article Number 8. Universiti Putra Malaya. 2002.

[17] Lau, V.L. 2004. Case study on the management of waste materials in Malaysia. Forum Geookol. 15(2). 2004.

[18] Piyaphant, S. and Prayong, K. 2011. Environmental impact on municipal solid waste management system in Chaiyaphum. in Proceedings of the 10th WSEAS international conference on System science and simulation in engineering, 2011, pp. 213-218.

[19] Sakawi, Z. 2011. Municipal solid waste management in Malaysia: Solution for sustainable waste management. Journal of Applied ciences in Environmental Sanitation. Vol 6. ISSN 0126-2807.

[20] Manaf, L.A., Samah, M.A.A,. 2009. Municipal Solid Waste Management in Malaysia: Practices and Challenges . Waste Management. 29:2902-2906

[21] Nazech E.M, Zaldi D., and Trigunarsyah B. 2008. Identification of Construction Waste in Road and Highway Construction Projects. Eleventh East AsiaPacific Conference on Structural Engineering & Construction (EASEC-11), Taiwan, (2008), pp.19-21.

[22] Serpell, A., Venturi, A., and Contreras, J. 1995. Characterization of waste in building construction projects. Presented on the 3rd workshop on lead construction, Albuquerque, pp. 67-77.

[23] Senaratne, S., and Wijesiri, D. 2008. Lean Construction as a Strategic Option : Testing its Suitability and Acceptability in Sri Lanka. Lean Construction Journal, pp. 34-48.

[24] Wahab, A. B., and Lawal, A. F. 2011. An evaluation of waste control measures in construction industry in Nigeria. African Journal of Environmental Science and Technology, Volume 5, pp. 246-254.

[25] Ekanayake, L. L., and Ofori, G. Construction material waste source evaluation. Proceedings of Strategies for a Sustainable Built Environment, Pretoria, (2000).

[26] Alwi, S., Hampson, K., and Mohamed, S. 2002. Non Value-Adding Activities in Australian Construction Projects.

[27] Zhao, Y., and Chua, D. K. H. 2003. Relationship between productivity and non value-adding activities.

[28] Faniran, O. O., and Caban, G. 2007. Minimizing waste on construction project sites. Engineering Construction and Architectural Management Journal, Volume 5(1), pp.182-188.

[29] Polat, G., and Ballard, G. 2004. Waste in Turkish Construction: Need for Lean Construction Techniques. Proceeding 12th Annual Conference of the International Group for Lean Construction (IGLC-12), Elsinore.

[30] Formoso C. T., Isatto E. L. and Hirota E. H. 1999. "Method for Waste Control in the Building Industry", Proceedings of the Seventh Annual Conference of the International Group for Lean Construction, Berkeley.

[31] Alwi, S., Hampson, K. and Mohamed, S. 2002. Waste in the Indonesian construction projects. Proceedings of International Conference of CIB W107 - Creating a sustainable Construction

Industry in Developing Countries, South Africa, pp. 305-315.

[32] Gavilan, R. M. and Bernold, L. E. 1994. Source evaluation of solid waste in building construction. Journal of Construction and Management, pp. 536–552.

[33] Lu, W., Yuan, H., Li, J., Hao, J. J. L., Mi, X., and Ding, Z. 2011. An empirical investigation of construction and demolition waste generation rates in Shenzhen city, South China. Waste management, Volume 31(4), pp. 680-687.

[34] Wang, J. Y., Kang, X. P., and Tam, V. W. Y. 2008. An investigation of construction wastes: an empirical study in Shenzhen. Journal of Engineering, Design and Technology, Volume 6(3).

[35] Wan, K. M. S., Kumaraswamy, M. M. and Liu, D. T. C. 2009. Contributors to Construction Debris from Electrical and Mechanical Work in Hong Kong Infrastructure Projects. Journal of Construction Engineering and Management, Volume 135(7), pp. 637-646.

[36] Osmani, M., Glass, J., Price A.D.F. 2008. Architects' perspectives on construction waste reduction by design. Journal of Waste Management, Volume 28, pp. 1147–1158.

[37] Innes, S. 2004. Developing tools for designing out waste pre-site and on-site. Proceeding of minimizing construction waste conference: Developing Resources Efficiency and Waste Minimisation in Design Construction. London.

[38] Jaillon, L., Poon, C.S., and Chiang, Y.H. 2009. Quantifying the waste reduction potential of using prefabrication in building construction in Hong Kong. 29(1), 309-320.

[39] Poon, C.S., Yu, A.T.W., and Ng, L.H. 2001. On-site sorting of construction and demolition waste in Hong Kong. 32(2),157-172.

[40] Hao, J.L., Hills, M.J., and Shen, L.Y. 2008. Managing construction waste on site through system dynamic modeling: The case of Hong Kong. 15(2), 103-113.

[41] Esin, T., and Cosgun, N. 2007. A study to conducted reduce construction waste generation in Turkey. Build Environment. 42(4) 1667-1674.

[42] Yuan, H.P., and Shen, L.Y. 2011. Trend of the research on construction and demolition waste management. Waste manage 31, 670-679.

[43] Lingard, H., Graham, P., and Smithers, G. 2000. Employee perception of the solid waste management system operation in a large Australia contracting organisation implication for company

# The Effectiveness EM Mudball and Banana Peels for Textile Wastewater Treatment

Mohd Ariff bin Ahmad Nazri[a] and Nurul Liyana binti Ghazali

*Department of Water and Environmental Engineering, University of Tun Hussein Onn Malaysia, Johor Bahru.*

**Abstract.** Textile industrial activities known as a higher consumer of water that involved several processed daily to maintain the quality of fabric that contribute to the water pollution. Wastewater produced from the textiles could lead to the presence of heavy metal contaminant. Meanwhile, heavy metal is one of the contaminants in wastewater and must be overcome with the appropriate treatment process. To overcome the heavy metal issue, the absorption method by organic waste could be used. Effective microorganism (EM) mudballs and banana peels is a method for threating wastewater quality from textiles effluent. EM is a medium that is able to improve water quality without altering the design area while the banana peels act as a nutrient to microorganism growth and have a potential to removed heavy metal contaminant that produced from textiles wastewater. Method that used to carry out this study are through the laboratory experimental for all consider parameters. The result was obtained shows that EM mudball itself are capable to reduce the concentration of BOD and COD as much as 84% for BOD with the value reduced to 37 mg/L and 90% for COD with the value reduced to 89 mg/L. The effectiveness of banana peels added to the EM mudball could be seen by the 70% of banana peels are well mixed with the EM mudball have the potential to reduce the concentration of BOD, COD and heavy metal contaminant in textiles wastewater as much as 81% for BOD (43 mg/L), 90% of COD (87 mg/L), and 86% of Zink ($Zn^{2+}$) with the value is 0.065 mg/L. From the comparison of result with the standard

## 1 INTRODUCTION

Textile industry has become the most important and rapidly developing sector in Malaysia and generated more wastewater that will affect the quality and quantity of the water [1]. Textile industrial activities known as a higher consumer of water that involved several processed daily to maintain the quality of fabric [2]. Thus, wastewater from textile industries will produced different pollutant which is dyes, surfactants, acids or bases, salts, heavy metals and suspended solid and the contaminant is highly contained BOD, COD, pH and will cause the color of the effluent change into black because of dyes processes.

Due to the severity of heavy metal contamination and potential adverse health impact on the public, efforts must be taken to purify water containing toxic metal ions [2]. The presence of heavy metal in textiles wastewater caused by the chemical that has been used in the printing, coloring, dyeing and bleaching process and contain synthetic chemical which are generally metal based and many of the metal are harmful for human body and aquatic life it is not complied with the permissible limit [3].

Therefore, by carried out both treatment of wastewater that included the treatment of water pollution and heavy metal in water, combination of this method of effective microorganism (EM) and banana peels were used. The EM and banana peels are material that easier to get and as initiative method of treating contaminated water. The combination of these material could be worked to improve the water quality and efficiently to reduce the cost of treatment.

## 2 LITERATURE REVIEW

### 2.1 Textile Wastewater

Textiles industries could generate millions of gallon dye wastewater daily and the percentage of water usage are different by the types of textiles product and between the processes of textiles production. The textile production processed need to consumes high amount of water, the water are mixed well with the chemical substance so that, it will produced high volume of wastewater that will contribute to water pollution [4].

The Printing and dyeing processed for textiles product is often rich in color and will produced wastewater that contained residues of reactive dyes, aromatic compounds, color, heavy metal, salt, surfactant and chemicals such as aerosol, high value of COD and BOD concentration and it could affect the pH value to become too acidity or alkalinity [5].

Effluent from textile dyeing and printing processed contain dyes, bleaching agents, salts, acids and heavy metals like Cr, Cu, Pb, Zn, Fe are discharged continuously without a proper treatment, thus the heavy metal contaminant was analyzed and has a value of Pb is 1.098 mg/L, Fe is 0.191 mg/L, Cu 4.66 mg/L, Cd 1.98 mg/L, Zn 3.29 mg/L, Ni 0.076 mg/L and Cr 3.96 mg/L [3].

[a] Corresponding author: ariffn@uthm.edu.my

### 2.2 Effective Microorganism Mudballs and Banana peels

EM is effective technology that contains useful microorganism and has a potential to degraded ecosystem that full of harmful microbes, to eliminate the foul smelling odors caused by high levels of ammonia, hydrogen sulphide and methane, and capable to reduce the COD and BOD contaminant and increased the DO for the contaminant water. COD and BOD showed large variations in concentrations during storms [6]. Effective microorganism were able to reduce the amount of BOD as much 93% while the value of COD was reduced to 20% and the SS was 94% (EMRO, 2014).

Banana peels are the product from food waste that have potential as absorbent of removal heavy metal such as $Cu^{2+}$, $Zn^{2+}$, $Co^{2+}$, $Ni^{2+}$, $Pb^{2+}$ , $Fe^{2+}$ from wastewater and converted as adsorbent because they have large surface area, high swelling capacities due to SEM images result with initial condition 5 mg/L of Fe and $Zn^{2+}$ [7].

## 3 METHODOLOGY

### 3.1 Preparation for Banana peels

Banana peels that have been used in this studies were obtained from the stall at Parit Raja, Batu Pahat. Banana peels were washed repeatedly by using distilled water to removes soluble impurities from their surface of banana peels and banana peels was dried naturally within 24 hours. Afterwards, banana peels was grinded until it become refined and was stored in airtight polyethylene containers.

### 3.2 Preparation for EM mudballs

50 ml of EM.1 solution are well mixed with 50 ml molasses and 900 ml distilled water in airtight bottle. The bottle was closed tightly and the cap were loosen regularly when gas formed after couple days to release the gas. The fragmentation processed takes a weeks to activate the EM solution with the pH value of EM is 4.0.

Firstly, Mixed the EM Bokashi with the top soil well. Secondly, activate EM are poured into the mixture of top soil and EM Bokashi until the compound become damp and easily formed. Thirdly, mixture of soil, activate EM and EM Bokashi are rolled until it turns to the shape of tennis ball. EM mudballs were placed into the box that covered with paper.

EM mudballs are kept at a low temperature to encourage the growth of fungi and avoid exposed directly to the sunlight or wind to avoid EM mudballs dries quickly. EM mudballs are ready to use when the entire surface of mudballs is filled fungi. FIGURE 1 show molasses as a solution for activate the EM 1 and EM mudballs preparation.

(a)

(b)

**Figure 1**: Activate effective microorganism and EM mudballs preparation

### 3.3 Wastewater Sampling

This study has carried out 5 wastewater sampling at Sri Gading with different percentage of banana peels and EM mudballs such as followed Table 1 shows the percentage of EM mudball and banana peels mixture. EM mudball was thrown into the wastewater tank for sample 2, 3 and 4 once a week to get the percentage of removal BOD, COD and heavy metal. Sample 1 is a blank sample that supplied with the oxygen while sample 5 is a blank sample. The oxygen that has been supplied was affect the percentage of oxygen needed in each sample.

Both of this sample were act as a variable sample to obtained oxygen needed for microorganism growth. TABLE 1 shows percentage of EM mudball and banana peels mixture.

i. Sample 2: Sample 2, 70% of EM are well mixed with the 30% of banana peels and the oxygen were supplied into the sample
ii. Sample 3: Sample 3 are supplied with the oxygen to help the microorganism in EM mudballs to breakdown the sludge and treated the water. For this sample 100% of EM mudballs is used without banana peels.
iii. Sample 4: Sample 4 was supplied with the oxygen, and the mixture of 30% of EM, and 70% of banana peels.

**TABLE 1:** percentage of EM mudball and banana peels mixture.

| Sample | EM mudballs | Banana peels |
|---|---|---|
| 2 | 70% | 30% |
| 3 | 100% | 0% |
| 4 | 30% | 70% |

### 3.4 Data Analysis Methods

Data analysis method are mostly used the procedures described in the APHA 2005 to obtain the average value for the consider parameters through the laboratory test conducted. The type of sample will divided by the number of samples tested and the result will be compared according to the standard of IWK of the

DOI: 10.1051/matecconf/20178701009

Environmental Quality Act 1974. All data presented through graphs, charts and analysed using Microsoft excel software.

## 4 LABORATORY TEST

For this studies, AAS test were conducted to carry out the heavy metal contaminant of textiles wastewater at Sri Gading such as $Zn^{2+}$. The concentration of $Zn^{2+}$ are low in range and the textiles wastewater are not necessary to prepare the dilution before the test. Others parameter are considered in this studies such as BOD, COD, DO, pH and temperature.

## 5 RESULTS AND DISCUSSIONS

Wastewater from the textiles industry were taken showed the characteristic of parameter such as pH, temperature, DO, $BOD_5$, COD and heavy metal such as Zink ($Zn^{2+}$).

Base on the characteristic table, obtained that the pH having an average value of 7.95 and the temperature is 25 °C which is permissible. The average concentration of DO for the textile wastewater was showed value of 3.80 mg/L while the $BOD_5$ is 222 mg/L and the COD parameter is 1325 mg/L.

The average concentration of the heavy metal obtained in textiles wastewater for Zink ($Zn^{2+}$) is 0.568 mg/L. The characteristic of textile wastewater shown in TABLE 2.

**TABLE 2:** Characteristic of Textile wastewater

| Parameter | Concentration of textiles wastewater | | |
|---|---|---|---|
| | Minimum | Maximum | Average |
| pH | 7.14 | 8.76 | 7.95 |
| Temperature ℃ | 24 | 26 | 25 °C |
| DO | 2.95 | 4.65 | 3.80 |
| $BOD_5$ | 187 | 256 | 222 |
| COD | 1000 | 1650 | 1325 |
| Zink ($Zn^{2+}$) | 0.450 | 0.686 | 0.568 |

*all units are in mg/L except the pH and Temperature

### 5.1 Dissolved Oxygen Analysis Result

Dissolved oxygen in the water are important for microorganism growth to capable of oxidizing the biodegradable organic matter in the sample and DO level are depends on the physical, chemical and biochemical activities in water bodies as the oxygen is considered poorly soluble in water [8]. Therefore, air is supplied by air pump to the water sample 2, 3, and 4 meanwhile for the sample 1 oxygen is supplied to the blank sample to comparing the percentage of oxygen needed in sample and act as the variable sample.

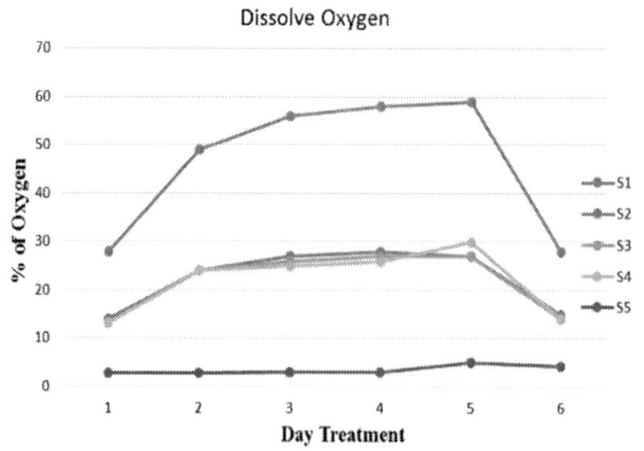


**Figure 2:** Total dissolved oxygen for textile wastewater treatment by using EM mudballs and banana peels

Sample 1 is representing of blank sample that has been supplied with the oxygen for comparison with the sample 5 (blank sample). The total of oxygen needed for each sample was calculated and compared through the percentage from the blank and supplied oxygen sample (sample 1). It is about 50% oxygen of all textile wastewater sampling needed for the microorganism to growth and biodegrade the organic matter in the wastewater.

FINGURE 2 shows sample 1 is the blank sample that supplied oxygen shows that increasing trend to the DO level from the 28% on day one supplied to 49% of DO level on the second day which the value is 4.05 mg/L to 5.65 mg/L respectively. The result of third day of oxygen supplied was obtained slowly increasing of DO level from the 56% with the value is 6.86 mg/L to the optimum level of DO is 59% with the value is 7.33 mg/L and this is because of the oxygen that supplied were enough for microorganism pollutant growth.

### 5.2 Biochemical Oxygen Demand Removal

FIGURE 3 shows that the total BOD removal by using EM mudball and banana peels as textiles wastewater treatment. Sample 2 has the optimum effectiveness removal by 81% as much as 43 mg/L and sample 3 are capable to reduce of BOD with the value decreased to 37 mg/L. While the sample 4 could reduce the BOD as much as 80% of removal with the value of removal is 47% mg/L.

The mixture of banana act as a food to microorganism and the effectiveness of the EM mudball in treated textiles wastewater were affected about 50% by oxygen needed that have been supplied all the time for microorganism to growth.

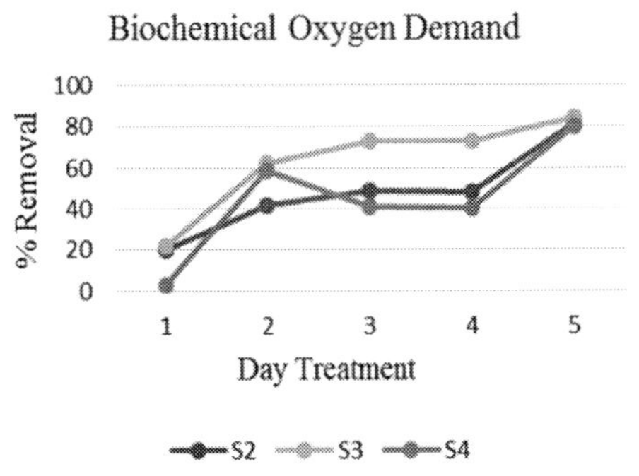


**FIGURE 3:** Total BOD removal by using EM mudballs and banana peels

### 5.3 Chemical Oxygen Demand Removal

The percentage of COD removal for sample 2 have reached the optimum percentage removal on the fifth day treatment as much as 90% with the value 87 mg/L. Sample 3 shows the optimum percentage of removal is 90% with the value of COD is reduced to 89 mg/L. Sample 4 could remove about 72% of COD in day two of treatment rather than the sample 2 only 31% of total removal which is 325 mg/L and 687 mg/L respectively.

The reducing of COD value is inversely proportional to the increasing value of DO which is contributed by neutral pH and this strengthened with statements made by previous studies[9] while the different percentage of banana peels that mixed in EM mudball also affect to the effectiveness of COD removal. FIGURE 4 shows the Total removal of COD by using EM mudball and banana peels.

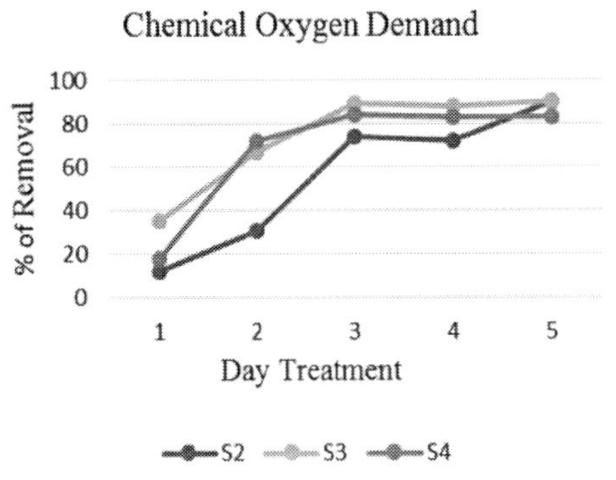


**FIGURE 4:** Total removal of COD by using EM mudball and banana peels

### 5.4 Heavy Metal Zink (Zn2+) Removal

From the treatment that have been done in this studies, showed that the banana peels has a potential as a medium to remove the heavy metal Zink ($Zn^{2+}$) contaminant in textiles wastewater. FIGURE 5 shows the total removal of Zink ($Zn^{2+}$) by using EM mudball and banana peels

Banana peels could contributed to reduce the heavy metal contaminant Zink ($Zn^{2+}$) about 86% of removal for sample 4 while the sample 2 has reached 80% optimum removal of the third day of treatment. The removal of Zink ($Zn^{2+}$) for sample 3 was increased to 60% and 78% of removal on the second and third day of treatment. Therefore, EM has a potential to remove about 89% of Zink ($Zn^{2+}$).

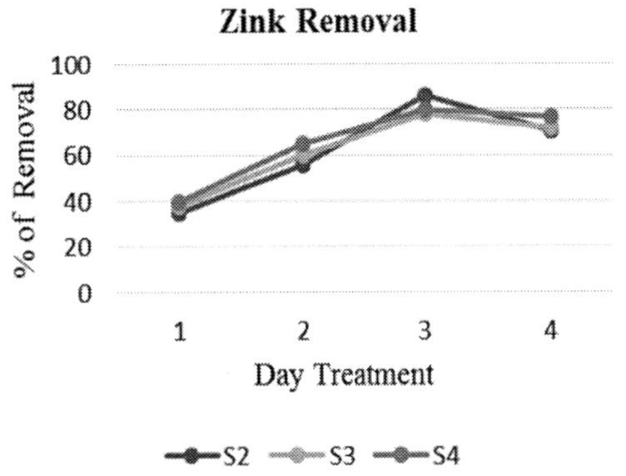


**FIGURE 5:** Total removal of Zink ($Zn^{2+}$) by using EM mudball and banana peels

## 6 CONCLUSIONS

The result was obtained from the parameter that have been carried out in this studies and it could be concluded that the EM mudball itself are capable to reduce the concentration of BOD, COD and Zink ($Zn^{2+}$) as much as 84%, 90% and 78% with the value reduced to 37 mg/L, 89 mg/L and 0.097 mg/L respectively.

The percentage of banana peels shows that 70% of banana peels well mixed with the EM mudball have the potential to reduce the concentration of BOD, COD and Zink ($Zn^{2+}$) contaminant in textiles wastewater as much as 81% (43 mg/L), 90% (87 mg/L) and 86% (0.065 mg/L) While the 30% of banana peels could reduce the BOD, COD and Zink ($Zn^{2+}$) as much as by 81%, 90% and 80% with the value is reduced to 43 mg/L, 87 mg/L and 0.09 mg/L respectively.

## ACKNOWLEDGEMENT

I would like to express my gratitude to University Tun Hussein Onn Malaysia for supporting STG grant no U527, Department of Water and Environmental Engineering and Micropollutant Research Centre for provide the most advanced laboratory facilities that available through time to fulfill this study.

## REFERENCES

[1] N. Sivri and İ. Toroz, "Pollutants of Textile Industry Wastewater and Assessment of its Discharge Limits by Water Quality Standards," vol. 103, pp. 97–103, 2007.

[2] A. A. Kadir, N. Othman, and N. A. M. Azmi. "Potential of Using Rosa Centifolia To Remove Iron and Manganese in Groundwater Treatment" International Journal of Sustainable Construction Engineering & Technology vol 3. pp. 70-82, 2012.

[3] M. A. Kamaruddin, M. S. Yusoff, H. A. Aziz, and C. O. Akinbile, "Review Paper Recent Developments of Textile Waste Water Treatment by Adsorption Process : A Review," vol. 1, no. 4, pp. 60–73, 2013.

[4] J. Rajasthan, "Monitoring of Heavy Metal in Textile Waste Water of," vol. 4, no. 3, pp. 3–6, 2014.

[5] A. Shaikh and T. Engineering, "Water conservation in textile industry," no. November, pp. 48–51, 2009.

[6] R. Nazahiyah, Z. Yusop and I. Abustan, "Stormwater Quality and Pollutant Loading from An Urban Residential Catchment in Johor, Malaysia", vol 56 No 7, pp. 1-9, 2007.

[7] Z. Wang, M. Xue, K. Huang, and Z. Liu, "Textile Dyeing Wastewater Treatment," 2010.

[8] G. A. and D. . L. Juang, Ruey-Shin, "Adsorption of Heavy Metals From Water Using Banana and Orange Peels," no. October, 2015.

[9] C. Principle, Patel H; & Vashi R. *Characterization of Textile Wastewater*, 2015. IBN:978-0-12-802326-6.

# Oil and grease (O&G) removal from commercial kitchen waste water using carbonised grass as a key media

Siti Nazahiyah Rahmat[1,a], Ahmad Zurisman Mohd Ali[1], Mohd Haziman Wan Ibrahim[1] and Nur Azrena Alias[1]

[1]*Faculty of Civil and Environmental Engineering, Universiti Tun Hussein Onn Malaysia (UTHM), Batu Pahat, Johor, Malaysia*

**Abstract.** Oil and grease (O&G) are usually found in kitchen waste water. O&G are poorly soluble in water and can cause serious problems during the wastewater treatment. Adsorption is a fundamental process in the treatment of kitchen waste water and very economical. Activated carbon is the most effective adsorbent for this application. Therefore, the aim of the current study is to determine the potential of four materials (i.e. sand, gravel, carbonised grass (CG) and clay powder) as filter media for O&G removal. The CGs were originated from "Elephant Grass" and it is a fast growing plant with significant potential as carbon. The clay acts as a magnet, drawing the oil molecules out of the water and causing them to attach to the surfaces of the clay. In the current study, two filters were developed with different media materials to get the best percentage removal. Filters 1 and 2 were filled with sand, gravel and CG, and sand, gravel, CG and powder clay, respectively. Three samples were taken during peak hour between April and May 2016. The initial O&G concentration varied from 101.37 mg/l to 248.30 mg/l and the final concentration varied from none to 22.57 mg/l for both filters. The percentage removal (%) of O&G were between 90.9 and 97.3 (Filter 1) and between 94.3 and 100 (Filter 2). Overall, both filters could efficiently remove O&G in the waste water and the quality of the carbonised grass proved to be as good as carbon produced from other traditional sources.

## 1 Introduction

Commercial kitchen waste water refers to any release to the sewerage system from the restaurants and other commercial food service facilities which varies significantly from residential waste water. When commercial kitchen reaches its peak operating hours, the waste water produced is much higher than the wastewater produced by the residential wastewater. This is due to the waste water consists higher level of oil and grease (O&G). Most of O&G are insoluble in water, thus will float on the water surface, due to the lower specific gravity than the water. Large amounts of oil and grease in the waste water will cause trouble in the collection system pipes. If not disposed of properly, O&G builds up in the sewer system constricting flow, which could cause sewer back-ups into homes. Additionally, it may hinder sewage treatment processes [1].

Many methods have been applied for the separation of O&G, including a variety of filters [2], reverse osmosis, gravity separation, biological processes [3], air flotation [4,5], membrane bioreactor, chemical coagulation, and electrocoagulation [6]. One commonly used technique and has gained wide attention for removing organics dissolved in water is the process of adsorption. It involves the separation of substances from one phase to the surface of another. The adsorbing phase is the adsorbent, and the material concentrated or adsorbed at the surface of that phase is the adsorbate [7]. The process is efficient and there are many waste materials that can be used as adsorbents.

Clay and carbon are among several adsorbents used in the adsorption method [8]. Carbons are used extensively as adsorbent for separation of contaminants due to its simplicity and has proven to be technically feasible [2]. Clays have been successful in removing oil-grease from wastewater streams [9]. CG media is produced by crushing the raw material to the correct size. The carbon-based material is converted to carbon by thermal decomposition in a furnace using a controlled atmosphere and heat. It is realised that wastes are unutilized resources. A number of studies have been carried out to produce CG such as bagasse [10], sugarcane bagasse [11], rice husks [12, 13] and coconut shells [14]. Goorahoo et al. [15] proposed that Elephant grass appears to have significant potential for scavenging excess soil, nitrogen and phosphorus and can be very useful in a filtration system. However, studies on the utilisation of grass as a source of CG are still very limited and so the current work may likely to get attention from researchers working on environmentalists.

Clay has been widely used because of their strong sorption and complexation ability [16]. It can remove approximately 70% of the waste of water in a water

[a] Corresponding author: nazahiya@uthm.edu.my

removal system. Therefore, the objectives of this study are threefold: (i) to investigate the quality of O&G released from the commercial kitchen, (ii) to propose filters that can treat and remove O&G using AC as a principal media, and (iii) to measure the efficiency of the proposed filters.

## 2 Materials and methods

### 2.1 Study Area

Parit Raja is a town in Batu Pahat District, Johor. It is located approximately 23 km away from Bandar Penggaram, Batu Pahat (the capital of the district) (see Figure 1). Along to Parit Raja, there are many villages and most of them are name after a *Parit*, as a result of relatively flat terrain and abundant waterways. In recent years, due to the development and rapid expansion of Universiti Tun Hussein Onn Malaysia (UTHM), the town has seen substantial growth. There is increasing number of commercial areas including restaurants and food stalls along the way to the city. In the current study, the location of sampling point was in Parit Daun, Parit Raja.

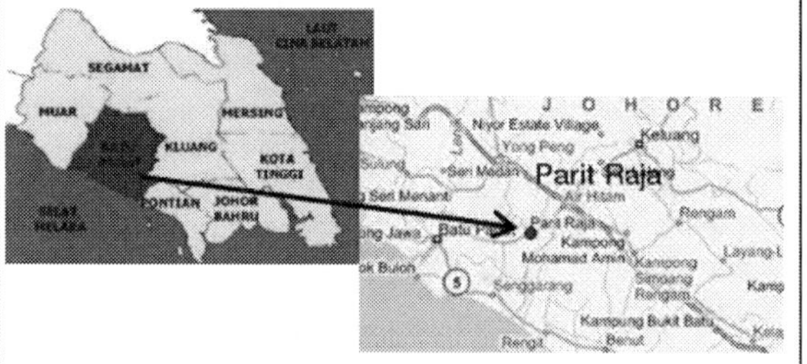


**Figure 1.** Location of the study site

### 2.2 Water Sampling

The water quality of the Parit Raja, one of the most densely populated areas within the region, is significantly degraded due to human activities as well as urbanization. In the current study, the waste water was released from a single discharge of the commercial kitchen at Parit Daun, thus, it is known as point source of pollution (see Figure 2).

The sampling was carried out between April 2016 and May 2016 (Table 1). The sampling days were scheduled such that they covered rainy and non-rainy days. A total of 3 samples were investigated. Samples were collected manually using a polyethylene bottle. To analyse, a volume of 1.5 L water sample for each time of sampling was filled into the bottle.

**Table 1.** Characteristics of monitored samples

| Sample | Sampling Date | No. of dry days | Rainfall (mm) |
|---|---|---|---|
| Sample 1 | 28/4/2016 | 2 | 0 |
| Sample 2 | 2/5/2016 | - | 5 |
| Sample 3 | 11/5/2016 | - | 13 |

**Figure 2.** Image of the drainage at the study area.

### 2.3 Filters

Two filters with different types of media were developed. Apart of sand and gravel for filter media, CG and clay were also used. The preparation of these media are as follows:

#### *2.3.1 Carbonised grass*

Carbonised grass was selected as the principal filter media. These carbons were originated from Pennisetum Purpureum (Figure 3) or also known as 'Elephant Grass'. Elephant grass requires very little supplementary nutrients for growth and can be harvested up to four times a year, which makes this plant one of the most prospective crops for energy use [15].

Elephant grass leaves were washed with deionized water to remove adhering dirt particles from the surface and dried at 110°C for 24 h, Dried leaves were cut and loading into a furnace and heated up to 700°C for 6 hours. The furnace lid was left closed as much as possible during the process to avoid the leaves turn into ash.

**Figure 3.** Carbon from Elephant Grass.

#### *2.3.2 Clay*

Apart from CG, clay is another adsorbent used in the adsorption technique [9]. In the current study, the clay sample was obtained from a factory located in Ayer Hitam, Johor. The clay was broken up into smaller pieces and loading into a crusher machine to get the clay powder (see Figure 4).

 DOI: 10.1051/matecconf/20178701010

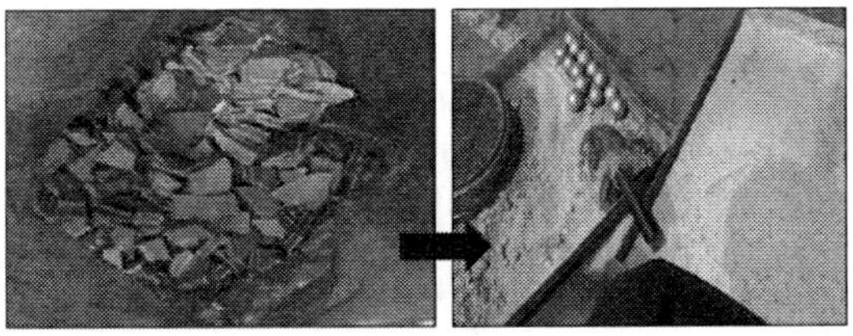

**Figure 4.** Powder clay used in the filter.

### *2.3.3 Dimensions*

Two buckets were used to construct the frame of the filter. They are identical and measure 0.5 m in height and 0.3 m in diameter. The following descriptions of the design begin with the left bucket continue to the right and in the same direction as the flow of waste water (Figure 5). For Filter 1, there are 90 mm of sand and beneath the sand, there are 75 mm of gravel. Waste water passes through the holes and into the second bucket. In the second bucket, there are 20 mm of CG, 85 mm of sand and 50 mm of gravel (Figure 5a). For Filter 2, in the first bucket there are 90 mm of sand 75 mm of gravel. In the second bucket, there are 20 mm of CG, 20 mm of clay, 65 mm of sand and 50 mm of gravel (Figure 5b). A plastic spigot is attached to the outer side of the bucket to release the water.

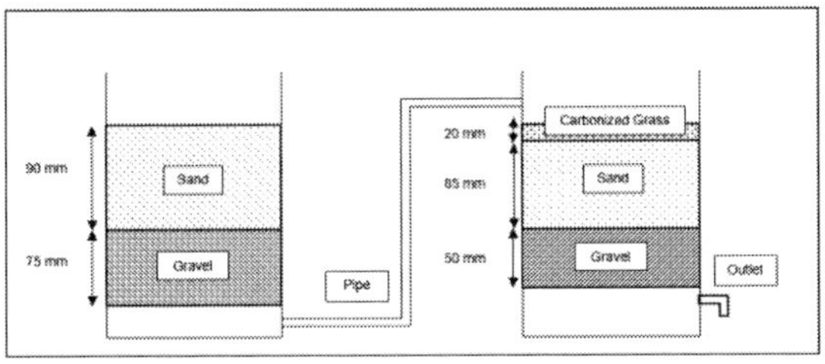


(a) Filter 1

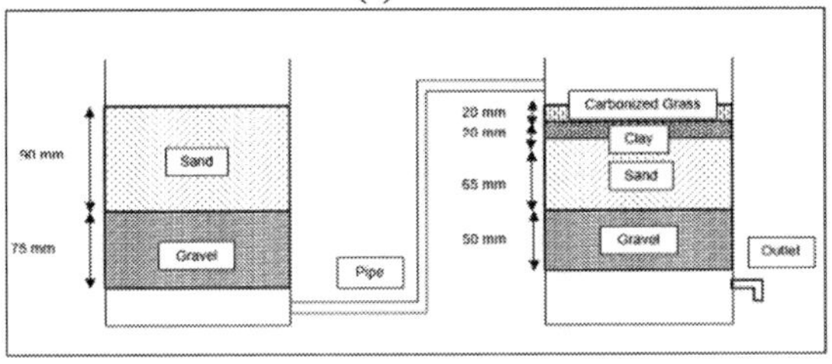


(a) Filter 2

**Figure 5.** Dimensions of the filters.

For each run, 1500 ml of sample was poured into the filter and the amount of time required for the samples to infiltrate was measured to determine the flow rate.

### 2.4 Waste water quality analysis

Samples were collected prior to and after infiltration to compare the concentration of O&G that have passed through the medium. Water samples were sent to the Micro-pollutant Research Centre (MPRC) and Waste Water Engineering Laboratory, UTHM for analysis within 24 hours after sampling. The laboratory analysis followed the Standard Methods for the examination of water and wastewater [17].

50 mL of acidified waste water was transferred to a separatory funnel. 50 mL of petroleum ether were then added to the funnel and the funnel was shaken vigorously for 2 minutes. The lighter waste water layer will be on the top and heavier petroleum ether layer will be on the bottom. The bottom layer of petroleum ether was drained into a clean and dried flask. The contents of the flask were then heated so that the petroleum ether was distilled into another container. The flask (containing the extracted O&G) (*A*) was reweighed. The original weight of the flask (*B*) was subtracted and the total O&G was calculated (see Equation 1).

$$\text{Value of O\&G} = \frac{(A - B)x\ 1000}{ml\ sample} \qquad (1)$$

*A* is the weight of flask plus with residue (*mg*) and *B* is the weight of flask (*mg*).

## 3 Results and discussion

### 3.1 Flow measurement

Flow rate is the time taken for the solvent to flow through the natural sorbent. The longer the contact time, the more the amount of organic material can be exposed to the adsorbent, causing in a more effective adsorption process.

The flow rate for each filter was calculated by measuring the volume of wastewater samples and time taken for the waste water flows through the media to the outlet. The measurement was determined using the bucket and stopwatch method. The volume of the bucket divided by the time taken to fill up the bucket produces the value of the flow. For Filters 1 and 2, the flow rates measured were 0.3 L/s and 0.001 L/s, respectively. Flow rate in Filter 2 was slower than Filter 1 due to the clay powder in the filter which had been slowing down the process. It is noted that the lower the flow rate, the more time the O&G be adsorbed.

### 3.2 O&G Removal Efficiency

Water samples were taken on 28th April 2016, 2nd May 2016 and 5th May 2016 by applying the grab technique. Table 2 shows the concentration values before and after the treatment for both filters. Sample 1 gives the highest initial concentration of O&G of 248.3 mg/l due to the sampling during dry day. Referring to Table 1, the observed antecedent dry days is 2 days. The concentration of O&G in the waste water is at its highest due to the accumulated pollutants including O&G since the preceding event. While for Samples 2 and 3, the initial concentrations are 101.37 and 205.48, respectively. These results could be associated with the

O&G that retained deposited in the drains during dry days and become diluted during storms.

For all samples, the O&G concentrations reduced from 248.3 mg/l to 22.57 mg/l (Sample 1), 101.37 mg/l to 2.74 mg/l (Sample 2), 205.48 mg/l to 9.45 mg/l (Sample 3) after being filtered through Filter 1. While for Filter 2, the concentrations were reduced tremendously for all samples from 248.3 mg/l to 3.72 mg/l (Sample 1), 101.37 mg/l to none (Sample 2) and 205.48 mg/l to 1.44 mg/l (Sample 3). These results show evidence the ability of the medium to remove the O&G.

pH of individual sample was also measured immediately after its collection. The initial pH of wastewater samples is generally towards acidic side. It is as low as 5.76 and as high as 6.55 before treatment (see Table 2). For both filters, the pH values after treatment ranging from 5.97 to 7.32. The media used in the filters (i.e. activated carbon and clay) absorbed some minerals, thus causing the changes in pH. The carbon itself can act as an ion exchange type media and contribute to pH rise.

**Table 2.** Concentrations of O&G and pH values.

| Parameter | Sample | Unfiltered | Filtered | |
|---|---|---|---|---|
| | | | Filter 1 | Filter 2 |
| O&G (mg/l) | Sample 1 | 248.30 | 22.57 | 3.72 |
| | Sample 2 | 101.37 | 2.74 | 0.00 |
| | Sample 3 | 205.48 | 9.45 | 1.44 |
| *pH* | Sample 1 | 5.87 | 6.34 | 7.13 |
| | Sample 2 | 6.55 | 7.14 | 7.32 |
| | Sample 3 | 5.76 | 5.97 | 6.39 |

From the observation, when the water sample was poured into the filter, there is still a slight sheen of oil and grease at the bottom of the first bucket. After flowing through the second bucket, it can be seen that the O&G retained at the carbonised grass layer and hence resulted in the removal of O&G. Image of the samples before and after the treatment is shown in Figure 6. It can be clearly seen that the sample after filtration using Filter 1 is very clear. In contrast, the water in Filter 2 is brownish in colour and very turbid due to the presence of clay powder.

**Figure 6.** Waste water samples (0) untreated, and treated using (1) Filter 1 and (2) Filter 2.

The percentage removal was also calculated to get the efficiency of the filters (see Figure 7). The results show that for all samples, Filter 1 could remove from 90.9% to 97.3% of O&G. While for Filter 2, for all samples, the percentage removal ranges from 94.3% to 100%. It is observed that for Sample 2, after passing the filter, the water is free from O&G.

From these results, it can be concluded that the materials tested can all be used as filter media material. Sand, gravel, CG and clay powder provided the best removal efficiency. Clay powder may be best used as an additive, however, in the future it is recommended not to use it in a powder form. This will slightly reduce the turbidity of water.

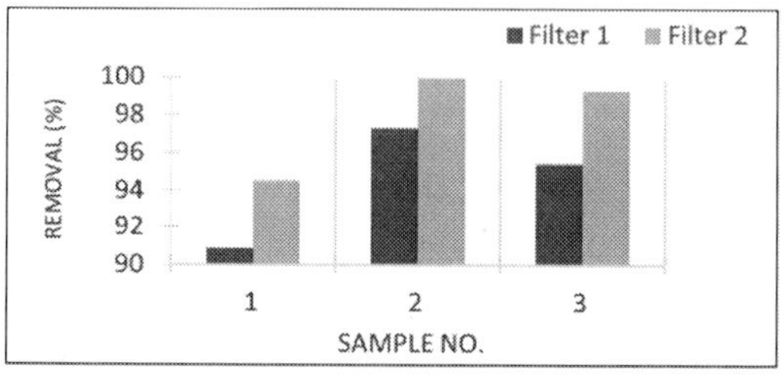


**Figure 7.** Percentages of O&G removal for both filters.

## 4 Conclusions and Recommendations

The current study determined the potential of four media for O&G removal which were sand, gravel, carbon and clay. Using carbon and clay could benefit us in terms of cost and also reassuring the sustainable use of waste materials. The work shows how environmental wastes could be a useful source of obtaining novel engineering material like carbon. In the present work grasses are considered as a safer, cheaper and greener way of utilizing them as a source of preparing carbon.

The initial concentrations of O&G in waste water were quite high ranged from 101.37 mg/l to 248.3 mg/l. The results showed that the waste water was discharged directly into the drainage system. In the current study, both filters could assist with O&G removal as the percentage removal recorded was between 90.9% and 100%.

Based on the findings of this study, several areas are identified that can be improved. Recommendations for future studies are summarised as follows:

1. A more intensive waste water monitoring is recommended to get more accurate and reliable results.
2. An important issue to be addressed is the influence of length of dry period and rainfall intensity on the water quality.
3. The clay used should not be in a powder form as the clay will disperse thus turning the water into brownish colour and slowing down the process of filtration.
4. Further analysis is carried out to study on the amount of water treated between maintenance intervals if the filter is going to be developed. This filter will be installed under the kitchen sink and the water quality will be monitored. This filter is expected to be a potential option to treat commercial kitchen waste water.

## Acknowledgements

The authors are grateful for the funding provided by the Universiti Tun Hussein Onn Malaysia (Vote U418) in support of this research.

## References

1. M.A. Fulazzaky and R. Omar, Removal of oil and grease contamination from stream water using the granular activated carbon block filter, Clean Techn Environ Policy, **14**, 965-971, (2012).
2. K. Okiel, M. El-Sayed and M. Y. El-Kady, Treatment of oil-water emulsions by adsorption onto activated carbon, bentonite and deposited carbon, Egyptian Journal of Petroleum, **20** (2), 9-15, (2011).
3. J.C. Campos, R.M.H. Borges, A.M. Filha and N.R. Oliveira, Water Research, **36**, 95-104, (2002).
4. B. Tansel, and B. Pascual, Removal of emulsified fuel oils from brackish and pond water by dissolved air flotation with and without polyelectrolyte use: pilot scale investigation for estuarine and near shore applications, Chemosphere, **85** (7), 1182-1186, (2011).
5. A.A. Sharmani, A. James and H. Xiao, Destabilisation of oil-water emulsions and separation by dissolved air flotation, Water Research, **36**, 1503-1512, (2002).
6. W. Scholz and W. Fuchs, Treatment of oil contaminated wastewater in a membrane bioreactor, Water Research, **34** (14) 3621-3629, (2000).
7. L. Frank Slejko, Adsorption Technology a Step-by-Step Approach to Process Evaluation and Application, Marcel Dekker, Inc., New York and Basel, (1985).
8. R.M. Zakaria, I. Hassan, M.Z. El-Abd, Y.A. El-Tawil, Lactic acid removal from wastewater by using different types of activated clay. In: Thirteenth international water technology conference (IWTC), Hurghada, 13, 403-416, (2009).
9. A.O. Alade, A.T. Jameel, A. S. Muyubi, M. I. Abdul Karim & M. D. Z. Alam, Removal of oil and grease as emerging pollutants of concern (EPC) in wastewater stream. *Iranian Journal of Oil & Gas Science and Technology*, **2**(4), 01-11, (2011).
10. V. Minkova, S.P. Marinov, R. Zanzi, E. Bjornbom, T. Budinova, M. Stefanova, et al., Thermochemical treatment of biomass in a flow of steam or in a mixture of steam and carbon dioxide, *Fuel Proc. Technol.* **62**, 45, (2000).
11. E. Cetin, B. Moghtaderi, R. Gupta, T.F. Wall, Influence of pyrolysis conditions on the structure and gasification reactivity of biomass chars, *Fuel*, **83**, 2139, (2004).
12. P.K. Malik, Use of activated carbons prepared from sawdust and rice-husk for adsorption of acid dyes: a case study of Acid Yellow 36, *Dyes Pigments*, **56**, 239, (2003).
13. N. Yalcin, V. Sevinc, Studies of the surface area and porosity of activated carbons prepared from rice husks, *Carbon*, **38,** 1943, (2008).
14. M. Osava, Elephant Grass for Biomass. *Inter Press Service News Agency*, October 10, (2007).
15. D. Goorahoo, F. Cassel, D, Adhikari and M. Rothberg, Update on elephant grass research and its potential as a forage crop, *California Alfalfa and Forage Symposium*. UC Cooperative Extension, Agronomy Research and Extension Center, Plant Sciences Department, University of California, Davis, Visalia, CA, (2005).
16. G. Sheng, H. Dong, Y. Li, Characterization of diatomite and its application for the retention of radiocobalt: role of environmental parameters, J. Environ. Radioactiv. **113**, 108-115, (2012).
17. American Public Health Association, APHA, Standard methods for the examination of water and wastewater, *19th edn. APHA*, Washington DC, (1995).

DOI: 10.1051/matecconf/20178701011

# Production and Characterisation of Microfine Sized Palm Oil Fuel Ash (POFA) Originated from Bau, Lundu Palm Oil Mill

R. Ahmadi [1,a], M.S. Saiful [1], D.F Zawawi [1], S.Z.A. Rahman [1], I. Ismail [1], A.B. Mannan [1] and A.S.Z. Abidin [2]

[1] *Department of Civil Engineering, University Malaysia Sarawak, 94300 Samarahan, Sarawak, Malaysia*
[2] *Department of Mechanical and Manufacturing Engineering, University Malaysia Sarawak, 94300 Samarahan, Sarawak, Malaysia*

**Abstract.** This paper investigates an effective and economical way for laboratory scale production of micro fine sized palm oil fuel ash (POFA) using an electric powder grinder. The raw POFA obtained from the palm oil mill is initially grinded by using Los Angeles abrasion machine, and then sieved using 150 µm sieve before it is burned in a furnace at 500°C. The burned POFA is then grinded using electric powder grinder to obtain the targeted micro fine sized. The physical, morphological and chemical properties of the micro fine sized POFA produced are analysed in the form of cement paste using Particle Size Analyzer (PSA), nitrogen sorption by using BET method, Fourier Transform Infrared Spectroscopy (FT-IR) and Scanning Electron Microscopy (SEM) with Energy Dispersive Spectroscopy (EDS). The results show that 96% micro fine sized POFA is produced when using the optimum grinding process. The microstructural analyses of cement paste with 20% micro fine sized POFA replacement give the optimum results that contribute to higher compressive strength. The overall results of this research show that the optimum grinding process by using electric powder grinder is relevant and can be used as pioneering work in the concrete production industry.

## 1 Introduction

In countries like Malaysia, Thailand and Indonesia, palm oil is one of the main agro-industries which produces a large amount of waste in the form of kernel shell, empty fruit bunches and fibers [2]. These by-products are used as fuel to heat up boiler for generation of electricity in the palm oil mill. As the second largest palm oil producer in the world, Malaysia has generated million tonnes of palm oil wastes each year [1,3,12]. The extracted product from the boiling process is known as palm oil fuel ash (POFA) which comprises a large amount of silica, thus, possesses pozzolanic properties suitable as cement replacement material.

Previous studies have showed that POFA can be used as a partial cement replacement material to produce high strength concrete with higher impermeability and durability [8]. The inclusion of up to 40% micro fine POFA tends to increase the workability and viscosity of high strength concrete and could improve the corrosion resistance properties of steel inside concrete [4,5,6,7,9,13,14]. Furthermore, utilization of POFA as partial cement replacement materials contributes to the development of concrete with resistance to adverse environment condition [10,11].

Until now, the best method of producing finer size of POFA particle size (1-10 $\mu$m) is still unavailable. Although previous studies are able to produce micro fine sized POFA, the cost of production is still too expensive and the methods used are considered not effective [3,13]. Due to the significance contribution of micro fine sized POFA as a new potential cement replacement material, an effective and economical method to produce micro

[a] Corresponding author araudhah@unimas.my

fine sized POFA is investigated. This research focuses on the simple and optimum grinding process and further investigates the physical and morphological properties of micro fine sized POFA which can affect the strength, workability and hardened concrete properties.

## 2 Materials and experimental procedures

### 2.1 Materials

The cement used in this experiment is Ordinary Portland Cement Type 1 whereas the Palm oil fuel ash (POFA) is obtained from Bau, Lundu Palm Oil Mill, Sarawak, Malaysia. The POFA is subjected to the $1^{st}$ stage of grinding by using Los Angeles abrasion machine and sieved through 150 μm sieve to remove coarser particles. Next, POFA is burned at 500°C for 1 hour in a furnace to remove excessive unburned carbon. Then, it is subjected to $2^{nd}$ stage of grinding by using electric powder grinder with 25000rpm to obtain micro fine sized which is ranging from 1-10 μm.

### 2.2 Experimental procedures

In order to determine the effectiveness of electric powder grinder, 2 samples were used which are; Sample A, 500 g of burned POFA and Sample B, 1000 g of burned POFA. All cement paste specimens were prepared with water to cement ratio of 0.35. One control sample and four samples were casted by replacing the cement with 10%, 20%, 30% and 40% of micro fine POFA (sample A). The test specimens were cured in water for 28 days.

The Particle Size Analyzer (PSA) was used to determine the physical size of burned POFA after $2^{nd}$ stage of grinding. The morphology of cement paste was investigated by using Scanning Electron Microscopy (SEM) with Energy Dispersive Spectroscopy (EDS). Fourier Transforms Infrared Spectroscopy (FT-IR) analysis is carried out to identify the pozzolanic characteristic in terms of hydration and identification of the composite group. Nitrogen sorption by using BET method is conducted to obtain the precise specific area of cement paste powder. Ordinary Portland Cement (OPC) and fly ash (FA) are also used as control samples.

## 3 Results and discussions

### 3.1 Particle Size Analysis (PSA)

3.1.1 Sample A

In this process, 500 g of burned POFA is grinded for 10 cycles which takes a duration of 30 minutes in total. After the $10^{th}$ cycles of grinding are completed, the final amount of POFA collected is 490g. Therefore, 10g POFA loss is equivalent to 2% of the total amount of sample used. Micro fine particles of POFA is seen to slightly escaped from the tighten lid of the grinder during the grinding and while opening the lid after the grinding process.

3.1.2 Sample B

In this process, 1000 g of burned POFA is used and grinded for 15 cycles which takes a duration of 45 minutes in total. After the $15^{th}$ cycles of grinding are completed, the final amount of POFA collected is 960 g. Therefore, 40 g POFA loss is equivalent to 4% of the total amount of sample used. The particle losses is also due to the same reason as in sample A. The $d_{50}$ (μm) particle sizes of OPC and FA, 18.29 μm and 8.64 μm respectively are also determined to compare the different sizes of POFA, OPC and FA. Table 1.1 shows the particle sizes result for every cycle while, Figure 1.1 shows the Particle size $d_{50}$ (μm) versus amount of cycles.

 DOI: 10.1051/matecconf/20178701011

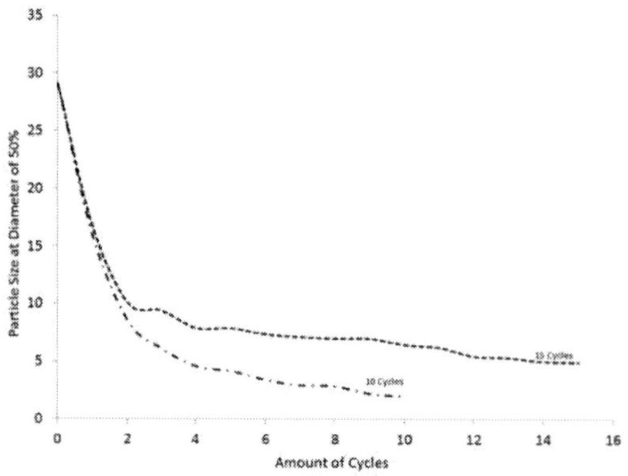


Figure 1.1: Particles Size at Diameter 50% vs Amount of cycles

Table 1.1 Particles Size Result for Samples A & B

| Samples | A | | | B | | |
|---|---|---|---|---|---|---|
| Cycles | Particle Size $d_{50}$ (μm) | Mean Diameter (μm) | Specific surface ($m^2/g$) | Particle Size $d_{50}$ (μm) | Mean Diameter (μm) | Specific surface ($m^2/g$) |
| 0 | 29.09 | 34.51 | 1.117 | 29.09 | 34.51 | 1.117 |
| 1 | 15.97 | 21.38 | 2.296 | 16.78 | 19.67 | 1.555 |
| 2 | 8.49 | 12.10 | 5.782 | 10.04 | 14.24 | 5.058 |
| 3 | 6.10 | 9.07 | 6.415 | 9.39 | 12.30 | 1.705 |
| 4 | 4.58 | 7.04 | 8.627 | 7.88 | 11.59 | 4.766 |
| 5 | 4.16 | 6.27 | 7.135 | 7.86 | 11.72 | 3.736 |
| 6 | **3.41** | **4.35** | **6.071** | 7.36 | 11.47 | 5.093 |
| 7 | 2.94 | 3.70 | 13.288 | 7.13 | 11.03 | 3.125 |
| 8 | 2.87 | 3.15 | 5.798 | 6.98 | 11.74 | 2.775 |
| 9 | 2.67 | 2.90 | 17.893 | 6.98 | 10.14 | 3.927 |
| 10 | 2.23 | 2.41 | 15.871 | 6.46 | 10.77 | 5.286 |
| 11 | | | | 6.22 | 9.37 | 3.700 |
| 12 | | | | **5.46** | **9.32** | **4.138** |
| 13 | | | | 5.34 | 7.95 | 7.834 |
| 14 | | | | 5.02 | 7.52 | 4.530 |
| 15 | | | | 4.94 | 7.38 | 2.583 |

Micro fine POFA which is ranging from 1-10 μm is obtained in both samples. For both Sample A and Sample B, the optimum cycle of grinding is at $6^{th}$ and $12^{th}$ cycles, respectively. This is because the diameter at 10%, 50%, 90% and mean diameter is already below 10 μm. By comparing the POFA diameter at 50% with OPC and FA, POFA with the highest amount of cycles has the smallest diameter. This factor influenced the surface area of the particle. Smaller particle size contributes to larger surface area and lead to more pore capillaries that can be filled in concrete. Hence, the strength of the hardened concrete will be higher.

### 3.2 Fourier Transforms Infrared Spectroscopy (FT-IR)

Fourier transform infrared spectroscopy (FT-IR) analysis is conducted to identify the pozzolanic characteristics of harden cement paste

DOI: 10.1051/matecconf/20178701011

through absorption of infrared spectrum by the formation of main hydration products calcium hydroxide (CH) and calcium silica hydrate (C-S-H). A strong band ranging from 965 $cm^{-1}$ to 975 $cm^{-1}$ is due to Si-O asymmetric stretching vibration of tricalcium silicate, $C_3S$ and /or dicalcium silicate, $C_2S$ which shows the formation of C-S-H [10]. Figure 1.2 shows the FTIR spectra of cement paste with POFA inclusion at (A) 0% replacement, (B) 10% replacement, (C) 20% replacement, (D) 30% replacement, (E) 40% replacement.

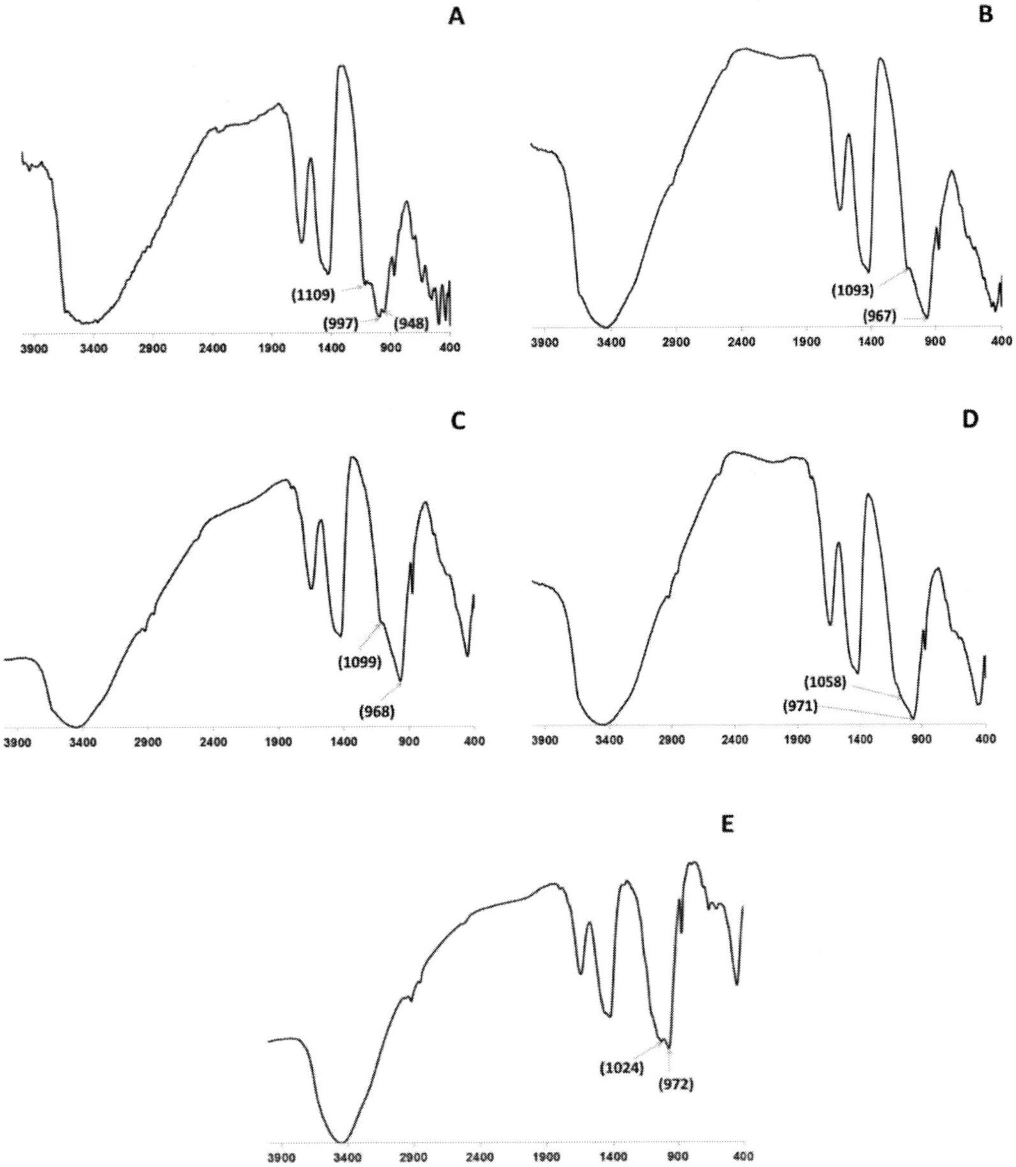


Figure 1.2: FTIR spectra of cement paste with POFA inclusion at (A) 0% replacement, (B) 10% replacement, (C) 20% replacement, (D) 30% replacement, (E) 40% replacement.

From Figure 1.2, the strongest absorption band centered at 972 $cm^{-1}$ to 997 $cm^{-1}$. Cement paste with 0% replacement of POFA shows the peak at 948 $cm^{-1}$ and 997 $cm^{-1}$ which indicates the paste starts producing the C-S-H gel. As mention earlier the strong band for C-S-H gel is at 965 $cm^{-1}$ to 975 $cm^{-1}$. Thus, the spectra show the paste is producing a weak band of C-S-H. Cement paste with 10% replacement of POFA shows the peak at 967 $cm^{-1}$ which shows the paste is producing a strong band of C-S-H gel. Compare to the 0% replacement, the peak of 1109 $cm^{-1}$ decrease to 1093 $cm^{-1}$. This indicates the extra amount of silica was added from the 10% of POFA replacement. The similar pattern was observed for 20% and 30% replacement. The absorption intensity of the corresponding peak increases with the percentage of replacement indicating the formation of more C-S-H in the cement paste due to the hydration of $C_3S$. Cement paste with 40% replacement shows a slight increment in peak which is 972 $cm^{-1}$. This is because of the large amount of POFA powder which takes longer time to contribute in hydration at later stage. Overall, cement paste with 10%, 20% and 30% replacement shows the strong band in the formation of the C-S-H gel.

### 3.3 Scanning Electron Microscopy (SEM) with Energy Dispersive Spectroscopy (EDS)

Scanning electron microscopy (SEM) is conducted to obtain the surface morphology of the cement paste while Energy Dispersive Spectroscopy (EDS) is carried out to obtain the chemical composition in the material. Figure 1.3 shows the surface morphology of the cement paste with POFA inclusion at (A) 0% replacement, (B) 10% replacement, (C) 20% replacement, (D) 30% replacement and (E) 40% replacement.

Figure 1.3: Surface morphology of the cement paste at (A) 0% replacement, (B) 10% replacement, (C) 20% replacement, (D) 30% replacement and (E) 40% replacement.

DOI: 10.1051/matecconf/20178701011

EDS analysis results can be obtained through mapping from the same point used for SEM. Table 1.2 shows the comparison of the Ca/Si ratio of cement paste with 0%, 10%, 20%, 30% and 40% of POFA replacement.

The results in Table 1.2 show that POFA tend to decrease in Ca/Si ratio which indicates pozzolanicity characteristic. The results show that Ca/Si ratio is highest in 20% replacement of POFA, with 4.22 ratios. Ca/Si ratio will decrease with the increment of silica present in C-S-H gel [11]. Comparison with compressive strength is also determined. Figure 1.4 shows the compressive strength versus Ca/Si ratio for each paste.

Table 1.2: Comparison of Ca/Si ratio for each cement paste

| Sample (28 Days) | Element (atomic %) | | Ca/Si |
|---|---|---|---|
| | Ca | Si | |
| 0% | 37.19 | 10.03 | 3.71 |
| 10% | 36.7 | 11.47 | 3.20 |
| 20% | **44.67** | **10.59** | **4.22** |
| 30% | 35.66 | 12.235 | 2.91 |
| 40% | 31.6 | 18.613 | 1.70 |

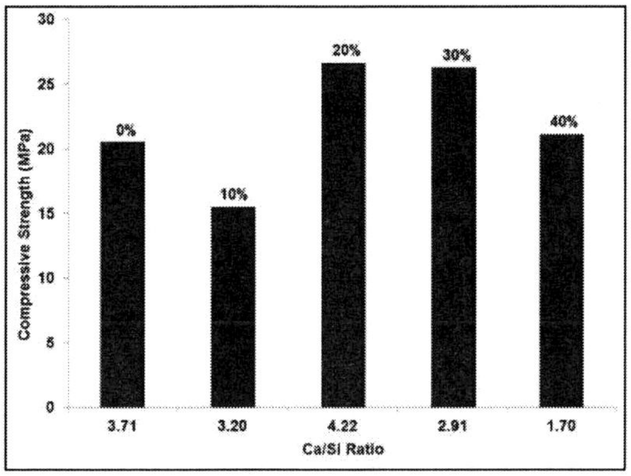


Figure 1.4: Compressive strength vs Ca/Si ratio

From Figure 1.4, it is shown that cement paste with 20% replacement is high in term of compressive strength together with Ca/Si ratio. This indicates that there is suitable amount of calcium in cement which contributes to higher initial strengths of concrete. However, as the chemical reactions take place, cracks and shrinkage in concrete might be produced due to the greater heat developed. Therefore, too much Calcium oxide, CaO is one of the reasons that may lead to failure in concrete. POFA has less calcium content but high in silica. Only with the presence of sufficient calcium and moisture, pozzolanic reaction will takes place resulting in formation of C-S-H gel [6]. This reaction helps to generate secondary C-S-H gel thus producing high strength concrete and makes concrete less destructible compare to OPC concrete. Hence, findings from EDS, supports findings from compressive test and FT-IR analysis.

### 3.4 Nitrogen sorption using $BET_{N2}$ analysis

Nitrogen sorption $BET_{N2}$ is conducted to obtain the precise specific area of cement paste. As shown in Table 1.3, the specific surface areas of POFA paste with inclusion of 10%, 20% and 30% are higher than 0% POFA replacement. This is because the micro fine sized POFA is used. Micro fine sized POFA has smaller particle sizes compare with OPC which causes it to fill up the capillary pores in cement. Hence, this results in denser internal structure of concrete which could sustain higher capacity load. Since the specific area is greater, the contact surface between pozzolan materials and water is also larger. This results to a better hydration process and increase the workability of concrete. The gaps between the particles of aggregates are able to be filled up with micro fine sized POFA which provides lubrication for aggregates to settle.

Table 1.3: Size of sample from $BET_{N2}$ method

| | 0% | 10% | 20% | 30% |
|---|---|---|---|---|
| Average Pore Radius (nm) | 7.05 | 7.59 | 7.33 | 7.18 |
| Total Pore volume (cc/g) | 0.02 | 0.03 | 0.03 | 0.02 |
| Specific Surface area (m2/g) | 52.87 | 69.39 | 65.00 | 63.97 |

## 4 Conclusion and Recommendation

Based on the results obtained from the experimental works on the production and characterization of micro fine sized palm oil fuel ash (POFA), the following conclusions can be made:

I. A simple yet effective method for laboratory scale production of micro fine sized POFA has been developed using a 25000-rpm electric powder grinder. The efficiency of this method achieved up to 96% production of micro fine sized POFA ranging between 1-10 μm of median particle size ($d_{50}$) when the optimum grinding process is used.

II. The microstructural analyses of the cement paste show formation of C-S-H gel as hydrated product is with the strongest band when 20% of the micro fine sized POFA replacement is utilized. Highest compressive strength together with Ca/Si ratio is also achieved indicating pozzolanic characteristics. $BET_{N2}$ test gives average pore radius of 7.3nm with specific surface area of $65m^2/g$.

The results obtained indicate that the new approach using an electric powder grinder is successful in producing micro fine size POFA and this approach is envisaged applicable for used in the industry provided further modification is carried out in future work to increase the efficiency if this device.

## Acknowledgement

This research is funded by Osaka Gas Foundation in Cultural Exchange Research Grant (OGFICE) session 2014/2015.

## References

1. Usman, J., Zaky, M., & Arifin, Z. (2014). Effects of Palm Oil Fuel Ash Composition on the Properties and Morphology of Porcelain-palm Oil Fuel Ash Composite. *Jurnal Teknologi* Full paper *5*, 5–10.
2. Abdul Awal, a. S. M., & Hussin, M. (2011). Effect of palm oil fuel ash in controlling heat of hydration of concrete. *Procedia Engineering*, *14*, 2650–2657.
3. Ahmadi, R., Zainudin, N., Ismail, I., Mannan, M.A. & Abidin, A.S.Z. (2016). Micro fine sized Palm Oil Fuel Ash (POFA) produced using a wind tunnel production system. *Journal of Advances in Materials Science and Engineering*. Volume 2016, Article ID 6792731
4. Abdullah, K., & Hussin, M. (2010). Fire Resistance Properties of Palm Oil Fuel Ash Cement Based Aerated Concrete, *1*(September), 107–114.
5. Usman, J., Zaky, M., & Arifin, Z. (2014). Jurnal Teknologi Full paper Effects of Palm Oil Fuel Ash Composition on the Properties and Morphology of Porcelain-palm Oil Fuel Ash Composite, *5*, 5–10.
6. Muthusamy, K., & Azzimah, Z. N. (2014). Exploratory Study of Palm Oil Fuel Ash as Partial Cement Replacement in Oil Palm

Shell Lightweight Aggregate Concrete, *8*(2), 150–152.
7. Kroehong, W., Sinsiri, T., & Jaturapitakkul, C. (2011). Effect of palm oil fuel ash fineness on packing effect and pozzolanic reaction of blended cement paste. *Procedia Engineering*, *14*, 361–369.
8. Olalekan, M., Universiti, Y., & Tebal, N. (2014). Improving the Engineering and Fluid Transport Properties of Ultra-High Strength Concrete Utilizing Ultrafine Palm Oil Fuel Ash Improving the Engineering and Fluid Transport Properties of Ultra-High Strength Concrete Utilizing Ultrafine Palm Oil Fuel, (October 2015).
9. Oss, H.G., & Padovani, A.C. (2003). Cement manufacture and the Environment – Part II: Environmental challenges and opportunities. *Journal of Industrial Ecology*, 7(1), 93-126.
10. Bensted, J. & Barnes, P. (2008). Structure and Performance of Cements, 2nd edition, Taylor and Francis Group, London and Newyork.
11. Awal, A.S.M. and Hussin, M.W. (1997) Some Aspects of Durability Performances of Concrete Incorporating Palm Oil Fuel Ash, Proceedings of the Fifth International Conference on Structural Failure, Durability and Retrofitting, Singapore, 210-218.
12. Mannan, A. Asrah, H., & Mirasa, A. K. (2015). The Performance of Ultrafine Palm Oil Fuel Ash in Suppressing the Alkali Silica Reaction in Mortar Bar, International Journal of Engineering and Applied Sciences (IJEAS) (9), 60–66.
13. Johari, M. A. M., Zeyad, A. M., Yusuf, M. O. (2014). Improving the Engineering and Fluid Transport Properties of Ultra-High Strength Concrete Utilizing Ultrafine Palm Oil Fuel Ash. *Journal of Advanced Concrete Technology*. 12, 127-137.
14. Yahaya, F. M., Muthusamy, K., & Sulaiman, N. (2014). Corrosion Resistance of High Strength Concrete Containing Palm Oil Fuel Ash as Partial Cement Replacement, *7*(22), 4720–4722.

MATEC Web of Conferences **87**, 01012 (2017) DOI: 10.1051/matecconf/20178701012
*ENCON 2016*

# A Review on Implication of Material Management to Project Performance

Zairra Mat Jusoh[1,a], Narimah Kasim[1]

[1]*Faculty of Technology Management and Business, Universiti Tun Hussein Onn Malaysia, 86400 Batu Pahat, Johor, Malaysia*

**Abstract.** Material management is an important element in project management as materials contribute a major portion to total project cost. It also plays a key role because of the successes of every construction project rely on having proper resources. Thus, it gives implication on project performance. Despite the importance of material management to project performance, there are limited numbers of research available related to this topic. Therefore, the aim of this paper is to identify the effects factors of material management on project performance. Literature from books, journal articles and conferences papers related to this topic are reviewed. In conclusion, exploring the implication of material management towards project performance will benefit construction players in improving the efficiency of material management in order to minimize the impacts on construction projects performance.

## 1 Introduction

Construction material is a critical element in the construction project as it contributes a major portion to the cost of projects. The contribution of construction materials may account 50-60 % of the project cost [1]. Due to its role as the major contributor to project cost, managing construction material becomes essential function in the construction project. Therefore, material management is an important element in project management.

Business Roundtable in Modern Management System defined material management as 'the process of planning and controlling all necessary efforts to ensure that the right quality and quantity of materials and equipment are specified in a timely manner, obtained at reasonable cost and available when needed [2, 3]. Some authors interpreted it directly to a concept related to integrated process of materials procurement that are combined under one management function [4, 5]. While others defined it as a combination process of planning, identification, inventory control, receiving and distribution, material handling and storage of materials [6, 7]. In essence, material management is a planned process combining identification, planning, purchasing, delivery, handling, storing, controlling and distributing with the purpose of ensuring the availability of sufficient quantities, appropriate quality and reasonable cost of construction materials for project's need.

Material management plays important roles in the success of construction projects as highlighted in previous studies [6, 8, 9]. It is important, because the outcome of every construction projects rely on having all proper resources such as materials, labours, plants and equipment delivered to site at appropriate time [10]. Without proper material management in construction projects, construction projects suffer delays, cost overrun, construction waste and low productivity [11-16]. As prior literature reports, among the factors that contribute to poor project performance is due to shortage of materials occur in construction projects [1, 12, 17]. Meaning that, the materials are not available when needed. Thus, existing papers acknowledged that material management has significant effects on project performance.

Despite the importance of material management to project performance. The material management area most often overlooked by the construction players and academicians [18, 19]. This claim seems to be true as only a few papers cover material management topic in local context [20-22]. The existing papers only briefly mentioned that material management influence project performance with limited exploration and empirical study specifically related to relationship of material management on project performance [1, 6, 23]. Moreover, it appears there is lacking of detailed discussion especially about the effect of material management on project performance. The effect of material management needs to be explored to gain better insight and understanding towards improving projects performance through effective material management. Hence, the purpose of this paper is to identify the effect of material management on project performance.

## 2 Material Management Processes and Functions

Material management processes and functions consist of multi-organizations and integrated processes of material definitions, acquisition and field disposition [24]. In this study, these processes and functions are project planning, material take-off, suppliers' enquiry, purchasing, material control, warehousing, expediting and shipping [2, 3].

[a] Corresponding author: zairramatjusoh@gmail.com

### 2.1 Project Planning

In material management, planning is the most important parts of the overall material management process [7, 9]. The planning process comprises the task of identification and determination of required materials, set up and maintain the materials records, determines target inventory levels, delivery frequency including materials logistic planning such as temporary facilities, access roads and storage area [3, 25-27]. Usually, the materials plan are developed based on detailed project design, bills of quantities, procurement plans, resource plan and project schedule [7]. Apparently, the planning process is also integrated with other processes.

### 2.2 Material Take-off

The purpose of material take-off is to obtain an actual total quantity of materials used in the construction projects [4]. Other than quantities of materials, this process is able to ascertain the materials specifications [2]. Clearly, this process is related to identification of what types of material are required and how much the quantities are needed.

### 2.3 Supplier Enquiry

The supplier enquiry process is related to suppliers' evaluation [2]. The criteria need to be considered in evaluating suppliers are price, quality, time of delivery, inventory held by suppliers, and flexibility [5, 9, 28]. However, the priority of these criteria depends on the strategies of each organization [28].

### 2.4 Purchasing

The purpose of purchasing process is to provide the materials at the right time, place, quality and an agreed budget [25]. Basically, this purchasing process starts by receiving the requisition form from the person in charge, followed by purchasing [29]. During this process, the current stockpile, availability of site space and storage areas need to be considered in avoiding material inventory problems [26].

### 2.5 Material Control

The aim of material control is to avoid shortage and surplus of materials occurs at construction sites [7]. This process related to handling, distribution and usage of materials [30, 31]. Within this task, material control is affected by the degree of the information technology used as the information technology has the ability to optimize this process [2, 24].

### 2.6 Warehousing

Primarily, warehousing related to material storage. Storage at the construction site is essential so that the materials already available when needed [18]. However, the stored materials need to be monitored in order to avoid waste, loss and damaged materials due to disorganized stock, improperly sorted, misallocate or deteriorate [25].

### 2.7 Material Take-off

The aim of expediting and shipping process is to ensure selected suppliers deliver the materials, technical specification and equipment to construction site accordingly as per schedule, terms and date specified in the purchase order [3]. This process also related to provide timely information related to materials deliveries [2]. Apparently, this process involves transfer of materials and equipment from suppliers to the construction site including deliveries information when necessary [26].

## 3 Influence of Material Management to Project Performance

As previously mentioned, existing studies on construction project performance have briefly mentioned the effect of material management on project performance. Therefore, reviews of literature summarize that the key criteria of project performance affected due to material management as in Table 1.

A summary is presented in Table 1 for 28 previous papers that mentioned the effect of material management to project performance. Generally, the criteria of project performance are identified both from a positive and negative perspective effects. From positive effects, effective material management has a positive impact on time optimization, cost saving, quality maximization, productivity improvement and waste minimization, [1-3, 32-34]. On the other hand, ineffective material management has negative effects on project performance such as time delay, cost overrun, poor quality, loss of productivity and excessive waste generation [2, 8, 11, 14, 35-37]. However, in the context of this paper, the criteria are termed in impartial form namely; time, cost, quality, productivity and waste. Discussion of material management effect on 5 criteria of project performance is as in the following.

### 3.1 Time

Time is defined as a degree that general conditions promoted completions of a project within allocated duration [38]. It is an essence of a construction contract, hence most of the contract have a similar provision about completion time and clearly stated in it [30]. Therefore, it is important for construction project completed on time.

Furthermore, time is widely recognized as minimum performance measured and one of the primary criteria of project success [13, 39-40]. As shown in table 1, total of 24 papers mentioned that material management affect time performance.

The insufficient stock of materials lead to idling time as workers try not to exhaust the stockpile or it is worsened by work stoppage [11, 41]. Due this shortage, materials need to be reordered and causes longer idling

DOI: 10.1051/matecconf/20178701012

time. Consequently, the work progress will be delayed. Therefore, the availability and sufficient quantity of materials affect the time performance.

**Table 1.** Summary of Material Management Affecting Project Performance

| No. | Authors | Criteria of Project Performance | | | | |
|---|---|---|---|---|---|---|
| | | **Time** | **Cost** | **Quality** | **Productivity** | **Waste** |
| 1 | Gulghane & Khandve (2015) | √ | √ | √ | √ | √ |
| 2 | Alanjari et al. (2014) | √ | √ | | √ | √ |
| 3 | Barry et al. (2014) | √ | √ | √ | | |
| 4 | Caldas et al. (2014) | √ | √ | √ | | |
| 5 | El-Gohary (2014) | | | | √ | |
| 6 | Hughes & Thorpe (2014) | √ | √ | | √ | |
| 7 | Othman et al. (2014) | √ | √ | √ | | |
| 8 | Safa et al. (2014) | √ | √ | | | √ |
| 9 | Abdul Rahman et al. (2013) | √ | √ | | √ | |
| 10 | Aziz (2013) | | √ | | | |
| 11 | Patil & Pataskar (2013) | √ | √ | √ | √ | |
| 12 | Doloi et al. (2012) | √ | | | | |
| 13 | Meng (2012) | √ | √ | | | |
| 14 | Nagapan (2012) | | | | | √ |
| 15 | Sardroud (2012) | √ | √ | | √ | |
| 16 | Ameh & Osegbo (2011) | √ | | | √ | |
| 17 | Haseeb et al. (2011) | √ | | | | |
| 18 | Ren et al. (2011) | √ | √ | | √ | √ |
| 19 | Kasim & Ern (2010) | √ | √ | √ | | |
| 20 | Nwachukwa & Emoh (2010) | √ | √ | | | |
| 21 | Wan & Kumaraswamy (2009) | | | | | √ |
| 22 | Kazaz et al. (2008) | √ | √ | | √ | |
| 23 | Enhassi et al. (2007) | √ | | √ | √ | √ |
| 24 | Caldas et al. (2006) | √ | √ | | | |
| 25 | Novan & Berkhovich (2006) | √ | √ | | | √ |
| 26 | Kasim, Anumba & Dainty | √ | √ | √ | √ | |
| 27 | Song et al. (2005) | √ | √ | | √ | |
| 28 | Thomas et al. (2005) | √ | √ | | √ | |
| | **Total Frequency** | **24** | **21** | **8** | **14** | **8** |

Material inventory affects the construction time. The systematic inventory control and documentation reduce time consumption for labour to retrieve materials [41]. Other effects are on time delivery of material increases work progress, used appropriate quality material speed up finishing time, on time material procurement reduces idling time and reasonable changes reduces extra time for adjustment of resources [35-37, 42].

### 3.2 Cost

Cost of construction project is defined as a price to a client including consultant's fees, main contractors' and trade contractors' charges including their overhead and profit [43]. Cost is one of the major consideration in entire cycle of construction projects. Unfortunately, majority of the projects failed to achieve project completion according to the contract cost [44].

Table 1 shows 21 existing papers stated that material management has effect on cost performance. With effective material management in construction projects, it is able to reduce overall cost of material [45]. For example, in purchasing process, discounts and bulk order may be economical as it reduced the transportation and ordering cost [29, 46]. It has been estimated that a saving of 2% material cost possibly will increase profits by 14.6% [7]. Thus, by minimizing the procurement cost of materials, the higher chances for reducing the overall project cost and concurrently increasing company profit.

However, the reasonable time needs to consider as if the materials ordered too early, it may affect the company capital, interest charges and storage charges [45]. The total quantity required also vital to be considered as the main reason of cost variance in construction projects is due to overstocked materials on site [6]. Because of overstocked materials, organizations need to spend extra expenses on cost for labour wages and transportation charges to move out the materials from site.

Lack of standard or poor quality control in material management contributes to increase of construction cost. This is because the chances to loss material during handling and implementation stages is higher and it requires a material replacement [46]. For instance, the materials with poor quality will damage due to lack of durability, strength and density [36].

The planning and procurement is claimed as an equally important process that control total project cost [6]. However, material control and expediting also important as the former process is to avoid shortage and surplus of materials occurred on site. While the later process ensures the material being delivered in a timely manner by suppliers [7, 46]. In sum, all mentioned processes are important to reduce the escalation of construction cost.

### 3.3 Quality

Quality is defined as the specific characteristic, features or nature of products, execution or installations of a task on construction project as stipulated in the contract document [47]. However, the term quality has sometimes become a problem as it is a subjective matter and understood differently by different people and organizations [39, 48]. Therefore, it is important to consultant be able to identify clients requirements and specify clearly in contract document.

As shown in Table 1, 8 papers mentioned that material management affects quality performance. Availability of resources such as materials and equipment as planned during project duration is one of the factors contributing to quality performance [49]. The available equipment also needs to be in good condition and in sufficient quantity. Without the proper and sufficient equipment, it will jeopardize the satisfactory quality of work done [50]. Similarly, the materials itself also need to be in appropriate quality according to the specification [11, 42]. Using the appropriate quality of materials can lead to the satisfactory quality of work, subsequently will be accepted by consultant [36]. In sum, the available resources must be in accordance to the specification, sufficient quantity and functional.

### 3.4 Productivity

Productivity can be interpreted in many ways. In context of construction, productivity is related to labour productivity. It constitutes units of work placed or produced per staff [15]. The productivity is measured in terms of unit completely accomplished during given period and the related costs in terms of man-hours or money [51]. In short, productivity can be defined as ratio of earned either completed units or associated cost to actual hours [15].

A study on factors affecting Australian construction productivity reported that lack of materials is a primary factors that had a moderate to strong effect on productivity [50]. Similarly, studies conducted in Indonesia and Thailand also revealed that lack of materials affect construction productivity [52, 53]. More importantly, it has been accumulated that disruptions in material management resulted in 9% of loss of labour productivity [54].

Material storage also had a significant impact on labour productivity. For example, inappropriate storage location increases unproductive inputs. This occurs as the workers need longer time and effort to retrieve materials. Due to this, it only wasting workers' energy and may end up physically fatigue [55]. On the contrary, with appropriate site storage, workers energy can be productively used to perform task under progress.

This situation also similar with the efficient site layout for material movement. With efficient movement of materials, it increases productivity whilst reduces material travel time [19, 32]. The accessibility of site and unconfined working space also affects workers' productivity [55]. Besides, the availability of material and equipment motivated workers to improve work productivity [10]. Indeed, these studies demonstrated the interrelationship between the material management and productivity performance in construction projects as 14 papers mentioned it (Table 1).

### 3.5 Waste

Construction waste refers to any material except earth material that needs to be moved out from the site due to damage, excess either non-use or cannot be used due to non-compliance with specification and surplus or debris by-product of construction process [56]. In simple word, waste is a product or material that is unwanted and required to transport out from site [11].

Inappropriate material storage contributes to waste generation in Singaporean construction sites [56]. Likewise, previous studies in Malaysia also revealed that poor material storage and poor material handling are the highest cause to waste generation in local construction industry [11, 57]. The above mentioned study supported 8 prior papers about the effect of material management to waste performance.

Strategies for waste minimization are stock control for minimization of over or duplicate ordering, good practices of material handling, systematic inventory process and proper material storage [6, 18, 58]. Hence, the efficient material management practice throughout entire construction process will minimize the waste generation [1]. Additionally, minimization of material wastage in construction project is important to avoid loss of profit for the contractors [18, 25, 56]. Evidently, it has been revealed that 2.5% of total budget can be saved due to implementation of waste minimization strategies [59].

## 4 Conclusion

This paper had identified the effect of material management to 5 criteria of project performance. In conclusion:-

i) The availability and sufficient materials and equipment have effect on time, quality, productivity and performance.
ii) Appropriate quality material has effect on time, cost and quality performance.
iii) On time and reasonable time of material procurement have effect on time and cost performance.
iv) Efficient inventory system and documentation have effect on time and waste performance.
v) Reasonable changes has effect on time performance.
vi) On time delivery has effect on time performance.
vii) Minimizing procurement cost has effect on cost performance.
viii) Appropriate site storage has effect on productivity and waste performance.
ix) Efficient site layout has effect on productivity performance.
x) Easy site access has effect on productivity performance.
xi) Unconfined working space has effect on productivity performance.
xii) Efficient material controlling has effect on waste performance.
xiii) Appropriate handling has effect on waste performance.

Further examination of these effects of material management to project performance will be conducted with practitioners to confirm the severity of each effect on criteria of project performance.

## Acknowledgement

The authors would like to express the profound gratitude to their supervisor for her invaluable guidance, encouragement and supervision. The gratitude also to UTHM as this paper was partly sponsored by Centre for Graduate Studies UTHM.

## References

1. M. Safa, A. Shahi, C.T. Haas, K.W. Hipel, Automation in Construction **48**, 64–73 (2004)
2. L.C. Bell, G. Stukhart, Journal of Construction Engineering and Management **112**, 14–21 (1986)
3. C.H. Caldas, C.L. Menches, P.M. Reyes, L. Navarro, D.M. Vargas, Practice Periodical on Structural Design and Construction **20**, 1–8 (2014)
4. K.K Chitkara, *Construction Project Management: Planning, Scheduling and Controlling.* (1998)
5. W.C. Benton, L. McHenry, *Construction Purchasing & Supply Chain Management.* (2010)
6. A.A. Gulghane, P.V. Khandve, International Journal of Engineering Research and Application **5**, 59-64 (2015)
7. Z. Ren, C.J. Anumba, J. Tah, Advanced Engineering Informatics **25**, 198-207 (2011)
8. J.M. Sardroud, Scientia Iranica **19**, 381-392 (2012)
9. J.H. Berka, W.D. Conn, *AACE International Transactions 1994*, SI2-1(1994)
10. W. Barry, F. Leite, W.J. O'Brien, In *Construction Research Congress 2014*, 2296–2305 (2014)
11. S. Nagapan, I. Abdul Rahman, A. Asmi, International Journal of Advances in Applied Sciences **1**,1-10 (2012)
12. H. Doloi, A. Sawhney, K.C Iyer, S. Rentala, International Journal of Project Management, **30**, 479–489 (2012)
13. X. Meng, International Journal of Project Management **30**, 188–198 (2012)
14. I. Abdul Rahman, A.H. Memon, A. Azis, A. Asmi, N.H Abdullah, Research Journal of Applied Sciences, Engineering and Technology **5**, 1963-1972 (2013)
15. K.M. El-Gohary, R.F. Aziz, Journal of Management in Engineering **30**, 1–9 (2014)
16. Z. Shehu, G.D. Holt, I.R. Endut, A. Akintoye, Built Environment Project and Asset Management **5**, 52–68 (2015)
17. R.A. Rivas, J.D. Borcherding, V. González, L.F. Alarcón, Journal of Construction Engineering and Management **137**, 312–320 (2010)
18. D.U. Kini, Journal of Management in Engineering **15**, 30-34 (1999)
19. S. Donyavi, R. Flanagan, R. *Proceedings of the 25th Annual ARCOM Conference*, 11-20 (2009)
20. H. Abdul Rahman, M.N. Alidrisyi, Construction Management and Economics **12**, 413–422 (1994)
21. N. Kasim, P.A.S. Ern, Int. Journal of Computer and Communication Technology **2**, 1–10 (2010)
22. N. Kasim, KICEM Journal of Construction Engineering and Project Management, 31–36 (2011)
23. S. Meeampol, S.O. Ogunlan, Journal of Financial Management of property and Construction **11**, 3-20 (2006)
24. W.E. Back, L.C. Bell, Cost Engineering-Morgantown **38**, 13-17 (1996)
25. N. Kasim, C.J. Anumba, A.R.J. Dainty, *Proceedings Twenty First Annual Association of Researchers in Construction Management (ARCOM) Conference,* 793-802 (2005)
26. H. Said, K. El-Rayes, *Construction Research Congress 2012*, 1580–1589 (2012)
27. C.H. Caldas, D.G. Torrent, C.T. Haas, Journal of Construction Engineering and Management **132**, 741-749 (2006)
28. G. Ali, Advances in Management **7**, 17 (2014)
29. T.P. Madhavi, S.V. Mathew, R. Sasidharan, International Journal of Research in Engineering and Technology, 400-403 (2013)
30. B. Cooke, P. Williams, *Construction Planning, Programming and Control*, (2009)
31. W.R. Mincks, H. Johnston, *Construction Jobsite Management*, (2011)
32. P. Alanjari, S. Razavialavi, S. AbouRizk, Automation in Construction **40**, 1-8 (2014)
33. S.K. Wan, M.M. Kumaraswamy, Engineering, Construction and Architectural Management **16**, 208–223 (2009)
34. H.R. Thomas, D.R. Riley, J.I Messner, Journal of Construction Engineering and Management **131**, 808–815 (2005)
35. A. Kazaz, E. Manisali, S. Ulubeyli, Journal of Civil Engineering and Management **14**, 95–106 (2008)
36. A. Enshassi, S. Mohamed, Z. Abu Mustafa, P.E Mayer, Journal of Civil Engineering and Management **13**, 245–254 (2007)
37. R. Navon, O. Berkovich, Construction Management and Economics **24**, 635–646 (2006)
38. A.P.C. Chan, D. Scott, E.W.M. Lam, Journal of Management in Engineering **18**, 120–128 (2002)
39. A.S. Ali, I. Rahmat, Journal of Retail & Leisure Property **9**, 25–35 (2010)
40. H.S. Cha, C.K. Kim, KSCE Journal of Civil Engineering **15**, 1319–1328 (2011)
41. M.R. Abdul Kadir, W.P. Lee, M.S. Jaafar, S.M. Sapuan, A. Ali, Structural Survey **23**, 42–54 (2005)
42. C.C. Nwachukwu, F.I. Emoh, Interdisciplinary Journal of Contemporary Research in Business **2**, 90–105 (2010)
43. D. Towey, *Cost Management of Construction Projects* (2013)

44. N.J. Smith, T. Merna, P. Jobling, *Managing Risk in Construction Projects* (2014)
45. A.R. Patil, S.V. Pataskar, International Journal of Engineering and Innovation Technology **3**, 96-100 (2013)
46. R.F. Aziz, Alexandria Engineering Journal **52**, 51-66 (2013)
47. W.J. Del Pico, *Project Control-Integrating Cost and Schedule in Construction* (2013)
48. A.A. Kwakye, *Construction Project Administration In Practice* (1997)
49. K.N. Jha, K.C. Iyer, Total Quality Management & Business Excellence **17**, 1155–1170 (2006)
50. R. Hughes, D. Thorpe, Construction Innovation: Information, Process, Management **14**, 210–228 (2014)
51. R.F. Cox, R.R. Issa, D. Ahrens, Journal of construction engineering and management 129, 142-151 (2003)
52. P.F. Kaming, P.O. Olomolaiye, G.D. Holt, F.C. Harris, International Journal of Project Management **15**, 21–30 (1997)
53. A. Makulsawatudom, M. Emsley, *17th Annual ARCOM Conference*, 281–290 (2001)
54. H.R. Thomas, D.R. Riley, V.E. Sanvido, Journal of Construction Engineering and Management **125**, 39-46 (1999)
55. A.M. Jarkas, C.G. Bitar, Journal of Construction Engineering and Management **138**, 811–820 (2011)
56. L.L. Ekanayake, G. Ofori, Building and Environment **39**, 851–861 (2004)
57. S.H. Hassan, N. Ahzahar, M.A. Fauzi, J. Eman, Procedia - Social and Behavioral Sciences **42**, 175–181 (2012)
58. R.A. Begum, C. Siwar, J.J. Pereira, A.H. Jaafar, Resources, Conservation and Recycling **51**, 190–202 (2007)
59. R.A. Begum, C. Siwar, J.J. Pereira, A.H. Jaafar Resources, Conservation and Recycling **48**, 86–98 (2006)

# Screening of Geomechanical Risks for Malaysian Development Field

Syed Muhammad Syafiq Syed Najmuddin and Muhammad Aslam Md Yusof[a]

*Peroleum Engineering Department, Universiti Teknologi PETRONAS, 32610 Seri Iskandar, Perak, Malaysia*

**Abstract.** Deeper drilling and exploitation of difficult reservoir is the new trend in oil and gas industry. Geomechanics study has, therefore, become a new requirement particularly for oil and gas field development. However, a complete geomechanics study is limited with the number of experts, time consuming and not a straightforward task. Therefore, there is an urgent need of a quick geomechanics screening criterion to be used as a standard guideline to evaluate the high level geomechanical risks and suitable analysis can be recommended for the identified development fields. The aim of this paper is to propose a screening criterion for geomechanical risks study based on four key parameters, drilling, depletion, injection and storage and sand production. The screening approach is designed based on Risk Assessment Matrix (RAM) risk screening where the likelihood is based on a set of scores developed to specific questions. The consequence for each failure scenarios is assessed based on educated estimation of the impact towards people, asset, environment and reputation. Recommendations for geomechanical study are made based on the severity of each failure category on the RAM risk matrix. Fourteen development fields in offshore Peninsular Malaysia, offshore Sarawak and offshore Sabah are selected for the assessment. Based on results, fields in offshore Sarawak and Sabah have higher potential for geomechacnical issues mainly because of their geological settings and formation characteristics. A set of geomechanical study is proposed for each individual field for prudent management of potential geomechanics risk associated with the depletion and EOR injection scheme planned for the fields.

## 1 Introduction

Geomechanics is related to the study of rock and soil behavior. There are two major types of geomechanics which are rock mechanics and soil mechanics [1]. Geomechanics deals with soil behavior starting from a small scale to a landslide scale. Presently, it deals with geosciences issues related to mass rock mechanics and mass rock characterization like things applies to high depths or petroleum industry, rock breakage, drilling of rock and tunnel design. As challenging conditions are more frequently encountered because of deeper drilling, exploitation of more difficult reservoirs, more focus on heavy oil, and other factors associated with the gradual depletion of easily recovered conventional oil, geomechanics has become more important to the oil and gas industries [2]. In the past years, geomechanics studies have only being done if it is requested by the field. Though, this practice is not reliable enough in the current world. Thus, a screening criteria spreadsheets need to be developed. It is essential to be used as a standard guideline to evaluate the high level geomechanical risks and suitable analysis can be recommended for every development fields.

To achieve geomechanics study goals, it is very crucial to reduce the risks and costs related to refining drilling programs, making a better predictions of production rate and optimizing the development plans of field. The key project objectives are to utilize a set of geomechanics screening criteria and process, assess in drilling, depletion, injection and storage also the sand production. Moreover, this is to assess the geomechanical risk for each individual development field in consideration of RAM risk matrix and rank the field that is affected geomechanically based on the development plans and their geological settings.

## 2 Screening Criteria

### 2.1 Drilling

A big changes in the fracture gradients and pressure of pore in drilling may result in stuck pipe, washouts, unplanned sidetracks and loss of circulation [3]. Therefore, the study of geomechanics helps to enabling either avoidance, identifying geohazards and predicting an operational windows that is safer. The operation of drilling has given an impact on wellbore instability, pack off, stuck pipe, mud loss and drilling induced fault reactivation. Borehole instability or also known as wellbore stability is a condition that is critical for an open hole interval. The causes are the failure mechanically causes by in-situ stresses, erosion causes by circulation of

[a] Corresponding author: aslam.myusof@petronas.com.my

fluid, chemical causes by borehole fluid interaction with the formation. When a pipe cannot be removed unless causing any pipe damages and without exceeding the maximum load off the hook of the drilling rigs. This phenomenon can be classified into differential pressure pipe sticking and mechanical pipe sticking.

Moreover, whenever the drilling fluid does not returning to the annulus but flows into the geological formation, it is called the mud loss. During the drilling operation in any well, the existence of lost circulation is considered as a serious problem. The presence of mud loss may cause the loss of high amount of drilling fluid costs and result in loss of humans life [4]. Therefore, a high concern should be put closely on the pits, tanks and flow from the well to ensure a fast assess lost control circulation is being taught well and practiced every time. In addition, the movement along formerly inactive faults is called fault reactivation. This will contribute to alleviate strain within the upper mantle and crust. There are 3 factors that has contributing much which are pore pressure being elevated locally because of hydrocarbon reservoir containing natural gas on the footwall side of the fault, an orientation of fault that is oriented optimally for frictional slip in today's stress field and a compressional stress that increase recently in the postglacial rebound area.

## 2.2 Depletion

Knowledge of deformation of stress within and surrounding the reservoir depletion is important in a multi-disciplinary approach to reservoir characterization, reservoir management, and enhanced oil recovery projects. The most common problems are related with surface subsidence and casing collapse that create substantial difficulties in some reservoirs due to compaction in weak formation [5]. The significant stress changes that occur in high reservoir being depleted can result in drilling new well into deeper target reservoir problematic due to the need to lower mud weights in depleted formations to avoid mud lost circulation. Induce faulting also is part of the possibility for depletion either within or outside the reservoirs depending on the well geological setting.

Excessive subsidence is the sinking of soil due to a major problem in the developing world without any enforcement and regulation and the extraction from groundwater. The pressurized fluid within pore spaces and the rock matrix are the one who support the overburden sediments weight above formation produced. More load is being transferred to the rock matrix especially when the oil and gas production reduces the fluid pressure [6]. Caprock shear failure will happen when pore pressure increases. This is mainly due to the change of phase in the water that results in the reduction of effective-stress.

Reactivation of fault has a lot of effects and consequences. It shears the wellbore which occur when the fault zone of slipped portions is drilled through, the stored elastic energy is being released, the permeability of formation and production-depletion strategies being changed and seismicity is induced [7]. Usually, four thresholds are defined for instability in wellbore, and drilling program design is partly based on trying to keep the mud weight within the safe mud window. Kick, break down, losses and breakout are the common thresholds.

## 2.3 Injection and Storage

Fluid injection induces an increase in the pore pressure and gradient of the pressure at the borehole. The pressure variations affect the materials mechanical equilibrium around the boreholes and changes the in-situ stresses. In theory, the poro-elastic effect does improve the total stresses [8]. Additionally, fluid injection in reservoirs also modifies the thermal and hydraulic equilibrium around wellbore and cause strains to the reservoir as well to the entire sedimentary column. In assessing geomechanical influences during injection and storage for reservoirs, it is essential to evaluate based on the containment characteristic of the formations in terms of risks of leakage through fault, well and top seal. Similarly, the evaluation of the induced subsurface and surface deformation in terms of risk of induced seismicity and ground movement are equally important.

Caprock shear failure is mainly because of steam injection in high pressure within the reservoir is an advantage [9]. However, if these fractures extend and grow to the overlying layers of shale, this may result in failure of shear at the interface of shale which compromise on the ability of containment. Besides that, the propagation fracture of the fluid-driven relates with the process in multiple physical such as the flow in porous aquifer and fracture and also mechanical fractures [10]. In addition, the rate-limiting process for propagation fractures will happen due to the low permeability of aquifer which result in a slow fluid outflow.

An analysis of slip tendency is used to assess the potential of shear reactivation and fractures dilatational in a deep geothermal reservoir which is based on the notion that slip on faults is controlled by the ratio of shear to normal stress acting on the plane of weakness in the in situ stress field [11]. The initiation of fracture coupled with micro seismic events is important to produce extra fractured flow paths that improve the productivity and permeability. However, an undesirable seismicity is caused by a fluid injection that are not being change to the strength of rock conditions and to the in-situ stress field [12].

With the slip tendency analysis the potential for slip along any fault orientation with respect to the ambient stress field is investigated and therefore it is possible to assess the fault reactivation potential. This technique has been used for seismic-risk and fault-rupture-risk assessment in earthquake-prone areas and to understand the relative importance of shearing versus dilation behaviors along faults and bedding planes during deformation.

## 2.4 Sand Production

The reservoir fluid flows causes the formation sand migration [13]. Commonly, the sand production is not desirable and if severe, sand production may require some remedial action to control or prevent production. Basically, unconsolidated sandstone reservoirs are the most prone to sand production where most probably initiated during initial flow or later when reservoir pressure decline or water breakthrough. Continuous sand production is sometimes can be neglected or considered tolerable. However, this way might cause the well to be extremely damaged, the equipment at the surface cannot function very well and the production operation can be killed.

For perforated well, it is also believed that sand production may be initiated from perforation tip, which may be assumed to have a hemispherical geometry after the perforation shape stabilizes through some time of oil or gas production [14]. They studied stress state near a sand arch by considering steady state and incompressible fluid flow. This is done in terms of wellbore pressure, drawdown, frictional angle and formation rock cohesion [15].

It is undeniable that when there is a presence of sand, it will result in a high cost which occur whenever drag forces on sand particles, induced by the flow of fluid and/or solution-gas drive, exceed the forces of inter-granular which result in the mechanical integrity loss. Quite a number of authors have realized and point out the problem from both control and prediction points of view. Sand production is considered one of the challenging issues in various steps of field development including borehole instability during production, drilling and IOR/EOR stages is a natural impact of flow of fluid into a borehole from the reservoir [16]. The ordinary production engineer still need to consider relying on the rudimentary predictive techniques being widely used in the world even though there are an existence of sophisticated proprietary computer programs, such as Schlumberger's Impact and Conoco's three dimensional FE models [17].

## 3 Methodology

The framework of the screening criteria and approach comprises of the following:

1. Evaluation aspect category: drilling, depletion, injection and storage, sand production
2. Questionnaire: question, weightage, scoring
3. Risk evaluation: category, impact quantification
4. Risk assessment matrix (RAM)
5. Recommendations

The screening evaluation is categorized into 4 aspects which are drilling, depletion, injection and storage, and sand production. The first 3 aspects directly change the stress state of the formation and thus, closely related to hydrocarbon containment. Meanwhile, the last aspect, sand production is the effect of stress changes in the formation and is not the primary focus of the project.

The screening criteria approach for this project focused mainly on reservoir geomechanics. It is essential to identify the potential issues or problems that might occur for each life cycle of the reservoir so that proper mitigation can take place. There are several main components in the screening spreadsheet that are crucial and needed for evaluation of the geomechanical risks in development fields. These components include major activities, initiating event, threat, hazard and barrier category, weight factor and question response. All of these components are decided after a thorough and several session of discussion between the project team based on team's field experiences and research studies.

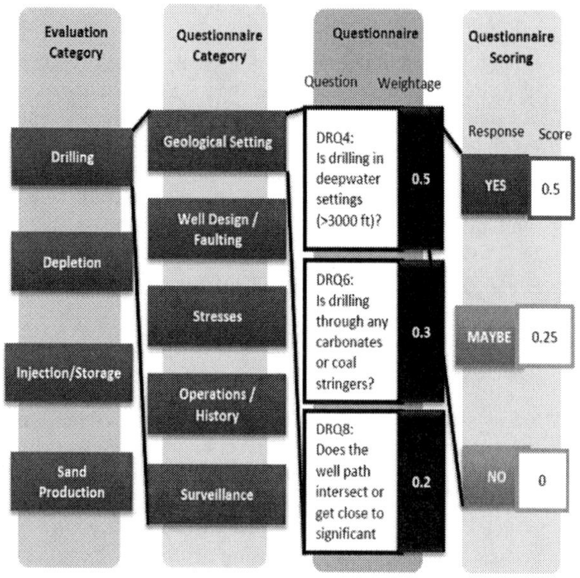


**Figure 1**: Questionnaire Classifications

There are four major activities being consider in the spreadsheet that includes drilling, depletion, injection and storage as well as sand production (Figure 1). Under each activities, there are number of questions, which also known as the initiating event that have been considered to have effects on geomechanics issues. In order to complete the list of initiating event, there are some deliberations done. For instance, there is a need to ensure that the questions are understandable. Apart from that, it need to make certain that the questions are answerable whereby the information needed is within the asset team scope of work. Last but not least, the number of the questions must be appropriate and not too time consuming for them to answer. Since the questions are closed-ended question, it is decided to use intensity scales i.e. yes, maybe or no as the answering mode for the screening criteria.

In addition, this method is believed to ease the answering process by the asset team. However, if the answer provided is vague the asset team need to explain further in the spreadsheet so that it is able to understand the issues for the specific field that will further help the evaluation process of the geomechanical risk.

 DOI: 10.1051/matecconf/20178701013

Threat is one of the main components in the spreadsheet. In this project, threat has the definition of issues that are under consideration that might cause by the main activities. For instance in the drilling section, some of the threats are wellbore instability, pack off and stuck pipe, mud loss and drilling induced fault reactivation. This is because while drilling, some of these issues might occur due to changes in the rock formation and in-situ stresses. The purpose of the threat section is to calculate and determine the possibility for the issues to occur for each field.

Hazard and barrier category is included in the screening criteria as well. There are six categories that vary for each of the major activities. The purpose of this section is to ensure that the initiating event developed covers the entire category. It is also to ensure the questions are balanced and all the aspects are being covered. On top of that, weight factor column has been created to accommodate the tracking of the hazard itself. It means that the hazard category and the weight factor are inter-related.

| Injection / Storage Threats Under Consideration: | | Normalized Score |
|---|---|---|
| IST-1 | Caprock Shear Failure | 1.00 |
| IST-2 | Fracture Propagation into Caprock | 1.00 |
| IST-3 | Fault Reactivation - stress-induced | 1.00 |
| IST-4 | Fault Reactivation - Induced Fracture Interception | 1.00 |
| Injection / Storage: Final Score | | 1.00 |

Recommendation on geomechanics study is based on the final score

**Figure 2**: Normalize score in specific issue

Under each threat, there is also one main component which is the influence factor. The influence factor is to indicate the presence of relation between the initiating event and the threat itself. By having this component, it will be much easier to evaluate the risk for the geomechanical issues for each field. As the score for the influence factor will eventually sum up and a normalized score is being calculated for each threat.

From the normalized score, the risk will be rank using the risk ranking table (Figure 2). The risk ranking is based on the likelihood and the consequences of the event to four main traits which are people, asset, environment and reputation (Figure 3). The final risk index will be automatically calculated whereby the recommended action can be generated based on the risk index itself. This will help to give some idea to the project team on the way forward for all of the fields as well as to categorize the need for geomechanical studies for all the fields.

**Figure 3**: Risk ranking score limit

Risk Assessment Matrix (RAM) is a matrix that is used during the assessment of risk study to define various levels of risk as the product of harm probability categories and harm severity categories (Figure 4). This is a simple mechanism to increase visibility of risks and assist management in decision making. Risk matrices in certain cases experience several problematic mathematical features making it harder to assess the risks. Furthermore, there is existence of error using this matrix. This is because the risk matrices can mistakenly assign higher qualitative ratings to quantitatively smaller risks. For risks with negatively correlated frequencies and severities, they can be "worse than useless," leading to worse-than-random decisions.

In addition, there is also a suboptimal resource allocation problem arises where effective allocation of resources to risk-reducing countermeasures cannot be based on the categories provided by risk matrices. Besides that, it is an ambiguous inputs and outputs problem. Categorizations of severity cannot be made objectively for uncertain consequences. Inputs to risk matrices (e.g., frequency and severity categorizations) and resulting outputs (i.e., risk ratings) require subjective interpretation, and different users may obtain opposite

| SEVERITY | CONSEQUENCES | | | | INCREASING LIKELIHOOD (SCORE LIMIT) | | | | |
|---|---|---|---|---|---|---|---|---|---|
| | People | Assets | Environment | Reputation | A | B | C | D | E |
| | | | | | >0 | >0.2 | >0.4 | >0.6 | >0.8 |
| 0 | No injury or health effect | No damage | No effect | No impact | | | | | |
| 1 | Slight injury or health effect | Slight damage | Slight effect | Slight impact | | | | | |
| 2 | Minor injury or health effect | Minor damage | Minor effect | Minor impact | | | | | |
| 3 | Major injury or health effect | Moderate damage | Moderate effect | Moderate impact | | | | | |
| 4 | PTD or up to 3 fatalities | Major damage | Major effect | Major impact | | | | | |
| 5 | More than 3 fatalities | Massive damage | Massive effect | Massive impact | | | | | |

**Figure 4**: Risk Assessment Matrix (RAM)

ratings of the same quantitative risks. These limitations suggest that risk matrices should be used with caution, and only with careful explanations of embedded judgments.

The probability of harm occurring might be categorized as 'Certain', 'Likely', 'Possible', 'Unlikely' and 'Rare'. However it must be considered that very low probabilities may not be very reliable. The asset team will only need to answer the spreadsheet by input yes, maybe or no and could add some justification so that the project team able to understand the issue in a correct manner. The evaluation part will be done by the project team based on past experiences, fields' history data and the research studies done.

## 4 Results & Discussion

Figure 5 shows the drilling score for all fields that have been studied. The final score is the generic results from four drilling threats that were given in the questionnaire for every field team. The four threats are general wellbore instability (DRT-1), packoff and stuck pipe (DRT-2), mud loss (DRT-3) and drilling induced fault reactivation (DRT-4). The score for each threat are gained from the influence factor that vary based on their impact towards each field. The bigger the impact, the higher the weightage it carry where the normalize score value towards '1' means that the field has a higher risk for the respective threat compared to the normalize score value towards '0'. Final score is obtained by calculating the average score from all threat present. This is to rank the field risk accordingly.

In Figure 5, Field L has the highest final score which is '0.6'. This is mainly due to the high effect of drilling induced fault reactivation. Thus, it is critical to conduct details geomechanics study in order to ensure smooth drilling and production operation in the future. The risk index for this field is '2'. The value is given based on the probability of harm occurring that has been discussed in the methodology section. Therefore, the recommended actions for this risk index is to review offset wellbore stability work. Update the formations stresses (based on any depletions, etc.) and formation mechanical properties (with any new data), and perform appropriate mud weight prediction for the new well trajectorie.

For depletion issue, as shown in Figure 6, Field B has the highest final score which is '0.76'. All fields were ranked based on their final mark in depletion issue. There are four threats that have been considered in depletion issue which are excessive surface subsidence (DPI-1), caprock shear failure (DPI-2), fault reactivation (DPI-3) and compaction/well failure (DPI-4). Based on RAM risk assessment, the risk index for Field B is '3'. Therefore, the recommended actions for this risk index is to perform comprehensive 3-D geomechanical modelling to evaluate the compaction/subsidence magnitudes, top seal integrity and potential of fault reactivation for various depletion scenarios and stages. It is also recommended to request GPS monitoring for subsidence prediction and model updates throughout the field life.

The next geomechanic issue is injection and storage. We evaluated the field based on their risk possibility in four types of threats namely caprock shear failure (IST-1), fracture propagation into caprock (IST-2), fault reactivation – stress-induced (IST-3) and fault reactivation – induced fracture interception (IST-4). Based on results shown in Figure 7, Field D has the highest final score which is '0.74'. This is mainly due to the high effect of caprock shear failure and fault reactivation – stress-induced. Meaning that, Field D will be the most prioritize field to be done research on geomechanics compared to other fields. The risk index for this field is '3'.

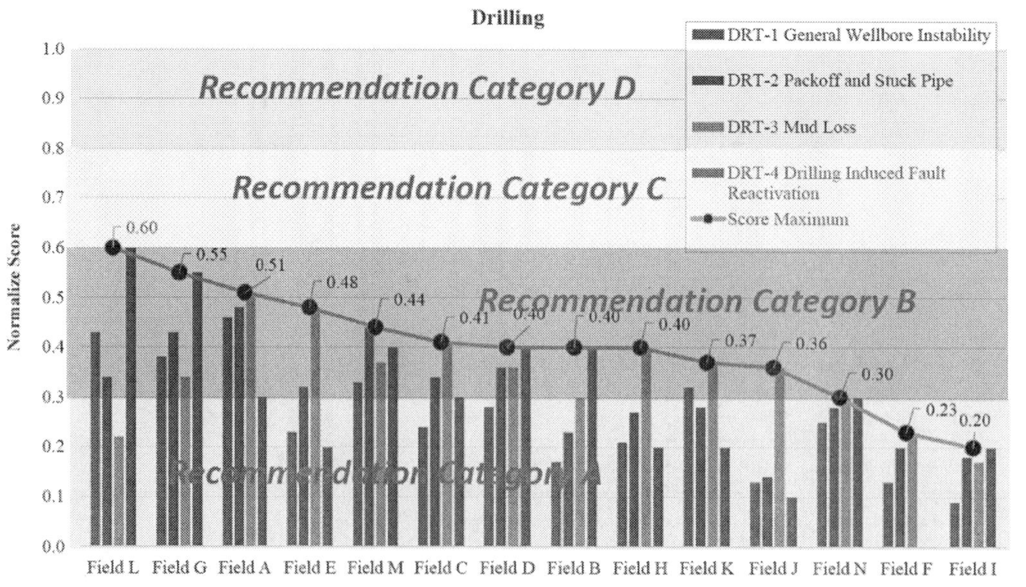


**Figure 5**: Drilling score for all fields

DOI: 10.1051/matecconf/20178701013

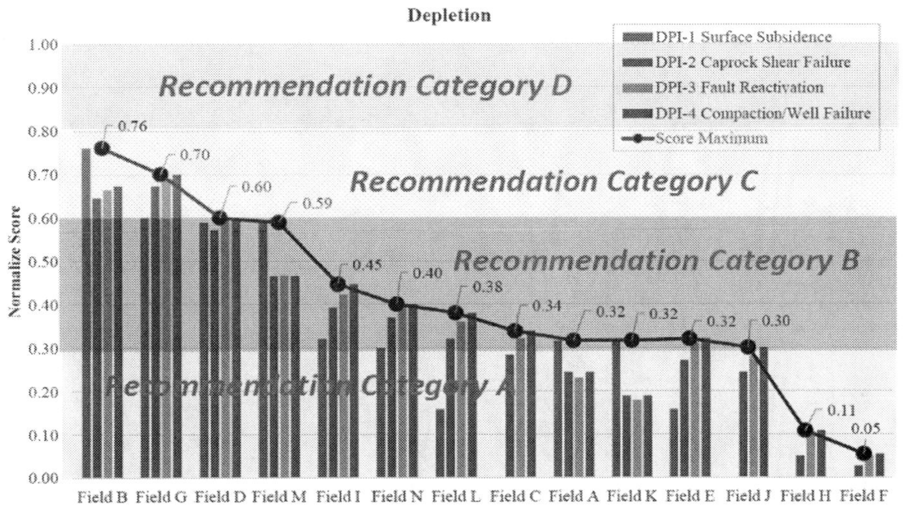


**Figure 6**: Depletion score for all fields

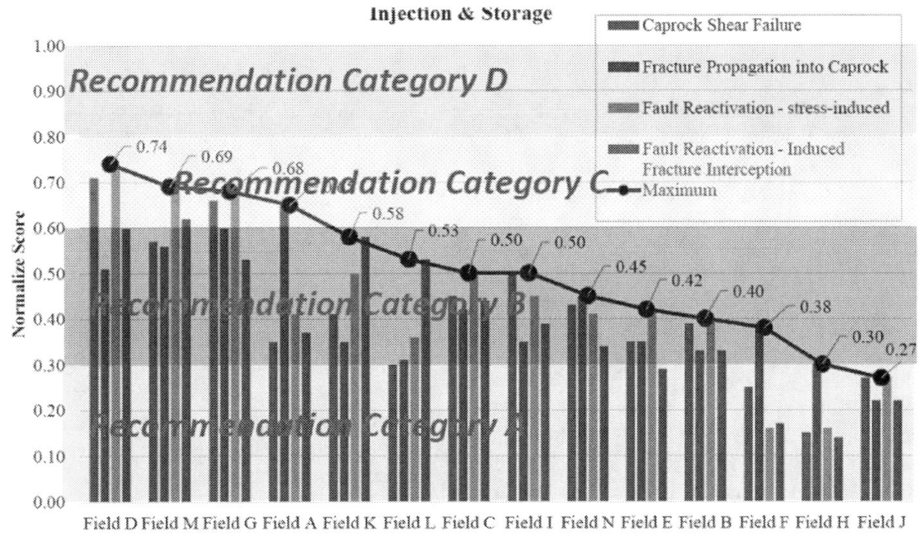


**Figure 7**: Injection and Storage score for all fields

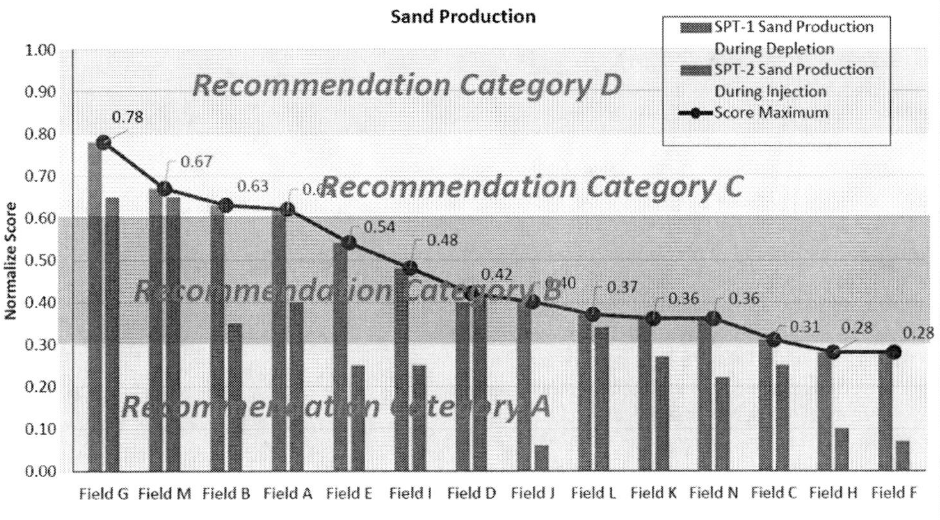


**Figure 8**: Sand Production score for all fields

 DOI: 10.1051/matecconf/20178701013

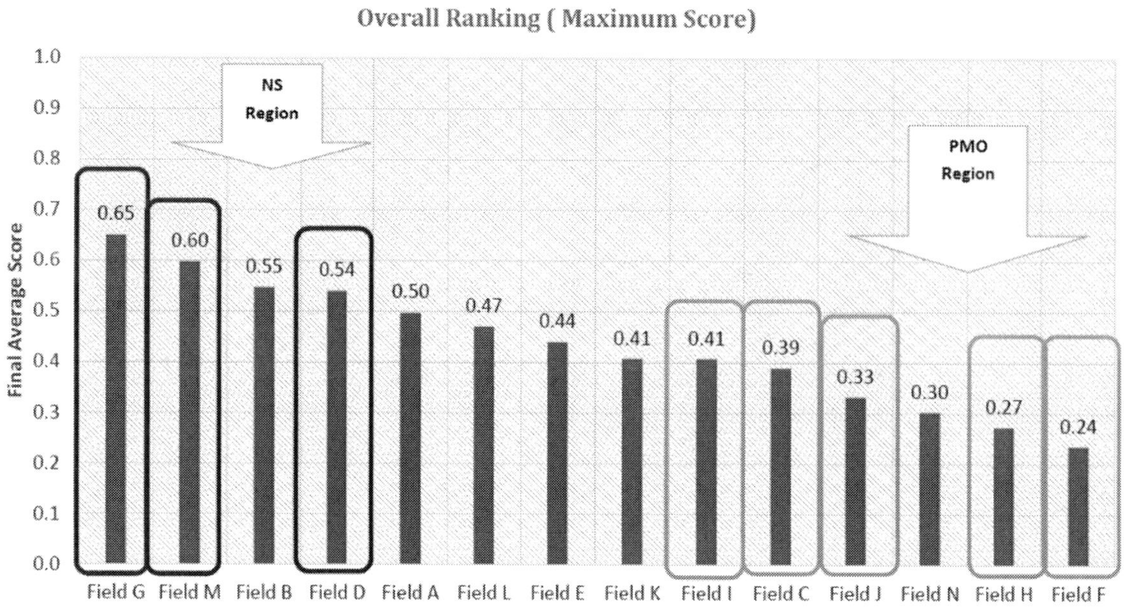


**Figure 9**: Overall risk ranking for all fields

Therefore, the recommended actions for this risk index is to perform comprehensive 3-D geomechanical modelling to evaluate top seal integrity and fault reactivation potential for various injection scenarios and stages. If the project team is planning to do injection under fracturing, it is highly recommended to perform fracture prediction (e.g., with REVEAL, PWRI-Frac or Frac-IT). In addition, the project team for Field D is advised to monitor for the critical locations based on the geomechanical model predictions.

Figure 8 shows the ranking for the sand problem issue for all fields. Only two threats that have been considered for this issue which are sand production during depletion (SPT-1) and sand production during injection (SPT-2). As a result, Field G has the highest final score which is '0.78'. This is mainly due to the high effect sand production during depletion. The risk index for this field is '3'. Therefore, the recommended actions for this risk index is to perform comprehensive sand production modelling, including formation characterization, rock mechanical characterization (UCS, TWC tests, etc.), and sanding estimation, e.g. with FIST.

Furthermore, Figure 9 summarizes the overall risk ranking for all field and thier regions. It is found that, North Sabah (NS) region i shaving higher geomechanical risk compared to Sarawak and Peninsular Malaysia (PMO). This is due to fact that majority of the oil and gas commercial accumulations were found in the Middle Miocene or younger clastic reservoirs in the Inboard Belt, Outboard Belt and East Baram Delta. The Inboard Belt is characterized by intense compressional features of the Sabah ridges separated by wide, deep synclines, particularly well-developed towards the western margin of the Belt. In the south, the main tectonic lineaments trend N-S and form a series of parallel, narrow wrench-related anticlines, with steep flanks and intensely faulted crest. The central part of the belt is tectonically more complex with both WNW-ESE and N-S trending tectonic features forming domal faulted fold interference patterns.

## 5 Conclusion

The questionnaires of screening criteria for geomechanical risk on development fields is successfully developed. This screening criteria can be used as a standard guideline for geomechanical study for all development fields. The result and discussion in the project is based on the geomechanical theory from the literature studied and the real assessment done towards relevant fields. Fourteen (14) identified development fields are selected for the assessment. These fields are from Baram Delta Operations (BDO) – 6 fields (Field E, A, L, K, B, N), North Sabah (NS) – 3 fields (Field G, D, M) and Peninsular Malaysia Operations (PMO) – 5 fields (Field J, I, F, H, C). Based on the assessment results, BDO and NS fields, comparing to PMO fields, have generally higher potential for geomechanics issues mainly due to the geological settings and formation characteristics.

Based on this study, the consequences towards people, asset, environment and reputation can be determined. This is very important due to the changes that occur to every field will affect all the 4 main aspects. Then, the level of effect is judged based on their severity. Moreover, there are also recommendation actions being suggested which was believed to help in reducing the risk of geomechanics. This mitigation are made differently for each threat which are drilling, depletion, injection & storage and sand production. Also, the recommended actions are rank based on the level of impact towards each field. Final result indicate that fields in offshore Sabah is more critical or should be prioritize for geomechanics study compared to peninsular Malaysia and Sarawak regions because of the geological complexity in that area.

## References

1. Sivakugan, N., S.K. Shukla, and B.M. Das, *Rock Mechanics: an introduction*. 2013: Crc Press.
2. Mody, F. and G. Wang. *Application of geomechanics technology in borehole stability reduces well construction costs*. in *The 42nd US Rock Mechanics Symposium (USRMS)*. 2008. American Rock Mechanics Association.
3. Naterstad, T.L., *Utilizing Managed Pressure Casing Drilling in Depleted Reservoir Zones*. 2014.
4. Engle, C.J. and S.J. Marwil, *Extracting and analyzing gas from well drilling mud*. 1960, Google Patents.
5. Zoback, M.D., *Reservoir geomechanics*. 2010: Cambridge University Press.
6. Beck, E., et al., *Concepts for Drilling and Excavating in and below the Ocean Bottom*. 1971, DTIC Document.
7. Germanovich, L.N., et al. *Poroelastic mechanism of fault reactivation as a result of reservoir depletion*. in *Pro& ceedings of the Biot conference on poromechanics*. 1998.
8. Nauroy, J.-F., *Geomechanics applied to the petroleum industry*. 2011: Editions Technip.
9. Collins, P.M. *Geomechanical effects on the SAGD process*. in *SPE International Thermal Operations and Heavy Oil Symposium*. 2005. Society of Petroleum Engineers.
10. Gor, G.Y., H.A. Stone, and J.H. Prévost, *Fracture propagation driven by fluid outflow from a low-permeability aquifer*. Transport in porous media, 2013. **100**(1): p. 69-82.
11. Moeck, I., G. Kwiatek, and G. Zimmermann, *Slip tendency analysis, fault reactivation potential and induced seismicity in a deep geothermal reservoir*. Journal of Structural Geology, 2009. **31**(10): p. 1174-1182.
12. Deichmann, N. and D. Giardini, *Earthquakes induced by the stimulation of an enhanced geothermal system below Basel (Switzerland)*. Seismological Research Letters, 2009. **80**(5): p. 784-798.
13. Muecke, T.W., *Formation fines and factors controlling their movement in porous media*. Journal of Petroleum Technology, 1979. **31**(02): p. 144-150.
14. Bratli, R.K. and R. Risnes, *Stability and failure of sand arches*. Society of Petroleum Engineers Journal, 1981. **21**(02): p. 236-248.
15. Nouri, A., et al., *A comparison of two sanding criteria in physical and numerical modeling of sand production*. Journal of petroleum science and engineering, 2006. **50**(1): p. 55-70.
16. Alipour Tabrizy, V. and Y. Mirzaahmadian. *Investigation of Sand Production Onset: A New Approach Based on Petrophysical Logs*. in *SPE International Symposium and Exhibition on Formation Damage Control*. 2012. Society of Petroleum Engineers.
17. Morita, N., et al., *Realistic sand-production prediction: numerical approach*. SPE Production Engineering, 1989. **4**(01): p. 15-24.

DOI: 10.1051/matecconf/20178701014

# Effectiveness of Mangrove Forest as Coastal Protection along the West Coast of Northern Peninsular Malaysia

Ahmad Mustafa Hashim[a] and Diana Bazila Shahruzzaman

*Department of Civil and Environmental Engineering, Universiti Teknologi PETRONAS, 32610 Seri Iskandar, Perak.*

**Abstract.** The effectiveness of mangrove forest as coastal protection was evidenced during Indian Ocean Tsunami (IOT) on $24^{th}$ December 2004. This paper presents the assessment on the characteristics and distribution of the mangrove forest along three northern states (Perlis, Kedah and Perak) affected by IOT 2004. At each site, the characteristics of individual tree were recorded including their species, height and diameter at breast height (DBH). A total of 52 matured trees were sampled that encompassing four species; *A. marina* (n=21), *A. alba* (n=7), *R. mucronata* (n=20) and *R. apiculata* (n=4) were found with *A. marina* as the dominant species. The zonation of mangrove forest along the northern states revealed that Avicenniacea inhabits the frontal area. The field survey also discovered that mangrove forest along the states was generally insufficient to protect the coastal area from future tsunami. There were several rehabilitation programs that had been implemented, but not all replanting efforts were successful. Outcomes from this research provided invaluable findings for the optimum replanting scheme to ensure acceptable level of protection along Malaysia's coastline.

## 1 Introduction

### 1.1 Indian Ocean Tsunami 2004

Tsunami is one of the disastrous natural events occurred and mainly generated by the sliding of the earth's surface, namely earthquakes. Generally, tsunami is triggered by the earthquake of magnitude 6.5 and above [1]. Up to 2012, Japan was recorded as the highest frequency of tsunami occurrences with 48 events [2].

The most catastrophic event recorded in tsunami history was the Indian Ocean Tsunami which occurred on $24^{th}$ December 2004 (IOT 2004). The tsunami waves reached 13 countries including Indonesia, Sri Lanka, India, Thailand, Somalia, Burma, Maldives, Malaysia, Tanzania, Bangladesh and Kenya [3]. The epicentre of the event occurred at about 30 km off the West Coast of Northern Sumatra, Indonesia. The 20 m wave height led to about 226,000 deaths and approximately USD 9.93 billion of economic losses [4,5,6]. The tsunami wave had propagated to the West Coast of Peninsular Malaysia and affected five states including Perlis, Pulau Pinang, Kedah, Perak and Selangor with 63 numbers of total deaths and missing people reported [7].

### 1.2 Mangrove Forest as Coastal Protection

The effectiveness of mangrove forest as coastal protection in front of the coastal area was evidenced during the IOT 2004 [8]. The coastal properties without any protection were completely destroyed by the tsunami wave. Figure 1 illustrates how mangrove forest helps to dissipate the wave energy, hence, reducing the impact along the coastal area [9]. The importance and benefits offered by mangroves have encouraged increasing number of related research. There had been numerous mangrove rehabilitation programs and replanting initiatives for protection at affected area organized by various agencies.

However, the optimum approach or technical strategy for the mangrove replanting scheme is still under active research. Several factors govern the performance of mangrove as coastal protection which include species of the tree, characteristics of tree (height, age, diameter of stem and root, diameter at breast height), basal area and also wave characteristics [8,10]. Thus, this paper aim to assess the characteristics and distribution of the mangrove forest along the West Coast of Peninsular Malaysia.

[a] Corresponding author: mustafa_hashim@utp.edu.my

DOI: 10.1051/matecconf/20178701014

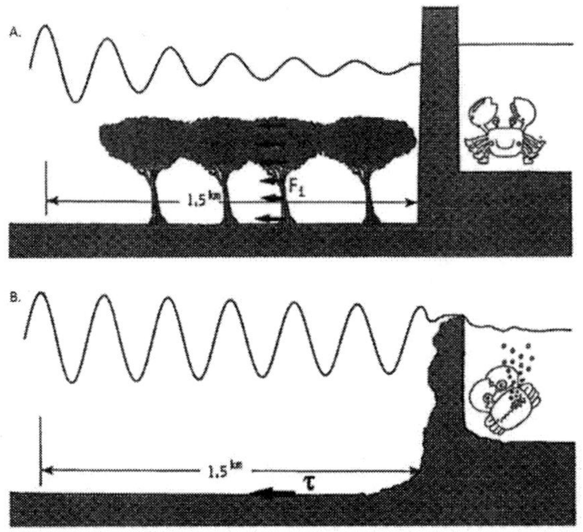


**Figure 1.** Wave energy and its impact at coastal area for A. With mangrove, B. Without mangrove [9].

## 2 Methodology

### 2.1 Study Area

West Coast of Peninsular Malaysia had been hit with IOT 2004. The area is bordered by the calm sea of Malacca Straits, which enabled the small particles of sediment (mud) to settle down at the coastal area [11]. Hence, the conditions contributed to the growth of mangrove forest in this part of coastline rather than the East Coast of Peninsular Malaysia [12]. However, these areas affected by the IOT 2004 are not fully recovered and potentially at high risk for future tsunami. In order to explore the current level of protection available at the previously affected IOT 2004 area, 10 sampling locations were assessed (Fig 2). These involved sites covering the three states, namely Perak (Parit Haji Dollah, Tg Kepah, Pantai Remis, Kuala Sepetang and Kuala Gula), Kedah (Pantai Merdeka, Kuala Sg Limau, Sg Ibus, Kuala Teriang) and Perlis (Kuala Sanglang).

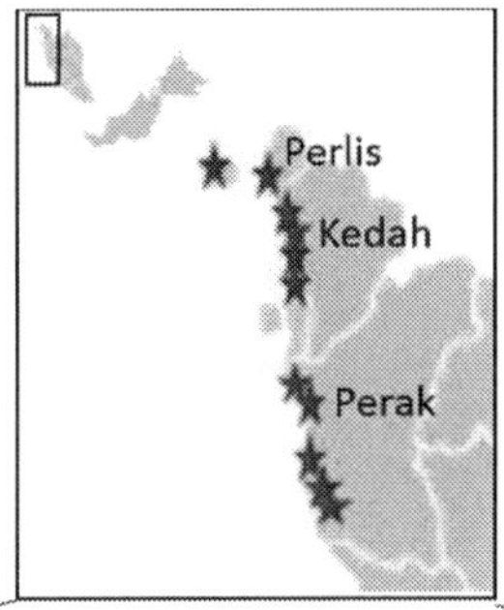

**Figure 2.** Locations of study area located at the coastline of Perlis, Kedah and Perak (Image adapted from Google Earth, 2016)

### 2.2 Data Collection

The mangrove forest at each site was assessed and the time, date, coordinate and tide level were recorded. The random sampling plots (30 x 30 m) were established at each site. The sampling plots were chosen as a representative of general mangrove condition for the accessed area. The selected matured trees found within the plot were identified. Measurement were carried out covering chosen trees representing different age groups. The data from selected trees were gathered include species of mangrove, basal area in $m^2$, height of tree in m, and diameter at breast height (DBH) in cm.

### 2.3 Statistical Analysis

#### *2.3.1. Diameter at Breast Height (DBH)*

To determine the DBH, the circumference of the mangrove tree at 1.3 m height was measured at site. The DBH was calculated by using the relationship (Eq 1) stated in [13].

$$DBH = \frac{Circumference_{1.3}}{\pi} \quad (1)$$

Where;

$Circumference_{1.3}$ = Circumference at 1.3 m height

#### *2.3.2 Species diversity*

The ecological approach was used to determine the diversity of the mangrove forest. The relationship proposed by Shannon-Weiner (Eq 2) was used to derive the diversity index.

$$H' = -\sum_{i=1}^{n} P_i \ln P_i \quad (2)$$

Where;

P = n/N

n = No of individuals

N = Total number of mangrove tree

## 3 Results and Discussion

### 3.1. Mangrove Distribution

Mangrove forest along the coastline of Perlis, Kedah and Perak was assessed. A total of 52 matured trees from four species and two families were identified. The species included *Rhizophora mucronata, Rhizophora apiculata, Avicennia alba* and *Avicennia marina*. Table 1 shows number of the inspected mangrove trees for the respective species. They were basically divided into three age groups; (1) three to six years, (2) seven to ten years, and (3) more than 10 years. The age was categorized based on the average height of the trees. The height of the tree greatly influenced their capabilities to attenuate wave. However, young mangrove trees below two years old were not included in this study. They are easily being washed away and not able to withstand the wave action [8].

*A. marina* (40%) was the dominant species found followed by *R. mucronata* (38%), *A. alba* (14%) and *R. apiculata* (8%) (Figure 3). The types of mangrove found relate well with general zonation of mangrove forest. Typical zonation of the mangrove forest in Malaysia is made up of *Avicennia* sp. or *Sonneratia* sp. on the seawards most and followed by *Rhizophora* sp., *Brugueira* sp., *Xylocarpus* sp., and transitional dryland (Fig 4) [14, 15]. The field investigation found high number of *Avicennia* sp. in the visited coastal area. *Avicennia* sp. are able to grow seaward as they may adapt to the high salinity range from 20.9 to 31.25 ppt [16].

**Table 1.** Mangrove species found at sampling plots.

| Location | Rhizophoraceae | | Avicenniaceae | |
|---|---|---|---|---|
| | ***R. mucronata*** | ***R. apiculata*** | ***A. alba*** | ***A. marina*** |
| Parit Haji Dollah | 3 | 0 | 1 | 0 |
| Tg Kepah | 0 | 0 | 0 | 4 |
| Kg Pantai Remis | 0 | 0 | 2 | 2 |
| Kuala Sepetang | 2 | 3 | 0 | 2 |
| Kg Kuala Gula | 4 | 0 | 0 | 0 |
| Pantai Merdeka | 3 | 0 | 0 | 1 |
| Kuala Sg Limau | 0 | 0 | 3 | 1 |
| Sg Ibus | 0 | 0 | 1 | 3 |
| Kuala Teriang | 4 | 1 | 0 | 8 |
| Kuala Sanglang | 4 | 0 | 0 | 0 |

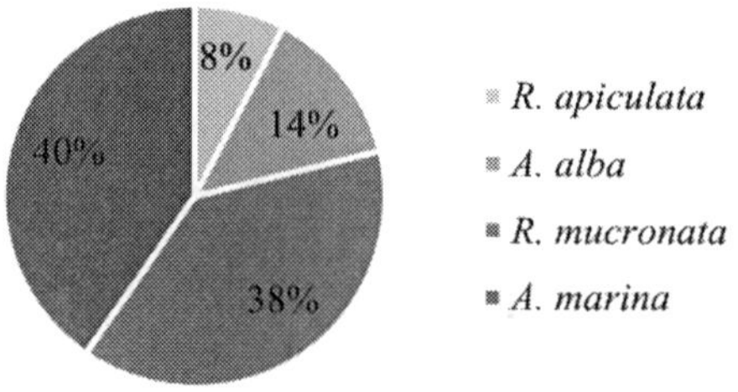


**Figure 3.** Distribution of mangrove forest in the surveyed coastlines.

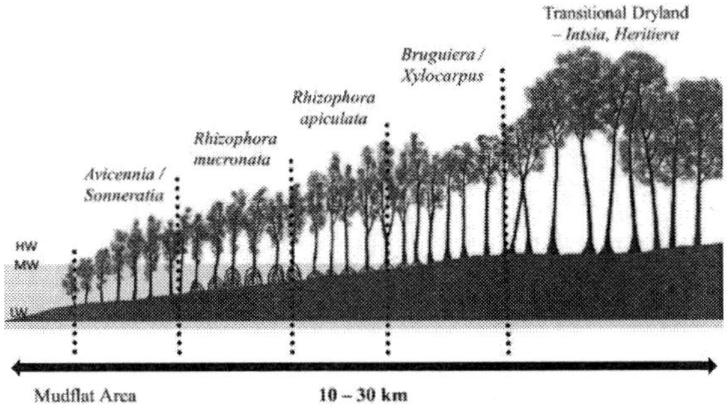


**Figure 4.** Typical mangrove forest zonation in Peninsular Malaysia [15].

### 3.2 Condition of Mangrove Forest

The diversity index is one of the approaches used to measure the diversity of the mangrove forest [17]. The derivation attributed from two facets including species richness and species evenness. The species diversity, evenness and richness are shown in Figure 5.

The species diversity along the coastlines was estimated to range from 0 to 1.08. The highest species diversity found was in Kuala Sepetang with a value of 1.08. The value attributed from 3 species including *R. mucronata, R. apiculata* and *A. marina*. Meanwhile, the area with zero diversity indicated only one species at the representative area. The diversity of the species indicates how the area is able to attenuate wave. Different species have different wave attenuation capability [10]. *Rhizophora* sp. have larger and complex root and pneumatophores which enable them to create greater friction to wave as compared to other species [18].

DBH indicates as the best height to determine the tree growth. The DBH is also an approach to predict the dimension of the large tree [19,20]. The derivation from the relationship between DBH and height of the mangrove tree can be used to estimate the size of the tree [13,19]; and it indicated positive correlation as height increases with DBH (Fig 6). It is important to note that the DBH of the tree should increase with height as it grows [19].

The Pearson correlation analysis is used to reveal how strong the relationship between the variables. The value for the correlation ranges from -1 to +1 ($-1 \leq r \leq 1$). Values close to +1 indicate the variables having strong positive relationship, while values close to -1 show strong negative relationship. However, value of zero implies no relationship.

It is revealed that *R. apiculata* recorded a strong correlation (n=4; r=0.96) as compared to other species. However, analysis concluded that the growth of mangrove forest synchronises well with DBH. All the correlation coefficient values (r) are above 0.5 which indicate moderate correlation. Hence, the growth of mangrove trees are considered satisfactory.

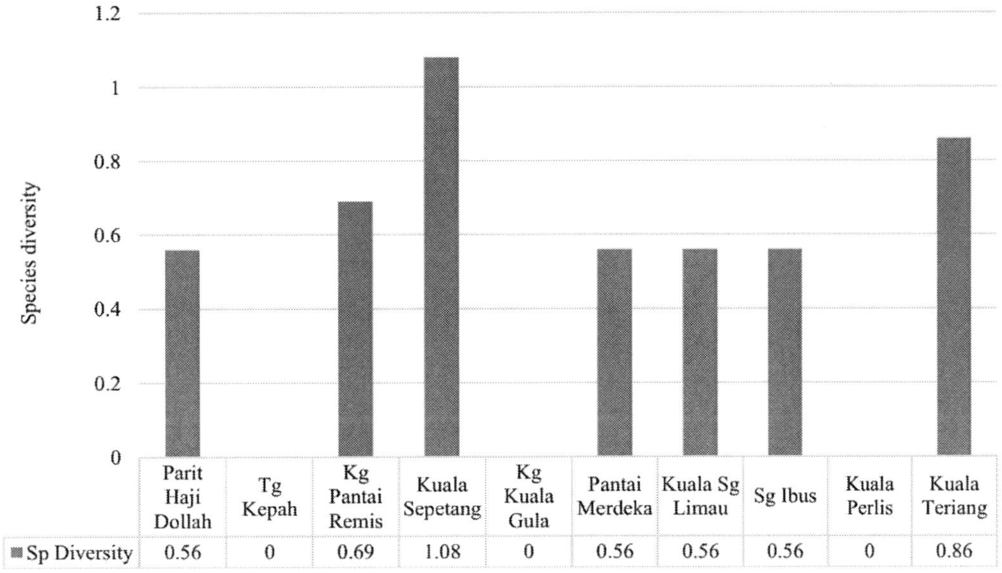


**Figure 5.** Species diversity of mangrove forest in the surveyed coastlines

DOI: 10.1051/matecconf/20178701014

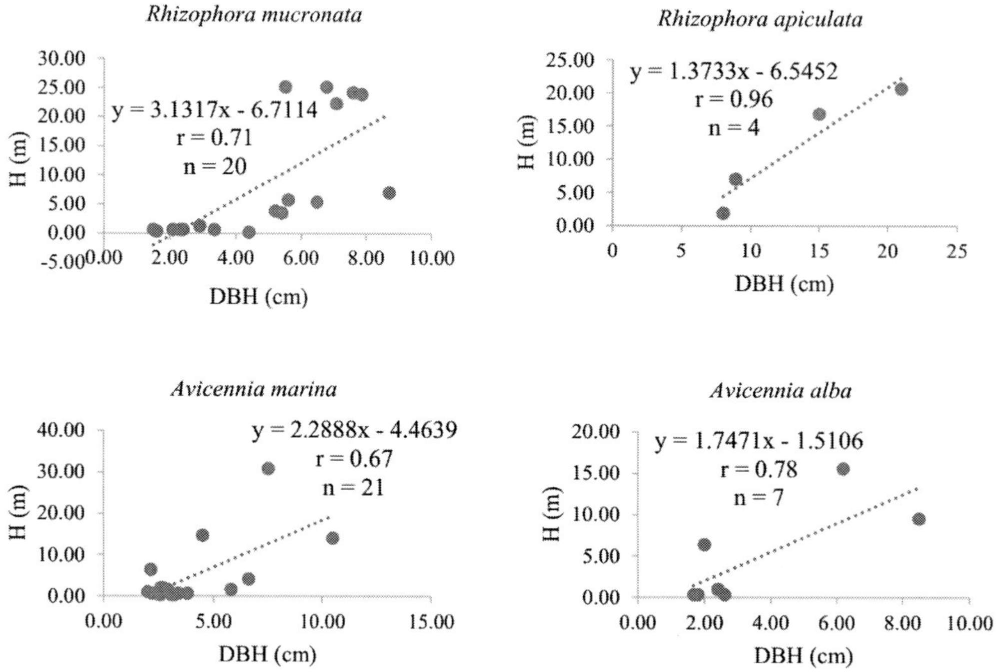


**Figure 6.** Correlation analysis between the height of the tree, H (m) and diameter at breast height, DBH (cm).

### 3.3 Mangrove as Coastal Protection

Mangrove forest provides countless benefits, not only to human but for ecosystem as well [21]. Unfortunately, despite the benefits offered, mangrove forest had been exploited extensively. The major factor of decreasing mangrove was due to land conversion [22]. The mangrove forest had been explored for shrimp aquaculture due to the suitability of its surrounding soil and water quality.

Analysis shows that Malaysia is exposed to some level of risk of tsunami especially at Sabah and West Coast of Peninsular Malaysia including Perlis, Penang, Kedah, Perak and Selangor (Fig 7) [15].

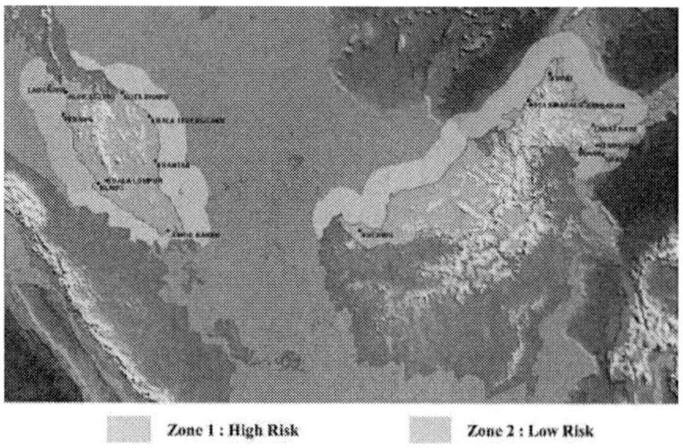


**Figure 7.** Map showing the tsunami risk area in Malaysia [15].

After the IOT 2004, about 1295.80 ha (4,395,472 number of mangrove trees) had been replanted in Peninsular Malaysia by the relevant authorities from 2005 to 2010 (Table 2) [15]. The replantation is hoped to protect the coastal properties from wave actions.

**Table 2.** Locations for the rehabilitation programme in Peninsular Malaysia after IOT 2004 [15].

| State | Location |
|---|---|
| Perlis | Kuala Sanglang |
| Kedah | Kg. Huma, Tg. Dawai<br>Kota Kuala Muda |
| P. Pinang | Pantai Malindo |
| Perak | Lekir, Manjung |
| Selangor | Bagan Nakhoda Omar |

However, recent visit in April 2016 revealed that some of the replantation programmes were not successful. As an example, the mangroves that were replanted in Lekir, Perak in 2008 showed a retarded growth. The villagers commented that the mangrove saplings had been planted for two times, but failed. The saplings only survived for a year and later washed away after the subsequent monsoon season at the end of the year. Currently, the third attempt with two years old *R. mucronata* were planted in early 2015. However, only few trees were found to still survive, while most of the replanted trees were destroyed by the monsoon.

Remnants of the destroyed matured trees were also observed at Sg. Haji Dollah which is located south of Lekir (Fig 8). These mangroves were struggling in front of the rock revetment. It is suggested that the presence of revetment has led to the degradation of these trees. Mangrove trees planted behind the revetment was found to survive well.

Among the successful replanting efforts include those in Kuala Teriang, Langkawi, Kedah. The area was one of the most affected area hit by the tsunami in Peninsular Malaysia. Healthy mangrove forest grows along the coastal area from Sg Melaka to Sg Kuala Teriang (Fig 9). In front of the mangrove forest, there was an approximately 100 m long geotextile fabric tube placed in front of beach. The geotubes had been laid as another line of coastal protection measure for this area. Interview with local villagers described further details on the mangrove forest at site. According to one of the senior citizens residing close to the beach, the mangrove replanting was done since 2005, after the IOT 2004. The three years old *R. mucronata* were planted with about 1.3 m spacing. The villagers were also briefed on how to grow the mangrove tree. They were advised to plant the mangrove seedlings from the mature tree at the nursery area. This awareness campaign by the authorities allowed the mangrove to grow well at the Kuala Teriang.

**Figure 8.** Replantation program implemented at Lekir, Perak, A. The replantation area in Tg Kepah. B. Fall-out of matured mangrove tree. (Image adapted from Google Earth, 2016; [15])

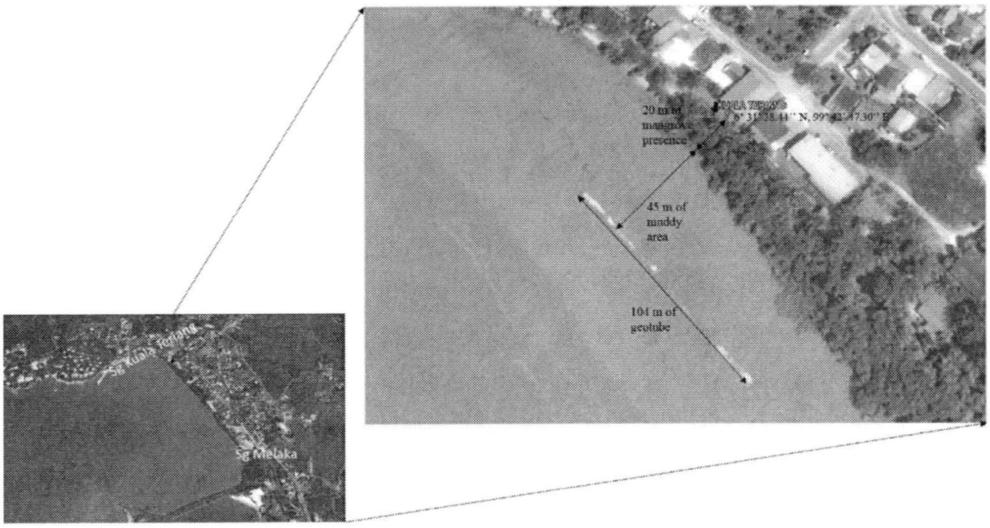


**Figure 9.** Geotube in front of mangrove forest at Kuala Teriang. (Image adapted from Google Earth, 2016)

## 4 Conclusion

The IOT 2004 was the most outrageous event occurred in Malaysia with 63 numbers of total deaths. It was evidenced from the incident that the mangrove forest had contributed significantly in reducing the impacts from the wave. The unique configuration of mangrove trees enabled them to dissipate wave energy.

The field investigation shows that *A. marina* was found as the dominant species along the West Coast of Peninsular Malaysia. The zonation of mangrove forest with Avicenniaceae as frontal species contributed to this result.

However, recent field investigation also discovered that there was no sufficient mangrove belt to provide effective protection especially at the South of Perak. The replantation of mangrove forest by the authorities had experienced failures twice due to the monsoons. However, further north of Perak, the distribution of mangrove forest was found to be increasing. It is discovered that the area is one of the tourist areas under the jurisdiction of the local government. Therefore, the area was given extra attention as compared to the South of Perak. Analysis recommended that future focus should be given to the area with less mangrove forest distribution. The optimum replanting scheme should be implemented immediately to ensure safety of Malaysia's coastline from future natural disaster. This field study provides useful database which can be used for future comprehensive assessment with the aid of GIS technology and remote sensing analysis.

## Acknowledgement

This research is fully supported by FRGS grant (FRGS/2/2013/TK03/UTP/02/1). The authors also wish to express special thanks to the Centre of Post Graduate Studies, Sustainable Resources MOR and Department of Civil and Environmental Engineering, Universiti Teknologi PETRONAS for their support and encouragement.

## References

[1] P. Whitmore, U. t. Brink, M. Carpolo, V. Huerfano-Moreno, W. Knight, W. Sammler and A. Sandrik, NOAA/West Coast Alaska tsunami warning center atlantic ocean response criteria, *Science of Tsunami Hazards,* 86-107 (2009).

[2] A. Suppasri, T. Futami, S. Tabuchi and F. Imamura, Mapping of historical tsunamis in the Indian and Southwest Pacific Oceans, *International Journal of Disaster Risk Reduction,* 62-71 (2012)

[3] M. Z. Asmawi and A. N. Ibrahim, The impacts of tsunami on the well-being of the affected community in Kuala Muda, Kedah, Malaysia, *Journal of Clean Energy Technologies,* **1**, 246-250 (2013)

[4] I. Fehr, P. Grossi, S. Hernandez, T. Krebs, S. McKay, R. Muir-Wood, A. Pomonis, D. C. Souch, D. Windeler and Y. Xie, Managing tsunami risk in the aftermath of the 2004 Indian Ocean earthquake and tsunami, Risk Management Solutions, Newark (2006).

[5] A. D. Hawkes, M. Bird, S. Cowie, C. Grundy-Warr, B. P. Horton, A. T. S. Hwai, C. Macgregor, J. Nott, J. Rigg, J. E. Ong, L. Law, R. Robinson, M. Tan-Mullins, T. T. Sa, Z. Yasin and L. W. Aik, Sediments deposited by the 2004 Indian Ocean tsunami along Malaysia-Thailand Peninsula, *Marine Geology,* 169-190 (2007)

[6] P. C. Athukorala, Disaster, generosity and recovery: Indian Ocean Tsunami, Australian national University, Canberra (2012)

[7] K. Rostam, I. Herman and A. B. M. Nor, Kawasan penempatan semula mangsa tsunami di Malaysia: Analisis kualiti kejiranan dan kesejahteraan isi rumah, *Malaysian Journal of Society and Space,* **5**, 33-43 (2009)

[8] D. M. Patel, V. M. Patel, K. Bhupesh and K. A. Patel, Peformance of mangrove in tsunami resistance, *International Journal of Emerging Technology and Research,* **1**, 29-32 (2014)

[9] Y. Mazda, M. Magi, M. Kogo and P. N. Hong, Mangroves as a coastal protection from waves in the Tong King delta, Vietnam, *Mangroves and Salt Marshes,* **1**, 127-135 (1997)

[10] Y. Mazda and M. Magi, Wave reduction in mangrove forest dominated by Sonneratia sp., *Wetlands Ecology and Management,* **14**, 365-378 (2006)

[11] F. Alemayehu, O. Richard, K. M. James and O. Wasonga, Assessment of Mangrove Covers Change and Biomass in Mida Creek, Kenya, *Journal of Forestry,* **4**, 398-413 (2014)

[12] N. A. Kamaruddin, S. Fujii and B. R. Shivakoti, Applicability of satellite monitoring on mangrove forests in Malaysia: A review of potential benefits and challenges, *International Journal of Research,* **2**, 185-204 (2015)

[13] T. Buba, Relationships between Stem Diameter at Breast Height (DBH), Tree Height, Crown Length, and Crown Ratio of V. paradoxa C.F. Gaertn in the Nigerian Guinea Savanna, *African Journal of Biotechnology,* **12**, 3441-3446 (2013)

[14] K. S. Sinfuego and I. E. Buot Jr, Mangrove zonation and utilization by the local people in Ajuy and Pedada Bays, Panay Island, Philippines, *Journal of Marine and Island Cultures,* **3** (2014)

[15] Y. Y. Hwai and S. Rosli, Management of mangrove forests: Policies and prescriptions, Jabatan Perhutanan Semenanjung Malaysia, Putrajaya (2012)

[16] H. Marisa and Sarno, Three Species Zonation of Sonneratia; based on Salinity in River Calik, South Sumatera, in *International Conference of PLant, Marine and Environmental Sceinces (PMES-2015),* Kuala Lumpur (2015)

[17] S. Das, P. S. Lyla and S. A. Khan, Marine microbial biodiversity: Present status and advanced statistical paradigms, *Conservation Biology in Asia,* 368-385 (2006)

[18] K. Jusoff, Malaysian mangrove forests and their significance to the coastal marine environment, *Pollution Journal Environmental Study,* **22**, 979-1005 (2013).

[19] A. Sumida, T. Miyaura and H. Torii, Relationships between stem diameter at breast height revisited: Analyses of stem growth using 20-year data of an even-aged Chamaecyparis obtusa stand, *Tree Physiology Advance,* 1-13 (2013)

[20] D. W. Aheto, A. A. O. Aduomih and E. A. Obodai, Structural parameters and above-ground biomass of mangrove tree species around the Kakum river estuary of Ghana, *Annals of Biological Research,* **2**, 504-514, (2011)

[21] A. M. Hashim and S. M. P. Catherine, A laboratory study on wave reduction by mangrove forests, *APCBEE Procedia,* 27-32, (2013)

[22] M. Simard, V. H. Rivera-Monroy, J. E. Mancera-Pineda, E. Castaneda-Moya and R. R. Twilley, A systematic method for 3D mapping of mangrove forests based on Shutle Radar Topography Mission elevation data, ICEsat/GLAS wavefoms and field data: Application to Cienaga Gande de Santa Marta, Colombia, *Remote Sensing of Environment,* **112**, 2131-2144, (2008)

# Heat Lump in Different Pavement Layer Using Ethylene Glycol as A Solar Heat Collector

Siti Hidayah Abu Talib[1, a], Syarifah Intan Najla Syed Hashim[2], Salmia Beddu[3], Anis Farhah Maidin[4], Muhammad Salleh Abustan[1]

[1]*Lecturer, Faculty of Civil and Environmental Engineering, Universiti Tun Hussein Onn Malaysia, 86400 Parit Raja, Batu Pahat, Johor, Malaysia.*
[2]*Postgraduate Student, Faculty of Civil and Environmental Engineering, Universiti Tun Hussein Onn Malaysia, 86400 Parit Raja, Batu Pahat, Johor, Malaysia.*
[3]*Senior Lecturer, Civil Engineering Department, College of Engineering, Universiti Tenaga Nasional, Putrajaya Campus, Jalan IKRAM-UNITEN, 43000 Kajang, Selangor, Malaysia.*
[4]*Undergraduate Student, Faculty of Civil and Environmental Engineering, Universiti Tun Hussein Onn Malaysia, 86400 Parit Raja, Batu Pahat, Johor, Malaysia.*

**Abstract.** Due to the hot weather in Malaysia, several studies being carried out regarding solar energy harvesting and use as a noble renewable energy. Therefore, researchers start developing various methods to utilize sun's heat as a renewable source of energy. Solar energy has been always concern researchers to develop most effective, durable and cheap methods and materials to be used as solar energy collectors in road pavements. Therefore, this study was conducted to determine the maximum heat area using ethylene glycol. The study was involved the field monitoring of the six samples of asphalt slab with stainless steel pipe with ethylene glycol and empty pipe act as the controlled sample. The pipe was arranged in three different depths of 50mm, 100mm and 150mm and keeping controlled the pipe diameter and the size of slab of 20mm and 300mm x 300mm respectively. These results show that the maximum heat extraction by the ethylene glycol was at a depth 150mm which was 51.2°C. The average percentage increments of the pipe with ethylene glycol for three days for pipe depth 50mm was 71%, while for pipe at depth 100mm and 150mm were 78% and 62% respectively. The ethylene glycol effect was more noticeable compare to the pipe without ethylene glycol and can store more heat. For the strength of pipe, the highest maximum load bearing capacity that can be supported by the stainless steel pipe was at depth 50mm and the second highest was at a depth 100mm while the third highest was at depth 150mm. So, the suitable depth to install the pipe with the consideration of both cases which was maximum heat gain and the maximum load that can be supported was at depth 50mm.

## 1 Introduction

Asphalt pavement as a solar collector was believed to be one of the most efficient techniques of harvesting solar energy by collecting solar radiations that hit the pavement throughout all day long and cause the pavement to contain amounts of heat energy which can be utilized and used for many purposes. In European countries, asphalt pavement as a solar collector is used for heating and melting the snow. The implementation of natural source of energy is more favourable due to its availability, environmental friendly and cost effective. Besides, asphalt collector is used for heating and cooling the buildings. However, high temperature in asphalt pavement also may lead to the global warming.

Solar energy is the primary energy source for all other forms of energy source distributed around the world. Road pavement is one of the largest solar thermal collector and storage system on land. It receives solar radiation all day and stores some of the energy from it. This energy is completely or partially dissipated to atmosphere by night time. An average temperature of an asphalt pavement can reach up to 60–70°C on hot day because of its excellent heat absorbing characteristic. The solar thermal energy collected by the asphalt pavement can be harvested by circulating fluid through it. A system that is designed for this purpose is called asphalt solar collector (ASC).

Renewable energy technologies are likely to become more important as other energy sources become depleted and the cost of power generation using fossil fuels rises. However, in most cases, renewable energy technologies require high initial investments that may deter potential users. Asphalt Solar Collector potentially offer a low-cost solution to sustainable and clean energy generation by using pavement infrastructure.

The water solvent in the pipes can absorb/reject heat to the pavement and store energy to in the pavement structure. A significant advantage of the system is that the pipes are installed within pavements that are already needed for structural reasons and need not to be specially constructed as the conventional thermal energy utilization systems such as solar panels. The ability of pavement to collect and/or store solar energy is a relatively new subject that is still at an early age of development. This research is in general to determine the maximum heat

[a] Corresponding author: hidayaht@uthm.edu.my

DOI: 10.1051/matecconf/20178701015

area in asphalt pavement by using steel pipe with ethylene glycol.

2. To determine the strength of steel pipe to withstand loading from vehicles.

### 1.1 Problem Statement

In recent years, many researchers have studied the use of road pavements as an energy collector by presenting several technologies and methods. These studies focus on the use of heat collected from pavements to produce energy. Through asphalt, the aim is to develop a solar collector that can make the roads and parking lots a source of electricity or hot water as the pavement layers stay hot at night and can continue generating energy after the sun goes down, unlike traditional solar electric cells. In this method there is needed to occupy lands for solar farms and the solar system can be invisible if the researchers can convince the industry to change the way they are building roads pavements.

To harvest the maximum amount of solar heat, several studies have been carried out on road pavement as a solar collector system. Multi-layered pavement with a high porous middle layer to be used instead of solar collector with an installed pipe network [1]. However the thermal efficiency results were excellent, this result contrasted with the low flow rate level recorded during the experiment.

Therefore, solar collector in pavement is an alternative of green energy harvesting while solving the problem of renewable energy for fossil fuels and reducing the heat island effect on asphalt pavement. Figure 1 shows the proposed distribution of heat energy from solar pavement to the housing areas.

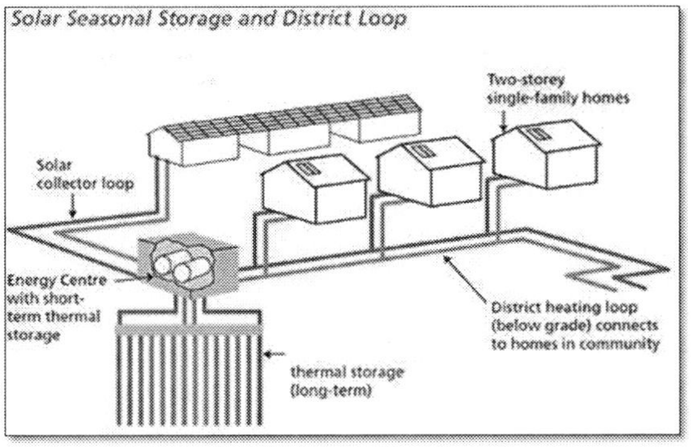


**Figure 1:** Solar pavement storage and distribution.

### 1.2 Objectives of Study

The aim of this study is to determine the optimum heat lump area in the pavement. There are two objectives of this study.

1. To determine the optimum depth using ethylene glycol in order to get maximum heat extraction.

## 2 Literature Review

Several authors have developed methods to predict the asphalt temperatures because they will significantly affect the pavements where high temperatures can cause plastic deformations while lower temperatures can leads to damage by hardening [2].

Asphalt solar collectors consist of pipes embedded in the pavement with a circulating fluid inside. Solar radiation causes an increase in pavement temperature. Due to the temperature gradient between the fluid circulating through pipes and the pavement, a heat transfer process occurs from pavement to fluid which leads to a drop in pavement temperature and an increase in fluid temperature [3].

The temperature distribution of an asphalt pavement is affected directly by the thermal environment conditions to which it is exposed. The primary modes of heat transfer include incident solar radiation, thermal and long wave radiation between the asphalt pavement surface and the sky, convection due to heat transfer between the pavement surface and the fluid (air or water) that is in contact with the surface, conduction inside the pavement, and the radiation heat loss from surface Figure 2 shows the various heat transfer modes in a heat conducting asphalt pavement exposed to solar radiation.

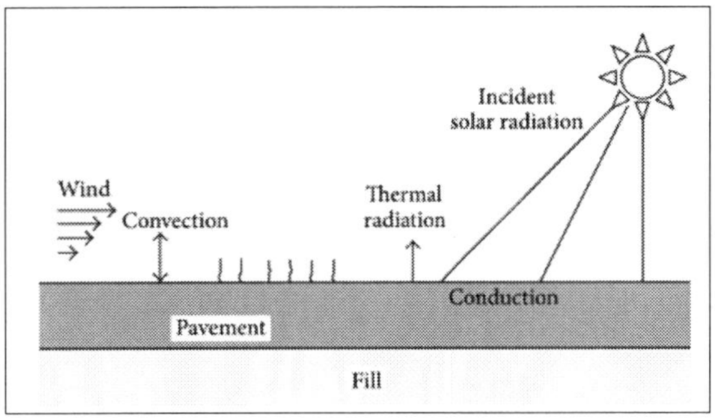


**Figure 2**: Schematic presentation of heat transfer mode of asphalt pavement [4].

A numerical simulation was developed by Wang et al. in which a model of an asphalt collector with a net of rubber pipes in its middle layer was used to predict the thermal response of these devices in order to determine which parameters were critical in the behavior of an asphalt collector [5]. Besides that, Wu et al. had investigated the process of extracting thermal energy from asphalt slabs with u-shaped metal tubes inside. It was proven that the circulating water can reduce the surface temperature of the asphalt pavement thus reducing the risk of permanent deformation and the urban heat island effect [6].

Heat exchangers are normally used only for the transfer and useful elimination or recovery of heat without an accompanying phase change. The fluids on either side of the barrier are usually liquids, but they may also be gases such as steam, air, or hydrocarbon vapors; or they may be liquid metals such as sodium or mercury. Fused salts are also used as heat-exchanger fluids in some applications. The exchangers are used to recover heat between two streams in a process; these fluids may be single compounds or mixtures [7].

The characteristics required for the circulating fluid are: high specific heat, stability in the collector operating temperature range, compatibility with pipes, abundance and low cost. Water is one of the best fluids for systems operating at low temperatures (range from 25°C to 90°C) such as asphalt solar collectors [8], but the solidification temperature must be less than the minimum temperature expected at the collector. For this reason mixtures of water and antifreeze are generally used.

## 3 Materials and Methods

Experiments began with the preparation of samples that was done according to the standard specifications of pavement from public work department (JKR) before the collection and analysis of data. This chapter will describes the process throughout the research and necessary steps to be studied from the beginning of the research until the laboratory testing to analyze the expected results of the study.

After identifying the materials and apparatus needed, field experiment and laboratory testing will be conducted in Highway Laboratory in UTHM. The results obtained from the observation and laboratory testing will be recorded and analyzed. Total 6 models of asphalt pavement are designed where the heat exchanger which is steel pipes are installed in different depth. Different depths of models with steel pipes analyzed in this study are shown in Table 1.

**Table 1:** Number of samples analysed in the study

| No. | Size of model (mm) | Depth (mm) | Steel Pipe |
|---|---|---|---|
| 1. | 300 x 300 | 50mm | Empty |
| 2. | | 100mm | |
| 3. | | 150mm | |
| 4. | | 50mm | Ethylene glycol |
| 5. | | 100mm | |
| 6. | | 150mm | |

### 3.1 Sample preparation

The sample preparation was conducted in Highway Laboratory at Universiti Tun Hussein Onn Malaysia. Six (6) physical models (pavement) are prepared in different pipe depth. 3 models were installed with empty steel pipe at different depths while the other 3 models were installed with steel pipes and ethylene glycol as a heat exchanger fluid at different depths.

The preparation of pavement is following Public Work Department (JKR) standard specifications. Steel pipes are inserted in the pavement as the heat exchanger pipe. In order to find out the temperature variation within the asphalt pavement and the amount of energy extracted by pavement, the models are placed outdoor with direct sunlight and connected to the data logger. The pavement structure has a dimension of 300 mm × 300 mm with a total depth of 150 mm. In the asphalt pavement structure, studies are conducted by varying the pipe depth (50 mm, 100 mm and 150 mm) from the bottom of pavement layers, keeping the pipe diameter fixed at 15 mm diameter as shown in Figure 3, 4 and 5.

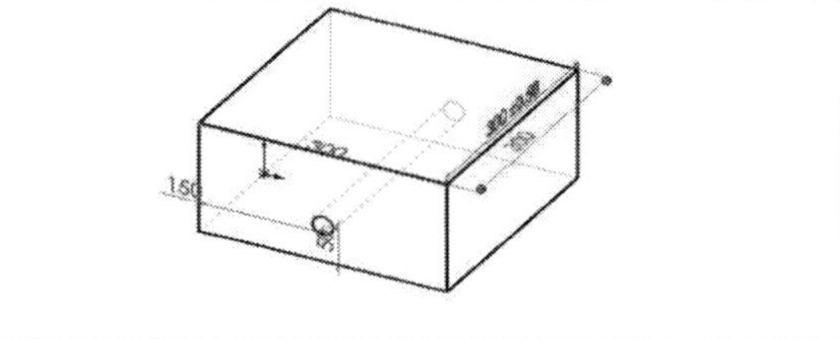

**Figure 3** : Steel pipe at depth of 50mm

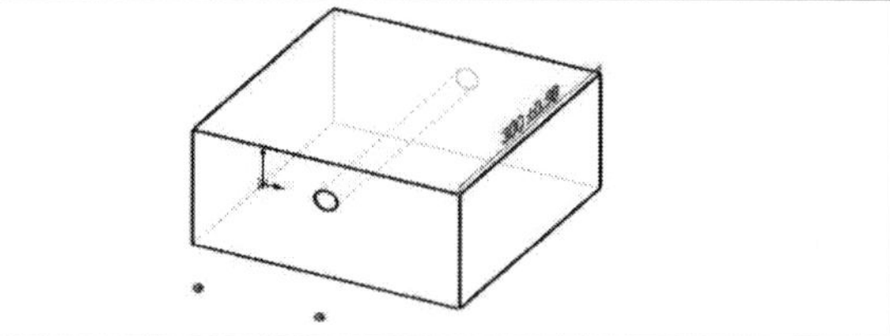

**Figure 4** : Steel pipe at depth of 100mm

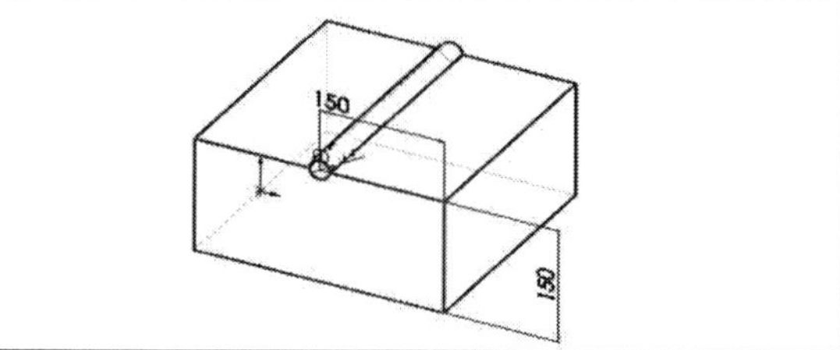

**Figure 5** : Steel pipe at depth of 150mm

### 3.2 Field experiment

This research was conducting a small-scale field test under real atmospheric conditions. The field test setup consists of asphalt pavement, heat exchanger pipe which is empty steel pipes and steel pipes filled with ethylene glycol, source of heat from sun, thermocouples and a data logger for data collection.

Thermocouples will be inserted in the pavement and connected to the data logger to measure the temperature of heat absorbed in pavement. The samples will be left under the sun for 24 hours in 3 days. The data will be recorded every 5 minutes and transmitted directly to the data logger. Figure 6 shows the arrangement of asphalt pavement solar collector for field testing in

UTHM with empty pipe and pipe filled with water solvent.

**Figure 6 :** Arrangement of models connected with data logger in field experiment

### 3.3 Laboratory testing

Laboratory testing for structural analysis is done to determine the maximum strength of steel pipe in pavement to withstand loading. Loading will be applied on samples by using California Bearing Ratio (CBR) machine. Three (3) samples were prepared by following the JKR standard specifications and compacted in mould before conducting the strength test. Samples will consist of steel pipe at 3 different depths which are at depth of 50mm, 100mm and 150mm. Figure 7 shows the samples of asphalt pavement prepared for laboratory testing.

**Figure 7 :** Laboratory samples of asphalt pavement

The California Bearing Ratio (CBR) test is a simple strength test by applying load to a small penetration piston at a rate of 1.3mm per minute and recording the total load of penetration applied on samples. This test was first developed by the California Division of Highway as a method for evaluating the strength of soil subgrade and other flexible pavement materials.

This test can be performed in the laboratory on prepared samples or in-situ on location. The CBR test is essentially used to evaluate the strength of pavement. The loading device on CBR machine having a capacity of 50kN and equipped with a movable head or base that travels at a uniform rate of 1.3mm per minute for use in forcing the penetration plunger to the sample. The machine's digital display provides the ability to monitor test data as well as the ability to see a test peak value and the maximum loading. Figure 8 shows the California Bearing Ratio (CBR) machine in Highway Laboratory, UTHM.

**Figure 8 :** California Bearing Ratio (CBR) machine

## 4 Results and Discussion

The field experiments and laboratory testing had been conducted and the data had been recorded for analysis. The result had been analyzed in order to achieve the research objectives which are to obtain maximum heat extraction of solar collector in different pipe depth and to determine the maximum strength of steel pipe to withstand loading.

The experimental setup was placed outside for exposure to the sun. In this experiment, the pavement is exposed to the sun and left outside for 24 hours from approximately 12 AM to 12 AM the next day. Data for the heat absorption in asphalt pavement was recorded every 5 minutes by using data logger. Then, the result were analyzed and graph were plotted to observe at which depth of pavement (50mm,100mm,150mm) have the maximum heat extraction. The field experiment was conducted under Batu Pahat, Johor climate condition and the location of study is in UTHM area.

### 4.1 Thermal performance of asphalt layer

The asphalt behavior for surface temperature is shown in Figure 9. The data were collected for 3 days continuously. Temperature was measured on both sunny and rainy days, which represent the real weather condition. The data logger was neither distracted nor shut off during the experiment.

The temperature was noticed to be more affected at the period between 12pm to 4 pm everyday where the maximum sun, heat radiations are more likely to hit the surface of the pavement.

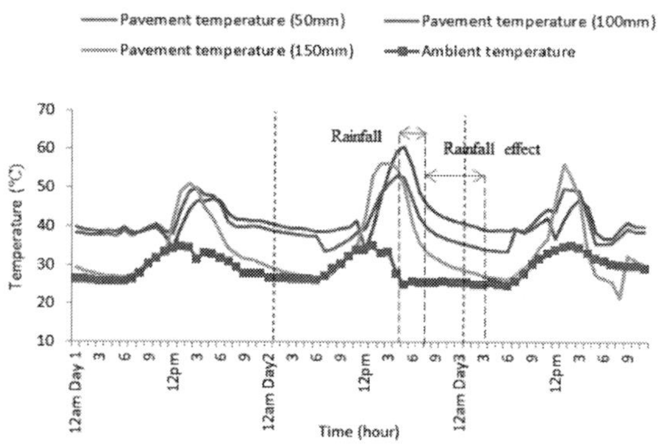


**Figure 9** : Thermal performance of pavement temperature and ambient temperature in different depths for 3 days

Temperature was low in the morning due to low solar radiation and resulting in less energy being absorbed by ethylene glycol in steel pipe. During daytime, solar radiation is high, and the ethylene glycol in steel pipe could absorb a large amount of thermal energy, resulting in a large degree of temperature for heat extraction.

The occurrence of rainfall during the experiment on Day 2 had affected the temperature of pavement surface. Further investigation showed that maximum pavement temperature is lowered until the next day after rainfall and the rainfall effect had altered the temperature of pavement surface which is up to 58.9°C (after rainfall) from 42.5°C (before rainfall). It shows that rainfall plays a major role as a cooling agent to reduce the surface temperature of asphalt pavement.

## 4.2 Efficiency of ethylene glycol as a heat collector

Table 2 shows the temperature difference of maximum temperature at pavement in different layers with ethylene glycol, empty pipe and ambient temperature for 3 days.

**Table 2 :** Percentage of heat increment in pavement surfaces

| Day | Depth (mm) | Temperature (°C) | | | | % Increment | |
|---|---|---|---|---|---|---|---|
| | | Ambient | | Pavement | | Pipe with ethylene glycol | Pipe without ethylene glycol |
| | | Pipe with ethylene glycol | Pipe without ethylene glycol | Pipe with ethylene glycol | Pipe without ethylene glycol | | |
| 1 | 50 | 31.5 | 31.5 | 49.7 | 53.2 | 58 | 41 |
| | 100 | 29.1 | 31.5 | 41.7 | 47.6 | 43 | 51 |
| | 150 | 33 | 31.7 | 50.9 | 50.1 | 54 | 58 |
| 2 | 50 | 25.1 | 32.5 | 52.7 | 63.3 | 110 | 95 |
| | 100 | 25.1 | 32.5 | 60.8 | 59.2 | 142 | 82 |
| | 150 | 33 | 32.5 | 56.3 | 57.8 | 71 | 79 |
| 3 | 50 | 34.6 | 34.6 | 49.8 | 53.6 | 44 | 55 |
| | 100 | 31.5 | 34.5 | 46.9 | 48.1 | 49 | 39 |
| | 150 | 34.6 | 31.5 | 56.1 | 51.2 | 62 | 63 |

At depth of 50mm, the percentage of increment recorded for 3 days for the pipe with ethylene glycol was in range of 44%-110% while at depth 100mm, the percentage of heat increment recorded was in range of 43%-142%. The increment was recorded for the pipe located at 150mm from the pavement base was in range of 62%-71%. The average percentage for the heat increment of the pipe with ethylene glycol for three days for pipe depth 50mm was 71% while for pipe at depth 100mm and 150mm was 78% and 62% respectively. The percentage increment of the pipe without ethylene glycol for three days for pipe depth 50mm was 64% while for pipe 100mm and 150mm was 57% and 67% respectively.

Not only that, the ethylene glycol effects were more noticeable at the period between 12 to 3pm where the maximum sun heat radiations are more likely to hit the surface of the pavement. The ethylene glycol also tend to cool faster during night at the pipes located closer to the pavement surface due to the reduction in surrounding temperature which affected the pavement surface at night. It can be conclude that, the ethylene glycol effect was more noticeable compared to the pipe without ethylene glycol.

A significant increment in temperature can be achieved when more heat is applied to the asphalt surface. Thus, the accumulation of heat which causes heat lump can be increase with the performance of asphalt pavement in absorbing heat from radiation.

## 4.3 Load bearing capacity of pipe

The load by the vehicle that passes through the road in daily life would be the effect to the strength of the pavement. The installed pipe in the pavement also can be affected by the loading of the vehicle. Figure 10 shows the maximum load can be supported by the steel pipe at different depth after laboratory testing using California Bearing Ratio (CBR) machine. The load for the AASHTO standard truck for 1 axle was 15.2kN and for 3 axle was 38.8kN while for automobile traffic the load was 9.9kN.

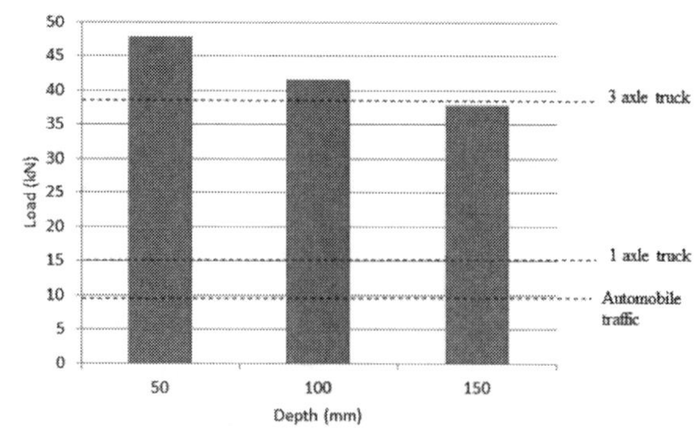


**Figure 10 :** Strength of steel pipe in different depths

The data in figure 10 shows that the higher the pipe installed in pavement from pavement base, the lower the maximum load that can sustain on it. At all depths, the pavement can sustain the load for automobile traffic and 1 axle truck but for 3 axle truck, only at depth 150mm it cannot support the load. So that it means that pipes must be installed deeper to the pavement to sustain the load that attack on the pavement without considering the load of any vehicle.

## 5 Conclusion

In conclusion, the highest heat that obtained by the ethylene glycol was at depth 150mm and the second highest was at depth 50mm while the third highest was at depth 100mm. The pipes located 50mm from pavement base have achieved the highest temperature comparing to 100mm and 150mm while those located 100mm above pavement base have achieved the lowest recorded temperature.

Due to the release of storing temperature during night, pipes located 150mm from pavement base have recorded the highest temperature as the asphalt layers will release the storing heat to the upper layer of the pavement where 50mm pipes were placed. The ethylene glycol effect was more noticeable for the pipes located closer to the pavement surface as the highest average temperature recorded was 51.2°C at depth 150mm. Not only that, the ethylene glycol was also noticed to be more affected at the period between 12pm to 4pm where the maximum sun heat radiations are more likely to hit the surface of the pavement. The ethylene glycol also tend to cool faster during night at the pipes located closer to the pavement surface this is due to the reduction in surrounding temperature which affect the pavement surface first at night.

For the strength of pipe, the highest maximum load bearing capacity that can be supported by the stainless steel pipe was at depth 50mm and the second highest was at depth 100mm while the third highest was at depth 150mm. So, the suitable depth to install the pipe with the consideration of both cases which was maximum heat gain and the maximum load that can be supported was at depth 50mm. We choose the second highest temperature for maximum heat and the first highest for the maximum load that can be supported.

## Acknowledgement

The funding for this project were provided by "The Composition and Distribution of Sediment in Sembrong Dam Reservoir by Using Coring and Subbottom Profiling Techniques" (U422), Special Research Contract Grant 2015 (UTHM) and "Green Pemeable Pavement Solar Hybrid Energy Collector" UNITEN Internal Grant 2015. We would also like to offer our sincere thanks to Department of Meteorology Malaysia for the data provided and all Laboratory staff of Faculty of Civil and Environmental Engineering, UTHM for their continuous cooperation and encouragement.

## References

1. Pascual-Muñoz, P., Castro-Fresno, D., Serrano-Br avo, P., & Alonso-Estébanez, a. (2013). Thermal and hydraulic analysis of multilayered asphalt pavements as active solar collectors. Applied Energy, 111, 324–332.
2. Solaimanian, M., and Kennedy, T., W., (1993). Predicting maximm pavement surface temperature using maximum air temperature and hourly solar radiation. Transportation Research Record, 1417.
3. Jonathan J. C., (2010). Design and analysis of an embedded pipe network in asphalt pavements to reduce the urban heat island effect. Master of Science in Worcester Polytechnic Institute.
4. Carolyn Roos (2008). Principles of Heat Transfer. The EnergyIdeas Clearinghouse. Energy Efficiency Factsheet. 1- 4.
5. Wang, H., Wu, S., Chen, M., and Zhang, Y., (2010). Numerical simulation on the thermal response of heat-conducting asphalt pavements. Physica Scripta.
6. Wu, S., Chen, M., and Zhang, Y., (2011). Laboratory investigation into thermal response of asphalt pavements as solar collector by application of small scale slabs. Journal of Applied Thermal Engineering, Vol, 31(10), 1582-1587.
7. Liu, X., Xiao, Y., Inthavong, K., and Tu, J., (2015). Experimental and numerical investigation on a new type of heat exchanger in ground source heat pump system. Journal of Energy Efficiency, 1-13.
8. Hasnain. S., M., (2011). Review on sustainable thermal energy storage technologies, Part I: Heat storage materials and techniques. Journal of Energy Conversion and Management. Vol, 39(11), 1127-1138.

MATEC Web of Conferences **87**, 01016 (2017) DOI: 10.1051/matecconf/20178701016
*ENCON 2016*

# 2 Dimensional Hydrodynamic Flood Routing Analysis on Flood Forecasting Modelling for Kelantan River Basin

Wan Hazdy Azad[1, 2,a], Lariyah Mohd Sidek[1], Hidayah Basri[1], Chow Ming Fai[1], Suhani Saidin[1] and Abd. Jalil Hassan[3]

[1]*Centre for Sustainable Technology & Environment (Centre for Energy Infrastructure) Universiti Tenaga Nasional, Jalan IKRAM- UNTEN 43000 Kajang, Selangor, Malaysia*
[2]*Department of Irrigation and Drainage, KM7 Jalan Ampang, 68000 Kuala Lumpur, Malaysia*
[3]*River Net Consulting Sdn. Bhd., Seksyen 32, 40460 Shah Alam, Selangor, Malaysia*

**Abstract.** Flood disaster occurs quite frequently in Malaysia and has been categorized as the most threatening natural disaster compared to landslides, hurricanes, tsunami, haze and others. A study by Department of Irrigation and Drainage (DID) show that 9% of land areas in Malaysia are prone to flood which may affect approximately 4.9 million of the population. 2 Dimensional floods routing modelling demonstrate is turning out to be broadly utilized for flood plain display and is an extremely viable device for evaluating flood. Flood propagations can be better understood by simulating the flow and water level by using hydrodynamic modelling. The hydrodynamic flood routing can be recognized by the spatial complexity of the schematization such as 1D model and 2D model. It was found that most of available hydrological models for flood forecasting are more focus on short duration as compared to long duration hydrological model using the Probabilistic Distribution Moisture Model (PDM). The aim of this paper is to discuss preliminary findings on development of flood forecasting model using Probabilistic Distribution Moisture Model (PDM) for Kelantan river basin. Among the findings discuss in this paper includes preliminary calibrated PDM model, which performed reasonably for the Dec 2014, but underestimated the peak flows. Apart from that, this paper also discusses findings on Soil Moisture Deficit (SMD) and flood plain analysis. Flood forecasting is the complex process that begins with an understanding of the geographical makeup of the catchment and knowledge of the preferential regions of heavy rainfall and flood behaviour for the area of responsibility. Therefore, to decreases the uncertainty in the model output, so it is important to increase the complexity of the model.

## 1 Introduction

A flood can be defined as the inundation of a normally dry area caused by an increased water level in a watercourse or in a body of water. Broadly, flooding can be classified as river flooding, coastal flooding, and urban flooding. This study is limited to the floods in rivers, which can be characterised by the sudden or gradual increase in water levels, normally beyond the bank level of the river watercourse and subsequently inundating its surrounding.

Flood disaster occurs quite frequently in Malaysia and has been categorized as the most threatening natural disaster compared to landslides, hurricanes, tsunami, haze and others. A study by JPS entitled National Register of River Basin, 2003 shows that 9% of land areas in Malaysia are prone to flood which may affect approximately 4.9 million of the population. This may result in a great loss of RM915 million per annum due to economic depletion of approximately RM2 billion in a year. Massive floods have been recorded since 1926 followed by 1967, 1971, 1973, 1970 and 1988. Recently, flood has occurred more frequently particularly in Pulau Pinang (1998/ 2003), Kuantan (2001/ 2003/ 2013), Kuala Lumpur (2003/ 2007/ 2011), Perlis/ Kedah (2005), Pahang/ Terengganu/ Kelantan (2004/ 2014), Johor Bahru (2004/ 2014), Shah Alam (2006/ 2014), Johor/ Pahang/ N. Sembilan/ Melaka (2006/ 2007/ 2011). In 5 years between 2011 and 2015, a total of 714 floods have been recorded comprising flash flood (556), monsoon flood (147), mud flood (4), flood due to dam release (7) and more than 200 flash floods particularly in 2015. The extreme flood disaster which happened from 14th December 2014 to 10th January 2015 affected 6 states in Malaysia namely Pahang, Terengganu, Kelantan, Perak, Perlis and Johor has led to loss of assets and lives. Further, the flood in Kelantan destroyed 2000 houses with a total loss of assets at RM2.9 billion. It also reported 25 deaths and 500,000 house evacuations.

The Red Flood which occurred in Kelantan back in 1926 recorded with continuous rainfall for 10 days which led to river overflow from Sg. Kelantan flooding Kita Bharu, Pasir Mas and Tanah Merah. After 40 years, heavy rain for 5 consecutive days from 2nd to 6th January 1967 has resulted in a massive flood of the same magnitude. It led to 38 deaths and affected approximately

[a] Corresponding author: wanhazdyazad@gmail.com

DOI: 10.1051/matecconf/20178701016

537,000 (84%) people in Kelantan. In 2004, another major flood disaster occurred in Kelantan due to heavy rain at Sg. Kelantan river basin affecting Kota Bharu, Pasir Mas and Tanah Merah. Only 10 years later, the 'Yellow Flood' returned to Kelantan and destroyed houses at Kuala Krai and Tanah Merah. The flood overflowed above two-storey schools and covered Kuala Krai town beyond 5 metres. It resulted in 14 deaths, and more than 339, 703 people or 87,024 families moved to flood evacuation centres or stayed with nearby family members.

## 2 Forecasting Sites and Model

The two sub-catchments selected in this study which is River Lebir and River Galas located upper of River Kelantan Catchment. Figure 1 gives an overview of the selected sites.

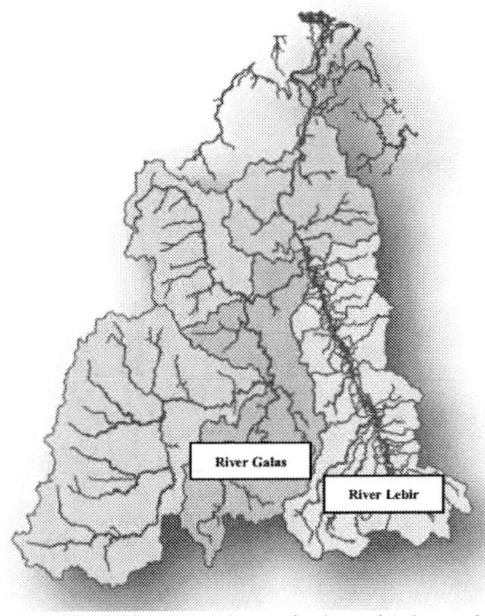


**Figure 1.** Detail of the locations of selected sub-catchments

### 2.1 River Lebir Sub-Catchment

The River Lebir flows generally in a general north from its main source near Kuala Koh National Park to its confluence with the Kelantan River near Kuala Krai. River Lebir Sub-Catchment is long and narrow, approximately 170km and with average of slope 0.09m/m. The catchment area is 3,302km$^2$. The percentage of the imperviousness and pervious area for this catchment area is 0.26% and 99.74%.

### 2.2 River Galas Sub-Catchment

The main river in this sub-catchment is River Galas with approximately 200km flows from Mount Tahan National Park to its confluence with Kelantan River near Kuala Krai. The percentage of the imperviousness and pervious area for this catchment area is 0.87% and 99.13%. The catchment area is 2,519km$^2$.

### 2.3 Rainfall-Runoff Model

Rainfall runoff models are characterized by a different level of complexity and data requirement. Thus, selection of rainfall runoff model are depends on how accurate the output is desired. Brocca, Melone, & Moramarco (2011) stated three main concerns in selecting rainfall runoff model as follows:

(i) Parameter identifiability (uncertainty) where the parameters incorporated in the rainfall-runoff model might increase the model uncertainty
(ii) Physical process description where reliability of the physical processes represented by the model structure is difficult to test and, also, different structures can produce similar results (as with parameters)
(iii) Applicability domain, where many hydrologists suggest choosing models with flexible structures that can be tailored to the 'local' observed conditions.

### 2.4 Probabilistic Distribution Moisture Model (PDM)

Response of the catchments to the rainfall has been modelled using Probabilistic Distribution Moisture Model (PDM). The PDM is a well-established rainfall-runoff model (Moore, 2007). A schematic of its basic processes is provided in Figure 2 and a more thorough description of its formulae and origins is available in (Moore, 2007). The PDM model essentially distributes rainfall between runoff and recharge according to a soil moisture store. The runoff and recharge is routed via stores to the catchment outflow. One of the main advantages of the model is the use of a probability distribution rather than a single value for the soil moisture store. This represents the spatial variability in soil storage across the catchment and prevents threshold type behaviour. The model's short computational run time and continuous soil moisture accounting model makes it suited to the continuous simulation of incoming data for flood forecasting. The model can also use incoming flow gauge telemetry to update its internal soil moisture values in a process known as state correction. This is a powerful tool in ensuring model outputs are tuned to observations where these are available. The PDM model has been integrated into InfoWorks ICM and this includes a calibration suite which allows rapid and user friendly calibration using automated and manual methods with the aid of a number of objective functions and graphs for visual inspection of fit.

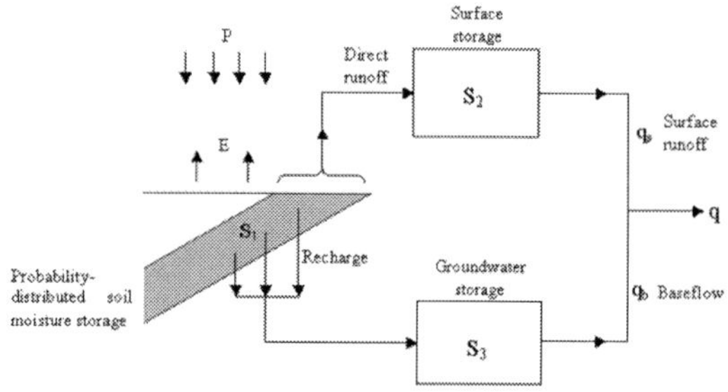


**Figure 2.** The PDM Rainfall Runoff Model (Moore, 2007)

PDM model was used in this study due to its less complexity. PDM model only considers heterogeneity in the top soil layer has been shown to improve subsequent river discharge when fed into a river routing scheme (Clark et al., 2011). The model is largely used in real-time river modelling systems to properly simulate the catchment hydrologic parameters over long periods of time, which take into account the variation in the depth of soil moisture storage capacity across a catchment, and that the catchment response to rainfall will change as these stores become saturated (Innovyze, 2012)

### 2.4 The 2 Dimensional Flood Routing Model

InfoWorks ICM (Integrated Catchment Modelling) is the first truly integrated modelling platform to incorporate both urban and river catchments. With full integration of 1D and 2D modelling techniques both the above- and below-ground elements of catchments can be modelled as never before. InfoWorks ICM enables the hydraulics and hydrology of natural and manmade environments to be incorporated into a single model (Innovyze, 2015).

## 3 Approaches and Methodology

Main components for the development flood forecasting model are:

- Hydrological component
- Hydrodynamic component

### 3.1 Hydrological Component

The Lebir and Galas catchment has a humid tropical climate and the average annual rainfall for Peninsular Malaysia is 3,000 mm (DID, 2014), although seasonal distribution of rainfall varies widely from place to place. Lebir and Galas catchment is highly forested with a mixture of some remaining areas of primary rainforest and managed agricultural plantations. Historically, the catchment was almost completely forested, until significant conversion to rubber plantations and palm oil plantations from the 1960s onwards. Since that time, flooding has become more common and severe. Furthermore, urbanisation is spreading, and in combination with the effects of necessary deforestation, soil erosion and flash flooding are becoming more prevalent.

#### *3.1.1 Hydrological Model*

The Soil Conservation Service Unit Hydrograph Method (SCS UHM) hydrological model is an event-based model and is unsuitable for flood forecasting where the soil moisture must be tracked continuously. Suitable method for flood forecasting hydrology using a combination of these models:

- Probability Distributed Model (PDM)
- Simple Runoff Model (SRM)

For larger sub-catchments, having reasonable amounts and quality of historical calibration data (concurrent rainfall and flow data), the PDM is the best choice to be run within InfoWorks ICM. For smaller sub-catchments, with fewer calibration data, the SRM is the best choice. Minimal amounts of historical flow data were available for model calibration. Historical level data were available, but these require rating equations to convert the level data to flow data.

Figure 3 to 5 shows the available hydrological data.

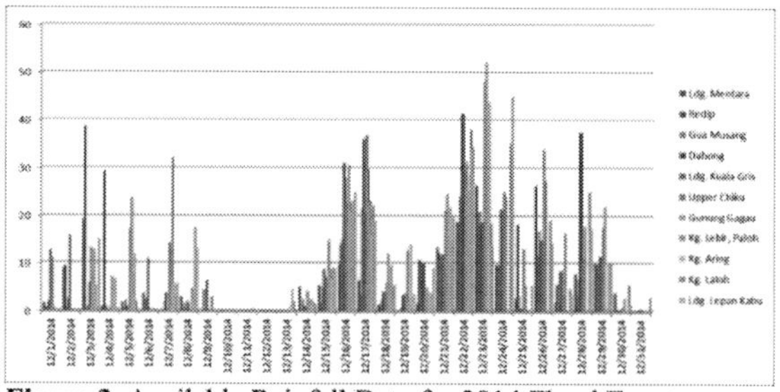

**Figure 3.** Available Rainfall Data for 2014 Flood Event

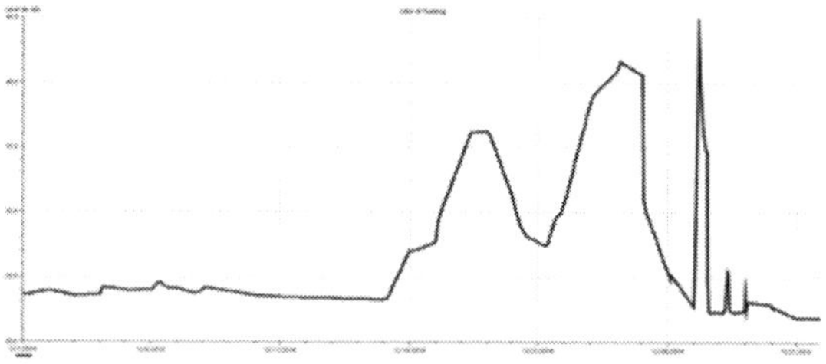

**Figure 4.** Overview of gauged hourly water level time series data for Lebir water level station (December 2014)

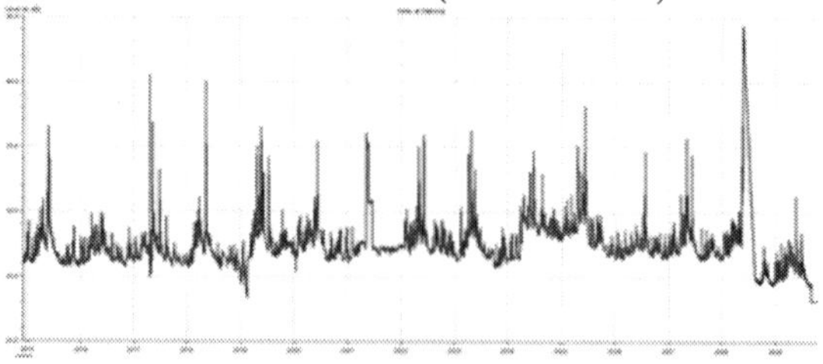

**Figure 5.** Overview of gauged hourly water level time series data for Galas water level station (December 2014)

### 3.2 Hydrodynamic Component

The purpose of carrying out the flood routing analysis is to simulate flood parameters such as flood water depths, flow velocities, flood extent and flow propagation.

In preparing the hydrodynamic flood forecasting model, integrated use of 1 dimensional (1D) and 2 dimensional (2D) hydrodynamic models is utilised to simulate the river and the flood plain interaction. The 1D model is used for water level assessment along the axis of the water body. In generating flood flows at the floodplains, which requires more complex computations, a 2D model is used.

#### *3.2.1 Governing Equations*

The basic equations used in most of the 1D Hydrodynamic Models are based on the one-dimensional unsteady state gradually varied flow equations, which are termed as the 'St. Venant equations';

- the mass conservation / continuity equation:

$$\frac{\partial Q}{\partial x}+\frac{\partial A}{\partial t}=c \qquad (1)$$

- the momentum conservation or dynamic equation:

$$\underbrace{\frac{\partial Q}{\partial t}}_{(I)}+\underbrace{\frac{\partial}{\partial x}\left[\frac{\beta Q^2}{A}\right]}_{(II)}+\underbrace{gA\left[\frac{\partial h}{\partial x}-S_0\right]}_{(III)}+\underbrace{g\frac{AQ|Q|}{K^2}}_{(IV)}=0 \qquad (2)$$

Where:

$Q(x,t)$ = discharge ($m^3/s$)
$t$ = time (s)
$x$ = streamwise direction (m)
$c$ = lateral inflow per unit length of flow
$A(x,t)$ = cross-sectional area ($m^2$)
$g$ = gravitational acceleration ($m/s^2$)
$h$ = water level (m)
$So$ = bed slope (m/m)
$K$ = conveyance ($m^3/s$)
$\beta$ = Boussinesq coefficient

and

*I.* local acceleration term
*II.* convective term (responsible for non-linearity of equation)
*III.* pressure term due to change in depth over reach – if *So* is neglected, then *dh/dx* approximates the friction slope based on the change in water level
*IV.* source/gravity term causes water to flow

In some instances, Equation 1 is set equal to c(*x*, *t*) $m^2/s$, which is equivalent to specifying lateral inflows from small rivers. Underground sources and ground water can influence the lateral inflow and this directly influences the calculation.

The assumptions inherent in the application of Equation 1 and Equation 2 are:

i. the flow is one-dimensional i.e. a single velocity and elevation can be used to describe the state of the water body in a cross-section;
ii. the water is incompressible with a constant density (=1000 $kg/m^3$) uniformly distributed;
iii. the bed slope is small;
iv. the streamline curvature is small and vertical accelerations are negligible, hence the pressure is hydrostatic;
v. the effects of boundary friction and turbulence can be accounted for by representations of channel conveyance derived for steady state flow; and
vi. all functions and variables are continuous and differentiable;

In the modelling of floods, flows often take short cuts through flood plains where the 1D description may become quite inaccurate. This is even more the case for dam or embankment failures, where the flow may leave the flood plain completely and inundate natural terrains. For this reason the 2D shallow water equations are introduced. Following the same principles as for 1D flow, the mass conservation / continuity equation reads as below (Equation 3):

$$\frac{\partial h}{\partial t}+\frac{\partial (uh)}{\partial x}+\frac{\partial (vh)}{\partial y}=0 \qquad (3)$$

Where, the ***y*** axis, orthogonal to the ***x*** axis, is introduced with its flow velocity ***v*** (m/s) associated to it. The convective momentum terms are subject to the same principles as discussed for the 1D approximation.

#### *3.2.2 Numerical Solution*

Numerical solution in a 2D model depends on the selection of the grid types. The example of grids are flexible mesh, rectilinear grid and curvilinear as shown in *Figure 6.*

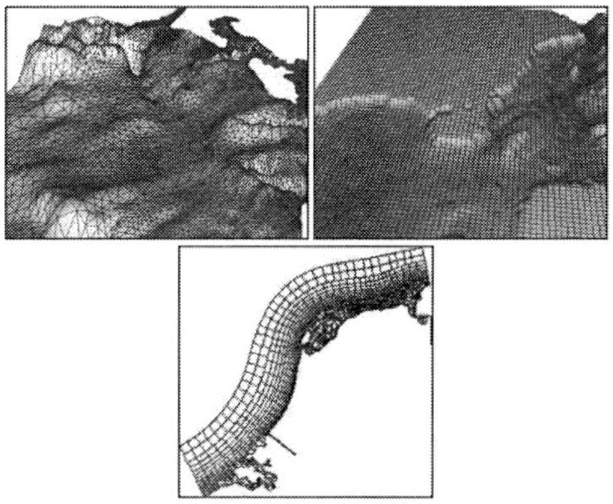

**Figure 6.** Example of types of grids for numerical solution in 2D model (flexible mesh, rectilinear grid and curvilinear)

Flexible mesh and rectilinear grid are usually applied for floodplain modelling. The general characteristics of these grids are as follows:

i. Flexible Mesh - Refer to *Figure 7*
   - Cell-centred finite volume method
   - Unstructured mesh
   - Triangular and quadrilateral elements
   - Spherical and Cartesian coordinates
   - Explicit up winding scheme limits time step for stability

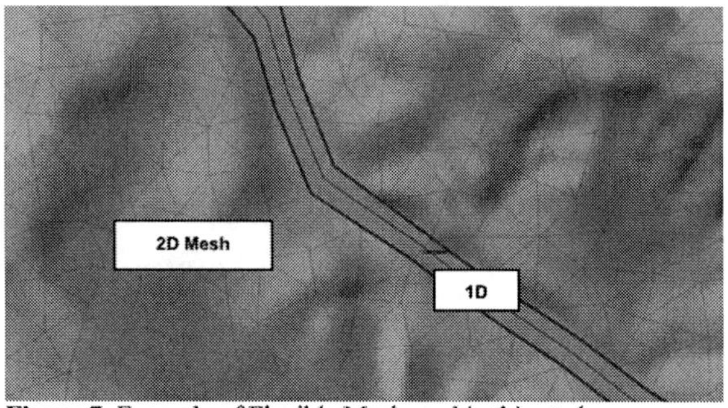

**Figure 7.** Example of Flexible Mesh used in this study

### *3.2.3 1D/2D Hydrodynamic Model Development*

The flood plain shall be modelled using 2D modelling with mesh technique. For 2D modelling, flow is not constrained to follow the direction specified by the modeller as in 1D modelling, but is allowed to spread freely over the ground surface under the influence of gravity. The modelled area is represented by a 2D mesh of calculation elements. The InfoWorks ICM engine uses a finite volume technique which takes advantage of unstructured meshes. The engine is fully integrated within the InfoWorks GUI and so benefits from the full range of data import, validation, version control and audit trail functionality.

2D finite volume methods are used to solve the shallow water flow equations, using the Riemann solver with a TVD shock capturing model that makes it particularly suitable for rapidly varying flood flows associated with bank overtopping or breaching. The 2D engine runs simultaneously with the 1D engine allow for a time step by time step exchange of water between the each zone. This ensures that there is feedback between the two simulations. Links between the 1D and 2D zones can either be lateral i.e. 1D channel and 2D floodplain or in-line, i.e. a 1D reach that feeds in to a 2D reach or vice versa. Each model network can have multiple 1D and 2D zones. Key to the efficiency and ease of use of the 2D engine is the automatic generation of the computational mesh. The mesh generator includes functionality to control the resolution by both maximum triage area and or minimum triangle angles. These can also be varied for differing parts of the mesh to enable high resolution to be maintained around key features, while using lower resolution for flatter, featureless, less important regions. In addition the 2D engine offers thematic mapping of the velocity vectors using either direction arrows whose length indicates the magnitude of the velocity coloured contouring.

## 4 Results and Discussion

### 4.1 PDM Calibration

The upstream boundary of the InfoWorks ICM model is at the location of the Lebir at Tualang water level gauge. A PDM model was defined for this sub-catchment. This PDM model will have a substantial impact on the overall performance of the flood forecasting system; it was therefore important to achieve the best calibration possible with the available data. The following steps were undertaken:

i. Quality control of gauged rainfall and gauged water level data;
ii. Analysis of correlations between rainfall and river water levels;
iii. Data selection for the PDM calibration; and
iv. Calibration of the PDM model.

From the quality control process of the data it was clear that there were continuous level time series available for model calibration. In addition, there were data for 11 rainfall gauges which might not all be equally relevant to the calibration of selected events due to the spatial variation of the rainfall. In order to get a good PDM calibration, there would need to be a high correlation between level and rainfall. In Dec 2014 there was a high correlation between river level and rainfall on a monthly basis. The PDM model was calibrated for this time period. The Lebir catchment is a fast responding system. The calibration approach was therefore to consider the following parameters and develop appropriate values for them given the catchment characteristics and the likely hydrological response. The parameters adjusted during the calibration are described below.

a) Rainfall and evaporation
   i. rainfac - Rainfall factor, from observed rainfall to effective rainfall.
   ii. be - Exponent in actual to potential evaporation rate function, affects variation between seasons.

b) Moisture Storage Distribution
   i. cmin - Minimum moisture store capacity.
   ii. cmax - Maximum moisture store capacity.
   iii. b - Exponent of store capacity distribution function controlling spatial variability of store capacity.

c) Recharge and Runoff
   i. Kg – Groundwater recharge time constant, controls rate of aquifer recharge.

DOI: 10.1051/matecconf/20178701016

ii. St – Soil tension storage capacity, increase to prevent complete drainage of soil moisture store.
iii. Bg – Exponent of recharge function, increase to magnify sensitivity of recharge rate to soil dryness.

d) Surface flow
   i. k1 – Time constant of surface flow storage for first linear reservoir.
   ii. k2 – Time constant of surface flow storage for second linear reservoir.

e) Baseflow
   i. kb – Base flow time constant, controls length of recession.

f) Time and flow
   i. tdly – Time delay, shifts hydrograph horizontally along time axis.
   ii. qconst – Constant flow representing return/abstractions, shifts hydrograph vertically along the flow axis.

The preliminary calibrated PDM model performed reasonably for the Dec 2014, however, it underestimated the peak flows. Result of the PDM calibration is presented in Figure 8. In this figure, the blue trace is the observed flow and the green trace is the PDM modelled flow. This result showing a reasonable goodness-of-fit (R2) between the model results and observed flows of 0.68. The R2 values vary considerably, and this reflects the nature of the data and the spatial location of the gauges relative to the spatial coverage of the flood event. For example, one event may be well recorded be a particular gauge and less well recorded by another gauge situated further away.

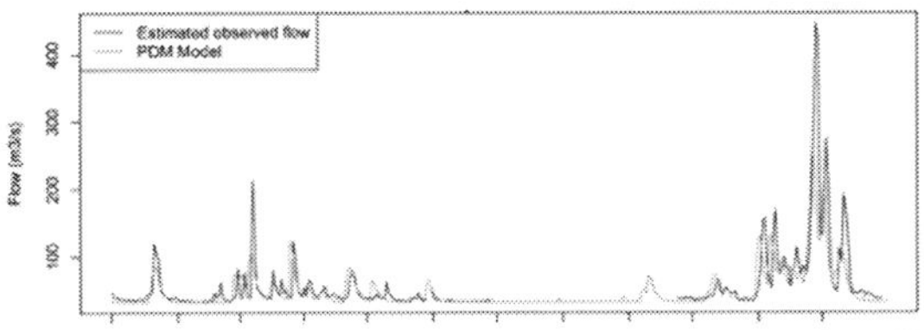


**Figure 8.** Preliminary PDM Calibration Result December 2014 ($R^2$=0.68)

### 4.2 Soil Moisture Deficit (SMD)

The soil moisture deficit (SMD) time series has been calculated using PDM. SMD is the depth of rainfall which would be required to bring the catchment to saturated conditions. Figure 9 shows the SMD and Runoff for December 2014 flood event. The orange trace is the SMD, the green trace is the runoff, and the black bar is the hyetograph. From this result soil moisture storage are set to zero and produce more runoff. During the SMD at zero, the soils became saturated and cause the flooding.

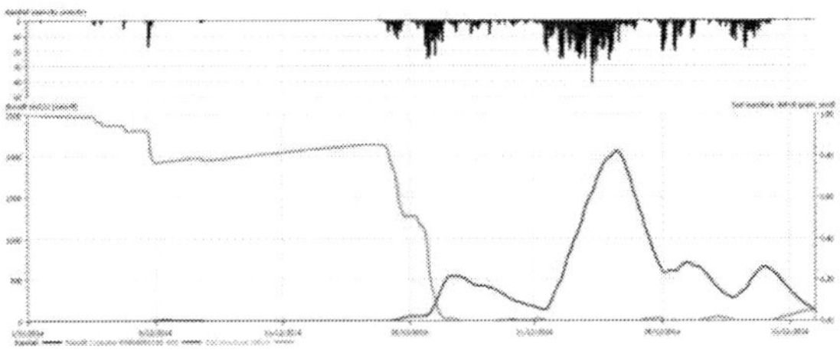

**Figure 9.** SMD and Runoff for Lebir Catchment

### 4.3 2D Flood Plain Representations – Mesh Polygon

Flood plain is represented by 2D polygon contained the definition of the mesh. The information includes the mesh size and roughness. The elevation for each mesh is taken from the ground model either from IFSAR. The connection between the main river and the flood plain is through the bank line. If the geometry is complex such as valley, then another polygon can be specify with finer mesh. As a start for this study, River Galas and River Lebir maximum mesh size of 20,000m2 and minimum mesh size of 4,000m2 has been applied. The preliminary Flood inundation area is shown in Figure 10.

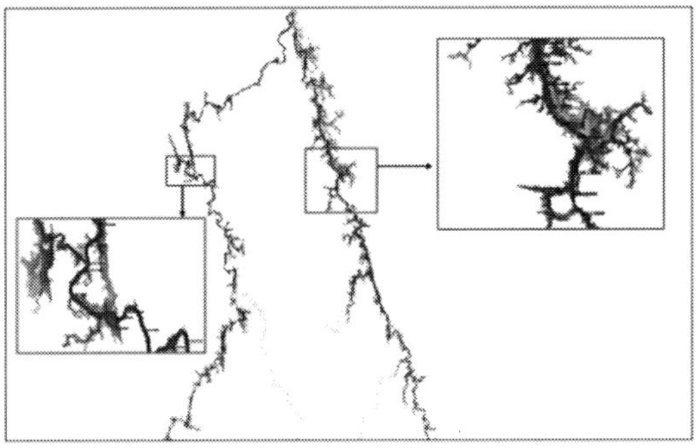

**Figure 10.** Preliminary Flood Hazard Map for Dec 2014

## 5 Future Works

Results presented in this paper are still at early stage. The ultimate contribution of this study is flood disaster warning dissemination to related agencies and community at risk. In order to achieve the study objectives, the following are planned to be performed near future:

(i) Production of flood trend analysis based on one selected event at particular area of interest.
(ii) Identify flood forecast lead time for particular area of interest in River Kelantan Catchment.
(iii) Development of Flood hazard map for River Kelantan Catchment using 2D mesh method to identify the critical area.

DOI: 10.1051/matecconf/20178701016

## 6 Conclusions and Recommendation

Calibration of the PDM for this study is not a straightforward process. The Pareto distribution of soil storage depths has been found to provide a simple yet flexible description of soil moisture storage for Lebir and Galas Catchment. SMD is essential to understand the correlation between the rainfall and runoff. This would give an advantage to the flood forecasting model using PDM which is the long duration hydrological model should be the first choice.

Two dimensional flow models have been progressively utilized as a part of flood modelling and are giving an important device in the appraisal of stream ways and along these lines the receptors of flooding. In this case, 2D hydrodynamic flood modelling is impressively produce the flood inundation area and flood propagation and this made modeller to have better understanding flood behaviour in this study area. This prompts an exchange off between the run time of the 2D model and precision of the model at speaking to the 2D surface. Bigger size of mesh components permit models to run all the more rapidly however may not accurately speak to every one of the elements of the surface. Frequently the full detail of the hidden geography is not representing in the hydrodynamic flood model. Generally 2D hydrodynamic flood model contain either a structured grid or an unstructured mesh to represent the ground topography. Unstructured mesh have the favourable position that triangle sizes can differ within the mesh, permitting the modeller to create a better work in territories obliging point by point examination and a coarse mesh in general to accomplish sensible keep running times. The ideal 2D mesh will represent the 2D topography with adequate precision to give trust in model results while keeping up a sensible run time. The exploration in this proposition speaks to a preparatory study to focus the impact that the discretization of the geography can have on the hydrodynamic flood model results gave by flood analysis model.

Due to complexity of overland flow analysis, 2D hydrodynamic flood model provides better understanding on the flood behaviour of the study area. This study aims to produce of flood trend analysis and provide information on flood forecast lead time for particular area of interest in River Kelantan Catchment. Finally, the model output will provide flood early warning dissemination to the public at risk, including identification of critical area through flood hazard map for River Kelantan Catchment using 2D mesh method.

## References

1. Brocca, L., Melone, F., & Moramarco, T. (2011). Distributed rainfall-runoff modelling for flood frequency estimation and flood forecasting. Hydrological Processes, 25(18), 2801–2813. http://doi.org/10.1002/hyp.8042
2. Garcia-Navarro P. and Brufau P., Numerical methods for the shallow water equations: 2D approach, in Donald W Knight and Asaad Y. Shamseldin (Eds.), River basin modeling for flood risk mitigation, (pp. 409-428), Taylor & Francis. (2006).
3. Clark, D. B., Mercado, L. M., Sitch, S., Jones, C. D., Gedney, N., Best, M. J., … Cox, P. M. (2011). The Joint UK Land Environment Simulator (JULES), model description – Part 2: Carbon fluxes and vegetation dynamics. *Geoscientific Model Development*, *4*(3), 701–722. http://doi.org/10.5194/gmd-4-701-2011
4. Clark, D. B., Mercado, L. M., Sitch, S., Jones, C. D., Gedney, N., Best, M. J., … Cox, P. M. (2011). The Joint UK Land Environment Simulator (JULES), model description – Part 2: Carbon fluxes and vegetation dynamics. Geoscientific Model Development, 4(3), 701–722. http://doi.org/10.5194/gmd-4-701-2011
5. Department of Irrigation and Drainage. Laporan Banjir Tahunan Jabatan Pengairan dan Saliran, Malaysia. (2015).
6. Environmental Agency. Comparison of Rainfall-Runoff Models for Flood Forecasting. R&DTechnical Report W241. (2001).
7. Gutierrez, J. et al. in prep. Testing and application of a practical new 2D hydrodynamic. Flood Risk 2008 Conference, Oxford, 30th September-2nd October 2008.
8. Hunter, N. M., Bates, P. D., Neelz, S., Pender, G., Villanueva, I., Wright, N. G., Liang, D., Falconer, R. A., Lin, B., Walle, S., Crossley, A. J., Mason, D. C.Benchmarking 2D hydraulic models for urban flooding, Proceedings of the ICE - Water Management, Volume 161, Issue 1, pages 13 -30. (2008).
9. Innovyze. Technical Paper. Innovyze Ltd Howbery Park, Wallingford, Oxfordshire, OX10 8BA, United Kingdom. (2015).
10. Innovyze. (2012). Real - Time Operational Forecasting , Warning & Reporting FloodWorks - Technical Review.
11. Price R.K. and Vojinovic, Z., Urban Hydroinformatics: Data, Model and Decision Support for Integrated Urban Water Management, IWA Publishing. (2011).
12. R.J. Moore., The PDM Rainfall Run-off Model. Journal of Hydrology and Earth System Science, 11 (1), 483-499. (2007).
13. Rosenthal, U., Hart, P. and Beyuzen, M. Flood Response and disaster management: a comparative perspective. In Flood Response and Crisis Management in Western Europe, Rosenthal, U. and Hart, P. (eds.), Springer-Verlag Berlin Heidelber, pp.1-13. (1998).
14. Schmitt T.G., Thomas M. and Ettrich N. Analysis and modeling of flooding in urban drainage systems. *J. Hydrology*, 299(3-4), 300-311. (2004).
15. Vojinovic Z., Bonillo J., Kaushik C. and Price R. Modelling flow transitions at street junctions with 1D and 2D models. 7 th Int. Conf. on Hydroinformatics, Nice, France, Vol. I, 377-384. (2006).

# Workability and Compressive Strength for Concrete With Coconut Shell Aggregate

Alif Syazani Leman[1], Shahiron Shahidan[1,a], Mohd. Yazid Yusuf[1], Sharifah Salwa Mohd Zuki[1], and Noor Aina Misnon[2]

[1]*Faculty of Civil & Environmental Engineering, Universiti Tun Hussein Onn Malaysia 86400 Batu Pahat, Johor, Malaysia*
[2]*Department Of Civil Engineering, Universiti Pertahanan Nasional Malaysia, 57000 Kuala Lumpur, Malaysia*

**Abstract.** This study was conducted to investigate the compressive strength and workability of concrete added with coconut shells. Comparisons were made between conventional concrete with concrete mix coconut shell. In this study, the concretes were mixes with coconut shell by percentage of weight concrete which is 0%, 5%, and 10%. The coconut shell has been crushed first, then it was sieved, to get the optimum size which, that retained on the 5mm sieve and passing 10mm sieve. Experimental tests conducted in this study are slump test and compressive test. The results from this study are workability of concrete added with 0% and 5% of coconut shell has medium degree of workability compared to concrete added with 10% that has low workability. For the compressive strength, the concrete added with 5% and 10% of coconut shell has lower strength compared with normal concrete.

## 1 Introduction

Concrete is a material produced from several materials such as cement, coarse and fine aggregates, and water mixture [1][2]. The application of concrete material in construction industry was highly demand which growing rapidly demonstrates the need for various types and properties of concrete to meet the diverse needs of the user [3]. However, due to the influx of construction projects that use a lot of concrete as the main material of construction, huge amounts of natural resources are required to make concrete [4][5]. As an alternative solution, the agricultural waste is the best way to reduce the raw materials used in construction industry and thus helping to reduce pollution caused by disposal of agricultural waste [6][7].

Coconut grown in over 86 countries worldwide and generates more than 54 billion nuts per annum [8]. However, after the coconut is scraped out, the shell is usually discarded as waste [9]. With the use of coconut shell as a replacement material in the construction industry, indirectly reduce the costs production of concrete and the disposal of waste [10][11]. The used of coconut shell in the manufacturing of concrete, indirectly creating new inventions in order to improve the concrete manufacturing industry as well as save costs and reduce solid waste disposal. Coconut shell is often used as a composite in concrete because of the characteristics found in it better than material that commonly used in production of concrete. Besides, coconut shell is potential material for the development of new composite material in concrete mix design because of their high strength and modulus properties [12]. However, further research is needed for better understanding of the behavior of coconut shell in concrete. Thus, the aim of this work is to provide more data on the strength of concrete with coconut shell at different percentage coconut shell used. Furthermore, in this study the workability of concrete with coconut shell was also investigated.

## 2 Coconut Shell Properties

Coconut shell may offer itself as a potential construction material in the field of construction industries and would solve the environmental problem of reducing the generation of solid wastes simultaneously. Coconut shell is often used as a composite in concrete because of the characteristics found in it better than material that commonly used in production of concrete. Coconut shell has high strength and modulus properties along with the added advantage of high lignin content. The shell also absorbs less moisture due to its low cellulose content [13]. Besides, the surface texture of the coconut shell was fairly smooth on concave and rough on concave faces. In addition, coconut shells are used in the production of activated carbon due to hardness and high carbon content [14]. Thus, coconut shells are fairly enough to get workability same or better than normal material used in concrete. The properties of coconut shells were compared with crushed granite and oil palm shell (Table 1).

[a] Corresponding author: shahironshahidan@gmail.com

**Table 1:** Properties of coconut shell, oil palm shell, crushed granite and river sand (Gunasekaran, Kumar, & Lakshmipathy, 2011)

| SI.No | Physical and mechanical properties | Coconut shell | Oil palm shells | Crushed granite | River sand |
|---|---|---|---|---|---|
| 1 | Maximum size (mm) | 12.5 | 12.5 | 12.5 | - |
| 2 | Moisture content (%) | 4.20 | - | - | - |
| 3 | Water absorption (24h) (%) | 24.00 | 23.32 | 0.50 | - |
| 4 | Specific gravity | 1.05-1.20 | 1.17 | 2.82 | 2.57 |
| 5 | SSD[a] apparent | 1.40-1.50 | - | 2.86 | - |
| 6 | Impact value (%) | 8.15 | 7.86 | 12.40 | - |
| 7 | Crushing value (%) | 2.58 | - | 6.30 | - |
| 8 | Abrasion value (%) | 1.63 | 4.80 | 1.85 | - |
| 9 | Bulk density (kg/m$^3$) | 650 | 590 | 1650 | - |
| 10 | Compacted loose | 550 | - | 1450 | - |
| 11 | Fineness modulus | 6.26 | 6.24 | 6.94 | 2.56 |
| 12 | Shell thickness (mm) | 2-8 | 1.5-2.5 | - | - |

## 3 Behaviour Of Concrete With Coconut Shell

Recently efforts have been made on the use of agricultural waste, coconut shell as composite materials to produce structural concrete [9]. In addition, the use of this shell has brought benefits in term of economic and environmental. Apart from that the ultimate bond strength of coconut shell as aggregate concrete under all type of curing conditions was much higher compared to the theoretical bond strength as per BS 8110 and IS 456 [10]. Besides, bonding between cement paste and coconut shell shows that the bond appears to be better [15]. In addition, the use of coconut shell in the production of concrete has better workability, compressive strength, flexural strength, slip tensile strength, impact resistance, bonding property of coconut shell have been studies, discussed and published already; all these properties are comparable with normal and other lightweight concrete [16]. Apart from that, from shell concrete beam behaviors under flexural, shear and torsion show that improvement and better result compare to normal concrete [17][18]. However, not all parameters of concrete experiments conducted using coconut shells can rival the result of normal concrete.

## 4 Experimental Works

In this research, it consists of 2 major part of testing in fresh concrete. There are slump test for workability and compressive test for the concrete strength. This investigation were prepare of 27 concrete cubes size of 150mm x 150mm x 150mm, with a grade C35. The detail percentage of coconut shell as presented in the next subtopic.

### 4.1 Material Preparation

The freshly discarded coconut shell was obtained from local processing oil mills were close to the study area. Coconut shell obtained has been cleaned from impurities. It was obtained in the already cracked and oil extracted form. Coconut shell was broken into a small chips manually using hammer before crushed it into small pieces. The fibrous outer parts of the nut were needed to be removed first, and then coconut shell is crushed by a crusher. Before being put into mix, crushed coconut shell were sieved by size to be used in concrete. The crushed of coconut shell was graded by using shaker machine to get the optimum size which retained on the 5mm sieve and passing 10mm sieve. In this research, percentage of coconut shell used in concrete was 0%, 5% and 10%. The percentage was calculated based on the weight of concrete. For other concrete mix constituents, such as cement, water, fine and coarse aggregates were obtained from Fkaas Laboratory, University Tun Hussien Onn Malaysia. The cement used as the binding agent was Ordinary Portland Cement. For coarse aggregates, the size used was which retained on the No.4 (4.75mm) sieve where fine aggregate are aggregate passing No.4 (4.75mm) sieve and predominately retained on the No.200 (700μm). The potable water available in Fkaas Lab premises was used for mixing and curing

### 4.2 Mix Proportions

There are three mixes were employed. First, control mix which is concrete without coconut shell was made by using a basic material such as Ordinary Portland cement, fine and coarse aggregates and water. Normal concrete sample made are as comparative experiments between coconut shells concrete. Then, mix of concrete added

DOI: 10.1051/matecconf/20178701017

with 5 and 10 percentages of coconut shell was made. The details of the mixture proportions used for the concrete are shown in Table 2. Every percent of coconut shell used, curing period and also coconut shell size used requires at least three cubes to get the average value for compressive strength test. Since all material has been prepared, the concrete work can be carried out by using mixer. Then, after several minute and the concrete were mixed well, the mixture will pour into a mould with prescribed size.

**Table 2:** Mixture proportions

| Mixes | Cement (kg) | Coarse aggregate (kg) | Fine aggregates (kg) | Water (kg/L) | Coconut shell (kg) |
|---|---|---|---|---|---|
| Control | 13.67 | 37.67 | 16.1 | 5.47 | 0.00 |
| 5% coconut shell | 13.67 | 37.67 | 16.1 | 5.47 | 3.75 |
| 10% coconut shell | 13.67 | 37.67 | 16.1 | 5.47 | 7.50 |

### 4.3 Testing Method

This research was conducted of two major test, which is workability and compressive strength test. Workability Test; this test was performed according to procedures established by the BS 1881: Part 102: 1983. Firstly, the cone were cleaned and damped. Next, the mold placed on a smooth and flat surface. Then, the mold was placed on flat tray with a part of the sampling cone diameter of 200 mm at the top and concrete mix was filled into the mold of three layers. Each layer that was filled in must be compacted 25 times using tapping rod. Once completed, the top of the cone mold is flattened using spatula. Mold slowly lifted vertically upward and the slump value is recorded. The workability of the concrete can be determined by measuring the slump drop.

Compressive Strength Test; this test was performed according to procedures established by the BS 1881: Part 102: 1983. Details about concrete such as the size of the cube should be recorded first so that the result of tests conducted compactible with size of the cubes were tested. After the materials and machines available, this test was carried out. The test was conducted on 150mm x 150mm x 150mm cube specimen at 7, 14, and 28 days. The sample cubes were placed in the testing machine. Loading were applied on the sample until the sample fails, where the reading started to decline. The test was repeated conducted on another two more cubes and the value was taken as the average.

## 5 Results And Discussions

A comprehensive summary of the workability and strength properties of normal concrete and concrete with coconut shell added are represented in Table 3 and Figure 4. Result were presented through univariate analyzes. For this univariate analyzes, it include demographic data consisting of the strength and workability of concrete cube with crushed coconut shells.

### 5.1 Workability Studies

For concrete mix with 0% (Figure 1) and 5% (Figure 2) of coconut shell that had 60.5mm and 50.0 mm slump, these mixtures had medium of degree of workability. For concrete mix with 10% (Figure 3) of coconut shell that had 25.0 mm slump, this mixture had low degree of workability. From the result, show that if the percentage of coconut shell added increase, the degree of workability of concrete mix will decreased. This result may be happen because of coconut shell have relatively high water absorption compared to the conventional aggregates.

**Figure 1:** Concrete with 0% coconut shell

**Figure 2:** Concrete with 5% added of crushed coconut shell

DOI: 10.1051/matecconf/20178701017

**Figure 3:** Concrete with 10% added of crushed coconut shell

**Figure 4:** Comparison of slump test result with practical standard referring to BS 1881: Part 102: 1983

**Table 3**: Data of workability test

| Percentage Of Coconut Shell | Slump Reading |
|---|---|
| 0% | 60.5 mm |
| 5% | 50.0 mm |
| 10% | 25.0 mm |

### 5.2 Compressive Strength

The data of compressive strength and density for each sample were recorded and were representing in graph below in Figure 5 and Figure 6. This graph was shown all the data collected including the age of concrete tested and also by percentages of coconut shell added

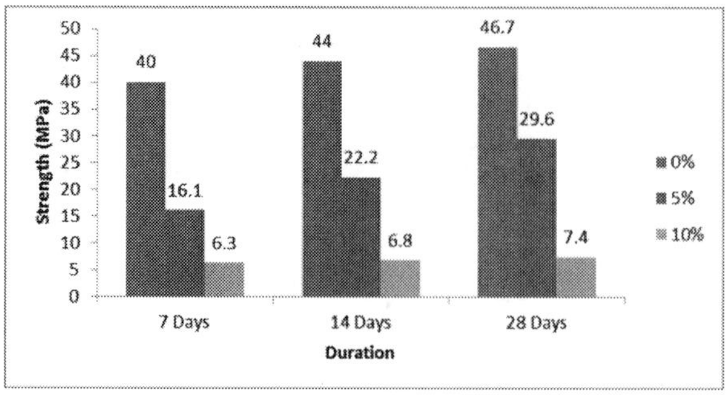


**Figure 5:** Graph of strength versus duration

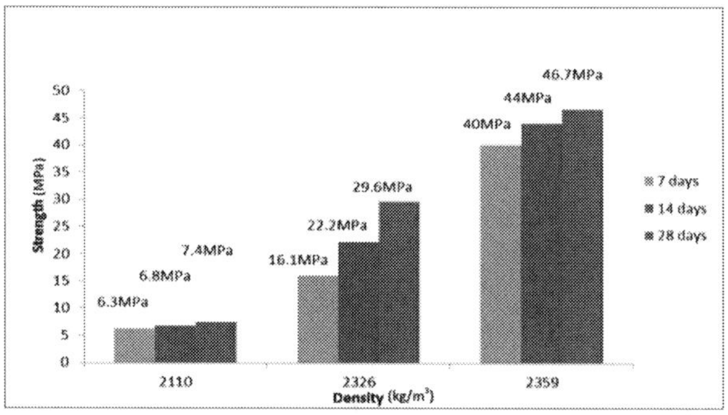


**Figure 6:** Graph of strength versus density

From the data collected, there was good relationship between the parameter. For concrete with 0% of coconut shell added or control concrete, the density of this sample is 2359 kg/m$^3$. In term of strength, the strength of concrete is increased follow by the age of concrete which is 40MPa for 7days, 44MPa for 14days and 46.7MPa for 28 days. For concrete added with 5% and 10% of coconut shells, the pattern of strength versus density is same with control concrete, which is the compressive strength of concrete increased when the density of concrete is increased. However, the strength of concrete specimen decreased when the percent of coconut shell added is increased. This may happen when the percent of coconut shell added are higher will decreasing the density of concrete. The decreasing value of the concrete strength while the percentage of coconut shell added into concrete increased happen maybe related to the workability of each concrete mix. The percentages of coconut shell added will decreased the degree of workability of concrete. Then, the lower degree of workability happens because of high water absorption in concrete. Therefore, while the level of water absorption of concrete is high, it will affect the strength of concrete.

 DOI: 10.1051/matecconf/20178701017

### 5.3 Relationship Between Compressive Strength and Density

In this study, the data of compressive strength and density for each sample were recorded. The data were representing in graph below in Figure 7(a) and Figure 7(b). This graph was shown all the data collected including the age of concrete tested and also by percentages of coconut shell added.

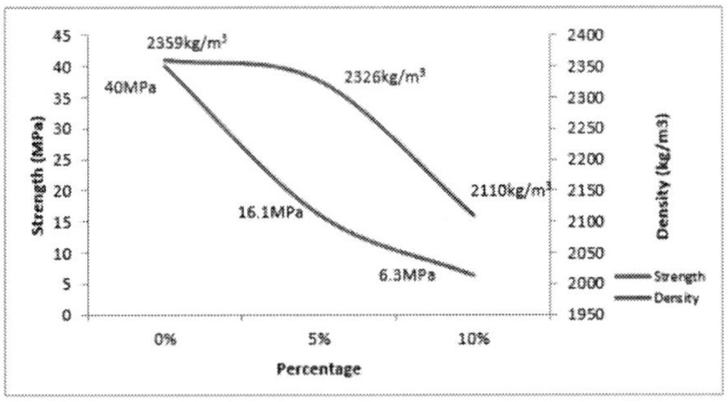


**Figure 7(a):** Graph of strength versus density for 7days

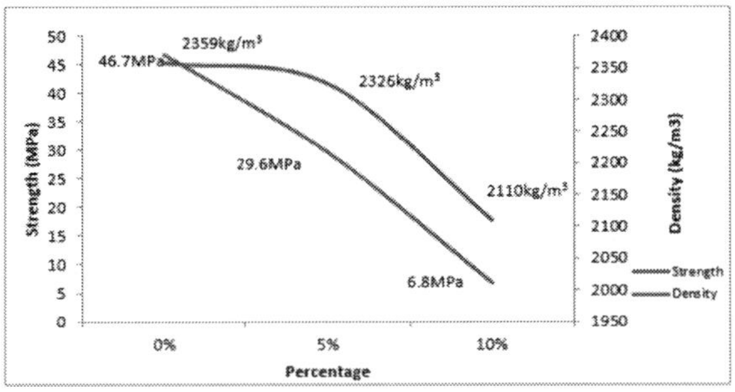


**Figure 7(b):** Graph of strength versus density for 28days

From the data collected, there was good relationship between the parameter. The compressive strength of concrete is increased when the density of concrete is increased. However, the strength of concrete specimen decreased when the percent of coconut shell added is increased. This may happen when the percent of coconut shell added are higher will decreasing the density of concrete. Although the densities of the concrete are more than 2000 kg/m$^3$, the compressive strength obtained is lower due to the high water absorption of the material. It shows that the hardened cement paste is less well-structured and more porous when hydration proceeds at faster rate.

## 6 Conclusions

A comparative study of workability and strength of concrete added with coconut shell and compare it with normal concrete has been carried out. Generally, the compressive strength of concrete added with coconut shell decreased as the percentage of coconut shell added is increased. Same result goes to the workability test, the workability of concrete decreased as percent of coconut shell added in concrete is increased. Based on the result, some suggestions for improvement should be given in this study.

This is to ensure that the research can have more appropriate result and better data than before. Here are some of the recommendations:

a) The coconut shell must be cleaned first, remove it from anything that can affect the properties or strength of the coconut shell

b) The coconut shell may could be dry or soaked into water first to changes the properties of coconut shell

c) Coconut shell may be treated first by any kind of treated method before use in concrete

d) Changes the shape or size of coconut shell use in concrete mix

e) Use the different type of curing process

f) Added some of admixture that can improve the bonding between coconut shell and concrete

g) Mix the coconut shell with the different type of cement used.

## Acknowledgements

The authors would like to thank Universiti Tun Hussein Onn Malaysia for Grant GPPS U449 and Grant Vot U523

## References

1. M. Abdul Rahim, N. M. Ibrahim, Z. Idris, Z. M. Ghazaly, S. Shahidan, N. L. Rahim, L. A. Sofri, and N. F. Isa, "Properties of Concrete with Different Percentange of the Rice Husk Ash (RHA) as Partial Cement Replacement," *Mater. Sci. Forum*, vol. **803**, pp. 288–293, (2014).
2. N. Md Nor, N. Muhamad Bunnori, A. Ibrahim, S. Shahidan, and S. N. M. Saliah, "An investigation on acoustic wave velocity of reinforced concrete beam in-plane source," in Proceedings - 2011 IEEE 7th International Colloquium on Signal Processing and Its Applications, CSPA 2011, 2011, pp. 19–22.
3. S. A. Kudus, N. M. Bunnori, S. R. Basri, S. Shahidan, M. N. M. Jamil, and N. M. Noor, "An Overview Current Application of Artificial Neural Network in Concrete," Adv. Mater. Res., vol. 626, pp. 372–375, 2012.
4. S. R. Abdullah, W. R. Wan Zainal Abidin, and S. Shahidan, "Strength of Concrete Containing Rubber Particle As Partial Cement Replacement," MATEC Web Conf., vol. **47**, pp. 2–5, 2016.
5. M. A. Rahim, Z. Ghazaly, R. Nurazira, R. Mamat, M. A. Azizan, N. F. Isa, and S. Shahidan, "Experimental Study of Slurry Infiltrated Fiber

DOI: 10.1051/matecconf/20178701017

Reinforced Concrete," *Mater. Sci. Forum*, vol. **857**, pp. 363–366, (2016).

6. Shafigh, P., Mahmud, H. Bin, Jumaat, M. Z., & Zargar, M. (2014). Agricultural wastes as aggregate in concrete mixtures – A review. Construction and Building Materials, 53,110–117. doi:10.1016/j.conbuildmat.2013.11.074
7. S. Shahidan, I. Isham, and N. Jamaluddin, "A Review on Waste Minimization by Adopting in Self Compacting Concrete," MATEC Web Conf., vol. **47**, pp. 1–7, (2016).
8. Yerramala, A. (2012). Properties of Concrete with Coconut Shells as Aggregate Replacement, 1(6), 21–31.
9. Gunasekaran, K., Annadurai, R., & Kumar, P. S. (2013). Study on reinforced lightweight coconut shell concrete beam behavior under flexure. Materials & Design, 46, 157–167. doi:10.1016/j.matdes.2012.09.044
10. Gunasekaran, K., Annadurai, R., & Kumar, P. S. (2012). Long term study on compressive and bond strength of coconut shell aggregate concrete.
11. I. Ismail, N. Jamaluddin, and S. Shahidan, "A Review on Performance of Waste Materials in Self Compacting Concrete (SCC)," *J. Teknol.*, vol. **5**, no. 78, pp. 29–35, (2016).
12. Kambli, P. S., & Mathapati, S. R. (2014). Application of Coconut Shell as Coarse Aggregate in Concrete : A Technical Review, 4(3), 498–501.
13. Chanap, R. (2012). Study of Mechanical and Flexural Properties of Coconut Shell Ash Reinforced Epoxy Composites Bachelor of Technology in Mechanical Engineering BY, (May), 1–41
14. T. U. Ganiron, "Sustainable management of waste coconut shells as aggregates in concrete mixture," *J. Eng. Sci. Technol. Rev.*, vol. **6**, no. 5, pp. 7–14, (2013).
15. S. Shahidan, H. B. Koh, A. M. S. Alansi, and L. Y. Loon, "Strength Development and Water Permeability of Engineered Biomass Aggregate Pervious Concrete," MATEC Web Conf., vol.**47**, pp. 2–7, (2016).
16. Gunasekaran, K., Kumar, P.S. & Lakshmipathy, M., Mechanical and bond properties of coconut shell concrete. Construction and Building Materials, 25(1), pp.92–98 (2011).
17. Gunasekaran, K. et al., Study on reinforced lightweight coconut shell concrete beam behavior under torsion. *Materials and Design*, 57, pp.374–382 (2014).
18. S. Shahidan, R. Pullin, K. M. Holford, M. B. N, and N. Nor, "Quantitative Evaluation of the Relationship between Tensile Crack and Shear Movement in Concrete beams," Adv. Mater. Res., vol. **626**, pp. 355–359, (2013).

# Properties of Concrete Mixes with Carwash Wastewater

Shahiron Shahidan[a], Mohamad Syamir Senin, Aeslina Binti Abdul Kadir, Lau Hai Yee, Noorwirdawati Ali

*Department of Structural and Material Engineering,University Tun Hussein Onn Malaysia, Batu Pahat, 86400 Johor, Malaysia.*

**Abstract.** The rapid growth of the car wash industry today results in the need for wastewater reclamation. Thus, this paper aims to investigate the effect of using car wash wastewater on concrete properties in terms of mechanical properties. The basic characteristics of wastewater were investigated according to USEPA (Method 150.1 & 3 00.0) while the mechanical properties of concrete with car wash wastewater were compared according to ASTM C1602 and BS EN 1008 standards. In this research, the compressive strength, modulus of elasticity and tensile strength were studied. The percentages of wastewater replaced in the concrete mix ranged from 0% up to 40%. In addition, the results also suggest that the concrete with 20% car wash wastewater achieved the highest compressive strength and modulus of elasticity compared to other compositions of wastewater. Moreover, the results also recommended that concrete mixed with car wash wastewater has better compressive strength compared to conventional concrete.

## 1 Introduction

Malaysia is located in South East Asia with a population of approximately 31 million people. This nation receives an average annual rainfall of 3000 mm. It means that there should not be any problems related to water shortage or scarcity [1-4]. Unfortunately, the water situation in Malaysia has changed from one of relative abundance to one of scarcity [5-6]. This is because of rapid population growth, urbanization, industrialization and the expansion of irrigated agriculture.

Generally, car wash stations are common in Malaysia. Therefore, car wash wastewater can be easily collected. Wastewater is normally generated by the process of washing private or public vehicles [3][4]. However, not much attention is being paid on reusing wastewater from the car wash industry. This may be due to the limited number of articles and journals emphasizing on the applications of reused car wash wastewater [5-13].

On the other hand, the lack of knowledge regarding the car wash wastewater management may be due to individual opinion. People may think that car wash wastewater is not as critically contaminated as other types of industrial wastewater [3]. Based on the evaluation of car wash effluent quality in Malaysia, the content of chemical oxygen demand (COD) in wastewater is above the tolerance limit specified by Malaysia standard A [3][11][14]. Table 1 shows the comparison between the typical characteristics of car wash effluent in Malaysia and the tolerance limit set by Malaysia standard A. According to the latest statistics of the car industry in Malaysia, the total vehicle sales in 2015 between January to September is 485,388 units and is predicted to hit another new record of a total of 670,000 units at the end of 2015 compared to 655,793 units and 666,465 units in 2013 and 2014 respectively [15-19]. The growth of land transports together with the current number of cars would definitely lead to increasing demand for car wash services especially in highly populated housing areas located in towns [3].

Besides that, car wash stations which consume larger capacities of fresh water on a daily basis could benefit from recycling programs [8]. According to Karakulski & Morawski (2003) and Panizza&Cerisola (2010), an average of 150 to 600 litres of car wash wastewater is produced from every car wash. Consequently, it is estimated that 2.5 million gallons of fresh water are consumed daily at car wash stations [7].

Apart from that, car wash wastewater can be harmful to humans and the environment if released untreated to surface water-bodies because it may contain many pollutants such as detergents that can be poisonous to fish, oil, grease, sand, dust, chemicals and solvent-based solutions, heavy metals and organic matter [8][13][14][20].Subsequently, the Environmental Pollution Agency raised public awareness on the impurities found in car wash wastewater such as heavy metals, polycyclic aromatic hydrocarbons (PAH) and volatile organic compounds (VOC) [17][20-23]. Considering the gigantic amounts of effluent generated and its complex characteristics, reusing car wash wastewater is not only a process to manage environmental problems but is also a way to recover a significant amount of wastewater directly discharged into rivers and oceans [24-29]. Therefore, reusing car wash wastewater in concrete mixes aims to create a sustainable

[a] Corresponding author: shahironshahidan@gmail.com

solution in this industry towards green technology in construction.

Al-Jabri, et al (2011) from Oman used car wash wastewater to manufacture concrete. Wastewater characteristics including pH value, chloride, alkalinity and sulphate content in wastewater were determined. The results indicated that the characteristics of wastewater were within the ASTM standard limits. The mechanical test of concrete included compressive strength, tensile strength and a flexural test on day 28 [30-32]. In addition, the percentages of car wash wastewater in the concrete mix ranged from 25% to 100% [33]. The experimental results suggested that the strength of concrete containing car wash wastewater was comparable with the control mix [29]. Therefore, this research will mainly focus on characterization of car wash wastewater and the effect of wastewater on concrete properties. The presence of chloride and sulphate in car wash wastewater were analysed and compared with the ASTM standards [33]. The tests for concrete properties which included compressive strength, tensile strength and modulus of elasticity were performed according to the ASTM standards [34-37].

## 2 Materials and Experimental Work

In this research, the replacement percentages of car wash wastewater in concrete mixes were at 10%, 20%, 30%, and 40% of tap water. The water cement ratio w/c is 0.45 and all the concrete specimens were designed to have a target mean strength of 35 MPa according to the DOE method [30]. 30 concrete specimens measured 150 x 150 X 150 mm while 45 measured 300 mm x 150mm in diameter. The tests related to this research are compressive strength, tensile strength and modulus of elasticity (MOE).

### 2.1 Materials

#### *2.1.1 Water*

Car wash wastewater was collected from two different car wash stations in Batu Pahat, Johor, Malaysia which are MAHAJU (MJ) car wash enterprise and Bandar U car wash station. Before the wastewater was collected, the washout water from vehicles was discharged into a pond through a drainage system. The chemical properties of car wash wastewater were analysed according to USEPA standard method and compared with the standard specifications [22][38-40].

#### *2.1.2 Cement*

A local Ordinary Portland cement type CEM 1 42.5N was used in this research and it complied with the requirements of MS EN 197-1[32]. The physical and chemical properties of cement shown in Table 1 and Table 2 respectively.

**Table 1.** Physical properties of cement

| Property | OPC |
|---|---|
| Specific gravity | 3.15 |
| Passing grain size | 78 % |
| Median grain size | 18 μm |
| Blaine specific surface | 300 $m^2$/kg |
| Initial setting time | 145 Min |
| Final setting time | 270 Min |

**Table 2.** Chemical properties of cement

| Component | % |
|---|---|
| Silicon dioxide, $SiO_2$ | 21.2 |
| Aluminium oxide, $Al_2O_3$ | 5.5 |
| Iron oxide, $Fe_2O_3$ | 3.1 |
| Calcium oxide, CaO | 63.7 |
| Magnesium oxide, MgO | 1.5 |
| Sodium oxide, $Na_2O$ | 0.18 |
| Potassium oxide, $K_2O$ | 0.71 |
| Sulfur trioxide, $SO_3$ | 2.63 |
| Loss on ignition | 0.96 |

#### *2.1.3 Aggregates*

Fine and coarse aggregates according to the ASTM C33 requirements [33], ASTM C33 and BS 882 standards were used to control the aggregates sizes using a sieve analysis [34-36].

### 2.2 Mix Proportion of Concrete

All concrete specimens were designed to have a target mean strength of 35MPa according to the Department of Environment (DOE) method [30]. Table 3 shows the mix proportions of concrete.

**Table 3.** Mix proportion of concrete

| Mix % | Cement (kg/m$^3$) | Tap Water (kg/m$^3$) | CW (kg/m$^3$) | F. A (kg/m$^3$) | C.A (kg/m$^3$) |
|---|---|---|---|---|---|
| CM 0 | 414 | 180 | 0 | 535 | 1245 |
| CW10 | 414 | 162 | 18 | 535 | 1245 |
| CW20 | 414 | 144 | 36 | 535 | 1245 |
| CW30 | 414 | 126 | 54 | 535 | 1245 |
| CW40 | 414 | 108 | 72 | 535 | 1245 |

*F.A (Fine Aggregate); C.A (Coarse Aggregate); CW (Car Wash Wastewater)

## 2.3 Experimental Work

### *2.3.1 Wastewater Analysis*

The car wash wastewater pH value was measured using Hanna HI 8424 digital pH meter. The sulphate and chloride content in car wash wastewater were analysed according to the United States Environmental Protection Agency USEPA method 300.0.

### *2.3.2 Concrete Testing*

The manufacturing and curing process of concrete were according to ASTM standards. The concrete slump tests were performed according to ASTM C 143.The compressive strength and tensile strength of concrete were evaluated in accordance with ASTM C39 and ASTM C496 respectively and modulus of elasticity (MOE) of concrete was tested according to ASTM C469[31-36].

# 3 Result and Discussion

## 3.1 Car Wash Wastewater Characterization

Table 4 and Table 5 show the results of car wash wastewater characterization and tolerance limit of water mixed in concrete respectively.

**Table 4.** Wastewater characteristics

| Parameters/ Sources | Parameters | | |
|---|---|---|---|
| | pH | Chloride $Cl^-$, ppm | Sulphate $SO_4^{2-}$, ppm |
| *Car Wash Wastewater* | | | |
| MAHAJU Car Wash Station | 8.8 – 9.5 | 19.4 – 33.0 | 12.6 – 115.5 |
| Bandar U Car Wash Station | 9.5 – 10.6 | 18.2 – 25.4 | 16.1 – 100.4 |
| *Tap Water* | | | |
| Tap Water | 7.6 – 8.1 | 19.4 – 19.7 | 13.3 – 13.4 |

**Table 5.** Tolerance limit of wastewater in concrete

| Standards | Tolerance limit, ppm | | |
|---|---|---|---|
| | Chloride $Cl^-$, ppm | Sulphate $SO_4^{2-}$, ppm | pH |
| ASTM C1602 | <500 | <3000 | - |
| ASTM C685 | <500 | <3000 | - |
| BS EN 1008 | <500 | <2000 | >4 |
| BS 3148 | <500 | <1000 | - |
| *Previous Researchers* | | | |
| [3][6][7][8][14][15][19][31] | 9 – 74.2 | 21.3 – 188.0 | 6.5 – 8.75 |

Referring to Table 4 and Table 5, the car wash wastewater characteristics were completely in line with the standards as well as the typical values suggested by previous researchers except the pH value for car wash wastewater. Even though this research obtained higher pH values compared to previous studies, the range of pH value is still in the range of typical car wash wastewater according to the Standard A-1978 which varies from 6.0 to 9.0. In addition, the higher pH value in car wash wastewater may be due to chemicals used in pre-soak stages during the car wash process as supported by previous studies by Zaneti et al., (2011) [37]. Furthermore, BS EN 1008, (2002) mentioned that the typical pH value of water for mixing concrete should be at least higher than four (>4). Therefore, car wash wastewater can potentially be used in concrete mixes due to its reasonable characteristics.

## 3.2 Properties of Fresh Concrete

### *3.2.1 Slump Test*

In this study, the concrete slump test was conducted according to ASTM C143 and the results were listed in Table 6.

**Table 6.** Slump Test

| Wastewater Type | Slump Value (mm) |
|---|---|
| Control Mix, CM | 32 |
| Car Wash Wastewater, CW 10% | 42 |
| Car Wash Wastewater, CW 20% | 38 |
| Car Wash Wastewater, CW 30% | 43 |
| Car Wash Wastewater, CW 40% | 45 |

As presented in Table 6, the slump values for car wash wastewater were comparable to the control mix. As mentioned in sub topic 2, the target mean slump value varied from 30 mm to 60 mm. This proved that the slump values of concrete with car wash wastewater were in the designated range. In addition, the results revealed that the slump value corresponds with the percentage of car wash wastewater in concrete.

## 3.3 Mechanical Properties of Concrete

### *3.3.1 Compressive Strength*

Table 7 show the comparison between car wash wastewater mix and control mix in terms of compressive strength.

DOI: 10.1051/matecconf/20178701018

**Table 7.** Compressive strength of car wash waste water mix

| Concrete Type | Strength on 7 Days (MPa) | Strength on 28 Days (MPa) | ASTM C1602 Minimum Wastewater Concrete At 7 days (>90%) of Control Strength |
|---|---|---|---|
| CM | 44.14 | 50.01 | 100.00% |
| CW 10% | 39.14 | 47.98 | 88.67% |
| CW 20% | 44.74 | 52.70 | 101.36% |
| CW 30% | 46.11 | 51.17 | 104.46% |
| CW 40% | 41.73 | 49.19 | 94.54% |

According to Table 7, the compressive strength of car wash wastewater concrete was in the range of 88% to 104% of normal concrete. This range was beyond the minimum limit of 90% specified in ASTM C1602 except for 10% car wash wastewater replacement. However, at 28 days all the specimens of concrete mix with car wash wastewater were not only above the limits of the standard, but also indicated a slight improvement especially for concrete with 20% and 30% car wash wastewater. This may be due to the fine solid particles present in the car wash wastewater which filled the pores or voids between fine aggregates and coarse aggregates. This may improve the strength of concrete but it would depend on suitable replacement percentages of car wash wastewater in concrete mixes [31][41-42].

In addition, the presence of chloride in car wash wastewater should be taken into account because the concentration of chloride in car wash wastewater is higher than normal concrete. This concentration may lead to corrosion on the steel reinforcement which could damage the bonding between rebars and concrete [40-45].

### *3.3.2 Tensile Strength*

Table 8 and Figure 1 demonstrated the comparison of tensile strength between car wash wastewater mix and control mix.

**Table 8.** Tensile strength of car wash wastewater mix

| Concrete Type | 7 Days Strength (MPa) | 28 Days Strength (MPa) |
|---|---|---|
| CM | 2.74 | 2.90 |
| CW 10% | 2.16 | 2.96 |
| CW 20% | 2.30 | 3.13 |
| CW 30% | 2.83 | 3.17 |
| CW 40% | 2.77 | 3.00 |

Referring to Table 8, all the results of the tensile strength of car wash wastewater were above the mean tensile strength value specified by BS EN 1992 [34]. The range of tensile strength was between 2.96 MPa to 3.17 MPa compared to the standard value of 2.8 MPa. These results concluded that the replacement of car wash wastewater in concrete mixes met the specifications from BS EN 1992. The presence of fine particles in wastewater may have also filled the voids and pores in concrete and improved the interconnection between concrete materials [19].

On the other hand, the existence of chloride concentration in car wash wastewater could affect the tensile strength of concrete [19][36]. Based on the data in Table 2, the concentration of chloride in car wash wastewater is slightly higher than tap with a range between 19.4 ppm to 33.0 ppm compared to tap water which contains chloride concentration ranging from 19.4 ppm to 19.7ppm. Therefore, the correlation between car wash wastewater and other concrete materials need to be further studied. This is because the presence of chloride content in wastewater may reduce the tensile strength of concrete [37][38][39].

### *3.3.3 Modulus of Elasticity (MOE)*

Figure 1 demonstrated the procedure of modulus of elasticity (MOE) testing. However, Table 9 present the comparison of modulus of elasticity between car wash wastewater mix and control mix.

**Figure 1.** Procedure of modulus of elasticity testing

**Table 9.** Modulus of elasticity of car wash wastewater mix

| Concrete Type | 28 Days MOE (GPa) |
|---|---|
| CM 0% | 29.38 |
| CW 10% | 32.96 |
| CW 20% | 33.62 |
| CW 30% | 28.73 |
| CW 40% | 28.62 |

As mentioned in BS EN 1992, the typical range of secant modulus of elasticity of concrete (Ecm) grade 35 was 32 $\pm$ 6 GPa [34]. So, the results in Table 9 completely achieved the typical range of MOE for concrete with a mean strength of 35MPa. According to the information in Table 8, the concrete with 20% of car wash wastewater replacement demonstrated the highest MOE value compared to other percentages as well as control mix with only 29.4GPa. These findings may correlate to the existence of fine solid particles in wastewater which could fill the pores and voids in between the fine and coarse aggregates in concrete structures. According to previous researchers, the factors affecting the MOE were concrete strength and concrete density [42-45]. The concrete strength effect was related to the concrete compressive strength while the concrete density was related to the porosity of concrete. Based on the results in Table 8, the sulphate content in car wash wastewater was between 12.6 ppm to 115.5 ppm compared to tap water which had a sulphate content of 13.3 ppm to 13.4 ppm. Previous researchers revealed that a higher sulphate concentration may result secondary gypsum is formed with calcium hydroxide and developing crystallise causing expansion. Thus it is recommended to use sulphate resistant cement to make sure the concrete impermeable as possible.

## 4 Conclusions

Based on the results from the experimental work the following conclusions for the use of car wash wastewater in concrete mixes can be drawn:

- The pH value for car wash wastewater was in the range of 8.8 to 10.6 which is slightly higher than tap water. However, the sulphate and chloride content in car wash wastewater were within the ASTM and BS standards requirement.
- The used of 10% car wash wastewater result the lowers value of compressive strength, tensile strength and modulus of elasticity, however it show increasing while 20% and 30% but beyond the 30% all the result decreasing.
- The results recommended that a higher compressive strength value may lead to a higher MOE value.
- According to the results for compressive strength and modulus of elasticity, the optimum percentage of car wash wastewater as a fresh water replacement in concrete mixes is 20%.

## Acknowledgement

The authors would like to thank UTHM fellowship, VOT U523 and Grant GPPS U572. This paper was partly sponsored by the Centre for Graduate Studies, UTHM.

## References

1. M. Y. Chong, "Malaysia Faces Looming Water Crisis: Sustainable Water Management," The Star Newspaper, Malaysia, (2011).
2. Z. B. Zaharaton, "Water Resources Management in Malaysia - The Way Forward," Asian Water, Kuala Lumpur, Malaysia, (2004).
3. S. Shahidan, I. Isham, and N. Jamaluddin, "A Review on Waste Minimization by Adopting in Self Compacting Concrete," MATEC Web Conf., vol. 47, pp. 1 7, 2016.
4. W. J. Lau, A. F. Ismail and S. Firdaus, "Treatment of car wash effluent using ultrafiltration and nanofiltration membranes," Separation and Purification Technology, vol. 104, pp. 26-31, (2013).
5. M. A. Wahid and H. Tanaka, "Potential Implementation of Wastewater Reclamation and Reuse in Malaysian Urban Area," International Sustainability and Civil Engineering Journal, vol. 1, no. 1, pp. 75-83, (2012).
6. K. Boussu, G. Van Baelan, W. Colen , D. Eelen, S. Vanassche, C. Vandecasteele and B. Van der Bruggen, "Technical and economical evaluation of water recycling in the carwash industry with membrane processes," Water Science and Technology, vol. 57, no. 7, pp. 1131-1135, (2008).
7. K. Boussu, C. Kindts, C. Vandecasteele and B. Van der Bruggen, "Applicability of nanofiltration in the carwash industry," Seperation Purification Technology, vol. 54, pp. 139-146, (2007).
8. A. Al-Odwani, A. Ahmed and S. Bou-Hamad, "Carwash water reclamation in Kuwait," Desalination, vol. 206, pp. 17-28, (2007).
9. N. Md Nor, N. Muhamad Bunnori, A. Ibrahim, S. Shahidan, and S. N. M. Saliah, "An investigation on acoustic wave velocity of reinforced concrete beam in-plane source," in Proceedings - 2011 IEEE 7th International Colloquium on Signal Processing and Its Applications, CSPA 2011, 2011, pp. 19–22.
10. T. Hamada and Y. Miyazaki, "Reuse of carwash water with a cellulose acetate ultrafiltration membrane aided by flocculation and activated carbon treatments," Journal of Desalination. Volume, vol. 169, pp. 257-267, (2004).
11. K. Karakulski and A. W. Morawski, "Treatment of wastewater from wastewater from car washes by ultrafiltration," Fresenius Environment Bulletin, vol. 12, pp. 343-348, (2003).
12. C. Jonsson and A. S. Jonsson, "The influence of degreasing agents used at car washes on the performance of ultrafiltration membranes," Desalination, vol. 100, pp. 115-123, (1995).
13. Malaysia Standard A, Environmental Quality Regulations: Sewage and Industrial Effluents, cyberjaya, Selangor: Government of Malaysia, (1978).
14. S. Shahidan, R. Pullin, K. M. Holford, M. B. N, and N. Nor, "Quantitative Evaluation of the Relationship between Tensile Crack and Shear Movement in Concrete beams," Adv. Mater. Res., vol. 626, pp. 355–359, 2013.
15. M. Panniza and G. Cerisola, "Applicability of electrochemical methods to carwash wastewaters for reuse. Part 1: Anodic oxidation with diamond and lead dioxide anodes," Journal of Electronic and Chemical, vol. 638, pp. 28-32, (2010).
16. E. Bazrafshan, F. K. Mostafapoor, M. M. Soori and A. H. Mahvi, "Application of Combined Chemical Coagulation and Electro-Coagulation Process for Carwash Wastewater Treatment," Environmental Bulletin, vol. 21, no. 9, pp. 2694-2701, (2012).
17. M. Abdul Rahim, N. M. Ibrahim, Z. Idris, Z. M. Ghazaly, S. Shahidan, N. L. Rahim, L. A. Sofri, and

N. F. Isa, "Properties of Concrete with Different Percentange of the Rice Husk Ash (RHA) as Partial Cement Replacement," Mater. Sci. Forum, vol. 803, pp. 288–293, 2014.

18. S. Shahidan, N. M. Bunnori, N. Md Nor, and S. R. Basri, "Damage severity evaluation on reinforced concrete beam by means of acoustic emission signal and intensity analysis," in 2011 IEEE Symposium on Industrial Electronics and Applications, 2011, pp. 337–341.
19. Z. A. Bhatti, Q. Mahmood, I. A. Raja, A. H. Malik and D. Wu, "Chemical Oxidation of Carwash Industry Wastewater as an Effort to Decrease Water Pollution," Journal of Physics and Chemistry of the Earth, vol. 36, pp. 465-469, (2011).
20. B. O. Kwach and J. O. Lalah, "High Concentration of Polycyclic aromatic Hydrocarbons found in Water and Sediments of Car Wash and Kisat areas of Winam Gulf, Lake Vctoria Kenya," Bulletin of Environmental Contamination and Toxicology, vol. 83, no. 5, pp. 727-733, (2009).
21. Z. Nadzirah, N. H. Haslina and H. Rafidah, "Removal of Important Parameter from Car Wash Wastewater," Applied Mechanics and Materials, vol. 773, pp. 1153-1157, (2015).
22. USEPA, "EPA's Polluted brochure," United States Environmental Protection Agency, United States, (1994).
23. K. S. Al-Jabri, A. H. Al-Saidy, R. Taha and A. J. Al-Kemyani, "Effect of using Wastewater on the Properties of High Strength Concrete," Procedia Engineering, vol. 14, pp. 370-376, (2011).
24. ASTM C1602, Specification for Mixing Water Used in the Production of Hydraulic Cement Concrete, Pennsylvania, United States: American Society for Testing and Materials, (2006).
25. ASTM C39, Test Method for Compressive Strength of Cylindrical Concrete Specimens, Pennsylvania, United States: American Society for Testing and Materials, (2005).
26. ASTM C496, Test Method for Splitting Tensile Strength of Cylindrical Concrete Specimens, Pennsylvania, United States: American Society for Testing and Materials, (2004).
27. ASTM C469, Test Method for Static Modulus of Elasticity and Poisson Ratio of Concrete in Compression, Pennsylvania, United States: American Society for Testing and Materials, (2002).
28. S. A. Kudus, N. M. Bunnori, S. R. Basri, S. Shahidan, M. N. M. Jamil, and N. M. Noor, "An Overview Current Application of Artificial Neural Network in Concrete," Adv. Mater. Res., vol. 626, pp. 372–375, 2012.
29. S. Shahidan, H. B. Koh, A. M. S. Alansi, and L. Y. Loon, "Strength Development and Water Permeability of Engineered Biomass Aggregate Pervious Concrete," MATEC Web Conf., vol. 47, pp. 2–7, 2016
30. DOE 1988, Design of Normal Concrete Mixes. Building Research Establishment, Watford, United Kingdom: Department of the Environmental, (1988).
31. BS EN 1008, Mixing Water for Concrete-Specification for Sampling, Testing and Assessing the Suitability of Water, Including Water Recovered from Processes in the Concrete Industry, as Mixing Water in Concrete, London: British Standards Institution, (2002).
32. MS EN 197-1, Composition, specifications and conformity criteria for common cements, Cyberjaya, Selangor: Department of Standards Malaysia , (2014).
33. ASTM C33, Specification for Concrete aggregates, Pennsylvania, United States: American Society for Testing and Materials, (2003).
34. BS 882, Specification for Aggregates from Natural Sources for Concrete, London: British Standards Institution, (1992).
35. ASTM C192, Practice for Making and Curing Concrete Test Specimens in the Laboratory, Pennsylvania, United States: American Society for Testing and Materials, (2006).
36. ASTM C143, Test Method for Slump of Hydraulic Cement Concrete, Pennsylvania, United States: American Society for Testing and Materials, (2005).
37. R. Zaneti, R. Etchepare and J. Rubio, "Car wash wastewater reclamation. Full- scale application and upcoming features," Resources, Conservation and Recycling, vol. 55, no. 11, pp. 953-959, (2011).
38. A. M. Neville, Properties of Concrete, 4th ed., London: Pearson, (1995).
39. L. Tang, N. Lars-Olof and B. P. Muhammed, Resistance of Concrete to Chloride Ingress, London: Spom Press, (2012).
40. C. Martin, "nfluence of Supplementary Cementitious Materials (SCMs) on concrete durability," Eco-efficient concrete, pp. 153-197, (2013).
41. BS EN 1992, Design of Concrete Structure, London: British Standards Institution, (2004).
42. O. S. Lee, M. R. Salim, M. Ismail and M. I. Ali, "Reusing treated effluent in concrete technology," Jurnal Teknologi, vol. 34, pp. 1-10, (2001).
43. G. Murali, C. M. Vardhan, S. Raju, C. Mahalakshmi, G. Srinidhi and D. S. Zachariag, "Influence of various industrial effluents on concrete structures," International Journal of Engineering Research and Application (IJERA), vol. 2, no. 2, pp. 704-709, (2012).
44. Z. Z. Ismail and E. A. Al-Hashmi, "Assessing the recycling potential of industrial wastewater to replace fresh water in concrete mixes: application of polyvinyl acetate resin wastewater," Journal of Cleaner Production, vol. 19, no. 2-3, pp. 197-203, (2011).
45. A. H. Noruzman, B. Muhammad, M. Ismail and A. Z. Majid, "Characteristics of treated effluents and their potential applications for producing concrete," Journal of Environmental Management, vol. 110, pp. 27-32, (2012).

# The effect of aggregate condition during mixing on the mechanical properties of oil palm shell (OPS) concrete

Delsye Teo Ching Lee[1,a] and Tan Suet Lee[1]

[1]Faculty of Engineering, Universiti Malaysia Sarawak, 94300 Kota Samarahan, Sarawak, Malaysia.

**Abstract.** One of the key characteristics affecting the properties of OPS concrete is the saturation condition of the coarse aggregates used during mixing. In this current study, the effect of different aggregate saturation conditions during mixing on the slump, compressive, flexural and splitting tensile strength of OPS concrete was investigated. Three aggregate saturation conditions were employed, namely saturated-surface-dry (SSD), air-dry (AD) and sun-dry (SD). The results revealed that the SD mix had a higher slump value than AD mix and SSD mix. On the other hand, SSD mix registered better compressive, flexural and splitting tensile strength compared to the AD and SD mix. This suggests that the use of SSD aggregates is more effective in producing higher strength concretes which could be credited to the internal curing provided by the SSD aggregates.

## 1 Introduction

The re-use of waste materials in many modern construction industries as an alternative approach to create a more sustainable and cleaner environment has been gaining increased attention. In the concrete industry, the making of concrete consumes a substantial amount of natural resources. To ensure the continuous availability of these resources in the future, one feasible approach is to replace natural stone aggregates with alternative coarse aggregates.

In Malaysia, the palm oil industry is one of the largest agricultural industries. Malaysia contributes about 57.6% of the total supply of palm oil in the world [1]. Large quantities of agricultural wastes have been produced due to increased agricultural production and development of agro-based industries [2]. Oil palm shell (OPS) is a solid waste material generated during the extraction of palm oil. Past studies have shown that OPS can be successfully used as full coarse aggregate replacement to produce structural lightweight concrete [3-5].

The characteristics of natural stone aggregates and OPS aggregates are very different. One distinct difference of OPS aggregates is the high water absorption capacity and this can affect the properties of the resulting concrete produced. Consequently, the current paper presents an investigation on the effects of different OPS aggregate condition during mixing on the properties of the resulting concrete. The aggregate conditions used include saturated-surface-dry (SSD), air-dry (AD) and sun-dry (SD) conditions.

## 2 Materials Used for Concrete Production and Mix Proportion

For the production of OPS concrete, Ordinary Portland Cement (ASTM Type 1), river sand as fine aggregate, OPS as coarse aggregate and Glenium C380 Type F superplasticizer were used. The OPS aggregates were obtained locally and have water absorption value of about 21%. The particle size distribution of the OPS aggregate and river sand used are shown in Figure 1 and Figure 2 respectively.

The three saturation conditions of the OPS aggregate used consisted of saturated-surface-dry (SSD), air-dry (AD) and sun-dry (SD) conditions. The SSD condition was obtained by immersing the OPS aggregates in potable water for 24 ± 2 hours. Then, the aggregates were dried using a large cloth until all the water on the aggregate surface was removed so that the aggregates are in SSD condition prior to concrete casting. For the AD condition, the OPS aggregates were left to air-dry in ambient laboratory conditions until the moisture content of the aggregates is in the range of 2% to 6%. For the SD condition, the aggregates were left out in the open to sun-dry for around two weeks until the moisture content of the aggregates fall below 2%.

[a] Corresponding author: tdelsye@unimas.my

DOI: 10.1051/matecconf/20178701019

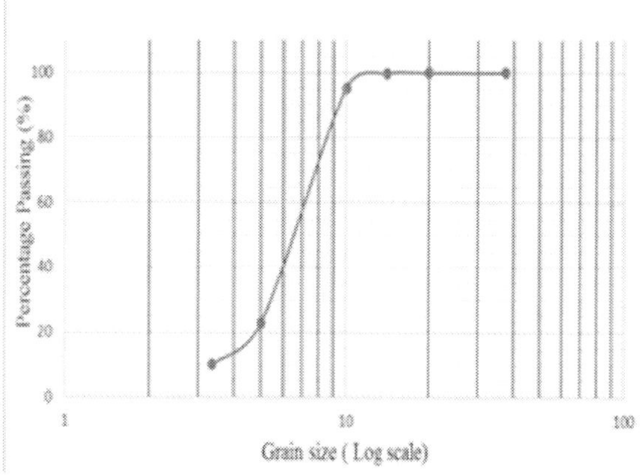


**Figure 1.** Particle size distribution for OPS aggregates

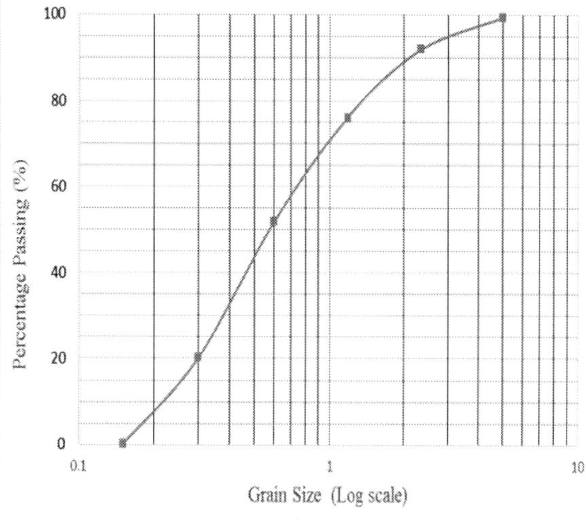


**Figure 2.** Particle size distribution for river sand

The mix proportion was produced using OPS as full coarse aggregate replacement and the mix was first determined using aggregates which were in SSD condition. Water adjustments were made to mixes which contained AD and SD aggregates as a means to ensure that the aggregates in these conditions do not absorb additional water from the mix. The water adjustments were made based on the moisture content of the aggregates. Table 1 shows the concrete mix proportions containing OPS aggregates with different saturation conditions. The OPS concrete mixes are called SSD mix, AD mix and SD mix hereinafter.

**Table 1.** Mix proportion of OPS concrete sample under different aggregate conditions.

| **Aggregate conditions** | **SSD** | **AD** | **SD** |
|---|---|---|---|
| Water to cement ratio | 0.42 | 0.42 | 0.42 |
| OPS aggregate (kg/m$^3$) | 384 | 384 | 384 |
| Fine aggregate (kg/m$^3$) | 749 | 749 | 749 |
| Cement (kg/m$^3$) | 480 | 480 | 480 |
| Water (kg/m$^3$) | 202 | 202 | 202 |
| Additional mixing water (water adjustment) (kg/m$^3$) | - | 54 | 71 |
| Superplasticiser (*l*/100kg of cement) | 1.2 | 1.2 | 1.2 |

## 3 Experimental Program

For each OPS concrete mix, the workability of the fresh concrete was determined using the slump test in accordance with BS 1881: Part 102 [6].

For the hardened concrete, the tests conducted included the compressive strength [7], flexural strength [8] and tensile splitting strength [9]. Samples were tested at 3, 7, 28 and 56 days after being subjected to full water curing and the results were reported as an average of three tested samples. For the compressive strength test, the samples used were 100mm cubes, while 100mm x 100mm x 500mm beam samples and 150mm (dia.) x 300mm cylindrical samples were used for the flexural strength test and split tensile test respectively.

## 4 Results and Discussion

### 4.1. Slump

The slump values for the fresh OPS concrete under different OPS aggregate conditions are illustrated in Figure 3. From Figure 3, it can be observed that the SD mix recorded the highest slump value followed by the AD mix, while the SSD mix had the lowest slump. This suggests that the drier aggregates did not managed to fully absorb the additional water from the water adjustment added for aggregate absorption during mixing which resulted in higher initial free water content in the mix containing drier aggregates. These findings are consistent with a previous study which also showed that the concrete mix containing dry aggregates had a slightly higher slump value as compared to the concrete mix with wet aggregates [10].

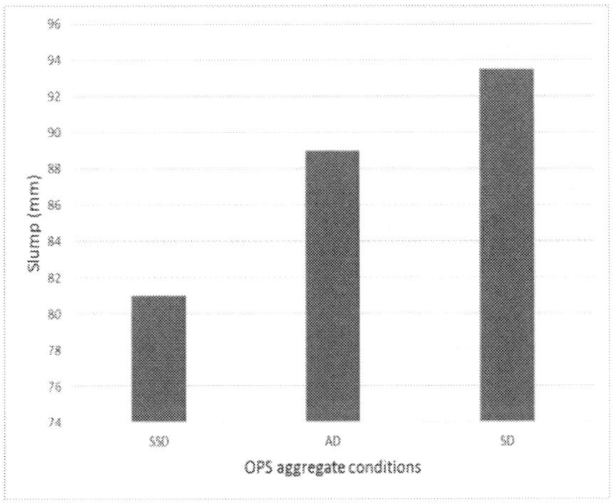


**Figure 3.** Slump of OPS concrete containing different aggregate conditions

## 4.2 Compressive Strength

Figure 4 presents the compressive strength development of the OPS concrete containing different aggregate conditions. The compressive strengths of the SSD mix were higher at all ages as compared to the AD and SD mix. More specifically, at an age of 28 days, the compressive strength for the SSD mix, AD mix and SD mix were 30.7 MPa, 24.7 MPa and 22.7 MPa respectively. The compressive strength of the AD mix and SD mix were approximately 19.5% and 26.1% lower compared to the compressive strength of the SSD mix at 28 days.

In general, concrete strength is dependent on the amount of water contained in the concrete for hydration process. Sufficient amount of water for hydration would lead towards higher concrete strength. From this study, it can be observed that although additional water is added to the AD and SD mix during mixing, the compressive strength is still much lower compared to the mix which contained SSD aggregates. This suggests that the use of SSD aggregates is more effective in producing higher strength concretes which could be attributed to the internal curing provided by the SSD aggregates. Past studies have also shown that the cementitious hydration can be enhanced due to the process of internal curing from the internal reservoir of water absorbed by the lightweight aggregate [11,12]. Furthermore, the high absorption capacity of OPS aggregates may also strengthen the contact zone between the cement paste and the OPS through the formation of new hydrated compounds at the interface zone when SSD aggregates are used during mixing [13].

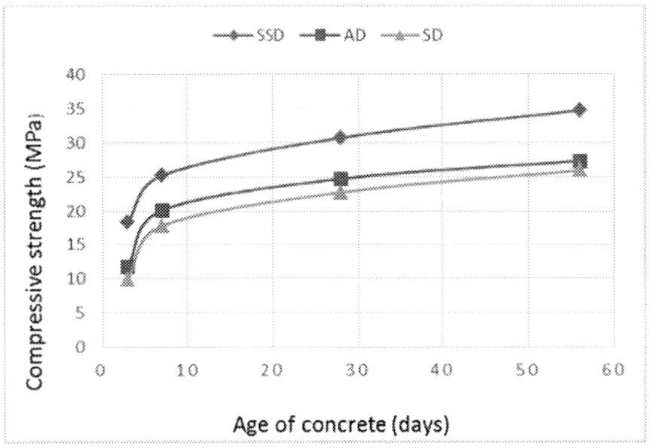


**Figure 4.** Compressive strength development of OPS concrete containing different aggregate conditions

## 4.3 Flexural Strength

The flexural strength of the OPS concrete containing different aggregate conditions is illustrated in Figure 5. From Figure 5, it can be observed that the flexural strength of all OPS concrete mixes continues to increase with age. As with the compressive strength results, the SSD mix registered the highest flexural strength for all ages followed by the AD and SD mix which again could be due to the effect of internal curing provided by the saturated aggregates.

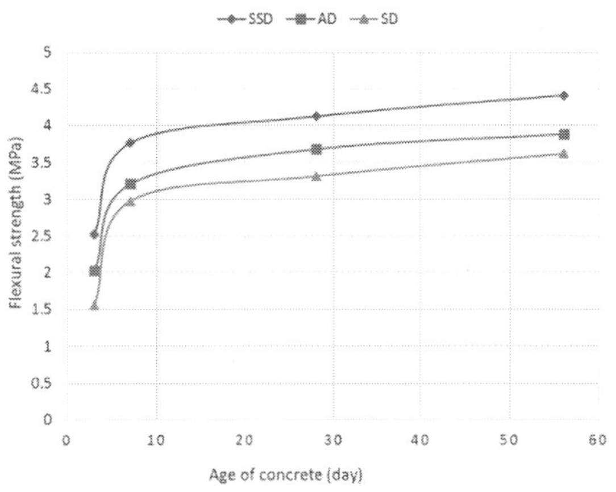


**Figure 5.** Flexural strength development of OPS concrete containing different aggregate conditions

In general, the flexural strength is about 10% to 20% of compressive strength of the concrete [14]. From Table 2, it can be observed that the ratio of flexural to compressive strength for all OPS concrete containing different aggregate saturation is in the range of 12.7% to 16.9% which is in close agreement with that of conventional concrete.

**Table 2.** Percentage of flexural strength to compressive strength

| Aggregate condition | Testing age (days) | Compressive strength (MPa) | Flexural strength (MPa) | Flexural to compressive strength (%) |
|---|---|---|---|---|
| SSD | 3 | 18.3 | 2.5 | 13.7 |
| | 7 | 25.2 | 3.8 | 15.1 |
| | 28 | 30.7 | 4.1 | 13.4 |
| | 56 | 34.7 | 4.4 | 12.7 |
| AD | 3 | 11.8 | 2.0 | 16.9 |
| | 7 | 20.1 | 3.2 | 15.9 |
| | 28 | 24.7 | 3.7 | 15.0 |
| | 56 | 27.3 | 3.9 | 14.3 |
| SD | 3 | 9.9 | 1.6 | 16.2 |
| | 7 | 17.8 | 3.0 | 16.9 |
| | 28 | 22.7 | 3.3 | 14.5 |
| | 56 | 25.9 | 3.6 | 13.9 |

### 4.3 Splitting Tensile Strength

Figure 6 shows the splitting tensile strength of OPS concrete at different ages under different aggregate conditions. Again, it can be observed that the concrete mix containing SSD aggregates recorded the highest splitting tensile strength followed by the mixes containing AD and SD aggregates respectively at all ages of testing.

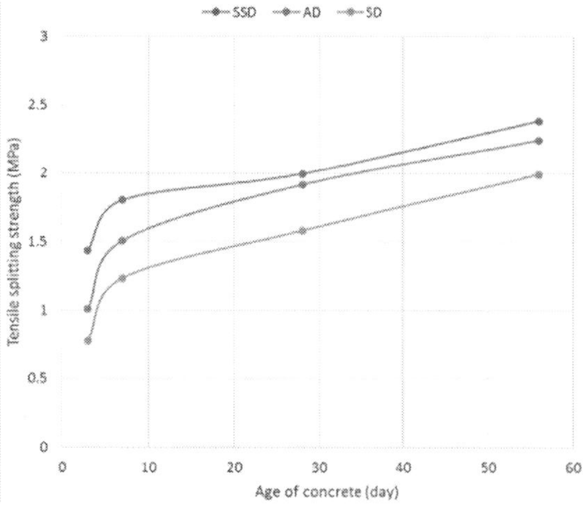


**Figure 6.** Splitting tensile strength development of OPS concrete containing different aggregate conditions

Table 3 shows the ratio of tensile splitting strength to compressive strength. Generally, for conventional concretes, the tensile splitting strength is about 8% to 12% of the compressive strength [15]. From Table 3, it can be seen that the splitting tensile strength is in the range of about 6.5% to 8.5%, which is slightly lower compared to that of conventional concretes.

**Table 3.** Percentage of splitting tensile strength to compressive strength

| Aggregate condition | Testing age (days) | Compressive strength (MPa) | Splitting tensile strength (MPa) | Splitting tensile strength to compressive strength (%) |
|---|---|---|---|---|
| SSD | 3 | 18.3 | 1.4 | 7.7 |
| | 7 | 25.2 | 1.8 | 7.1 |
| | 28 | 30.7 | 2.0 | 6.5 |
| | 56 | 34.7 | 2.4 | 6.9 |
| AD | 3 | 11.8 | 1.0 | 8.5 |
| | 7 | 20.1 | 1.5 | 7.5 |
| | 28 | 24.7 | 1.9 | 7.7 |
| | 56 | 27.3 | 2.2 | 8.1 |
| SD | 3 | 9.9 | 0.8 | 8.1 |
| | 7 | 17.8 | 1.2 | 6.7 |
| | 28 | 22.7 | 1.6 | 7.0 |
| | 56 | 25.9 | 2.0 | 7.7 |

## 5 Conclusions

In the current study, the effects of OPS aggregate conditions used during mixing on the workability and mechanical strength of concrete were investigated. Three aggregate saturation conditions were used namely saturated-surface-dry (SSD), sun-dry (SD) and air-dry (AD). Based on the results obtained from this experimental study, the main conclusions drawn are:

1. The mix containing OPS aggregates under the SD state exhibited the highest slump value followed by the mix containing AD and SSD aggregates respectively. This suggests that the drier aggregates did not manage to fully absorb the additional water from the water adjustment added for aggregate absorption during mixing which resulted in higher initial free water content in the mix containing drier aggregates and therefore resulted in higher slump values.

2. The compressive strength of the SSD, AD and SD mix at the age of 28 days were 30.7MPa, 24.7MPa and 22.7MPa respectively. This suggests that the use of SSD aggregates is more effective in producing higher strength concretes which could be attributed to the internal curing provided by the SSD aggregates.

3. In terms of the flexural strength, the SSD mix also registered the highest strength followed by the AD and SD mix with values of 4.1MPa, 3.7MPa and 3.3MPa respectively at 28 days. It was determined that the ratio of flexural strength to compressive strength of concrete is

about 12.7% to 16.9% which is in close agreement with that of conventional concrete.

4. The splitting tensile strength developed in the OPS concrete with SSD, AD and SD aggregates were 2MPa, 1.9MPa and 1.6MPa respectively at 28 days. It was also found that the splitting tensile strengths of the OPS concrete were about 6.5% to 8.5% of the respective compressive strengths.

## Acknowledgement

The authors would like to thank Universiti Malaysia Sarawak (UNIMAS) for providing financial support to present these findings.

## References

1. Z. Ahmad, H.M. Saman, P.M. Tahir, Oil palm trunk fiber as a bio-waste resource for concrete reinforcement, International Journal of Mechanical and Materials Engineering, **5**(2), 199–207 (2010)
2. S. Chandra, *Waste materials used in concrete manufacturing* (Noyes Publications, U.S.A, 1997)
3. M.A. Mannan, C. Ganapathy, Long-term strengths of concrete with oil palm shell as coarse aggregate, Cement and Concrete Research, **31**(9), 1319–1321 (2001)
4. D.C.L. Teo, M.A. Mannan, V.J. Kurian, Structural concrete using oil palm shell (OPS) as lightweight aggregate. Turkish Journal of Engineering and Environmental Sciences, **30**(4), 251–257 (2006)
5. D.C.L. Teo, M.A. Mannan, V.J. Kurian, Lightweight concrete made from oil palm shell (OPS): Structural bond and durability properties, Building and Environment, **42**(7), 2614–2621 (2007)
6. BS 1881: Part 102, *Testing Concrete.* Method for determination of slump (British Standard Institution, London, 1983)
7. BS 1881: Part 116, *Testing Concrete.* Method for determination of compressive strength of concrete cubes (British Standard Institution, London, 1983)
8. BS EN 12390: Part 5, *Testing Hardened Concrete.* Flexural strength of test specimens (British Standard Institution, London, 2009)
9. BS 1881: Part 117, *Testing Concrete.* Method for determination of tensile splitting strength (British Standard Institution, London, 1983)
10. A.M. Alhozaimy, Effect of Absorption of Limestone Aggregates on Strength and Slump Loss of Concrete, Cement and Concrete Research, **31**(7), 470–473 (2009)
11. A. Bentur, S. Igarashi, K. Kovler, Prevention of Autogenous Shrinkage in High Strength Concrete by Internal Curing using Wet Lightweight Aggregates, Cement and Concrete Research, **31**(11), 1587–1591 (2001)
12. H. Al-Khaiat, M.N. Haque, Effect of Initial Curing on Early Strength and Physical Properties of a Lightweight Concrete, Cement and Concrete Research, **28**(6), 859–866 (1998)
13. D.C.L. Teo, M.A. Mannan, V.J. Kurian, Durability of lightweight OPS concrete under different curing conditions, Materials and Structures, **43(**1), 1–13 (2010)
14. National Ready Mixed Concrete Association (2000), Flexural strength concrete. Retrieved 5 March 2015 from http://www.nrmca.org/aboutconcrete/cips/16p.pdf
15. S.H. Kosmatka, B. Kerkhoff, W.C. Panarese, Design and control of concrete mixtures. 14th ed. USA: Portland Cement Association (2002)

# Web Content Analysis On Sustainable Campus Operation (SCO) Initiatives

Ruzaimah Razman[1], Abd Halid Abdullah[1], Abdul Zaki Abd Wahid[1] and Rahmat Muslim[1]

[1]*Faculty of Civil and Environmental Engineering, Universiti Tun Hussien Onn Malaysia*

**Abstract.** The purpose of this paper is to identify and analyse the current practices implemented in global universities for achieving sustainability throughout campus operations. This study adopted a web content analysis method where 30 international green universities' websites have been thoroughly examined to identify common initiatives implemented to achieve sustainability through campus operations. The findings are ranked based on the implementation of these initiatives by participating universities. From the websites reviewed, as much as 31 initiatives have been identified as common initiatives frequently implemented by green universities to achieve sustainability in campus operations. It was found that the common initiatives frequently implemented by most of the universities include 'Provide bin with clearly marked signs to increase the number of recycling items', and 'Generate electricity on campus by establishing power generation plants' with 87% and 83% respectively. This paper fills the gap by presenting the investigation of sustainability initiatives from some of the major green universities internationally. It is suggested that higher education institutions, particularly Malaysian universities, initiate or manage their implementation of sustainable campus operation (SCO) initiatives based on the findings of this research.

## 1 Introduction

Universities are generally known for their wide area with large population, and carrying out complex operations and activities which comprise various teaching and learning, research and development, and provision of support services. They can generate significant environmental impacts because of their huge use of energy, extensive transportation, massive waste, high consumption of materials, and extensive development of built facilities [1]. Most researchers believe that to overcome these issues, elements of campus sustainability should be incorporated into all aspects of university's governance, operation, education, research, and outreach [2][3][4][7][8]. However, dimensions of governance and operations have been given more focus in assessing the sustainability at university as compared to other areas [7].

In Malaysia, although there have been many campus sustainability initiatives being implemented by local universities to achieve Sustainable Campus Operations (SCO), it seems that the efforts are still divergent and not systematically applied. The knowledge on the current and actual campus sustainability practices is still limited as there is currently very few published articles provide in depth discussion on the conceptual or practical frameworks of campus sustainability [1]. Therefore, further studies are still needed in order to share information, understand the issues and concepts, and develop action plans to be used as guidelines that are appropriate to the local universities. This paper fills the gap by presenting the investigation of the real sustainability practices from some major green universities in the world. Malaysian universities could use the findings whether to initiate or to manage their implementation of SCO-initiatives.

[a] Corresponding author: emmaradzman@gmail.com

## 2 Sustainable Campus Operations Background

About 400 university leaders from over 52 countries have signed Talloires Declaration to show their commitment and support to environmental sustainability in Higher Education Institutions (HEIs) [9]. The declaration urges universities to carry out a more sustainable campus operations, as well as incorporating sustainability and environmental literacy within universities' approach as a whole by explicitly linking research, educational and operational activities. The universities should be the role model of environmental responsibility by establishing institutional ecology policies and practices of resource conservation, recycling, waste reduction, and environmentally sound operations [10]. Kyoto Declaration also encourages universities to review their operations to reflect the sustainable development practices. Both declarations are often referred by HEIs to develop and implement SCO through centralized programs to promote green practices in achieving campus sustainability objectives. It is not surprising that most of the university's sustainability policies and sustainability initiatives focus more on operations rather than other areas [11].

Campus Operations is responsible for the provision of all buildings and facilities within universities to maintain a safe, comfortable and attractive campus environment that supports excellence in teaching, learning and research. Sustainable development benefits can be achieved through consuming fewer resources in construction and operation, reduce negative impacts on environment, using efficient building material energy, minimize construction waste, and reduce planning risk, extended building life and viability [5] [6].

In adopting sustainability in campus operations, various focus areas and goals are being set by universities around the world. Based on [7], campus operation and environmental performance are related to energy and water consumption, and waste and hazardous materials management. While, [12] mentioned that Environmental Management System (EMS) covers two areas, namely (i) environmental management and improvement, which relates to waste minimization, energy efficiency, and environmental conservation, and (ii) green campus, which aims to promote construction of green buildings and transportation facilities. Sustainable operations emphasize features such as saving, environmental quality, and efficiency and effectiveness of resources for university. For instant, the operations must be able to lower the carbon emission, as well as able to reduce the consumption of non-renewable sources, and the resources are selected based on life cycle assessment.

On a personal level, by doing recycling, limit energy use, buy products with low carbon footprints, and buy products from sustainable sources are the several methods adopted to go green. In turn, universities can do their bit too, such as developing their own sustainability policies and implementing the initiatives towards SCO. As universities are unique due to its topography, population size, culture and climate where they exist, and etc., therefore the selection of areas and initiatives that can be effectively conducted at a particular university are important to be identified.

## 3 Method

A total of 30 green universities have been identified as samples for this study. Table 1 outlines the list of universities obtained through Google search using several keywords including 'Campus Sustainability Initiatives', 'Campus Green Practices', 'Members of International Alliance of Research Universities (IARU)', 'University Listed in UI GreenMetric', and 'The 39 Greenest Universities'. The investigation of the websites is only focused on web menus and content of web pages consist of the campus operations practices or initiatives adopted at the universities. From the investigation of the 30 green universities' websites, as much as 31 initiatives have been identified as common initiatives frequently implemented to achieve SCO.

## 4 Findings

Table 2 presents the findings of this study. Most of the green universities have implemented initiatives to achieve sustainability in campus operations by 'Provide bin with clearly marked signs to increase the number of recycling items' and 'Generating electricity in campus through a large-scale plants or power generation plants e.g. solar energy to reduce nonrenewable fuels'. The top 2 initiatives mentioned earlier can be considered as the most significant initiatives implemented by most of the international universities to achieve SCO with involvement percentages of 87% and 83%, respectively.

Including the two initiatives mentioned above, the top 5 list of initiatives that are frequently implemented by universities towards achieving SCO also consist of 'Design, build and renovate university buildings according to green building certification/standard/index', 'Collect food waste from dining halls and convert it into a high grade compost which can be used in the gardens and landscape in the campus', 'Use of energy efficient equipment and eco-friendly products in campus to enhance performance and to reduce carbon footprint' and 'Use unprocessed water supply such as portable water, rainwater harvesting or on-site recycling of water which can be used in operational activities and reduce the water bills'. Other implemented initiatives such as 'Providing secure bicycle storage area', 'Provide parking facilities for carpooling, bicycling and shuttle bus purpose', 'Providing safe cycling & walking paths', 'placing the bin at convenient place' and 'Buying foods from local producers to reduce the distance in transporting foods to consumer' are also among the significant initiatives frequently implemented on campus encompassing 60-80% of universities considered in this study.

However, several others; such as 'Create native and drought plants landscape as significantly reduced the need for irrigation and maintenance', 'Provide in-campus workshop to maintain and repair bicycles', 'Collect and process garden/landscape waste into compost for a campus gardening and landscape', 'Consider the full life-cycle of building from the extraction of resources during building construction, maintenance, demolition and material recycling during renovation' and 'Provide real time shuttle-bus schedules', are less adopted as only less than 40% of the green universities implemented such initiatives. Whereas, the only initiative that can be considered as insignificant is 'providing showers for cyclist' since it is rarely implemented by the green universities towards achieving SCO.

**Table 1.** Samples of Green Universities

| Green Universities Identified | Search Key words | | | | |
|---|---|---|---|---|---|
| | Campus Sustainability Initiatives | Campus Green Practices | Members of International Alliance of Research Universities (IARU) | Top 15 University Listed in UI Green Metric 2015 | The 39 Greenest Universities of 2015 |
| University of Nottingham [13] | X | | | X | |
| University College Cork [14] | X | | | X | |
| University of California, Davis [15] | X | | | X | |
| University of Oxford [16] | X | | X | X | |
| University of Connecticut [17] | X | | | X | |
| Colorado State University [18] | X | | | | X |
| Harvard University [19] | X | | | | |
| Duke University [20] | X | | | | |
| Thompson Rivers University [21] | X | | | | |
| Bloomsburg University [22] | X | | | | |
| Emory University [23] | X | | | | X |
| Colgate University [24] | X | | | | X |
| University of California, Berkeley [25] | X | | X | X | |
| Cornell University [26] | X | | | | X |
| University of Alberta [27] | X | | | | |
| Gonzaga University [28] | X | | | | |
| University of Massachusetts Amherst [29] | X | X | | | X |
| Lewis and Clark Institute [30] | | X | | | |
| University of California, Irvine [31] | | X | | | |
| Carnegie Mellon University [32] | | X | | | |
| Australian National University [33] | | | X | | |
| National University of Singapore [34] | | | X | | |
| University of Cambridge [35] | | | X | | |
| University of Cape Town [36] | | | X | | |
| University of Copenhagen [37] | | | X | | |
| Yale University [38] | | | X | | |
| Northeastern University [39] | | | | X | |
| University of Limerick [40] | | | | X | |
| University of North Carolina [41] | | | | X | X |
| University of Plymouth [42] | | | | X | |

**Table 2.** Ranking of SCO Initiatives

| SCO-initiatives | | 30 green universities websites | | |
|---|---|---|---|---|
| | | Frequency | Percentage | Average |
| *1) Energy* | | | | |
| **E1** | Generate electricity on campus by establishing large-scale plants or power generation plants such as solar energy systems to reduce/replace total consumption of nonrenewable fuels (e.g.: oil & natural gas) | 25 | 83% | 76.5% |
| **E2** | Use of energy efficient equipment and eco-friendly products within the campus to enhance performance and reduce carbon footprint | 21 | 70% | |
| *2) Building* | | | | |
| **B3** | Install building management systems to monitor real-time electricity, water and chilled-water usage | 14 | 47% | 52% |
| **B4** | Consider the full life-cycle of building from extraction of resources to demolition and recycling when building or renovating a building | 11 | 37% | |
| **B5** | Design, build, and renovate building according to green certification standard | 23 | 77% | |
| **B6** | Design the building for high performance | 14 | 47% | |

| | | | | |
|---|---|---|---|---|
| ***3) Green space*** | | | | |
| **G7** | Reserve green area to reduce heat-island effect by provide effective shading, oxygen, and isolating carbon emissions generated by campus activities | 17 | 57% | 49% |
| **G8** | Create green area for storm water management to reduce the risk of flooding | 15 | 50% | |
| **G9** | Create native and drought plants landscape as significantly reduced the need for irrigation and maintenance | 12 | 40% | |
| ***4) Water*** | | | | |
| **Wr10** | Use of water efficient equipment which have water conservation and water saving features | 17 | 57% | 63.5% |
| **Wr11** | Use unprocessed water supply such as portable water, rainwater harvesting or on-site recycling of water which can be used in operational activities | 21 | 70% | |
| ***5) Purchasing & procurement*** | | | | |
| **P12** | Establish central information medium or a portal to assembled materials for re-use or exchange | 18 | 60% | 55% |
| **P13** | Run programs for students or department to donate or sale unwanted items to local community and charities | 13 | 43% | |
| **P14** | Set technical specifications or purchasing guidelines with include environmental preferable criteria | 18 | 60% | |
| **P15** | Include procedures of treatment or proper handling of solid waste (inorganic waste), hazardous waste and sewage disposal, without polluting the environment and public health | 17 | 57% | |
| ***6) Transport*** | | | | |
| **T16** | Use renewable fuels such as bio-diesel or use of energy efficient vehicles such as hybrid or electric vehicle for university transportation (official vehicle or even bus) | 17 | 57% | 49.9% |
| **T17** | Offer staff and students to purchase public transport season tickets | 18 | 60% | |
| **T18** | Provide real time shuttle-bus schedules | 10 | 33% | |
| **T19** | Provide a centralized parking facilities for carpooling, bicycling, walking, and shuttle bus purpose | 19 | 63% | |
| **T20** | Provide secure bicycle storage area | 20 | 67% | |
| **T21** | Provide showers for cyclist | 4 | 13% | |
| **T22** | Offer access to free or cheap bicycles or provide bicycle charter | 16 | 53% | |
| **T23** | Provide safe cycling & walking paths | 19 | 63% | |
| **T24** | Provide in-campus workshop to maintain and repair bicycles | 12 | 40% | |
| ***7) Waste*** | | | | |
| **Ws25** | Re-use or recycle construction and demolition waste (e.g.: unused metal, concrete, plywood etc.) in different project applications | 13 | 43% | 61.2% |
| **Ws26** | Provide bin with clearly marked signs to increase the number of recycling items | 26 | 87% | |
| **Ws27** | Place the bin in areas both convenient and close to the waste source and allow easy access by the community members | 19 | 63% | |
| **Ws28** | Collect and turned garden/landscaping waste into compost for re-used on campus gardens and landscape | 12 | 40% | |
| **Ws29** | Collect food waste from dining halls and convert it into a high grade compost which can be used in the gardens and landscape in the campus | 22 | 73% | |
| ***8) Food*** | | | | |
| **F30** | Buying foods from local producers to reduce the distance that food is transported from product consumer | 19 | 63% | 61.5% |
| **F31** | Choose the agricultural products that using organic fertilizer or minimal chemical fertilizer | 18 | 60% | |

However, based on the average score of each element of SCO initiatives, it can be seen clearly from Table 2 that the Energy sector contributes the most initiatives implemented amongst the green universities with 76.5% followed by Water 63.5%, Foods 61.5% and Waste 61.2%. Meanwhile, element of Purchasing and Procurement has received less attention with only 55% of the initiatives implemented by the green universities as well as other elements of SCO initiatives including Building area 52%, Transport 49.9%, and Green space 49%, respectively.

## 5 Conclusions and Recommendation

The concept of sustainable campus is relatively new in Malaysian, and several local public universities are still at the infant stage in adopting and implementing the SCO initiatives towards sustainability. Obviously, the planning and implementation of these initiatives should be phased and prioritized accordingly so that they can affect optimally on sustainability in campus operations.

There are various initiatives and sustainable practices that can be adopted and implemented by the top management to ensure that universities operate in a sustainable manner. However, more focus and attention should be given to the initiatives that can give positive impact on the environment, economic and social aspects of the campus community. In the early stages, the local universities may implement only the top 5 lists (e.g. Provide bin with clearly marked, Generating electricity in campus through power generation plants, Design, build and renovate building according to green building certification/standard/index, Compost foods wastes, and Use of energy efficient equipment and eco-friendly products) to ensure more focused and manageable implementation of the initiatives for successful results. Otherwise, top management can also implement sustainability initiatives according to the focus areas such as energy and water sectors, as this focus area can give greater impact on sustainable development at the university although it requires higher initial investment for a long term benefits.

However, it is not denying that factors such as self-awareness, knowledge, finance, leadership, organization, technology, climate, and etc. also affect the selection of sustainability initiatives to be implemented. The initiatives discussed earlier are for reference purposes only and unnecessarily be implemented fully by local universities. Finally, the SCO initiatives identified from website review will be considered for further investigation on their level of significance through questionnaire survey should they be implemented at Malaysian universities.

## Acknowledgements

This paper is a part of the research work funded by the Ministry of Higher Education Malaysian under Short Term Grant Scheme (Vote U345). We are also grateful to those individuals who have rendered their assistance throughout this study.

## References

1. Gunawan, Tarigan, E., Prayogo, D.N., and Mardiono, L. (2012). "Eco-sustainable Campus Initiatives: A Web Content Analysis," Proceedings of The 3rd International Conference on Technology and Operations Management: Sustaining Competitiveness through Green Technology Management, Bandung–Indonesia (July 4-6), pp. 59-65. ISBN: 978-979-15458-4-6.
2. Cortese AD. (2003) "The critical role of higher education in creating a sustainable future", Planning for Higher Education, vol 31, no 3, pp. 15-22
3. Velazquez, L., Munguia, N.P.A. and Taddei, J. (2004), "A Sustainable University: What can be the matter?" *Journal of Cleaner Production*, vol. 14, iss.9-11, pp. 810-819
4. Waas, T. et al (2012), "Sustainable Higher Education: Understanding and Moving Forward", Retrieved on August 6, 2015 from https://lirias.kuleuven.be/handle/123456789/374087.
5. Azis A.A.A.et al (2012), "Challenges faced By Construction Industry in Accomplishing Sustainability Goals", *IEEE Symposium on Business, Engineering and Industrial Applications*, pp. 630-634. ISBN: 978-1-4577-1632-4.
6. Yunus R. & Jay Y. (2011), "Sustainability Criteria for Industrialised Building Systems (IBS) in Malaysia", *The Twelfth East Asia-Pacific Conference on Structural Engineering and Construction*, Procedia Engineering 14 (2011) 1590–1598.
7. Yarime & Tanaka (2012), "The Issues and Methodologies in Sustainability Assessment Tools for Higher Education Institutions: A Review of Recent Trends and Future Challenges", Journal of Education for Sustainable Education, Vol 6(1): 63–77.
8. Norhayati, N. (2014), "Kajian Terhadap Model Kampus Lestari Universiti Awam di Malaysia: Suatu Pendekatan Terintegratif", Doctoral Thesis
9. University Leaders for a Sustainable Future, Talloires Declaration, Washington, DC. (2010), in Benjamin O. (2011), "Investigating Sustainability Professors' Conceptualizations of Sustainability and Sustainable Universities at Dalhousie University", Retrived on January 7, 2016 from http://www.dal.ca/content/dam/dalhousie/pdf/science/environmental-science-program/Honours%20Theses/BenOjoleck.pdf
10. University Leaders for a Sustainable Future, (2001), "The Talloires Declaration", Retrived on January 7, 2016 from http://www.ulsf.org/programs_talloires_td.html
11. Wright, T.S.A. (2002) "Definitions and Frameworks for the Environment Sustainability in Higher Education". Higher Eduction Policy, vol 15, no 2, pp.105-20.
12. Alshuwaikhat, H.M. and Abubakar, I. (2008) "An integrated approach to achieving campus sustainability: assessment of the current campus environmental management practices", Journal of Cleaner Production, vol. 16, pp. 1777-1785.
13. Nottingham Trent University, Retrieved on September 20, 2015 from https://www.ntu.ac.uk/sustainability/ecocampus/index.html
14. University College Cork, Retrieved on August 8, 2015 from https://www.ucc.ie/en/greencampus/
15. University of California, Davis Retrieved on October 18, 2015 from http://sustainability.ucdavis.edu/action/index.html

16. University of Oxford, Retrieved on August 8, 2015 from http://www.sustainia.me/solutions/
17. University of Connecticut, Retrieved on October 20, 2015 from http://ecohusky.uconn.edu/development/tour.html
18. Colorado State University Retrieved on August 13, 2015 from https://www.fm.colostate.edu/sustain/index.cfm?page=projects/energy
19. Harvard University, Retrieved on October 13, 2015 from http://green.harvard.edu
20. Duke University Retrieved on August 13, 2015 from http://sustainability.duke.edu/campus_initiatives/index.html
21. Thompson Rivers University Retrieved on August 13, 2015 from http://www.tru.ca/sustain/initiatives.html
22. Bloomsburg University Retrieved on August 15, 2015 from http://www.bloomu.edu/sustainability
23. Emory University Retrieved on August 15, 2015 from http://sustainability.emory.edu
24. Colgate University Retrieved on August 15, 2015 from http://www.colgate.edu/distinctly-colgate/sustainability/climate-action-planning/alternative-transportation
25. University of California, Berkeley Retrieved on September 10, 2015 from http://sustainability.berkeley.edu
26. Cornell University (2013), "Sustainability: Today and Tomorrow", Retrieved on August 07, 2015 from http://cscproduction.s3.amazonaws.com/2014/04/06/21/20/07/271/PSCC2013Sustainability_Today_Tomorrow_FINAL.pdf.
27. University of Alberta Retrieved on August 18, 2015 from http://www.sustainability.ualberta.ca/en/Campus Initiatives.aspx
28. Gonzaga University Retrieved on August 18, 2015 from http://www.gonzaga.edu/campus-resources/sustainability/initiatives.asp
29. University of Massachusetts Amherst Retrieved on August 18, 2015 from https://www.umass.edu/sustainability/green-campus
30. Lewis and Clark Institute Retrieved on August 18, 2015 from https://www.lclark.edu/about/sustainability/campus/
31. University of California, Irvine Retrieved on August 19, 2015 from http://sustainability.uci.edu/sustainablecampus/
32. Carnegie Mellon University Retrieved on August 19, 2015 from http://www.cmu.edu/environment/recycling-waste/index.html
33. Australian National University Retrieved on August 25, 2015 from http://sustainability.anu.edu.au/themes/energy-and-climate
34. National University of Singapore Retrieved on August 25, 2015 from http://www.nus.edu.sg/oes/
35. University of Cambridge Retrieved on August 25, 2015 from http://www.environment.admin.cam.ac.uk
36. University of Cape Town Retrieved on August 26, 2015 from http://www.greening.uct.ac.za/action/
37. University of Copenhagen Retrieved on September 3, 2015 from http://greencampus.ku.dk/sustainable_campus_development/sustainability_initiatives/
38. Yale University Retrieved on September 3, 2015 from http://sustainability.yale.edu
39. Northeasten University Retrieved on August 8, 2015 from http://www.northeastern.edu/sustainability/
40. University of Limerick Retrieved on September 5, 2015 from http://www3.ul.ie/envirocom/Green-Guide/
41. University of North Carolina, Chapel Hill Retrieved on September 5, 2015 from http://sustainability.unc.edu/initiatives/
42. University of Plymouth Retrieved on September 5, 2015 from https://www.plymouth.ac.uk/your-university/sustainability/our-sustainable-campus

 DOI: 10.1051/matecconf/20178701021

# Fire resistance of ultra-high performance fibre reinforced concrete due to heating and cooling

Fadzli Mohamed Nazri[1,a], Ramadhansyah Putra Jaya[2], Badorul Hisham Abu Bakar[1], Raudhah Ahmadi[3]

[1]*School of Civil Engineering, Engineering Campus, Universiti Sains Malaysia, 14300 Nibong Tebal, Penang.*
[2]*Faculty of Civil Engineering, Universiti Teknologi Malaysia, 81310 Skudai, Johor.*
[3]*Department of Civil Engineering, Universiti Malaysia Sarawak, 94300 Kota Samarahan, Sarawak.*

**Abstract.** This study investigated the performance of ultra-high performance fibre reinforced concrete (UHPFRC) at elevated temperatures. The specimens were exposed to high temperatures, specifically 200, 400, and 600 °C, for 2 h.The fire resistance performance of the specimens was classified on the basis of their compressive strength, spalling, and weight loss; residual strength after heating was also examined. Results showed that UHPFRC processes excellent fire resistance in terms of flame spread and fire growth. While strength loss was not significant at low temperatures, the specimen subjected to high temperature spalled severly and showed deterioration because of heat.

## 1 Introduction

Ultra-high-performance fiber-reinforced concrete (UHPFRC) is a breakthrough in modern concrete mix design with compressive strengths benchmarking 150 MPa and above and tensile strengths of over 10 MPa. Previous studies clearly show the advantages of using high-performance fiber-reinforced cementitious composites, such as UHPFRC, engineered cementitious composites, and slurry-infiltrated concrete, in structural members under static loading conditions [1-3]. In a related study, [4] found that UHPFRC has outstanding material characteristics, such as self-consolidating workability, very high mechanical properties, and low permeability, all of which result in excellent environmental resistance. Millon et al. [5] reported that UHPFRC can significantly improve the impact resistance of cladding panels and walls while maintaining its standard thickness and appearance. UHPFRC is a cementitious composite reinforced by fibers with characteristic values exceeding 150 $N/mm^2$ in compressive strength, 5 $N/mm^2$ in tensile strength, and 4 $N/mm^2$ in first cracking strength [6]. This concrete also shows compressive strengths over seven times and tensile strengths greater than three times those of conventional concrete [4]. The fibers in UHPC provide tensile capacity across cracks, resulting in high shear capacity in bending members. These fibers improve tensile strength. Parsekian et al. [7] reported that small brass-coated steel fibers with a diameter of 0.185 mm and a length of 14 mm are commonly used as reinforcements in UHPC. Synthetic fiber and poly-vinyl alcohol have also been used [8]. The high compressive strength of UHPFRC is achieved by the densely packed state of the cement matrix, and its tensile strength is attributed to steel or polypropene fibers embedded in the matrix. The superior characteristics of UHPRFC allow its use in different applications that demand high strength and durability, including bridges, tunnels, and high-rise buildings. Massive structures may be at risk and endanger lives if UHPFRC is not resistant to fire exposure. Conventional fiber-reinforced concrete exhibits good capacity to absorb impact energy [9].

Despite the positive characteristics of these structures, they are still susceptible to fire. Fire exposure induces temperatures of up to 1000 °C, which could be detrimental to the structural integrity of UHPFRC. Water stored in the fine pores of the dense matrix evaporates under temperature extremities, and pressure builds up internally. When stresses cannot be withstood, explosion of concrete follows, a phenomenon known as spalling. Unfortunately, spalling is unpredictable; it can occur during the heating or cooling of UHPFRC. The behavior of UHPFRC during and after fire exposure requires further study to fully understand the mechanism of failure and risks in using this concrete [10].

The present research principally focuses on the fire reactions of UHPFRC. In this work, two types of UHPFRC mixes are prepared and subsequently subjected to high temperatures and fire resistance tests to observe their physical characteristics, spalling, and strength loss. Considering that the behavior of the concrete under fire is unknown, this research assists in understanding how the product behaves during fire. Use of UHPFRC not only reduces the amount of concrete required but also increases the serviceability and durability of the resulting structures. Understanding fire resistance is crucial to gain

[a] Corresponding author: cefmn@usm.my

confidence in the reliability of the product as a future replacement for conventional OPC.

## 2 Methodology

UHPFRC was prepared in accordance with the mix design shown in Table 1.

**Table 1.** Mix proportions of UHPFRC

| Materials | Kg/m$^3$ |
|---|---|
| Cement | 768 |
| Silica Fume | 192 |
| Steel Fiber (vol%) | 2% |
| White sand | 1140 |
| Superplasticizer (vol%) | 1.9% |
| Water | 145 |
| Water/binder ratio | 0.15 |
| Total binder amount | 960 |

The materials were mixed in a drum mixer in the following order: cement, silica fume, and sand. These materials were dry mixed until they were homogeneous. Pre-mixed superplasticizer and water were added and mixing was continued until no dry materials remained. Finally, the steel fibers were added and mixed. The mixing process took 15 min.

Finished batches were poured into a concrete bucket and then placed into a wheelbarrow. Cubic molds were filled using hand shovels. No vibration was included during placement because the concrete is self-consolidating. The specimens were cured by steam curing for 48 h at a temperature of 90 °C and a relative humidity of 100%. Then, the specimens were cured in water maintained at a room temperature of 27 ± 2 °C until the testing day. The specimens were cast on the basis of the four tests. For the first test (i.e., a furnace fire test), 24 cube specimens were prepared. Two specimens from the two mix designs were exposed to a temperature of 200 °C. Then, the test was repeated at 400 and 600 °C. This test was repeated once more. The three tests related to BS476 fire testing of building materials and structures comprised the sizes and quantities described in Table 2.

**Table 2.** Quantity of specimens required for BS 476 fire reaction tests

| BS 476 Fire Reaction Tests | Size of Specimen | Quantity |
|---|---|---|
| Part 4 (Non-Combustibility test) | 40 mm × 40 mm × 50 mm (h) | 3 |
| Part 6 (Method of test for fire propagation of products) | 225 mm × 225 mm × 50 mm (thk) | 3 |
| Part 7 (Method for classification of the surface spread of flame of products) | 885 mm × 225 mm × 50 mm (thk) | 6 |

The mix quality of self-consolidating concretes is frequently assessed through a slump cone flow test., The newer BS EN 12350-8:2010 test is commonly used with UHPFRC. This test measures self-compacting concrete. An initial flow reading and a dynamic flow reading were recorded. This test was completed immediately after mixing to assess the consistency between mixes and appropriateness for casting. Compressive strength is arguably the most readily captured and used property of concrete. Standard concrete compression testing methods (i.e., BS 1881-119:2011) are applicable to UHPFRC. However, these test methods may benefit from slight modification to facilitate efficient use. A loading rate of 1 MPa/s is acceptable and allows for individual tests to be completed within a reasonable timeframe.

At the high strength levels achieved with UHPFRC, cube compressive tests are an appropriate substitute for cylinder compression tests. Companion cylinder and cube strength results tend to be within 5% of one another, thereby allowing for direct substitution of results. The high compressive strengths of UHPC may necessitate the use of high-capacity compression testing plates and machines. The furnace test is a method in which UHPFRC samples are heated at specific temperatures to observe their physical properties and behavior as a function of heat. The furnace used was a ThermConcept KC80/14, as illustrated in Figure 1. The furnace specifications are tabulated in Table 3.

**Figure 1.** ThermConcept KC 80/14 furnace used in this work

**Table 3.** Specifications of the ThermConcept KC 80/14 furnace

| Model | KC 80/14 |
|---|---|
| T max (°C) | 1400 |
| Inside dimensions (w x d x h) | 430 x 430 x 430 |
| Liter (L) | 80 |
| External dimensions (w x d x h) | 985 x 1066 x 1540 |
| Power (kW) | 13 |
| Weight (kg) | 280 |
| Volatege (V) | 400 3/N |

The specimens were placed in the furnace for 2 h at varying temperatures, specifically 200, 400, and 600 °C. Four specimens were used for every temperature. The experiment was repeated twice with a different batch mix. After heating, the specimens were allowed to cool to

DOI: 10.1051/matecconf/20178701021

below 70 °C and subsequently removed from the furnace. The spalling behavior, residual mass, and residual compressive strength of the samples were then observed, noted, and tested accordingly.

## 2.1 Fire reaction tests in BS 476: Fire tests on building materials and structures

Materials used in the construction and finishing of buildings or structures are classified as "non-combustible" or "combustible" on the basis of their behavior in the non-combustibility test. To begin, the size of specimens illustrated in Figure 2 was heated, and its temperature was stabilized at 750 ± 10 °C for a minimum period of 10 min.

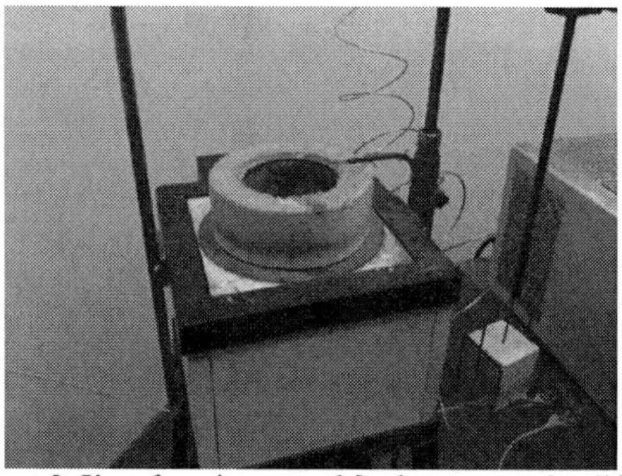

**Figure 2.** Size of specimens used for the non-conmbustibility test : 40 mm (l) x 40 mm (w) x 50 mm (h).

The specimen was then inserted in the furnace. The whole operation required no more than 5 s. A record of the temperature of the two thermocouples was made over the next 20 min using a continuous recorder. The occurrence and duration of any flaming in the furnace were noted. The stabilized heating current remained unchanged throughout the duration of the test. The material was deemed non-combustible if, during the test, none of the three specimens either:

i. caused the temperature reading from either of the two thermocouples to rise by 50 °C or more above the initial furnace temperature, or
ii. continuously flamed for 10 s or more inside the furnace. Otherwise, the material was deemed combustible.

Before the test, specimens were dried in a ventilated oven at 60 ± 5 °C for 24 h and then cooled to ambient temperature in a desiccator containing anhydrous calcium chloride.

Results are expressed as a fire propagation index, which provides a comparative measure of the contribution to the growth of fire made by an essentially flat material, composite, or assembly. This index is primarily intended for assessing the performance of internal wall and ceiling linings. The test apparatus is illustrated in Figure 3.

**Figure 3.** Fire propagation test for a specimen measuring 225 mm x 225 mm x 50 mm

The thermocouple output indicator was turned on, and the initial output Ei (expressed in mV) was recorded, as measured by the thermocouples inside the cowl. After allowing any residual gas within the combustion chamber to disperse, the gas supply was turned on and the jets were ignited using the gas igniter. The test was timed from the time of ignition. After 2 min and 45 s, the electrical supply was turned on to produce an indicated input of 1800 W. About 5 min after the time of ignition, the input was reduced to 1500 W. The output was recorded from the thermocouples, Er, in mV, at the following points:

a) 0.5 min intervals, up to and including 3 min from the time at which the gas was ignited;
b) 1 min intervals, up to 10 min from the time at which the gas was ignited; and
c) 2 min intervals, up to 20 min from the time at which the gas was ignited.

BS 476 Part 7 specifies a test method for measuring the lateral spread of flame along the surface of a specimen of a product orientated in the vertical position, as illustrated in Figure 4, and provides a classification system based on the rate and extent of flame spread.

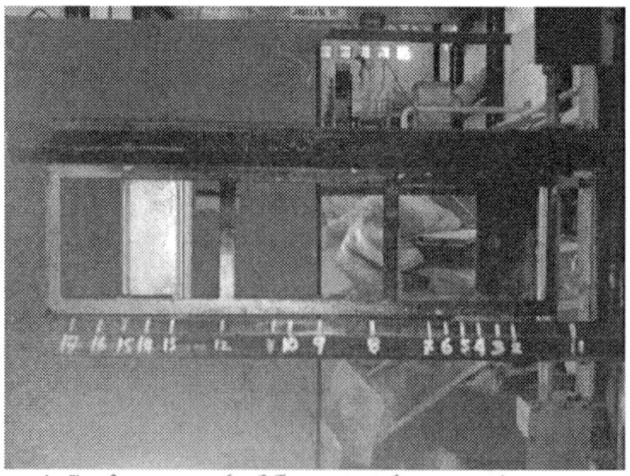

**Figure 4.** Surface spread of flame test for a specimen measuring 825 mm x 225 mm x 50 mm

This standard provides data suitable for comparing the performances of essentially flat materials, composites, or assemblies that are primarily used as the exposed surfaces of walls or ceilings.

In practice, standard testing methods are used to determine the fire performance of materials and building or structural elements. These methods replicate the conditions of typical fires, either on a smaller scale (e.g., in a specially built oven/furnace) or in a full-scale test (i.e., on a part or whole mock-up of a building).

i. ISO 834 or BS 476:1987: Fire tests on building materials and structures.
ii. BS EN 1363-1:2012: Fire resistance tests
iii. BS 1881:1983: Testing Concrete.
iv. ASTM C150 / C150M - 12 Standard Specification for Portland Cement.

# 3 Results and discussion

## 3.1 Development of optimum mix design

Park et al. [11] explained that the strength of UHPFRC depends on the interface failures between the cement paste and aggregates. A low water-to-cement ratio reduces the size of inter-transition zones between the cement particles and aggregates and subsequently maximizes the strength development of the UHPFRC. The mix designs are specified in Table 4 below.

**Table 4.** Mix design of UHPFRC samples.

| Materials | Mix A kg/m$^3$ | Mix B kg/m$^3$ |
|---|---|---|
| **Cement** | 768 | 875 |
| **Silica Fume** | 192 | 190 |
| **Steel Fiber (vol%)** | 2% | 2.5% |
| **White Sand** | 1140 | 1272 |
| **Superplasticizer (vol%)** | 1.9% | 1.9% |
| **Water** | 145 | 182 |
| **Water/binder ratio** | 0.15 | 0.17 |
| **Total binder amount** | 960 | 1065 |

## 3.2 Compressive strength

The average values obtained are illustrated in Table 5.

**Table 5.** Average compressive strength of UHPFRC cubes.

| Time (Days) | Mix A (MPa) | Mix B (MPa) |
|---|---|---|
| 7 | 131 | 127 |
| 28 | 152 | 161 |

The compressive strength of UHPFRC, at above 150 MPa, is satisfactory. Mix A showed a compressive strength of 152 MPa, while Mix B showed a strength of 161 MPa after 28 d. The higher strength development at 28 d in Mix B is most likely due to the higher cement content packed per cube of Mix B samples. Mix B has a higher binder amount than Mix A. Silica fume was maintained at equal levels for both mixes because of the physical and chemical aspects of the silica fume itself. First, silica fume undergoes a pozzolanic reaction in the presence of calcium hydroxide during cement hydration. Addition of silica fume when the calcium hydroxide required for the pozzolanic reaction is relatively fixed could be redundant and decrease the compressive strength. The particle size of the silica fume is very small, and the silica can increase the packing density of the material when the cement is substituted. Packing effects can increase the compressive strength of UHPFRC, lower its permeability, and increase its durability. Steel fibers of 0.2 mm diameter at 2% and 2.5% of the total volume were used in Mixes A and B, respectively. The percentages of steel fiber were adopted from Yu et al. [12] in their study on optimized UHPFRC. The 2% steel fiber amount is an optimized value practiced at the industry and research levels.

Steel fibers play an important role in the 28 d compressive strength of UHPFRC. Steel fibers also contribute to the tensile strength of the concrete. Ultimately, the main contributors to UHPFRC strength are a low water-to-cement ratio and the highly dense state of the cement.

## 3.3 Spalling

UHPFRC cubes measuring 100 mm × 100 mm × 100 mm were placed in a furnace at 200, 400, and 600 °C for 120 min at a rate of 15 °C/min. The furnace test was performed with both mixes A and B (two samples for each mix) and repeated twice at three temperatures, yielding a total of 24 samples. All samples were allowed to cool until ambient temperature before removal from the furnace. At a low temperature of 200 °C, the concrete was observed to be in perfect condition despite some discoloration. This sample appeared greyish-white after being exposed at 200 °C for 2 h. However, at very high temperatures of 400 and 600 °C, the concrete underwent severe spalling. The concrete cubes became unrecognizable and completely deteriorated at high temperatures. All physical behaviors described above are illustrated in Figures 5 to 9.

**Figure 5.** UHPFRC before and after exposure to 200 °C temperature.

DOI: 10.1051/matecconf/20178701021

**Figure 6.** UHPFRC before and after exposure to 200 °C temperature.

**Figure 7.** UHPFRC before and after exposure to 400 °C temperature.

**Figure 8.** UHPFRC before and after exposure to 600 °C temperature.

**Figure 9.** Core sample of UHPFRC recovered after exposure to 600 °C.

The low porosity of UHPFRC induces great internal stresses. When pores completely closed, water vapor cannot escape during heating. Increases in pressure within the material cause spalling. Polypropylene fibers are effective in reducing spalling because these fibers melt and provide escape routes for trapped steam. These fibers create additional capillary pores in the matrix. However, a different phenomenon was observed in the UHPFRC meshed with steel fibers. The concrete underwent severe spalling at 400 and 600 °C until the specimen was destroyed beyond recognition. This result shows that UHPFRC fire resistance is unfavorable when subjected to high fire loadings primarily because of its very dense and brittle material. Steel fibers remained in good condition without much distortion from the heat. The color of these fibers was a dark rusty brown. This color can be attributed to Fe oxidation in the steel fibers to FeO.

### 3.4 Mass and strength loss

Mass loss was apparent after subjecting the samples to high heat loadings. Two representative samples, P and Q, were weighed before the samples were heated. The test was repeated twice. Figure 10 shows the standard weighing process of the cubic samples.

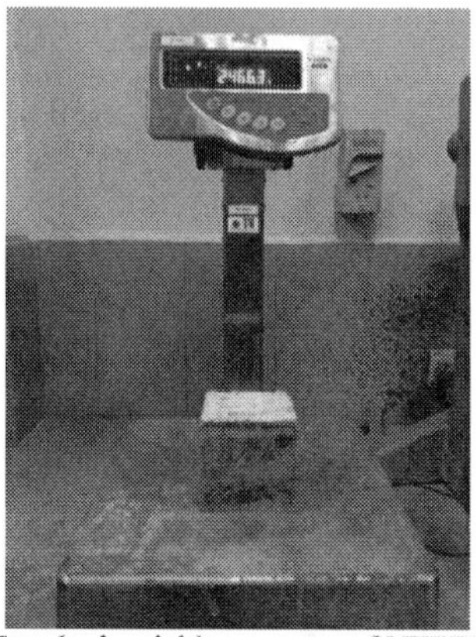

**Figure 10.** Standard weighing process of UHPFRC samples.

Table 6 summarizes the mass loss data before and after heating.

**Table 6.** Mass of various UHPFRC samples before and after the furnace test.

| Mix Type | | | Temperature | | | | | |
|---|---|---|---|---|---|---|---|---|
| | | | 200 °C | | 400 °C | | 600 °C | |
| | | | Before (g) | After (g) | Before (g) | After* (g) | Before (g) | After* (g)' |
| A | Test 1 | P | 2433.5 | 2359.4 | 2441.1 | - | 2465.5 | - |
| | | Q | 2432.4 | 2353.3 | 2457.2 | - | 2443.6 | - |
| | | | 2466.3 | 2380.3 | 2456.5 | - | 2449.5 | - |
| | | | 2455.4 | 2354.5 | 2453.5 | - | 2438.3 | - |
| | Test 2 | P | | | | | | |
| | | Q | | | | | | |
| B | Test 1 | P | 2465.5 | 2369.3 | 2463.8 | - | 2467.4 | - |
| | | Q | 2465.4 | 2368.7 | 2446.3 | - | 2398.7 | - |
| | | | 2338.4 | 2269.2 | 2368.7 | - | 2346.8 | - |
| | | | 2348.4 | 2279.5 | 2351.4 | - | 2348.8 | - |
| | Test 2 | P | | | | | | |
| | | Q | | | | | | |

* samples were destroyed beyond recovery

Figure 11 shows that the concrete lost some mass at 200 °C.

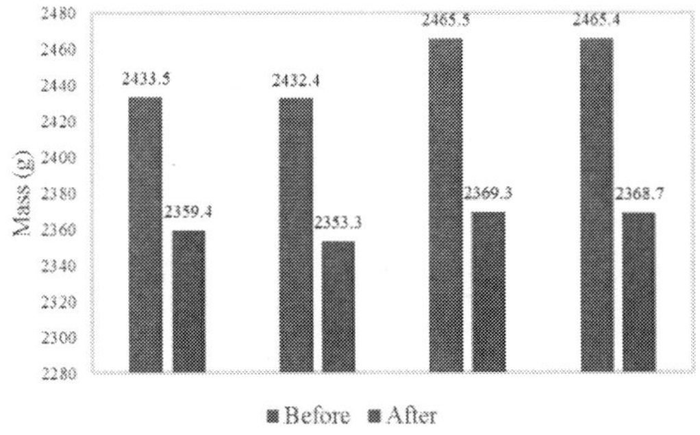


**Figure 11.** Residual mass of UHPFRC before and after 200 °C furnace test 1.

This result is attributed to losses in the water content of the concrete. This behavior is similar to test 2. Mass loss determination could not be completed at 400 and 600 °C because of complete deterioration of the specimens under high temperature.

Representative samples of specific batch mixes were taken for compressive strength determination before the test. The average strength was taken from two samples. Strength loss was not apparent when subjected to a low temperature loading of 200 °C. However, severe spalling behavior at high temperatures of 400 and 600 °C made a compressive strength test impossible to perform. Figure 12 shows a cubic sample subjected to compressive strength after heating at 200 °C.

**Figure 12.** UHPFRC subjected to compressive strength testing after heating at 200 °C.

The figure 12 shows the failure mode of the crack due to the compressive test. The sample fared well in compressive strength test, and nearly no strength loss was detected. Table 7 summarizes the strength loss before and after the furnace test. No compressive strength was obtained after 400 and 600 °C because of sample deterioration.

**Table 7.** Strength loss of UHPFRC of various mix designs before and after furnace test.

| Mix Type | | Temperature | | | | | |
|---|---|---|---|---|---|---|---|
| | | 200 °C | | 400 °C | | 600 °C | |
| | | Before (MPa) | After (MPa) | Before (MPa) | After* (MPa) | Before (MPa) | After* (MPa) |
| A | Test 1 | 157<br>154 | 155<br>151 | 157<br>154 | -<br>- | 157<br>154 | -<br>- |
| | Test 2 | | | | | | |
| B | Test 1 | 162<br>163 | 161<br>161 | 162<br>163 | -<br>- | 162<br>163 | -<br>- |
| | Test 2 | | | | | | |

* samples were destroyed beyond recovery

### 3.5 BS 476 fire tests on building materials and structures: Part 4

Table 8 shows a summary of the results. The test results were inconclusive because all samples exploded within a few minutes.

**Table 8.** Test results of non-combustibility test.

| Parameters | Units | 1 | | 2 | | 3 | |
|---|---|---|---|---|---|---|---|
| Specimen Reference | | | | | | | |
| Weight | g | 209.65 | | 201.70 | | 204.60 | |
| Width & Breadth | mm | 40 × 40 | | 40 × 40 | | 40 × 40 | |
| Height | mm | 50 | | 50 | | 50 | |
| Density | kg/m$^3$ | 2620.63 | | 2521.25 | | 2557.50 | |
| Average Density | kg/m$^3$ | 2566.46 | | | | | |
| Temp. Measurement* | | CH 1 | CH 2 | CH 1 | CH 2 | CH 1 | CH 2 |
| Max. Furnace Temp. | °C | * | * | * | * | * | * |
| Stabilized Temp. | °C | 750 | | 750 | | 750 | |
| Temp. Difference | °C | - | - | - | - | - | - |

CH1: maximum temperature of the furnace
CH2: maximum temperature at the centre of specimen
* - unable to record

The non-combustibility test was deemed invalid. The sample totally disintegrated in the furnace within a few minutes and the test could not be completed. Similar to the furnace test performed previously, UHPFRC did not perform well under high temperatures. The specimen underwent explosive spalling and could not be recognized. Before the test, specimens was dried in a ventilated oven at 60 ± 5 °C for 24 h and then cooled to ambient temperature in a desiccator. Explosive spalling was attributed solely to the build-up of pressure from steam that had no exit route because of the low porosity of UHPFRC. This problem must be addressed to commercialize UHPFRC in the industry for public safety. To overcome this problem, the bond between the matrix and steel fibers as well as quartz sand particles must be exceptionally strong. Polypropylene fibers could enhance the fire resistibility of UHPFRC and may be used alongside steel fibers.

### 3.6 BS 476 fire tests on building materials and structures: Part 6

The purpose of this test is to obtain a fire propagation index, which provides a comparative measure of the contribution to the growth of fire. Comparison is performed against a calibration sheet, in this case, an asbestos sheet, and the temperature rise was recorded accordingly. Three specimens measuring 225 mm × 225 mm × 50 mm were tested, and their average density was 2316 kg/m$^3$. The test apparatus consisted of a combustion chamber with a specimen holder fixed to one face. The combustion chamber contained a horizontal gas burner tube and two electrical heating elements, and was surmounted by a removable steel chimney and cowl. The rise in temperature of the specimen was recorded using a thermocouple located on the cowl. During heating, three sub-indices, namely, $S_1$, $S_2$, and $S_3$ were calculated. These sub-indices are defined as follows:

$$S_1 = \sum_{t=0.5}^{t=3} \frac{\theta_s - \theta_c}{10t}; \quad S_2 = \sum_{t=4}^{t=10} \frac{\theta_s - \theta_c}{10t}; \quad S_3 = \sum_{t=12}^{t=20} \frac{\theta_s - \theta_c}{10t} \qquad (1)$$

where

$\theta_s$ is the actual temperature rise of the specimen to the nearest °C

$\theta_c$ is the actual temperature rise of the calibration material to the nearest °C

$t$ is the time, in min, of the specified interval

Once the sub-indices were calculated, the index of overall performance, also known as the fire propagation index, $i$, was calculated from these sub-indices. The relevant formula for $i$ is as follows:

$$I = i_1 + i_2 + i_3 \qquad (2)$$

where

$$i_1 = \tfrac{1}{3}[(S_1)_A + (S_1)_B + (S_1)_C]$$
$$i_2 = \tfrac{1}{3}[(S_2)_A + (S_2)_B + (S_2)_C]$$
$$i_3 = \tfrac{1}{3}[(S_3)_A + (S_3)_B + (S_3)_C]$$

where A, B, and C represent individual specimens yielding valid test results, and $S_1$, $S_2$, and $S_3$ are the sub-indices.

As shown in Table 9, the sub-indices for all three specimens were 0.0 except sub-index 2 of specimen C, which was 0.1 or negligible in reality. The $i$ is also 0.0.

**Table 9.** Results of the fire propagation test.

| Time (min) | Calibration | Specimen A | Specimen B | Specimen C | Specimen Average |
|---|---|---|---|---|---|
| 0.5 | 14 | 11 | 9 | 8 | 9 |
| 1.0 | 17 | 15 | 12 | 12 | 13 |
| 1.5 | 22 | 18 | 16 | 15 | 16 |
| 2.0 | 25 | 20 | 18 | 17 | 18 |
| 2.5 | 29 | 22 | 19 | 20 | 20 |
| 3.0 | 34 | 25 | 25 | 29 | 26 |
| 4.0 | 68 | 57 | 64 | 70 | 64 |
| 5.0 | 104 | 87 | 96 | 99 | 94 |
| 6.0 | 129 | 108 | 116 | 123 | 115 |
| 7.0 | 153 | 128 | 133 | 143 | 134 |
| 8.0 | 170 | 143 | 151 | 158 | 151 |
| 9.0 | 187 | 157 | 163 | 174 | 164 |
| 10.0 | 203 | 167 | 173 | 183 | 174 |
| 12.0 | 222 | 191 | 193 | 200 | 194 |
| 14.0 | 232 | 201 | 204 | 214 | 206 |
| 16.0 | 245 | 206 | 214 | 225 | 215 |
| 18.0 | 253 | 214 | 220 | 229 | 221 |
| 20.0 | 258 | 218 | 220 | 236 | 225 |
| Subindex 1 | | 0.0 | 0.0 | 0.0 | 0.0 |
| Subindex 2 | | 0.0 | 0.0 | 0.1 | 0.0 |
| Subindex 3 | | 0.0 | 0.0 | 0.0 | 0.0 |
| Index of Performance | | 0.0 | 0.0 | 0.1 | 0.0 |

As shown in Figure 13, the temperature of UHPFRC does not rise as quickly as that in the calibration material.

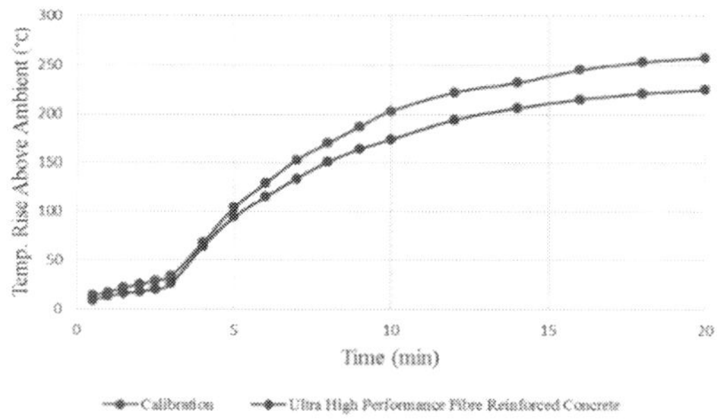


**Figure 13.** Graph of temperature rise of UHPFRC vs. calibration.

Asbestos sheeting is a common material used in wall and ceiling claddings. Thus, using UHPFRC as a structural or non-structural component resists the growth of fire even if the claddings are in flames. The highest average temperature rise of UHPFRC above the ambient temperature (25 °C) recorded over a 20 min time frame is 225 °C, which is lower than the calibration material's 258 °C. The specimens performed exceptionally well. No spalling was observed. An $i$ of 0.0 is the best index possible given the conditions of this test. Although the behavior of UHPFRC toward fire in this test was good, the results should not be the sole criterion for assessing the potential fire hazard of the product in use.

### 3.7 BS 476 fire tests on building materials and structures: Part 7

This test measures the lateral spread of flame along the surface of UHPFRC specimens and enables classification of the sample based on the rate and extent of the flame. This classification would allow UHPFRC to be compared with other products on the basis of their fire performance when used as exposed surfaces of walls or ceilings. Four classes, Classes 1–4, have been specified and Class 1 is the class with the best fire performance. The classes are solely based on the spread of flame. Table 10 demonstrates how the code classifies the spread of flame. Six specimens measuring 885 mm × 225 mm × 50 mm were used. The average density of the specimens was 2313 kg/m$^3$. Test results indicated that the specimens may be classified under Class 1, which features the shortest distance travelled and slowest flame spread. Table 10 shows a summary of the results of the flame test.

**Table 10.** Classification of spread of flame.

| Classification | Spread of flame at 1.5 min | | Final spread of flame (10 min) | |
|---|---|---|---|---|
| | Limit (mm) | Limit for one specimen in sample (mm) | Limit (mm) | Limit for one specimen in sample (mm) |
| Class 1 | 165 | 165+25 | 165 | 165+25 |
| Class 2 | 215 | 215+25 | 455 | 455+45 |
| Class 3 | 265 | 265+25 | 710 | 710+75 |
| Class 4 | Exceeding the limits for class 3 | | | |

Under a Class 1 fire rating, UHPFRC can reasonably protect against the spread of flame; this rating is the best class possible under the provisions of the BS476 code. The specimens exhibited exceptionally behavior in resisting flame. This result indicates that the specimens show no signs of fire spread. After the first 90 s of testing, no flame ignition was observed on the specimens. This phenomenon continued until the end of the 10 min test. The resistance of UHPFRC to fire spread is highly typical of the natural fire resistance concrete.

## 4 Conclusions

UHPFRC performed exceptionally well at 200 °C. Its compressive strength was not compromised and the material did not spall, although some mass loss attributed to water loss was observed. Explosive spalling was observed at temperatures beginning from 400 °C and above. The presence of steel fibers did not contribute to prevent explosive spalling.

Strength loss was imminent when UHPFRC underwent explosive spalling at temperatures above 400 °C. The material disintegrated, leaving no core sample to be tested. No compressive strength was left to be recovered.

UHPFRC can neither be classified as a combustible nor non-combustible material. The non-combustibility test had to be stopped because the material was no longer suitable to carry out the test because of explosive spalling. However, UHPFRC showed excellent performance in resisting heat when heated at low temperatures. The temperature rise was not prominent, which indicates that UHPFRC is an excellent material for resisting the growth of fire. UHPFRC scored a Fire Propagation Index *I* of 0.0, the best index available under BS476 Part 6, and showed excellent resistance to flame spread. No flame was observed to spread on the material when the pilot flame was ignited. Thus, the material was given a Class 1 fire rating under BS476 Part 7.

## Acknowledgement

This research was supported by Universiti Sains Malaysia under Research University Individual grant (814190).

## References

1. N. Krstulovic-Opara, E. Dogan, C.M. Uang, A.R. Haghayeghi, Flexural behavior of composite RC-Slurry infiltrated mat concrete (SIMCON) members, ACI Structure Journal, 94 (1997).
2. J.H.J. Kim, Y.M. Lim, J.P. Won, H.G. Park, K.M. Lee, *Shear capacity and failure behavior of DFRCC repaired RC beams at tensile region*, Engineering Structure 29 **(1)** 121-31 (2007).
3. K. Habel, E. Denarié, E. Brühwiler, Experimental investigation of composite UHPFRC concretre members, ACI Structure journal, 104 **(1)** 93-101 (2007).
4. B.A. Graybeal, Compressive behaviour of Ultra-High-Performance Fibre-Reinforced Concrete, ACI materials journal, **40**, 104-107 (2007).
5. O. Million, W. Riedel, C. Mayrhofer, K. Thoma, *Fibre-reinforced ultra-high performance concrete-a material with potential for protective structures*, Proceeding of the first international conference of protective structures, Manchester, 103 (2010).
6. Y. Uchida, J. Niwa, Y. Tanaka, M. Katagiri, G. Fischer, V. Li, Recommendations for design and construction of ultra high strength fibre reinforced concrete structures, Concrete committee of Japan Society of Civil Engineers (JSCE), 345-351 (2006).
7. G.A. Parsekian, N.G. Shrive, T.G. Brown, J. Kroman, P.J. Seibert, V.H. Perry, A. Boucher, Innovative ultra-high performance concrete structures, Tailor Made Concrete Structures, Taylor&Francis Group, London, 325-330 (2008).
8. B.A. Graybeal, MAterial property characterization of ultra-high performance concrete FHWA-HRT-06-103, Faderal highway administration, Washington, DC (2006).
9. N. Wang, S. Mindess, K. Ko, *Fibre reinforced concrete beams under impact loading*, Cement & Concrete research, **26**, 363-376 (1996).
10. A.H. Akca, N.O. Zihnioğlu, *High performance concrete under elevated temperatures*, Construction and building materials, **44**, 317-328 (2013).

DOI: 10.1051/matecconf/20178701021

11. J. J, Park, D. Yoo, S.W. Kim, Y.S. Yoon, *Drying shrinkage cracking characteristics of ultra-high performance fibre reinforced concrete with expansive and shrinkage reducing agents.* Magazine of concrete research 65, 248-256 (2013).
12. R. Yu, P. Spiesz, H.J.H. Brouwers, *Mix design and properties assessment of Ultra-high performance fibre reinforced concrete (UHPFRC),* Cemented concrete research, **56**, 29-39 (2013).

 DOI: 10.1051/matecconf/20178702001

# Numerical Study of Frequency-dependent Seismoelectric Coupling in Partially-saturated Porous Media

Ulugbek Djuraev[1], Shiferaw Regassa Jufar[1,a] and Pandian Vasant[2]

[1]*Department of Petroleum Engineering, Universiti Teknologi Petronas, 32610 Seri Iskandar, Perak, Malaysia*
[2]*Department of Fundamental & Applied Sciences, Universiti Teknologi Petronas, 32610 Seri Iskandar, Perak, Malaysia*

**Abstract.** The seismoelectric phenomenon associated with propagation of seismic waves in fluid-saturated porous media has been studied for many decades. The method has a great potential to monitor subsurface fluid saturation changes associated with production of hydrocarbons. Frequency of the seismic source has a significant impact on measurement of the seismoelectric effects. In this paper, the effects of seismic wave frequency and water saturation on the seismoelectric response of a partially-saturated porous media is studied numerically. The conversion of seismic wave to electromagnetic wave was modelled by extending the theoretically developed seismoelectric coupling coefficient equation. We assumed constant values of pore radius and zeta-potential of 80 micrometers and 48 microvolts, respectively. Our calculations of the coupling coefficient were conducted at various water saturation values in the frequency range of 10 kHz to 150 kHz. The results show that the seismoelectric coupling is frequency-dependent and decreases exponentially when frequency increases. Similar trend is seen when water saturation is varied at different frequencies. However, when water saturation is less than about 0.6, the effect of frequency is significant. On the other hand, when the water saturation is greater than 0.6, the coupling coefficient shows monotonous trend when water saturation is increased at constant frequency.

## 1 Introduction

The seismoelectric phenomenon has been attracting attention of many researchers all over the world. The first interest to study this phenomenon was shown the early 1940s [21]. Ever since, numerous works have been conducted to show and justify the potential of the method. Technologically, the seismoelectric method combines a high resolution of the seismic surveys and the sensitivity of the electrical methods to hydrological properties of subsurface formations [9, 23] that are commonly used in Oil and Gas Industry. In fact, it has been shown that the seismoelectric method can have very promising applications in various sectors of the Petroleum Engineering, especially, in Reservoir Monitoring. As such, the seismoelectric method can be applied in Reservoir Monitoring because it is able to detect subsurface fluid saturation changes. Subsequently, with sufficient knowledge of the subsurface geology, the distribution of reservoir fluids can be mapped.

As it was said earlier, the interest to explore the seismoelectric phenomena began in the twentieth century. In 1944, Frenkel investigated how streaming potentials could be generated by seismic waves in fluid-saturated porous media [7]. Thereafter, he developed a theoretical framework similar to Biot's [1] wave theory that allowed Frenkel to estimate the amount of relative fluid motion triggered by a seismic wave. Later on, he was able to calculate the magnitude of electric potentials by adopting the Helmholtz-Smoluchowski equation for streaming potentials [14].

Despite the aforementioned early breakthroughs, further investigations on the seismoelectric phenomenon were on hiatus until recently, because conducting field measurements were nearly impractical due to very low signal-to-noise ratio of the seismoelectric signals. Nevertheless, last couple of decades have been productive to deliver advances on the phenomenon thanks to theoretical breakthroughs [11, 13, 14], a series of successful practical laboratory [24, 27, 29] and field experiments [4, 5]; as well as to recent numerical modeling developments [16, 19, 23].

Frequency excitation of the seismic source has a significant impact on the seismoelectric effects. Moreover, dissipation of a seismic wave travelling through a fluid-saturated porous medium is affected by the relaxation time and effect of nonlinearities that vary with fluid saturation. Therefore, this study intends to investigate the effects of seismic frequency and water saturation affecting the phenomenon via the seismoelectric coupling in two-phase saturated porous media.

[a] Corresponding author: shiferaw.jufar@petronas.com.my

DOI: 10.1051/matecconf/20178702001

## 2 Theory of seismoelectric phenomenon

### 2.1 Electrical double layer

Seismic energy transforms into electromagnetic energy and vice versa due to the reversible electrokinetic conversion. The electrokinetic conversion occurs due to the presence of another phenomenon which is formed in a fluid-containing porous medium – electrical double layer.

As shown in Figure 1, the surface of porous medium grain, which is in contact with pore fluids, gains a chemically bound charge due to deprotonization and ion adsorption between surface and fluid reactions [18]. In a thin layer surrounding the grains, mobile counter ions tend to balance the bound charge. While the bound charge remains immobile, the transportation of counter ions is done by the flowing pore fluid. In fact, electrostatic forces and thermal diffusivity balance determine the distribution of mobile ions [20]. As a result, an electrical potential also known as the "zeta-potential" is defined at the interface between immobile and counter-ions. Zeta-potential varies exponentially when it moves away from the interface. The Debye length, the corresponding characteristic decay length, is on the order of several nanometers for typical solid grain-electrolyte combinations [11]. Essentially, the adsorbed and diffuse charged layers make up the ensemble, which is referred to as the *Electrical Double Layer* (EDL). Hence, due to the existence of the double layer, escalation and propagation of coupled seismic and electromagnetic waves occur simultaneously.

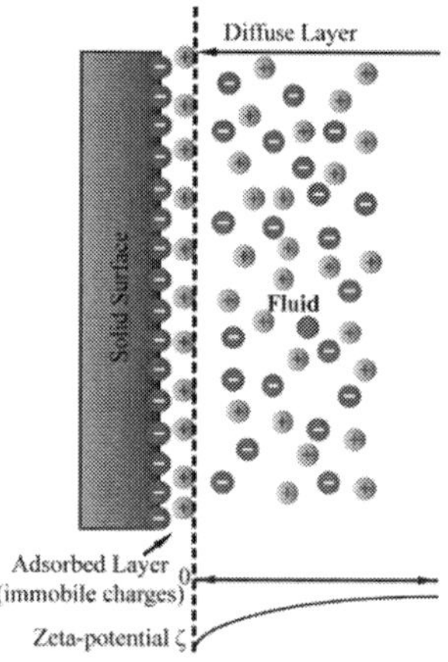


**Figure 1.** Electric double layer

Normally, petroleum reservoirs are predominantly water-wet. Hence the double layer is formed throughout the grain surface, and has a thickness of a few nanometers. Nonetheless, formation of the EDL on an oil-wet solid surface is also possible, though further investigation is required.

The magnitude of zeta-potential has a direct effect on the seismoelectric signals. The lower the magnitude of zeta-potential is, the weaker seismoelectric response is observed. A number of researchers have conducted experiments to measure zeta-potential for brine-saturated porous media, although the measurement procedure remains quite complex. From their experimental results they confirm that with increasing salinity, the value of zeta-potential reduces. [2, 10, 17, 22]. Moreover, they suggest that for numerical studies it is sufficient to use the empirically-derived Eq. 1, where $C_f$ is brine concentration in mol/l and $\zeta$ is zeta-potential in mV. The equaions is presented in a general format, however, various researchers report different equation coefficients. As such, one must consider all assumptions prior using a certain model.

$$\zeta = a\log_{10} C_f + b \tag{1}$$

For our numerical investigation, we adopt the model presented by Vinogradov, et al. [22] that is shown below in Eq. 2. The equation remains valid for brine concentrations are up to 0.4 mol/l.

$$\zeta = 19.02\log_{10} C_f - 9.67 \tag{2}$$

### 2.2 The seismoelectric effects in fluid-saturated porous medium

When a seismic wave travels through a fluid-saturated porous medium, it induces a fluid-to-solid relative motion that leads to charge separation in the EDL. As result, an electric field is produced [8, 11, 25-27, 29]. This phenomenon is referred to as "coseismic" electrical field.

A compressional wave (P-wave) travelling through a homogeneous fluid-saturated porous medium is shown in Figure 2 (left), where the wave is represented by vertical lines. There exist pressure peaks when the lines are closer together, and there exist pressure troughs when they are far apart. Hence, the counter-ions of the double layer will be transported by the fluid flow in relation to immobile, fixed charges. In this fashion, the pressure troughs are accumulated by the counter-ions (+), whereby the bound charge (-) is exposed to in pressure peaks, leading to creation of the coseismic electrical field on the wavelength scale. Thus, the exact balance of the hydraulic current flow is driven by a conduction current of that electric field. Subsequently, there is not net electric current, because there is no support upon the coseismic field outside the wave [12, 13, 18]. However, the instant a seismic wave that crosses a boundary with a contrast in mechanical and/or electrical properties, it produces an electromagnetic signal referred to as the "interface response field" as shown in Fig. 5 (right).

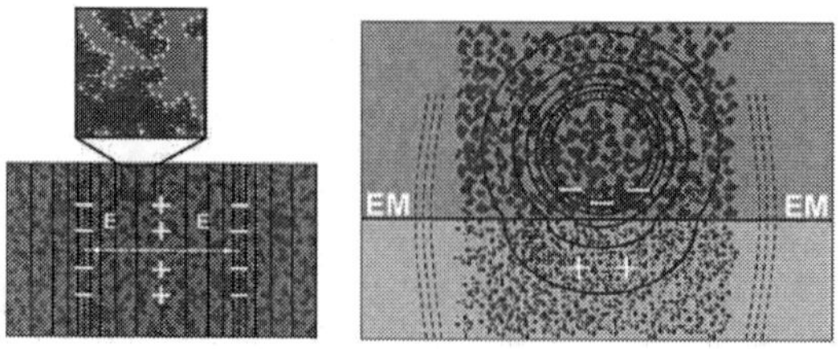


**Figure 2.** Seismoelectric effects: creation of the coseismic electric field (left). The interface response (electromagnetic) field is produced (right) [18]

### 2.3 Seismoelectric coupling: numerical and experimental studies

Zhu, et al. [28] recorded conversion of seismic wave to electromagnetic wave in the laboratory borehole model at ultrasonic frequency, and showed that the seismic wave induced localized seismoelectric field and seismomagnetic field within a porous medium only. That verifies that the induced fields can only exist within the area disturbed by the travelling wave. Additionally, the they confirmed the sensitivity of the seismoelectric conversion to salinity of the pore fluid, as the amplitude of the converted electromagnetic signals decreased with increasing brine salinity [25].

Zhang, et al. [25] also point out that there is a relation between the seismoelectric coupling coefficient and rock properties. They show that the coupling coefficient increases when porosity and permeability increase too. Furthermore, Zhu, et al. [26] show that the converted electrical signal are mostly sensitive to varying permeability at a constant porosity of the formation and same acoustic source. Additionally, Warden, et al. [23] show that the seismoelectric signals are directly proportional to the pressure gradient, so the higher production rate (the higher pressure gradient) yields the more distinct seismoelectric signals.

### 2.4 Frequency-dependent seismoelectric signals in fully-saturated porous media

Zhu, et al. [29] performed a series of seismoelectric measurements in different rock samples submerged in a water tank. The recorded converted seismoelectric signals were triggered by a power hydrophone, used as an acoustic source. The source was excited with a single sine burst pulse of 100 kHz. The other hydrophone was placed near the rock samples to record acoustic field around the rock sample. Additionally, they employed a pair of electrodes to record voltage across the sample induced by the electric current due to the seismoelectric conversions. Consequently, they were able to record both acoustic disturbances and seismoelectric signals, where the conversions occurred only in the porous rock samples.

Afterwards, in order to validate that the electric signals were generated by the seismic wave exactly in the rock samples, Zhu, et al. [29] further proceeded by conducting measurements specifically over the Berea sandstone alone. Likewise, they recorded electric signals induced by the seismoelectric coupling in the frequency range 15 kHz – 150 kHz. As they compared the measured seismoelectric responses with theoretical calculations, they concluded that the seismoelectric coupling decreases when seismic frequency increases.

## 3 Methodology

In our study, we extended the numerical study presented by Zhu, et al. [29], where we performed calculation of the theoretical seismoelectric coupling coefficient designed for partially-saturated porous medium. Hence, we adopted the analytical coupling coefficient formula developed by Reppert, et al. [15] shown in Eq. 3 and Eq. 4. Moreover, in order to satisfy partially-saturated conditions of the variables, we also introduced additional analytical formulations developed by other researchers. As such, the density and viscosity of the fluid displayed by Eq. 5 and Eq. 6, respectively, are presented by Revil, et al. [16]. Furthermore, the formula of the dielectric constant for two-phased fluid mixture was adopted from Fabbri, et al. [6].

$$L = \frac{\Delta V(\omega)}{\Delta P(\omega)} = \frac{a}{2\varepsilon_m \zeta \kappa} \frac{\left( \frac{2}{\kappa a} \frac{J_1(\kappa a)}{J_0(\kappa a)} - 1 \right)}{\left( \frac{J_1(\kappa a)}{J_0(\kappa a)} \right)} \quad (3)$$

where $$\kappa = \sqrt{\frac{-i\rho_f \omega}{\eta_f}} \quad (4)$$

$$\rho_f = S_w \rho_w + (1 - S_w)\rho_g \quad (5)$$

$$\eta_f = \eta_g \left( \frac{\eta_w}{\eta_g} \right)^{S_w} \quad (6)$$

$$\varepsilon_m = (S_w \varepsilon_w^\alpha + (1 - S_w)\varepsilon_g^\alpha)^{1/\alpha} \quad (7)$$

where $\Delta V(\omega)$ and $\Delta P(\omega)$ are voltage and seismic pressure at angular frequency $\omega$; $J_1$ and $J_0$ are Bessel functions of first and zero orders respectively; $\zeta$ is the zeta-potential; $\varepsilon_m$ is the dielectric constant of fluid mixture; $S_w$ is water saturation; $\eta$ and $\rho$ are viscosity and density of a fluid mixture respectively (w – water, water; g – gas, air).

In our calculations, we assumed constant value for pore radius of 80 μm. Zeta-potential of 48 mV was calculated using Eq. 2 at a constant salinity with an assumption that the EDL is not affected by changing water saturation. Values for other remaining variables were obtained from engineering tables at the temperature of 25 $^0$C. Furthermore, we performed calculations of the seismoelectric coupling at various water saturation values within the frequency range of 10 kHz to 150 kHz.

## 4 Results and Discussion

The seismoelectric coupling was calculated for a two-phased saturated porous medium. The phases considered are water and air. The calculations were performed at the range of seismic frequency and water saturation – 10 kHz to 150 kHz and 10 % to 100 %, respectively. The normalized coupling coefficients against normalized frequencies are shown in Fig. 3. The coupling coefficient was calculated for every water saturation value at the given frequency range. Subsequently, we normalized the values of each coupling coefficient with respect to the largest coupling coefficient, which was obtained at the frequency of 10 kHz for each water saturation. Furthermore, each frequency value was then normalized by the highest frequency, i.e. 150 kHz. The results show that the normalized curves at various water saturations coalesce into a single curve, indicating that the relative strength of the coupling coefficient is the same at the given frequency regardless of water saturation. Apparently, Fig. 3 illustrates the frequency-dependence

 DOI: 10.1051/matecconf/20178702001

of the coupling coefficient, where the normalized curves at each water saturation values decreases exponentially when normalized frequency increases. The decline trend of the coupling can be divided into two regions depending on the normalized frequency. When the normalized frequency $\omega_n$ is low, i.e. less than about 0.4 (60 kHz), the coupling coefficient decreases fast. On the other hand, when the normalized frequency $\omega_n$ is high, i.e. greater than about 0.4, the coupling coefficient decreases slowly.

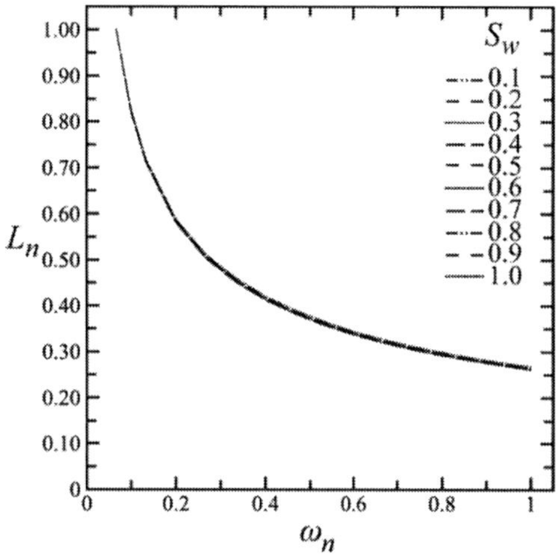


**Figure 3.** Normalized seismoelectric coupling ($L_n$) as a function of normalized seismic frequency ($\omega_n$)

Apparently, at high frequency, the seismoelectric conversion is less efficient due to interference and nonlinearities and as such, the coupling coefficient is low at higher frequencies. Figure 4 shows variation of the normalized coupling coefficient with water saturation at various seismic frequencies. As it can be observed from the figure, the frequency-dependence of the coupling appears to be significant in the lower frequency range, i.e. less than 60 kHz. However, frequencies greater than 60 kHz does not affect the coupling coefficient that much. Moreover, Fig. 3 and Fig. 4 imply that the seismoelectric coupling curves would exhibit identical trends, irrespective of water saturation.

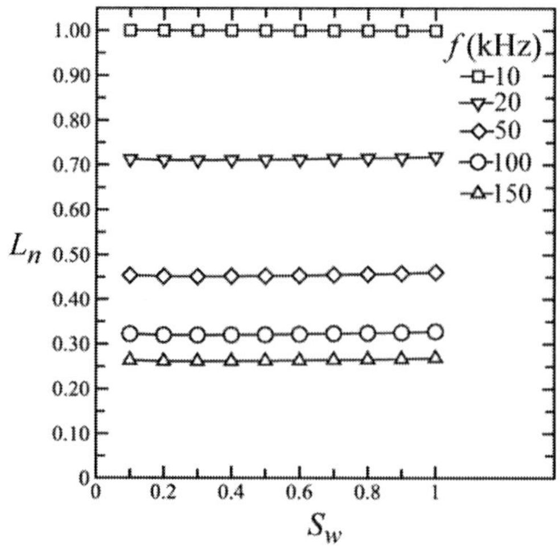


**Figure 4.** Normalized seismoelectric ($L_n$) as a function of water saturation ($S_w$)

Figure 5 shows effect of water saturation on the seismoelectric coupling at various frequencies. Brine concentration was fixed at 0.01 mol/l. When water saturation is 0.1, the magnitude of the coupling coefficient reduces drastically by 380 %, i.e. from 0.19 nV/Pa to 0.05 nV/Pa, when frequency increases from 10 kHz to 150 kHz. Although subsequent reduction in the coupling coefficient is seen when water saturation is fixed and frequency increases, the decline rate lessens when water saturation increases from 0.1 to 0.6. When water saturation is greater than 0.6, the rate of reduction of the coupling coefficient remains almost constant as frequency increases at fixed water saturation suggesting that frequency has negligible effect on the coupling coefficient at high water saturation.

In a porous medium saturated with water and gas, the seismoelectric coupling is rather enhanced at low water saturations. Our results obtained from investigating the effect of water saturation on the coupling coefficient show consistence with experimental and theoretical findings obtained by Bordes, et al. [3]. Additionally, our results are in full agreement with findings of Zhu, et al. [29], where they show frequency-dependence of the coupling in fluid-saturated porous media.

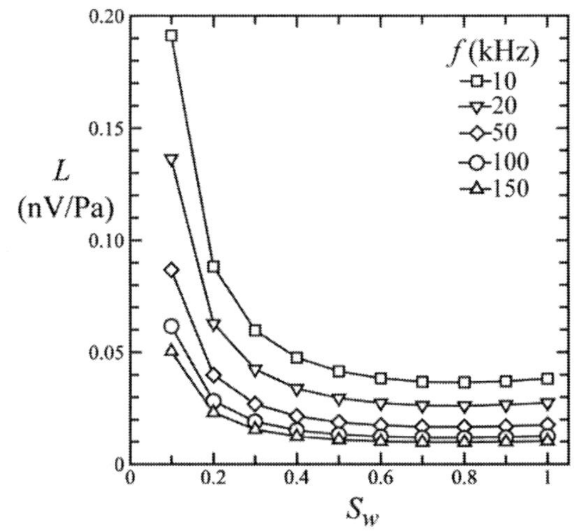


**Figure 5.** Seismoelectric coupling as a function of water saturation ($S_w$) and frequency ($f$)

## 5 Conclusion

In this paper, we extended the calculation of theoretical seismoelectric coupling coefficient to partially-saturated conditions, whereby the ratio was a function of seismic frequency and water saturation. Our results suggest that regardless of fully- or partially-water saturated conditions, the seismoelectric coupling coefficient reduces when frequency increases. At lower water saturation, i.e. less than 0.6, the magnitude of the coupling coefficient significantly drops when frequency increases. On the contrary, the magnitude of the coupling coefficient remains rather constant when water saturation increases at constant frequency. When water saturation is greater than about 0.6, regardless of frequency the rate of decline of the coupling coefficient remains almost constant. The results observed in our study are consistent with experimental and numerical findings of other researchers.

## Acknowledgements

The authors express their gratitude to UTP for providing financial support under cost-centre No. 0153AB-DA5. We thank the editor and the referees for valuable comments and their time.

## References

1. M.A. Biot. "Theory of propagation of elastic waves in a fluid-saturated porous solid. II. Higher frequency range," the Journal of the Acoustical Society of America Vol. 28, No. 2, (1956), pp. 179-191.
2. A. Boleve, A. Crespy, A. Revil, F. Janod, and J.-L. Mattiuzzo. "Streaming potentials of granular media: Influence of the Dukhin and Reynolds numbers," Journal of Geophysical Research: Solid Earth Vol. 112, No. B8, (2007).
3. C. Bordes, P. Sénéchal, J. Barrière, D. Brito, E. Normandin, and D. Jougnot. "Impact of water saturation on seismoelectric transfer functions: a laboratory study of coseismic phenomenon," Geophysical Journal International Vol. 200, No. 3, (2015), pp. 1317-1335.
4. J.C. Dupuis, K.E. Butler, and A.W. Kepic. "Seismoelectric imaging of the vadose zone of a sand aquifer," Geophysics Vol. 72, No. 6, (2007), pp. A81-A85.
5. J.C. Dupuis, K.E. Butler, A.W. Kepic, and B.D. Harris. "Anatomy of a seismoelectric conversion: Measurements and conceptual modeling in boreholes penetrating a sandy aquifer," Journal of Geophysical Research Vol. 114, No. B10, (2009).
doi: 10.1029/2008jb005939
6. A. Fabbri, T. Fen-Chong, and O. Coussy. "Dielectric capacity, liquid water content, and pore structure of thawing–freezing materials," Cold Regions Science and Technology Vol. 44, No. 1, (2006), pp. 52-66.
7. J. Frenkel. "On the theory of seismic and seismoelectric phenomena in a moist soil," J. Phys. J. Phys. (Soviet) Vol. 8, No. 4, (1944), pp. 230-241.
8. N. Grobbe, J. Hunziker, and E. Slob. "Seismoelectric wave propagation modeling for typical laboratory configurations: A numerical validation," (2014), pp. 2072-2077. doi: 10.1190/segam2014-0689.1
9. S.S. Haines, S.R. Pride, S.L. Klemperer, and B. Biondi. "Seismoelectric imaging of shallow targets," Geophysics Vol. 72, No. 2, (2007), pp. G9-G20.
10. M. Jaafar, J. Vinogradov, and M. Jackson. "Measurement of streaming potential coupling coefficient in sandstones saturated with high salinity NaCl brine," Geophysical Research Letters Vol. 36, No. 21, (2009).
11. S.R. Pride. "Governing equations for the coupled electromagnetics and acoustics of porous media," Physical Review B Vol. 50, No. 21, (1994), p. 15678.
12. S.R. Pride, and S. Garambois. "The role of Biot slow waves in electroseismic wave phenomena," The Journal of the Acoustical Society of America Vol. 111, No. 2, (2002), pp. 697-706.
13. S.R. Pride, and M.W. Haartsen. "Electroseismic wave properties," The Journal of the Acoustical Society of America Vol. 100, No. 3, (1996), pp. 1301-1315.
14. S.R. Pride, and F. Morgan. "Electrokinetic dissipation induced by seismic waves," Geophysics Vol. 56, No. 7, (1991), pp. 914-925.
15. P.M. Reppert, and F.D. Morgan. "Frequency-dependent electroosmosis," Journal of Colloid and interface Science Vol. 254, No. 2, (2002), pp. 372-383.
16. A. Revil, G. Barnier, M. Karaoulis, P. Sava, A. Jardani, and B. Kulessa. "Seismoelectric coupling in unsaturated porous media: theory, petrophysics, and saturation front localization using an electroacoustic approach," Geophysical Journal International, (2013), p. ggt440.
17. A. Revil, P. Pezard, and P. Glover. "Streaming potential in porous media: 1. Theory of the zeta

potential," Journal of Geophysical Research: Solid Earth Vol. 104, No. B9, (1999), pp. 20021-20031.
18. M.D. Schakel. Coupled seismic and electromagnetic wave propagation: TU Delft, Delft University of Technology, (2011).
19. M.D. Schakel, Z. Zhu, and E.C. Slob. "Full-waveform model and measurements of electromagnetic to seismic conversion," 2011 SEG Annual Meeting. Society of Exploration Geophysicists, (2011).
20. J.J. Sheng. "Foams and Their Applications in Enhanced Oil Recovery," Enhanced Oil Recovery Field Cases Elsevier, (2013), pp. 251-276.
21. R. Thompson. "The seismic electric effect," Geophysics Vol. 1, No. 3, (1936), pp. 327-335.
22. J. Vinogradov, M. Jaafar, and M. Jackson. "Measurement of streaming potential coupling coefficient in sandstones saturated with natural and artificial brines at high salinity," Journal of Geophysical Research: Solid Earth Vol. 115, No. B12, (2010).
23. S. Warden, S. Garambois, L. Jouniaux, D. Brito, P. Sailhac, and C. Bordes. "Seismoelectric wave propagation numerical modelling in partially saturated materials," Geophysical Journal International Vol. 194, No. 3, (2013), pp. 1498-1513.
doi: 10.1093/gji/ggt198
24. X. Zhan, Z. Zhu, and M.N. Toksoz. "Quantitative DC And High Frequency AC Seismoelectric Measurement On Berea Sandstone," 2011 SEG Annual Meeting. Society of Exploration Geophysicists, (2011).
25. Y. Zhang, and J. Zhang. "An Experimental Study of Seismoelectric Conversion in Model Borehole," 2003 SEG Annual Meeting. Society of Exploration Geophysicists, (2003).
26. Z. Zhu, S. Chi, X. Zhan, and M.N. Toksöz. "Theoretical and experimental studies of seismoelectric conversions in boreholes," Commun. Comput. Phys Vol. 3, No. 1, (2008), pp. 109-120.
27. Z. Zhu, and M.N. Toksöz. "Experimental Studies Of Electrokinetic Conversions In A Fluid-Saturated Porous Medium." Massachusetts Institute of Technology. Earth Resources Laboratory, (1996).
28. Z. Zhu, and M.N. Toksöz. "Crosshole seismoelectric measurements in borehole models with fractures," The 7th International Symposium on Recent Advances in Exploration Geophysics (RAEG 2003). (2003).
29. Z. Zhu, M.N. Toksöz, and D.R. Burns. "Electroseismic and seismoelectric measurements of rock samples in a water tank," Geophysics Vol. 73, No. 5, (2008), pp. E153-E164.

MATEC Web of Conferences 87, 02002 (2017) DOI: 10.1051/matecconf/20178702002
*ENCON 2016*

# Effects of Cigarettes Smoking on Pulmonary Function among University Students

Azian Hariri[1,a] and Wan Mohd Mazdey Wan Mansor[1]

[1] *Industrial and Indoor Environment Research Group (IIERG)*
*Centre for Energy and Industrial Environment Studies (CEIES), Faculty of Mechanical and Manufacturing Engineering, Universiti Tun Hussein Onn Malaysia, 86400, Parit Raja, Batu Pahat, Johor, Malaysia*

**Abstract.** Pulmonary function testing is a physiological test that measures how an individual inhales or exhales volumes of air as a function of time. Smoking is greatly associated with reduction of pulmonary function. This research is aimed to estimate the values of peak expiratory flow (PEF), forced expiratory volume in first second (FEV1), forced vital capacity (FVC) and ratio between FEV1/FVC among smoking and non-smoking students in Universiti Tun Hussein Onn Malaysia. Smoking is often related to obstructive disorder with low value of FVC, FEV1 and FEV1/FVC. These pulmonary functions were analyzed based on several variables such as; the number of cigarette smoked per day, duration of smoking, age, and body mass index (BMI) values. 70 healthy volunteers consist of smoking and non- smoking students was selected through several sessions. Students were interviewed to answer questionnaire on demographic, lifestyles and their smoking habit. The pulmonary function tests were conducted according to American Thoracic Society (ATS) standards. The results of the pulmonary functions were analyzed by using SPSS software to compare the pulmonary functions between the smoker and the non-smoker students. The results of the studies showed that the number of cigarettes smoked by respondent and the BMI values were the significant predictors of the decrease in FEV1/FVC values among university students

## 1 Introduction

The use of tobacco by adolescents remains a major public health concern worldwide. There are 1.2 billion smokers globally, of which more than 50% are young people. There are nearly 600 million tobacco users within the Southeast Asian countries, with the region contributing to about 50% of the global burden of tobacco users [1]. It is estimated that tobacco-related deaths will be the most important cause of deaths in developing countries by 2020 [2]. According to Global Adult Tobacco Survey 2011, 32.5% of the respondents in Malaysia where tobacco smoker with the age between 15-24 years [3].

It is estimated that tobacco related death will become 1 billion in the 21st century compared to 100 million in the 20th century. It is also estimated 1 person die every 6 seconds from a tobacco related diseases [4]. Cigarette smoking is the major cause of many chronic diseases, such as stroke, heart disease, chronic obstructive pulmonary disease (COPD), periodontal disease, peripheral vascular disease, pneumonia, lung and oral cancer [2], [5]. Lung is a directly affected organ by cigarette smoking. Various respiratory diseases including lung cancer, chronic obstructive pulmonary disease, interstitial lung diseases, bronchial asthma, are caused and worsen by cigarette smoking [6].

The degradation of lung function and the severity of respiratory disease can be identified by pulmonary function tests (PFTs) [7–9]. PFTs is a tools to measure the volumes of air during individual inhalation or exhalation process. The main parameters measured by PFT are the peak expiratory flow (PEF), forced expiratory volume in first second (FEV1), forced vital capacity (FVC) and ratio between FEV1/FVC.

Study conducted by Bano et. al in rural area of India concluded that smoker had 17.3 times more risk of having impaired pulmonary function as compared to non-smoker and obstructive pulmonary impairment was the commonest in smoker[10- 11]. Study by Boskabady et. al in Iran also concluded smoking leads to increased respiratory symptom (cough, wheezing and tightness) and reduction of PFTs value [12]. Study done among Thailand youth between 15 to 18 years old lead to the conclusion of FVC value of nonsmoker were significantly greater than smoker. It is also highlighted that early effect of cigarette in youth can lead to problem with respiratory system [13].

There are limited study done to assess the pulmonary functions of young people especially that relate to young tobacco smoker mainly because the effects of smoking is believe to be significant for longer duration of tobacco smoking years, and may insignificant in young people with shorter duration of tobacco smoking years. However, in the last five decades, the anthropometric data of young adults

[a] Corresponding author: azian@uthm.edu.my

DOI: 10.1051/matecconf/20178702002

have changed significantly owing to factors such as increasing body height, increasing body mass index, increasing air pollution level and decreased in physical activities [7]. Therefore, new pulmonary function data regarding young people, particularly young smokers and non-smokers are needed.

## 2 Methodology

### 2.1 Questionnaire Interview

The questionnaire interview questions were based on the following topics:

i. Demographic and health status (age, race, height, weight, previous pulmonary illness, asthma)
ii. Lifestyles (physical activities)
iii. Smoking habit (duration of smoking, the number of cigarette per day)

The questionnaire interviews were made to collect required variables that may correlate significantly with the pulmonary function results. It also conducted as screening tools to select healthy young university students that suitable to participate in this study. Students who exercised regularly/athletic or who had a history of or currently had obstructive or restrictive types of respiratory diseases and were taking treatment were excluded from this study. Student less than 19 years old or exceeding 29 years old or with excessive body mass index (BMI) was excluded. Only male students were asked to participate in this research to avoid inaccurate results due to limited sample of female student smokers. 70 students were divided into 2 main group of; smoker and non-smoker with the same sample size.

### 2.2 Pulmonary Function Tests (PFTs)

There are three distinct phases to the spirometry maneuver, as follows:

i. Maximal inspiration
ii. A "blast" of exhalation
iii. Continued complete exhalation to the end of test

The subject should inhale rapidly and completely from functional residual capacity, and the breathing tube should be inserted into the subject's mouth, making sure the lips are sealed around the mouthpiece and that the tongue does not occlude it. It is important that the preceding inspiration is fast and any pause at full inspiration be minimal (i.e., only for 1 - 2 seconds). The test assumes a full inhalation before beginning the forced exhalation, and it is imperative that the subject takes a complete inhalation before beginning the maneuver [7]. Table 1 shows the pulmonary function parameters investigated in this study with definition. Each of these parameters was compared with the predicted value data from the spirometry software database (CareFusion Version 1.08) based on sex, age and height of the welders. The predicted value data were based on the healthy nonsmoking reference data taken during Pneumobile Project, Indonesia [14]. The results of pulmonary function were calculated in percentage predicted (% predicted) as the following;

$$\%\,predicted = \frac{value\ recorded}{predicted\ value} \times 100 \qquad (1)$$

**Table 1** The definitions of forced expiratory (FEV1), peak expiratory flow (PEF) and forced vital capacity (FVC) [7].

| Parameter | Definitions |
|---|---|
| Forced Vital Capacity (FVC) | FVC is the total maximum exhaled of individuals, which is the ability maximum force exhaled, expressed in litres / sec. |
| Forced Expiratory (FEV1) | The volume delivered in the first second of an FVC manoeuvre |
| Peak Expiratory Flow (PEF) | The maximum expiratory flow achieved from a maximum forced expiration, starting without hesitation from full inspiration |

Figure 1 shows the acceptability and repeatability criteria that need to be fulfilled for pulmonary function test. Flow should be measured from a series of at least three force expiratory curves that have an acceptable start of the test. The spirometer device needs to be calibrated daily with a 3-L volume calibrator syringe (Kahntact 3L Volume Calibration Syringe Type 5570) for low, medium, and high flow for accurate results.

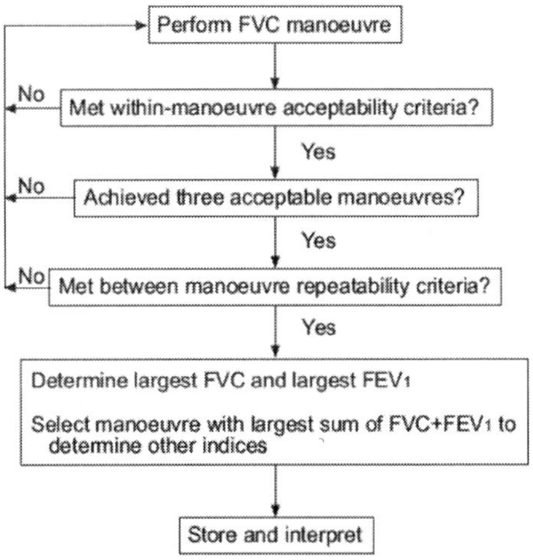


**Figure 1** Acceptability and repeatability criteria for spirometric test [7]

### 2.3 Statistical Analysis

Statistical analysis was conducted to correlate and recognize the significant variables for the pulmonary functions. Statistical analysis was conducted by the SPSS software version 18 (SPSS Inc., Chicago). T-test analysis was used to compare mean pulmonary function parameters between smoker and non-smoker. Kendall's Tau-b correlation analysis was done to get association between smoking duration and smoking status (smoker or non-smoker) with pulmonary function parameters. Further analysis using multiple regressions were conducted to confirm the predictors of the pulmonary functions. The level of significance is taken as $p < 0.005$.

## 3 Results and Discussion

The questionnaire is divided into four parts, A, B, C and D. Part A is a question about the demographics of the respondents, while part B is a question about their smoking activity information. In section C are questions about their health status information, and the next part D is about the status of their lifestyle information. A total of 70 respondents (35 smoker and 35 non-smoker) were taken into consideration in this study.

Data of the lung function tests (Mean + standard deviation (SD)) for 70 students are shown in Table 2. The mean age is 24.49 years for overall respondents with SD of 1.78.

**Table 2** The mean age of the respondents and the results of pulmonary function tests.

| Criteria | N=70 (Mean ± SD) |
|---|---|
| Age (Years) | 24.49 ± 1.78 |
| FEV1 (% predicted) | 81.23 ± 8.56 |
| FVC (% predicted) | 79.09 ± 10.47 |
| PEF (% predicted) | 79.83 ± 9.01 |
| FEV1/FVC | 105.17 ± 6.99 |

Table 3 shows the T-test analysis to compare the mean value of investigated pulmonary functions parameters of 35 smoking students and 35 non-smoking students. All investigated pulmonary functions parameters of FEV1, FVC, PEF and FEV1/FVC mean values were higher for the non-smoker students compare to the smoker students as expected. In addition, there is a significant correlation between FEV1 and FVC with smoking and non-smoking status among students as shown in Table 3. The finding on FVC value were on the same agreement with Tantisuwat et. al [13] and Nawafleh et.al [15].

**Table 3** T-test results for different group

| Parameters | Group | N | Mean | SD | SE | Sig (2-tailed) |
|---|---|---|---|---|---|---|
| FEV1 | Smoker | 35 | 78.08 | 7.57 | 1.28 | .002* |
| | Non-smoker | 35 | 84.37 | 8.43 | 1.43 | |
| FVC | Smoker | 35 | 75.97 | 8.98 | 1.52 | .012* |
| | Non-smoker | 35 | 82.20 | 11.03 | 1.86 | |
| PEF | Smoker | 35 | 77.94 | 8.05 | 1.36 | .080 |
| | Non-smoker | 35 | 81.71 | 9.63 | 1.63 | |
| FEV1/FVC | Smoker | 35 | 104.4 | 6.53 | 1.10 | .360 |
| | Non-smoker | 35 | 105.9 | 7.43 | 1.26 | |

*p < .005

Table 4 shows Kendall's Tau-b correlation analysis to determine the correlation between lung function parameters with the number of cigarettes smoked per day, duration of smoking and BMI values. There was a significant correlation between FVC and the number of cigarettes smoked per day (r = -.256, two tailed p <.001). In addition, there is also a significant correlation between FEV1 with the number of cigarettes smoked per day (r = -.288, two tailed p <.001).

There were also significant correlation between duration of smoking with FVC values (r = -.125, two tailed p <.001) and FEV1 (r = -.151, two tailed p <.001). A significant correlation was also found between the ratio of FEV1 /FVC with BMI values (r = -.137, two tailed p <.001). Results of the study showed a decrease in lung function of FVC and FEV1 values was correlated with the increased number of cigarettes smoked per day and duration of smoking among students while increase BMI values correlated with a decrease in FEV1 / FVC. These results clearly shows the obstructive lung symptoms were detected in smoker group of students which significantly correlates with the number of cigarettes smoked per day, duration of smoking and BMI values.

**Table 4** The Kendall's tau-b correlation between lung functions parameters with the number of cigarettes smoked, duration of smoking and BMI.

| Types of questions | FVC | FEV1 | PEF | FEV1/ FVC |
|---|---|---|---|---|
| Number of cigarettes smoked per day | -.256** | -.288** | .051 | .116 |
| Duration of smoking | -.125** | -.151** | .072 | .058 |
| Body mass index *(BMI)* | .053 | .066 | .030 | -.137** |

***p< .001*

Table 5 shows the multiple regression analysis for the FEV1. The multiple regression analysis was performed to determine the significant predictor for the decrease in lung function. The multiple regression analysis showed that only the number of cigarettes smoked per day is the significant predictor for the decrease value in FEV1. Duration of smoking and BMI did not show a significant predictor for the decrease in FEV1.

**Table 5** Analysis multiple regression for FEV1

| Types of questions | B | SE | β |
|---|---|---|---|
| **Step 1** | | | |
| Constant | 89.303 | 6.963 | |
| Number of cigarettes smoked per day | -.5.401 | 1.743 | -.644** |
| Duration of smoking | 1.07 | .571 | .391 |
| Body mass index, BMI | -.262 | .289 | .103 |
| **Step 2** | | | |
| Constant | 83.090 | 1.296 | |
| Number of cigarettes smoked per day | -5.352 | 1.74 | -.638** |
| Duration of smoking | 1.027 | .568 | .375 |
| **Step 3** | | | |
| Constant | 83.593 | 1.287 | |
| Number of cigarettes smoked per day | -2.713 | .963 | -.323** |

Dependent variable: FEV1, **p<.001

Table 6 shows results of the multiple regression analysis for the FVC. Results of the analysis showed only the number of cigarettes smoked per day is the significant predictor for the decrease values in FVC. Duration of smoking and BMI did not showed any significant predictor for the decrease in the FVC values.

**Table 6** Analysis multiple regression for FVC.

| Types of questions | B | SE | β |
|---|---|---|---|
| **Step 1** | | | |
| Constant | 78.672 | 8.441 | |
| Number of cigarettes smoked | -7.501 | 2.113 | -.751** |
| Duration of smoking | 1.685 | .692 | .503 |
| Body mass index, BMI | .098 | .350 | .032 |
| **Step 2** | | | |
| Constant | 80.996 | 1.562 | |
| Number of cigarettes smoked per day | -7.519 | 2.097 | -.733** |
| Duration of smoking | 1.701 | .685 | .508 |
| **Step 3** | | | |
| Constant | 81.830 | 1.583 | |
| Number of cigarettes smoked per day | -3.149 | 1.184 | -.307** |

Dependent variable: FVC, **p<.001

Table 7 shows the results of multiple regressions analysis for the ratio of FEV1 / FVC. The multiple regression analysis showed that only BMI is a significant predictor for the decrease in the value of FEV1 / FVC. Duration of smoking and number of cigarettes smoked per day did not show any significant predictor for the decrease in the FEV1 / FVC values.

**Table 7** Analysis multiple regression for FEV1/FVC.

| Types of questions | B | SE | β |
|---|---|---|---|
| **Step 1** | | | |
| Constant | 116.734 | 6.013 | |
| Number of cigarettes smoked per day | .384 | 1.505 | .076 |
| Duration of smoking | -.075 | .493 | .034 |
| Body mass index, BMI | -.487 | .249 | -.235** |
| **Step 2** | | | |
| Constant | 105.180 | 1.144 | |
| Number of cigarettes smoked per day | .476 | 1.536 | .069 |
| Duration of smoking | - .155 | .501 | .069 |
| **Step 3** | | | |
| Constant | 105.104 | 1.11 | |
| Number of cigarettes smoked per day | .077 | .830 | .011 |

Dependent variable: FEV1/FVC, **p <.001

The significant relation between BMI and lung function parameters were also had been highlighted by Nawafleh et. al in study conducted in Jordan [15].

The results of pulmonary functions of investigated students cannot be compared directly with previous study from other researcher mainly due to the predicted set used were differs from one study to another. There were also difference existed in age and sex of the investigated subjects which had limit the comparison to be done between this study and previous study by other researcher

## 4 Conclusion

Investigation of pulmonary function among university students shows that the pulmonary function value of smokers was decreased compared to non-smokers. In addition, a significant correlation was found between the number of cigarettes smoked per day and duration of smoking with the decrease value of FVC and FEV1 of the smoking students. Besides that, the increase BMI values had shown a significant correlation with the decrease value of FEV1/FVC.

The results of the multiple regression analysis showed that the significant predictor of decrease in FEV1 and FVC values was the number of cigarettes smoked per day by respondent, while the decrease in value of FEV1/FVC was the influence of the BMI values.

In conclusion, smoking students had the lower pulmonary function values compared to non-smoking students.

## Acknowledgement

This work was supported by the Fundamental Research Grant Scheme (FRGS) vote 1539 under the Malaysian Ministry of Higher Education (MOHE). The authors acknowledged Office for Research, Innovation, Commercialization and Consultancy Management (ORICC), Universiti Tun Hussein Onn Malaysia for the opportunities given. The authors gratefully acknowledged the technical support from Mr. Mohd Azizi bin Mohd Afandi during the operation of equipments used in this study.

## References

1 N. Al-Sadat, a Y. Misau, Z. Zarihah, D. Maznah, and Tin Tin Su, *Asia-Pacific journal of public health / Asia-Pacific Academic Consortium for Public Health*, **22**, p. 175S–180S, (2010).

2 R. A. Al-Naggar, S. A. R. Al-Dubai, T. Hamoud, R. Chen, and K. Al-Jashamy, *Asian Pacific J Cancer Prev*, **12**, pp. 619–624, (2011).

3 Malaysia Ministry of Health, *Report of the Global Adult Tobacco Survey (GATS) Malaysia 2011*, (2011).

4 Action on Smoking and Health, *Tobacco Statistics and Facts*, (2015).

5 D. B. Abrams, *The 50th Anniversary SG Report*, (2014).

6 Y. Ishii, *Nihon Rinsho*, pp. 416–420, (2013).

7 A. Bandyopadhyay, I. Bhattacharjee, R. Dalui, and S. Pal, *Malaysian Journal of Medical Sciences*, **20**, pp. 17–25, (2013).

8 J. Macleod, R. Robertson, L. Copeland, J. Mckenzie, R. Elton, and P. Reid, *British Journal of General Practice* pp. 89–95, (2015).

9 A. Bandyopadhyay, *Indian J Med Res*, **134**, pp. 653–657, (2011).

10 R. Bano, M. Am, K. Nb, N. Ahmad, and S. Nighute, *World Health*, **4**, no. 1, (2009).

11 R. Bano, N. Ahmad, and A. M. Mahagaonkar, *Walawalkar International Medical Journal*, no 1, pp. 33–38, (2014).

12 M. H. Boskabady, M. Mahmoodinia, M. Boskabady, and G. R. Heydari, *Revista Portuguesa de Pneumologia*, **17**, no. 5, pp. 199–204, (2011).

13 A. Tantisuwat and P. Thaveeratitham, *Journal of physical therapy science*, **26**, no. 2, pp. 167–70, (2014).

14 Ingelheim Boehringer, *Pneumobile Project, Indonesia.*,(1992).

15 H. A. Nawafleh and S. A.-S. A. Z. D. F. Al-Maghaireh, *Health Science Journal*, vol. **6**, no. 2, pp. 773–783, (2012).

DOI: 10.1051/matecconf/20178702003

# Energy Consumption Analysis Based on Energy Efficiency Approach: A Case of Suburban Area

K.A Rahman[1,a], A.M Leman[2], M. Faris Mubin[1], M.Z.M Yusof[1], Azian Hariri[1], and M.N.M Salleh[3]

[1]*Faculty of Mechanical & Manufacturing Engineering, Universiti Tun Hussein Onn Malaysia, 86400 Batu Pahat, Johor, Malaysia*
[2]*Faculty of Engineering Technology, Universiti Tun Hussein Onn Malaysia, 86400 Batu Pahat, Johor, Malaysia*
[3]*Faculty of Computer Science & Information Technology, Universiti Tun Hussein Onn Malaysia, 86400 Batu Pahat, Johor, Malaysia*

**Abstract.** Sufficient data about electricity consumption over large periods of time was accumulated and analysed in order to develop appropriate electricity-saving measures. An important first step was to analyse and identify electrical appliances that had energy saving potential. Different behavioural consumption profiles were analysed using information from two sources: 1) technical data about electricity consumption (electricity bill) and 2) data about household electrical appliance usage and consumer awareness obtained from a questionnaire survey. The questionnaire consisted of four sections of questions which concerned residents' backgrounds, residential information, consumers' awareness about energy efficiency (EE) and the energy usage of each house. The results showed that the electricity consumption profile of an individual household could be most related to the EE approach. It indicated that consumers' behaviour (awareness and practice) resulted in one of the important factors related to high electricity consumption. Another important conclusion of our questionnaire-based analysis was that the implementation of the EE approach could help the consumer to reduce their monthly energy consumption.

## 1 Introduction

Energy consumption in the residential sector represents an important part of total electricity demand. Most consumers are not alert about how to use energy efficiently in their daily lives. In other words, consumers use electric energy inefficiently because they use inefficient electrical appliances. Saving electricity requires either conserving or improving efficiency. Conserving means performing fewer activities that use electricity. Improving energy efficiency (EE) is about choosing the right appliances which use less energy for the same tasks. From the questionnaire, it can be seen that consumers did not understand the high power consumption that occurs via their electrical appliances [1]. Based on a 2013 report by the Energy Commission of Malaysia, 20.6% of electricity usage was by domestic users [2]. There are about 7.8 million registered electricity consumers with Tenaga Nasional Berhad (TNB), the largest electricity utility in Malaysia. Effective from 1st January 2014, TNB's tariff remained unchanged at 21.8 cents/kWh for the first 200kWh of consumption per month, therefore there are nearly 50.4% (or 3.25 million) of all domestic users who pay RM43.60 or less monthly [3]. On average, domestic customers will experience an increase of 10.6% (3.03 cent/kWh). Figure 1 shows the distribution of electricity users in Peninsular Malaysia and of the total 7.8 million, 82% are domestic users (6.4 million). The rest of the users are thus divided up: 17 % commercial, 0.4% industrial, and 0.6% other types [4].

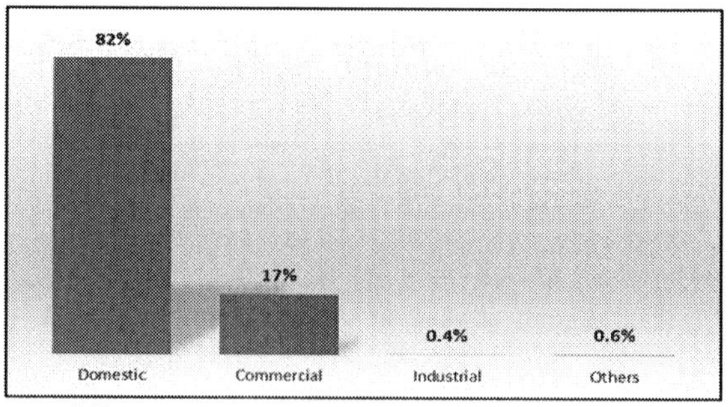


**Figure 1.** Distribution of Electricity Users in Peninsular Malaysia

The increase in the number of residential area development projects has had a great impact on national development, but it has also increased energy demand. By making households more energy efficient, the negative impacts on the environment can be reduced.

The internationally recognized energy star symbol is a simple way for consumers to identify the most energy-efficient electrical products on the market. Devices carrying the energy star service mark, such as computer

[a] Corresponding author: khairunnisa7480@gmail.com

products and peripherals, kitchen appliances, buildings and other products, generally use 20–30% less energy than required by federal standards [5]. In this study, the energy consumption of electrical appliances for a household will be analysed in order to identify the energy consumption pattern for residential area involved. Furthermore, this study aims to identify the relationship between two components of energy awareness: awareness and practice towards the increased energy bill at home. The recommendation and solution will be proposed at the end of this study to help consumers to reduce the energy consumption.

## 2 Energy Efficiency Standards & Labelling (S&L)

EE standards are a set of procedures and regulations that prescribe the energy performance of manufactured products, sometimes prohibiting the sales of products that are less energy efficient than the minimum standards [6]. This actually means an efficiency level set of standards must be met by the appliance manufacturer in order to sell the products. EE standards can be mandatory or voluntary, which means they can be in the form of a minimum allowable EE or a maximum allowable energy use. Prescriptive standards, minimum energy-performance standards (MEPS), and class-average standards are types of energy standards [7]. Energy labels as shown in Figure 2 are informative labels to describe the manufactured product's energy performance in the form of energy use, efficiency, or energy cost. [8]. The labels properly furnish consumers with the knowledge to buy more efficient appliances. The EE label is meant to be issued by the Energy Commission to manufacturers of electrical appliances who comply with the standards and requirements of the energy performance test for a star rating established by the Energy Commission [9].

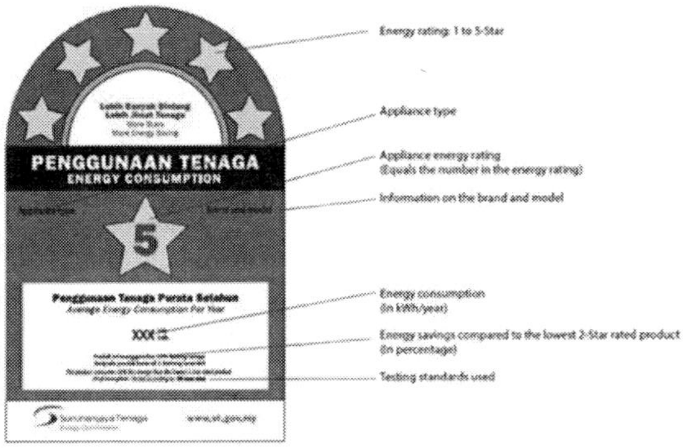


**Figure 2.** Energy Star Rating in Malaysia

EE labels encourage people to buy more efficient products by helping them to understand and compare how much energy is used. Mahlia [10] highlighted that the purpose of introducing energy star labels is to convince consumers to buy EE appliances. At the same time, it is also to influence manufacturers to produce more EE products into the market. A label is a mandatory sticker that is affixed to products or their packaging containing information about the EE or EC of the product. The three types of labels that are currently in use in various countries are endorsement, comparison and a combination of both [11,8]. In Malaysia, with the regulations in place, the five domestic appliances (domestic fans, domestic lamps, air-conditioners, refrigerators, and televisions) will be issued with a Certificate of Approval (COA) by the Energy Commission Malaysia. The star rating ranges from 2-star to 5-star, with 2-star representing minimal efficiency and maximum efficiency represented by the 5-star label [8].

### 2.1 Energy Star Rating

The demand for energy in Malaysia is influenced by many factors such as population and economic growth. Energy consumption usually varies from one sector to another. In the domestic sector, household electrical appliances are one of the biggest energy consumers in Malaysia. Based on the ninth Malaysia Plan, residential and commercial sectors contributed about 12.8% of the total energy demand in 2012 [12], which shows that the authorities are already taking action to reduce energy consumption in this country [12, 13]. Implementing the Energy Star Rating will help consumers to purchase efficient electrical home appliances, which will bring many more advantages to the country, manufacturers, consumers, and environment [14, 15].

The Energy Star Rating is a program that was first developed in 1992 by the US Environmental Protection Agency (EPA) as a method to identify and promote products that are energy efficient. According to TNB [4], the benefits of 5-star appliances include reduced domestic electricity consumption and electricity bills. This will in turn produce increased disposable household income in the long run, which will improve the quality of life. Advantages of using 5-star appliances include:

i. Better savings in the long run.
ii. Reduced carbon footprint from electricity generation.
iii. Improved quality of life with more efficient appliances.

Research and development are still ongoing, as manufacturers strive to deliver low-cost energy efficient products. A major appliance which sports the energy star label is not necessarily a better product than a comparable model, but in order to be rated with an energy star, it must meet very strict energy efficiency guidelines as set out by the program.

Energy guide ratings for appliances will show their typical usage in kWh per year - the lower the energy number, the more cost efficient it is to run. Appliances carrying the Energy Star Rating are typically 10 to 20% more energy efficient than non-rated models [16].

## 3 Methodology

For this study, the level of energy consumption among the households was found by obtaining electricity bills and distributing a questionnaire. Figure 3 shows the flow chart for this study.

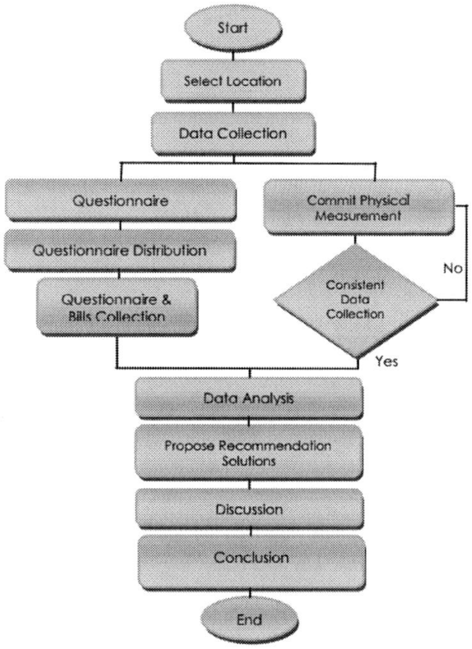


**Figure 3.** Flowchart of the Study

The process started with location selection, followed by data collection, which was divided into two parts – the distribution of a questionnaire and the collection of three months worth of household electricity bills. The questionnaire was developed. Questions developed based on a literature review and verified by experts. After data collection was successfully carried out, the data were analysed using the SPSS (Statistical Package for Social Science). The next step was the discussion and proposal of a solution. The last part of the study was the final discussion and conclusion.

### 3.1. Study Case Location

This study was located in Parit Raja, Batu Pahat, Johor which is a suburban area. The purpose of the study was to analyze the level of energy consumption among the residents in the area. The reason a suburban area was selected for this study was to help the residents to manage their energy consumption by using the EE approach. This application will help them to reduce their monthly electricity bills and also raise the environmental awareness level among consumers.

### 3.2 Questionnaire Survey and Electricity Bills

A questionnaire was distributed to survey the consumers' awareness of EE. Meanwhile, electricity bills were collected to obtain, analyse, compare and classify the total energy consumption of every household studied according to their use of Energy Star Rating appliances. Houses involved were divided into three main groups of energy consumption (kWh). Electricity bills of all houses studied were taken over three continuous months. The Pearson's Correlation and mean score analysis were used to survey consumer awareness of the EE approach.

### 3.3 Electricity for Domestic Consumers

The voltage used in Malaysia is 230/240 volts - 50 Hz. In Malaysia, most households use a single-phase system rather than a three phase one. Households receive their power supply through a two-wire single-phase 240V from three-phase distribution system provides 415V through the connection at neutral wire and one wire from three wires phases. Figure 4 shows the simple flow of electricity in a domestic household.

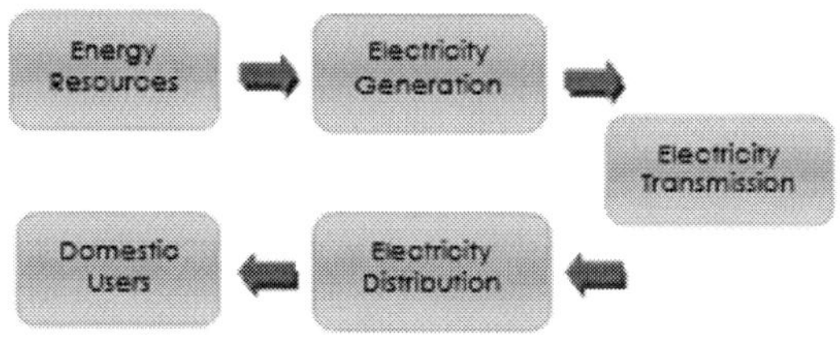


**Figure 4.** Flow of Electricity in a Domestic Household

#### *3.3.1 Electricity Tariff*

The cost of electricity from generation to distribution before reaching the end user will be translated into a tariff. The current tariff for domestic consumers is shown in Table 1 [3]:

**Table 1.** TNB's Electricity Tariff for Domestic Households

| Tariff Category | Unit | Current Rate (1 Jan 2014) (Malaysian Ringgit - MYR) |
|---|---|---|
| Tariff A - Domestic Tariff | | |
| For the first 200 kWh (1 - 200 kWh) per month | cent/kWh | 21.80 |
| For the next 100 kWh (201 - 300 kWh) per month | cent /kWh | 33.40 |
| For the next 300 kWh (301 - 600 kWh) per month | cent /kWh | 51.60 |
| For the next 300 kWh (601 - 900 kWh) per month | cent /kWh | 54.60 |
| For the next kWh (901 kWh onwards) per month | cent /kWh | 57.10 |
| ***The minimum monthly charge is MYR3.00* | | |

The monthly electricity usage was based on actual meter readings performed at the households. The average consumption was then multiplied by the billing period

DOI: 10.1051/matecconf/20178702003

and the applicable tariff rates to determine the total bill amount. Multiply the rate depends on unit of energy use.

#### *3.3.2 Utility Bill Analysis*

The electricity consumption cost per household depends on several variables, including family size, economic background, type of housing, number and age of electrical appliances and hours of usage. Customers can calculate the estimated electricity cost for different appliances using the power rating of the electrical appliance and its efficiency, usage by number of hours and the domestic tariff rate per kilowatt hour (kWh) [4].

#### *3.3.3 Measurement of Energy Consumption*

The measurement of energy consumption can be calculated by multiplying the power (watt) and hour of operations together with 30 days. The result will be divided by 1000. The equation is as below:

Electricity Consumption (kWh) :

$$\frac{Power\ (Watt) \times Hours\ of\ Operation \times 30\ days}{1000} \tag{1}$$

** *kW = kilowatt*
*h = Hour*

## 4 Result and Discussion

### 4.1 Energy Consumption

Energy consumption categories for this study were divided into three main groups; energy consumption between 0-200 (kWh) (MYR0-MYR43.60), between 201-300 (kWh) (MYR43.90-MYR77), and above 300 (kWh) (above MYR77). Figure 5 shows the distribution of energy consumption by group.

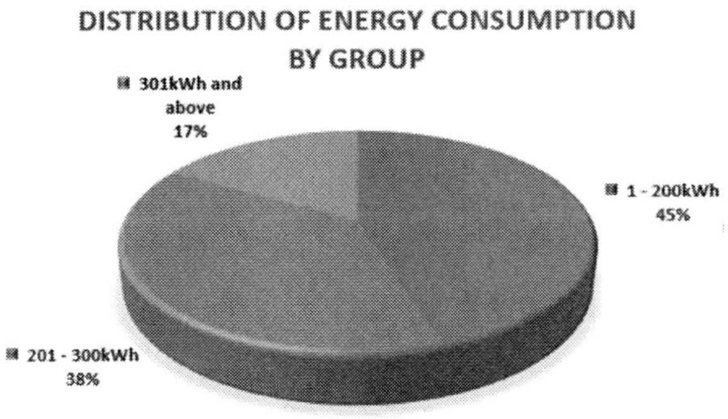


**Figure 5.** The Distribution of Energy Consumption by Group for All Residents

The graph shows that the energy consumption of 45% (32 respondents) of the 71 respondents was between 1-200kWh per month, followed by 201-300kWh per month used by 27 respondents, and finally, above 300kWh per month used by 12 respondents.

From the calculations, it was found that the average energy consumed by each household was about 237kWh per month. According to the report by the Association of Water & Energy Research Malaysia (AWER) [17], about 50% of consumers in Malaysia use less than 200 kWh of electricity per month. This statistic has increased in line with the improved economic status of each family. According to the Ministry of Energy, Green Technology and Water (KeTTHA) [18], 75% or 4.4 million domestic consumers used electricity at the rate of 300 kWh per month or less and of the total, 900,000 were low-income earners, while 200,000 comprised the poor population. Therefore, EE could be one of the solutions to help consumers to reduce their energy consumption every month [17]. In this study, the data was analysed by graph to show the energy consumption over three months and the difference in energy consumption between months for each group.

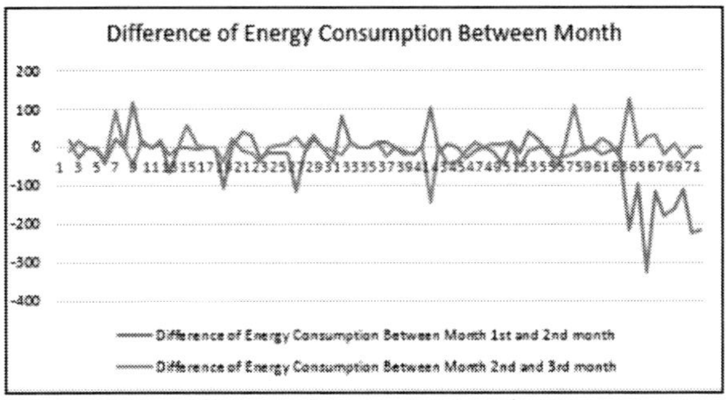


**Figure 6.** The Difference of Energy Consumption Between Months for All Residents

The graph in Figure 6 shows the difference in energy consumption between months for all participating households. The positive value in this graph means a decrease in energy consumption while the negative value means an increase in energy consumption between months. The results show the highest increase in energy consumption was 325kWh while the highest decrease of energy consumption was 115kWh in between the first and second months. For the second and third months, the results show the highest increase was 143kWh while the highest decrease was 126kWh.

#### *4.1.1 Group of Energy Consumption 1-200 (kWh)*

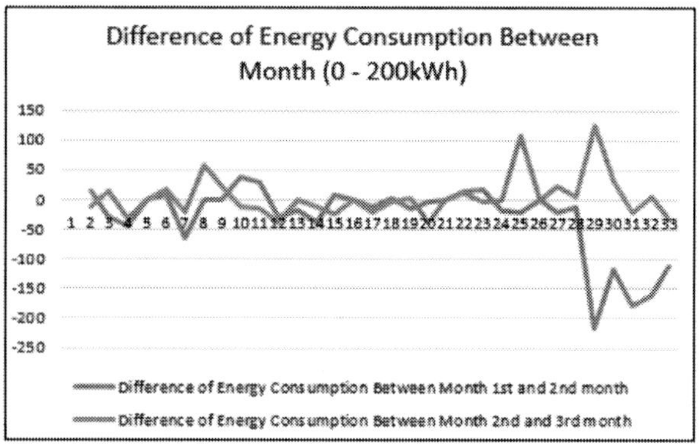


**Figure 7.** The Difference in Energy Consumption Between Months for Group of Energy Consumption 1-200 (kWh)

Figure 7 shows 32 residents had a total energy consumption of between 1-200kWh (costing from MYR3 to MYR43.60) This group represented almost half of the total homes participating in this research. Examining the difference in energy consumption between the first and second months, the highest increase was 215kWh while the highest decrease was 38kWh. The highest increase between the second and third months was 38kWh and the highest decrease was 126kWh.

#### *4.1.2 Group of Energy Consumption 201-300 (kWh)*

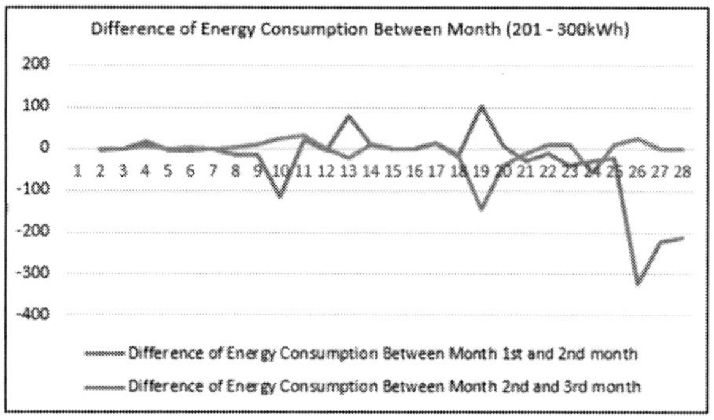


**Figure 8.** The Difference of Energy Consumption Between Months for Group of Energy Consumption 201-300 (kWh)

Figure 8 shows the results for the group with the energy consumption of between 200-300 (kWh). 33.8% of the total 71 residents who participated fell into this group. The total energy demand for all residents in this group for three months was 20,328kWh. The highest increase of energy usage in this group was 325kWh. This reading was dramatically increased from the previous bill (65kWh to 390kWh). This situation may be due to the installation of new high energy electrical appliances such as air conditioners or refrigerators. Meanwhile, between the first and second months' energy consumption, the highest decrease amongst respondents was only 18kWh which was the lowest decrease among all the groups. On the other hand, between the second and third months, the highest increase was 143kWh and the highest decrease was 25kWh, and the lowest energy usage was in this group (3.41%). In this group, three residents did not show a significant change in energy consumption between months. This is because their energy usage was consistent throughout the months.

#### *4.1.3 Group of Energy Consumption 301 (kWh) and above*

Figure 9 shows the difference between the energy consumption in the first to second months and the second to third months for households with energy consumption of 301 (kWh) and above. The highest increment from the first to the second month was 105kWh where for the first month the energy usage was 428kWh and in the second month it was 533kWh.For the second to third month, the highest increase was 50kwh while seven residents from this group decreased their energy usage from the previous month with the highest decrease of 95kWh. The lowest energy cost in this group was RM 77.00 and this consumer had to spend more money on their electric bill. In Malaysia, if the energy consumption is more than 300kWh per month, the consumer must include 6% of Goods and Service Tax (GST) [3].

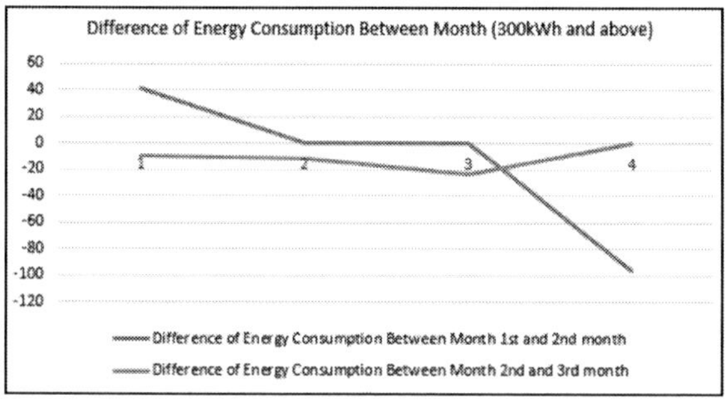


**Figure 9** The Difference of Energy Consumption Between Months for Group of Energy Consumption 301 (kWh) and Above

#### *4.1.4 Energy Consumption of Electrical Appliances*

This section will discuss the analysis of energy consumption by electrical appliances. Figure 10 shows the most commonly used electrical appliances in households. Of the 20 items listed, lamps (71), fans (71) and refrigerators (70) were the most popular electrical appliances.

Currently in Malaysia, EE Standards and Labelling are only implemented on five types of electrical appliances including refrigerators, air conditioners, televisions, fans, and lamps. The number will increase gradually [19] over the years. Based on the results, the aforementioned electrical appliances had the highest usage compared with the others. Thus, the usage of high EE electrical appliances can help consumers to reduce their monthly energy consumption.

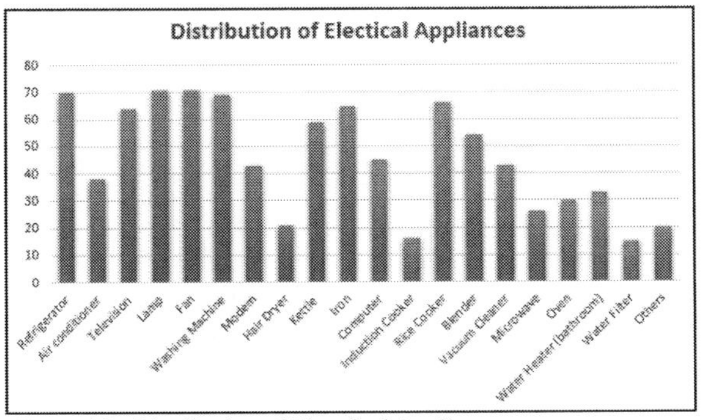


**Figure 10.** Distribution of Electrical Appliances

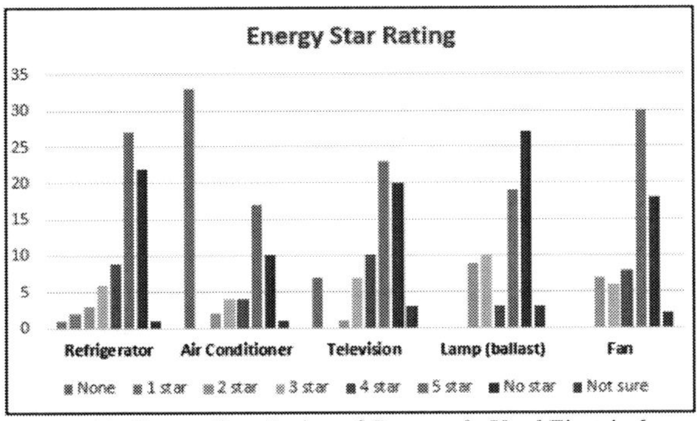


**Figure 11.** Energy Star Rating of Commonly Used Electrical Appliances

Figure 11 shows the statistic of Energy Star Rating for those five types of electrical appliances. The results show that 33% of consumers were aware of EE as they used 5 star EE products, followed by 10% who used 4 star products, 9% (3 star), and 6% (2 star), compared with 27% of consumers who did not use any energy star rating products. Of the total, about 4% of consumers were not sure about Energy Star Rating products. This was because they had not been exposed to the importance of the EE approach and implementation.

## 4.2 Consumer Awareness and Practice

A Pearson product-moment correlation or Pearson's r was run to determine the relationship between awareness and the implementation of the energy efficiency approach (practice) in the residents' daily lives.

**Table 2.** Correlation between Awareness and Practice

**Correlations**

| | | Practice | Awareness |
|---|---|---|---|
| Practice | Pearson Correlation | 1 | .316** |
| | Sig. (2-tailed) | | .007 |
| | N | 71 | 71 |
| Awareness | Pearson Correlation | .316** | 1 |
| | Sig. (2-tailed) | .007 | |
| | N | 71 | 71 |

**. Correlation is significant at the 0.01 level (2-tailed).

The data showed no violation of normality and linearity. There was a positive correlation between knowledge and awareness, which was statistically significant ($r = 0.316$, $n = 71$, $p < 0.01$) as shown in Table 2. This correlation means an increase in awareness was increasing the practice and both variables were important for the consumer to implement the energy efficient approach for electrical appliances in residential areas.

**Table 3.** Mean Score for Awareness

**Awareness**

| | Mean | Std. Deviation |
|---|---|---|
| Know about energy crisis | 3.58 | .750 |
| Need energy saving for home | 3.92 | .368 |
| Electricity consumption can be reduced | 3.80 | .576 |
| Reducing energy consumption to implement green technology | 3.35 | .958 |
| Exposure to energy saving by implementation of energy efficiency | 3.65 | .719 |
| Impact of increasing electricity tariff | 3.77 | .566 |
| Understands the energy efficiency concept | 3.23 | .974 |
| Needs guidance, a service advisor, or campaign | 3.82 | .516 |
| Knows energy saving is important | 3.63 | .741 |
| Has opportunity to save energy | 3.59 | .729 |
| Valid N (listwise) | | |

**Table 4** Mean Score for Practice

| Practice | Mean | Std. Deviation |
|---|---|---|
| Uses high energy consumption of electrical appliances | 3.56 | .788 |
| Unplugs the socket after use | 2.94 | 1.040 |
| Irons clothes every day | 3.08 | 1.052 |
| Electrical appliances help to do work at home | 2.55 | 1.025 |
| Uses high energy efficiency electrical appliances (Energy Efficiency/Energy Star) | 2.97 | 1.000 |
| Shows concern about the electricity consumption of electrical appliances before purchasing | 3.42 | .936 |
| Washes clothes at the maximum level | 3.90 | .345 |
| Boils water at the maximum level | 3.25 | 1.010 |
| Switches off the lamp before leaving | 3.73 | .632 |
| Valid N (listwise) | | |

Table 3 shows that the average mean value of the awareness was between 3.20 to 3.95. However, the results showed that the average mean value for nine items of practice was between 2.50 to 3.90 as shown in Table 4. Robinson [20] suggested that for the four-point Likert scale, the mean scores should be 1 for very low priority, 2 for low priority, 3 for high priority, and 4 for very high priority. This means that the majority of respondents had a high awareness about EE while consumers held a low priority for a certain item in practice and the rest had already been implementing the EE practice in their daily lives.

In conclusion, awareness and practice can cause the residents to implement energy efficiency, although they place a low priority on practising it. This means that although they know and are aware of EE, they do not practice it, or practice it minimally in their daily lives. Most of the consumers had less awareness about EE at home.

## 5 Conclusion and Recommendation

Based on the analysis, it can conclude that there is an increase in energy consumption at home by the consumer in three months observation. This scenario caused by many factors, including the use of electrical appliances that are not consistent (time of use), the lack of awareness of the use of efficient energy and the use of electrical appliances that are not energy efficient. The consumer should use electrical appliances which are high energy efficiency based on the energy star rating to save more energy for their home.

Furthermore, from this study, it has been proven that awareness and practice towards energy efficiency affect the energy consumption of consumers. Consumers had a high awareness about EE but at the same time consumers held a low priority for a certain item in practice and this can increase their energy consumption at home.

Meanwhile, the Energy Standard and Energy Star Rating label help the consumer choose more efficient electrical appliances. Some recommendations would also be useful to improve the study, including benefits, the method of measurements and consultancy in the renovation of homes or new home construction for energy efficiency.

EE approaches such as MEPS and Energy Star Rating can help consumers choose more efficient electrical appliances and start to use energy wisely. The overall result of this study successfully showed that the EE approach is important in energy consumption. In addition, consumers should use electrical appliances which have high EE based on the Energy Star Rating to reduce energy consumption.

In other hands, the findings from this study recommend a number of aspects that should be considered for future works. The recommendation are as follows :

i. Equipment to create questionnaire and analyse difficult to learn and should be provided with complete instruction so that many types of measurements can be carried out.
ii. The development of energy management system that can help consumer to estimate the energy consumption (by appliances) easily at home and this can raise their awareness the importance of good energy management.
iii. To raise awareness and stimulate motivation and the need for action, a broad public campaign may also be necessary. Such a campaign should comprehend the general public, similar to the campaigns of energy suppliers who have created a clear and long-lasting presence in the public.

## Acknowledgement

The author would like to express her thanks and gratitude to University Tun Hussein Onn Malaysia (UTHM) and the Ministry of Education Malaysia for HLP (*Hadiah Latihan Persekutuan*) Scholarship for the study (author 1).

## References

[1] National Renewable Energy Laboratory (NREL). [Online]. From: http://www.nrel.gov/ [Acessed on 20 February 2016].

[2] Energy Commission, "Peninsular Malaysia Electricity Supply Industry Outlook 2013," Putrajaya, 2013].

[3] Tenaga Nasional Berhad (TNB). Electricity Tariff. [Online]. From: https://www.tnb.com.my/residential/pricing-tariffs/. [Acessed on 20 February 2016].

[4] TNB. Tenaga Nasional Berhad. [Online]. From: www.tnb.com.my/. [Acessed on 20 February 2016]

[5] C. S. Ward, D. O., Clark, C. D., Jensen, K. L., Yen, S. T., & Russell. 2011. Factors influencing willingness-to-pay for the ENERGY STAR® label. *Energy Policy*. 39(3): 1450–1458.

[6] I. Richardson, M. Thomson, D. Infield, and C. Clifford. 2010. Domestic electricity use: A high-resolution energy demand model. *Energy Build.* 42(10): 1878–1887.

[7] M. Varman, H. H. Masjuki, and T. M. I. Mahlia. 2005. Electricity savings from implementation of minimum energy efficiency standard for TVs in Malaysia. *Energy Build.* 37(6): 685–689.

[8] Energy Commission. Energy Commission. 2014. [Online]. From: http://www.st.gov.my/. [Acessed on 20 February 2016].

[9] J. E. (2003). Wiel, S., & McMahon. 2003. Governments should implement energy-efficiency standards and labels - cautiously. *Energy Policy*. 31(13): 1403–1415, 2003.

[10] T. M. I. Mahlia, H. H. Masjuki, and I. a. Choudhury. 2002. Theory of energy efficiency standards and labels. *Energy Convers. Manag.* 43(6): 743–761.

[11] A. Thøgersen, J., & Grønhøj. 2010. Electricity saving in households-A social cognitive approach. *Energy Policy*. 38(12): 7732–7743.

[12] The Economic Planning Unit, "Ninth Malaysia Plan 2006-2010," 2006.

[13] T. M. . Mahlia, H. . Masjuki, R. Saidur, and M. . Amalina. 2004. Cost–benefit analysis of implementing minimum energy efficiency standards for household refrigerator-freezers in Malaysia. *Energy Policy*. 32(16): 1819–1824.

[14] K. A Rahman, M. Zainal, M. Najib, and A. M. Leman. 2015. Implementation of Energy Efficiency Standards and Labelling For Household Electrical Appliances : A Comparison among Asian Countries. *3rd Postgraduate Colloquium for Environmental Research (POCER).* Sarawak, Malaysia. August 2015.

[15] K. A. Rahman, M. Z. . Yusof, M. N. . Salleh, and A. M. Leman. 2015. A Review on Implementation of MEPS as A Standard and Labelling (S&L) Program in Malaysia. *Technology and Innovation National Conference Proceedings (TECHON 2015)*, Sarawak, Malaysia, June 2015. 551–562.

[16] L. Reeder. 2010. *Guide to Green Building Rating Systems: Understanding LEED, Green Globes, Energy Star, the National Green Building Standard, and More*. 2010.

[17] Association of Water & Energy Research Malaysia (AWER), "KECEKAPAN TENAGA DI MALAYSIA Pembuatan dan Penggunaan Lestari : Menghentikan Penggunaan Produk-produk Tidak Cekap Tenaga Secara Berperingkat di Malaysia," 2012.

[18] "Tarif Baru Elektrik," *Berita Harian*, 2011.

[19] Z. Umar. 2014. Implementation and Enforcement of Minimum Energy Performance Standards (MEPS) in Malaysia. Kuala Lumpur, Malaysia.

[20] P. Robinson and R. Shepard. 2011. Outreach, applied research, and management needs for Wisconsin's Great Lakes Freshwater Estuaries: A cooperative extension needs assessment model. *J. Ext.*49: 1–13, 2011.

MATEC Web of Conferences **87**, 02004 (2017)
*ENCON 2016*

DOI: 10.1051/matecconf/20178702004

# Applied TRIZ in Improving Productivity in Textile Industry

Aminah Ahmad[1,a], Nor Hamizah Miswan[1], Rahaini Mohd Said[1] and Rohana Abdullah[1]

[1] *Universiti Teknikal Malaysia Melaka, UTeM, 71600, Durian Tunggal, Melaka, Malaysia*

**Abstract.** TRIZ is a methodology and a collection of problem solving tools and strategies that has been used in many other fields. Therefore, this paper proposes TRIZ method for improving the productivity in a textile industry. It focuses at the packing department in a textile company situated in Malaysia. The process was monitored and the problem was observed. TRIZ method is applied in this problem using Functional Analysis and trimming method. A comparison between before and after implementation is done in order to evaluate the productivity effectiveness.

## 1 Introduction

Profitability of a firm much relies on productivity improvements. Furthermore, it is a common knowledge that the productivity improvement is one of the most important ways to increase competitiveness and especially in price competitiveness. Proper planning on productivity is possible to reduce the unit's costs of products and services.

The study is performed at a textile industry where productivity evaluation initiatives had been carried out for the purpose of increasing current productivity. The study will focus at the packaging department since it is one of the important areas in the company. There are two key aspects are considered in this department; packaging and labeling process.

Most of the packaging processes at the packaging area have a weighing step. Two types of weighing equipment are used, one is manual weight and the other one is digital weight. The main concern in this study is to define which one is more efficient to be used and more accurate. During the load process the operators load the items manually into the plastic bag and into a dritz box. This manual step is seen as waste because it is time consuming and causes the operators to easily get tired.

The entire dritz box used in packing needed a label at bottom of the box. This step again is done manually by pulling out the label from a piece of paper and then sticking the label to the dritz box. The label sticker material is made of plastic and it is difficult for the operators to separate the plastic label from the paper. Therefore, the operator had to use their nail to pull out the label and it is also another time consuming task.

## 2 The method of finding and solving the problem using functional analysis.

Functional Analysis is one of the concepts in TRIZ, which assist in building a comprehensive model of the Engineering System. Functional analysis consists of three steps, which are Component Analysis, Interaction Analysis and Function Model. The components analysis identifies components of the Engineering system and its supersystem, whereas interaction analysis identifies the interaction between components. Then, function model identifies and evaluate the functions performed by the components. Figure 1 shows the function model for this problem.

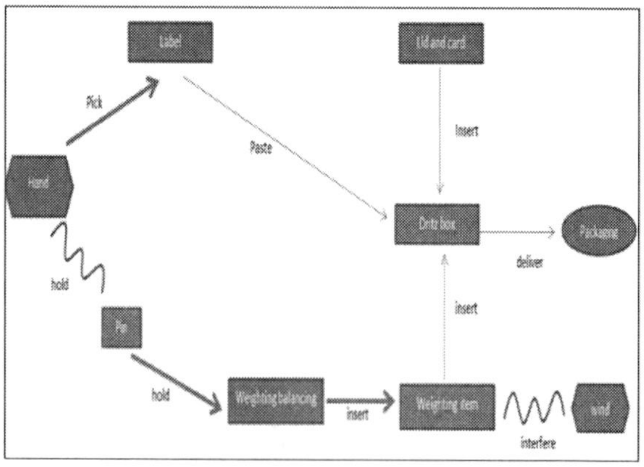


**Figure 1.** Functional Analysis before implementation.

[a] Corresponding author: aminah@utem.edu.my

 DOI: 10.1051/matecconf/20178702004

From Figure 1, the process starts with the operator's hand holding the pin. The interaction between hand and pin are harmful since the operator might injure their hands while performing the process. Next, the operator needs to weigh and balance the pins to ensure the weight meets the required weight specification before inserting the pin into a dritz box. During the weighing process, harmful supersystem such as wind is identified since it can affect the measurement accuracy.

Next, the dritz box will need to be labeled using a sticker. The labeling process is an exhaustive process where it requires high focus because the label needs to be put at the specific position at the bottom of the box and therefore, proper adjustment needed to be made before the label can be attached. This step is done manually by separating the plastic label from the paper, observing, adjusting the position of the sticker and finally sticking the label to the dritz box. Separating the label from the paper is very tedious and time consuming since the operator will need to use their nails to pull out the label from the paper. Finally, once the weighting and labeling process are done, a card is inserted and the dritz box's lid is closed and the part is ready to be delivered to the packaging department.

## 3 Results and Discussion.

Based on the analysis done, several improvements are proposed. The Functional Analysis results proposed is shown in Figure 2 below.

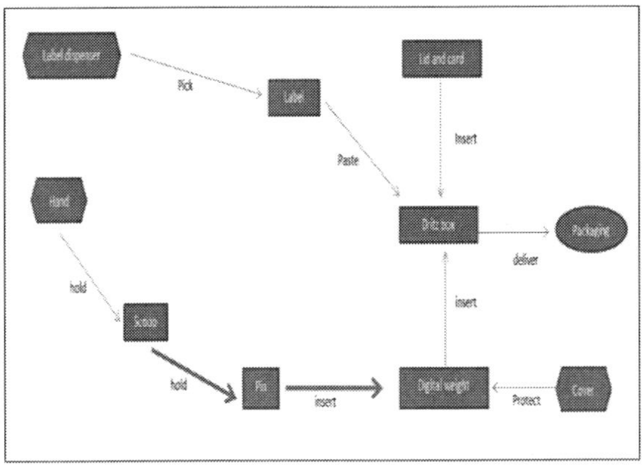


**Figure 2.** Functional Analysis after implementation.

Based on Figure 2, some improvements have been suggested at three process including weighting process, load bulk pack and stick label to dritz box by using trimming process in TRIZ.

### 3.1. Improvement At Weighting Process

The improvement that is suggested in this process is to change all the weighing equipment from balancing weight to digital weight. Figure 3 shows the balancing weight that has been used before improvement. Figure 4 shows the improvement of weighing process by using digital weight. The new time taken to weigh using digital weight is compared with the previous method. The results showed that using digital weight reduces the time taken to perform the weighing activity. In addition, kaizen improvement is also applied in this step where a cover is located around the digital weight to prevent the wind from affecting the reading of the weight and to ensure more accurate reading.

In Figure 5, the histogram shows the comparison between before and after improvement in term of average time. There is an improvement where the average time is reduced from 6.99 seconds to 5.92 seconds. It is about 16% improvement from the previous method.

**Figure 3.** Before Improvement.

**Figure 4.** After Improvement.

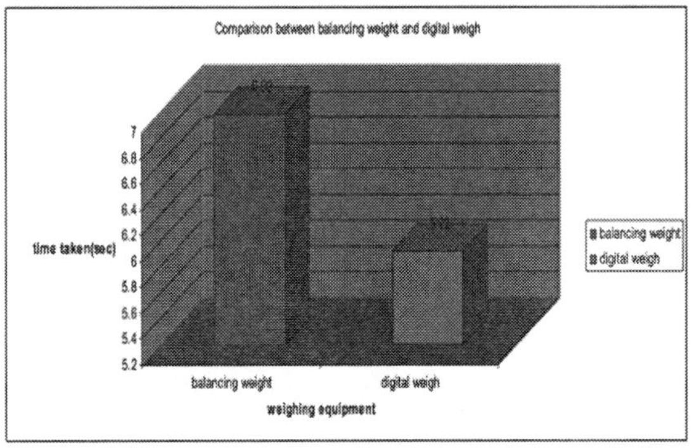


**Figure 5.** Histogram for comparison between before and after improvement at the weighing method.

DOI: 10.1051/matecconf/20178702004

### 3.2 Improvement At Load Bulk Pack

As mentioned earlier and is shown in Figure 4, the weighing activity is done manually where it requires the operator to pick the pins using their bare fingers and this will cause the pins to prick and injure the fingers. The activity of picking, estimating, weighing and loading the pins is termed as the load bulk pack process. In this process, the improvement that is suggested is loading the item into pollybag using a scoop. The prototype of the scoop is designed and made using the stainless steel. From the observation, it is more efficient compared to the manual load since it's safer for the fingers and faster because the capacity of the scoop is already designed to meet the packaging weight. This means when using the scoop, it is much easier for the operator to estimate the item's weight requirement and lesser time is spent on this operation. Figure 8 shows the difference between previous method and the new methods. The average time is reduced from 23.24 seconds to only 9.03 seconds (62% improvement). This shows that using the measurement scoop is a much more efficient and safer method than using the bare fingers.

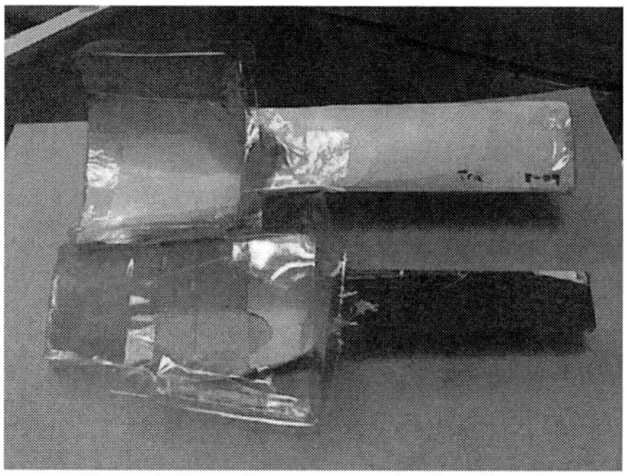

**Figure 6.** Scoop Prototype using hard cardboard.

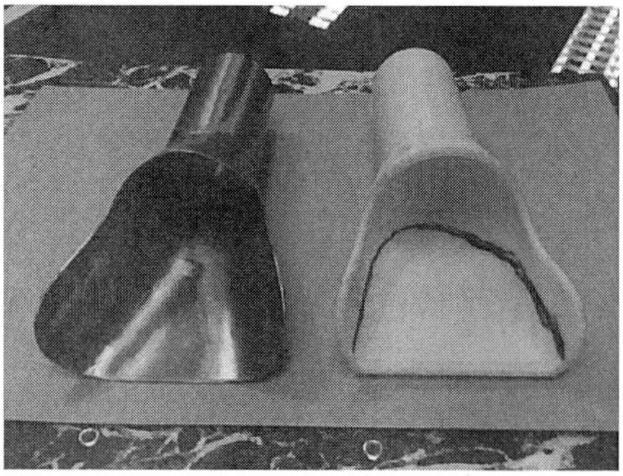

**Figure 7.** Scoop Prototype using stainless steel and polymer.

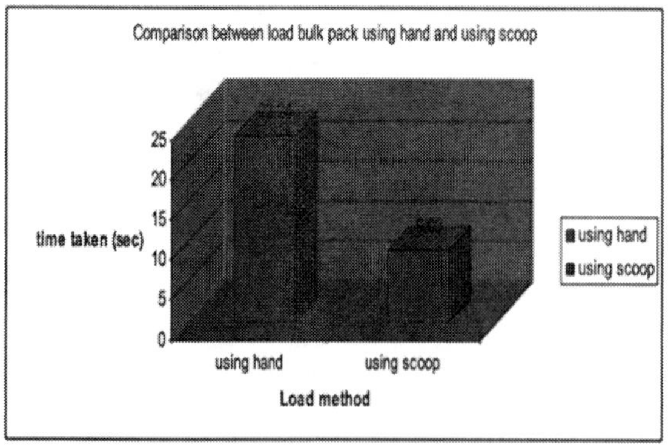


**Figure 8.** Histogram of comparison between load bulk pack using hand and using scoop.

### 3.3 Improvement At Load Bulk Pack

In this process, the main concern is to separate the plastic label from the paper and stick to the dritz box. The improvement suggested is to use a dispenser to remove the label from the paper instead of using the nails. This suggestion is made after reviewing several case studies. Many companies have implemented using the electric dispenser to replace the manual activity. However, for this study, a simple manual sticker dispenser was designed and fabricated based on the case studies' design of dispensers. The dispenser is designed using solid work software. The prototype is built using stainless steel material. From the time study, by using this dispenser, the average time is reduced from 3.23 seconds to 2.61 seconds (20% improvement). Figure 10 shows the improvement of the stick label activity between using the nails and using the dispenser.

**Figure 9.** The label/sticker Dispenser Usage

DOI: 10.1051/matecconf/20178702004

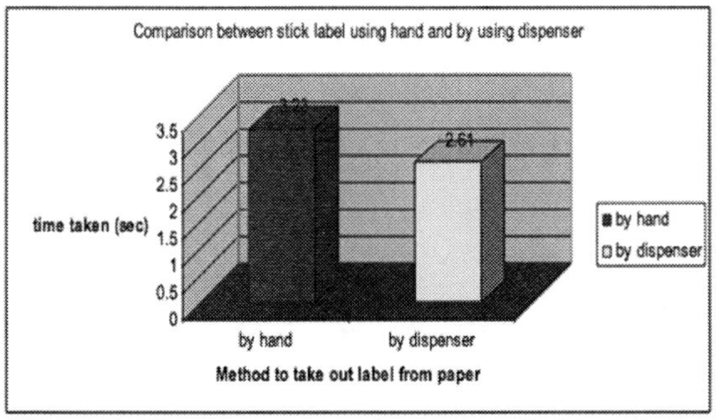


**Figure 10.** Histogram of comparison between before and after improvement of stick label method

## 4 Conclusions

The aim to study and improve the efficiency of the packaging department process at the textile company is achieved. The Functional Analysis approach in TRIZ is proven to be a very effective method to solve a problem where a problem is first analyzed using a functional analysis model and then the trimming process is done in order to obtain the improved Functional Analysis. After improvement activities has been implemented, comparisons in terms of the time taken between the before and after are done in order to measure the effectiveness of the productivity between before and after improvement.

## References

1. Yeoh, T.S.,Yeoh, T.J and Song, C.L, *TRIZ : Systematic Innovation in Manufacturing. Malaysia: Firstfruits*, (2009).
2. D.Daniel Sheu and Chuan Ting Hou, *TRIZ-based problem solving for process-machine improvements: Slit-valve Innovative redesign, Proceeding of the 41st International Conference on Computer & Industrial Engineering* (2013).
3. Valery Krasnoslobodtsev and Richard Langevin, *Applied TRIZ in High Tech Industry, The TRIZ Journal*, (2006).
4. Zhang Ruihong, Tan Runhua, and Cao Guazhong, *Case study in ad and TRIZ: A paper Machine, The* TRIZ Journal, (2004).
5. Nuralesya Ramlee, *Productivity Improvement in Textile Industry. Malaysia: UTeM library*, (2010).
6. Meyers F. E. and Stephens M. P., *Manufacturing Facilities Design and Material Handling 3rd Ed. New Jersey: Prentice Hall*, (2005).
7. Amrine H. T., Ritchey J. A., Moodie C. L. and Kmec J. F., *Manufacturing Organization and Management 6th Ed. New Jersey: Prentice Hall.(1993)*
8. Longenecker, C.O. and Stansfield, T.C., *Why plant managers fail: causes and consequences, Industrial Management*, (2000).
9. Stevenson J. W., *Operations Management. New York: Mc Graw Hill*(2007).
10. Ali M.M, *Implementation of the DMAIC Analytical Method on Industrial Machinery Repair Service Company in Indonesia*, (2008)

# Virtual Mass of Any Shape of Body by using General Mapping Function

Norio ARAI[1]

[1]*Department of Mechanical Precision Engineering, MJIIT, Universiti Teknologi Malaysia, Kuala Lumpur, 54100 Malaysia*

**Abstract.** When an object is moving in acceleration with fluid interaction from surrounding, it possesses virtual mass (added mass) in which the total mass of object is higher than the original mass. The exact conformal mapping method was proposed to determine the virtual mass. However, it is difficult to determine the virtual mass for the body of random shape by using conformal mapping. Therefore the new concept of mapping function is proposed. It is based on Laurent series that has been proposed to map the body of random shape to a circle. The function of a body shape is expressed in Fourier series which is essential to determine the complex coefficients of mapping function (Laurent series). Proposed method is compared with exact conformal method (equilateral polygon) and the validity is confirmed.

## 1 Introduction

When the body is accelerated by fluid for instance, the body motion is expressed by the equation of motion like $m\ddot{x}+c\dot{x}+kx=f(t)$. Here *m*(mass) is composed by both the pure mass of body and the 'virtual' mass. Especially, when the acceleration is large or the mass ratio is almost larger than unity, the effect of virtual mass should not be ignored. Although the virtual mass effect is very important term in motion, a little special cases could have been investigated because of the difficulty of calculating it analytically and/or exactly. While special cases are gotten from the famous text (Lamb[1], Milne-Thomson[2], Landau and Lifshitz[3]), the body geometries are restricted in the flat plate, circle, ellipse, and so on.

Here we need to specify the conformal mapping function from the target geometry (*z*-plane) to the circle (ζ-plane), while it is very difficult to specify a general function for the conformal mapping analytically.

In this article the objective is to show the virtual mass of the any shape of body which geometry is used in architecture sometimes.

The conformal mapping function is given by the infinite series like Laurent Series $\sum \frac{a_n}{z^n}$, in which the body shape is expressed by using Fourier Series. Finally in order to validate the present method, some examples are shown.

## 2 Analysys

### 2.1 Virtual mass and kinetic energy

The flow field around a body at angle of attack 'α' is expressed by using complex velocity potential '*f*' as follows,

$$f = Uze^{-i\alpha} + w \tag{1}$$

where *U*: speed of uniform flow, *w*: 'disturbance' from 'body'.

For example, the flow around a circle is expressed

$$f = U\left(ze^{-i\alpha} + \frac{a^2}{z}e^{i\alpha}\right) \tag{2}$$

$$= Uze^{-i\alpha} + w \tag{3}$$

$$= \text{'Uniform flow'} + \text{' Disturbance'} \tag{4}$$

As shown in Milne-Thomson[2], the total kinetic energy is derived by using 'disturbance *w*' as follows,

$$T = \frac{1}{2}\rho \oint_s \frac{\partial}{\partial z}\left(w\frac{d\bar{w}}{dz}\right) dS \tag{5}$$

This equation is reduced to Eq.(6) by using 'Area Rule'

$$T = -\frac{1}{4}i\rho \oint_{c1} w d\bar{w} + \frac{1}{4}i\rho \oint_{c2} w d\bar{w}, \tag{6}$$

where *c1* : inner boundary, *c2* : outer boundary.

When *c2* tends to infinity, the second term is vanished, then

$$\begin{aligned} T &= -\frac{1}{4}i\rho \oint_{c1} w d\bar{w} = \frac{1}{2}U^2 M' \\ &= \text{'Total Kinetic Energy'}, \end{aligned} \tag{7}$$

where$M'$ : virtual mass.

[a] Corresponding author: arai@utm.my or arai@cc.tuat.ac.jp

DOI: 10.1051/matecconf/20178702005

Therfore, we can calculate the virtual mass by using Eq(7).

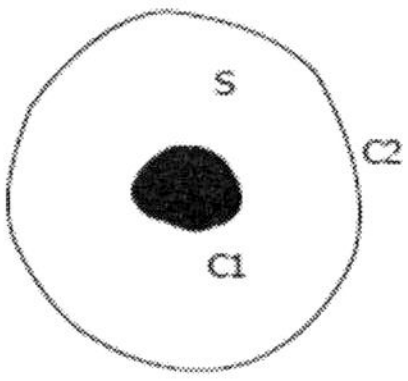


**Figure 1.** Flow field *S*, inner and outer boundaries *C1*, *C2*.

### 2.2 General mapping function

The body shape is expressed by using Fourier series as follows

$$\text{real part} \quad x_n = x_n(\theta)$$
$$\text{imaginary part} \quad y_n = y_n(\theta)$$

$$x_n(\theta), y_n(\theta) = \frac{1}{2}a_0 + \sum_{n=1}^{\infty}\left\{a_n \cos\left(\frac{n\pi\theta}{l}\right) + b_n \sin\left(\frac{n\pi\theta}{l}\right)\right\} \tag{8}$$

While the body shape is expressed by Eq.(8), the general mapping function is defined (assumed) by the considerations in fluid dynamics as follows

$$z = \sum_{n=0}^{\infty} A_n\zeta^n + \sum_{n=1}^{\infty}\frac{B_n}{\zeta^n} \tag{9}$$

$$= A_0 + A_1\zeta + A_2\zeta^2 + A_3\zeta^3 + \cdots\cdots + \frac{B_1}{\zeta} + \frac{B_2}{\zeta^2} + \frac{B_3}{\zeta^3} + \cdots\cdots \tag{10}$$

where $A_n = {}^{\mathrm{R}}A_n + i\,{}^{\mathrm{I}}A_n \quad (n = 0, 1, 2, 3, \cdots)$

$B_n = {}^{\mathrm{R}}B_n + i\,{}^{\mathrm{I}}B_n \quad (n = 1, 2, 3, \cdots)$

The body shape (Fourier series) is compared with the mapping function ($z$). Then,

$$n = 0 \;:\; {}^{\mathrm{R}}A_0 = \frac{1}{2}{}^{x}a_0, \quad {}^{\mathrm{I}}A_0 = \frac{1}{2}{}^{y}a_0$$

$$\vdots$$

$$n = n \;:\; a^{n\,\mathrm{R}}A_n + \frac{1}{a^n}{}^{\mathrm{R}}B_n = {}^{x}a_n, \quad -a^{n\,\mathrm{I}}A_n + \frac{1}{a^n}{}^{\mathrm{I}}B_n = {}^{x}b_n$$
$$a^{n\,\mathrm{I}}A_n + \frac{1}{a^n}{}^{\mathrm{I}}B_n = {}^{y}a_n, \quad a^{n\,\mathrm{R}}A_n - \frac{1}{a^n}{}^{\mathrm{R}}B_n = {}^{y}b_n$$

These algebraic equations are solved easily for ${}^{\mathrm{R}}A_n$, ${}^{\mathrm{I}}A_n$, ${}^{\mathrm{R}}B_n$, ${}^{\mathrm{I}}B_n$, then coefficients of general mapping function (Eq.9) are determined as follows,

$$\begin{aligned} {}^{\mathrm{R}}A_n &= \frac{1}{2a^n}({}^{x}a_n + {}^{y}b_n), \quad {}^{\mathrm{R}}B_n = \frac{a^n}{2}({}^{x}a_n - {}^{y}b_n) \\ {}^{\mathrm{I}}A_n &= \frac{1}{2a^n}({}^{y}a_n - {}^{x}b_n), \quad {}^{\mathrm{I}}B_n = \frac{a^n}{2}({}^{y}a_n + {}^{x}b_n) \end{aligned} \tag{11}$$

The "disturbance *w*" is expressed using the complex velocity potential *f* as follows,

$$\begin{aligned} w &= f - Uze^{-i\alpha} \\ &= U\left(\zeta e^{-i\alpha} + \frac{a^2}{\zeta}e^{i\alpha} - ze^{-i\alpha}\right) \end{aligned} \tag{12}$$

Eq.9 is substituted in Eq.12. Then we calculate the virtual mass *M'* as follows,

$$\begin{aligned} M' &= -\frac{1}{2}\frac{i\rho}{U^2}\oint_{c1} w\,d\bar{w} \\ &= -\rho\pi\left\{(1 - A_1 - \bar{A}_1)a^2 + A_1\bar{A}_1a^2 + 2A_2\bar{A}_2a^4 + \cdots\cdots \right. \\ &\quad + nA_n\bar{A}_na^{2n} + \cdots\cdots \\ &\quad - a^2 + B_1e^{-2i\alpha} + \bar{B}_1e^{2i\alpha} - \frac{B_1\bar{B}_1}{a^2} - \cdots\cdots \\ &\quad \left. - \frac{nB_n\bar{B}_n}{a^{2n}} - \cdots\cdots\right\} \end{aligned} \tag{13}$$

## 3 Results and discussion

### 3.1 Comparison between equilateral polygon

The comparison amongst different polygons is shown in Table 1 for their respective mapping functions and diagrams.

**Table 1.** Comparison between equilateral polygons.

| | General mapping function, *z* | Mapping diagram |
|---|---|---|
| Triangle | $0.6288\,\zeta + (0.0865i)\zeta^4 - 0.000005\zeta^5 + \frac{0.0328}{\zeta^5}$ | 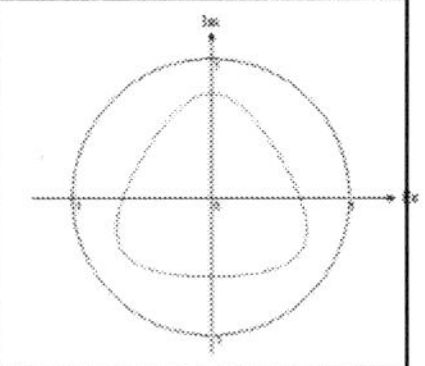 |
| Pentagon | $0.8695\,\zeta + (0.0369i)\zeta^6 + \frac{0.0382i}{\zeta^4}$ | 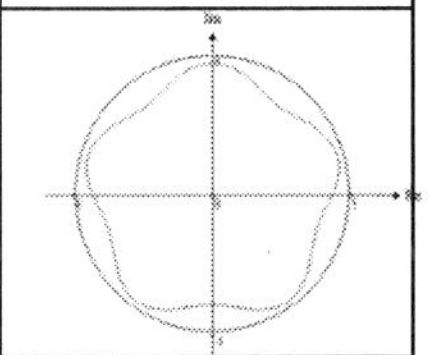 |
| Heptagon | $0.935\,\zeta - \frac{0.02i}{\zeta^6}$ | 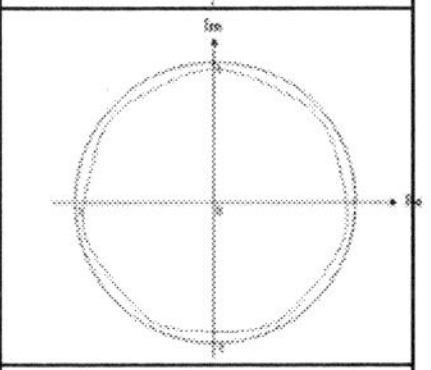 |
| ......... | ......... | ......... |

In this calculation, number of terms of Laurent series is limited to only four Laurent series coefficients. Therefore descrepancies between 'Geometry (triangle, square,...)'

DOI: 10.1051/matecconf/20178702005

and 'Mapping diagram' are appeared. However when more terms are applied to calculate it, it is expected to improve it.

### 3.2 Coefficient of virtual mass

The virtual mass is calculated using exact conformal method[4] so as to make comparison, in which it is more proper for the respective virtual mass to be expressed in the form of coefficient. The coefficient of virtual mass is defined as follows,

$$C = \frac{M_{body}}{M_{circumscribed\,circle}} \tag{14}$$

Based on Eq.(14), the coefficient can be obtained by dividing calculated virtual mass of a body with its circumscribed circle. For example, Fig. 2 shows the square and hexagon inscribed in their respective circle.

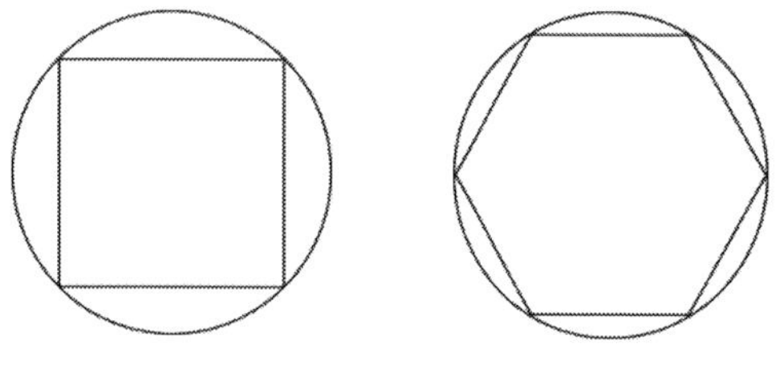

**Figure 2.** Inscribed square and hexagon.

Table 2 shows the comparison $C_G$ (present method) with $C_E$ (conformal mapping[4]) .

**Table 2.** Comparison of virtual mass coefficient for equilateral polygon.

| | $C_G$ (present) | $C_E$ [4] | error, $\Delta C$ |
|---|---|---|---|
| Triangle | 0.898 | 0.661 | 0.237. |
| Pentagon | 0.981 | 0.833 | 0.148 |
| Heptagon | 0.998 | 0.925 | 0.073 |
| ......... | ......... | ......... | ......... |

It is shown that the differences decrease as the body shape reaches a circle, in which this is because the coefficient will be equal to unity when it is a circle.

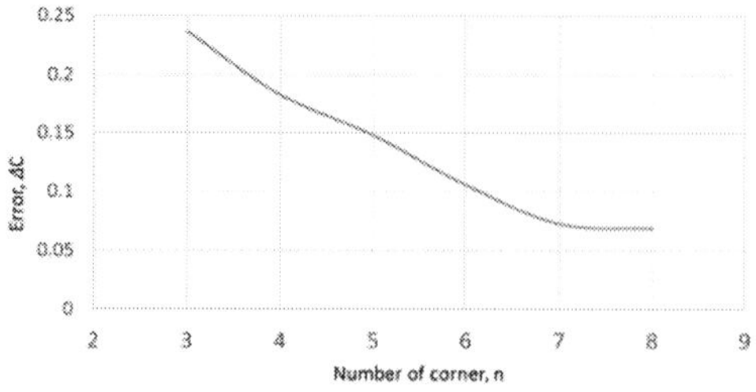


**Figure 3.** Error.

Figure 3 shows errors which decreases when the body reaches a circle. In this comparison only 4 terms of general mapping function at maximum are used. When much more terms are applied, it is expected that errors are decreasing.

This method is applied for the random shape body. The essence of general mapping method is that it can determine the general mapping function which can be used to map from any shape body in $z$-place to a circle $\zeta$-plane. The concept of Laurent series is the backbone of this newly proposed general mapping method as it can be used complex functions. The general mapping function can be expressed in Eq.(15).

$$\begin{aligned} z(\zeta) = & -0.0075 + (0.88 - 0.0604i)\zeta - 0.005\zeta^2 \\ & +0.0062\zeta^4 - 0.069\zeta^5 - 0.089\zeta^6 + \cdots\cdots \\ & + \frac{0.074i}{\zeta} + \frac{0.005}{\zeta^2} + \frac{0.026i}{\zeta^3} + \frac{0.0062}{\zeta^4} + \cdots\cdots \end{aligned} \tag{15}$$

The general mapping function from Eq.(15) is applied to map the random shape body to a circle shown in Fig.4, for instance.

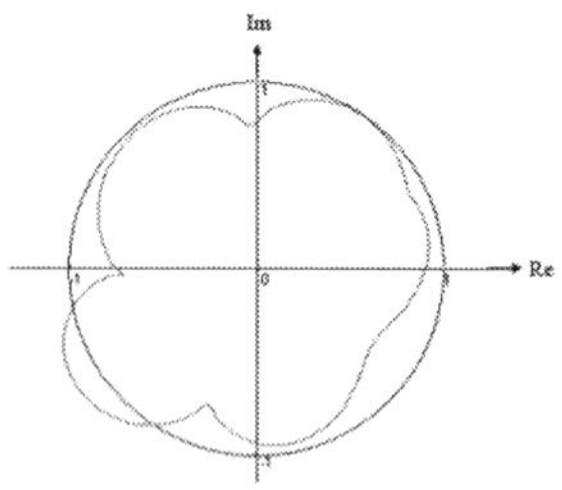


**Figure 4.** Formation of any shape body with general mapping function.

Note that some coefficients of Eq.(15) are neglected due to very small values which are insignificant to the characteristic of mapping function.

## 4 Conclusion and recommendations

The analytical solution of virtual mass of any shape body with angle of attack is shown by using complex velocity potential, in which the body shape is expressed by Fourier Series and the conformal mapping function is expressed by Laurent Series. The validation of the analysis is confirmed by comparing with other analytical and exact results (circle, flat plate, regular polygon and so on).

There are some recommendations that are important to be considered in near future in order to improve results.

1. Number of terms of Fourier and Laurent series :In order to obtain the specific general mapping function, Fourier coefficients are required to be determined. In this calculation, number of terms of Fourier series is limited to only five Fourier coefficients for both $a_n$ and $b_n$ . When more coefficients are used to calculate it, it is expected that results are improved to show more accurate results.

2. Convergency of series : In this sample calculation, first 4 or 5 terms of series are applied to calculate the virtual mass. If the convergency is so slow, then more terms should be applied to calculate it. However, when more terms are applied to calculate it, the oscillation llike Gibbs phenomenon may appear. So it is very important to validate the convergency.

## References

1. H. Lamb, *Hydrodynamics* (Cambridge University Press, 6th ed., 1932)
2. L. M. Milne-Thomson, *Theoretical Hydrodynamics* (Macmillan, 5th ed., 1968)
3. L. D. Landau and E. M. Lifshitz, *Fluid Mechanics* (Pergamon Press, 1959)
4. N. Arai, *Proc. Of IsySM'2015* International Symposium in Science and Mathematics, Bandung, Indonesia, Nov.24-26, 2015.

# Level of Bus Performance Based On the Relationship Between Distance and Travel Time of Universiti Tun Hussein Onn Malaysia (UTHM) Bus Services

Joewono Prasetijo[1,a], Wan Zahidah Musa[1], Zaffan Farhana Zainal[1], Kamarudin Ambak[1] and Mohd Erwan Sanik

[1]*Smart Driving Research Center (SDRC), Universiti Tun Hussein Onn Malaysia, 86400 Parit Raja, Batu Pahat, Johor*

**Abstract.** UTHM bus service is an important transport mode for most students at Universiti Tun Hussein Onn because it such primary public vehicle to support students movements around the campus, within and outside student apartments. This service is often associated with the quality of service in terms of time, stops and comforts. Therefore, this following study is focused on investigation on several UTHM bas routes which was based on bas operations such as distance and travel time/travel distance that would determine the level of service provided. Primary data were collected mainly to obtain information relate to speed, bas position, time movement/circulations of bus and time headway. These were obtained by installing GPS-Slute Gear i-trail along bus travels. In addition, additional data were collected by exploring previous studies regarding to the bus services such as Highway Capacity Manual (HCM 2010). The approach used was based on assessment of progress used to determine the level of quest service and such obstacle frequently occurs with results on bus delays, volume of unequal that can be overcome according to the assessments that have been provided.

## 1 Introduction

Bus is one of the most important transport modes and contributes some aspects in daily life. The mode would be an alternative for users to travel [1]. Generally bus would be use for public transport services instead of use private vehicles or other public transports/modes such as taxis, carpooling or chartered bus, which is not for sharing [2]. Furthermore, most public transport is on the way of progressing on punctuality with service scheduled for the user's convenience.

Public transport system in universities is often used for the purpose to facilitate the students/staff movement from one place to another within and out of the campus. The Office of Academic Management (PPA) Universiti Tun Hussein Onn Malaysia (UTHM) provides UTHM bus service [3]. Therefore, bus quality services must be emphasized on comfort and safe for all students/university members, especially for students who stay in the residential colleges and they do not have private vehicles. In addition, the bus functioned as transport services to transfer students from a class to another, from lecture hall to other destinations (halls, class or café). Therefore, these movements of buses would show the level of bus service surrounding university area in term of speed, time travel, passenger taking and alighting from bus, and the impact of stops. The following study shows how the level of bus services can be measured and the quality of services can be predicted [4].

## 2 Literature Review

### 2.1 Characteristic Transport Services

One of the process stages of determining the network selection and option in the perception of transport services/quality is based on an assessment of each option in the nature of services related to the transport modes. Characteristics of transport services (when/time, where/place, why, type of modes and the route) would affect on the results of the tendency of transport users to travel. Therefore, all transport services should be well considered [5].

#### *2.1.1 Travel Time*

Generally, travel time is defined as the time taken to complete the journey. As speed increases, the travel time required to reach the destination which also decreases and vice versa. Therefore, time travel is inversely proportional to the speed. It usually measured by the average local time and distance [6]. Referring to the User Guide of Public Transportation (2012), journey times for the local road to the bus stop; a must not be more than 30

[a] Corresponding author: joewono@uthm.edu.my

minutes in one trip and the following figure 1 shows its characteristics [2].

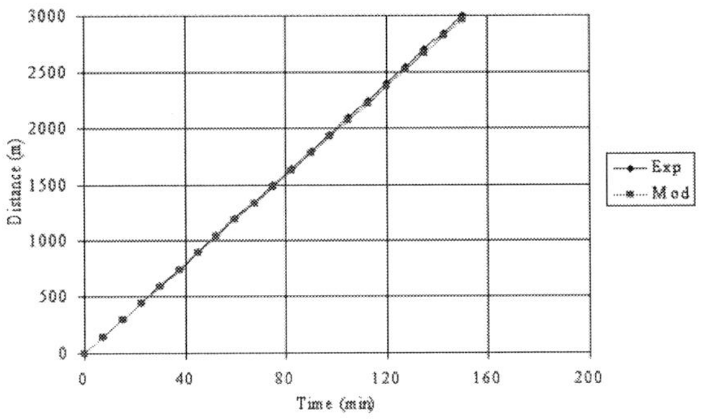


**Figure 1.** Graph plotted show the distance vs time is vice versa

### *2.1.2 Time Headway*

It is considered that vehicles microscopic characterization related to volume is the advancement of time or just progress. Generally, headways/progress of time is defined as the time difference between any two consecutive vehicles when they cross a point. In practice, it involves in measuring the time between the passages of the rear-bumper that passes a given point. Normally, headways for the convenience of public transport bus design gap would be between 5 to 20 minutes.

### *2.1.3 Headway Distance*

Another relevant parameter is the distance progress (space). It is defined as the distance between the same two consecutive vehicles at any time. It involves the measurement of the distance measured from the rear-bumper to the rear-bumper of the successive vehicles pass at a point within time interval. It is also known as the 'spacing'. The following figure 2 shows the concept of progress/headways.

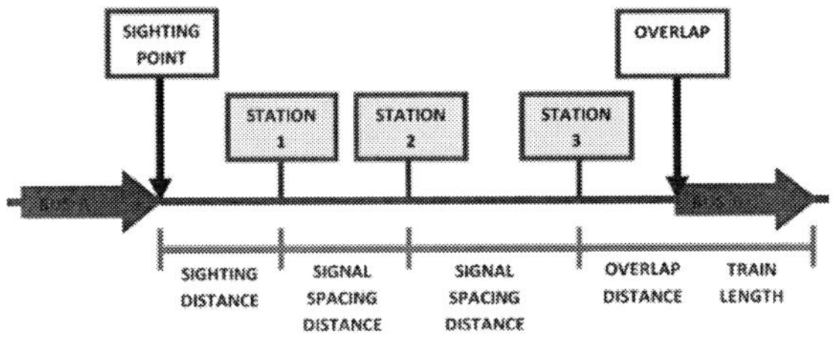


**Figure 2.** Concept of headway distance

### *2.1.4 Trip Chaining*

Travel chain can be defined as bus travelling through an area from a stop to others, which every stop has been defined. Chaining travel is often used to change the direction of bus travelling as a result of bus route changes through certain area. It also provide a huge influence towards a bus journey times and quality of the services.

### *2.1.5 The Period Used In the Course (Time Travel)*

Regardless of passenger income and mode of transit, all individual eventually faces the same problem in as 24 hours a day of traffic operations. There was the time spent in which is mainly focused on affects the scheduling of buses and it closely related to the travel distance and travel speed used while transit travelling from one destination to another. There is in common that the amount of time travel of vehicle used by different modes; the time spent of each vehicle per day is constant in average.

### *2.1.6 Rates of Excessive Waiting Period*

Excessive waiting time is defines as the difference between the scheduled time of arrival and the actual to the bus arrived in such interval/section of road divided by the number of observations. Early arrival without departing early is not counted and assumed as time errors. However, travel at the normal departure time with the early arrival would be counted as "excessive' for the purpose of calculating the average exceeds waiting time. Rate of excessive waiting times are excessive waiting time (in minutes) divided by the mean length of bus passengers while travelling route (s) in section of the road.

### *2.1.7 Rates of Stop Time at the Bus Stops (Dwell Time)*

Stop time is the time that the bus stop and the measuring instrument are in seconds. This includes the time for the bus to stop, time of opening and closing the door of the bus, the time for passenger to entrance and exit at the bus stop before further bus movements.

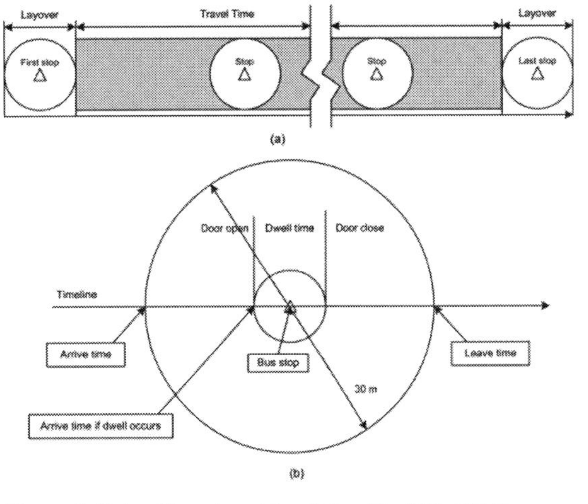


**Figure 3.** Rate of bus stop at the station (Sunif F., 2011)

## 3 Materials and Methods

To conduct the current study and observing the data, authors have chosen the common method of observations

DOI: 10.1051/matecconf/20178702006

by using data logger, (four) 4 digits tally counter and a stopwatch, which is considered more simple and easier to control and monitor. Data logger is the device that easy to use, it is placed in the bus during the study/data collection and data recording; speed, bus positions/the bus stops which are shown in longitude and latitude coordinates as well as the time of bus stopped. Furthermore, there was also considered more accurate and easily operated data. The stopwatch is simply used to measure time (headway) of a bus through the interval/sections of road. In addition, 4-digit tally counter data are used to observe the distribution of students/passenger on boarding and alighting.

Further data collection on bus velocity and distance is to use data logger, 4-digits tally counter and stopwatch with following steps; selecting buses, no bus parking and bus must be stop at the shelter/bus stops. Data was generated in accordance to the sections selected. Bus stops were selected that may represent the high number passengers.

Excessive time of bus movements through the study areas were observed which a stop as a constant value; distance was measured by distance in minutes (measured by bumper-bumper).

Numbers of passengers/students boarding and alighting were observed by a 4-digit tally, which focused on the bus travelling within campus routes. Further analysis used the common standard of Manual 2010, which data observation was taken from GPS measurements [7, 8].

# 4 Results and Conclusion

## 4.1 Analyze the level of bus service

**Table 4.1.** Result analyse the level of bus service.

| Track Selection | Average Time Travel | Distance Travel (meter) | Average Headway (minute) | Evaluation Study by Area Line, $I_{t,seg}$ |
|---|---|---|---|---|
| Bus travelling in university type Etika Delta (morning) | 15 minute 03second | 5480 | 6 | C |
| Bus travelling in university type Etika Delta (evening) | 18 minute 15second | 5610 | 7 | C |
| Bus travelling in university type Sikun Jaya (morning) | 18 minute 28second | 6900 | 8 | B |
| Bus travelling in university type Sikun Jaya (evening) | 35 minute 50second | 6870 | 5 | D |

Refer to Table 4.1 the highest ratings is Sikun Jaya movement in the university area in the morning and get assessment score is B while the lowest scoring is movement in the university for Sikun Jaya bus in the evening is D. The average travel time to the time difference can also be seen in turn gives a bad impression on consumers who use the service. Scoring can be seen through the data required and analysed in accordance with the format of the existing formula in the Highway Capacity Manual 2010 (HCM 2010) [9].

## 4.2 Analyze the impact of stop on travel time

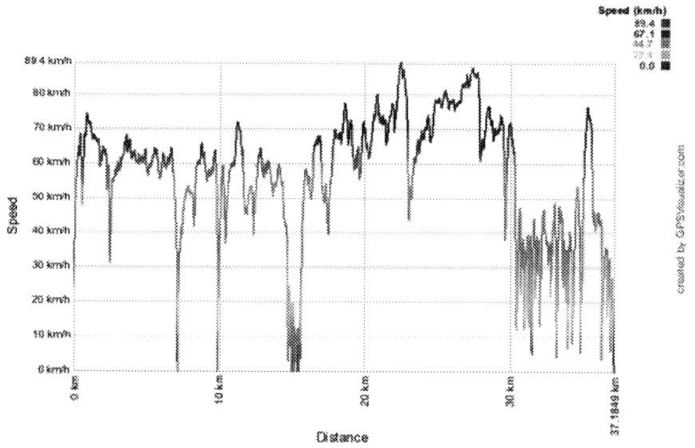


**Figure 4.** Example graph plotted speed versus time at the morning time

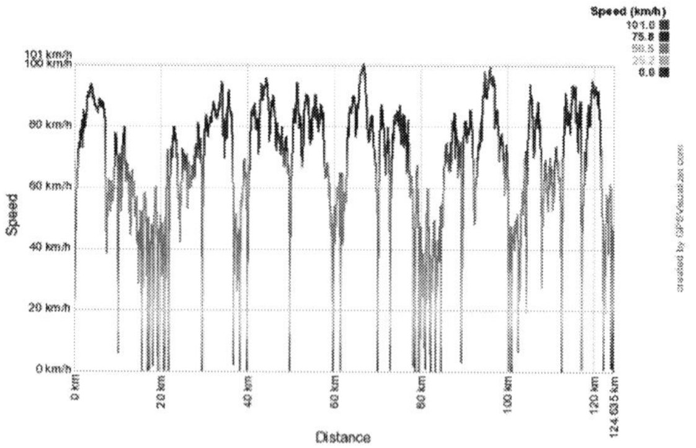


**Figure 5.** Example graph plotted speed versus time at the evening time

Refer to the figure 4 and figure 5, there are differences in the pattern of the data speed which is plotted against the distance travelled. There are differences in the speed at point 0 because of differences in the volume of traffic is different in the morning and in the evening there are differences cause velocity of 0 km/h. there was also a difference that can be seen is the travel time taken by the travel time of a trip in the morning and evening. The average difference in velocity also varies due cause deceleration stops occur every time you want to stop. Figure 4 and figure 5 also were as an example of speed profile against distance from the study measurement.

### 4.3 Analyze of potential stop area

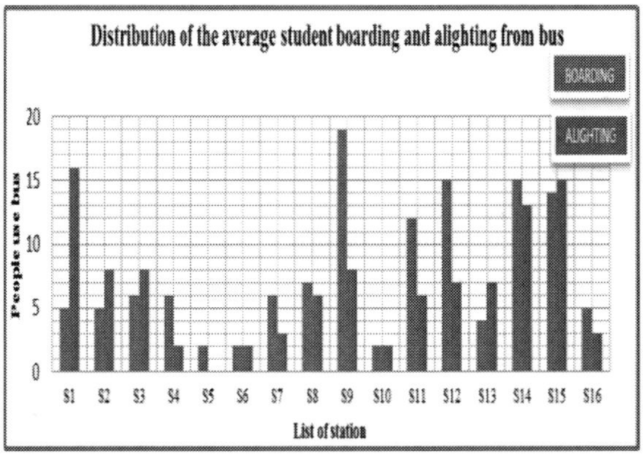


**Figure 6.** Graph show distribution of the average student

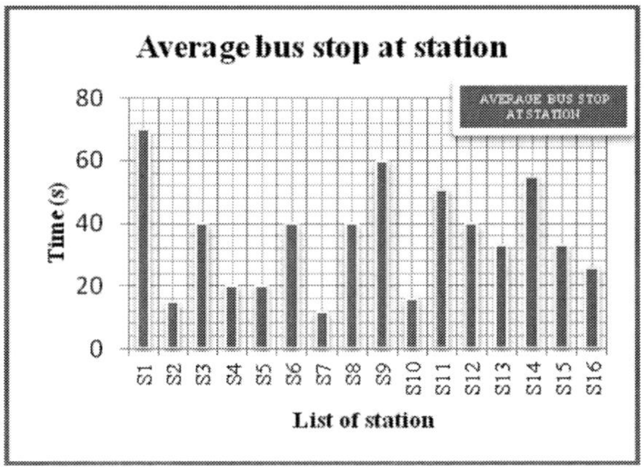


**Figure 7.** Graph show the average bus stop at the station

Figure 6 shows the average distribution of students boarding and alighting from the bus. This graph shows the frequency of the highest student boarding is at the stop bus in front S9 which is library Tun Aminah. While the distribution alighting is at S14 which is in front of Tun Fatimah College. Besides that, figure 7 shows the average time bus stop at the station. Figure 7 show the highest reading is S1, which is ATM station.

**Table 4.2.** Different of road taken by bus and average time travel bus taken each road and time

| Travel Direction | Average Time Travel | Distance (meter) | Average Speed (km/h) |
|---|---|---|---|
| Perwira Residential College (morning) | 17minute 30seconds | 5610 | 17.83 |
| Perwira Residential College (evening) | 42minute 50seconds | 5610 | 15.98 |
| Melewar Residential College (morning) | 14minute 05seconds | 5250 | 23.81 |
| Melewar Residential College (evening) | 19minute 29seconds | 5250 | 20.15 |
| Taman University Residential College (morning) | 37minute 01seconds | 5800 | 24.50 |
| Taman University Residential College (evening) | 17minute 30seconds | 5800 | 10.91 |
| Kelisa Residential College (morning) | 17minute 30seconds | 9670 | 29.14 |
| Kelisa Residential College (evening) | 17minute 30seconds | 9680 | 17.52 |
| Etika Delta bus activity in campus (morning) | 17minute 30seconds | 5480 | 20.41 |
| Etika Delta bus activity in campus (evening) | 17minute 30seconds | 5610 | 17.38 |
| Sikun Jaya bus activity in campus (morning) | 17minute 30seconds | 6900 | 18.15 |
| Sikun Jaya bus activity in campus (evening) | 17minute 30seconds | 6800 | 11.16 |

## 5 Conclusions

The following study found that results of the current existing bus services that should be improved in the university area or changed to an alternatives type of buses [10] because these services are often used by students who do not have their own vehicles to travel. It also the main transit for students/passengers to travel from campus to outside places. The study also found the locations of bus stops that provide the potential for students to use this service. The results also shows that stop area used by students should further extend the service because the bus ride would be faster and more efficient because of less waiting time at the stop and also shorter the journey time to complete a cycle. In addition, it shows that bus services should be improved and provides convenience bus stops for a comfortable which can accommodate a large number of students.

The bus facilities provide huge impact on student's comfort because it may reduces the feeling such a long wait that may not satisfy the students/passengers. The final objective is to study the effect of the number of stops on the time travel. The results of this study demonstrate the effect on number of stops that and travel time by linking the greater number of stops made by a longer bus time travel to complete a full circle. In addition, the results of this study may also explain the existing volume of bus traffic and it can be used to predict expectedly travel time accurately.

## Acknowledgement

The authors thank to the Universiti Tun Hussein Onn Malaysia for supporting the study through U395 contract grant. Appreciations are also presented to the Smart Driving Research Center, Faculty of Civil and Environmental Engineering and Registrar Office Universiti Tun Hussein Onn Malaysia for facilitating the study and make possible the results to be published.

## References

1. Ambak, K., Kasvar, K. K., Daniel, B. D., Prasetijo, J., Ghani, A. R. A., MATEC Web of Conf. **47**, 03008 (2016)
2. Public Transportation Fact Book, 63rd Edition. American Public Transport Association (2012)
3. HEPA, Pelan UTHM. Dec 3, https://www.uthm.edu.my/HEPA (2014)
4. Ruslawati A. W., Muhamad N.B., Riza Atiq A. O. K. R., Jurnal Teknologi (Sciences & Engineering) 78(2): 157-163 (2016)
5. P. Bullock, Journal of Public Transportation, no 1. (vol8), ms21-40 (2005)
6. Aimee Flannery, First Edition, New Fettern Lane, London: Spon Press (2010)
7. Prasetijo, J. & Zainal, Z. F., MATEC Web of Conf. **47**, 03001 (2016)
8. GPS Online, GPS Online tracks Sydney buses: 1 Sept (2000)
9. P. Reus, HCM 2010 First Edition, Sydern Denmark:Spon Press, California and Ostorn Engineeting (2010)
10. Azhar Yusof, Bas Elektrik Mulakan Operasi. Melaka Hari Ini.4-5 (2014)

# Mapping of Tidal Energy Potential based on High and Low Tides for Sabah and Sarawak

Kamran Ahmed Samo[1,a], Andrew Ragai Henry Rigit[2], Azhaili Baharun[3]

[1,a]*Kamran Ahmed Samo, PhD student, Department of Mechanical and Manufacturing Engineering, Faculty of Engineering, Universiti Malaysia Sarawak, 94300 Kota Samarahan, Sarawak Malaysia*
[2]*Andrew Ragai Henry Rigit, Professor, Department of Mechanical and Manufacturing Engineering, Faculty of Engineering, Universiti Malaysia Sarawak, 94300 Kota Samarahan, Sarawak, Malaysia (arigit@unimas.my).*
[3]*Azhaili Baharun, Associat Professor/Director Centre of Renewable Energy, Faculty of Engineering, Universiti Malaysia Sarawak, 94300 Kota Samarahan, Sarawak, Malaysia (bazhaili@unimas.my).*

**Abstract.** Tidal energy is one of the best predictable and reliable source of renewable energy. Therefore, this paper aims to map extractable tidal energy, and to determine potential locations to generate electricity from tidal power along the coastline of Sabah and Sarawak states of Malaysia. The data was obtained from Sarawak Marine Department, and analyzed by using ArcGIS version 10.3. There are two potential locations, namely Tawau in Sabah, and Pending in Sarawak, where 67.0kW and 115.4kW of optimum power generated were obtained respectively.

## 1 Introduction

The oceans around the world are a bountiful and reliable source of renewable energy [1]. Electricity generation from the ocean power offers a numerous advantages over other sources of renewable energies [2]. The sea energy has a vast potential in its operation and developments [3]. Sea energy can be categorized broadly into two main types, namely tidal and wave power. The specific forms of sea current are surface waves, tides, salinity, thermal gradients and circulation [1]. In general, the waves were produced by the interface of wind with the sea surface, and potential to deliver a boundless source of renewable power [4]. As soon as the wind blows across huge starches of sea, then the ocean waves were created [5]. The wave power can be extracted and transformed into electricity through wave power technologies and devices [4] [6].

The tidal power is more predictable than wave energy and wind power. The tidal power activities ebb and flow of the coastal tidal waters affected by interface of gravitational fields of the moon, sun and the earth [4, 7]. The tidal water levels follows a periodic highs and lows which were related with variations of water levels, and hence the tidal currents. Stronger tides were form, due to either in water level height or tidal current velocities [8]. The tidal generator uses this phenomenon to produce power. Tidal power is a method of non-pollution energy production, which has a vast potential [9]. Besides, growing awareness in the enhancement of tidal power, potentially support to shift use of fossil fuel in the generation of electricity. However, tidal power is not a new idea and several barrage type plants were working, the biggest existence 240 MW in Larance, Farance [10]. The tidal power comprises both potential energy and kinetic energy components [6, 7]. Hence, either utilizing the kinetic energy for tidal stream energy or constructed a barrage to generate the electricity using the potential energy by storage of water [11].

Till the year 1990, for utilization of tidal power required a basin with an appropriate water level variation as comparative to the outside body of water, as an existing turbine required a significant hydraulic head to work [12]. The tidal energy involved ocean waters at high tide and low tide, although its waves show the movement of its flow horizontally [7]. The diurnal tides, semidiurnal tides and the mixed tides are main types of tide. These tides have varying tidal periods, for example the diurnal tides have about 24 hours and 48 minutes whereas a period of 12 hours and 24 minutes reported for semidiurnal tides. The combination of the diurnal and the semi diurnal tides were known as a mixed tide, exists in the most different locations. Some of mixed tides were dominant on the diurnal and the semidiurnal tides. When the diurnal tides were dominant, the maximum tidal current arises at the greater declination of the moon and last current at the zero declination. While the semidiurnal tides were dominant, then the maximum tidal current arises at spring tide and the last tidal current at neap tides [13, 14, and 15].

The principle of tidal energy usage as a renewable energy source has a key role in the improvement of the human society by serving it to control and adapt to the environment [16]. Its utilization depends on the state of the technology and local characteristics of potential regions. For example, some of the challenges were often found with the operating principles, status, and efficiency and cost of generating energy related with each technology [17].

The barrages were constructed through the river estuary. The water level would flow in and flow out through the channel by which turbine rotates as soon as the forces applied onto the turbine (example shown in figure 1) [7].

[a] corresponding author: kamransamo2@gmail.com

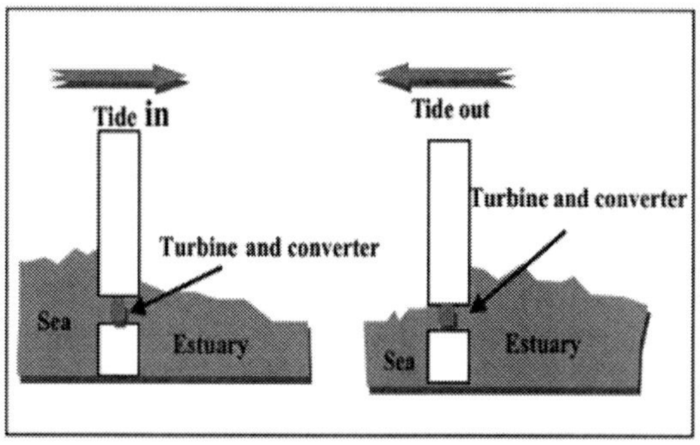


Figure 1: Basic method of operation for tidal barrage [7]

The tidal barrage is a dam, which blocks the incoming and outgoing tides and then builds a water head [18]. It makes use of potential energy due to its alteration in height between high and low tides [19]. The high tide and low tide of ocean level of water is effect by the gravitational force of the solar system. The scale of gravitational force for an object depends on its mass and distance. The moon has higher gravitational force compare to the sun, even though it is smaller than the sun. This force affected to the increment about 71% to the sea level. The greater movement of sea waters comprises largest amount of energy [7]. The tidal barrage were typically comprises turbines of either unidirectional or bi-directional [6]. The tidal barrages were planned to generate electric power by ebb or flood mood or can be both the ways [18]. In general, tidal barrage structure needs a very high investment cost [20] and thus, these schemes were not economically much appealing to the investor due to their delayed in time line of profit [21].

In contrary to the barrage technology, tidal current turbines extract kinetic energy by moving water to produce the electric power. The tidal technology is alike wind technology [6, 11]. The tidal turbines were mounted underwater at the sites in the ocean wherever great tidal stream occur and the tidal stream energy was an intermittent of horizontal water flow convoying increase and decrease of the tides [22] [23]. As the water flow stores a huge quantity of power, which can be extracted with the support of tidal in stream energy conversion (TISEC) plans [24]. The TISEC devices extracts power from the tidal stream as windmills gain power from the wind [25]. Tidal stream technology is extra suitable than the windmills as of its predictability and greater water density as related to air [26]. Tidal stream technology requires very low cost and has low environmental impact as compared to tidal barrage technology [27, 28].

## 2 Methodology

The data was collected from the Sarawak marine department (SMD), Malaysia. The data was acquired, hourly high and low tides tables and nautical charts were used for tidal energy resource valuation and mapping of potential locations. These tables illustrate the usage of hourly tidal energy flows according to the moon age (depicted by the symbols). Data was obtained from twelve locations of the year 2015 data as published by the directorate of marine, Sarawak, Malaysia. The Nautical charts were used to describe the physical features of the selected locations. The generated powers by the wave, dam of water tidal and tidal water energy were described in the following sections.

### 2.1 Power generated by the wave

The potential power from the dam determined through the volume of water stored. The amount of potential energy in water volume is given by Eq. (1):

$$E=hMg \tag{1}$$

Where; h = height of tidal water, M = mass of water (1025 kg/m$^3$), g = gravitational force (9.81 m/s$^2$).

This method was considered together with area of barrage and also ocean tidal 4 times in a day. This resultant method is used to estimate the produced energy [7].

$$W=\frac{(E\times(A\times H)\times 4)}{T} \tag{2}$$

Where; E = potential energy (Joule), A = area of barrage (m$^2$), H = difference of tidal height and T= duration in one day (Second).

For a most appropriate location to construct the barrage is at great amplitude tidal wave. Since, the tidal energy shown was 80% to variation energy of water to electricity generation [7].

### 2.2 Dam of water tidal

The barrage technology build to apply this method but construction of barrage involves capital cost and also have environmental impact to the huge area [7]. The potential energy of water head in the barrage facilitates to produce power from different tidal conditions [29]. A barrage was constructed through an estuary or bay so as to allow water to flow into basin as the tide comes in. The barrage has gates which let the water to permit through, thus filling basin. The gates are locked when the tide has stopped coming in, signifying the beginning of ebb. The level of water outside the basin continues to drop in ebb while the level of water inside the basin leftovers high. Alteration of water level produces a head. Water in the basin is then free into the sea through a set of turbines [30].

Power is generated by turbines, which happens during ebb and continues till tide floods and increasing water decreases head to the smallest operating point. The water is then taken again and repeats procedure. Power generation takes place during ebbs and then this method of procedure is called ebb generation. Through flood tide, gates and turbines are kept locked till a significant hydrostatic head is developed through the barrage. When adequate hydrostatic head is reached, turbine gates are opened permitting water to flow through them into the basin. Electricity is generated by flood tide method is termed as flood generation [30].

## 3 Result and discussion

### 3.1 Tidal water energy

The tidal energy mapping of potential sites was done after the analysis of tidal energy resources. The data of low and high tide energy extraction have been analysed and processed it with the help of an excel program. A maximum height of 6.2 m was identified at Pending location which has low and high tide among the other analysed locations. There are twelve identified locations to perform tidal water investigation, positions of these potential locations from the coastline of Sabah and Sarawak were as shown in figure 2.

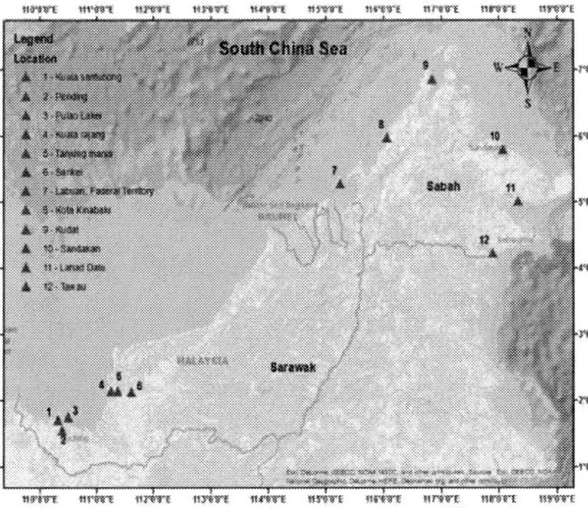


Figure 2: Map showing potential locations for tidal energy along Sabah and Sarawak coastline

The twelve locations of low and high tide data have been presented by utilizing ArcGIS software version 10.3. The energy mapping was carried out by using ArcGIS and the corresponding data as represented in excel spread sheets.

The river widths of these locations were assumed at about 200 meters, as proposed by Maulud *et.al* [7]. In this study, calculation was concluded using the tidal barrage technology and has been applied to produce tidal water to region. Tidal barrage area is about 200 m x 200 m. This was assumed according to department of irrigation and drainage, Malaysia [7].

The data in tide table (2015), for a height of either inward measure or outward measure of tide specified by time. Power was considered from the difference between height of inward and outward measure of tidal power. Graph shows the result from calculation water energy location have been plotted from the values shown, to categorize the areas where water tidal energy is maximum and reliable throughout the year. It is to determine maximum value realized at a certain locations. The parameters used for determining the potential locations for tidal energy were tabulated in table 1.

Table 1: The parameters used for locating tidal energy at various potential locations

| Locations | Lat N | Long M | Height (m) | E (Energy) | $P_{max}$ (kW) | $P_{min}$ (Kw) |
|---|---|---|---|---|---|---|
| Kuala Santubong | 01 43 | 110 19 | 5.0 | 50276.2 | 93.1 | 80.1 |
| Pending | 01 33 | 110 23 | 6.2 | 62342.5 | 115.4 | 100.6 |
| Palau Lakei | 01 45 | 110 30 | 4.8 | 48265.20 | 89.4 | 76.3 |
| Kuala Rajang | 02 09 | 111 15 | 5.7 | 57314.9 | 106.1 | 91.2 |
| Tanjung Manis | 02 09 | 111 22 | 5.8 | 58320.4 | 108.0 | 95.0 |
| Sarikei | 02 08 | 111 37 | 5.5 | 55303.8 | 102.4 | 91.2 |
| Labuan | 05 17 | 115 15 | 2.2 | 22121.5 | 41.0 | 27.9 |
| Kota Kinabalu | 05 59 | 116 04 | 1.9 | 19104.9 | 35.4 | 24.2 |
| Kudat | 06 52 | 116 50 | 2.0 | 20110.5 | 37.2 | 26.1 |
| Sandakan | 05 48 | 118 04 | 2.4 | 24132.6 | 44.7 | 33.5 |
| Lahad Datu | 05 01 | 118 20 | 2.3 | 23127.0 | 42.8 | 37.2 |
| Tawau | 04 14 | 117 53 | 3.6 | 36198.9 | 67.0 | 57.7 |

A graph of optimum height limit was generated to give a better result on focus locations, for generation of electricity. The highest energy generated at Pending in Sarawak was 115.4 kW, in the month of October, while the lowest energy of 100.6 kW was observed in June. Similarly, second highest energy of 67.0 kW was generated in the month of January at Tawau in Sabah, while the lowest energy was 57.7 kW in the month of June. A consistent energy could be generated at these areas.

The Figure 3 shows, the maximum energy generated at Pending in Sarawak among other analyzed areas. Therefore, Pending area is considered as the highest potential area of Sarawak coastline for tidal energy.

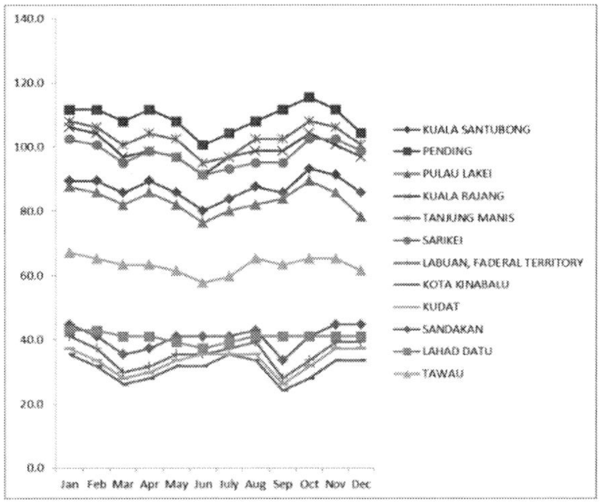


Figure 3: Graph of optimum height limit of power generated vs (month) of twelve locations at Sabah and Sarawak

DOI: 10.1051/matecconf/20178702007

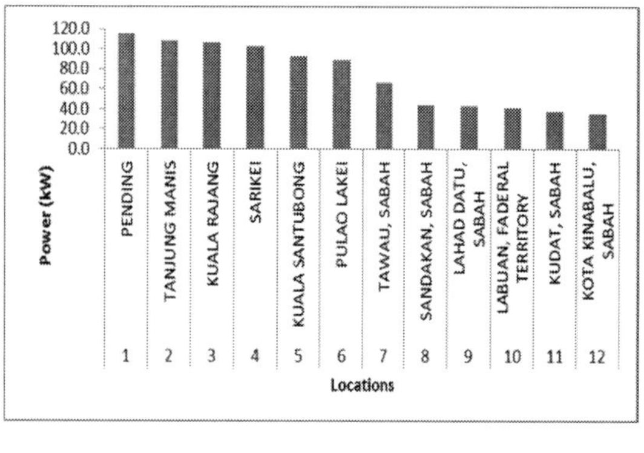


Figure 4: The tidal water energy generation

Figure 4 shows the generated energy between these twelve analyzed areas. The highest energy of 115.4 kW was generated from Pending in Sarawak. The second highest energy of 108.0 kW was generated from Tanjung Manis in the Sarawak area. The third highest energy generated was noted from Kuala Rajang in Sarawak area at 106.1 kW. An energy of 102.4 kW was reported as the fourth highest energy generated from Sarikei area. The fifth highest energy generated from Kuala Santubong area was 93.1 kW and the sixth highest energy generated from Pulau Lakei in the Sarawak area was 89.4 kW.

Further, the figure 4 shows highest energy of 67.0 kW was generated from Tawau in Sabah. . Subsequently 44.7 kW, 42.8 kW, 41.0 kW, 37.2 kW and 35.4 kW were generated respectively from Sandakan, Lahad Datu, Labuan, Kudat and Kota Kinabalu in Sabah area.

The graph shows the different of generated energy for six areas of Sarawak coast line involved in this research. Notably, Pending was observed as a potential area to generate electricity. The maximum generated power of 115.4kW for throughout year 2015 was observed. . While a lowest power of 89.4 kW was generated from the Pulau Lakei area as shown in figure 3.

Similarly, the graph shows the different ranges of energy generated for six areas of Sabah coast line involved in this research. From the results, Tawau area is a potential area to generate electricity. A maximum energy of 67.0 kW was generated power throughout year 2015. . While lowermost energy of 35.4 kW has been generated from Kota Kinabalu area as shown in figure 3.

There are two factors that affect tidal power. The first factor is the height difference between inward measure and outward measure. For Pending, the height differential is measured as 2.9 m where the highest inward movement is 6.2 m compare to others related areas which were not more than 5.8 m. The second factor in producing energy is the barrage area; barrage area that was used in the calculation has been standardized to estimate the comparison between each area. The generated energy will increase as the barrage area increases.

Among the data observed from 12 locations for high and low tides of east Malaysia, the two main locations identified as potential locations for generating maximum power of electricity are Pending in Sarawak coastline and Tawau in Sabah coastline. Figure 5 shows two potential locations of low and high tide east Malaysia.

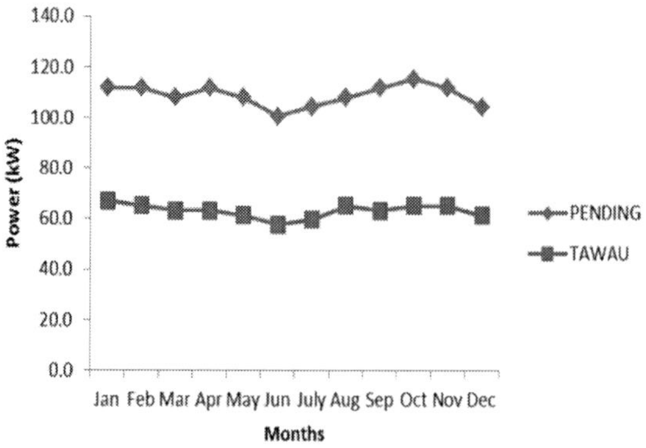


Figure 5: Shows two potential locations

## 4 Conclusions

Tidal energy is an environment friendly and potential renewable energy source Energy potential of tidal power can be utilized to generate electricity. The cost of increment and resource decrement of petroleum and natural gas give a way to improvement of renewable energy.This research has observed the potential of tidal barrage energy extraction for twelve locations at East Malaysia. The mapping of locations was carried out by ArcGIS ver. 10.3. In an effort to evaluate the potential of harnessing tidal barrage for electricity generation in east Malaysia two locations with great potential for tidal barrage energy extraction were identified. It was found from the study that the Pending location was identified as a potential location to generate the electricity with great amount of 115.4 kW among other examined locations of Sarawak coastline. The second potential location was identified at Tawau with amount of 67.0kW to generate electrify in Sabah coastline. However, the final work of the energy generated depends on the specification and design of the barrage size and turbine.

## 5 Recommendation for further work

For extraction of worthy electric power for utilization, it is recommended that suitable turbine will be identified for the chosen site. The turbine can be a vibrator or a spring type. A suitable low speed turbine will be identified for generation of electricity at Sabah and Sarawak coastline.

## Acknowledgement

The financial support for this work was provided by Universiti Malaysia Sarawak and data was provided by Sarawak Marine Department, Malaysia.

## References

1. A. S. Bahaj, Generating Electricity From the Oceans, *Renewable and Sustainable Energy Reviews*, Vol. 15, pp. 3399-3416, 2011.

2. L. Myers and A. S. Bahaj, Simulated Electrical Power Potential Harnessed by Marine Current Turbine Arrays in The Alderney Race, *Renewable Energy,* Vol. 30, pp. 1713-1731, 2005.
3. H.B. Lu, Y.L. Li, L.L. Ma, and X.L. Pao, The Design and Optimization for The Dome Of Tidal Turbine, *Energy Procedia,* Vol. 17, part B, pp. 1710-1716, 2012.
4. A. Q. Malik, Assessment of The Potential of Renewables for Brunei Darussalam, *Renewable and Sustainable Energy Reviews,* Vol. 15, pp. 427-437, 2011.
5. P. Lenee-Bluhm, R. Paasch, and H. T. Ozkan-Haller, Characterizing The Wave Energy Resource of The U.S Pacific Northwest, *Renewable Energy,* Vol. 36, pp. 2106-2119, 2011.
6. F. O. Rourke, F. Boyle, and A. Reynolds, Renewable Energy Resources and Technologies Applicable To Ireland, *Renewable and Sustainable Energy Reviews,* Vol. 13, pp. 1975-1984, 2010.
7. K. N. A. Maulud, O.A.Karim, K.Sopian and S.N.F.A.Aziz, Determination of Tidal Energy Resource Location in East Coast of Peninsular Malaysia Using Geographical Information System, 2009.
8. P. Meisen and A. Loiseau, Ocean Energy Technologies for Renewable Energy Generation, *Global Energy Network Institute,* 2009.
9. Z. Fan, Tidal Power Energy, *Renewable Energy In Future. Retrieved,* pp. 30-11, 2011.
10. J. F. Atwater and G. A. Lawrence, Power Potential of A Split Tidal Channel, *Renewable Energy,* Vol. 35, pp. 329-332, 2010.
11. A.R.H.Rigit, A. Q. Jakhrani, S.A.Kamboh, W. H. Kong and K. A. Samo, Mapping of Tidal Stream Energy Resources in The Coastline of Sarawak, pp. 1252-1261, 2013.
12. G. I. R. Carballo, A. Castro, Numerical Model Evaluation of Tidal Stream Energy Resources in The Ria De Muros (Nw Spain) *Renewable Energy,* pp. 1517-1524, 2009.
13. T. J. Hammons, Tidal Power, pp. 419-433, 1993.
14. G. Tom, *Oceanography: An Invitation to Marine Science (6th Edition Ed.)*,2007
15. P. B. H. George, Epri North American Tidal in Stream Power Feasibility Demonstration Project., 2006.
16. K. A. Maulud, W. W. Mohtar, and O. Karim, Spatial Multi Criteria Analysis for The Determination o Areas With High Potential Wave Energy, *Jurnal Teknologi,* Vol. 65, 2013.
17. J. K. N. P. Purbaa, R. Sandrob, S. Gibranb, R. A.I. Permatab, and M. K. M. F. Maulidab, Suitable Locations of Ocean Renewable Energy (ORE) in Indonesia Region – GIS Approached, *Energy Procedia,* pp. 230 – 238, 2015.
18. J. Xia, R. A. Falconer, B. Lin, and G. Tan, Estimation of Annual Energy Output from A Tidal Barrage Using Two Different Methods, *Applied Energy,* Vol. 93, pp. 327-336, 2012.
19. Z. Defne, K. A. Haas, H. M. Fritz, L. Jiang, S. P. French, X. Shi, B.T.Smith, V.S.Neary and K.M.Strewart , National Geodatabase of Tidal Stream Power Resource in U.S.A, *Renewable and Sustainable Energy Reviews,* Vol. 16, pp. 3326-3338, 2012.
20. E. Denny, The Economics of Tidal Energy, *Energy Policy* pp. 1914–1924, 2009.
21. R. H. Charlier, Sustainable Co-Generation from The Tides: A Review, *Renewable and Sustainable Energy Reviews,* pp. 187–213, 2003.
22. J. Hardisty, Assessment of Tidal Current Resources: Case Studies of Estuarine and Coastal Sites, *Energy & Environment,* Vol. 18, pp. 233-249, 2007.
23. S. Couch and I. Bryden, Tidal Current Energy Extraction: Hydrodynamic Resource Characteristics, *Proceedings of The Institution of Mechanical Engineers, Part M: Journal of Engineering for The Maritime Environment,* Vol. 220, pp. 185-194, 2006.
24. M. Previsic and R. Bedard, Methodology for Conceptual Level Design of Tidal in Stream Energy Conversion Power Plants, *Epri North American Tidal In Stream Power Feasibility Demonstration Project,* 2005.
25. I. Bryden, S. Naik, P. Fraenkel, and C. Bullen, Matching Tidal Current Plants to Local Flow Conditions, *Energy,* Vol. 23, pp. 699-709, 1998.
26. J. B. Blanchfield, The Extractable Power from Tidal Streams, Including A Case Study for Haida Gwaii, University of Victoria, 2007.
27. M.S. Guney and K.Kaygusuz, Hydrokinetic Energy Conversion Systems: A Technology Status Review, *Renewable and Sustainable Energy Reviews,* pp. 2996-3006, 2010.
28. C. M. Johnstone, D. Pratt, J. A. Clarke, and A. D. Grant, A Techno-Economic Analysis of Tidal Energy Technology, *Renewable Energy,* Vol. 49, pp. 101-106, 2013.
29. C. Baker, J. Walbancke and P. Leache, Tidal Lagoon Power Generation Scheme in Swansea Bay, A report on behalf of the Department of Trade and Industry and the Welsh Development Agency, 2006.
30. A. Q. Jakhrani,A.R.H.Rigit and S. R. Samo, Estimation of Tidal Stream Energy Resources at Sarawak Coastline and Their Potential Impact on Environment, *Australian Journal of Basic and Applied Sciences,* vol. (6), pp. 503-514,, 2013.

# Active Heterogeneous CaO Catalyst Synthesis from *Anadara granosa* (Kerang) Seashells for Jatropha Biodiesel Production

Reddy ANR[1], AA Saleh[1, a], M S Islam[2] and S. Hamdan[1]

[1]*Department of Mechanical and Manufacturing Engineering, Faculty of Engineering, Universiti Malaysia Sarawak, Malaysia*
[2]*Department of Chemistry, Bangladesh Army University of Engineering & Technology, Qadirabad, Cantonment, Natore-6431, Bangladesh*

**Abstract.** Heterogeneous catalysts are often used at large to produce biodiesel from non-edible vegetable crude oils such as *Jatropha curcas* oil (JCO). In this study, an active heterogeneous CaO catalyst was synthesized from a tropical biodiversity seashells *Anadara granosa* (*A.granosa)*. The catalytic efficiency of *A.granosa* CaO was investigated in transesterification of JCO as biodiesel. The *A.granosa* CaO catalyst was synthesized using 'Calcination – hydration – dehydration' protocol. The spectral characterization of the catalyst were investigated by employing FT-IR, SEM, BET and BJH spectrographic techniques. The experimental design was executed with four reaction parameters that include catalyst concentration (CC), methanol ratio (MR), transesterification time (TT) and reaction temperature (RT). The JCO transesterification reactions as well as impact of reaction parameters on the Jatropha biodiesel yield (JBY) were analyzed. The sufficiency of the experimental results conformed through sequential validation tests, as a result, an average of 96.2% JMY was noted at optimal parametric conditions, CC of 3wt. %, TT of 120 min, MR of 5 mol. and RT of 60°C at a constant agitation speed of 300rpm. An average JMY of 87.6% was resulted from the *A.granosa* CaO catalyst during their recycling and reuse studies up to third reuse cycle.

## 1 Introduction

Ever increasing energy needs for the world's comprehensive developments emphasis on intrinsic and sustainable approaches into multiphase coherent and seamless research for the future[1]. On the contrary, despite substantial policy and research initiatives, the global warming, pollution as caused by the fossil fuel combustion emissions besides depletion of natural petroleum deposits reiterates search for renewable and alternate fuel and energy sources. In early 1980's the biodiesels were commercially launched as an alternate fuel in the view of minimizing the greenhouse exhaust gas emissions[2]. The renewable feed stocks such as edible and non-edible vegetable oils, algae oils, animal fat wastes etc.[3,4] were prominent in biodiesel production by their transesterification with methanol and a catalyst. To minimize the demand for edible oils and to maintain sustainable human food cycle, non-edible feed stocks such as *Jatropha curcas*, a euphorbia family plant mainly grown in tropical and sub-tropical regions, has accomplished both commercial and researcher's attention, besides the plants supplementary benefits[4].

A large number of homogeneous and heterogeneous catalysts were experimentally utilized in transesterification of various vegetable oils and among which heterogeneous catalysts demonstrated higher catalysis, stability, reusability and ease in production mechanisms[5,6]. An adequate usage of catalysts as synthesized from diverse sources of wastes is crucial as part of the sustainable and strategic management of wastes and so as the environmental concerns[7,8]. Many researchers have reported on calcium based catalysts synthesized by make use of direct calcination over a temperature range of 800°C – 900°C from domestic waste shells such as cockle shells[9], *Pomacea* sp. shells[10] derived CaO from mussel shells[11], waste chicken-eggshell[12], ostrich-eggshell[12] etc. and utilized effectively in transesterification. These reports emphasizes CaO catalysts prominence in catalysis of triglycerides of oils to biodiesel. Since the domestic chicken eggshells are notably abundant, economic synthesis of catalysts from theses waste shells encompasses the recycling of wastes for sustainability and environmental advantages as well. *Anadara granosa* (*A.granosa)* seashells are prominent and economic edible protein source in the Indo-Pacific tropical region especially from wetlands of Kuching[13] and Asajaya[14] areas in Sarawak state and also in other coastal areas of Malaysia. *A.granosa* shells are largely available in the region and locally known as 'Kerang'[13,14]. According to our literature review, any research was published on synthesis of heterogeneous CaO catalyst.

The present research reports on CaO synthesis with higher surface area form *A.granosa* seashells to produce an active CaO heterogeneous catalyst. 'Calcination – hydration – dehydration' protocol was employed for

[a] Corresponding author: aasaleh@unimas.my

economic catalyst synthesis and the catalyst was fully characterized by spectrographic techniques. The synthesized CaO Catalytic activity was tested in transesterification of the *Jatropha curcas* oil (JCO). The Jatropha biodiesel yield (JBY) optimization was achieved through analysis of four principal reaction parameters include catalyst concentration, methanol ratio, transesterification time and reaction temperature. The *A.granosa* CaO catalyst reusability together with economic catalyst synthesis in comparison with a commercial CaO were investigated.

# 2 Experimental

## 2.1 Materials

The *Jatropha curcas* oil (JCO) used for biodiesel production was extracted from their seeds using oil expelling machine without purification and or refining. The JCO saponification (SV) and acid values (AV) were calculated by adopted a standard procedure[15] and noted as 210.04 $mg\,KOH/g$ and 29.46 $mg\,KOH/g$ respectively. The JCO molecular weight (M) = $(56.1 \times 1000 \times 3)/(SV - AV)$[15] resulted as 932 $g/mol$. All experiments were performed using laboratory grade chemicals that include potassium hydroxide 95%, methyl alcohol >99%, calcium oxide CAS.NO.0001305788 and sulfuric acid 99.9%. The experimental protocol was designed and executed at the energy lab, mechanical and manufacturing engineering department, faculty of engineering, Universiti Malaysia Sarawak (UNIMAS), Malaysia.

## 2.2 Catalyst synthesis

The *A.granosa* shells were collected from a local market then carefully rinsed simultaneously with tap, hot and distil water to clean up the edible organic fractions adhered to the eggshells as well as other impurities. The *A.granosa* shells were overnight dried at 105°C[16] in a hot-air oven then grounded using a domestic blender and sieved through 80µm mesh. The shells powders were calcined at 900°C for 2hrs utilizing a muffle furnace (KSL-1700X-A4, MTI Corporation, USA). The *A.granosa* shells derived CaO were refluxed in distilled water continuously for 6hrs at 60°C. The filtered calcium solid particles were overnight dried in a hot-air oven at 120°C further finely grounded for 3hrs by employing a ball mill with an operating speed of 200rpm. The fine calcium particles were recalcined for 3hrs at 600°C to dehydrate the hydroxides and to form calcium oxides. The powders were dehydrated for 1hr in a hot air oven at 100°C to make it free from working atmosphere effects, before being used as a catalyst. Thus, an active heterogeneous CaO catalyst from *A.granosa* shells was synthesized subjected to the 'Calcination–hydration–dehydration' synthesis protocol.

## 2.3 CaO catalyst characterization

To analyse the chemical features of synthesized CaO sample particles FT-IR: Fourier Transform Infrared, Spectrophotometer (Perkin Elmer Model 100 series) was employed and tested over a wavelength region of 4000 to 280 $cm^{-1}$. The particle morphology was studied using a SEM: scanning electron microscopy (Hitachi TM3030). Brunauer-Emmett-Teller (BET) and Barrett-Joyner-Halenda(BJH) (Quantachrome iQ-AG-C) were utilized to examine the catalyst surface area pore diameter and pore volumes respectively.

## 2.4 Catalyst economic synthesis study

The economic synthesis of CaO from *A.granosa* shells was carried out considering the raw *A.granosa* shells as procured from the market together with CaO synthesis costs at laboratory scale. The price of *A.granosa* shells synthesized CaO was compared with price of a commercial CaO (CAS number: 0001305788).

## 2.5 Experimental design for biodiesel production

Jatropha biodiesel was produced from JCO following a two-step transesterification based on many preliminary laboratory experiments to analyze all-important parameters such as catalyst concentration (CC), transesterification time (TT), methanol ratio (MR), and reaction temperature (RT) were ascertained and analyzed in contrast with previous studies reported in the literature[15,16]. An in-house laboratory protocol was followed for biodiesel synthesis, in brief free fatty acid (FFA) analysis, acid esterification and transesterification together with methanol and catalyst mixture, separation of glycerol, washing and drying. FFA of a vegetable oils impacts on their transesterification reactions during biodiesel production[17,18]. 29.3% of FFA was measured from the JCO samples by following a standard method as described our recent work published[19]. Since the oil samples have reported a very high FFA, which lead to lowering the biodiesel yield, especially, in solid heterogeneous catalyzed transesterification[18–20]. Acid esterification of JCO was carried out in the presence of $H_2SO_4$ and methyl alcohol to reduce the JCO FFA to 1% which is suitable for base catalyzed transesterification[21].

Transesterification reaction firstly transforms the triglycerides of JCO samples to diglycerides and then to monoglyceride and methyl esters (biodiesel) subsequently as a final product in three consecutive reactions which are equivalent and reversible in nature. *A.granosa* shells synthesized CaO catalyzed biodiesel reaction was carried out with esterified JCO. The catalyst and methanol mixture were prepared in accordance with oil ratio by volume. The transesterification reaction process operating ranges were comprehend with state of the art literature[18,19,22]. The reactant mixture contents were allowed to react over a course of transesterification four variables catalyst concentration (CC: 1% - 6 %wt.), transesterification time (TT: 30min. – 180 min), methanol ratio (MR: 1 – 6 mol.), and reaction temperature (RT: 35°C-65°C) at constant agitation speed of 300rpm with a magnetic stirrer. The catalyst was collected from the mixture with a centrifuge while glycerol and biodiesel layers were separated followed by warm deionized double distilled water wash and heating in order to obtain pure biodiesel.

## 2.6 Catalyst reusability study

Transesterified JCO and other mixtures were centrifuged at 3500rpm for 1hr to recover *A.granosa* shells CaO. The solid precipitate was thoroughly washed using n-hexane successively for four times to prepare an impurities free catalyst and oven dried for 24hrs. The recovered CaO was recalcined for 2hrs at 600°C before using in transesterification of JCO.

# 3 Results and discussion

## 3.1 Catalyst characterization analysis

The infrared (IR) spectrum of CaO catalyst as synthesized from *A.granosa* shells was graphically shown in Figure 1. It's evident that Ca-O group absorptions were strongly attributed over wavenumbers <600cm$^{-1}$, along with medium bands at 1427cm$^{-1}$ and 876cm$^{-1}$, as well as weak bands at 3740cm$^{-1}$, 3850cm$^{-1}$ and 978cm$^{-1}$ A Sharp absorption peak at 1427 cm$^{-1}$ corresponds a carbonyl group due to their asymmetrical stretching vibration[23]. The carbonyl and hydroxyl group traces were exist in the IR spectrograph due to absorption of carbon dioxide and moisture on to the calcined CaO samples from the atmosphere[10].

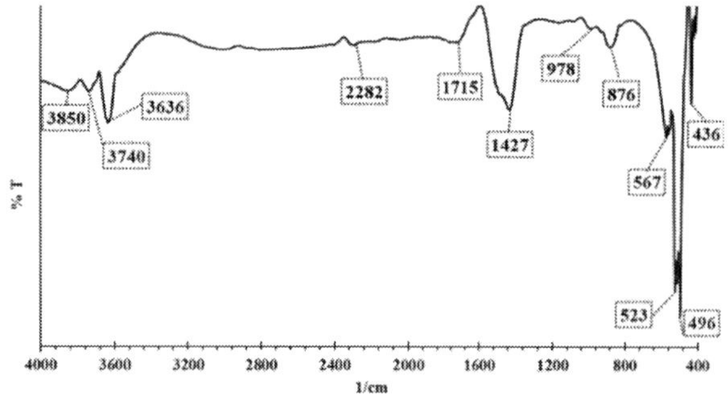


**Figure 1.** IR Spectrum of *A.granosa* shells synthesized CaO catalyst

The SEM monograms in Figure 2 (a-b) shows morphology and active sites of calcined CaO particles synthesized from *A.granosa* shells. The CaO particle monograms illustrates presence of uneven particle surfaces and hence formation of a porous catalyst structures. As a result, the calcined CaO agglomerations could provide highly active surfaces. The surface area of CaO particles synthesized from *A.granosa* shells together with their average pore diameter and pore volume were noted form the BET and BJH results 34.6 m$^2$/g, 9.151 nm and 0.012 cc/g respectively. The catalyst pore diameter represents mesopores it would show cases higher suitability for catalytic reactions[24]. Further, these results significantly agrees higher specific surface area of CaO particles by which catalyst catalytic activity improves in transesterification of triglycerides to methyl esters and hence the synthesized catalyst is stated as an active catalyst[24].

**Figure 2.** SEM spectral monograms of *A.granosa* shells CaO particles

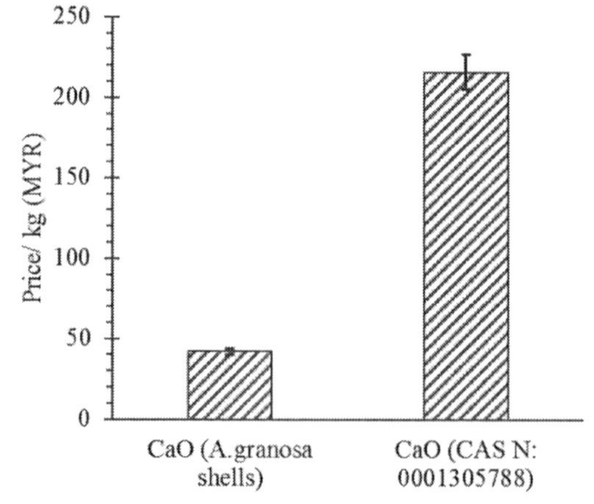


**Figure 3.** *A.granosa* shells CaO synthesis cost analysis in comparison with commercial CaO.

**Table 1.** Jatropha biodiesel yield % w.r.t. varation in transesterification reaction parameters

| Catalyst concentration (CC) | | Methanol ratio (MR) | | Transesterification time (TT) | | Reaction temperature (RT) | |
|---|---|---|---|---|---|---|---|
| CC (wt. %) | JBY % | MR (mol.) | JBY % | TT (min.) | JBY % | RT (°C) | JBY % |
| 1 | 78 | 1 | 57 | 30 | 81 | 40 | 75 |
| 2 | 93 | 2 | 60 | 60 | 90 | 45 | 83 |
| 3 | 97 | 3 | 70 | 90 | 93 | 50 | 89 |
| 4 | 88 | 4 | 84 | 120 | 96 | 55 | 92 |
| 5 | 74 | 5 | 96 | 150 | 88 | 60 | 97 |
| 6 | 68 | 6 | 94 | 180 | 82 | 65 | 93 |

## 3.2 *A.granosa* Catalyst synthesis economic analysis

The CaO synthesis cost form *A.granosa* shells was calculated based on the raw shells as well as approximate laboratory synthesis costs and compared with market price of a commercial CaO (CAS number: 0001305788) as shown in Figure 3. The *A.granosa* shells synthesized CaO was found to be more economic then the commercial CaO. The major cost difference may be due to usage of very low cost raw materials as well as customized synthesis protocols adopted at the laboratory.

## 3.3 Jatropha biodiesel production analysis

The biodiesel production metrics using *A.granosa* shells CaO from JCO were tabulated Table 1. The experimental results of Jatropha biodiesel yield (JBY) percentage with respect to each of transesterification reaction parameter performed were plotted in Figure 4. The catalyst concentration influences the biodiesel yield at large together with other factors. The JMY% was incremental within 1wt. % - 3wt. % of CC and resulted a maximum JBY of 97% at 3wt. %. However, further increase in CC resulted in lowering the JBY as a result of soap formation and secondary reactions[25].

Methanol ratio is a primary parameter in transesterification of vegetable oils. The results from both Table 1 and Figure 4 demonstrates the relevance of higher MR i.e. 5 mol. yields 96% of biodiesel. A reduction in JMY% was noted beyond 5 mol. of MR, which may be a result of reversible reaction due to the presence of excess over methanol[26]. The transesterification time for biodiesel production was a collective impact of other reaction parameters[27,28]. JMY% as obtained from the experiments shows, Figure 4, 120 min. of TT yields 96% biodiesel which is maximum. Experiments run 120 min. reveals a decrease in JMY as the excessive TT destabilizes the state of reaction equilibrium besides emergence of soap formation[27,29,30] and hence on the JMY as well.

The biodiesel production was testified over 35°C-65°C reaction temperature (RT). At 60°C of RT a maximum JMY of 97% was reported. Reaction kinetics are considered to be accelerated at elevated RT, however viscosity JCO predominantly minimizes which further leads to soap formation as well as lowering the JMY[3]. Furthermore, temperatures over 60°C were close to the methanol's boiling point 64.7°C, consequently expedites methanol evaporation which further curtails their availability for completing the reaction[27,31].

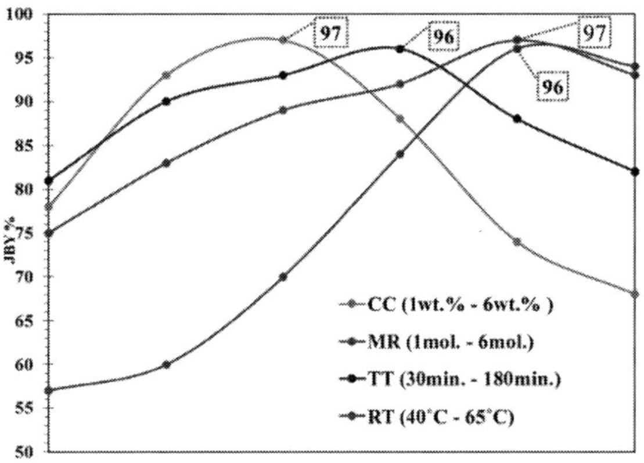


**Figure 4.** Graph showing Jatropha biodiesel yield % vs transesterification parameters

From the experimental result analysis, the optimal parameters contributed for maximum JMY% were noted. A series of four confirmatory experiments were performed simultaneously at CC of 3wt. %, TT of 120 min, MR of 5 mol. and RT of 60°C at a constant agitation speed of 300rpm, as a result, an average of 96.2% JMY was noted.

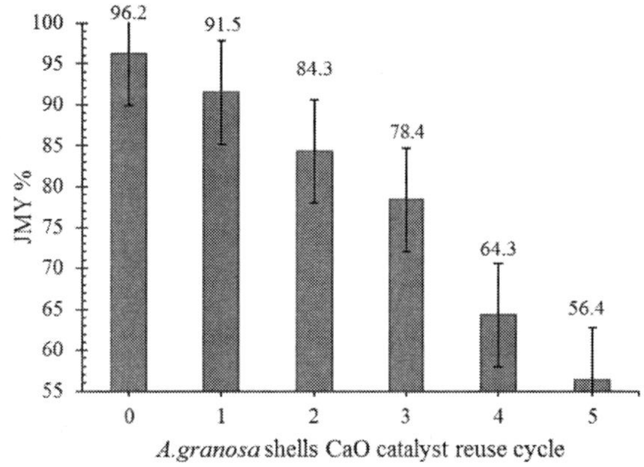


**Figure 5.** *A.granosa* shells synthesized CaO catalyst reusability study over five reuse cycles.

## 3.4 *A.granosa* Catalyst reusability analysis

The *A.granosa* shells synthesized CaO catalyst reusability in transesterification of JCO was studied over optimal reaction conditions that include CC of 3wt. %, TT of 120 min, MR of 5 mol. and RT of 60°C at a constant agitation speed of 300rpm. During five catalyst reuse cycles conducted successively, the JMY was reducing from each of their previous reuse cycle as depicted in Figure 5. This may be due to the fact that losing of active sites present in the *A.granosa* shells synthesized CaO as well as agglomerations formed out of catalyst particles and JCO. However, the *A.granosa* shells synthesized CaO catalyst was stable and yields 78.4% of biodiesel after third reuse cycle.

## 4 Conclusions

In this research, an active heterogeneous CaO catalyst was successfully synthesized from a tropical biodiversity seashells *Anadara granosa* (*A.granosa).* The *A.granosa* CaO catalyst was synthesized by following 'Calcination – hydration – dehydration' protocol. The spectral characterizations of the synthesized catalyst were investigated using FT-IR, SEM, BET and BJH spectrographic techniques. The SEM and BET results reveals presence of mesopores uneven particle surfaces with an average pore diameter and pore volume of 34.6 $m^2/g$, 9.151 nm and 0.012 cc/g respectively. The *A.granosa* CaO catalytic efficiency was studied over transesterification of JCO as biodiesel considering catalyst concentration (CC), methanol ratio (MR), transesterification time (TT) and reaction temperature (RT). An average of 96.2% JMY was noted as optimal, at CC of 3wt. %, TT of 120 min, MR of 5 mol. and RT of 60°C at a constant agitation speed of 300rpm. In the *A.granosa* CaO catalyst recycling and reuse an average JMY of 87.6% was resulted up to third reuse cycle.

## Acknowledgements

The authors would like to thank Faculty of Engineering, UNIMAS and all the staff members for providing research facilities and their continuous support during this work. The first author, ANR, would greatly acknowledge and express his highest gratitude to Ministry of Higher Education, Malaysia for awarding prestigious Malaysian International Scholarship MIS-2015; KPT.B.600-4/1/12 (60).

## References

1. G.B. Asheim, "WPS1302", Washington, DC 20433. (1994).
2. US-EPA, "RFS2: EPA-420-R-10-003", Washington DC (2010).
3. J. Qian, H. Shi, and Z. Yun, *Bioresour. Technol.*, **101**, 7025–7031 (2010).
4. L.C. Meher, C.P. Churamani, M. Arif, Z. Ahmed, and S.N. Naik, *Renew. Sustain. Energy Rev.*, **26**, 397–407 (2013).
5. E.F. Aransiola, T.V. Ojumu, O.O. Oyekola, T.F. Madzimbamuto, and D.I.O. Ikhu-Omoregbe, *Biomass and Bioenergy*, **61**, 276–297 (2014).
6. J.M. Marchetti, V.U. Miguel, and a. F. Errazu, *Renew. Sustain. Energy Rev.*, **11**, 1300–1311 (2007).
7. I. Abdulrahman, H.I. Tijani, B.A. Mohammed, H. Saidu, H. Yusuf, M.N. Jibrin, and S. Mohammed, *J. Mater.*, 6 (2014).
8. M.A. Martin-Luengo, M. Yates, M. Ramos, J.L. Salgado, R.M.M. Aranda, F. Plou, J.L. Sanz, R.L. Pirrongellil, E.S. Rojol, L.G. Gill, and A.M.M. Serranol, *IEEE WCST '11*, 19–22 (2011).
9. K.N. Islam, M.Z.B.A. Bakar, M.E. Ali, M.Z. Bin Hussein, M.M. Noordin, M.Y. Loqman, G. Miah, H. Wahid, and U. Hashim, *Powder Technol.*, **235**, 70–75 (2013).
10. Y.Y. Margaretha, H.S. Prastyo, A. Ayucitra, and S. Ismadji, *Int. J. Energy Environ. Eng.*, **3**, 33 (2012).
11. A. Buasri, N. Chaiyut, V. Loryuenyong, P. Worawanitchaphong, and S. Trongyong, *Sci. World J.*, **2013**, 1–7 (2013).
12. Y.H. Tan, M.O. Abdullah, C. Nolasco-Hipolito, and Y.H. Taufiq-Yap, *Appl. Energy*, **160**, 58–70 (2015).
13. H. Hamli, M.H. Idris, M.K. Abu Hena, and S.K. Wong, *Int. J. Zool. Res.*, **8**, 52–58 (2012).
14. F. Hossen, S. Hamdan, and R. Rahman, *Sci. World J.*, **2014**, 4 (2014).
15. X. Deng, Z. Fang, and Y.H. Liu, *Energy Convers. Manag.*, **51**, 2802–2807 (2010).
16. B. Yoosuk, P. Udomsap, B. Puttasawat, and P. Krasae, *Bioresour. Technol.*, **101**, 3784–3786 (2010).
17. H. a. Choudhury, P. Srivastava, and V.S. Moholkar, *AIChE J.*, **60**, 1572–1581 (2014).
18. X. Deng, Z. Fang, Y.H. Liu, and C.L. Yu, *Energy*, **36**, 777–784 (2011).
19. R. ANR, A.A. Saleh, M.D.S. Islam, S. Hamdan, and M.A. Maleque, *Energy & Fuels*, **30**, 334–343 (2016).
20. H. Jaliliannosrati, N.A.S. Amin, A. Talebian-Kiakalaieh, and I. Noshadi, *Bioresour. Technol.*, **136**, 565–573 (2013).
21. H.J. Berchmans and S. Hirata, *Bioresour. Technol.*, **99**, 1716–1721 (2008).
22. A. Okullo, a. K. Temu, J.W. Ntalikwa, and P. Ogwok, *Int. J. Eng. Res. Africa*, **3**, 62–73 (2010).
23. R. Ljupkovic, R. Micic, M. Tomic, N. Radulovic, A. Bojic, and A. Zarubica, *Hem. Ind.*, **68**, 399–412 (2014).
24. Y. Ren, Z. Ma, and P.G. Bruce, *Chem. Soc. Rev.*, **41**, 4909 (2012).
25. M. Sánchez, F. Bergamin, E. Peña, M. Martínez, and J. Aracil, *Fuel*, **143**, 183–188 (2015).
26. P. Verma, M.P. Sharma, and G. Dwivedi, *Renew.*

*Sustain. Energy Rev.*, **56**, 319–333 (2016).
27. D.Y.C. Leung, X. Wu, and M.K.H. Leung, *Appl. Energy*, **87**, 1083–1095 (2010).
28. D.Y.C. Leung and Y. Guo, *Fuel Process. Technol.*, **87**, 883–890 (2006).
29. K. Prasertsit, P. Phoosakul, and S. Sukmanee, *Songklanakarin J. Sci. Technol*, **36**, 195–200 (2014).
30. M. Kouzu, T. Kasuno, M. Tajika, S. Yamanaka, and J. Hidaka, *Appl. Catal. A Gen.*, **334**, 357–365 (2008).
31. M.T. Freedman B, Pryde EH, *J Am Oil Chem Soc*, **10**, 1638–1643 (1984).

DOI: 10.1051/matecconf/20178702009

# Development of Wind Tunnel for Ultrafine Palm Oil Fuel Ash Separator

Sylvester Kenneth Sedah[1], Ana Sakura Zainal Abidin[1,a], A. Rigit[1], Mohd Azlan Ismail[2], Raudhah Ahmadi[3], and Md Abdul Mannan[3]

[1] *Department of Mechanical and Manufacturing Engineering, Faculty of Engineering, Universiti Malaysia Sarawak, Kota Samarahan, Sarawak, Malaysia*
[2] *Faculty of Engineering, Universiti Malaysia Sabah, Kota Kinabalu, Sabah, Malaysia*
[3] *Department of Civil Engineering, Faculty of Engineering, Universiti Malaysia Sarawak, Kota Samarahan, Sarawak, Malaysia*

**Abstract.** Palm Oil Fuel Ash (POFA) is a pozzolanic material that has great potential in concrete production. POFA sizes range from 1 to 10 µm is preferable. Thus, there is a need for efficient method to separate the POFA according to the sizes needed. This study explores the idea of using a simple wind tunnel as a separator for ultrafine POFA. The existing operation of the wind tunnel is simulated using ANSYS-CFX, a numerical modelling software. The identified weakness of the wind tunnel design is located at the hopper. The wind tunnel is simulated with a different air inlet velocity in order to visualize the air flow profile as well as air flow energy at the hopper. Different air inlet velocities has a similar air flow profile but the air flow intensity is different. For further study the modifications of the air flow profile were conducted. By altering the hopper angle shows that the air flow profile is changed. Based on the simulation results, the hopper at 60° yields the best flow characteristic. At angle 60° of the hopper, the air flow is mostly directed downward into the tunnel and the air flow energy is low. Therefore, it allows the POFA to be directed into the tunnel and reduced the ultrafine POFA losses at the tip of the hopper. This practical method utilizes a simple open wind tunnel that enables the process to be more efficient and cost efficient thus the production of ultrafine POFA can be increased.

## 1 Introduction

Palm oil has become a significant product in the global market. Palm oil is obtained from a plant that is oil palm tree, scientifically known as Elaeis guineensis [1]. In tropical countries, palm oil is an important commodity. Malaysia is one of the producer and exporter in the world [2]. Malaysia has endeavoured to advance with the increasing demand for vegetable oil as palm oil industry is one of the biggest agricultural industries.[3].

As millions of tons of the by-product known as Palm oil fuel ash (POFA) have been disposed off annually; it indirectly contributed a negative impact to the environment. Researches had conducted many studies to convert this waste material into useful products. POFA has been identified as pozzolanic material [4] which is suitable to be used in concrete production in 20th century [5]. This waste material can be integrated as a partial cement replacement to produce various types of concrete which consume lesser amount of cement, more environmental friendly and economical [4]. As the fineness of the POFA influenced the strength of the concrete [3], it is essential to have a proper method to separate the ultrafine POFA in order to get ultrafine POFA sizes range from 1 to 10µm. [6].

Therefore, an application of wind tunnel concept in separating the ultrafine POFA had been applied. Fundamentally, at suitable velocity of air, the fine and light particles will be blown away and the heavy particle will drop. The wind tunnel; a tube-like-tunnel at varying cross sections is made to blow through them at certain speed [7]. It was classified as open wind tunnel with blowdown principle in which a fan was located at the inlet of the tunnel. The wind tunnel has five main components which are a settling chamber, contraction cone, test section, diffuser, and drive section [8].

The proposed wind tunnel for this analysis as compared to the existing wind tunnel by other researches is simpler in design with lower cost and safer to build [9]. The wind tunnel sensitive parameters that are commonly observed in these area of studies are pressures, forces, and direction of air flow impacting the tested models, such as spacecraft, cars, trucks, trains, or entire cities [7]. The behaviour of a model can be predicted visually.

In this study, further research on the design of the wind tunnel is carried out to extend the practical method of separating the ultrafine POFA. Previously, it was proven that ultrafine POFA was able to be collected at a certain distance in the tunnel [6]. Figure 1 shows the design of the wind tunnel with a length of 10 metres and it consists of three main components; diffuser, hopper and tunnel.

Application in using numerical modelling software, enables better understanding of the flow pattern, reducing

[a] Corresponding author: zaasakura@feng.unimas.my

 DOI: 10.1051/matecconf/20178702009

number of repetitive experiments that consequently cut down on experimental cost [10].

The research utilized POFA collected from Bau Palm Oil Mill (BAPOM) which is located in Bau, Sarawak. The study focused on improving the wind tunnel design that must be simple and lower construction cost, consequently allowing for low cost research applications in other faculty researches regarding ultrafine POFA. The wind tunnel is designed to cater for the separation of ultrafine POFA obtained from BAPOM only.

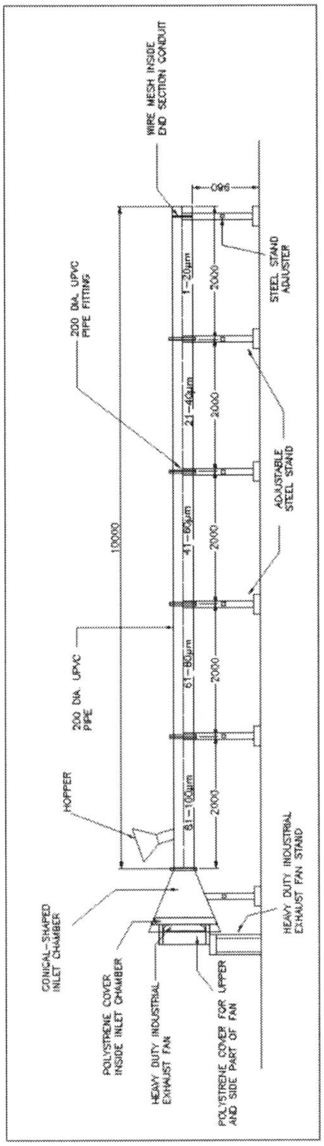


**Figure 1.** Schematic diagram of the wind tunnel

## 2 Methodology

In this study, the CFD model was used to simulate the air flow across the wind tunnel with a hopper positioned between the flow passages. The air flow can be simulated in the wind tunnel especially when the air passed through the hopper section. The simulation flow domains take into account the overall internal flow region of the wind tunnel, however, the observed region was the air passing through the hopper channel.

The meshing is set with element size of 0.01 metres, with high smoothing and fast transition setting as shown in Figure 2.

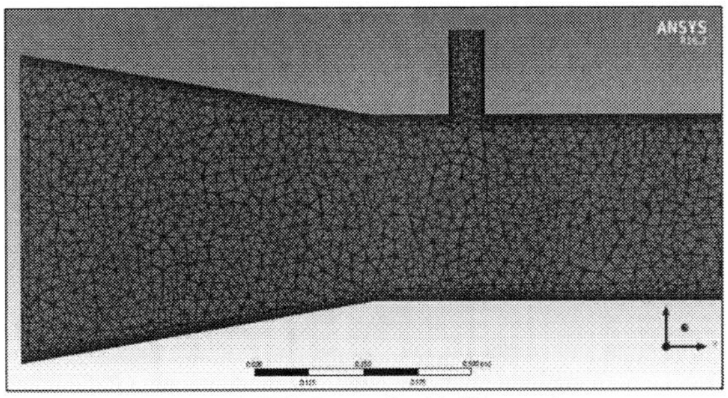


**Figure 2.** Wind tunnel with fine mesh

The inflation meshing setting was focused on the hopper in order to see the air flow changes when passing through the hopper as shown in Figure 3.

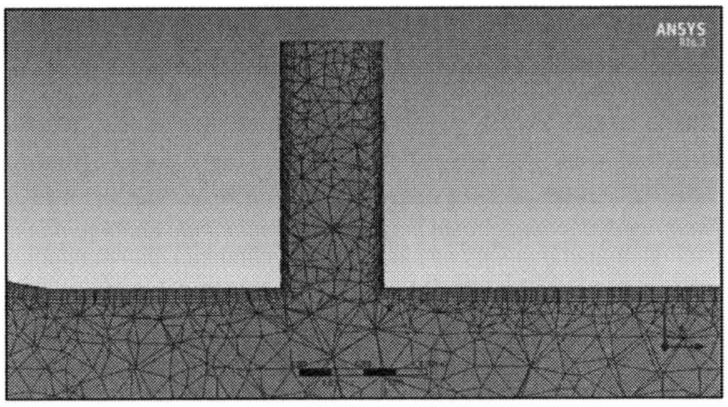


**Figure 3.** Inflation on the hopper

The simulation of air flow inside the wind tunnel was set on the basis of the following basic assumptions:

- Steady state condition
- Constant fluid properties
- Incompressible fluid flow
- Smooth wall surface

The boundary conditions were set as shown in Table 1. These conditions specified the flow on the boundaries of a physical model. The boundary condition would determines the behaviour of the solution given a set of differential equations at the boundary of its domain.

**Table 1.** Domain and boundary type

| Domain | Boundary Type |
|---|---|
| Air inlet | Inlet |
| Hopper opening | Opening |
| Outlet | Opening |
| Wind tunnel body | Wall |

 DOI: 10.1051/matecconf/20178702009

Advection scheme in high resolution technique was used to simulate the air flow profile in the wind tunnel. Turbulence numeric was first order. Air at 25°C was used as a working fluid. Number of iteration used for the simulation was set to 500. The solution parameters was chosen after conducting mesh independent analysis and convergence criteria analysis.

Simulation of the wind tunnel was divided into two objectives:

1. Visualize air flow profile at different air velocity
2. Modify the air flow profile to find effective separation effect

In order to achieve the simulation objectives, the variable is determined as shown as in the Table 2.

**Table 2.** Parameter settings

| Objectives | Manipulate variable | Constant variable |
|---|---|---|
| Objective 1 | Different velocity of air flow (5.0, 2.0, 1.0 , 0.5 m/s and 0.2 m/s) | i. Initial design of wind tunnel<br>i. Angle of hopper : 0° |
| Objective 2 | Different angle of hopper (0°, 30°, 45°, and 60°) | i. Initial design of wind tunnel<br>ii. Velocity of air : 0.2 m/s |

The simulation was using governing Navier-Stokes equations. The equation was based on the conservation laws for physical properties of fluids, which included mass, energy, and momentum. The mathematical equations that were applied to the mass, momentum and energy were expressed by the following equations [11].

Continuity Equation

$$\frac{\partial \rho}{\partial t} + \nabla \bullet (\rho U) = 0 \quad (1)$$

Momentum Equation

$$\frac{\partial \rho U}{\partial t} + \nabla \bullet (\rho U \otimes U) = -\nabla p + \nabla \bullet \tau + S_M \quad (2)$$

Continuity Equation

$$\frac{\partial (\rho h_{tot})}{\partial t} - \frac{\partial \rho}{\partial t} + \nabla \bullet (\rho U h_{hot}) = \nabla \bullet (\lambda \nabla T) + \nabla \bullet (U \bullet \tau) + U \bullet S_M + S_E \quad (3)$$

Where, fluid density, ρ, time, t, vector of velocity, U, pressure, p, viscosity, μ, temperature, T, specific enthalpy, h, thermal conductivity, λ, shear stress, τ, mass source, $S_M$, energy source, $S_E$.

## 3 Result And Discussion

### 3.1 Different air inlet velocity

Figure 4 shows the simulation of air flow in the wind tunnel with air inlet at maximum fan speed of 5 m/s. The maximum velocity of air at the hopper is 16.5 m/s, in which can be observed at right side of the hopper from the Figure 4. Due to the high velocity built up at the hopper, the simulation shows that the POFA was being pushed back out of the hopper and blown to the surrounding, causing a major loss in obtaining ultrafine POFA.

**Figure 4.** Air inlet velocity at 5m/s

The air inlet velocity of 5 m/s produced strong flow of energy at the hopper due to the sudden change at the hopper section as shown in Figure 4. Therefore, the simulation of different air inlet velocity was studied to prove that the flow energy is reduced when the air inlet velocity decreased.

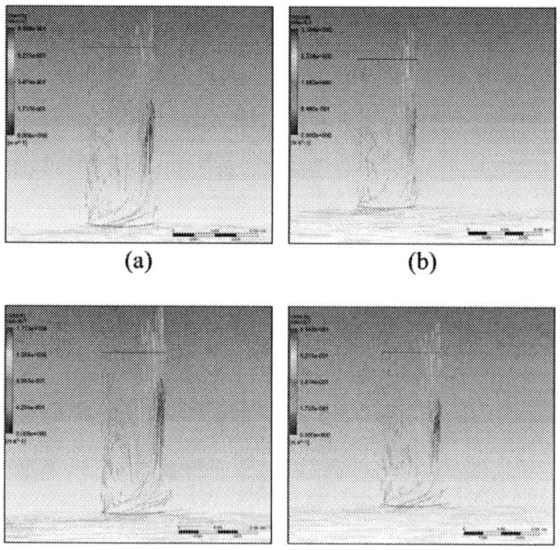

(a) (b)

(c) (d)

**Figure 5.** Air inlet at (a) 2.0 m/s, (b) 1.0 m/s, (c) 0.5m/s and (d) 0.2 m/s

Figure 5 showed the result for different velocity of air introduced into the tunnel. It was observed that the air flow out of the hopper was weakened when using a low velocity of air as shown in Figure 7. The flow of energy became less at the bottom of the hopper with the maximum velocity of 0.6948 m/s occurred at the very bottom of the hopper but with less intensity

The data was plotted in the graph by constructing line Y from the middle of the tunnel to the top of the hopper as shown in Figure 6. The line Y is placed near the hopper wall. Along the line Y, the velocity v is

DOI: 10.1051/matecconf/20178702009

measured and visualized using the graph function in the software.

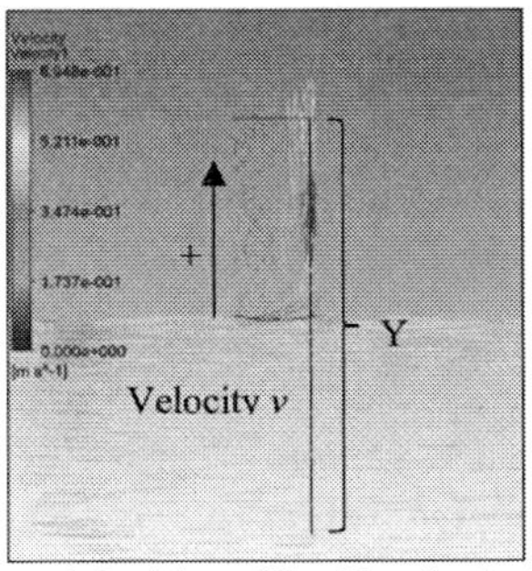


**Figure 6.** Position of line Y and velocity v

The graph as shown in Figure 7 confirmed the air flow profile for different velocity showed the same air flow profile. As a conclusion, air flow profile at different velocity produced different intensity of energy flow. Low velocity of air would make sure less or no POFA coming of the hopper. If the air velocity was too low, the POFA could not be separated effectively.

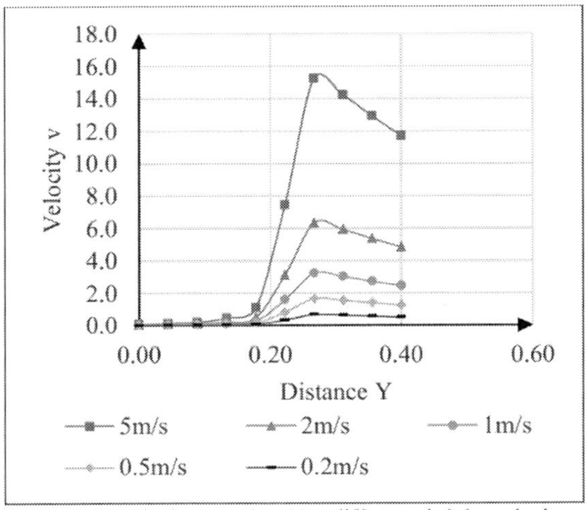


**Figure 7.** Velocity v against Y at different air inlet velocity

### 3.2 Different angle of hopper

The wind tunnel is simulated for a different angle of hopper to observe the air flow inside the tunnel. At certain degree of angle, it is claimed that the flow energy are varies when different angle of the hopper is installed.

Figure 8(a) shows the hopper is being slanted about 0° as in the initial design with velocity of air 0.2 m/s. The intensity of air flow at the bottom of the hopper shows the air flow as high as 0.6948 m/s is directed upwards in the hopper.

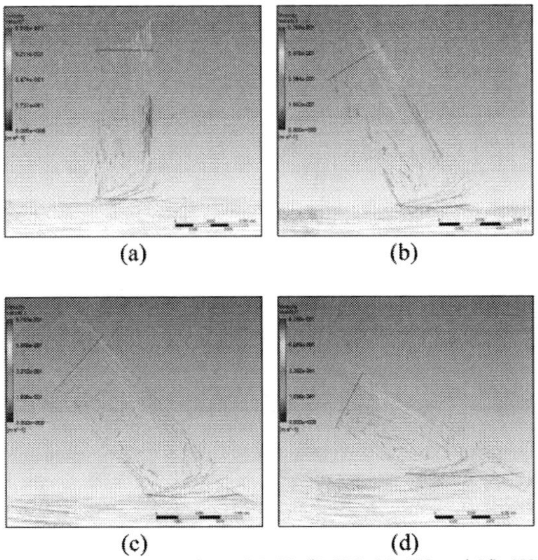


**Figure 8.** Hopper angles at (a) 0°, (b) 30°, (c) 45° and (d) 60°

The hopper at different angles were simulated as shown in Figure 8. Visibly, by modifying the angle of the hopper changed the air flow profile in the hopper.

At the intersection between the hopper and the tunnel body, greater turbulence was required in order to separate the ultrafine POFA from the fine POFA. In Figure 8(a) and 8(b), the flows were mostly directed out of the hopper. The flow profiles created by the angles of 0° and 30° were not effective since the air flow profile is totally flowing out of the hopper.

Figure 9 showed the summary of the results of simulation at different angles of the hopper. Likely, the angle of 60° exhibit the effective separation process. The negative value of the data shows the air flow was directed downward. Meanwhile, at 60° angle of hopper, obvious intensity of vortices happened at the bottom of the hopper and the flow was directed downward. Consequently, mixed sizes of POFA could be separated effectively at an angle of 60° of the hopper angle as well as easier to insert POFA to the hopper. If hopper angle was greater than 60°, there would be a tendency of the POFA to stick onto the hopper wall.

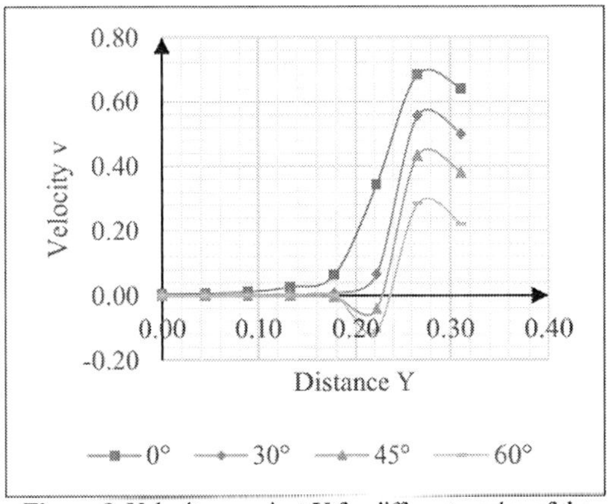


**Figure 9.** Velocity v against Y for different angles of the hopper

## 4 Conclusion

As a conclusion, the design for the wind tunnel had been successfully improved. Based on the result of the simulations, the best angle of the hopper at 60° was able to improve the effectiveness of the ultrafine POFA separation. The backflow of the POFA in the hopper could be minimized.

In simulation perspective, the mesh was generated successfully using ANSYS-CFX. The mesh quality was proven good as the skewness was low and the result of the simulations were converged.

The simulation results showed the presence of air backflow in the hopper, but had been reduced as the flow energy was low. Therefore a practical and economical design of a simple wind tunnel had been improved consequently would had increased the production of POFA. It was expected more research on POFA can be executed, where POFA size is concerned.

The simulations were done using a single phase flow without considering the POFA effect in the tunnel. For further research, it was recommended to use the multiphase flow in order to improve the simulation on the effect of air collision with the various sizes of POFA.

## References

[1] K. Poku, "Smal-scale palm oil processing in Africa," *FAO Agric. Serv. Bull. 148*, pp. 1–56, 2002.

[2] Y. Basiron and M. A. Simeh, "Vision 2020 – The Palm Oil Phenomenon," *Oil Palm Ind. Econ. J.*, vol. 5, no. 2, pp. 1–10, 2005.

[3] R. Karim, M. F. M. Zain, M. Jamil, and N. Islam, "Strength of concrete as influenced by palm oil fuel ash," *Aust. J. Basic Appl. Sci.*, vol. 5, no. 5, pp. 990–997, 2011.

[4] L. Y. Jong and D. C. L. Teo, "Concrete containing palm oil fuel ash ( POFA ) and oil palm shell ( OPS ) subjected to elevated temperatures," *UNIMAS e-Journal Civ. Eng.*, pp. 13–17, 2015.

[5] A. Munir, Abdullah, Huzaim, Sofyan, Irfandi, and Safwan, "Utilization of palm oil fuel ash ( POFA ) in producing lightweight foamed concrete for non-structural building material," *Procedia Eng.*, vol. 125, pp. 739–746, 2015.

[6] N. Zainudin, "Properties of micro fine size palm oilf fuel ash (POFA) using wind tunnel production system," *UNIMAS e-Journal Civ. Eng.*, 2015.

[7] M. Singh, N. Singh, and S. K. Yadav, "Review of design and construction of an open circuit low speed wind tunnel," *Glob. J. Res. Eng.*, vol. 13, no. 5, pp. 1–22, 2013.

[8] Y. Q. Nguyen, "Designing, constructing, and testing a low-speed open-jet wind tunnel," *Int. J. Eng. Res. Appl.*, vol. 4, no. 1, pp. 243–246, 2014.

[9] T. H. Yong and S. S. Dol, "Design and development of low-cost wind tunnel for educational purpose," *Mater. Sci. Eng.*, vol. 78, pp. 318–328, 2015.

[10] S. Selvi Rajan, M. Santhoshkumar, N. Lakshmanan, N. P. S, and M. Paramasivam, "CFD analysis and wind tunnel experiment on a typical launch vehicle model," *Tamkang J. Sci. Eng.*, vol. 12, no. 3, pp. 223–229, 2009.

[11] M. A. Ismail, A. K. Othman, H. Zen, and M. S. Misran, "CFD Modelling of pump as turbine with various number of blade for microhydro system," *J. Appl. Sci. Process Eng.*, vol. 3, no. 1, pp. 17–23, 2016.

# Sliding wear behaviour of steel carburized using $Na_2CO_3$-NaCl

Willey Y. H. Liew[a], Jester L.J. Ling, Nancy J. Siambun

*Faculty of Engineering, Universiti Malaysia Sabah, Jalan UMS, 88400 Kota Kinabalu, Sabah, Malaysia*

**Abstract.** Experiments have been carried out to investigate the effect of carburization process on the sliding wear resistance of mild steel. The carburization process was conducted in carbonate salts mixtures of $Na_2CO_3$-NaCl. Carburization followed by water quenching resulted in the formation of martensite with a hardness of 900 HV in the subsurface, up to the depth of 400 µm. This hardness value was substantially higher than the non-carburized steel which had a hardness of 520 HV. In the initial stage of sliding in air, abrasive wear and cluster of fine cavities due to adhesion were formed. This was followed by the formation of large-scale fracture at the cavities. The high hardness of the carburized steel reduced the severity of adhesive wear and thus the tendency of the worn surface to fracture.

## 1 Introduction

Carburization processes, involve the intentional diffusion of carbon into the subsurface of iron-based alloys at high temperature, are widely used to enhance the wear resistance of mild steel. The carbon rich layer at the subsurface will transform to other microstructure (such as martensite and bainite) depending on the rate of cooling used after the carburization process. Zhang et al. [1] used alcohol and kerosene as carburizer to produce 2.5 mm thick of bainite and martensite on the subsurface of steel specimens. The bainitic steel is produced by immerging the carburized specimens into a salt bath at 230°C for 48 hours before being air cooled to room temperature. Despite having lower hardness, it exhibited higher wear resistance than the martensitic steel produced by rapid cooling after the carburization process. This superior wear resistance of the bainitic steel could be attributed to (i) the high strength and toughness of the bainitic microstructure, (ii) the effectiveness of the carbon enriched film-like austenite formed between the ferrite in preventing crack propagation and (iii) the formation of extremely fine and hard ferrite-phase microstructure at the surface of the bainitic microstructure during sliding. Izciler and Tabur [2] and Ulutan et al. [3] reported that the formation of martensitic microstructure in the carburization process resulted in an increase in the abrasive wear resistance of steel. Bensely et al. [4] found that deep cryogenic treatment applied to the carburized steel after quenching resulted in higher conversion of austenite to martensite and induced precipitation of finer carbides, and thus an increased hardness and wear resistance.

We had demonstrated that the electro-carburization process utilizing non-toxic $Na_2CO_3$-NaCl carbonate salts mixture as the electrolyte was effective in enhancing the wear resistance of mild steel under reciprocating sliding condition [5]. The carburized layer, rich in retained austenite on the surface and dominated by martensite in the subsurface, increased the load carrying capacity, and adhesive and abrasive wear resistance of the steel. The superior wear resistance of the martensite, as compared to the austenite, could be attributed to its high cracking and adhesive wear resistance owing to its high hardness and tendency to form oxides. This paper reports the study carried out to investigate the effect of this carburization process on the sliding wear resistance of mild steel.

## 2 Experimental

### 2.1 Carburization process

Fig. 1 shows the set up of the carburization process. It had a retort fabricated using a 316 stainless steel tube. Gas inlet and outlet pipes which continuously supplied $CO_2$ into the sealed retort were also welded on to the lid. The central body of the stainless steel retort was placed in the middle of the furnace (hottest zone) and alumina wool was placed on top of the furnace to minimize heat losses. A crucible of alumina or stainless steel filled with carbonate salts was placed in the middle of the retort on a stainless steel stand before the lid was secured to the retort flange. The mild steel specimens were immersed in the molten salt.

[a] Corresponding author: wyhliew@ums.edu.my

During the carburization process, the temperature of the salt was maintained at 860° ± 10°C and the voltage was set to 4.5 V. The temperature of the $Na_2CO_3$-NaCl molten salt inside the crucible was measured at various location using several thermometers. Whereas the gas outlet pipe was connected to a dreschel bottle containing mineral oil to prevent the ingress of air to the system.

The $CO_2$ gas supplied into the furnace above the molten salt bath was the carbon source in the proposed electro-carburization process. When an external voltage was applied through the electrodes, the carbonate ions ($CO_3^{2-}$) decomposed into carbon atoms and oxide ions ($O^{2-}$) at the cathode (mild steel) according to reaction 1. The carbon atoms produced then diffused into the mild steel. The $O^{2-}$ ions would either react with the $CO_2$ at the molten salt/gas interface to reform the $CO_3^{2-}$ ion (reaction 2), or transport to, and discharge at the stainless steel plate as $O_2$ gas (reaction 3).

$$CO_3^{2-} + Fe + 4e \Rightarrow C - Fe + 3O^{2-} \quad \text{reaction 1}$$
$$O^{2-} + CO_2 \Rightarrow CO_3^{2-} \quad \text{reaction 2}$$
$$2O^{2-} \Rightarrow O_2 + 4e \quad \text{reaction 3}$$

where $C - Fe$ represents the carbon dissolved in the steel cathode.

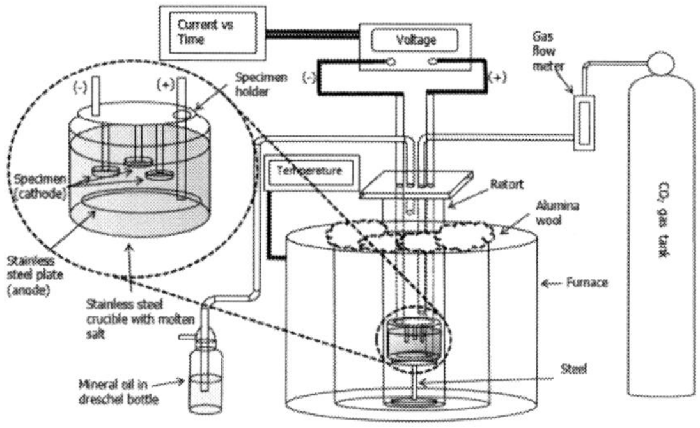


**Figure 1.** Set up of the carburization process.

## 2.2 Materials

Uncoated cemented carbide balls (6wt% Co and 94% WC; grade ISO K10) with a hardness 1600 HV and diameter of 6 mm were used to slide on the non-carburized and carburized mild steel discs. 2 types of carburized mild steel were used. They were (i) steel carburized at 860°C for 1 hour followed by quenching in water (referred as C1 specimen) and (ii) the non-carburized steel (referred as NC specimen). The NC specimens were produced by heating the specimens at 860° C in a furnace, followed by quenching. The mild steel had a nominal chemical compositions of 0.2 wt.% C, 0.19wt.% Si, 0.60 wt.% Mn, 0.014wt.% P and 0.018wt.% S, balance Fe.

## 2.3 Sliding tests

The frictional and wear behaviour of the non-carburized and carburized steel was investigated using a Ducom TR-20EV-M3 ball-on-disc tester. The steel specimens were carburized at 860°C for 1 hour followed by quenching in water. Tests were carried out at 10 m/min and different normal loads of 10, 15, 30, 50 and 90 N for a sliding distance of 100 m in ambient air and vacuum (1× 10-5 Pa). The coefficient of friction was continuously measured throughout the tests by a load cell. The morphology of the worn surface was examined using a scanning electron microscopy (SEM).

# 3 Results and discussion

## 3.1 Hardness measurement

Fig. 2 shows the variation of the hardness of the subsurface of theC1 specimen. It had a peak hardness of 910 HV at 45-150 μm and a case depth of 450 μm. Towards the core, the hardness decreased significantly to 520 HV, the hardness of the non-carburized steel.

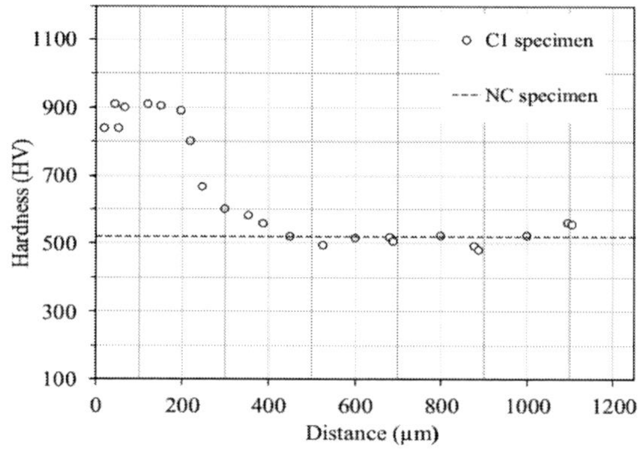


**Figure 2.** Variation of the subsurface hardness of the carburized and non-carburized steel

## 3.2 Microstructure of the carburized steel

The contaminant on the surface of the C1 specimen was removed by slight polishing using 1 μm size diamond paste before they were etched with nital to unveil the microstructure on the surfaces. Examination of the etched surfaces using an optical microscope showed that the surfaces the specimens had both martensite and retained austenite surrounded by grain boundaries (fig. 3(a)). XRD analysis on the slightly polished surface produced peaks at $2\theta$ = 44.9°, 50.8°, 65.1° and 82.2° (fig. 4). The peaks at $2\theta$ = 50.8° showed that the white grains were retained austenite. The presence of the peaks at $2\theta$ = 44.9°, 65.1° and 82.2° indicated that the dark grains were bcc martensite [6-8]. Towards the core, the amount of austenite reduced while the amount of martensite increased. The subsurface at 100 μm beneath the original surface with the highest hardness value was dominated by martensite with no grain boundaries (fig. 3(b)). This observation was consistent with the XRD results which showed the absence of the austenite peak.

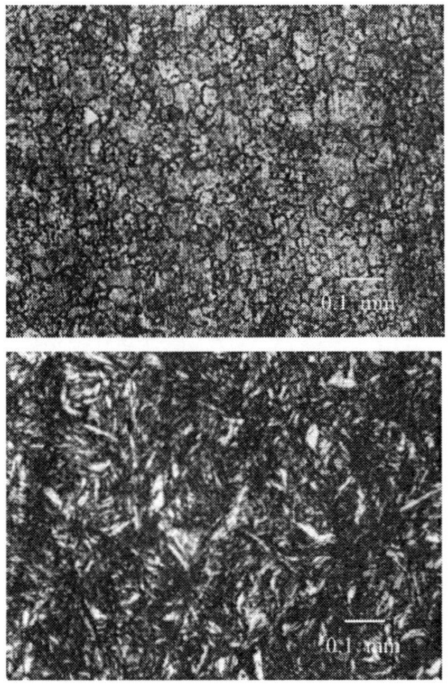


**Figure 3.** The microstructure at the surface and subsurface of the C1 specimen. (a) Grain boundaries, and austenite and martensite grains were evident on the surface (b) At the subsurface 100 μm beneath the surface, the microstructure was dominated by martensite.

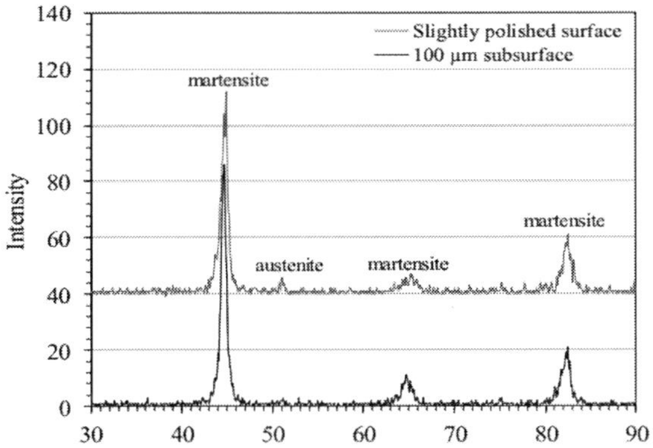


**Figure 4**. XRD analysis on the surface and sub-surface of the C1 specimen

### 3.3 Coefficient of friction

Figs. 5-8 show the variation of the COF (coefficient of friction) with sliding distance for the C1 and NC specimens in air and vacuum at various normal loads. A notable feature of the results obtained in air is the sharp increase followed by a rapid drop in the COF to a low prevailing steady-state value in the initial stage of sliding. This reflects the nature of the running-in process. During the running-in process, plastic deformation takes place on the subsurface results in an increase in the hardness of the material. Once the hardness reaches to a critical value to support a lubricant film, the friction coefficient drops to a low prevailing steady-state value. In vacuum, no oxide film was form after the running-in process. Therefore, an abrupt drop in COF either did not take place or was not evident. In sliding NC and C1 specimens at 10, 15 and 30 N, the COF produced in air was much lower than that produced in vacuum throughout the entire test duration (fig. 5). In sliding NC specimen at 50 N in air, a low prevailing COF (at least 0.1 lower than that in vacuum) was obtained in the initial stage of sliding. A marked increase in the COF was observed at the distance of 550 m. At the distance of between 550-1000 m, the value of the COF in air and vacuum was approximately the same and the COF in air showed larger fluctuations (fig. 6). The same phenomenon also occurred at 90 N. The tests conducted on the C1 specimens at 50 N in air produced either a low prevailing COF throughout the entire test duration or a gradually increased COF (figure 7) or a sudden increase in the COF to the vacuum value (similar to that observed in the NC tests). All tests conducted at 90N produced a sudden increase in the value and fluctuation of the COF (fig. 8), similar to that observed in the sliding of the NC specimen (fig. 6).

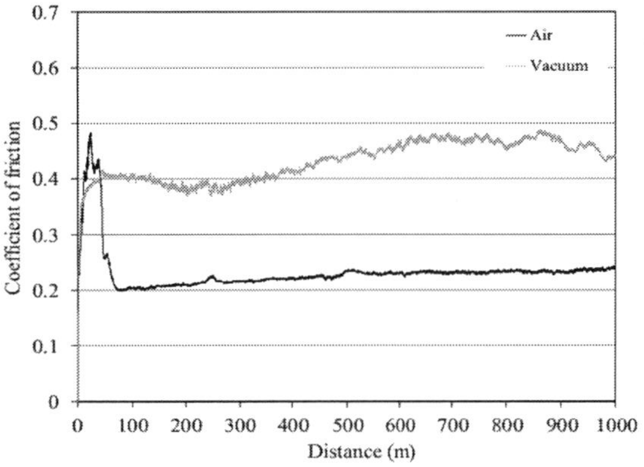


**Figure 5.** Variation of the coefficient of friction during sliding of C1 specimen at 30N.

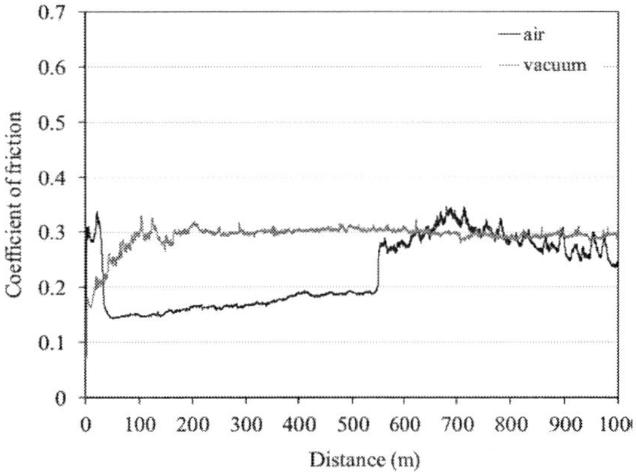


**Figure 6.** Variation of the coefficient of friction during sliding of NC specimen at 50N.

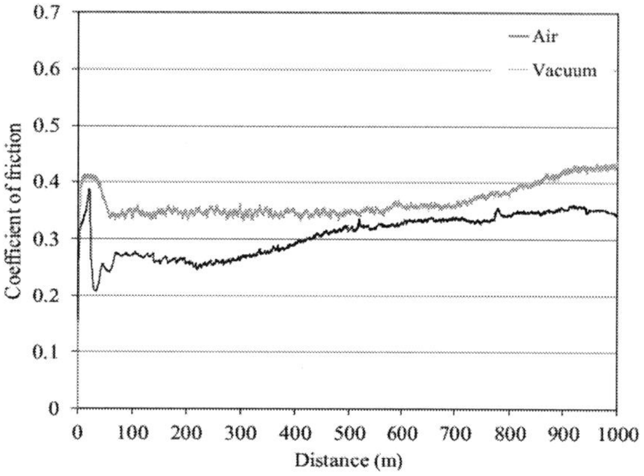


**Figure 7.** Variation of the coefficient of friction during sliding of C1 specimen at 50N.

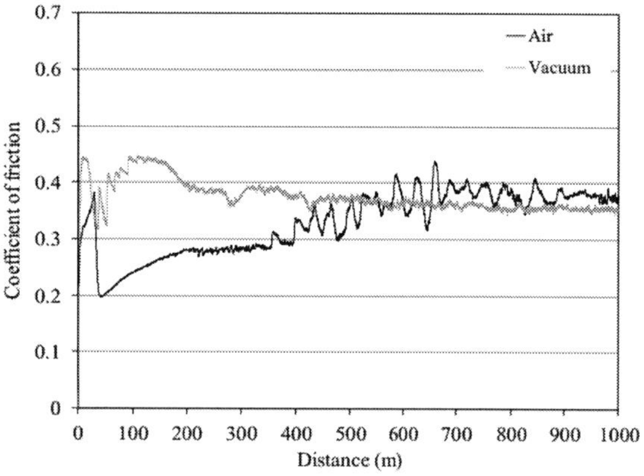


**Figure 8.** Variation of the coefficient of friction during sliding of C1 specimen at 90N.

### *3.4 Morphology of the worn surface*

The worn surfaces of C1 specimen after sliding at 10 and 15 and 30 N at 10 m/min in air had cavities due to the ploughing action of the debris (fig. 9(a)). This indicated that the dominant wear mechanism was abrasive wear. The NC specimen which had a lower hardness was subjected to a lower degree of abrasive wear but to a more severe adhesive wear as can be seen in fig. 9(b).

In the early stage of sliding of the NC specimen at 50 and 90 N using the speed of 10 m/min in air, grooves indicating of abrasive wear and clusters of fine cavities with the diameter of between 5-20 μm were formed on the worn surface (fig. 10). Formation of the fine cavities was due to welding and breaking of adhesive junctions. Such wear mechanism produced fine debris. Examination of the worn surface using the EDX-element-mapping showed that the cavities had much lower concentration of O element as compared to the groove surface (fig. 11). EDX examination showed that oxygen content in the groove surface was between 27.2-30.2 wt%. Whereas, the cavities had oxygen content of less than 1.0 wt%. Samples of the EDX results are shown in Table 1. This showed that the fine cavities formed on the worn surface were too small to make contact with the carbide ball in the subsequent sliding events to produce sufficient heat for high degree of oxidation to take place. The cavities would then merge to form a larger scale of nascent surface. Once the cavities were large enough to make contact with the carbide ball, severe adhesion between the contact surfaces occurred resulting in high frictional force. High frictional traction at the nascent contact surface caused fracture to take place (fig. 12(a)) and hence the release of flake-like delamination debris from the surface. Other observations further supported the hypothesis that the fracture was initiated at the fine cavities were (i) the distance between the clusters of the fine cavities was almost the same as that between the fractured surface, and (ii) the formation of fracture led to the diminish of the fine cavities. Due to a large amount of oxide film removed by fracture, the frictional condition and thus the COF became similar to that in vacuum which wear was dominated by fracture.

In the sliding of C1 specimen at 50 N, while clusters of fine cavities were generated, the severity of the adhesive wear was less than those observed on the NC specimen and they did not always develop into fracture. Only at the load of 90 N had caused fracture to take place in all tests which was reflected by the sudden increase in the frictional force (fig. 8). The size of the fracture surface was less as compared to that formed on the NC specimen (compare fig. 12 (a) and (b)). It could be concluded that the increase in the hardness brought about by the carburization process resulted in a lower degree of adhesive wear, making clusters of nascent fine cavities less likely to form. This in turn reduced the tendency of the worn surface to fracture.

(a)
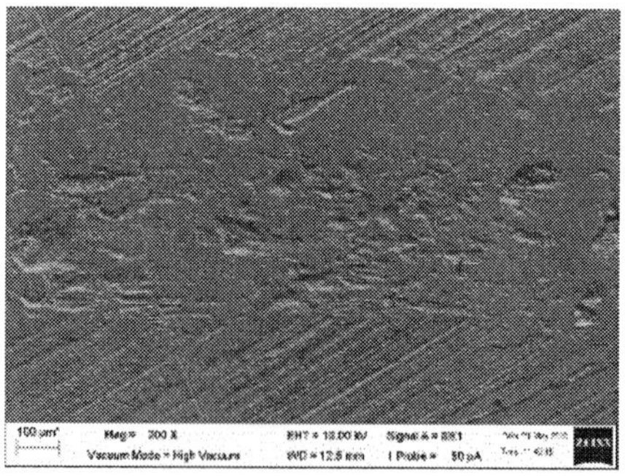

(b)
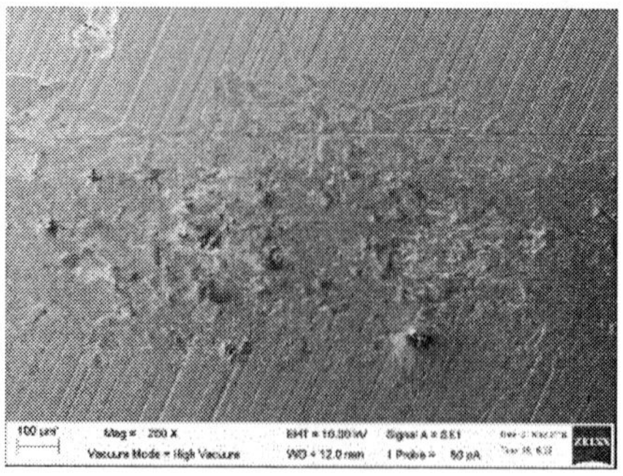

**Figure 9.** SEM images of the worn surfaces of the (a) C1 and (b) NC specimens after sliding for 100 m at 15 N in air.

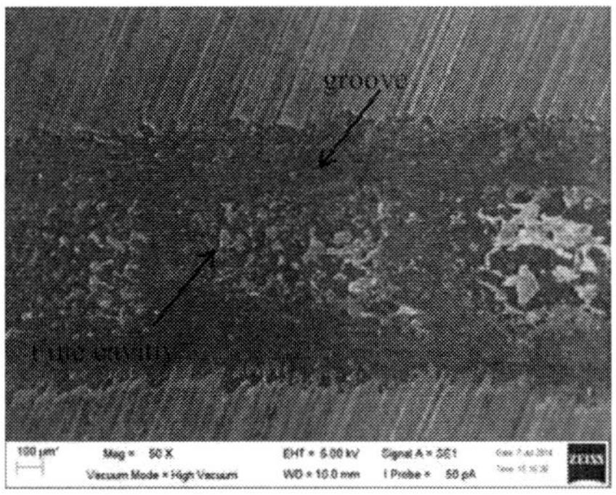


**Figure 10.** Worn surface of the NC steel in the initial stage sliding at 90 N in air. The worn surface consisted of clusters of fine cavities and enlarged cavities.

(a)

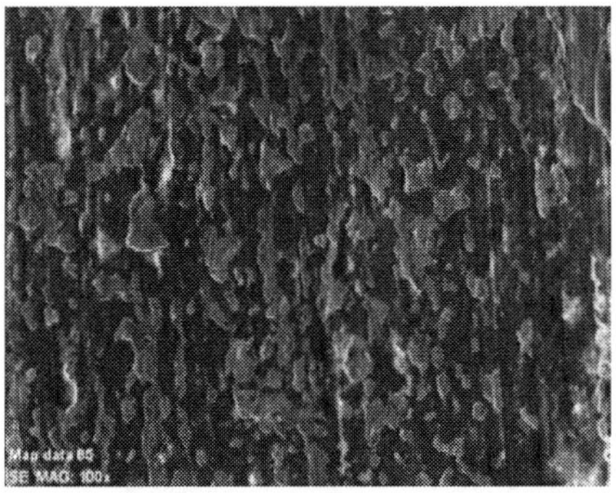

(b)

**Figure 11.** EDX-element-mappings of the worn surface of the NC which consist of numerous fine cavities. The cavities had lower concentration of O element.

**Table 1.** EDX analysis on the groove surface and fine cavities.

| Element | norm. [wt.%] at the groove surface | norm. [wt.%] at the fine cavities |
|---|---|---|
| Carbon | 3.67 | 5.63 |
| Oxygen | 0.79 | 28.91 |
| Iron | 94.56 | 62.53 |
| Sulfur | 0.00 | 0.00 |
| Phosphorus | 0.00 | 0.00 |
| Silicon | 0.15 | 2.30 |
| Manganese | 0.83 | 0.63 |

## 4 Conclusions

Experiments have been carried out to investigate the effect of carburization process utilizing $Na_2CO_3$-NaCl as the electrolyte on the microstructure and the sliding wear behaviour of mild steel. It resulted in an increased hardness of 900 HV in the subsurface, up to the depth of 400 μm. This hardness value was substantially higher than the non-carburized steel which had a hardness of 520 HV. The surfaces of the specimens had both martensite and retained austenite surrounded by grain boundaries. Towards the core, the amount of austenite reduced while the amount of martensite increased. In the initial stage of sliding in air, abrasive wear and cluster of fine cavities due to adhesion were formed. The five cavities acted as preference sites for the formation of large-scale fracture. The high hardness of the carburized steel reduced the severity of adhesive wear and thus the tendency of the worn surface to fracture.

(a)

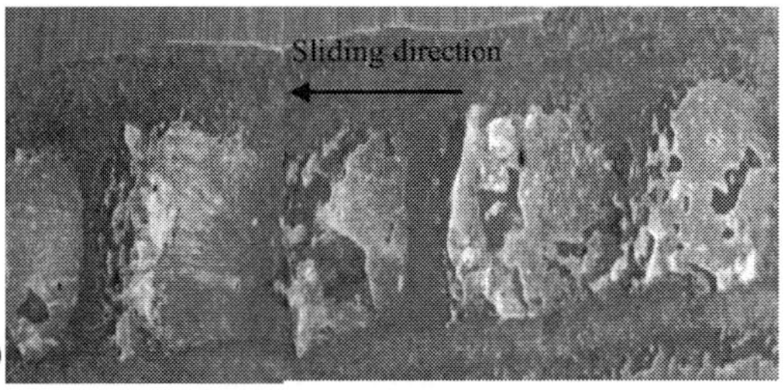


(b)

**Figure 12.** SEM image of the worn surface of the (a) NC and (b) C1 specimens after sliding for 100 m at 90 N in air.

## References

1. P. Zhang, F.C. Zhang, Z.G. Yan, T.S. Wang, L.H. Qian, "Wear property of low-temperature bainite in the surface layer of a carburized low carbon steel," Wear **271**, pp. 697-704 (2011).
2. M. Izciler and M. Tabur, "Abrasive wear behavior of different case depth gas carburized AISI 8620 gear steel," Wear **260,** pp. 90-98 (2006).
3. M. Ulutan, O. N. Celik, H. Gasan and U. Er, "Effect of Different Surface Treatment Methods on the Friction and Wear Behavior of AISI 4140 Steel," J. Mater. Sci. Technol. **26**, pp. 251-257 (2010).
4. A. Bensely, A. Prabhakaran, D. Mohan Lal and G. Nagarajan, "Enhancing the wear resistance of case carburized steel (En 353) by cryogenic treatment," Cryogenics **45**, pp.747-754 (2006).
5. Willey Y. H. Liew, Roonie Protasius, Jester L.J. Ling, Nancy J. Siambun and Noor-Ajian Mohd-Lair, "Reciprocating wear behavior of mild steel carburized using $Na_2CO_3$-NaCl," Tribology International **95**, pp. 406-418 (2016).

6. D. Frölich, B. Magyar, B. Sauer, P. Mayer, B. Kirsch, J.C. Aurich, R. Skorupski, M. Smaga, T. Beck, D. Eifler, "Investigation of wear resistance of dry and cryogenic turned metastable austenitic steel shafts and dry turned and ground carburized steel shafts in the radial shaft seal ring system," Wear **328-329**, pp. 123-131 (2015).
7. J.R. Tolchard, A. Somme, J.K. Solberg, K.G. Solheim, "On the measurement of austenite in supermartensitic stainless steel by X-ray diffraction," Materials Characterization **99**, pp. 238-242 (2015).
8. R. Bhambroo, S. Roychowdhury, V. Kain, V.S. Raja, "Effect of reverted austenite on mechanical properties of precipitation hardenable 17-4 stainless steel," Materials Science & Engineering A **568**, pp. 127-133 (2013).

# Physical and Mechanical Properties of Ni-Cr based composites with addition of solid lubricants produced through powder metallurgy process

Wan Farhana Mohamad[1,a], Amir Azam Khan[1], Faiz Ahmad[2] and Abdullah Yassin[1]

[1]*Department of Mechanical and Manufacturing, Faculty of Engineering, Universiti Malaysia Sarawak, 94300 Kota Samarahan, Sarawak, Malaysia*
[2]*Mechanical Engineering Department, University Technology PETRONAS, 31750 Tronoh, Perak, Malaysia*

**Abstract.** Ni-Cr based composites with and without the addition of solid lubricants ($MoS_2$, silver and $CaF_2$) were prepared by powder metallurgy method. The samples were sintered at two different temperatures, 1000 °C and 1200 °C. The physical properties such as shrinkage, sintered density and porosity were studied. The microstructures of the Ni-Cr based composites were observed by using SEM analysis while the mechanical properties of the composites were measured by Rockwell Hardness Tester. The results revealed that the increased in sintering temperature improved the shrinkage, sintered density and hardness of the composites while less porosity produced. Ni-Cr based composites with the addition of silver and $MoS_2$ exhibited better shrinkage, density and porosity. Besides, 5% of $MoS_2$ addition in the composites improves the hardness of the composites at sintering temperature 1200°C.
**Keywords:** solid lubricants, Ni-Cr composites, powder metallurgy, $MoS_2$, silver, $CaF_2$

## 1 Introduction

Giant industries such as automotive and aerospace are having an economical loss due to high maintenance cost for mechanical components. Failure of mechanical components such as bearings and bushings used in the advanced jet engines are caused by friction and wear as subjected to wide temperature range. As for example, the operation temperature of nozzles for turbojet propulsion system is reaching the temperature of 1650°C [1]. At this high temperature, the liquid lubricant is unstable and tends to lose its lubricating properties. Thus, solid lubricant is preferable compared to the liquid lubricant. Besides, solid lubricant also has advantages compared to liquid lubricants either under the extreme pressure conditions, radiation environment or cryogenic temperature [2].

The incorporation of the solid lubricant in the composites is called self-lubricating composites. The composite is able to form a lubricating film to reduce the effect of friction and wear. The examples of metal used in the self-lubricating composites are iron based, copper based and nickel based composites. Iron based self-lubricating composites are mostly used in the automotive application such as piston ring, clutch, brake system and engine liners [3]. While copper based composites are used for the application of thermal management application due to excellent properties thermal and heat conductivity [4]. Among the metal composites, nickel based composites have become the most attention of researchers for high temperature application. The high cost of refractory metals and complex manufacturing process also make the metal nickel as an option for high temperature application [5].

Nickel Chromium (Ni-Cr) has become one of the leading materials for high temperature application due to its excellent performance at high temperature. A series of nickel based composites has been developed in order to achieve the great need of high temperature solid lubricating system produced through powder metallurgy process [6]. Ni-Cr matrix also acted as a binder and offered excellent high temperature oxidation/corrosion resistance and essential mechanical strength [7], [8]. Nickel itself offers good mechanical properties and anti-oxidizing properties when exposed to air atmosphere at high temperature while chromium offers anti-wear and lubricating properties at high temperature [9] . The strength of the Cr particles can determine the strength and bonding between the matrix and Cr reinforcement.

Efforts have been made in order to enhance the performance of self-lubricating composites with solid lubricants addition. The work is continued in order to develop a perfect combination of matrix and solid lubricant for high temperature system to meet the requirement of advanced technology. In this research work, the author is working on Ni-Cr based composites with the addition of single, dual and multiple solid lubricants in order to obtain an excellent mechanical as

[a] Corresponding author: wanfarhana88@gmail.com

DOI: 10.1051/matecconf/20178702011

well as tribological properties. The composites are fabricated by powder metallurgy method.

## 2 Experimental

The Nickel –Chromium (Ni-Cr) composites are based on 80% Ni and 20% Cr. The solid lubricants added are molybdenum disulphide ($MoS_2$), silver (Ag) and calcium fluoride ($CaF_2$). All of the raw materials are supplied by Robert Scientific Sdn Bhd. The characterizations of powders were done using Scanning Electron Microscope (SEM) by Hitachi Tabletop TM3030 and X-Ray Fluorescences (XRF) by Bruker S4 Pioneer, USA.

### 2.1 Preparation of samples

#### *2.1.1 Samples preparations*

Ni-Cr based composites have been produced by powder metallurgy method which consists of mixing, compaction and sintering. NC denoted for Ni-Cr based composite (without solid lubricant), NCM (with $MoS_2$), NCA (with Ag), NCCf (with $CaF_2$), NCMA (with $MoS_2$ and Ag), NCMCf (with $MoS_2$ and $CaF_2$), NCACf (with Ag and $CaF_2$) and NCMACf (with $MoS_2$, Ag and $CaF_2$). The powder has been weighted by analytical balance based on composition in Table 1. The powder mixture was mixed homogeneously in a ball mill for 30 minutes with the speed of 300 rpm. A mixture of powder was compacted using compression machine with the pressure of 100kN at room temperature. The compacted samples were in a pellet shape with the dimension of 13mm x 5mm. The compacted samples were sintered in a high temperature furnace for 60 minutes. The sintering atmosphere was argon gas with a flow rate of 4 L/min. The compacted samples were sintered at the temperature of 1000 °C and 1200 °C with a heating rate of 10 °C/min.

**Table 1.** Materials compositions for Ni-Cr based composites.

| Samples | Designation | Composition (wt%) | | | |
|---|---|---|---|---|---|
| | | 80Ni-20Cr | $MoS_2$ | Ag | $CaF_2$ |
| 1 (Pure) | NC | Balance | 0 | 0 | 0 |
| 2 (Single) | NCM | Balance | 5 | 0 | 0 |
| 3 (Single) | NCA | Balance | 0 | 5 | 0 |
| 4 (Single) | NCCf | Balance | 0 | 0 | 5 |
| 5 (Dual) | NCMA | Balance | 5 | 5 | 0 |
| 6 (Dual) | NCMCf | Balance | 5 | 0 | 5 |
| 7 (Dual) | NCACf | Balance | 0 | 5 | 5 |
| 8 (Multiple) | NCMACf | Balance | 5 | 5 | 5 |

### 2.2 Physical Properties

The diameter and height of the sintered samples were measured by using vernier caliper while the mass of the samples was obtained by analytical balance. At least 3 measurements were taken and the average value was measured. The shrinkage, density, and porosity of the samples were calculated using the geometric method (formula).

#### *2.2.1 Shrinkage*

The volume of the green and sintered samples in pellet size are calculated by using the formula $V= \pi\ (d/2)^2h$. The percentage of shrinkage is measured by using formula :

$$\%\text{ of shrinkage} = (V_1 - V_2)/(V_1) \times 100$$

$V_1$= Volume before sintering

$V_2$= Volume after sintering

The measurements of height (h) and diameter (d) were taken 3 times to get the average value.

#### *2.2.2 Density*

The density of green and sintered samples is measured before and after sintering. The density, ρ formula is :

$$\rho = m / V$$

$m$ = mass of pellet (g)

$V$ = Volume of pellet (cm$^3$)

#### *2.2.3 Porosity*

The porosity of the composites was measured by formula of:

$$\text{Porosity}= (\rho_{theoritical} - \rho_{sintered}) / (\rho_{theoritical}) \times 100$$

$\rho_{theoritical}$ = theoretical density

$\rho_{sintered}$ = sintered density

### 2.3 Microstructure and Mechanical Properties

#### *2.3.1 Microstructure*

The microstructure of the sintered samples was observed and analyzed by using Scanning Electron Microscope (SEM). The samples were ground with silicon carbide paper and polished with diamond paste as a surface preparation before SEM analysis. The effect of sintering temperature and composition of the composites were studied.

#### *2.3.2 Hardness*

The hardness of Ni-Cr based composites was measured by using Mitutoyo Hardness Machine with the diamond indenter at an indentation load of 1471N. At least 10 measurements were taken and the average value is calculated.

## 3 Results and Discussions

### 3.1 Characterizations of raw materials

SEM analysis was done to observe the particle shape and distribution while XRF analysis was conducted to determine the purity of the powder. Figure 1 and 2 show the SEM micrograph of the nickel and chromium powder.

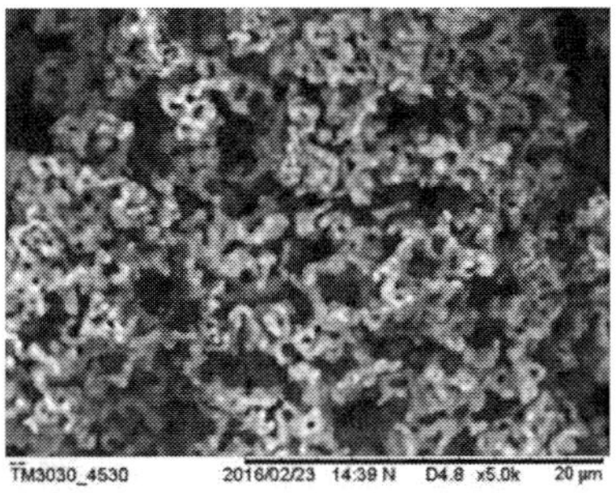


**Figure 1.** SEM microstructure of nickel powder at a magnification of 5000X.

In Figure 1, SEM analysis with a magnification of 5000X illustrates the nickel powder used in this research. The purity of the nickel powder is 99.1% determined by XRF analysis.

**Figure 2.** SEM microstructure of chromium powder at a magnification of 1000X.

Figure 2 shows the particle shape and distribution of chromium powder. The particle shape is in the irregular shape. XRF analysis determines the purity of chromium powder is 93.3%. Figure 3-5 shows the SEM microstructure of solid lubricants which are $MoS_2$, silver and $CaF_2$.

**Figure 3.** SEM microstructure of $MoS_2$ powder at a magnification of 10000X.

From the Figure 3, it demonstrates the particle shape of $MoS_2$ powder which is in flake shape. The purity of the powder is Mo 72% and S 25.1%.

**Figure 4.** SEM microstructure of silver powder at a magnification of 5000X.

SEM microstructure in Figure 4 shows the particle distribution of silver powder used in the research work.

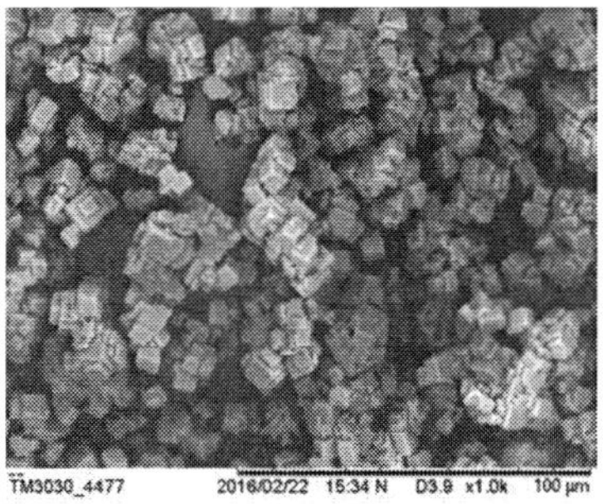


**Figure 5.** SEM microstructure of $CaF_2$ powder at a magnification of 1000X.

Based on Figure 5, the micrograph illustrates the stacked crystallize structure of $CaF_2$. The microstructure in cubic particle shape. The purity of powder is Ca 99.1%.

### 3.2 Characterization of mixed powders.

**Figure 6.** A mixture of Ni and Cr powders after 30 minutes of ball milling.

Figure 6 shows the distribution of Nickel and Chromium powder after mixing by using ball mill machine for 30 minutes. The bigger particles represent chromium powder while the smaller particles represent nickel powder. The mixture was homogeneously distributed throughout the process. Some smaller particles of nickel tend to agglomerate on the chromium surface. After that, the mixture was compacted by a hydraulic press into a pallet shape. The compacted samples were sintered in the tube furnace at the sintering temperature of 1000 °C and 1200 °C for 60 minutes in the argon atmosphere.

### 3.3 Shrinkage of Ni-Cr based composites

After sintering at high temperature, the compacted samples were subjected to shrinkage. There was a reduction in the volume of compacted samples. Table 2 demonstrates the percentage of shrinkage after sintering at different temperatures. As the sintering temperature increased from 1000 to 1200 °C, the shrinkage of the samples also increased.

**Table 2.** Percentage of Shrinkage after sintering at 1000 °C and 1200 °C.

| Samples | Shrinkage (%) | |
|---|---|---|
| | at 1000 °C | at 1200 °C |
| NC | 15.56 | 32.47 |
| NCM | 21.65 | 35.07 |
| NCA | 25.14 | 38.18 |
| NCCf | 24.90 | 34.77 |
| NCMA | 24.83 | 37.55 |
| NCMCf | 22.09 | 33.79 |
| NCACf | 24.35 | 33.61 |
| NCMACf | 20.92 | 37.50 |

Based on Table 2, the highest percentage of shrinkage for both temperatures is NCA composites for 25.14% (1000 °C) and 38.18% (at 1200 °C) while the lowest percentage of shrinkage for both temperatures is NC composites for 15.56% (1000 °C) and 32.47% (at 1200 °C). During sintering, the center to center distance between the powder particles is reducing and at the same time, the pore shrinks [10]. The addition of solid lubricant/s increased the percentage of shrinkage of the Ni-Cr based composites. The highest shrinkage were NCA composites at both temperatures. This was due to the smaller particles of silver which filled in the pores and reduced the size of pores. Shrinkage is one of the factors to the increment of sintered density.

### 3.4 Sintered Density of Ni-Cr based composites

Shrinkage of the composites leads to the densification of the composites. Sintered density is the measurement result for densification. Table 3 illustrates the sintered density of Ni-Cr based composites samples at sintering temperature of 1000 °C and 1200 °C. Sintered density was also enhanced from the sintering temperature of 1000 to 1200 °C.

**Table 3.** Sintered density of Ni-Cr based composites samples at sintering temperature of 1000 °C and 1200 °C.

| Samples | Sintered Density (g/cm$^3$) | |
|---|---|---|
| | at 1000 °C | at 1200 °C |
| NC | 5.59 | 6.99 |
| NCM | 5.67 | 7.08 |
| NCA | 6.07 | 7.36 |
| NCCf | 5.91 | 6.67 |
| NCMA | 6.15 | 7.23 |
| NCMCf | 5.58 | 6.54 |
| NCACf | 6.08 | 6.94 |
| NCMACf | 5.57 | 6.88 |

From Table 3, the highest sintered density is achieved by NCA composite followed by NCMA composites at sintering temperature 1200 °C. Increased in density demonstrate that the process of diffusion, densification, recrystallization and grain growth between the particles at the contact area. The result is supported by the percent density graph as shown in Figure 7.

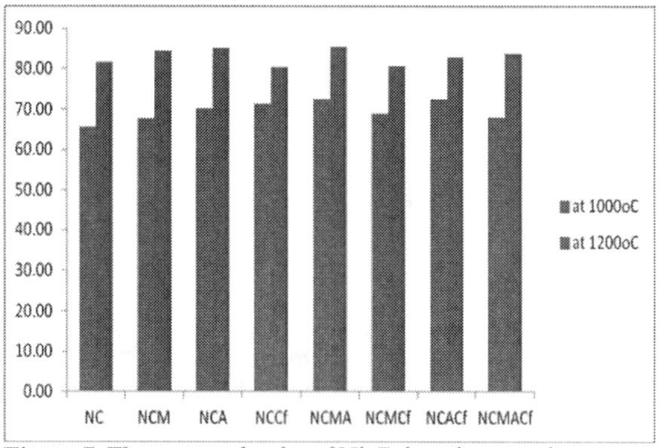


**Figure 7.** The percent density of Ni-Cr based composites at sintering temperature of 1000 and 1200 °C.

The percent density indicated the densification or pore shrinkage of the composites. During sintering, the particles are diffusing into each other as the temperature rise. Based on Figure 7, the highest percent density was achieved by NCMA followed by NCA and NCM composites at sintering temperature of 1200 °C. Small and fine size of solid lubricant which was silver and

 DOI: 10.1051/matecconf/20178702011

$MoS_2$ compared to $CaF_2$ help in the diffusion and densification process during sintering. Then, densification leads to decrease in porosity and increase the particle contact area.

### 3.5 Porosity in Ni-Cr based composites

As increasing the sintering temperature, the porosity of the Ni-Cr based composites is decreasing. The number of porosities is decreased and the size of pores are reduced. When a higher temperature is applied, the grain of the composite will grow bigger and the pores tend to shrink. Thus, the porosity of the composites is reduced [11]. The reason is the driving force of sintering is increasing when the sintering temperature rises. At the earlier stage of sintering, the driving force is produced by the surface energy which is associated with the internal surface area of the particles. This is where the grains of the composites grow bigger and the pore shrinks. Less porosity is desirable because it will contribute to the positive effect on mechanical properties. Table 4 shows the porosity in the Ni-Cr based composites.

**Table 4.** Porosity in the Ni-Cr based composites samples after sintering at a temperature of 1000 °C and 1200 °C.

| Samples | Porosity (%) | |
|---|---|---|
| | at 1000 °C | at 1200 °C |
| NC | 35 | 18 |
| NCM | 32 | 16 |
| NCA | 30 | 15 |
| NCCf | 29 | 19 |
| NCMA | 27 | 15 |
| NCMCf | 31 | 19 |
| NCACf | 27 | 17 |
| NCMACf | 32 | 16 |

From Table 4, the lowest porosity is achieved by NCMA composites at both sintering temperature. This is due to the particle size of $MoS_2$ and Ag which is smaller compared to $CaF_2$. The particles of $MoS_2$ and Ag tend to fill in the pores and eventually reduce the size of the pores. The porosity results can be supported by the SEM results in Figures 8-13. The black areas which represent porosity are decreasing as the sintering temperature increased from 1000 to 1200 °C.

### 3.6 Microstructure of Ni-Cr based composites

Figure 8 and 9 show the microstructure of Ni-Cr based composites without any addition of solid lubricants after sintering at 1000 °C and 1200 °C. The grey phase represents the Nickel matrix while the darker phase represents the Chromium phase. On the other hand, the darkest (black) phase represent the porosity in the composite. More and larger size of pores could be seen in Figure 8 compared to Figure 9. When sintering at a higher temperature, the percentage of shrinkage is bigger, thus, reduce the distance between the particles and shrink the size of pores.

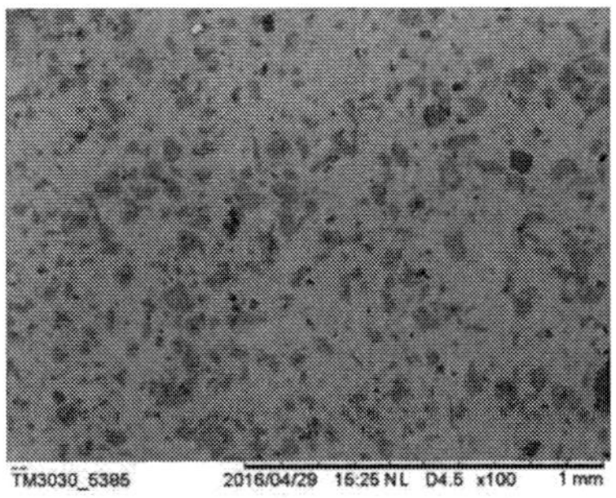


**Figure 8.** SEM micrograph of NC composite after sintering at 1000 °C.

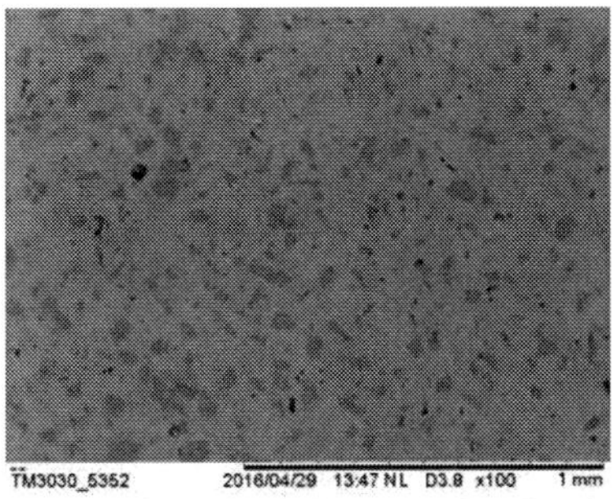


**Figure 9.** SEM micrograph of NC composite after sintering at 1200 °C.

Figure 10 and 11 show the Ni-Cr based composites with the addition of silver as a solid lubricant. The existence of white phase represents the silver addition in the composite.

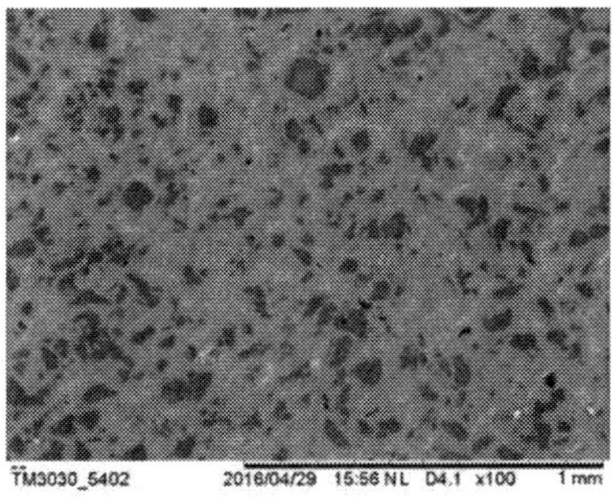


**Figure 10.** SEM micrograph of NCA composite after sintering at 1000 °C.

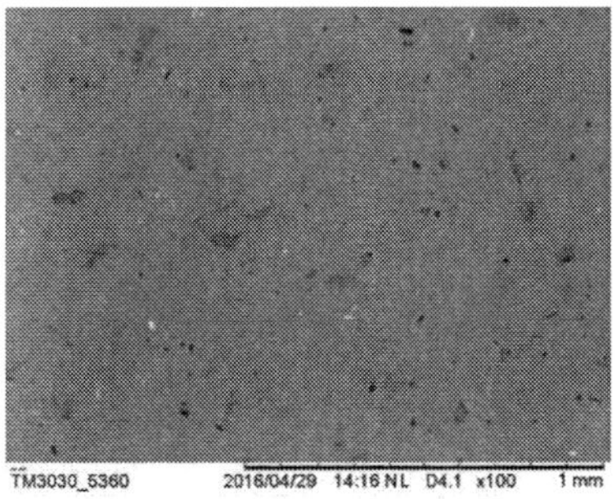


**Figure 11.** SEM micrograph of NCA composite after sintering at 1200 °C.

Figure 12 and 13 demonstrate the microstructure of Ni-Cr based composites with the addition of dual solid lubricant which are $MoS_2$ and silver.

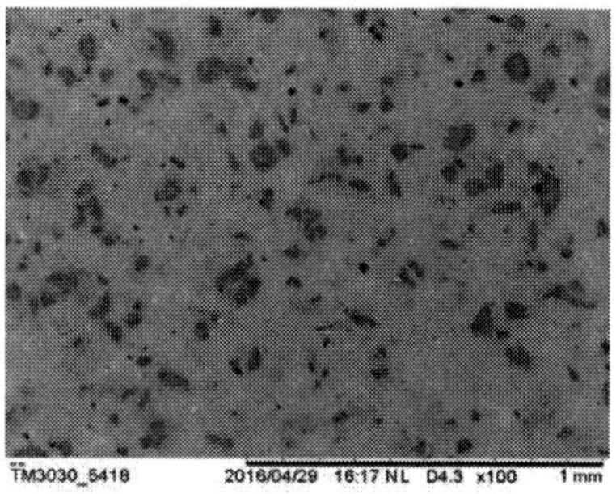


**Figure 12.** SEM microstructure of NCMA composites sintering at 1000 °C.

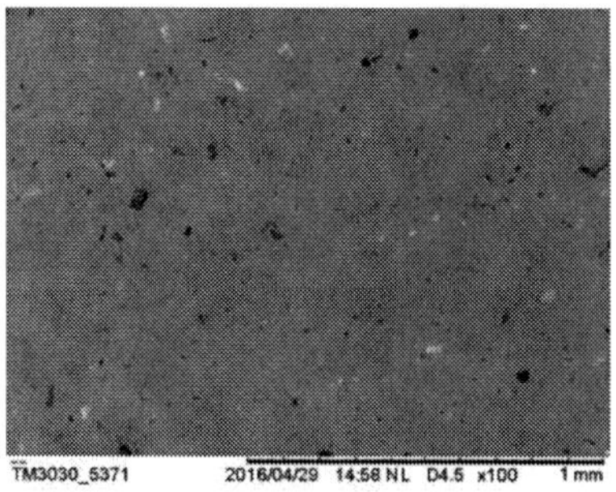


**Figure 13.** SEM microstructure of NCMA composites sintering at 1200 °C.

### 3.6 Hardness of Ni-Cr based composites

Higher shrinkage, enhancement of density and reduction of porosity will contribute to better mechanical properties. The hardness of the composites were listed in Table 5.

**Table 5.** The hardness of Ni-Cr based composites samples after sintering at temperature of 1000 °C and 1200 °C.

| Samples | Hardness (HRC) | |
|---|---|---|
| | at 1000 °C | at 1200 °C |
| NC | 42.6 | 51.5 |
| NCM | 36.8 | 52.1 |
| NCA | 35.1 | 48.5 |
| NCCf | 37.5 | 44.0 |
| NCMA | 41.9 | 57.1 |
| NCMCf | 35.8 | 46.4 |
| NCACf | 36.0 | 44.8 |
| NCMACf | 32.0 | 57.5 |

Based on Table 5, at sintering temperature of 1000 °C, the highest hardness is attained by the Ni-Cr composites. As we added the solid lubricants, the hardness of all Ni-Cr based composites is decreasing. However, at a higher sintering temperature of 1200 °C, the hardness of all of the composites containing $MoS_2$ are increasing (NCM, NCMA and NCMACf). 5% addition of $MoS_2$ as solid lubricant helps in improving the mechanical properties of the composite. A suitable amount of solid lubricant addition is important in order to enhance mechanical properties of the composites [12].

The highest hardness achieved by NCMACf composites after sintering at 1200 °C with a value of 57.5 HRC. The lowest hardness also achieved by NCMACf composites after sintering at 1000 °C with a value of 32 HRC. The main reason is the sintering temperature play the main role in the reinforcement and strengthening of the composites. At sintering temperature of 1000 °C, the addition of solid lubricant such as $MoS_2$, Ag and $CaF_2$ is not fully forming a good bonding of grain boundary. Thus, the grain growth between the particles is less thus the strengthening of the particles is also low compared to sintering at a higher temperature (1200 °C) [13].

## 4 Conclusions

1. The increment in sintering temperature from 1000 °C to 1200 °C enhanced the shrinkage, density and hardness while reduced the porosity of Ni-Cr based composites.

2. Ni-Cr based composites which contain silver and $MoS_2$ (NCA and NCMA) achieved better physical properties such as shrinkage, density and porosity due the fine size of these solid lubricants compared to $CaF_2$. The fine size of solid lubricant helped in filling the pores and thus increased the densification of the composites.

3. The mechanical properties of Ni-Cr based composites are improving with the increasing sintering temperature and addition of $MoS_2$. The addition of 5% $MoS_2$ in Ni-Cr based composites improved their hardness. A suitable amount of solid lubricant addition was helpful to enhance the mechanical properties.

4. Further research will be conducted on the wear properties of Ni-Cr based composites with the single, dual and multiple solid lubricants in order to achieve a composite with excellent mechanical as well as wear properties.

## Acknowledgement

The author would like to thank UNIMAS DPP grant for providing financial support of this research work.

## References

[1] M. B. Peterson, S. LI, and S. F. Murray, "Wear-resisting Oxide Films for 900°C," *J.Mater.Sci.Technol*, **13**, 99–106, (1997).

[2] K. Miyoshi, "Solid Lubricants and Coatings for Extreme Environments: State-of-the-Art Survey," (2007).
[3] A. Vadiraj, M. Kamaraj, and V. S. Sreenivasan, "Effect of solid lubricants on friction and wear behaviour of alloyed gray cast iron," **37**, 569–577, (2012).
[4] S. Huang, Y. Feng, K. Ding, G. Qian, H. Liu, and Y. Wang, "Friction and wear properties of Cu-based self-lubricating," *Acta Metall. Sin.*, **25**, 391–400,( 2012).
[5] M. Suzuki, M. Moriyama, M. Nishimura, and M. Hasegawa, "Friction and wear of self-lubricating composites at temperatures to 450 °C in vacuum," *Wear*, **162–164**, 471–479, ( 1993).
[6] Q. Bi, S. Zhu, and W. Liu, "High Temperature Self-Lubricating Materials," in *Tribology in Engineering*, (2013).
[7] M. Xue, "High temperature oxidation and wear behaviour of powder metallurgically developed Ni-Cr-W-Al-Ti-MoS2 composite," *Sci. York*, **16**, 111–115, (2009).
[8] C. H. Ding, P. L. Li, G. Ran, Y. W. Tian, and J. N. Zhou, "Tribological property of self-lubricating PM304 composite," *Wear*, **262**, 575–581, (2007).
[9] E. Y. Liu, Y. M. Gao, W. Z. Wang, X. L. Zhang, X. Wang, G. W. Yi, and J. H. Jia, "Effect of the synergetic action on tribological characteristics of Ni-based composites containing multiple-lubricants," *Tribol. Lett.*, **47**, 399–408, (2012).
[10] C. Suiyuan, W. Jing, L. Yijie, L. Jing, and L. Changsheng, "Synthesis of New Cu-based Self-lubricating Composites with Great Mechanical Properties," *J. Compos. Mater.*, **45**, 51–63, (2010).
[11] L. F. Güleryüz, S. Ozan, R. İpek, and D. Uzunsoy, "Production of B 4 C p reinforced magnesium metal matrix composites by powder metallurgy," **1**, 51–58, ( 2012).
[12] X. Dangsheng, "Lubrication behavior of Ni–Cr-based alloys containing MoS2 at high temperature," *Wear*, **251**, 1094–1099, (2001).
[13] F. Li, J. Cheng, Z. Qiao, J. Ma, S. Zhu, L. Fu, J. Yang, and W. Liu, "A Nickel-Alloy-Based High-Temperature Self-Lubricating Composite with Simultaneously Superior Lubricity and High Strength," *Tribol. Lett.*, **49**, 573–577, (2013).

# Optimisation of electrospinning parameter for Poly(*L*-lactic) acid (PLLA) electrospun nanofiber

*Marini* Sawawi[1] , *S.K* Sahari[1], *Nicholas* Kuan[1], *Mahshuri* Yusof[1], *Magdalene* Andew[1], *SF* Salleh[1], *Rohana* Sapawi[1]
[1]Faculty of Engineering, Universiti Malaysia Sarawak, 94300 Kota Samarahan,

**Abstract.** Poly (L-lactic) Acid (PLLA) ia one of the biodegradable polymer that normally used in the biomedical application. In this work it is shown that electrospinning of PLLA electrospun fibre was affected by electrospinning processing parameters such as the solvent systems selection, polymer solution concentration, flow rate and acceleratingvoltage. Optimisation of these parameters is necessary to obtain a beadless and submicron diameter fibre as desired in this research.

## 1 Introduction

Poly (L-lactic) Acid (PLLA) is a thermoplastic, semi crystalline polymer. It has become widely used recently due to its biocompatibility and biodegradability [8]. One of the methods to produce PLLA fibre is using electrospinning. Electrospinning has become an oft-reported technique for the fabrication of polymeric fibrous membranes that have been used for a variety of applications including membrane technologies, sensors and in particular for tissue engineering as scaffolds that mimic features of the extracellular matrix [3, 16-18, 20, 25] It is a method that makes use of electrical charge with polymer solutions to produce fine filaments. The basic concept dates from 1902 [5] and a series of patents were obtained by Formhals commencing in the 1930s [7] to produce polymer threads from various polymers using such an electrostatic force. This method has subsequently been recognised as a fast and simple process for making continuous, sub-micron to nano-size fibres, when compared with other conventional methods such as drawing, template synthesis, phase separation, and self-assembly. It also has been used to produce various types of fibre materials ranging from natural polymers, synthetic polymers, polymer solutions, polymer melts and biopolymers. One of the attractive aspects of the electrospinning technique is the flexibility of the process and that the morphology of the fibres can be readily varied by changing processing conditions, including producing fibres with solid, porous and core-shell morphologies. Solid and porous fibres can be obtained by changing the processing condition such as types of solvents and humidity whilst the core-shell morphology is obtained through setting up the inner and outer needle during electrospinning [2, 9, 10, 27]. In its most basic form, electrospun webs are produced from a relatively simple process that consists of a pipette to hold a dissolved polymer solution, a high voltage source and a collector. Briefly, the positive electric field induces charges in the polymer solution, and once this charge build-up exceeds that of the surface tension, a Taylor cone is formed [11]. At this critical voltage, a polymer jet erupts from the cone tip, accelerating through the electric field towards the collector. As the jet travels towards the collector, the solvent evaporates and a solidified polymer fibre is collected. There are a number of parameters that influence the morphology of electrospun nanofibres, some of which include: the concentration of polymer solution, applied voltage, solution feed rate, working distance and ambient parameters [3, 15, 19, 20].

Although electrospinning of PLLA polymers has been reported by others in a range of studies [4, 14, 20, 26], it is clear that using a different apparatus and working conditions such as environment will lead to different fibre morphologies, in addition to any changes in processing conditions. In this research, the PLLA polymer was electrospun and optimization of the processing parameters were performed so that the electospun fibres achieve the surface morphology that is beadless and is of a sub-micron diameter size.

## 2 Experimental

*Materials.* Poly(L-lactide) Acid (PLLA, Inherent viscosity = 0.9-1.2 dL/g) was purchased from Lactel, USA. It is in the form of pellets with a brownish, translucent colour. Dodecyl trimethyl ammonium bromide (DTAB), were purchased from Sigma Aldrich, Australia. Chloroform, dimethylformamide (DMF) and acetone from Merck Pty Ltd, Australia.

*Electrospinning.*
The PLLA electrospun web was prepared using the solvent systems and electrospinning condition as shown in Table 1. The PLLA-solvent mixture was dissolved using a magnetic stirrer overnight. The solution was then placed in a 10 mL syringe with the gauged needle. The electrospinning process was conducted using an electrospinner built in the laboratory which used a high voltage supply from Gamma High Voltage Research, Florida, whilst the syringe pump is from Razel Scientific Instruments, Inc. The fibres were collected onto a flat metallic collector with aluminium (Al) foil placed on top. After collection, the non-woven web was stored in a desiccator under vacuum before further use. This procedure was repeated for a range of concentrations, until beadless fibres with a submicron range diameter were obtained.

*Characterisation.*
The morphology of the as-spun fibres was characterized using FEI Phenom Desktop SEM. Prior to imaging, gold coated at 1 nm thickness with a sputter coater (Cressington 208HR, UK).

**Table 1 Electrospinning condition**

| Solvent system | Concentration (%)(w/v) | Feed rate (ml/hr) | Voltage (+kV) | Needle size (G) | Working Distance (cm) |
|---|---|---|---|---|---|
| chloroform | 6.5 | 2.4 | 20 | 18 | 15 |
| chloroform | 8 | 2.4 | 20 | 18 | 15 |
| chloroform (1mM DTAB) | 8 | 2.4 | 20 | 18 | 15 |
| chloroform (1mM DTAB) | 12 | 2.4 | 20 | 18 | 15 |
| chloroform (1mM DTAB) | 13.7 | 1.6 | 20 | 18 | 15 |
| chloroform (with 1mM DTAB) | 15.3 | 1.6 | 20 | 21 | 23 |
| chloroform+acetone (3:1)(with 1mM DTAB) | 13 | 1.2 | 20 | 21 | 23 |
| chloroform+acetone (3:1)(with 1mM DTAB) | 13.7 | 1.2 | 15 | 21 | 23 |
| chloroform+acetone (3:1)(with 1mM DTAB) | 15 | 1.2 | 15 | 21 | 21 |
| chloroform+acetone (3:1)(with 1mM DTAB) | 16 | 1.2 | 25 | 21 | 21 |

## 3 Result and discussion

The common solvents that have been used to electrospin PLLA include chloroform [22] and a mixture of chloroform and acetone [23]. We initially present results with chloroform alone and Figure 1 shows PLLA electrospun fibres spun from chloroform at 6.5 wt/v% and 8 wt/v% concentration, with a needle feed rate of 2.4 mL/h, through an 18 G gauge needle with a 20 kV voltage, and 15 cm working distance to the collector.

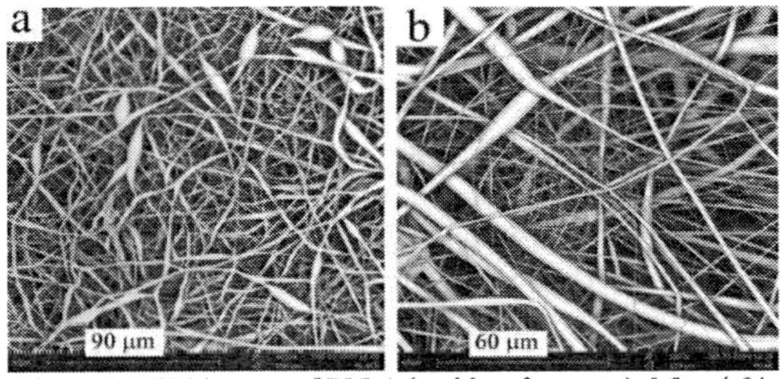

Figure 1 SEM images of PLLA in chloroform at a) 6.5 wt/v%, b) 8 wt/v%

It can be seen in Figure 1 that at 6.5 wt/v % polymer concentration, the fibres produced contained elongated beads. Increasing the concentration to 8 wt/v% lead to beadless, non uniform fibres, with two distinctive types of fibres' diameter. This bimodal size distribution is due to the secondary jet that erupts from the primary electrospinning jet, producing smaller fibre diameters which result from the fibre-solvent interaction and higher voltage used [20]. It has often been suggested that addition of ionic surfactant can improve the conductivity and reduce the surface tension of the polymer solution [20, 23] with the increased conductivity resulting in more charge carried by the electrospinning jet, leading to more fibre stretching and thus a thinner fibre [20] The results of incorporation of DTAB are shown in Figure 2, where the PLLA was electrospun in chloroform with 1 mM DTAB, with the other parameters the same as before for 8 wt/v%,13.7 wt/v% and 15.3 wt/v% polymer concentrations.

Figure 2 Electrospun PLLA fibres in with chloroform as the solvent, with 1mM DTAB at the polymer concentration of a) 8 wt/v% , b) 13.7 wt/v% and c)15.3 wt/v%.

It can be seen in Figure 2 that with the addition of the cationic surfactant, DTAB, electrospun fibres in chloroform at 8 wt/v% polymer concentration produced fibres with spherical beads, while at the same concentration without DTAB addition, the fibres obtained were beadless (Figure 1b). However, it is important to note that although without addition of DTAB, the fibres showed no beads, the non uniform fibre diameter is undesirable. At 8 wt/v% concentration, incorporating of DTAB in chloroform resulted in a difference in charge distribution on the Taylor cone and generated nanofibres with beads [12]. The addition of a cationic surfactant will increase the charge density, leading to a higher conductivity. If the conductivity level is too high, multijets will be produced and lead to fibres with beads [1, 21]. It was previously shown that without DTAB addition, 8 wt/v% results in a bimodal distribution in fibres resulting in distinct fibres diameter due to secondary jets. With the addition of DTAB, an increase in conductivity leads to an unstable jet and results in jet break-ups (electrospraying). It was also found by Kriegel et al [13] that the addition of DTAB reduced the viscosity of PEO solutions which also might contribute to the beads in our findings. This can be overcome by increasing the concentration at which chain entanglements increase to enable continuous stretching of the jet. This is particularly true when the polymer concentration increased to 12 wt/v% (not shown here), and 13.7 wt/v% (Figure 2b) where the beads changed from spindle-like shapes to elongated beads. . This shows that by increasing the concentration, the intermolecular force or chain entanglement of the polymer is increased and sufficient for jet stretching. Finally, the optimum fibre morphology for PLLA in chloroform with 1 mM DTAB was obtained at the concentration of 15.3 wt/v% with a feed rate of 1.6 mL/h , 20 kV accelerating voltage and 15 cm working distance which obtained average fibre diameter of 790 ± 390 nm where no beading was observed although the uniformity of the fibres was still not satisfactory, as shown in the high standard deviation of the fibres diameter.

In order to improve the uniformity and reduced the diameter of the electrospun fibres, PLLA was electrospun with a mixture of chloroform and acetone, with 1 mM DTAB. It has been reported that the addition of acetone to the PLLA in chloroform solution leads to improved continuity of the electrospinning jet [23, 24], likely due to the lower polarity of acetone compared to chloroform, which reduces the interactions between the solvent and polymer molecules [24]. This in turn reduces jet cohesive energy, contributing to a greater ease of spinning and thus more uniform fibre diameters. The electrospinning of PLLA with chloroform and acetone with 1mM DTAB addition was carried out at the concentration of 13 wt/v%, 13.7 wt/v%, 15 wt/v% and 16 wt/v%, with the other parameters, as stated in Table 1. The result showed that increasing the polymer concentration lead to a larger fibre diameter. This is because the greater concentration leads to a higher solution viscosity and an increased molecular entanglement that leads to high cohesive forces between the polymer molecules, thus reducing the stretching and thinning of the fibres [6]. For the same concentration of 13.7 wt/v% PLLA, reducing the accelerating voltage from 15 kV to 10 kV leads to a larger fibre diameter (not shown). This reduced electrostatic field leads to less polymer stretching and fibres with a larger diameter result [20]. For this system, the optimum fibre morphology was obtained at 13.7 wt/v% polymer concentration with the feed rate of 1.2 mL/h, 20 kV accelerating voltage, 18 G needle gauge and 16 cm working distance which obtained the average fibre diameter of 770 ± 240 nm. The larger fibre average diameter was obtained at 16 wt/v%, with an average fibre diameter of 970 ± 400 nm, as shown in Figure 3b.

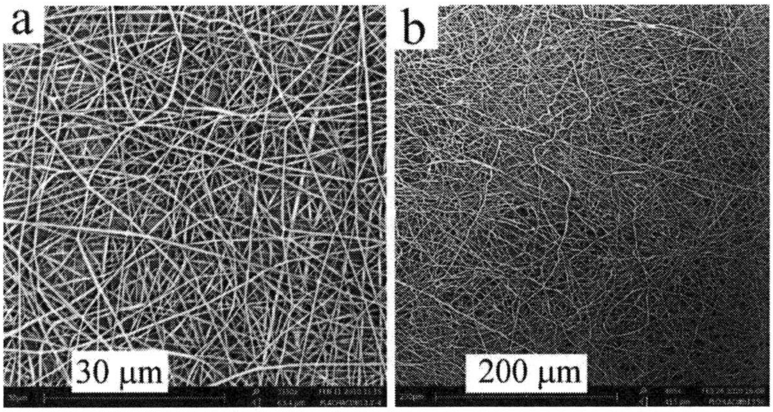


Figure 3 Electrospun PLLA in chloroform:acetone with 1 mMDTAB at a)13.7 wt/v% and b)16 wt/v%

Figure 3 demonstrates that the optimum fibre morphology obtained was more uniform for the 13.7 wt/v% PLLA concentration, where the standard deviation of the fibre diameter was about 200 nm. It was thus found that the addition of acetone did improve the electrospinning process of PLLA with the uniformity of the fibres achieved ( lower standard deviation of the fibre diameter showing the distribution of the fibres' diameter is more uniform) with the average fibre diameter within submicron range. Meanwhile, the 16 wt/v% PLLA concentration showed that the higher concentrations result in higher viscosities that make maintaining flow difficult and the resulting fibres had a greater average fibre diameter and were less uniform.

## 4 Conclusion

From this work, the optimization process of PLLA electrospun fibre were conducted by manipulating the processing parameters such as solvent systems selection, cationic surfactant addition, polymer concentration, polymer solution properties such as viscosity, accelerating voltage and flow rate. For instance, the polymer concentration affects the intermolecular force or chain entanglement that by increasing the concentration , chain entanglement of the polymer is increased and sufficient for jet stretching. Low concentration would lead to less chain entanglement and can cause electrospraying or jet break ups. The addition of DTAB would lead to a change in conductivity and viscosity of the polymer solution. At sufficient concentration and viscosity, this surfactant would produce more uniform fibre compared with no surfactant addition. Reducing the accelerating voltage leads to a larger fibre diameter due to reduced electrostatic field leads to less polymer stretching. By manipulating the parameters, we can produce the fibres at with certain surface morphology and diameter to suit our application.

## References

1. C.J Angammana and H.J Shesha, Industry Applications, IEEE Transactions on. **47**(3): p. 1109-1117, (2011)
2. A. Bazilevsky, A. L. Yarin, and C. M. Megaridis, Langmuir. **23**(5): p. 2311-2314, (2007)
3. N Bhardwaj, and C.K Subhas, Biotechnology Advances. **28**(3): p. 325-347,(2010)
4. Y.Z Chen et al., Journal of Polymer Science, Part B: Polymer Physics. **47**(12): p. 1211-1218, (2009)
5. J.F Cooley Apparatus for electrically dispersing fluids,: US Patents, (1902).
6. J.M Deitzel et al., Polymer. **42**(1): p. 261-272, (2001)
7. A. Formhals , Process and Apparatus for Preparing artificial threads, US. Patent, Editor (1934).
8. B.Gupta, R. Nilesh, and H.Jöns, Progress in Polymer Science. **32**(4): p. 455-482, (2007)
9. C. He et al., Gaojishu Tongxin/Chinese High Technology Letters. **16**(9): p. 934-938, (2006)
10. C.L He, Z.M. Huang, and X.J. Han, Journal of Biomedical Materials Research - Part A. **89**(1): p. 80-95, (2009)
11. J.H He, et al., Electrospun Nanofibers and Their Applications., Shawbury,UK: iSmithers (2008)

12. Z.M Huang et al., Composites Science and Technology. **63**(15): p. 2223-2253, (2003)
13. C.Kriegel et al., Food Biophysics. **4**(3): p. 213-228, (2009)
14. T. Lin et al., Nanotechnology. **15**(9): p. 1375, (2004)
15. Y.P Neo et al., Journal of Food Engineering. **109**(4): p. 645-651, (2012)
16. D.R Nisbet et al., Journal of Neural Engineering. **4**(2): p. 35, (2007)
17. D.R Nisbet et al., Journal of Biomaterials Applications. **24**(1): p. 7-29, (2009)
18. D.R Nisbet et al., Biomaterials. **30**(27): p. 4573-4580, (2009)
19. N.Okutan et al, Food Hydrocolloids. **39**: p. 19-26, (2014)
20. S. Ramakrishna et al., An Introduction to Electrospinning and Nanofibers, Singapore: World Scientific Pub, (2005)
21. C.Wang et al, Macromolecules. **39**(22): p. 7662-7672, (2006)
22. Y. You et al., Journal of Applied Polymer Science. **95**(2): p. 193-200, (2005)
23. J.Zeng et al., Journal of Controlled Release. **92**(3): p. 227-231, (2003)
24. J.Zeng_et al., Journal of Applied Polymer Science. **89**(4): p. 1085-1092, (2003)
25. Y.L Zhang et al. Electrospinning process of ethylene-co-vinyl alcohol nanofiber with DMAC/IPA solution system. in Advanced Materials Research.. Trans Tech Publ (2014)
26. Z.Zhou et al., Polymer. **50**(13): p. 2999-3006, (2009)
27. E.Zussman et al., Journal of Polymer Science, Part B: Polymer Physics. **44**(10): p. 1482-1489, (2006)

MATEC Web of Conferences 87, 02013 (2017) DOI: 10.1051/matecconf/20178702013
*ENCON 2016*

# A Study on Relationship Between Carrying Schoolbags and The Prevalence of Neck and Back Pain Among 7 – 9 Year Old Students

N. Kabilmiharbi[1] and T. Santhirasegaram[1]

[1]*Mechanical Engineering Department, Universiti Tenaga Nasional (UNITEN), Malaysia*

**Abstract.** There is a growing concern these days that children are carrying way too much weight on their backs. School students are at risk of developing Musculoskeletal Disorder or possibly long term back pain due to the frequent exposure of carrying heavy schoolbags. In Malaysia, this problem is currently an overlooked issue as not many are aware of the importance of maximum allowable load that should be carried by school students together with proper posture and how all these will affect the children in the future. An interview was conducted based on 90 school students of age 7 to 9 year old regarding the body part discomfort that they experience while carrying the schoolbag. Furthermore, the weight of the schoolbag and the student's height and weight was taken for this study. The responses were classified based on age and gender as the anthropometry differs on ages and gender. A statistical analysis is carried out with the results obtained. The finding of the statistical analysis showed that the prevalence of neck and back pain does exist among 7-9 year old school students and almost 90 of them are affected. Among the main problems are that the schoolbag design, the incorrect ways of carrying the schoolbag, the excessive weight of the schoolbag and the prolonged duration of constantly carrying the schoolbag. Necessary action should be taken to overcome these problems.

## 1 Introduction

School students usually caries their schoolbag to, from and around the school which inside the schoolbag contains their textbooks, sports attire, packed meals and stationaries. There has been a growing concern around the world these days regarding the school children that carrys way too much weight on their backs. Some countries are coming up with variety of ways on how to overcome this issue such as electronic based learning where students are provided with tablets or laptops and they only need to carry the laptops or tablets to school instead of the heavy school books or by making their study materials available online which will then reduce the burden of carrying heavy textbooks to school. Contrary, in Malaysia, the risk factors of carrying heavy schoolbag faced by school students are currently an overlooked issue as Malaysians are not well exposed on the ergonomics awareness where there are merely few studies and reports done in Malaysia regarding this issue [1, 2, 3].

According to Dutch Labour Inspectorate, to prevent work-related back disorders as a result of lifting, the maximum load mass to be lifted by one adult worker is to be 25 kg and if it is more than that, it should be lifted by more than one workers provided it is not more than 50 kg and loads which are more than 50kg should always be lifted mechanically [4]. Furthermore, according to the National Institute of Occupational Safety and Health, there are different weight limit among different countries and the mean carrying weight for boys and girls not more than 16 years old of age are 14kg and 8 kg [5]. Meanwhile, for primary school students, 10% of body weight (BW) has been commonly accepted as a recommended schoolbag weight as the physical capability of school children are different from the adults [6]. The National Back Pain Association recorded that the schoolbag weights for British children of 13-year olds is 10.4% of their body weight and 10.2% of their body weight for 16 year old children [7]. Besides that, an investigation was done to see the effect of backpacks where athletic bags were loaded with 17% of student's body weight and a conclusion was made that the carriage of schoolbags significantly altered the posture and gait of students and the result may leads to a variety of musculoskeletal complaints such as muscle soreness, back pain, numbness and shoulder pain [8]. Furthermore, the National Back Pain Association has created a guideline which suggested that in an ideal world, a schoolbag should weigh under 10% of the student's body weight and this weight should be held on both shoulders in a backpack style schoolbag [5]. All these research proves that school children carrying heavy school bags are subjected to MSDs in the long run.

School students are at risk of developing Musculoskeletal Disorder or possibly long term back pain due to the frequent exposure of carrying heavy schoolbags. A research done on the weight of schoolbags and prevalence of musculoskeletal symptoms amongst 140 students in New Zealand found that the students were carrying weights which exceeds the recommended guideline load limits for an adult industrial workers and carrying heavy schoolbags was suspected as the contributory [6]. The research shows that 77.1% of the students reported to suffer from musculoskeletal symptoms and the symptoms was most prevalent in the neck, shoulders, upper back and lower back. Furthermore, another recent study in Ireland also found that the prevalence of baseline musculoskeletal discomfort was as high as 63.4% on student of mean age 10.6, where schoolbag related discomfort was reported more frequently on the shoulders (27.3%) than on the back (15%) and time factor was also taken into account as most students carried their bags for almost less than 10 minutes [9]. This shows that students who carries heavy burden on their back more than the recommended weight will somehow develop pain on their neck shoulders and back which might leads to Musculoskeletal disorder at early age.

School students are responsible on carrying their own schoolbag and so it is important for each students to carry their schoolbag properly in order to avoid or minimize postural problems, back pain and musculoskeletal disorder. There are proper ways on carrying a load and this is also applied on carrying a luggage or a bag. Carrying a load in a lateral position, such as carrying a suitcase by hand, requires a greater muscular effort in the contralateral muscles compared to carriage in a backpack where it requires less muscular effort in the lower back [10]. Besides that, carrying a backpack was associated with a lower heart rate and oxygen uptake than with shoulder load carriage where carrying materials on the back seems to be an acceptable method of load carriage as long as it follows the safety guidelines in carrying which are the load should be kept as close as possible to the trunk and heavy load should not be carried using small muscle groups, and optimum use should be made of large muscle groups of the body in order to minimize fatigue and local muscle discomfort [11].

Hence, the objective of this study is to investigate the relationship between carrying schoolbag and the prevalence of neck and back pain among students aged 7 to 9 years old by interviewing the students on their discomfort level experienced while carrying the schoolbag. It is important to study on how carrying schoolbag affected the students and the risk factors which causes the occurrence of neck and back pain among them. Furthermore, the results can be used to help promoting the ergonomics awareness among the parents, teachers, educational institutions, medical professionals and the government.

## 2 Methodology

### 2.1 Focus of the Study

This study was conducted on 90 school students aged 7 to 9 years old at a primary school in Shah Alam, Selangor. There are a total of 43 male students and 47 female students and 30 students were taken for each age, 7, 8 and 9 years old. All the students were interviewed and the height and weight of each student were taken followed by measuring the weight of their respective schoolbags.

### 2.2 Observation on the students

An interview was done to all the students by modifying the Nordic questionnaire in order to gather information regarding the occurrence discomfort experienced by the students. The modified Nordic questionnaire is used together with a Visual Analogue Scale (VAS) as shown in Figure 1. This is to help the students provide a more accurate responses to the intended questions and the usage of VAS with a questionnaire is very common in research conducted among school children as it would be much easier for kids to understand since it is more graphical rather than words [6, 8, 10, 11]. Furthermore, the height and weight of the students were measured together with the weight of their respective schoolbags.

**Figure 1**. The Visual Analogue Scale (VAS) used to interview the students.

### 2.3 Analysis

After observation and data collection, the method used to analyze the data is the same as done in previous studies [6, 13] where the mean weight and height of each group of students by age and gender are being evaluated. This is then compared with the mean weight of this schoolbags to obtain the percentage of weight of their schoolbag to their body weight. The method is used as it is recommended that school children are only allowed to carry load which is 10% of their body weight. Necessary action and suggestion will be discussed based on the results obtain form the analysis done.

## 3 Results and Discussion

### 3.1 Respondent Profile

The response rate for this survey varied by age and gender group from 14.44% to 18.89%. The highest number of responses were from 9 year old female students (17 respondents) and the lowest were from 9 year old male students (13 respondents) as shown in Figure 2 below.

 DOI: 10.1051/matecconf/20178702013

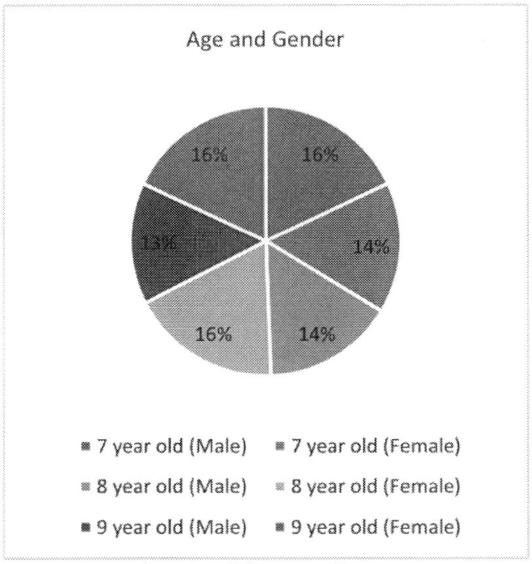


**Figure 2**. Age and gender of the respondents.

## 3.2 Statistical Analysis

### *3.2.1 7 year old male student*

There are 16 respondents from this group where the range of weight are from 22.1 kg to 32.4 kg and the range of height were from 118 cm to 132 cm while the range of schoolbag weight are from 3 kg to 4.6 kg. The mean weight for the students is 29.3 kg and the mean height is 125.9 cm while the mean weight of the schoolbag is 3.7 kg. The interview results regarding the occurrence of discomfort experienced by the male students age 7 is as shown in the Figure 3. The results shows that 7 year old male student from this school do experience neck and back pain especially they experience distress on the lower back part of their body. Meanwhile, the average weight of the schoolbag shows that it is 12.6% which is more than 10% of the student's weight.

**SCORE OF PAIN/ DISCOMFORT ON AFFECTED AREAS**

| | Low | Discomforting | Distressing | Horrible | Excruciating |
|---|---|---|---|---|---|
| **NECK** | 11 | 5 | | | |
| **SHOULDER** | 9 | 5 | 2 | | |
| **UPPER BACK** | 13 | 3 | | | |
| **LOWER BACK** | 3 | 5 | 7 | 1 | |

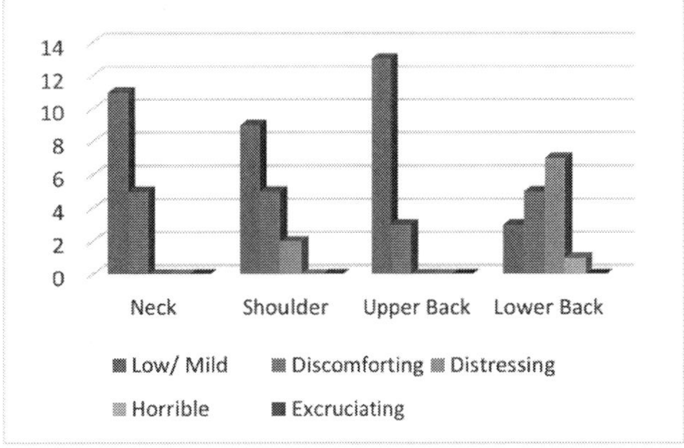


**Figure 3.** Results of discomfort level experience by male student age 7.

### *3.2.2 7 year old female student*

There are 14 respondents from this group where the range of weight are from 21.5 kg to 34.7 kg and the range of height are from 117 cm to 130 cm while the range of schoolbag weight are from 3.1 kg to 3.8 kg. The mean weight is 27.3 kg and the mean height is 123.9 cm while the mean weight of schoolbag is 3.5 kg. Meanwhile, the results of the interview among the 7 year old female students from this school regarding the occurrence of discomfort experience while carrying the schoolbag are as shown in Figure 4. From the result obtain, it shows that most of the students felt discomfort on the neck and the lower back area. Furthermore, the mean weight of the schoolbag for this group is 12.8% of their weight which shows that the heavy schoolbag has also taken toll on 7 year old female students.

DOI: 10.1051/matecconf/20178702013

**SCORE OF PAIN/ DISCOMFORT ON AFFECTED AREAS**

| | Low | Discomforting | Distressing | Horrible | Excruciating |
|---|---|---|---|---|---|
| **NECK** | 5 | 7 | 2 | | |
| **SHOULDER** | 12 | 2 | | | |
| **UPPER BACK** | 10 | 4 | | | |
| **LOWER BACK** | 1 | 6 | 7 | | |

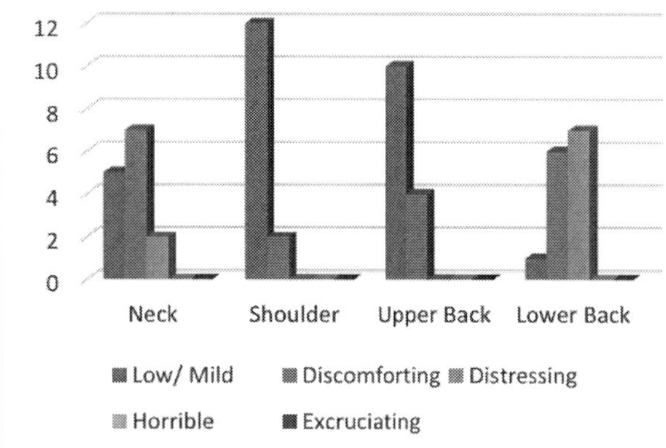


**Figure 4**. Results of discomfort level experience by female student age 7.

### *3.2.3 8 year old male student*

There are 14 respondents from this group where the range of weight are from 26.4 kg to 42.2 kg and the range of height are from 120 cm to 137 cm while the range of schoolbag weight are from 3.8 kg to 4.6 kg. The mean weight is 34.5 kg and the mean height is 129.8 cm while the mean weight of schoolbag is 4.3 kg. Based on the result shown in Figure 5, half of the student felt discomfort on their lower back area and some of them felt discomfort on their shoulder. Furthermore, their mean schoolbag weight is 12.5% of their mean weight. Neck and back pain does exists due to heavy schoolbags but the shoulder pain might be caused the strain of neck pain extending to the shoulders.

**SCORE OF PAIN/ DISCOMFORT ON AFFECTED AREAS**

| | Low | Discomforting | Distressing | Horrible | Excruciating |
|---|---|---|---|---|---|
| **NECK** | 10 | 4 | | | |
| **SHOULDER** | 8 | 6 | | | |
| **UPPER BACK** | 11 | 3 | | | |
| **LOWER BACK** | 5 | 2 | 7 | | |

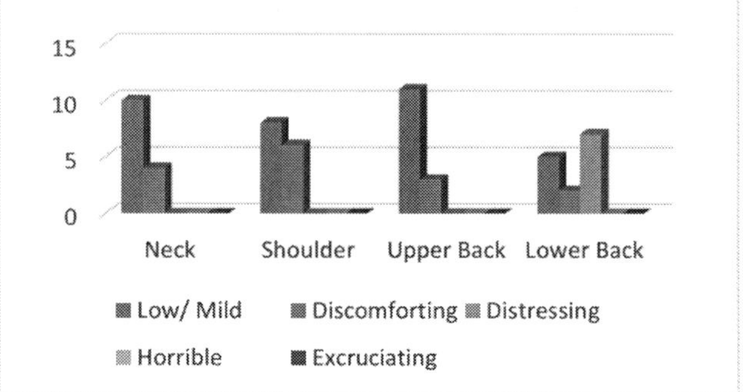


**Figure 5**. Results of discomfort level experience by male student age 8.

### *3.2.4 8 year old female student*

There are 16 respondents from this group where the range of weight were from 21.7 kg to 37.1 kg and the range of height were from 119 cm to 133 cm. The range of schoolbag weight was from 4.1 kg to 4.6 kg. The mean weight is 30.2 kg and the mean height is 126.8 cm. The mean weight of schoolbag is 4.3 kg. According to the result obtained based on Figure 6, 8 year old most of the female students from this school experience distress pain on their lower back while carrying their schoolbag. On the other hand, the mean weight of schoolbag is 14.2% of the mean body weight for this group which is at least 1.4% more than all the other groups before.

 DOI: 10.1051/matecconf/20178702013

**SCORE OF PAIN/ DISCOMFORT ON AFFECTED AREAS**

| | Low | Discomforting | Distressing | Horrible | Excruciating |
|---|---|---|---|---|---|
| **NECK** | 5 | 6 | 5 | | |
| **SHOULDER** | 9 | 7 | | | |
| **UPPER BACK** | 12 | 4 | | | |
| **LOWER BACK** | 1 | 3 | 9 | 3 | |

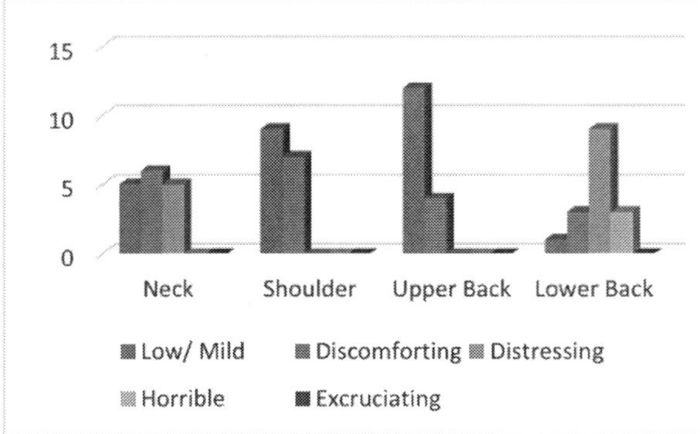


**Figure 6**. Results of discomfort level experience by female student age 8.

### *3.2.5 9 year old male student*

There are 13 respondents for this group where the range of weight are from 32.1 kg to 50 kg and the range of height are from 133 cm to 142 cm while the range of schoolbag weight are from 4.8 kg to 5.6 kg. The mean weight is 42.7 kg and the mean height is 137.8 cm while the mean weight of the schoolbag is 5.2 kg. The results shown in Figure 7 indicates that, most of them felt discomfort and distress pain on the lower back of their body. The mean schoolbag weight for this group is 12.2% of the mean body weight of this group which is clearly more than 10% of their BW.

**SCORE OF PAIN/ DISCOMFORT ON AFFECTED AREAS**

| | Low | Discomforting | Distressing | Horrible | Excruciating |
|---|---|---|---|---|---|
| **NECK** | 7 | 5 | 1 | | |
| **SHOULDER** | 11 | 1 | 1 | | |
| **UPPER BACK** | 10 | 3 | | | |
| **LOWER BACK** | 2 | 5 | 5 | 1 | |

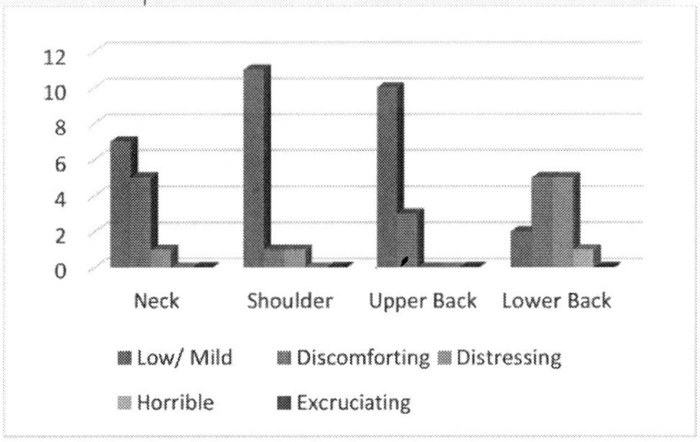


**Figure 7.** Results of discomfort level experience by male student age 9.

### *3.2.6 9 year old female student*

There are 17 respondents for this group where the range of weight are from 21.7 kg to 47.9 kg and the range of height are from 120 cm to 142 cm while the range of schoolbag weight are from 4.8 kg to 5.3 kg. The mean weight is 42.7 kg and the mean height is 137.8 cm while the mean weight of schoolbag is 5.2 kg. According to the results from Figure 8, it shows that nearly half of the students experience a distress pain on their lower back due to carrying the schoolbag and at least 4 of them experience discomfort pain on their neck, shoulder, lower back and upper back while carrying the schoolbag. The mean schoolbag weight is also 12.2% of their mean body weight which the occurrence of prevalence of back and neck pain can be seen based on the result.

DOI: 10.1051/matecconf/20178702013

**SCORE OF PAIN/ DISCOMFORT ON AFFECTED AREAS**

| | Low | Discomforting | Distressing | Horrible | Excruciating |
|---|---|---|---|---|---|
| **NECK** | 13 | 4 | | | |
| **SHOULDER** | 9 | 6 | 2 | | |
| **UPPER BACK** | 11 | 6 | | | |
| **LOWER BACK** | 1 | 4 | 8 | 4 | |

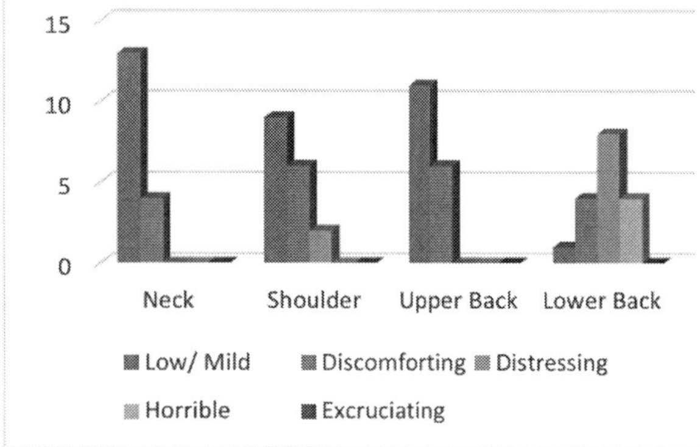


**Figure 8.** Results of discomfort level experience by male student age 9.

### 3.3 Discussion

According to the results and statistical analysis shown above, it is found that the prevalence of neck and back pain does exist among 7-9 year old school students and almost all students of all age and gender groups are affected. Comparing the results obtained above, it is found that most student age 7-9 of both gender felt distressing amount of discomfort especially on their lower back while most of them felt slight discomfort on other parts of thair body such as on neck, shoulder and upper back

Moreover, results obtained from this study shows that most of the student are carrying schoolbags that weighs around 12% to 14% more than their body weight. This means that the students carries 2%- 4% more than the recommended weight. This excessive weight of the schoolbags will create an imbalanced stance and will cause the students to lean forward which in time will disrupts both coordination and posture of the students and worst, it may distort the back's natural curves. Furthermore, heavy backpacks creates a forward trunk lean which causes a forward head posture with an extended neck, creating neck and shoulder pain and making it difficult for muscles and ligaments to hold up the body. This could eventually lead to chronic muscle imbalance. There could also be other external factors such as schoolbag design and schoolbag carrying postures which can alter the results yet the weight of the schoolbag would still be the main factor of the prevalence of neck and back pain among 7-9 year old students.

Additionally, it is found quite common among the primary school students to use a trolley bag since looks like it is easier for children to transport their belongings but overlooked the fact that most schools in Malaysia have classrooms which are located on an at least a 3 story high building with no elevator and only a staircase to excess from one floor to another. Therefore, students who used trolley bags are unable to pull their bags once they have reached the staircase. The students will then carry their bags from there on. What would make it worst is that, the weight of the trolley bag itself without any load or baggage inside is at least 1kg where this will further add up to the weight of the students belonging.

For future research, the amount of respondent for this study can be increased where this will improve the finding. The larger the number of respondent for the study, means more accurate data can be obtained and a clearer pattern of data can be acquired. Moreover, a comparative study could also be done to examine the effect of MSDs among students who carry their schoolbag in a proper posture with less loads on their schoolbags. This different grouping of research could help in determining a better posture of carrying schoolbag, ideal load that could be carried by a 7-9 year old student and improving the design of schoolbags in order to reduce MSDs among the school children.

## 4 Conclusion

In conclusion, this study showed the prevalence of neck, shoulder and back pain among 7-9 year old school students due to carrying heavy schoolbags. The weight of the schoolbags for each respective student is around 12% - 14% of the student's body weight. In addition, the analysis on the students height and weight and the weight of their schoolbag is used to assist in detecting the discomfort experienced among the students while carrying their schoolbag. Even though there are many factors exist contributing towards the prevalence of neck and back pain among 7-9 year old students such as the way of carrying the schoolbag or even the duration spent on carrying the schoolbag, but the main factor that contributes to this problem is overweight schoolbags will eventually leads on into many other small factors which later might leads to Musculoskeletal diseases in the future. It is hoped that the results from this study will help in suggesting and supporting improvements to be taken regarding this heavy schoolbag issue.

## References

1. S. T. Wei, *School Bags In Malaysia - A 'Heavy' Issue* (Malaysian Digest, 2014)
2. R. Lim, *A Load Too Heavy* (The Star, 2013)
3. H. N. Shasmin, N. A. Abu Osman, R. Razali, J. Usman and W. A. B. Wan Abas, *A Preliminary Study of Acceptable Load Carriage for Primary School Children* (Biomed 06, IFMBE 2006)
4. *Physical Work Demands in the Construction Industry* (Dutch Labour Inspectorate, 2012)
5. *Work practices guide for manual lifting*, (National Institute of Occupational Safety and Health, 1981)

6. J. Whittfield, S. J. Legg, and D. I. Hedderley, Schoolbag weight and musculoskeletal symptoms in New Zealand secondary schools. Appl. Ergono. **36,** 193 (2005)
7. *NBPA Schoolbag Survey '97* (National Back Pain Association, 1997)
8. D. D. Pascoe, D. E. Pascoe, Y.T. Wang, D. M. Shim, C.K. Kim. Influence of carrying book bags on gait cycle and posture of youths. Ergonomics. **40**, 631 (1997)
9. S. Dockrell, S. Ciaran, C. Blake. Schoolbag carriage and schoolbag-related musculoskeletal discomfort among primary school children. Appl. Ergon. **51**, 281 (2015)
10. T.M. Cook, D.A. Neumann, The effects of load placement on the EMG activity of the low back muscles during load carrying by men and women. Ergonomics. **30**, 1413 (1987)
11. S. J. Legg, T. Ramsey, D. J. Knowles, The metabolic cost of backpack and shoulder load carriage. Ergonomics **35**, 1063 (1992)
12. A. A James, M. R. Jafri, M. Abdul Rani, Back pain arising from schoolbag usage among primary schoolchildren. Int. J. Ind. Ergonom **44**, 590 (2014)
13. D. Iman, N. Sorkhi, A. Pourhossein, A. Alipour, M. Asghari-Jafarabadi, Neck, shoulder and low back pain in secondary schoolchildren in relation to schoolbag carriage: should the recommended weight limits be gender-specific?. Appl. Ergon. **45**, 437 (2014)

# Preparation of Nano-Scale Biopolymer Extracted from Coconut Residue and Its Performance as Drag Reducing Agent (DRA)

Muhammad Luqman Bin Hasan[1,a], Mohamad Naqib Bin Mohamed Khalid[1]

[1]*Department of Petroleum Engineering, Universiti Teknologi PETRONAS, Malaysia*

**Abstract.** Drag or frictional force is defined as force that acts opposite to the object's relative motion through a fluid which then will cause frictional pressure loss in the pipeline. Drag Reducing Agent (DRA) is used to solve this issue and most of the DRAs are synthetic polymers but has some environmental issues. Therefore for this study, biopolymer known as Coconut Residue (CR) is selected as the candidate to replace synthetic polymers DRA. The objective of this study is to evaluate the effectiveness of Nano-scale biopolymer DRA on the application of water injection system. Carboxymethyl cellulose (CMC) is extracted by synthesizing the cellulose extracted from CR under the alkali-catalyzed reaction using monochloroacetic acid. The synthesize process is held in controlled condition whereby the concentration of NaOH is kept at 60%wt, 60 °C temperature and the reaction time is 4 hours. For every 25 g of dried CR used, the mass of synthesized CMC yield is at an average of 23.8 g. The synthesized CMC is then grinded in controlled parameters using the ball milling machine to get the Nano-scale size. The particle size obtained from this is 43.32 Nm which is in range of Nano size. This study proved that Nano-size CMC has higher percentage of drag reduction (%DR) and flow increase (%FI) if compared to normal-size CMC when tested in high and low flow rate; 44% to 48% increase in %DR and %FI when tested in low flow rate, and 16% to 18% increase in %DR and %FI when tested in high flow rate. The success of this research shows that Nano-scale DRA can be considered to be used to have better performance in reducing drag.

## 1 Introduction

### 1.1 Drag

In oil and gas industry, drag typically occurs in the pathway of fluid inside the injection tubing or pipelines at high flow rate. Zakaria & Jasamai mentioned that drag, which is also known as friction, is defined as energy that oppose to an object's relative motion through a fluid [1]. In other words, drag will cause a downward force to the fluid and will create drop in pressure of the fluid. Drag are classified as one of the major problems in oil and gas industry as the presence of drag will reduce the flow capacity and subsequently will involve in installation of more pumps as the easiest way to increase the pressure of the fluid in order to overcome the friction pressure losses is by pumping [2], and hence, will contribute in higher operating cost.

As stated by Bailey W. J. & Peden J. M. (1997), there are two (2) types of flow regime; known as laminar flow and turbulent flow [3]. Drag are not usually occur during laminar flow as the movement of fluid's particles are in parallel or layer manner if compared with turbulent flow, which the fluid's particles are moving in random style, and hence, will have more contact with the pipeline wall to create drag or friction force. However, to show the type of flow regime of fluid in a pipeline, Avila et al. (2011) in their study indicate that flow regime is usually identified by using the dimensionless Reynolds number (NRE) which is defined in equation below [4]:

$$N_{Re} = \frac{\rho v d}{\mu} \quad (1)$$

Ikoku (1984) said that for flow in a circular pipe, the change from laminar to turbulent flow is typically estimated to happen at N_RE = 2100 [5]. The fluid behaviors are generally classified as follow:

**Table 1.** Types of Flow Regime

| | |
|---|---|
| $N_{Re} < 2100$ | Laminar flow |
| $2100 < N_{Re} < 4000$ | Transitional flow |
| $N_{Re} > 4000$ | Turbulent flow |

Transitional flow is where laminar flow and turbulent flow occurs at the same time.

[a] luqman.hasan@petronas.com.my

### 1.2 Drag Reduction

Douberet (2009) defined drag reduction as an occurrence where a high cutback (over 70%) of turbulent friction of a fluid happens due to the inclusion of just a few parts per million of additives [6]. According to Campbell & Jovancicevic (2001) in their research, the drag reduction can results in increased fluid flow and/or decreased in pressure loss [7]. Therefore, reducing drag in a pipeline is really important to ensure that the flow capacity of a pipeline will be high. Ibrahim & Braimoh (2005) reported that drag reduction occurs due to suppression of the energy dissipation by turbulent eddy currents near the pipe wall during turbulent flow based from the Drag Reduction Theory [8].

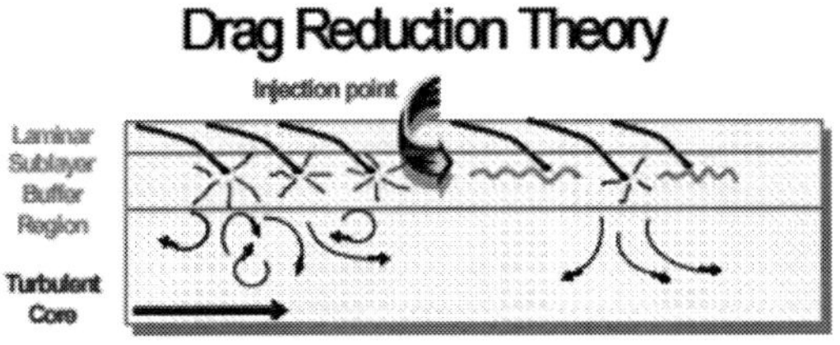


**Figure 1.** Drag Reduction Theory [8]

From **Figure 1**, part of laminar sub layer known as "turbulent streak" will move towards the buffer region and starts to oscillate and vortex. It will then tend to move rapidly as it gets closer to the turbulent core and eventually becomes unstable and burst out where it throws fluid into the turbulent core. This phenomenon is known as "turbulent burst" which causes wasted in energy.

Drag reducing agent will disturb the bursting process by dampen the burst and act as shock absorber, thus, reducing the subsequent turbulent bursts. This absorption of burst will result in less wasted energy and subsequently reduces the frictional pressure loss.

In 1906, Blatch conducted a study of "Water Filtrate" and observed that when some addictives were added to a flowing fluid, a significant pressure drop occurred [9]. The theory of his study was then further explained and proved in 1948 by Toms where he observed a decrease in wall friction and pressure drop in a flow with turbulent regime after polymers were added [10]. Tom's study has played an important and major role in introducing the term of drag reduction effect with extensive studies on various source of polymers.The calculation of drag reduction percentage (%DR) is as follow:

$$\%DR = \frac{|\Delta P_f - \Delta P_{fDRA}|}{\Delta P_f} \times 100 \qquad (2)$$

The formula is famously found in most research papers that have been used to check the relationship between the presence of DRA and resulting pressure drop [2].

### 1.3 Drag Reducing Agent

An additive that is made up of high molecular weight polymers which is use to improve the flow of oil in transportation pipeline as well as in water injection system by reducing the pressure drop is called Drag reducing agent, or DRA [9]. Whilst the term drag reduction refers to the decrease of friction in the turbulent flow inside the tubing causing the rise in flow capacity and the decline in pressure loss. Darbouret et. al (2009) stated that the DRA is able to reduce the pressure drop of the fluid if it is categorized as turbulent flow and DRA also helps to boost the tubing capacity [6].

Drag reducing agents which can be classified into three (3) main categories: polymers, surfactants and fibers, are majorly used in the industry due to high efficiency in reducing drag and economical. The polymers are divided into two which are synthetic and natural polymers. Synthetic polymers are made from numerous mixtures of chemicals, while the natural polymer (known as biopolymer) are extracted from natural waste materials. Synthetic polymers are widely used in industry due to their excellence performance in reducing pressure drop and reducing drag but recently there are environmental issues rise as the accumulation of synthetic polymers can seep into the soil and will cause harm to the ecosystem. Based from the issue arises, researchers around the world conducted studies on biopolymers as replacement of synthetic polymers. They conducted many experiments to test on the effectiveness of biopolymers DRA on different parameters of fluid.

The effectiveness of a DRA is depending on several factors, such as the polymer's solubility, molecular weight, cloud point, injection location, salinity of water, flow turbulence, temperature of fluid, degradation, concentration, shear rate etc. As reported by Salamone (1996), the two most important factors that decide the effectiveness of DRA is the linear flexible chain structure and ultra-high-molecular-weight (UHMW) polymers [11]. Percentage of drag reduction (%DR) can be related with the percent flow increase (%FI) by using the equation below:

$$\%FI = \left\{ \left( \frac{100}{100 - \%DR} \right)^{0.556} - 1 \right\} \times 100 \qquad (3)$$

In order to get the desired concentration, the volume of DRA ($V_{DRA}$) to be injected into the liquid can be determined by the following equation:

$$V_{DRA} = \frac{C_{DRA} \times V_{TOTAL}}{1 \times 10^6} \qquad (4)$$

The most used polymer as DRA in industries are Carboxymethyl cellulose (CMC), Polyethylene Oxide (PEO), Guar Gum (GGM) and Polyacrylamide (PAM) [20]. Carboxymethyl cellulose (CMC) is the result from the alteration of natural polymer known as cellulose and which is made from the derivation of cellulose resulting from the reaction of alkali (e.g. sodium hydroxide) and acid (e.g. monochloroacetic acid). Due to its abundance availability and cheaper price if compared to synthetic polymers, these natural polymers have been used in many studies to test on the best candidates to be the most effective natural polymers DRA as well as to test the effectiveness of natural polymers DRA on different fluid and pipeline parameters. Other than that, these polymers also contain larger molecular weight, which according to Kim (2009) is one of the characteristic of good drag reducers.

After the coconut milk is extracted from the flesh, the wastes produced are call coconut residue. Coconut residues are usually disposed as it cannot be eaten or used again. The reason why coconut is chosen as the DRA candidate is due to the availability of that substance. In 2011, Malaysia ranked the ninth coconut producing country in the world with the estimated total of coconut production of 30,000 MT [15]. It also contains high cellulose which indicates a high quality of DRA. From the studies made by Ng, Tan, Lai, Long, & Mirhosseini (2010), coconut residues has a total of 72.6% of cellulose [16].

In this project, Nano size particles will be used. As stated by Filipponi et al. (2012), material that is prepared in Nano particles will have alterations in both physical and chemical properties such as higher shear strength as compared to their bulk materials and more soluble in water since they were having in increased surface-to-volume ratio [12].

## 2 Experimental Procedure

### 2.1 Carboxymethyl cellulose (CMC) Extraction and Synthesization

CR is washed with tap water and rinsed completely. Then, it is dried in oven (Temperature: 60 °C, Duration: 24 hours). The dried CR is kept in air-tight container.
To eliminate any unwanted products and impurities, 40 g of CR is mixed with 1M of Sodium Hydroxide (NaOH) in a beaker. Then, they are cooked together by using magnetic stirrer hot plate (Temperature: 90 °C, Speed: 200 RPM, Duration: 1 hour). The solution is left to cool down (Duration: 20 - 40 minutes). Then, the mixture is filtered by using sieve and residue is accumulated. The residue is rinsed by using tap water until the dark reddish-purple color turns to light red. To ensure that the CR is dry enough for the extraction of CMC, it is dried down in oven (Temperature: 60 °C, Duration: 24 hours).

For alkalization, 100 ml with 60% concentration of NaOH and 900 ml isopropanol is prepared and mixed together with 25 g of CR. The mixture is stirred by using magnetic stirrer (Speed: 200 RPM, Duration: 30 minutes). To initiate carboxylmethylation reaction, 36 g of monochloroacetic acid is added into the solution. The solution is stirred until the mixture becomes clear. (Speed: 200 RPM, Duration: 30 minutes, Temperature: 60 °C). The temperature of the mixture is kept constant at 60 °C by gradually checking the thermometer. After 30 minutes of stirring, the polymers started to become thicker and the magnetic stirrer could not be able to stir it. The beaker is then transferred into oven. (Duration: 210 minutes, Temperature: 60 °C). Polymer is separated from the mixture by using sieve. It is then suspended into a beaker with 100 ml of methanol with 70% v/v concentration (Duration: 24 hours).

100 ml glacial acetic acid is poured into the beaker for neutralization and left for a while (Duration: 15 minutes). Then, the neutralized polymer is filtered from the solution by using sieve. To ensure any unwanted by products are removed, the polymer is suspended in 300 ml of ethanol with 70% v/v concentration (Duration: 10 minutes). Then, the polymer is separated by using sieve. It is then washed by using 300 ml of absolute methanol. The washing activity by using methanol is repeated for 3 times. Then, the polymer is separated by using sieve and then it is dried in the oven (Temperature: 60 °C, Duration: 24 hours). After 24 hours of drying, the extracted CMC sample is grinded by using mortar grinder until become powder size. CMC sample is stored in air-tight container.

### 2.2 Planetary Mill (Nano-scale Grinding)

**Table 2.** Planetary Mill Test Parameters Setup

| Parameters | Description |
|---|---|
| Rotation Speed | 330 rpm |
| Ball Weight : Powder Ratio | 10 : 1 |
| Grinding Time for 1 Cycle | 6 minutes operation and 4 minutes stop |
| Total Grinding Time Needed | 42 hours = 252 cycle (7 hours/day for 6 days) |

**Table 3.** Grinding Ball Calculation

| Diameter | Ball needed | Weight |
|---|---|---|
| 3 cm | 1 X 74 g | 74 g |
| 2 cm | 1 X 23 g | 23 g |
| 1 cm | 20 X 2.4 g | 48 g |
| 0.5 mm | ±300 X ±0.05 g | 15 g |
| **Total Weight** | | **160 g** |

This process involves 2 grinding jars. 16 g of CMC powder and 160 g of grinding balls are placed into each grinding jar. The hood is opened and the jars are placed at designated arms. The parameters on the machine is keyed in which includes the rotational speed, time per cycle and total cycle as stated in Table 2. The hood will be automatically locked upon the running of the machine. The main power of the machine is turned off after the grinding process is over.

### 2.3 Fluid Friction Apparatus Test

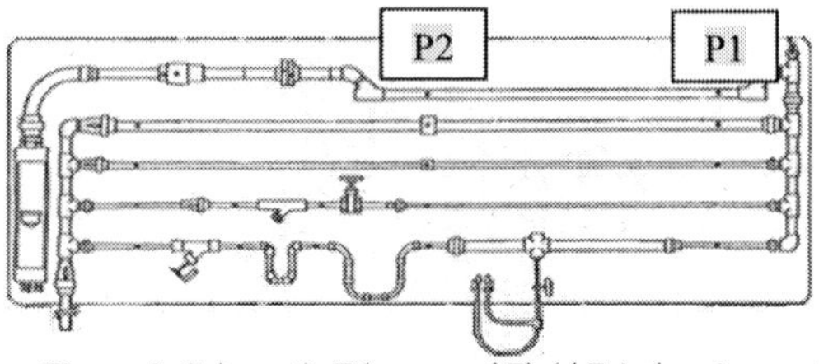


**Figure 2.** Schematic Diagram of Fluid Friction Apparatus

Flow loop tank is filled with 40 liters of tap water. The pump is switched on and the choke valve is opened until the flowmeter reading is stabilized at rate of 1 m3/h. The flow loop is left circulated for 15 minutes to ensure no air inside the pipe. The water manometer readings which are labeled as point P1 and P2 in **Figure 2** are observed and recorded. The experiment is continued by slowly increase the liquid flow rate to 2 $m^3/h$ by adjusting the flow control valve and monitoring the flowmeter. Again, the water manometer readings are observed and recorded after it is stabilized for 1 minute. The pump is then switched off and the drain valve at the side of the tank is fully opened to drain all water from the tank to prepare for the new run and then flushed once with clean water to remove any leftovers inside the system. The experiments are repeated for Nano-scale (Sample C1 and C2) and normal-scale DRA solution (Sample C1 and C2). Results are recorded and tabulated.

## 3 Result and Discussion

### 3.1 CMC Extraction and Synthesization

**Table 4.** Mass of CMC Obtained After Synthesization

| Batch No. | Materials | Yield |
|---|---|---|
| | CR Cellulose (g) | CMC (g) |
| 1 | 25 | 23.61 |
| 2 | 25 | 23.84 |
| 3 | 25 | 23.96 |
| **TOTAL** | | **71.41** |

Three batches of CMC sample have been prepared and the results are tabulated in the Table 4. Table 4 shows an average of 23.8 g of CMC powder are produced for every 25 g of CR cellulose used. The 4.79% loss of mass may occurred during the filtration process and transportation of the CMC. Despite of the losses, the synthesization has managed to achieve a total of 71.41 g of CMC powder which will then be used for the next procedure.

### 3.2 Preparation of Nano Size CMC

After grinding the CMC powder by using the Ball Milling Machine, the powder obtained is then tested using Particle Size Analyzer to determine whether the size reaches the Nano size (<100 nm) or not.

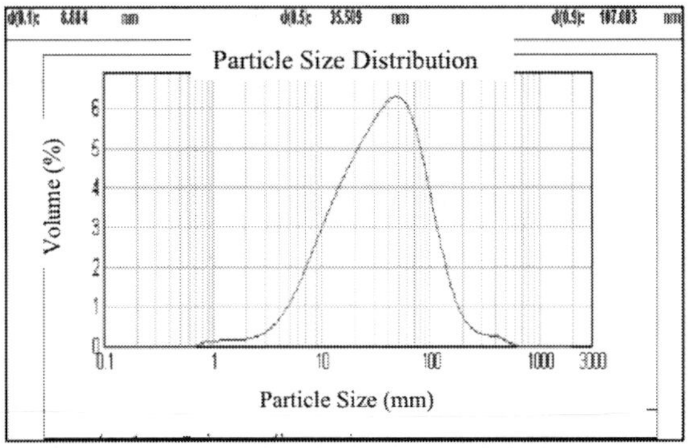


**Figure 3.** Size Distribution Graph (C1)

**Figure 3** shows the size distribution of sample C1. 10% of the sample has an average size of 8.884 Nm, while 50% of the sample has an average size of 35.509 Nm, 90% of the sample has an average size of 107.003 Nm. The smallest particle size is around 0.724 Nm, and the largest is 549.541 Nm. The average size of sample C1 particles is 47.541 Nm.

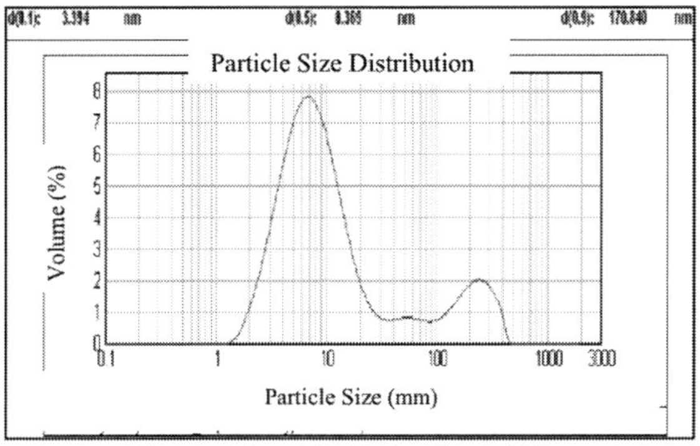


**Figure 4.** Size Distribution Graph (C2)

**Figure 4** shows the size distribution of sample C2. 10% of the sample has an average size of 3.394 Nm, while 50% of the sample has an average size of 8.369 Nm, 90% of the sample has an average size of 170.840 Nm. The smallest particle size is around 1.259 Nm and the largest is 416.869 Nm. The average size of sample C2 particles is 39.098 Nm.

From the result of the Size Particle Size test, it shows that both of the samples have the average size of 43.32 nm, thus proving that Nano-scale has been achieved from the ball milling of CMC.

### 3.3 Reynolds Number

DRA can only work effectively in a turbulent regime flow where the Reynolds number is greater than 4000 (NRe > 4000). To confirm the flow regime, the Reynolds number of the flow in the pipeline is calculated by using Equation 1.

DOI: 10.1051/matecconf/20178702014

The parameters used to calculate Reynolds number are as follow:

**Table 5.** Parameters to Calculate Reynolds Number

| | |
|---|---|
| Density of solution (kg/m$^3$) | **1017.5** |
| Flow rate (m$^3$/s) | **2.778 × 10$^{-4}$** |
| Diameter (m) | **0.0235** |
| Viscosity of solution (kg/m.s) | **1.09 × 10$^{-3}$** |
| Velocity (m/s) | **0.64048** |

The Reynolds number is as follow:

$$N_{Re} = \frac{1017.5 \times 0.64048 \times 0.0235}{1.09 \times 10^{-3}}$$
$$N_{Re} = 14050.16$$

Since $\mathbf{N_{Re} > 4000}$, thus the flow regime is turbulent.

### 3.3 Results from Fluid Friction Test

The fluid friction test uses manometer readings that give the value of water pressure at the entry of the pipe (P2) and the water pressure at 1 meter from the entry (P1) in the unit of centimeter (cm). Few assumptions have been made for the fluid friction test which are:

i. The solution prepared is homogeneous and the CMC powder has been mixed thoroughly with the water.
ii. The pipe used for the test is assumed to have smooth surface, thus, the effect of friction in the experiment is neglected.

The values of pressure that in centimeter are converted into psi unit by using the hydrostatic pressure formula as below:

$$\Delta P = (\rho g \Delta h) \times (1.45038 \times 10^{-4}) \quad (5)$$

The difference between the value of P2 and P1 represents the pressure drop inside the pipe. Table 6 below shows the value of pressure drop from the fluid friction test when there is no CMC (tap water) added:

**Table 6.** Pressure Drop Using Tap Water (Baseline)

| Flow rate (m$^3$/h) | Tap Water | | | |
|---|---|---|---|---|
| | P2 (cm) | P1 (cm) | ΔP (cm) | ΔP (psi) |
| 1 | 53.1 | 46.9 | 6.2 | 0.088215 |
| 2 | 55.7 | 44.3 | 11.4 | 0.162201 |

The value of the pressure drop when using tap water only will be used to calculate the drag reduction percentage (%DR) which is calculated by using Equation 2. Then, by using the %DR value, the flow increase percentage (%FI) can be calculated by using Equation 3.

The graph in **Figure 5** shows a significant increase of drag reduction percentage. It can be seen that the percentage of flow increased as the flow rate increases.

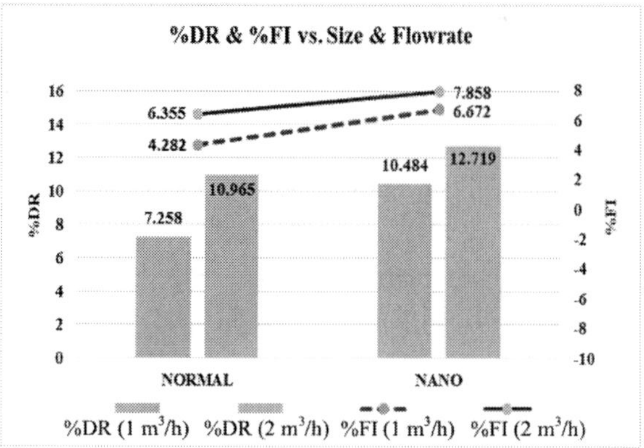


**Figure 5.** Summary Result for Fluid Friction Test

For the normal size biopolymer DRA, the increment is almost 51% increase in %DR and 56% increase in %FI when the flow rate is increased from 1 m$^3$/h to 2 m$^3$/h.For Nano size biopolymer, the %DR and %FI increases 21% and 24% respectively when the flow rate increases. This is due to the high friction force inside the pipeline that caused by the high flow rate that will lead to a high pressure drop. Therefore, CMC's effect in increasing the pressure drop is more visible in the case of higher flow rate compared with lower flow rate.

It can also be observed that the value of %DR and %FI is much higher when using Nano-scale CMC compared to normal scale CMC. In 1 m$^3$/h flow rate, normal scale CMC gives, %DR reading of 7.258%, and the %FI is 4.285%. The value of %DR and %FI increases to 10.484% and 6.355% respectively when using Nano size CMC. This shows an increment of 44% for %DR and increment of 48% for %FI. In 2 m$^3$/h flow rate case, the %DR is 10.965%, and the %FI is 6.672% when using normal scale CMC. When using Nano size CMC, the value of %DR and %FI increases to 12.719% and 7.858% respectively. This shows an increment of 16% for %DR and increment of 18% for %FI.

The higher percentage of drag reduction and flow increase in Nano-scale solution compared to normal-scale solution proves that the Nano-scale CMC has better ability to reduce drag. This is due to two reasons; the higher solubility and the better alterations in physical and chemical properties. Since the Nano size materials have larger surface area, it will allow the materials to be dissolved in solution much better and thus can increase the efficiency to decrease drag. Normal size materials will have difficulties to dissolve completely in the solution. The assumption that the powder has dissolved completely is only made by observation, whereby due to the solution is homogenous, it is assumed that the powder have completely dissolved. However realistically, there might be particles that are not fully dissolved when using the normal size materials. This will then affect the performance of the CMC to reduce drag.

Nano materials also have better ability to react with water in reducing drag. The bursting process created from turbulence flow can be effectively dampen by the Nano particles due to the large surface area of the materials.

## 4 Conclusion

Friction force or drag does exist in a turbulent flow of fluid in a pipeline. This issue is really critical for oil and gas industry especially during the transportation of oil and re-injection of water during water flooding, as it can cause drop in pressure and reduction of flow capacity. From previous researches that have been done, pressure of the fluid can be increased by using pumps, but, it will involve higher cost. So, drag reduction technique has been used to tackle the problem. Drag reduction is where the pressure loss is decreased and the flow performance will be increased by just adding additives called Drag Reducing Agents (DRA). This study shows that a biopolymer DRA can be extracted from coconut residue (CR) which is more environmental friendly.

The study shows that for every 25 g of dried CR used, it can yield an average of 23.8 g of synthesized CMC under controlled conditions. The CMC yield is grinded into Nano-scale particle using ball milling machine using controlled parameters. The process yields CMC powder with an average particles size of 43.32 Nm, which is in range of Nano particle size (<100 Nm). Both of Nano scale and normal scale CMC are tested by using the fluid friction apparatus to test on the ability to reduce pressure drop. The experiment shows that Nano-size CMC has higher percentage of drag reduction (%DR) and flow increase (%FI) if compare with normal-size CMC when tested in high and low flow rate; 44% to 48% increase in %DR and %FI when tested in low flow rate, and 16% to 18% increase in %DR and %FI when tested in high flow rate. The success of this research shows that the CMC extracted from coconut residue can have the potential to be used to replace the current synthetic DRAs and also Nano-scale DRA can be considered to be used to have better performance in reducing drag.

**NOMENCLATURE**

$N_{Re}$ = Reynolds Number
%DR = Drag Reduction Percentage
$\Delta P_f$ = Pressure Drop before injecting DRA
$\Delta P_{fdra}$ = Pressure Drop after injecting DRA
%FI = Flow Increase Percentage
$V_{DRA}$ = Volume of DRA to be injected
$C_{DRA}$ = Concentration of DRA solution
$V_{TOTAL}$ = Total volume solution
CMC = Carboxymethylcellulose
CR = Coconut Residue
DRA = Drag Reducing Agent
NaOH = Sodium Hydroxide
Wt% = Concentration solution on percentage by weight
v/v = volume to volume concentration

## References

[1] Zakaria, A. N. and Jasamai, M., "To Evaluate Effectiveness of Hibiscus leaves as Drag Reducing Agent in Water System," Universiti Teknologi PETRONAS, Malaysia.

[2] Karami, H. R. and Mowla, D., "Investigation of the effect of various parameters on pressure drop reduction in crude oil pipelines by drag reducing agents," *Journal of Non-Newtonian Fluid Mechanics*, pp. 37-45, 2012.

[3] Bailey, W. J. and Peden, J. M. "A Generalised and Consistent Pressure Drop and Flow Regime Transition Model for Drilling Hydraulics Suitable for Slimhole, Underbalanced and Horizontal Well." SPE 39281(paper presented at the SPE/IADC Middle East Drilling Technology Conference held in Bahrain, 23-25 November 1997)

[4] Avila, K., Moxey, D., de Lozar, A., Avila, M., Barkley, D., & Hof, B. (2011). The onset of turbulence in pipe flow. Science, 333(6039), 192-196.

[5] Ikoku, C. U. (1984). Natural gas production engineering: Wiley.

[6] Darbouret, M., Henaut, I., Palermo, T., Glenat, P., and Hurtevent, C., "Experimental Methodology To Evaluate DRAs: Effect of Water Content and Waxes on Their Efficiency," presented at the SPE International Symposium on Oilfield Chemistry, The Woodlands. Texas, 2009.

[7] Campbell, S. E., & Jovancicevic, V. (2001). *Performance Improvements From Chemical Drag Reducers.*

[8] Ibrahim, J. U. and Braimoh, L. A., "Drag Reducing Agent Test Result for Chevron Texaco, Eastern Operations, Nigeria (SPE 98819)," in Society of Petroleum Engineers, Abuja, Nigeria, 2005.

[9] Blatch, N. S., "Water Filtrate," ASCE, pp. 400-408, 1906.

[10] Toms, B. A., "Some observation on the flow of linear polymer solution through straight tubes at large Reynolds number," in Proceedings of the 1st International Congress on Rheology, North Holland, Amsterdam, 1948, pp. 135-141.

[11] Salamone, J. C. (1996). Polymeric Materials Encyclopedia, Twelve Volume Set: Taylor & Francis.

[12] Filipponi, L. and Sutherland, D. (2012). *NANOTECHNOLOGIES: Principles, Applications, Implications and Hands-on Activities*. Luxembourg: Publications Office of the European Union.

[14] Choi, H., Moin, P. and Kim, J., 1992, "Turbulent drag reduction: Studies of feedback control and flow over riblets," *Technical Report TF-55*, Stanford University

[15] United Nations Conference on Trade and Development (2011), *World Coconut Oil Production*. Retrieved on 18th June 2015 from http://www.unctad.info/en/Infocomm/AACP-Products/COMMODITY-PROFILE---Coconut2/

[16] Ng, S., Tan, C. P., Lai, O. M., Long, K., & Mirhosseini, H. (2010). *Extraction and characterization of dietary fiber from coconut residue. J Food Agric Environ, 8*, 172-177

# Simple Arm Muscle Model for Oil Palm Harvesting Process

*Aliff* Rahman[1], *Shahrol* Mohamaddan[1,a] , *Noor Aliah* Abdul Majid[1], *Magdalene* Andrew-Munot[1], *Nur Tahirah* Razali[1], *Baba* Md Deros[2] and *Syah Runniza* Ahmad Bakri[3]

[1]*Department of Mechanical and Manufacturing, Faculty of Engineering, Universiti Malaysia Sarawak (UNIMAS), 94300 Kota Samarahan, Sarawak, Malaysia*
[2]*Department of Mechanical and Materials Engineering, Faculty of Engineering, Universiti Kebangsaan Malaysia, 43600 UKM Bangi, Selangor, Malaysia*
[3]*Faculty of Computer Science and Mathematics, Universiti Teknologi MARA, 94300 Kota Samarahan, Sarawak, Malaysia*

**Abstract.** Arm are essential in order to perform manual material handling work that normally involves lifting, handling, placing, push and pull, carrying and moving heavy loads. When these work elements are performed over prolonged periods repeatedly, it will expose workers arm to awkward posture and possible ergonomic risk factor. For example, work element that requires repetitions frequently may lead the arm to face physical stress and mental fatigue. The situation can be extremely risky if the worker task requires higher focus or time consumable. These issues are unavoidable in palm oil harvesting process since the workers are still using manual handling when harvesting the fresh fruit bunch (FFB). The worker using a chisel to harvest the young palms and a sickle mounted on a bamboo or aluminum pole to harvest taller palms. When perform this work element combining with heavy physical workload, it may lead to work-related muscle disorders (WSMDs). This study was conducted to identify the force reaction and inverse dynamic analysis during oil palm harvesting process using ergonomics software called Anybody Technology. Inverse dynamic analysis is a technique for figuring strengths and/or moments of power (torques) taking into account the kinematics (movement) of a body and the body's inertial properties.

## 1 Introduction

Arm are essential in order to perform manual material handling work that normally involves lifting, handling, placing, push and pull, carrying and moving heavy loads [1]. Tasks that require forceful exertions place higher loads on the muscles, tendons and joints of arm. Increasing the force means increasing the body demands such as higher muscle exertion. The weight and height of a load that has to be lifted and the frequency of lifting are the factors that will contribute to the level of exertion on the muscles and joints of the arm.

In addition, when these work elements are performed over prolonged periods and repeatedly, it will expose the arm to awkward posture and possible ergonomics risk factor. For example, work element that requires repetitions frequently may cause physical stress on the arm and also mental fatigue. This situation can be extremely risky if the worker task requires higher focus and time consumable.

This issues are unavoidable in palm oil harvesting process due to the workers are still using manual handling when they harvest the fresh fruit bunch (FFB)[2]. In spite of the fact that there have been a few improvements and interventions have been done on the harvesting tools, the estate are still utilizing the manual tools and depending on the labor strength. The worker is still using a chisel to harvest the palms that are reachable and a sickle that is mounted to or aluminum pole to harvest taller palms.

This work element is causing heavy physical workload to the workers and it may lead to work-related muscle disorders (WSMDs). This may contribute to awkward posture that leads to fatigue discomfort and injury. There are some assessment that has been done to investigate the posture and possible risk of the worker when using the palm oil harvesting tools. Rapid Entire Body Assessment REBA [3], Rapid Upper Limb Analysis RULA [4] and Quick Exposure Check QEC [5] are some of the techniques used to determine the possible musculoskeletal risk.

REBA techniques are used to investigate the possible risk of the entire body sections for example trunk, wrists, arms and legs. It is a worksheet of survey for whole body movement based on the task of palm oil harvesting process [6]. RULA technique is focusing on the posture of worker.

RULA analysis focuses on time cycle, number of movements, static muscle force and the working posture. These factors will be analyses and resulted in the form of score with the range of 1-7. QEC is a technique used to determine the possible of musculoskeletal disorder. It focuses on four main section of human body which is back bone, shoulder/hand, wrist/arm and neck. This technique also include the psychosocial factor evaluation, the movement and it frequency.

The objective of this paper is focusing on identifying the length of contraction of arm muscle and the reaction force throughout the movement of arm through Inverse Dynamic Analysis. It is a technique for figuring strengths and/or moments of power (torques) taking into account the kinematics (movement) of a body and the body's inertial properties (mass and moment of inactivity).

In this research, the AnyBody Modelling System programming is utilized. AnyBody is an intense devices that immeasurably utilized by examination foundations to recreate the musculoskeletal. It empowers analysts and understudies to assemble and investigate musculoskeletal model, or to make adjustments of existing models.

[a] Corresponding author: mshahrol@unimas.my

 DOI: 10.1051/matecconf/20178702015

AnyBody programming is utilized for breaking down the responses inside of the human body or the cooperation of human body with a favoured situation.

The framework permits client to show and dissect any musculoskeletal framework either human or creature [7]. The framework are accompanied with full body demonstrate that can be adjustable as per clients inclinations and necessities. Body movements and external force can be adjusted in the AnyBody programming to ascertain the individual muscle reaction to certain development.

## 2 Methodology

### 2.1 Modeling Harvesting Tools

This study starts with modelling the tools for oil palm harvesting process using 3D design software. There are three main tools that have been applied in the harvesting process namely the chisel, pole and loading spike as shown in Figure 1. In the 3D design, the dimensions of the tools had followed the current/existing specification.

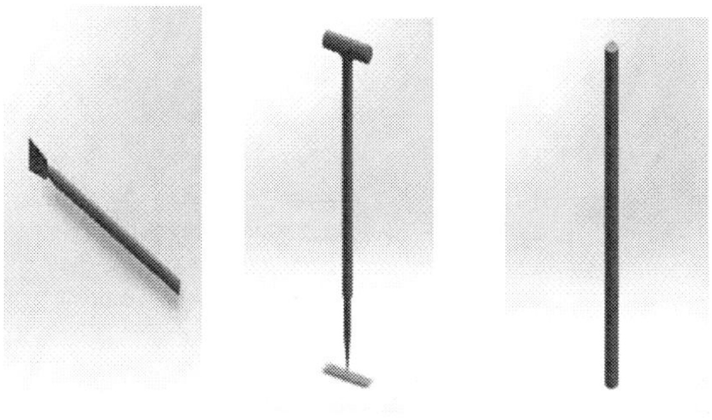

Chisel Loading Spike Aluminium Pole

**Figure 1.** Modelled harvesting tools using 3D modelling software

Chisel is used to cut the fresh fruit bunch (FFB). This is applicable for the small size tree. Pole is used to cut the FFB with long-armed sickle. This is aimed for high size tree. Lastly, the loading spike is used to stomp the FFB and lift it on the shoulder.

### 2.2 Modeling Simple Arm Muscle Model

Simple Arm Muscle was modeled using Anybody Technology Software. AnyScript Programming was used to design the segment of arm, muscle and the movement. The data had been input manually into the programming command. Two segments of arm namely the upper arm and forearm were formed. Upper arm is the extended segment from the shoulder to the elbow. The segment focused on lifting and pulling strength. Forearm is the segment from the elbow to the risk and focused on the wrist extension, flexion and rotation.

Elbow acts as the hinged between the upper arm and forearm. It allows the segment to rotate until 180 degrees; the normal arm full extension. The segments then are linked with the muscle. There are 4 muscles that linked the segment which is deltoid, triceps, biceps and brachialis. Figure 2 shows the simple arm muscle model based on actual anatomy of the arm [8].

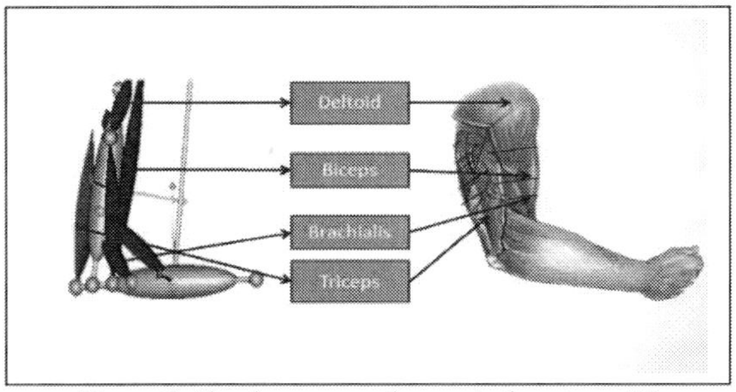


**Figure 2.** Simple arm muscle vs actual anatomy [8]

### 2.3 Modelling Simple Arm Muscle Model Using Harvesting Tools

The simple arm muscle model was then remodelled by adding a harvesting tool in the environment condition. The angle of posture and movement of arm during the work element had been input into the AnyScript programming manually according to the working condition. Table 1 show the model of simple arm using harvesting tools.

### 2.4 Data Analysis – Inverse Dynamics

Inverse dynamics analysis was depends on the stimulated muscle and joint strengths of the whole body that experience specific complex movements. It also took dynamic inertia forces into consideration and was not necessarily requiring the measured forces in the interface which was between the body and the environment. Inverse dynamic analysis had the advantages which it includes the consideration in investigation of extremely complex musculoskeletal frameworks that may including with several of muscles.

In this research, the focus was on using the inverse dynamic analysis to determine the reaction force of the muscle and the length of contraction. In order to understand and investigate the better posture during the harvesting process, it is crucial to determine how the muscle reacts through the movement of task.

In palm oil harvesting process the task such as pulling, pushing and lifting was required to be performed by the workers. The task involved a lot of forces and energy. Besides the forces or energy performed by the body, there is a need to consider others parameters such as the weight of the fresh fruit bunch (FFB) and the harvesting tools. In the simulation conducted, the unit for force was in Newton (N) while length was in meter (m).

**Table 1.** Simple arm muscle model using harvesting tools

**1. Chisel**

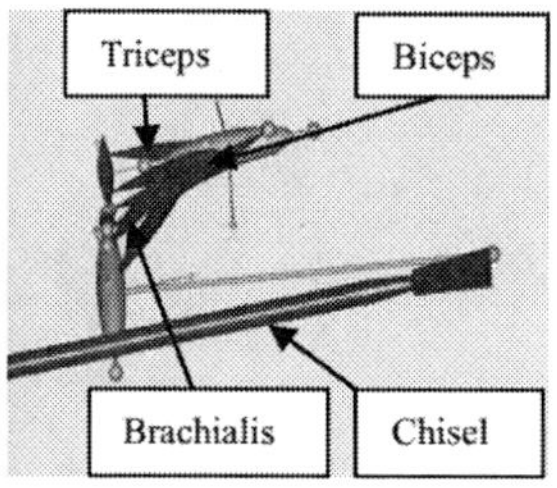


**2. Pole**

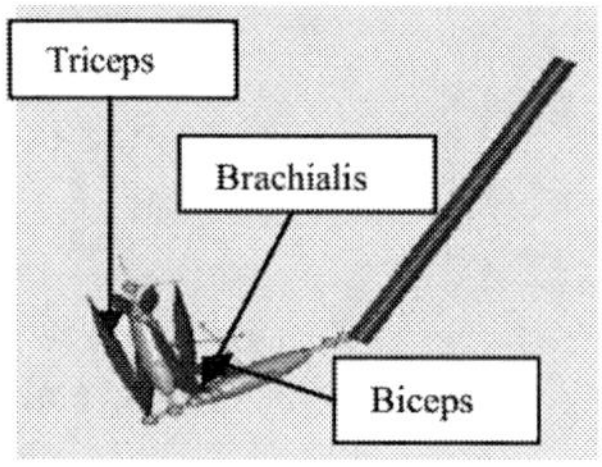


**3. Loading Spike**

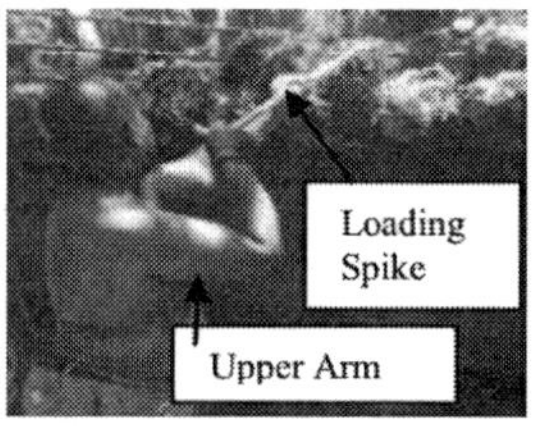


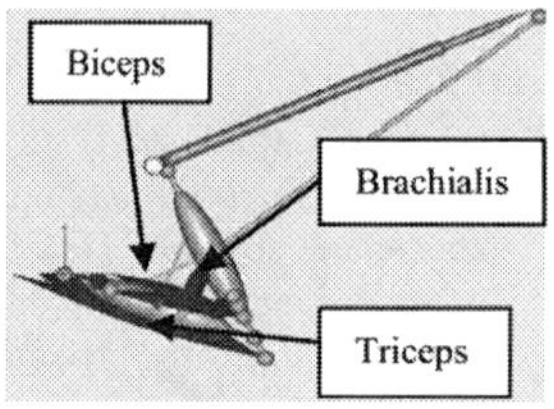


## 3 Results and Discussion

The analysis was performed to the simple arm muscle model based on the selected muscle. The analysis discussed the reaction forces exerted to the muscle and the contraction length of the muscle when using different harvesting tools. There were three selected muscles including Biceps, Triceps and Brachialis. Biceps is a 2-headed muscle that link to the upper arm between the elbow and shoulder. Triceps is the largest muscle in the upper limb. The muscle is responsible for the extension of elbow joint when straightening the arm. Lastly, Brachialis is a muscle at the upper arm that flexes elbow joint.

Figure 3 and 4 show the contraction force and length of the Biceps muscle corresponding to time based on the three harvesting tools. For pole, it shows that at 0.00 second the reaction force is 3.50E+02 N and contraction length is 3.89E-01 m. The value shows that in order to lift the aluminium pole, biceps muscle required 3.50E+02 N as the arm start to move vertically. At this point the arm is bending upwards at 75 degrees.

For chisel, it shows that at 0.00 second the initial force is 2.91E-01 N and the contraction length is 2.92E-01 m. The value shows that in order to start moving horizontally, biceps muscle required 2.91E-01 N to move the chisel. At this point the arm is bending horizontally at 25 degrees. For loading spike, it shows that at 0.00 second the initial force is 4.02E+03 N and contraction length is 4.68E-01 m. The value shows that in order to start lifting the loading spike vertically, biceps muscle required 4.02E+03 N to move upward. At this point the arm is bending downwards at 150 degrees.

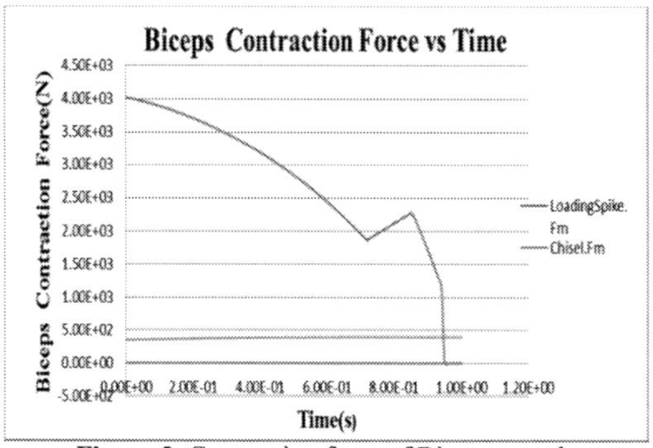


**Figure 3.** Contraction force of Biceps muscle

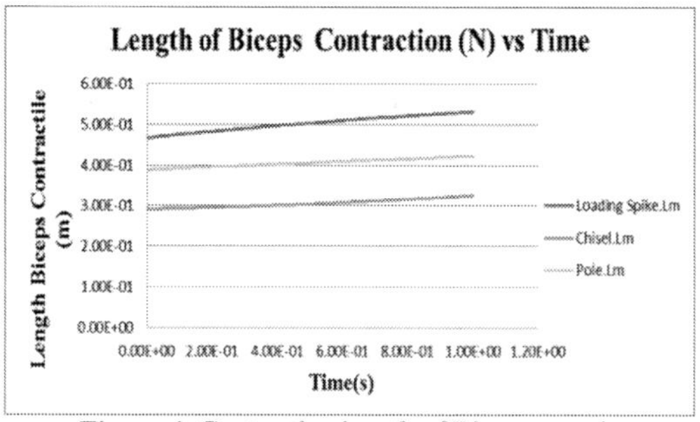


**Figure 4.** Contraction length of Biceps muscle

Figure 5 and 6 show the contraction force and length of the Triceps muscle corresponding to time based on three harvesting tools. For pole, it shows that at 2.00E-01 second the reaction force is 2.18E-04 N and contraction length is 4.83E-01 m. The value showed that in order to lift the aluminium pole, Triceps muscle required 2.18E-04 N as the arm start to move vertically. At this point the arm is bending upwards at 135 degrees.

For chisel, it shows that at 2.00E-01 second the initial force is 2.51E+03 N and the contraction length is 2.92E-01 m. The value shows that in order to start moving horizontally, Triceps muscle required 2.51E+03 N to move the chisel. At this point the arm is bending horizontally at 55 degrees. For loading spike, it shows that at 2.00E-01 second the initial force is 2.15E-07 N and contraction length is 3.96E-01 m. The value shows that in order to start lifting the loading spike vertically, triceps muscle 2.15E-07 N to move upward. At this point the arm is bending downwards at 110 degrees.

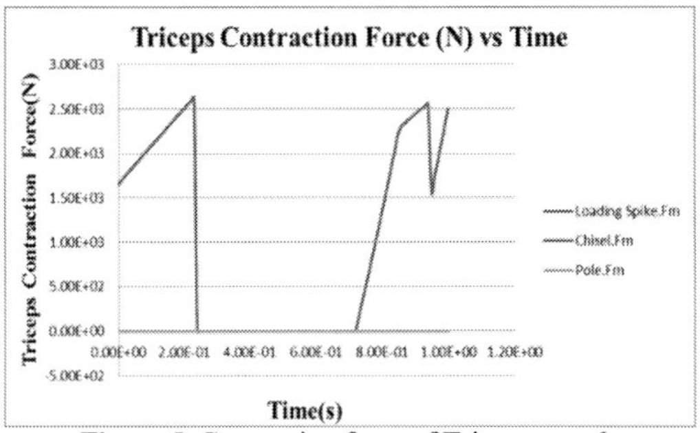


**Figure 5.** Contraction force of Triceps muscle

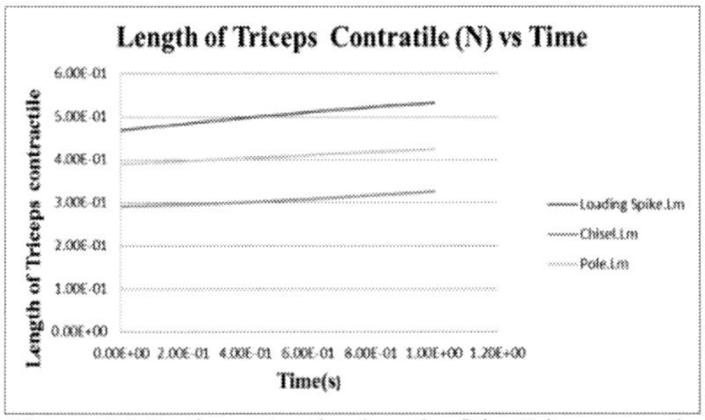


**Figure 6.** Muscle contraction length of the Triceps muscle

Figure 7 and 8 shows the contraction force and contraction length of the Brachialis muscle corresponding to time based on the three harvesting techniques. For pole, it shows that at 8.50E-01 second the reaction force is 8.14E-07 N and contraction length is 5.24E-01 m. The value shows that in order to lift the aluminium pole, Brachialis muscle required 8.14E-07 N as the arm start to move vertically.

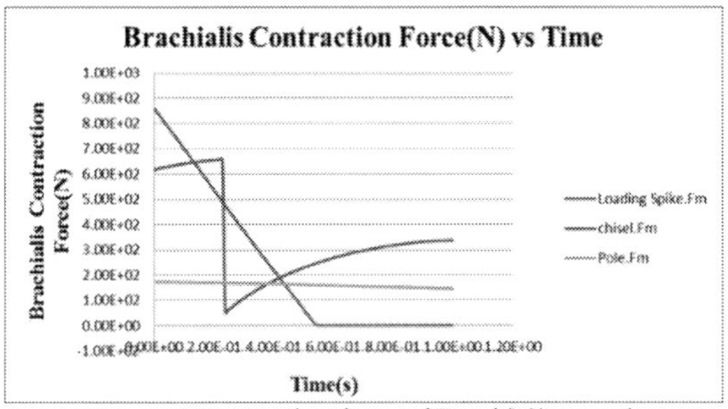


**Figure 7** Contraction force of Brachialis muscle

At this point the arm is bending upwards at 160 degrees. For chisel, it shows that at 2.00E-01 second the initial force is 3.26E+02 N and the contraction length is 3.19E-01m. The value shows that in order to start moving horizontally, Brachialis muscle requires 3.26E+02 N to move the chisel. At this point the arm is bending horizontally at 95 degrees. For loading spike, it shows that at 2.00E-01 second the initial force is 1.50E+02 N and contraction length is 4.18E-01m. The value shows that in order to start lifting the loading spike vertically, Brachialis muscle required 1.50E+02 N to move upward. At this point the arm is bending downwards at 85 degrees.

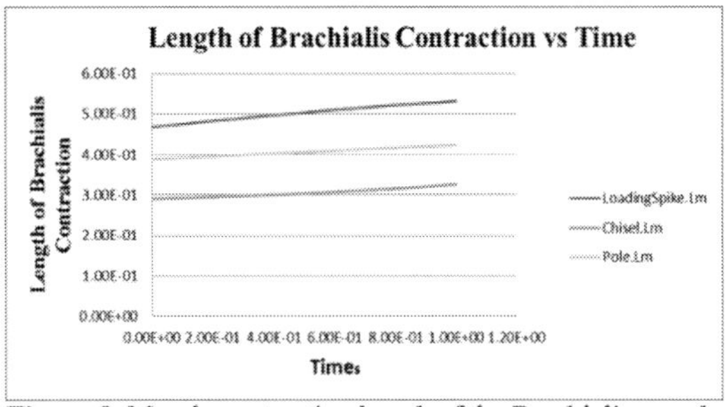


**Figure 8.** Muscle contraction length of the Brachialis muscle

## 4 Conclusions

Simple arm muscle model is the simplified stage of investigation to measure reaction exerted to the human arm when performing the harvesting process by using different tools. These work elements should be analysed further in order to assist reducing the potential injury due to work-related muscle disorders. The musculoskeletal of human are mechanically complex and needs a lot of interconnected tasks and consideration.

## Acknowledgement

This research is funded by UNIMAS MyRA Special Grant Schemes [Grant no: F02/SpSTG/1385/16/27]. The authors would like to thank Universiti Malaysia Sarawak (UNIMAS) for providing facilities for this research.

## References

1. Deros, B. M., Daruis, D. D. I., & Basir, I. M. A. S. Ergo A. M.M.H, **195**, 1666-1673(2015)
2. Nawi, N. S. M., Deros, B. M., & Norani, N. J. Oil Palm Res. **10**, pp. 122-127( 2013)
3. Hignett, S, & McAtamney, L. R.E.B.A Ergo, **31**, 201–5. (2000)
4. Deros, B. M., Khamis, N. K., Mohamad, D., Kabilmiharbi, N, & Daruis, D. D. I. I. Oil Palm R.U.L.A. 287-290. (2014)
5. G. Li, P. Buckle, A.P Wok. Mus. Q.E.C. 1351 -1355, (1998)
6. Nawi, N. S. M., Deros, B. M., & Norani, N. A. Oil Palm FFB. R.E.B.A, **10**, 122-127, (2013).
7. M.Damsgaard, J. Rasmussen, Anal. Mus. Syst. Any. Mod.System,(2006)
8. Moore, K. L., Dalley, A. F, A. M. Cli. O. Ana. (2013)

# Development of Control System for *Keropok Keping* Drying Machine

*Muhammad Naim* Leman[1], *Shahrol* Mohamaddan[2,a], *Mohamad Syazwan Zafwan* Mohamad Suffian[2], *Abang Mohamad Aizuddin* Abang Mohd Mohtar[2], *Nur Alia Athirah* Mohtadzar[2], *Nazreen* Junaidi[2] and *Satria Arjuna* Julaihi[3]

[1] *Mechatronic Technology, Institut Latihan Perindustrian (ILP) Kota Samarahan, 94300 Kota Samarahan, Sarawak, Malaysia*
[2] *Faculty of Engineering, Universiti Malaysia Sarawak (UNIMAS), 94300 Kota Samarahan, Sarawak, Malaysia*
[3] *Faculty of Computer Science and Mathematics, Universiti Teknologi MARA, 94300 Kota Samarahan, Sarawak, Malaysia*

**Abstract.** This research is focused on the development of a control system for the *keropok keping* drying machine by using programmable logic controller (PLC) as the controller. The control panel and human machine interface (HMI) were developed for the machine. Experimental study was conducted to validate the HMI speed data by the control panel speed display. The evaluation of the duty cycle (%) and current flow (amp) effect to the motor and fan speed were also conducted. The results showed that around 1 - 30 % and 1 - 3 % differences of the speed readings were recorded by the HMI for the motor and fan respectively. The percentage difference needs to be controlled as small as possible to ensure the HMI speed readings are more accurate. The linearity of the current flow curve to all duty cycle can be observed. The flow (amp) increases when the speed of dryer motor and fan increases. The development of the control system is expected to improve the operation of the *keropok keping* drying machine.

## 1 Introduction

*Keropok keping* is a dried crispy food product relatively popular in Malaysia and South-East Asian countries. Currently, its manufacturing is mostly practiced in small scale industry [1]. In Malaysia, the *keropok keping* industry is widely operated in coastal areas in the state of Terengganu, Kelantan, Johor, Kedah and some parts of Pahang due to high seafood supply, high temperature and windy area that contribute to the sustainability of the industries [3].

Ingredients for making of the *keropok keping* are starch or sago flour, fish, squid, prawn or shrimp and a little of seasoning such as pepper, garlic, salt, sugar and monosodium glutamate (MSG). All the ingredients are mixed by using a mixer to obtain dough. The dough is formed into a cylindrical shape with a diameter around 6 to 10 cm and then is cooked by boiling or steaming. The cooked dough is drained, cooled and cut into the thin slices with thickness around 3 mm and then is dried under the sunlight [2].

The dried *keropok keping* obtained is considered as a half-finished product or an intermediate product. Then, the *keropok keping* has to be fried in hot oil to obtain the edible puffed cracker. In *keropok keping* industries, most of the production processes are implemented by semi-automated machines such as a mixer, grinding and slicing machine. However, the drying process still uses the traditional method. In existing method, the *keropok* is arranged and dried on a drying board called *pemidai* under the sunlight. The *keropok keping* is exposed to the heat and surrounding wind to be dried [3]. In order to improve the drying process, new rotary type of *keropok keping* drying machine was invented as shown in Figure 1.

In general machine operation, *keropok* with high level moisture is arranged and hold by a special holder inside the rotary drum which is the drum and supported by four rubber tires. The gear motor is utilized to transmit power to rotate the drum to produce anti-clockwise rotation. Air generated by the fan is flowing to vaporize the water from the *keropok keping* inside the rotary drum. The process is continuously performed until the moisture level in the *keropok keping* is considered low and suitable for packaging [3]

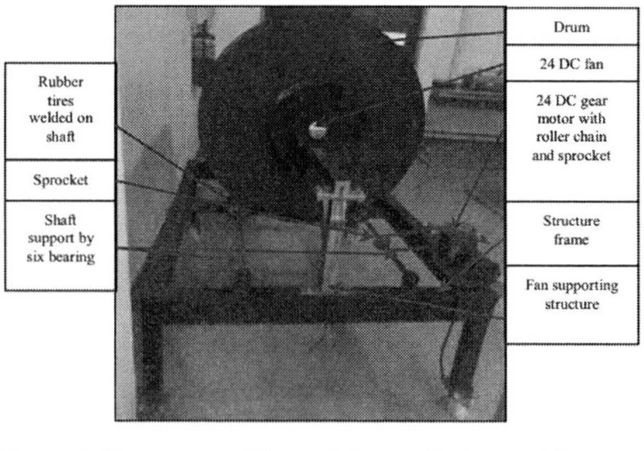


**Figure 1**: Rotary type of *keropok keping* drying machine [3]

However, the machine does not have any special system to control the dryer operation. All tools and machines need an appropriate control system to work. Otherwise it will be difficult to complete the task designation accurately. Control system is needed in order to guide, instruct and regulate tools and machines.

[a] Corresponding author: mshahrol@unimas.my

DOI: 10.1051/matecconf/20178702016

Common control system includes mechanical, electronic electrical and computer aided. A system usually contains three main parts which are input, process and output.

The objective of this paper is to develop a control system for the *keropok keping* drying machine by using PLC as a main controller. The purpose of the control system development is to improve the machine operation. The current machine design does not have a control system to instruct and regulate the machine during the drying process. Based on the machine design, the speeds (rpm) of motor and fan has significant effect to the effectiveness of the dryer. The drying rate of the *keropok keping* is depends on the variation of the rotation speed for both of components. However, in the current design, the control system for the machine is not developed. This situation may cause difficulty to the users to control the machine. It is necessary to develop a user friendly control system to ensure that the machine can work in a proper way.

## 2 Methods

### 2.1 Electrical design

**Design of Electrical Block Diagram**: A block diagram is a diagram of system components that are represented by blocks. It is connected by lines to show the relationships of the system. The block diagram is widely used in electrical hardware and software design. The block diagram is typically used for higher level, less detailed descriptions that are intended to clarify overall concepts without concern for the details of implementation. In this method, block diagram for the dryer control system was developed by using selected electrical components. The diagram can be used as a basic primary input reference before drawing the detail electrical wiring diagram for the dryer control system.

**Design of Control Panel:** The electrical control panel is a cabinet for electrical or electronic equipment such as mounted switches, knobs and displays. It is used to prevent electrical shock to the equipment, users and to protect the contents from the environment. The panel is also known as an electrical enclosure. Electrical enclosures are usually made from rigid plastics, metals, particular stainless steel, carbon steel and aluminum. Mechanical design using computer aided design software is used to represent the dimension of the control panel, components arrangement and type of the design. The process is conducted to ensure that all the electrical components and wiring connection can be fixed into the design during fabrication work.

### 2.2 Control design

**Controller Unit Selection**: The significant unit in intelligent systems is the controller unit which allows defining complex applications and also implementing practical interface [4]. There are several types of controllers which are being used in various engineering processes such as Programmable Logic Control (PLC), and micro controller (PIC). Almost any production line, machine function or process can be significantly enhanced using this type of the control system. For this research, OMRON PLC was selected as the main controller. The PLC is considered as an economical controller for the complex system. The benefit in using PLC is the ability to modify and replicate the operation or process while collecting and communicating vital information. Besides, the controller is also easy to program and the durability and reliability of the components have made PLCs likely to operate for several years [5].

**Design of Ladder Diagram:** The PLC ladder diagram was created by using CX-Programmer. CX- Programmer is the programming software for all Omron's PLC series, which is fully integrated into the CX-One software suite. CX-Programmer includes a wide variety of features to speed up the development of PLC programs. Advantages of using this type of software are new parameter setting dialogues to reduce setup time. The development of PLC programs is also simple with drag and drop configuration [6]. In the ladder logic development, the important part is to determine the dryer sequence of operation. After that, all external I/O devices for the dryer must be determined and assigned number corresponding to the I/O number before uploading into the PLC hardware.

**Simulation:** For the simulation purpose, CX-Designer software was utilized to test the functionality of the created ladder diagram. Similar to CX-programmer, this software is also fully integrated into CX-One software suite. In the simulation method, development of the Simulator Control Panel (SCP) is needed to simulate the created ladder diagram. The SCP design is used for the simulation purpose only. In the SCP, all I/O devices must be developed and represented in the software screen based on I/O created from the ladder diagram. The simulation is important to ensure that the created ladder diagram can fully function without any problem. Besides, the simulation is able to monitor the I/O progress and also can detect malfunctioning of the I/O. Through the simulation, the modification of the ladder diagram can be done if the diagram has some problem or mistake before it is uploaded into the PLC hardware.

**Design of wiring diagram for PLC connection:** A wiring diagram is a simple visual representation of the physical connections and physical layout of the electrical system or circuit. In this method, the wiring diagram is developed to show how the electrical wires are interconnected and where dryer fixtures and components may connect to the system. The idea of wiring diagram is based on the developed block diagram. The diagram is important to assist in developing the control system during fabrication work. For this purpose, an electrical computer aided design is used to develop the wiring diagram for the dryer control system.

DOI: 10.1051/matecconf/20178702016

**Design of Sensory Devices:** Sensors are common components for converting physical characteristics to electrical value. In the drying machine, two types of sensors were utilized for recording motor and fan speed and also temperature and humidity inside of rotary drum. Suitable sensor and controller must be selected properly in order to obtain an accurate data from the sensor during data acquisition (DAQ) process.

### 2.3 Design of Human Machine Interface (HMI)

**The design of Human Machine Interface (HMI)**: A Human Machine Interface (HMI) is an interface for providing the means of controlling, monitoring, managing and visualizing device processes [7]. Visual Basic 2008 was used to design the HMI for the control system. An external controller is needed as a medium to interface between HMI and PLC. Therefore, the controller has to be developed to ensure the HMI can fully function.

### 2.4 Prototyping and Testing

A prototype of the control system was fabricated base on all design methods. The fabrication includes mechanical, electrical and electronic work. Finally, the control system is tested to ensure the functionality of the overall system and to validate the data recorded by the sensors.

## 3 Results and discussion

### 3.1 Electrical design

**Electrical block diagram:** The developed blocks diagram for the dryer controller is shown in Figure 2. The system can be controlled by two types of methods. Firstly by using the Control Panel (CP) and secondly by using the Human Machine Interface (HMI) which is assisted by an external controller.

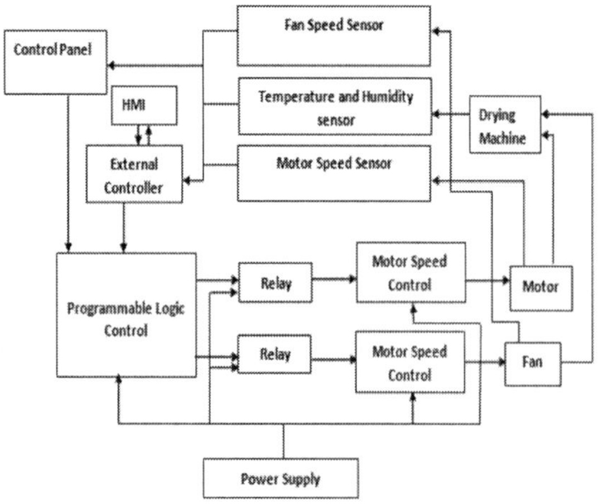


**Figure 2:** Block diagram of drying machine control system

The power supply was given to the PLC as well as to both motor speed controls. In the design, two relays were used as a switching component for the motor and fan. Two types of sensors were attached into the design. The sensors are motor speed and temperature/humidity sensor. The motor speed sensor is used to measure the speed of the motor and fan during operation while temperature sensor is used to measure temperature and humidity inside the rotary drum.

**Control Panel:** Figure 3 shows the 3D model of the control panel. The panel consists of two main parts namely, the electrical enclosure and the structural frame. Electrical enclosure was utilized to mount all switches, knobs, and displays that are used in the control panel. It is also used to protect the contents inside of the enclosure such as one unit PLC, two units 24VDC relays, two unit 24V DC power supplies and one unit miniature circuit breaker (MCB) from the environment. While the structural frame is used as a support stand for the enclosure to ensure that the panel is ergonomic and easy to control by the user.

**Figure 3:** CAD model of control panel for *keropok keping* drying machine

The panel can be divided into five sub controls including main switch, motor switch, fan switch, emergency and automatic switch. The main switch is used to activate or deactivate the overall system in the control panel. Both motor and fan control is used to switch ON or OFF the motor and fan respectively. For safety purposes, an emergency switch is allocated to the control panel to shut off the overall system if any undesirable condition happens during the machine operation. Automatic control is also designed for the machine. Therefore, automatic switch rotary type was placed at the right of the control panel to provide an auto signal to the controller.

Motor and fan speed are important parameters that need to be controlled in the drying machine. Pulse-Width Modulation (PWM) motor speed controls with potentiometer were installed for both motor and fan in the control panel. It is the feasible approach in motors' and speed control [8]. The PWM approach will assist in achieving different speed requirements for the motor and fan. It uses duty-cycle principle and average voltage for controlling the motor and fan speed smoothly. The different speed levels range between 0 to 100% of maximum speed [9].

### 3.2 Control design

**Controller Unit Selection**: OMRON PLC CJ series model CJ2M-CPU11 was selected as the main controller device for the system. The PLC unit can be categorized as a modular type PLC or rack mounted units due to the power supply, CPU and I/O interface comes separately [10]. The advantages of using this type of PLC are the PLC have far more memory and has the capability to store a higher volume of information. Besides, the PLC also has the capacity to accomplish more complex processes. The type of PLC was used in the control panel can be illustrated in Figure 4.

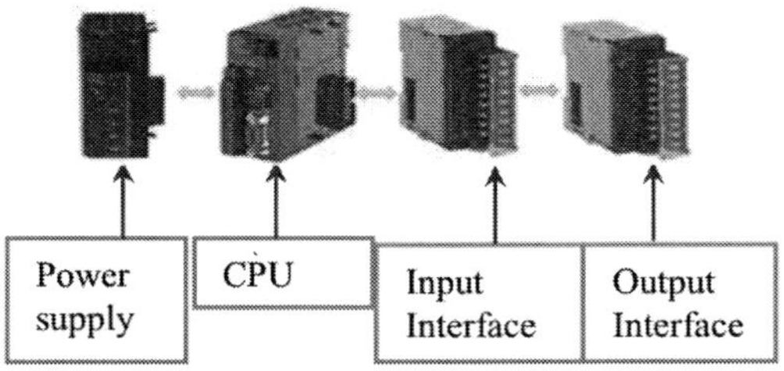


**Figure 4**: PLC OMRON CJ2M-CPU11 modular type

The controller (PLC) is used to give the switching commands to all input and output devices for the drying machine. PLC hardware setup is very important before creating a ladder diagram. The detail PLC setup can be illustrated in Figure 5. In the PLC, two types of cables can be used to upload the program into the PLC hardware. The cables are RS232 or USB cable. However, both of the cables need different setup before it can be used. For this research, the USB cable was selected as the interface cable between computer and PLC due to easy setup compared to the with RS232 cable.

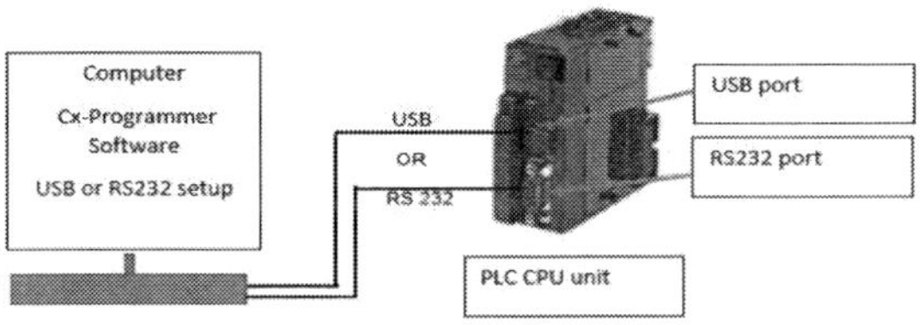


**Figure 5:** PLC hardware setup

**Dryer Control Sequence:** The algorithm used in *keropok keping* drying machine control system technique is shown in Figure 6. The control was programmed in either automatic or manual operation. When the automatic switch is turned on, the timer starts and the drying process begins. The motor and fan will rotate respectively and the duration of the rotations depends on the timer setup in the ladder diagram. Pilot lamps were utilized as an indicator to represent each output operation. After the timer time out, all system is stopped. In manual operation motor and fan can be controlled independently by the user during the drying process.

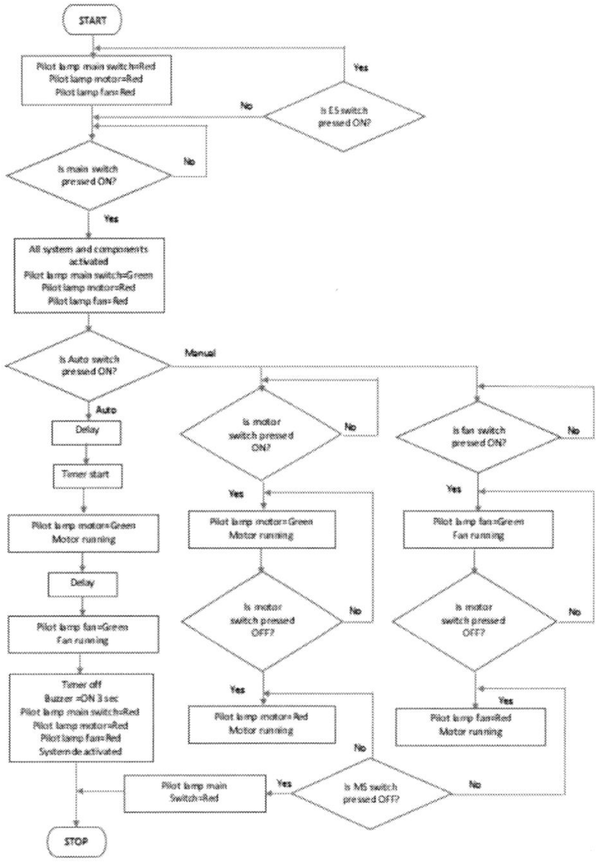


**Figure 6:** Algorithm for *keropok keping* drying machine control system

**Assignment of Inputs and Outputs:** After the system sequence of operation was determined, all the external input and output devices connected to the PLC must be assigned the addressing number corresponding to the input and output number. Table 1 shows the assignment of the inputs and outputs number to the PLC. The number is important before creating the detailed ladder diagram in the CX-Programmer software.

**Table 1**: Assignment number for Inputs and Outputs

| Input | Address | Output | Address |
|---|---|---|---|
| Main Switch ON | I:0.00 | PL Main switch ON | O:1.00 |

DOI: 10.1051/matecconf/20178702016

| | | | |
|---|---|---|---|
| Main Switch OFF | I:0.01 | PL Main switch OFF | O:1.01 |
| Motor On | I:0.02 | PL Motor switch ON | O:1.02 |
| Motor OFF | I:0.03 | PL Motor OFF | O:1.03 |
| Fan ON | I:0.04 | Motor | O:1.05 |
| Fan OFF | I:0.05 | PL Fan ON | O:1.06 |
| Emergen cy Switch | I:0.06 | PL Fan OFF | O:1.07 |
| Auto Switch | I:0.07 | Fan | O:1.09 |
| | | PL Emergency | O:1.10 |
| | | Buzzer | O:1.11 |

**Simulation:** Simulator control panel (SCP) was developed to simulate and test the functionality of the created ladder diagram. The panel can be shown in Figure 7. After the testing process completed, the diagram will be uploaded into the PLC hardware.

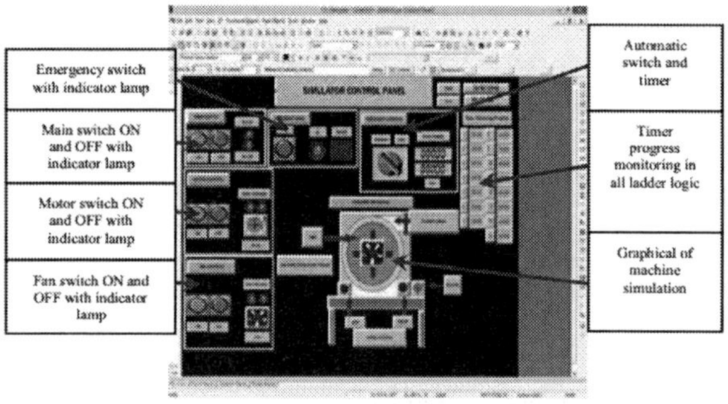


**Figure 7:** Simulator Control Panel (SCP)

**Sensory Device:** Architecture of sensory devices in the control system can be shown in Figure 8.

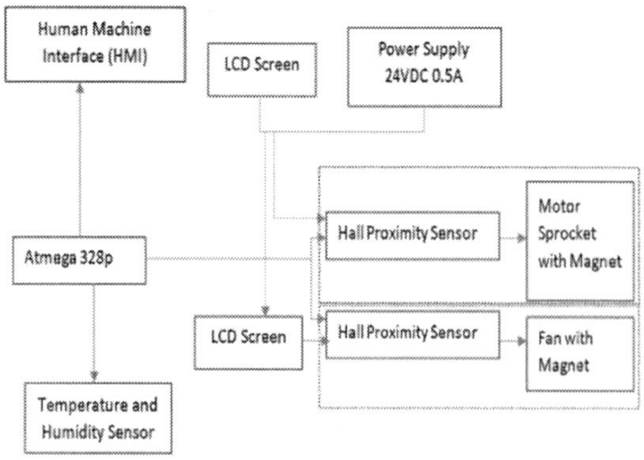


**Figure 8:** Architecture of sensor system

Atmega 328P microcontroller was used as a controller to all the sensors in the system. The controller also can be used as a data acquisition (DAQ) device through a serial communication port. The data then will be transferred to the HMI for recording process via receiving data command.

### 3.3 Human Machine Interface (HMI)

Human machine interface (HMI) for the drying machine can be illustrated in Figure 9. The HMI was created using Visual Basic Express 2008. However, the HMI needs to work together with an external controller.

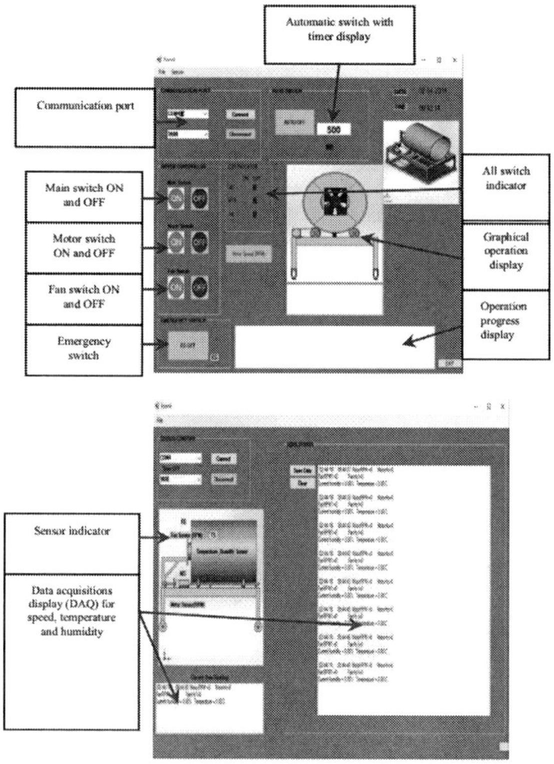


**Figure 9:** Human Machine Interface (HMI) for *keropok keping* drying machine with sensor data acquisition (DAQ)

The external controller was designed based on electronic relay module. The component consists of 12VDC relay, 10k Ohm resistor, IN 4007 diode, transistor 2N2222 and AT Mega328P as the controller. In the design, eight relay modules were developed.

### 3.4 Prototyping and Testing

The prototype of the control panel can be illustrated in Figure 10. The experimental study of the duty cycle (%)

effect to the dryer motor and fan speed was conducted. The testing was conducted without any load applied onto the motor. The study is conducted in order to know the relationship between the duty cycle (%) and current flow (amp) to the motor and fan speed (rpm). It is also done to investigate the minimum and maximum speed that can be achieved by the motor and fan without any load applied in the system. The speed of the control panel was used as the reference speed for the validation process due to the display is a real digital tachometer which was attached to the control panel but in the HMI, the speed reading is based on the equation written in C programming language.

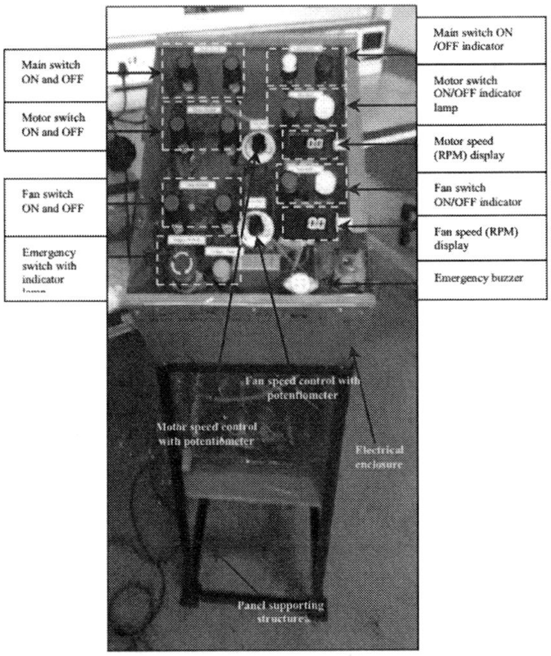


**Figure 10:** Control panel and HMI controller prototype

Measurement of the differences of the speed readings between both of the methods is very important to validate the HMI data reading. In order to obtain the current flow, a multimeter was fixed between the life wire of the motor and fan. The multimeter was setup in 10Amp measurement mode to obtain the current flow. The duty cycle 10, 20, 30, 40, 50, 60, 70, 80, 90 and 100% were tested on the motor and fan. Motor and fan were run for every duty cycle until they achieve steady state condition before the data were taken. The experiment results are represented in line graphs as shown in Figure 11 to 16.

Figure 11 shows the motor speed (rpm) versus PWM duty cycle based on the panel and HMI reading. The result shows that the motor started to rotate at 40% of the duty cycle. During that time, the speed is recorded at 23.95 rpm for the panel reading and 30 rpm for the HMI reading. When the motor is rotated at 100% of the duty cycle, the recorded motor speed is 134 rpm for the panel reading and 120 rpm for HMI reading. The percentage different between panel and HMI reading is around 22.43% during 40% of duty cycle and 11.02% for the 100% of duty cycle as shown in Figure 12. Overall, percentage difference for the motor is around 1 to 30%.

On the other hand, the fan is started to rotate at the 10% of duty cycle with 2946 rpm based on panel reading and 3015 rpm from HMI reading. When the rotation is at 100% of duty cycle, the fan is recorded to rotate at 3139 rpm for panel and 3225 for HMI reading as shown in Figure 14. The percentage difference between panel and HMI reading for fan speed is shown in Figure 15. It was recorded that the percentage difference of fan speed is around 1 to 3%.

In general, the percentage difference of duty cycle to rotate the motor and fan can be observed due to different range of speed needed based on equipment specification. At the same time, the different readings between panel and HMI reading may occurred due to the errors in the programming equations.

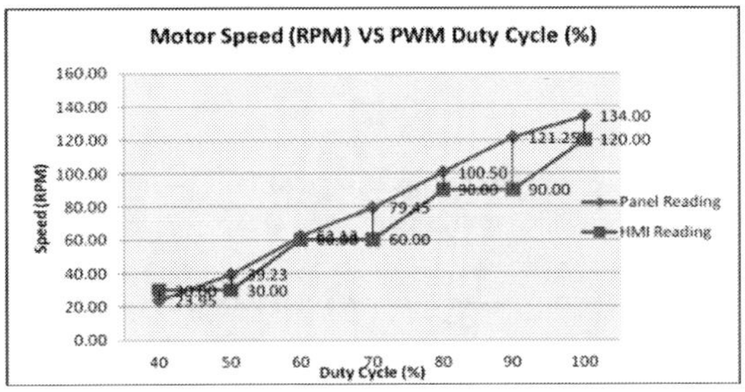


**Figure 11**: Motor speed (RPM) Versus PWM duty cycle

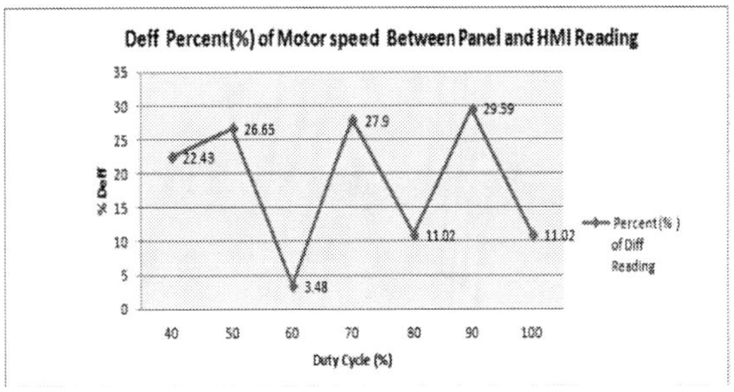


**Figure 12**: Difference percentage of motor speed between control panel and HMI

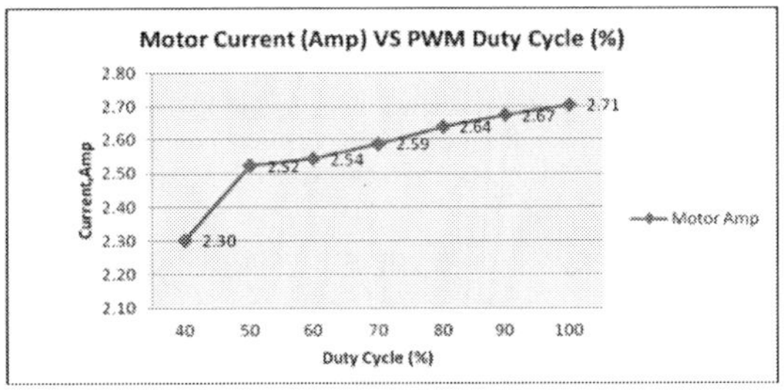


**Figure 13:** Motor current (amp) versus duty cycle

 DOI: 10.1051/matecconf/20178702016

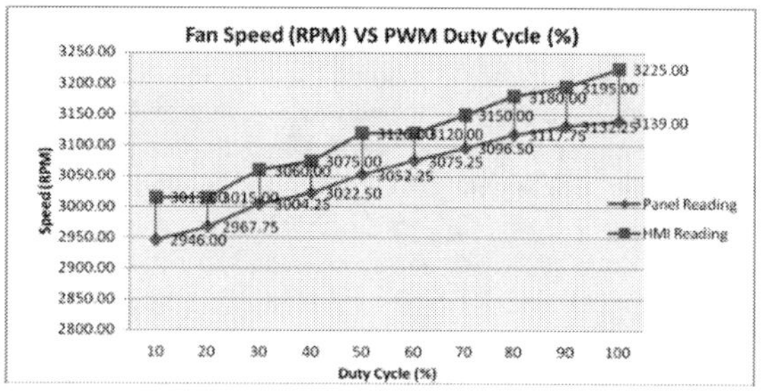


**Figure 14**: Fan Speed (RPM) versus PWM duty cycle

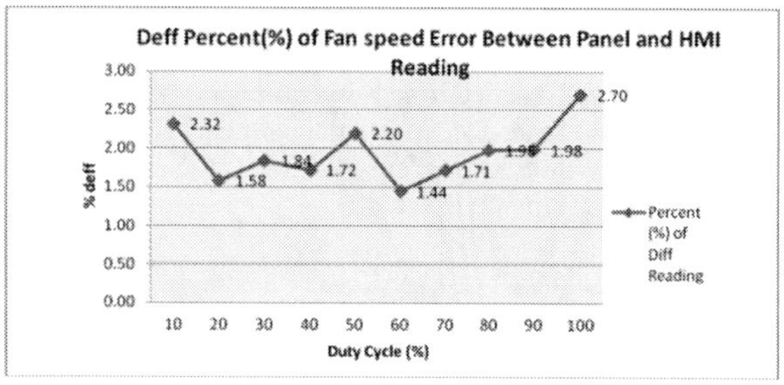


**Figure 15**: Different percentage of fan speed between control panel and HMI

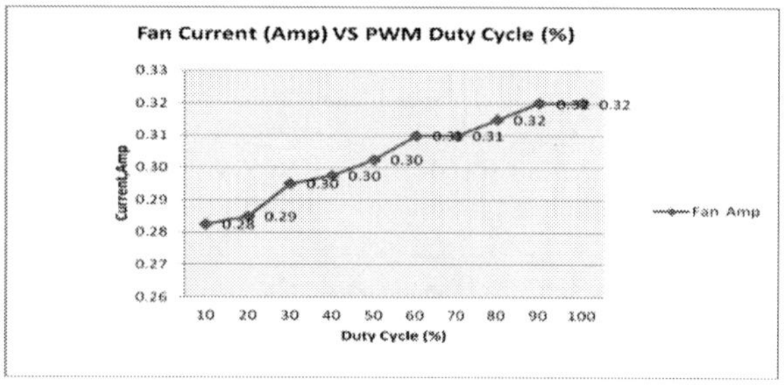


**Figure 16:** Fan current (amp) versus duty cycle

Lastly, Figure 13 and 16 shows the motor and fan current versus duty cycle. Both graph shows the linearity of the current flow curve to all duty cycles. The current flow consumption for motor and fan required to start the rotation are 2.30 amp and 0.28 amp respectively. The maximum current consumptions for the motor and fan when achieved 100% of the duty cycle are 2.71 amp and 0.32 amp. Speed (rpm) and current flow (amp) tends to be constant when the duty cycle is close to 100%. Based on the current consumption, the selected power supplies for the design is considered enough to support the overall dryer operation. The graphs also proved that PWM speed control worked based on the power supplied to the system via current flow (amp). When the current flow increases the speed of the dryer motor and fan also will increase.

## 4 Conclusion

In this research, a control system for the *keropok keping* drying machine was successfully developed. Programmable Logic Controller (OMRON) was utilized as a main controller and programming using ladder logic language. The inputs signal can be specified by the user to the controller through the control panel (CP) or human machine interface (HMI). Based on the signal, PLC will make motor, fan and all indicators to work correspondingly. Pulse-width Modulation (PWM) motor speed control was attached for both of motor and fan to control the speed. There are two types of sensors installed into the system which are speed and temperature sensor. The sensors were used to measure the speed, temperature and humidity of the machine during data acquisition (DAQ). A testing was conducted to the system in order to know the relationship between the duty cycle (%) and current flow (amp) to the motor and fan speed (rpm). The percentage difference readings between the control panel and HMI were also measured in order to validate both of the data readings. However some improvement can be taken as further work. The recommendations and suggestions for the improvement can be listed as follows;

a) The Human Machine Interface (HMI) is does not have a special function to control the speed of motor and fan. The system will be better if the function is added into the system software.
b) The wiring is too complicated and the best way is to simplify the components wiring connection.
c) The developed system still has differences readings between control panel and HMI, Some modification needs to be done to the microcontroller C programming language to ensure that the HMI reading will be more reliable.
d) Temperature reading of the dryer machine is still not covered in the testing due to the panel and HMI have not been assembled yet to the actual machine. The effect of the temperature and humidity to the motor and fan speed is another important parameter that needs to be measured in order to know the effectiveness of the dryer.

## Acknowledgement

This research is funded by Research Acculturation Grant Schemes [Grant no: RAGS/TK01(4)/1318/2015(12)]. The authors would like to thank Universiti Malaysia Sarawak (UNIMAS) for providing facilities for this research.

## References

1. T.K Tawee. *International Food Research* **18,** 855-866 (2011)
2. Info on from htpp//: www.dof.gov.my-crackers1.
3. S. Mohamaddan *et al. IOP Conf. Ser.: Mater. Sci. Eng.* **114** 012037 (2016)

4. M Javanmard, J Endan, KA Abbas, F Arvin.American Journal of Applied Sciences **7,** 2 160-166 (2010)
5. S. Mahapatra. A. Jain, D. Singh. IJITEE. **4** , 512-15 (2014)
6. Info on https://industrial.omron.eu/en/products/catalogue/automation_systems/software/configuration/cx-one/cx-programmer.html.
7. E, Husinu. F., Agyare. O.R. Journal of Emerging Trends in Computing and Information Sciences, **5,** 2 134-144 (2014)
8. Erikson, R.W., Dragon. M. Fundamental of Power Electronic.2nd Ed ,(2011)
9. Valentine, R. (1998). Motor Control Electronics HandBook. 1 st Ed,(1998)

MATEC Web of Conferences **87**, 02017 (2017)
*ENCON 2016*
DOI: 10.1051/matecconf/20178702017

# A Review on Development of Kek Lapis Sarawak's Machine

Mohamad Syamim Hamsawi[1], Ana Sakura Zainal Abidin[1,*], Shahrol Mohamaddan[1], Rasli Muslimen[1], Mohd Syahmi Jamaludin[1], Nurul 'Izzati Hashim[2], Nazreen Junaidi[2]

[1]*Department of Mechanical and Manufacturing Engineering, Faculty of Engineering, Universiti Malaysia Sarawak (UNIMAS), Malaysia*

[2]*Department of Electrical and Electronic Engineering, Faculty of Engineering, Universiti Malaysia Sarawak (UNIMAS), Malaysia*

**Abstract.** Kek Lapis Sarawak is a beautifully decorated and flavorful cake that can be considered as the artwork of the Malays in Sarawak. There is an overwhelming demand of the Kek Lapis Sarawak in the market that consequently need for automation. This paper summarizes comprehensive reviewed patents of the current machine and journals which have similar functions and related to the current function of manual process of producing Kek Lapis Sarawak. Conceptual idea of the Kek Lapis Sarawak machine consists of three modules namely Depositor Module, Cooling and Pressing Module and Baking Module. Therefore, patents and journals that have similar functions with these modules are reviewed that look into its advantages. The disadvantages of the design also considered for future design. Hence, the crucial advantages of each module considered are namely improved production rates, increased of hygienic rating, easy to maintenance and portability of a design. The reviewed patents and journals can be considered for the development of Kek lapis Sarawak's machine with some improvisations are needed to meet up the specific requirements of the machine.

## 1 Introduction

Kek Lapis Sarawak can be considered as the artwork of the Malay people in Sarawak. The uniqueness of the cake's pattern and flavour captures the world's attention. Kek Lapis Sarawak is a must menu during the festive seasons in Sarawak including birthdays and wedding ceremonies. The popularity of the cake creates an overwhelming market demand.

The current production of Kek Lapis Sarawak which is done manually is unable to meet the growing demand. Conventional method involves repetitive processes of depositing, layering, baking, cooling and pressing dependent on number of layered needed [1]. The tedious process is very time consuming and relying on the labour's skills. The issues consequently limit the productivity and quality of the cakes produced. The needs of producing compact and cheaper machine that can produce varieties of Kek Lapis Sarawak caused by the high dependable to a foreign supplier that costs more on the maintenance such as delay in repairing process and international expertise that contributes to the loss of production. Likely, automation of the Kek Lapis Sarawak machine is the most viable solution.

The objective of this paper is to present summary of reviewed any related patent, journal and proceeding that have similar functions to the current process namely depositing, baking and layering. Advantages of the reviewed publications can be considered for the development of the future machine. While, any drawback need to be improvised to suite with the specific needs of the Kek Lapis Sarawak machine.

* Corresponding author: zaasakura@unimas.my

## 2 Discussions

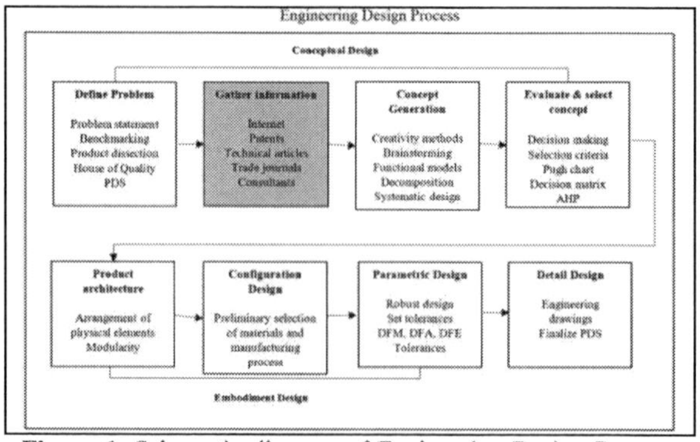


**Figure 1.** Schematic diagram of Engineering Design Process. [2]

Currently, this project is in the stage of gather information as illustrated in the higlighted box in **Figure 1**. Engineering Design Process is one of the method to do the design process [2]. The next stage will be development of conceptual design of the Kek Lapis Sarawak machine.

The conventional method of making Kek Lapis Sarawak can be divided into three main processes namely depositing, cooling and pressing, and baking [1]. The cycle of process Kek Lapis Sarawak production is shown in **Figure 2**. More over, the number of process cycle is depended on the desired production of the cake. This section presents summary of related publications followed by discussion according to the named processes.

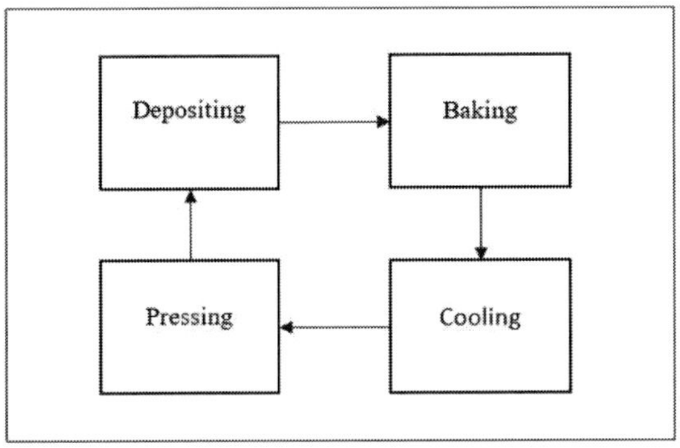


**Figure 2.** Flow process of cycle of a Kek Lapis Sarawak production. [1]

## 2.1 Depositor Module

Table 1 shows summary of reviewed patents for depositor [3, 11, 16, 20, 22]. "Depositing confectionary nozzle mass" is based on the slit and elongated nozzle mechanism [3-10]. This patent used to deposit chocolate that has quite similar behavior to the cake batter. The negative of de-aeration effect of this patent can be beneficial to the development of Kek Lapis Sarawak's machine because it will improve accuracy of the depositing of the cake batter.

"Volume division dough and implementing device" is based on the hopper design that applies gravitational force [11]. However, this patent claimed that the flow of the dough by gravity is utterly incapable to meet the customary rates of industrial production of bread–making products. Somehow, the idea of using gravitational force is good because it can eliminate additional force needed. Another benefit of this patent is the deformation of the medium that can be controlled. Controlling function is very crucial in the development of Kek Lapis Sarawak as consistency the cake thickness represents the quality of the cake produced [11-15].

"Food flavouring by using a metering roller and a blade" is based on the dispensed by plurality indentations [16]. Those distinctly "chunky" or "gluey" coatings generally tend to obstruct dishing out apparatus and are susceptible to intermittent, irregular flow. The mechanism potentially able to prevent it to occur [16]. Indirectly, the flow of the medium will be in steady flow. This patent can lessen excess of the batter used which can contribute to the optimum production rate [16-19]. Moreover, this patent is easy to maintain which is essential consideration to the development of Kek Lapis Sarawak machine.

"Producing sandwich cookie" having dissimilarly base cakes [20]. This patent is based on hopper filler with nozzle. The medium that has been used in this patent is slightly similar to the medium for Kek Lapis Sarawak's batter which is cream. This patent can controls the flow of the cream through the discharge orifices which it is good for optimum rate of depositing process of the medium. This patent also claimed that it can operates in high speed due to it processing system which is high speed capability [20-21]. This patent also provides continuous production which is good for mass production. Moreover, this patent also equipped changeable molder that make it more portable. Unfortunately, the rotary molder part is not suitable to dispense Newtonian fluid and Non-Newtonian fluid because it is built to deposit dough.

"High viscosity dough depositor machine" is based on hopper and a die concept [22]. It is mentioned that the dough transfer of the die could provide 50% greater than conventional dies for the identical size baked product which lead to increase the production rate. Moreover, this patent tolerate for enhanced deposit of the high viscosity of non-traditional dough and resolve readily spread to the preferred product form [22]. Besides, it could save time in depositing process and lead to improve the production rate which is essential for development of Kek Lapis Sarawak's machine.

Unfortunately, many of these patents are limited to solid state or dough characteristic [3, 11, 16, 20, 22]. The idea of the mechanisms are informative for the development of Kek lapis Sarawak's machine but different characteristic of the transported medium for the cake batter as Non-Newtonian fluid must be improvised.

## 2.2 Cooling and Pressing Module

Table 2 shows summary for cooling and pressing mechanism [23, 31, 38, 44]. Most of the available cooling mechanism applied fan structured. Cooling structure of electronic equipment rack is based on the ordinary fan structured [23]. The patent provides constant air speed for steady flow of air ventilation in order to achieve stable cooling effect [24-30]. Moreover, this patent produce less noise due to arrangement of the fan itself that enable the fans to operate with less noise. However, this patent is considered as expensive as more fans needed to provide stable cooling effect. At the same time, the patent has limited degree of freedom because of the arrangement of the fans are placed in vertical upwards.

The next cooling mechanism is vacuum cooling. The mechanism is able to cool down without damaging the cake [31]. This patent is recommended for hygienic and safety. This patent potentially meet the requirement for hygiene and safety criteria of the development of Kek Lapis Sarawak's machine. Moreover, this patent is also efficient for product with medium of slurries characteristic which is difficult to cool [32-35]. However, there is drawback of the vacuum cooling mechanism which is high in capital investment. Therefore, the mechanism is recommended for long term operation as upsetting for short term of operation can lead to loss. The mechanism can be considered as the development of the Kek Lapis Sarawak's machine is targeted for mass production.

In spite of that, the process is quite grimy when operated. The sauces splattered on the roofs of processing tunnel during vacuum cooling and strenuous to clean [36]. Moreover, the product aroma is disturbed due to vacuum cooling which might reduce quality of the product [37]. On top of that, if the vacuum cooling process is operate at low vacuum state, the process will not fully satisfied the objective. Thus, the mechanism must be ran at high pressure that consequently consume a lot of power. Therefore, the idea can be considered with some improvements to meet up the requirements needed.

Pressing mechanism has very limited design concepts as most of the pressing part applied in the current designs are stamper and roller presser. "Device for pressing the coffee powder within the filter holder" has applied concept of stamper presser [38]. This patent provides easy to maintenance since is easy to clean after being used. The reliability on the stamper presser is considered as good to the current development of Kek Lapis Sarawak's machine because the patent is able to provide equal pressure to the whole cake surface that potentially able to produce better quality cake that has firm texture [39]. Besides, pressure of the presser can be controlled to any part of the cake surface with specific amount of pressure either little or harder pressure [40-43].

"Bun machine" is based on the concept of roller presser [44]. This patent need less maintenance as it is easy to clean due to simple assembly. Moreover, this patent can be considered for the development of Kek Lapis Sarawak's machine because the pressure exert from the presser is able to spread evenly and more reliable [44-51]. This patent also consume less space that can be considered for compact design and less material usage.

## 2.3 Baking Module

Table 3 shows summary of reviewed patents for baking module [52, 58, 64, 68, 74]. "Convection oven" using steam as heating element [52]. This patent provides portability in accessible of electrical components which is essential for development Kek Lapis Sarawak's machine [52-57]. Moreover, variable in speed by the reversible blower during operation is one of the important consideration of the develop machine that allows for heat controlling. Thus, it could improve the production rate. Unfortunately, maintenance seems hard for the patent because it is not reliable. Besides, this patent is not eco-friendly because the heat emitted by the gas.

"Conveyor oven" is based on infrared radiation heating for heating element [58]. This patent offers moveable design in which needed for development of Kek Lapis Sarawak's machine by accessible of changing parts [59]. Thus, make it is easy to maintain [58-63]. Moreover, this patent also provides optimum baking capacity that consequently increase the production rate [58-63]. However, the heat emitted by this patent is not fully utilized because of the open space. Hence, this idea can be considered but need some improvements.

"Forced convection gas oven" is operated by gas [64]. The heat transfer is caused by the burning of the liquefied petroleum gas (LPG). Heat of the oven is fully utilized as the patent applied closed design concept [64-67]. However, this patent is not portable. Hence, it is hard to maintain because of inaccessibility of the compact design.

"Multi access microwave oven" applied microwave radiation for its heating element [68]. The emitted microwave radiation improved the efficiency of the cooking method because the radiation is able to cook with less power and faster as multiple unit of microwave ovens can be installed [68-73]. Unfortunately, the price is higher than a conventional microwave oven. Moreover, this patent is hard to maintenance due to its closed design.

"Bidirectional heating cooker" is based on infrared radiation as its heating element [74]. This patent capable in controlling heat emitted by the infrared radiation [74]. Thus, controllable heat transfer could improves the production rate which is good for development of Kek Lapis Sarawak's machine [74-78]. Furthermore, suitable amount of infrared radiation in this patent is able to improve the cooking efficiency [74-78]. Unfortunately, this patent is hard to maintain because of the complexity of the design.

**Table 1.** The reviews for Depositor Module.

| **Details** | **Mechanisms** |
|---|---|
| Name: Depositing confectionery mass (Slit and elongated nozzle). [3] | Depositing confectionery mass comprising at least one discharge passageway extending to at least one elongate discharge outlet, wherein at least one discharge passageway diverges in a direction towards the discharge outlet, and a machine for producing a confectionery product having at least one such apparatus for depositing. In a method of producing a confectionery product, the confectionery mass is deposited as at least one strip having a width in the range of 5 to 500 mm and/or thickness in the range of 0.5 to 100 mm. [3]<br><br>***Related mechanism for Slit and elongated nozzle.** [4-10] |
| Name: Volume division dough and implementing device (Hopper). [11] | It concerns a method for volume division of bread-making, bakery, and pastry dough and the like, contained in a hopper. The method includes allowing the dough to by gravity into a chamber capable of measuring an amount of dough to be divided and extending beneath an opening located at the base of the hopper. [11]<br><br>***Related mechanism Hopper.** [12-15] |
| Name: Food flavouring by using a metering roller and a blade (Dispensed by plurality indentations). [16] | An exterior axial surface of the metering roller may define a plurality of indentations, which may be configured to capture the food flavoring as the metering roller rotates. The system may also include a metering blade and a displacement; each may be coupled to the frame. The metering blade may be configured to level the food flavoring as the food flavoring is dispensed by the plurality of indentations of the metering roller as the metering roller rotates. The displacement blade may have a plurality of teeth that are configured to correspond to the plurality of indentations. The teeth may be configured to positively displace the food flavoring as the metering roller rotates. [16]<br><br>***Related mechanism Dispensed by plurality indentations.** [17-19] |
| Name: Method for producing sandwich cookie having dissimilarly-sized base cakes (Hopper filler and nozzle). [20] | Continuously producing a sandwich cookie having dissimilarly-sized base cakes and for producing filled deep-walled cookies or cookie cups. The production of the sandwich cookies includes at least one rotary molder for producing lower base cakes and upper base cakes and a filler cream depositor that deposits a layer of filler cream on each of the lower base cakes. [21]<br><br>***Related mechanism for Hopper filler and nozzle.** [20-21] |
| Name: High viscosity dough depositor machine (Hopper and a die). [22] | Hopper receive a supply of dough. A forming portion comprising a die having at least one dough transfer port through which the dough is extrude-able to form the dough into a predetermined shape, wherein the at least one dough transport comprises a cross sectional area not less than about 〖0.75 in〗^2 and a diameter of at least about 1 inch. The dough been delivered by the hopper and being transferred to roller. [22] |

**Table 2.** The reviews of Cooling and Pressing Module

| **Details** | **Mechanisms** |
|---|---|
| Name: Cooling Structure of Electronic Equipment Rack (Fan Structured). [23] | The fan unit of each cooling block generates a cooling wind which is passed linearly through the shelf unit from the bottom to the top of the shelf unit as well as introducing atmosphere through vent holes formed on side face portions of each fan unit to mix the atmosphere with the cooling wind. The flow rate of the cooling wind will be gradually increased as the cooling wind proceeds to the upper cooling blocks due to the addition of an upward blowing force caused by the fans of the fan units. [23]<br><br>***Related mechanism for Fan Structured.** [24-30] |
| Name : Vacuum cooling technology for the food processing industry. (Vacuum Cooling). [31] | Vacuum cooling is a batch process whereby moist products containing free water are cooled by evaporation of moisture under vacuum. Products to be cooled are loaded into a vacuum chamber and the system put into operation. [31]<br><br>***Related mechanism for Vacuum Cooling.** [32-37] |
| Name: Device for pressing the coffee powder within the filter holder (Stamper Presser). [38] | The presser will be pulled down by the lever holder if the powder coffee's cup already been placed under the presser. [38]<br><br>***Related mechanism for Stamper Presser.** [39-43] |
| Name: Bun machine (Roller Presser). [44] | The tank will deliver the fermented dough to the roller presser for forming a first layer and then forming a second layer when it deliver to the second roller presser. [44]<br><br>***Related mechanism for Roller Presser.** [45-51] |

**Table 3.** The reviews of Baking Module

| Details | Mechanisms |
|---|---|
| Name: Convection oven (Steam). [52] | The vapor collection system collects vapor from the cooking chamber during a cooking event, condenses the vapor, and drains the condensed vapor. The water injection system injects water for impact against a blower wheel for dispersion into the air circulating through the cooking chamber. The electrical components are housed Within a housing that in a closed position conceals the components and in a closed position exposes the components for easy access. The rotational speed and direction of the variable-speed, reversible blower is controlled during a cooking event according to predetermined speed curves which may include one or more reversal events to achieve more uniform cooking of food. A main controller is programmable via an operator input (e.g., liquid crystal display touch screen) to control operating parameters of the oven. [52]<br><br>***Related mechanism for Steam.** [53-57] |
| Name : Conveyor oven (Infrared Heating). [58] | The oven described herein comprises at least one conveyor for carrying in food products through a bake chamber to be bake chamber with openings for conveyor(s), an air heating distribution and return System a fan box that contains means for pressurizing heated air and a drive end that contains the fan motor, conveyor drive motor(s) and oven temperature controls. [58]<br><br>***Related mechanism for Infrared Heating.** [59-63] |
| Name: Forced convection gas oven (Gas). [64] | The invention provide a microwave oven having multiple (i.e., at least two or more) independently operable cooking chambers. Each of the multiple cooking chambers has a separate access door for opening and closing the corresponding cooking chamber. The microwave oven according to embodiments of the present invention further includes a control panel for separately controlling the operations of the at least two cooking chambers. [64]<br><br>***Related mechanism for Gas.** [65-67] |
| Name: Multi access microwave oven (Microwave Radiation). [68] | The invention provide a microwave oven having multiple (i.e., at least two or more) independently operable cooking chambers. Each of the multiple cooking chambers has a separate access door for opening and closing the corresponding cooking chamber. The microwave oven according to embodiments of the present invention further includes a control panel for separately controlling the operations of the at least two cooking chambers. [68]<br><br>***Related mechanism for Microwave Radiation.** [69-73] |
| Name: Bidirectional heating cooker (Infrared Radiation). [74] | Provided is a bidirectional heating cooker that supplements and improves a function of a conventional infrared radiation cooker, in which the bidirectional heating cooker further includes a lower heating unit that is placed at the lower end of a cooking pan that is heated by radiant heat irradiated by an infrared lamp and that directly heats the cooking pan, to thereby heat the cooking pan quickly by the lower heating unit to thus make the upper and lower portions of food such as none meat put on the cooking pan more quickly roasted than the conventional infrared radiation cooker and reduce a burden of electric charges based on a shortened heating time of the cooking pan, and to thereby make it possible to cook food rapidly by an upper heating unit with the infrared lamp and the lower heating unit that is placed on the cooking pan. [74]<br><br>***Related mechanism for Infrared Radiation.** [75-78] |

**Table 4.** Proposed Solution and Improvisation of the chosen patents

| Process | Mechanism | Major Improvisation of current patents |
|---|---|---|
| Depositor | Name: Method for producing sandwich cookie having dissimilarly-sized base cakes (Hopper filler and nozzle) | • The patent must be improvised to meet the requirement for characteristic of Kek Lapis Sarawak's batter which the batter can be controlled while the batter is being deposited. |
| Baking | Name: Conveyor oven (Infrared Heating) | • The conveyor oven is being improvised to meet the requirement of Kek Lapis Sarawak's baking time when the batter is being baked with optimum time. |
| Cooling | Name: Cooling Structure of Electronic Equipment Rack (Fan Structured) | • The fan structured cooling system is being improvised to meet the requirement for cooling the cake with optimum time. The main idea is being incorporated in which a fan is used to cool down the cake. |
| Pressing | Name: Bun machine (Roller Presser) | • The Roller Presser is improvised to meet the requirement for the quality of layering specification that is being controlled. |

## 3 Conclusions

Prior publications offer greater advantages. For instance, depositor module, showed on how the patent is able to improve the production rate, negative de-aeration effect, depositing by using simple gravity, controllable deformation, technique to prevent uneven flow, minimize waste of material, easy to maintain, improve high speed capabilities as well as delivery of the high viscosity and high withstand and rigidity of the design.

Moreover, the cooling and pressing module have its own advantages whilst reviewing the patents and the journals which are secured specified wind velocity, stable cooling effect, less noise, no mechanical damage, efficient cooling system, low running cost, increased productivity, easy to maintenance, evenly distributing the pressing force, consume less space area and improved the quality of the medium that has been used.

Furthermore, the baking module also has its own advantages while reviewing the patents and the journals which are portability of the design, controllable speed of the device, fast production, controllable heat transfer, improved cooking efficiency and easy to maintenance. Significantly, the improvisation of those ideas is needed to meet up the requirement for development of Kek Lapis Sarawak's machine. Summary of proposed solution and suggestion for improvisation of the chosen patents towards development of Sarawak Kek Lapis's Machine is shown in Table 4.

## References

1. Abidin, A. S. Z., Annisa, J., Tan, W. K., Rigit, A. R. H., Chong, J. S., Kiprawi, M. A., & Johnapi, P. R. (2014, November). Development of a prototype model for fully automated Sarawak layered cake machine. In *Engineering and Technology (BICET 2014), 5th Brunei International Conference on* (pp. 1-5). IET.
2. George. E. D., Linda. C. S., *Engineering Design*. 5th ed. New York: McGraw-Hill Higher Education, 2012.
3. Abylov, M., & Durco, J, "Apparatus for depositing a confectionery mass, and method of producing a confectionery product", U.S. Patent 8 658 233, Feb 25 2014.
4. Berger, T. G., & Shtilerman, M, "Dough Feeding System", U.S. Patent 13/098 736, Nov 3 2011.
5. Campbell, B. V, "Dough Pump and Developer", U.S. Patent 13/007,198, Jul 21 2011.
6. Bertuzzi, I., Zanetti, U., & Biondi, A, "FEEDER DEVICE FOR FEEDING A FOOD PRODUCT IN A MACHINE FOR MAKING FILLED PASTA AND MACHINE FOR MAKING FILLED PASTA COMPRISING SUCH A FEEDER DEVICE." U.S. Patent No. 20,150,366,222, Dec 24 2015.
7. Chiang, W. D, "Forming machine that quickly compacts hollow materials and automatically recycles the materials." U.S. Patent No. 8,057,212, Nov 15 2011.
8. Swami, Shrikant Baslingappa, B. Maiti, and S. K. Das. "Development of an extrusion system for Bori and force characteristics of its batter during extrusion." *Journal of food engineering*, vol. 73, no. 1, pp. 20-28, March 2006.
9. Gabriele, D., Baldino, N., Migliori, M., De Cindio, B., & Tricarico, C, "Modelling flow behaviour of dairy foams through a nozzle." *Journal of food engineering*, vol. 109, no. 2, pp. 218-229, Mar 2012.
10. Kusakari, T., Von Essen, K., & Hoisington, P. A, "Nozzle layout for fluid droplet ejecting." U.S. Patent No. 8,591,003, Nov 26 2013.
11. Voegtlin, C, "Method for volume division of dough and implementing device." U.S. Patent Application No. 13/444,468, Sep 20 2012.
12. Rahaman, M. F., Bari, S., & Veale, D, "Flow investigation of the product fill valve of filling machine for packaging liquid products." *Journal of food engineering*, vol. 85, no. 2, pp. 252-258, Mar 2008.
13. Alden, R. M, "High protein whipped collet extrusion crunchy snack product and manufacturing apparatus and process." U.S. Patent Application No. 13/999,015, Aug 7 2014.
14. Medina, F., Wicker, R., Palmer, J. A., Davis, D. W., Chavez, B. D., & Gallegos, P. L, "Methods and systems for integrating fluid dispensing technology with stereolithography." U.S. Patent No. 8,252,223, Aug 28 2012.
15. Thomas, R., Dufort, C., & Tence, J. F. M. A, "Method and apparatus for manufacturing decorated ice cream confectionery items." U.S. Patent No. 7,658,960, Feb 9 2010.
16. Villarreal, J. A, "System and method for flavoring food." U.S. Patent No. 9,254,052, Feb 9 2016.
17. Robinson, C. E., Janulis, T. N., & Mihalos, M. N, "Ribbon cutter apparatus and method for making sandwich baked goods." U.S. Patent No. 8,334,005, Dec 18 2012.
18. Groff, E. T., Zaleski Jr, J. S., & Zvoncheck, K. J, "Rotating nozzle die machine for extrusion of two or more types of dough." U.S. Patent No. 8,177,542, May 15 2012.
19. Wijshoff, H, "The dynamics of the piezo inkjet printhead operation."*Physics reports*, vol. 491, no. 4, pp. 77-177, June 2010.
20. Mihalos, M. N., Janulis, T. N., Robinson, C. E., & Wines, C, "Method for producing sandwich cookie having dissimilarly-sized base cakes." U.S. Patent No. 8,309,148, Nov 13 2012.
21. Wang, L., Smith, S., & Chessari, C, "Continuous-time model predictive control of food extruder." *Control Engineering Practice*, vol. 16, no. 10, pp. 1173-1183, Oct 2008.
22. Rodrigues, D, "High-viscosity dough depositor machine." U.S. Patent No. 8,926,307, Jan 6 2015.
23. Okuyama, K., Arai, S., Ishiwata, M., & Takada, H, "Cooling structure of electronic equipment rack." U.S. Patent No. 4,774,631, Sep 27 1988.
24. Cronin, J. E., Previti-Kelly, R. A., Ryan, J. G., & Sullivan, T. D, "Cooling microfan arrangements and process." U.S. Patent No. 5,326,430, Jul 5 1994.

25. Cronin, J. E., Previti-Kelly, R. A., Ryan, J. G., & Sullivan, T. D, "Cooling microfan arrangements and process." U.S. Patent No. 5,296,775, Mar 22 1994.
26. Canon, K. H, "Pulse-width modulation speed controllable DC brushless cooling fan." U.S. Patent No. 5,099,181, Mar 24 1992.
27. Kuo, C. R., Chang, C. Y., & Wey, T. W, "Axial flow type cooling fan with shrouded blades." U.S. Patent No. 7,192,258, Mar 20 2007.
28. Hsu, C. C, "Snap coupling cooling fan module." U.S. Patent No. 6,587,342 Jul 1 2003.
29. Laguerre, O., Duret, S., Hoang, H. M., Guillier, L., & Flick, D, "Simplified heat transfer modeling in a cold room filled with food products." *Journal of Food Engineering*, vol. 149, pp. 78-86, Mar 2015.
30. Hoang, H. M., Duret, S., Flick, D., & Laguerre, O, "Preliminary study of airflow and heat transfer in a cold room filled with apple pallets: Comparison between two modelling approaches and experimental results." *Applied Thermal Engineering*, vol. 76, pp. 367-381, Feb 2015.
31. McDonald, K., & Sun, D. W, "Vacuum cooling technology for the food processing industry: a review." *Journal of food engineering*, vol. 45, no. 2, pp. 55-65, Aug 2000.
32. Anon, "Bakery products cooled in minutes instead of hours. Modulated vacuum cooling is the key." *Food Engineering International*, vol. 3, pp. 33-34, 1978.
33. Anon, "Vacuum cooling for fruits and vegetables." Food Processing Industry, vol.12, pp. 24, 1981.
34. Anon, "Rapid vacuum cooling." *Food Processing Industry*, vol. 9, pp. 49, 1981.
35. Mellor, J. D, "Vacuum techniques in the food industry." *Food Technology Australia*, vol. 32, pp. 77-177, 400-401, 1980.
36. Kratochvil, J., & Holas, J, "Effect of vacuum cooling on the content of aroma substances and sensory properties of bread." *Sobornik UVTIZ Potravinarske Vedy*, vol. 2, pp. 241-251, 1984.
37. James, S. J, "Cooling systems for ready meals and cooked products." *Process engineering in the food industry*, vol. 2, pp. 88-97, 1990.
38. Mazzer, G, "Device for pressing the coffee powder within the filter holder." U.S. Patent No. 20,160,095,463, Apr 7 2016.
39. Bimman, L, "Clamp for pressing food in pan during cooking." U.S. Patent No. 20,160,095,463, Nov 27 1984.
40. "Bun toasting machine." U.S. Patent No. 20,160,095,463, Dec 17 1940.
41. Buckley, C, "Adjustable food press." U.S. Patent No. 20,160,095,463, Jan 6 2015.
42. Errera, R. B, "Food embossing stamper device." U.S. Patent No. 7,331,776, Feb 19 2008.
43. Kovalevich, P. R., & Boni, L. M, "Handheld hamburger patty forming system." U.S. Patent No. 8,202,073, Jun 19 2012.
44. Huang, P. Z, "Bun machine." U.S. Patent No. 20,160,113,291, Apr 28 2016.
45. Linneman, R. H, "Bun making machine." U.S. Patent No. 2,354,681, May 6 1947.
46. Emerson, D. B., & Narwankar, S, "Dough rolling apparatus, method and product." U.S. Patent No. 8,668,485, Mar 11 2014.
47. Kolodesh, M. S., Peters, E. L., & Madgett, R. J, "Dough rolling apparatus for laminated cookies." U.S. Patent No. 4,511,318, Apr 16 1985.
48. Quinlan, M. J, "Method and apparatus for producing thin sheets of starch containing food material." U.S. Patent No. 5,268,187, Dec 7 1993.
49. Hayashi, T, "Method for processing bread dough." U.S. Patent No. 4,276,317, Jun 30 1981.
50. Hayashi, Torahiko, "Apparatus and method for rolling croissant dough pieces." U.S. Patent No. 4,905,583, Mar 6 1990.
51. Rhodes, H. C, "Method for distributing moisture in bread dough." U.S. Patent No. 2,479,864, Aug 23 1949.
52. Greenwood, J., & Reese, R. J, "Convection oven." U.S. Patent No. 8,378,265, Feb 19 2013.
53. Knost, D, "Baking oven." U.S. Patent No. 5,875,705, Mar 2 1999.
54. Karabin, T., & Johnson, L. B, "Chambered flame oven." U.S. Patent No. 8,464,701, Jun 18 2013.
55. Klobucar, J. M., Pakkala, J. L., & Yu, G, "Convection combustion oven." U.S. Patent No. 7,905,723, Mar 15 2011.
56. Baik, O. D., Marcotte, M., & Castaigne, F, "Cake baking in tunnel type multi-zone industrial ovens Part I. Characterization of baking conditions." *Food Research International*, vol. 33, no. 7, pp. 587-598, Aug 2000.
57. Baik, O. D., Marcotte, M., & Castaigne, F, "Cake baking in tunnel type multi-zone industrial ovens Part II. Evaluation of quality parameters." *Food Research International*, vol. 33, no. 7, pp. 599-607, Aug 2000.
58. Hardenburger, P. A, "Conveyor oven." U.S. Patent No. 6,526,961, Mar 4 2003.
59. Shukla, K. C., Hurley, J. R., Orcheski, C. J., & Grimanis, M. P, "Conveyor oven." U.S. Patent No. 4,951,648, Aug 28 1990.
60. Smith, S. R, "Conveyorized baking oven with steam tunnel." U.S. Patent No. 8,881,646, Nov 11 2014.
61. Alden, L. B, "Infra-red baking oven." U.S. Patent No. 4,960,977, Oct 2 1990.
62. Lee, J. H, "Infrared radiation cooker." U.S. Patent No. 8,948,579, Feb 3 2015.
63. Westerberg, E. R., & Robert, I. B. I, "Visible light and infra-red cooking apparatus." U.S. Patent No. 5,517,005, May 14 1996.
64. Violi, R, "Forced convection gas oven." U.S. Patent No. 7,100,596, Sep 5 2006.
65. Onuschak, A. D, "Dual mode convection oven." U.S. Patent No. 6,116,895, Sep 12 2000.
66. Hwang, Y. I, "Overheated steam oven." U.S. Patent No. 7,143,761, Dec 5 2006.
67. Bonuso, J. L., & Swayne, S. M, "Priority controlled multi-fan convection oven." U.S. Patent No. 8,304,695, Nov 6 2012.
68. Blazevich, J. Z." U.S. Patent Application No. 12/365,137, Aug 5 2010.

69. Dobie, M. J., Norris, J. R., Cooper, N. S., Bacigalupe, C., & Foreman, R. W, "Multi-shelved convection microwave oven." U.S. Patent No. 7,087,872. Aug 8 2006.
70. Fang, L, "Microwave oven with rotary cooking apparatus." U.S. Patent No. 8,847,131, Sep 30 2014.
71. Lee, S. G, "Built-in microwave oven." U.S. Patent No. 6,528,774, Mar 4 2003.
72. Keskin, S. O., Sumnu, G., & Sahin, S, "Bread baking in halogen lamp–microwave combination oven." *Food Research International*, vol. 37, no. 5, pp. 489-495, Jun 2004.
73. Megahey, E. K., McMinn, W. A. M., & Magee, T. R. A, "Experimental study of microwave baking of Madeira cake batter." *Food and Bioproducts Processing*, vol. 83, no. 4, pp. 277-287, Dec 2005.
74. Lee, J. H, "Bidirectional heating cooker." U.S. Patent No. 8,761,588, Jun 24 2014.
75. Moreth, R. E, "High efficiency carousel infrared oven." U.S. Patent No. 6,250,210, Jun 26 2001.
76. Jovanovic, D, "Infrared apparatus for baking pastries and pizzas." U.S. Patent No. 5,378,872, Jan 3 1995.
77. Ploteau, J. P., Glouannec, P., Nicolas, V., & Magueresse, A, "Experimental investigation of French bread baking under conventional conditions or short infrared emitters." Applied Thermal Engineering, vol. 75, pp. 461-467, Jan 2015.
78. Sumnu, G., Sahin, S., & Sevimli, M, "Microwave, infrared and infrared-microwave combination baking of cakes." Journal of food engineering, vol. 71, no. 2, pp. 150-155, Nov 2005.

 DOI: 10.1051/matecconf/20178702018

# Burner Characteristics for Activated Carbon Production

Supaat zakaria[1,a], A.M. Leman[1], Dafit Feriyanto[1] ,Azmarini A. Nazri[2], N.M. Sunar[1] and Mohd Najib Mohd Salleh[3]

[1]*Faculty of Engineering Technology, Universiti Tun Hussein Onn Malaysia (UTHM) 86400, Parit Raja, Batu Pahat, Johor, Malaysia*
[2]*Department of Mechanical Engineering, Politeknik Ungku Omar, Jalan Raja Musa Mahadi, 31400 Ipoh, Perak, Malaysia*
[3]*Faculty of Computer Science and Information Technology, Universiti Tun Hussein Onn Malaysia 86400, Batu Pahat,Johor, Malaysia*

**Abstract.** Carbonization process has become an important stage in developing activated carbon. However, existing burner are not efficient in time production which take 24 hours to15 days for charcoal production. Therefore, new design of burner/kilns is quite needed in order to produce larger number of charcoal in short time production, to improve charcoal quality regarding to the smooth surface area and pore volume. This research proposed new design burner which divided into two types which are vertical and horizontal types. Vertical is not completed by auto-rotating system while horizontal type is complete by auto-rotating and fume handling system. It developed using several equipment such as welding, oxy-cutting, drilling grinding and cutting machine. From the result of carbonization process shows that coconut shell charcoal need shorter time of 30 minutes as compared to palm shell charcoal of 2 h to completely carbonized. This result claim that the new design better than existing kiln that need longer time up to 24 h. The result of the palm and coconut shell charcoal believe will produce better properties of activated carbon in large surface area and higher total volume of pores. Therefore, this burner is high recommended for producing palm and coconut shell charcoal as well as other bio-based material.

## 1 Introduction

Activated carbon (AC) is the carbon which activated using several methods like physical and chemical activation. There are many sources for AC development such as palm shell, coconut shell, woods, fish bones and bamboo. Prior to developing AC, raw materials need to be charcoaled and commonly called as carbonization process. Producing AC was conducted using several equipment such as using burner, drum kiln or conventional burning process. In the carbonization process, medium to high temperature has been investigated by previous researchers i.e. 500 to 1200 $^0$C with the total energy produced from the charcoal is listed in Table 1. According to Seifrtiz (1993) [1] that ±50% of burned biomass is extracted become charcoal form. Meanwhile, refer to Glaser (2002) [2] average 49.9% carbon yield which produced from various types of kilns including laboratory furnace. That phenomenon means that approximately of 50% from total carbon inside of raw material emitted to the air with half-carbonized matter form [2].

**Table 1.** Product of charcoal carbonization [3]

| Product of carbonization | % Dry base |
|---|---|
| Charcoal (80% fixed carbon) | 33 |
| Pyroligneous acid | 35.5 |
| Acetic acid | 0.5 |
| Methanol | 0.2 |
| Soluble tar | 5 |
| Water and other | 23.5 |
| Insoluble tar | 6.5 |
| Non condensable gases (HCG) | 25 |
| Hydrogen-0.63% | 0.16 |
| CO-34% | 8.5 |
| $CO_2$ – 62% | 15.5 |
| Methane – 2.43% | 0.61 |
| Ethane – 013% | 0.03 |
| Others – 0.81% | 0.2 |

[a] Corresponding author: paat77@gmail.com

There are many products resulted from the carbonization process as mentioned in Table 1. It respective to the energy loses and it is improved by increasing charcoal yield and make more efficient palm tree plantation in Malaysia since it very abundant (Figure 1) [4].

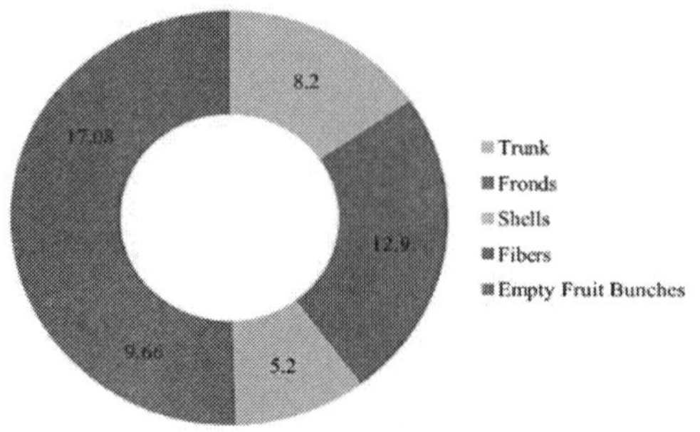


**Figure 1** Palm biomass production in Malaysia (million tons per year) [4]

DOI: 10.1051/matecconf/20178702018

According to previous researchers [5-11] they have produce activated carbon for air and water filter using laboratory burner and conventional burner which is not based on air flow, thermal flow calculation and not completed by fume handling system. Therefore, the conventional activated carbon is produced by high cost operation and longtime production as well as not environmental friendly. Some limitation of the existing burner that fine activated carbon size is dominated in this production and also high ash volume. Fine activated carbon is only suitable for lab scale and it caused by slower diffusion through particles and high pressure loss across the bed. High ash volume is indicated that the designed burner is not effective to control the temperature inside the burner, it will cause over heated and reduce the properties of the activated carbon. Therefore, this research is aimed to develop new burner for producing activated carbon where designed burner based on charcoal size before and after burned, air circulation, heat circulation and ash volume analysis.

## 2 Types and Characteristic of Kilns

There are several types of kilns which have been investigated by previous researchers. It has unique characteristics for burning and producing charcoal regarding to the time and temperature.

**Figure 2** Hot tail kilns

This kiln has maximum diameter, height and charcoal/load of 2.8 m, 2.8 m and 4 $m^3$, respectively. It is producing low yield and it also completed by smoke and tar emission. This kiln need 10 to 12 days for producing charcoal [3].

**Figure 3** Surface kilns

Surface kiln available for producing 20 $m^3$ charcoal/cycle, diameter of 5 m and need 10 to 12 days for producing charcoal [3].

**Figure 4** Rectangular kilns

Rectangular kiln has dimension of 13 x 4 x 3.5 m (l x w x h) with maximum capacity of 200-700 $m^3$ and this kiln need 15 days operation times for producing charcoal [3].

**Figure 5** Slope kilns

Work principle of slop kilns is similar with hot tail kilns with low cost and low yield not more than 25% [3].

DOI: 10.1051/matecconf/20178702018

**Figure 6** DPC process

This kiln is completed by gas flow system which used as thermal fluid and it need 60 h for each cycle [3].

**Figure 7** CML France

This kiln available for producing 2500 t/years with low yield of 20-25%. Every cycle, this kilns can prodding at least 2 $m^3$ which need 24 h for production time [3].

**Figure 8** Continuous retort

Continuous retort conduct the burning process by hot gases from combustion system. Raw material at condition approximately of 20% humidity and feeding material in the top of the kilns while discharging cold charcoal is located at the bottom with tar recovering. This kiln is available for producing charcoal of 2,000 to 10,000 t/year [3].

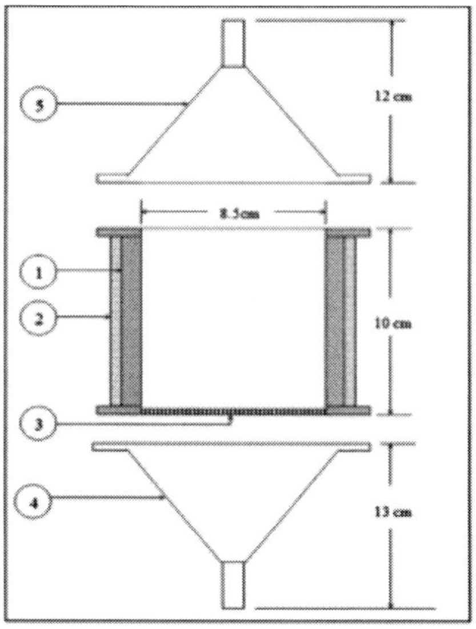


**Figure 9** The layout and dimensions of burner housing, inlet chamber and top cover. 1. Stainless steel body (thickness = 3mm), 2. Silica insulation (thickness = 5 mm), 3. Cast iron grill (plat form for porous layers), 4. Inlet chamber (stainless steel, diameter 8.5cm), 5. Top cover (for exhaust gas sampling) [12].

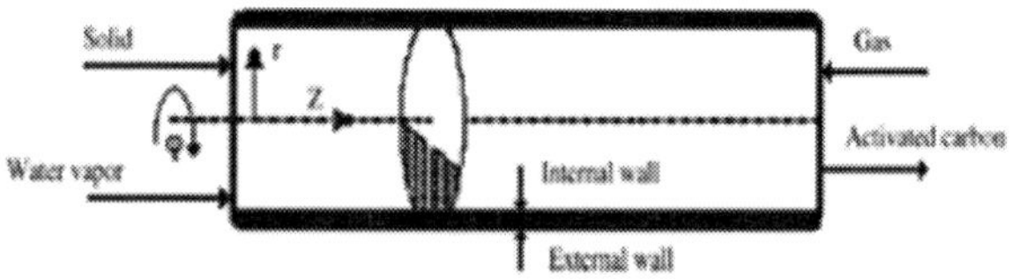


**Figure 10** A pilot rotary kiln [13]

Working principle of a pilot rotary kiln is it rotates in longitudinal axis and it operates as heat exchanger. System of this kiln i.e. hot gas is produced by combustion of natural gas which originated from central burner; supply the necessary energy for the activation reaction. In this system, water vapor is injected in concurrent mode as activation agent. The moisture content of raw material approximately of 5-10% and impurity content of the raw material is negligible. It is supposed that the solid bed moves as a pseudo fluid with axial displacement and without retro-mixing and it rolls or slides in transverse direction as the cylinder rotates [13].

**Table 2** Specification of pilot rotary kiln [13]

| | |
|---|---|
| Length | 3.7 m |
| Internal diameter | 0.3 m |
| External diameter | 0.6 m |
| Raw material | Solid carbonaceous material |
| Reaction temperature | 800 - 1000°C |
| Rotary kiln pressure | Atmospheric |
| Rotation speed | 1 to 3 rpm |
| Heating medium | Natural gas combustion |

 DOI: 10.1051/matecconf/20178702018

In real production of AC, the rotary kiln plan was able to convert 2000 kg/hr capacity of olive pits into char fated for the production of activated carbon in Italy. The pyrolysis process temperature between 50 - 750°C at fixed biomass with flow rate of 1700 kg/hr. operating temperature of 300 - 400 °C and higher operating temperature is 800 to 900 °C. The length of the kiln is 20 m and internal diameter is 1.6 m [14]. Rotary kiln is also used for recycling of waste composite material (thermoset – based polymer composites) collected from the products such as automobiles, wind turbines and aircrafts [15].

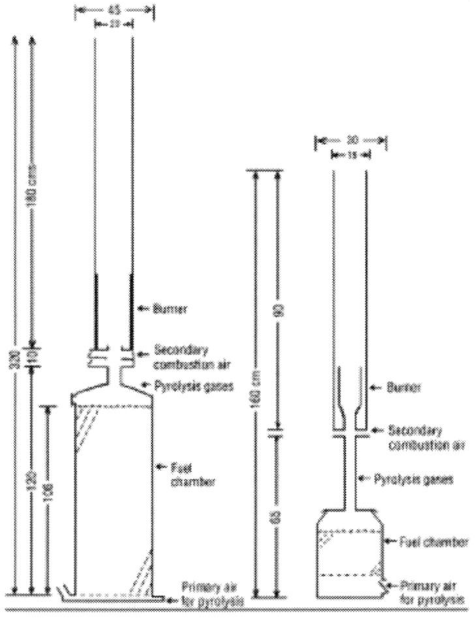


**Figure 11** Large pyrolyser/burner on the left and small pyrolyser/burner on the right (individually to scale) [16]

According to Figure 11 that those design have basic operation mode i.e.

1. Fuel chamber is filled with raw material.
2. Pellets are lit on top.
3. Pellets are allowed to burn for a few minutes until pyrolysis has moved down into the fuel and spread over the entire top surface of the fuel.
4. Gas and air mixture is lit in the burner.
5. Flame is sustained until fuel is completely turned to charcoal.
6. Fuel chamber is sealed and allowed to cool.
7. Charcoal is removed.

There are differences between small and large burner regarding to the fuel chamber and the burner. Both burners are ignited by pilot flames up to gas production can support sell-ignition in the combustion chamber. Primary air supply into the bottom of the fuel chamber is left fully open at the beginning. Primary air was not adjusted for large burner and it different with small burner. Small burner need operator to control the primary air to moderate burning gas production while secondary air was adjusted for both of burner. In small burning process, the operator adjust air supply level up to appropriate results is achieved which based on stack gases, visual and auditory characteristics of the flame. Testing process of large burner controlled by plug as the function to stop the air inlet to the fuel chamber prior to reaction zone is reached by fuel column bottom in order to allowing an inspection of reaction zone.

## 3 Proposed Kilns

There are several types of the proposed kilns such as vertical kilns and horizontal kilns. It developed using mild steel hollow, mild steel plate and mild steel angle bar and it fabricated using several equipment such as welding, oxy-cutting, drilling grinding and cutting machine. New burner is completed by several equations regarding to the air and thermal flow analysis.

**Volume calculation**

$$Air\ Change\,/\,hour = \frac{CFMx60}{Cubicfeet}$$
$$CFM = Room\ Volume\ x\,\frac{Air\ Change\,/\,hour}{60} \tag{1}$$

**Air flow formula**

$$CFM = C_V x \sqrt{\Delta P(inches)H_2O}$$
$$C_V = \sqrt[3]{\frac{CFM}{\Delta P(inches)H_2O}} \tag{2}$$

Where;
$CFM$ = Duct area sq. ft. x Velocity
$\Delta P$ = Pressure in - pressure out

**Air thermal calculation**

$$BTUH = M\ x\ sp.ht\ x\ \Delta T \tag{3}$$

Where;
BTUH = Air thermal
M = Mass flow rate (Lbs./h)
Sp.ht = Specific heat (BTU/Lb./Degree)

Proposed burner is designed with environmentally concern in air pollution. Therefore, it completed by fume handling system. Dimension of vertical kilns is 15 cm and 120 cm for diameter and height. It completed by several holes in the bottom side purposed to air circulation and discharge hole for cold charcoal. Meanwhile, horizontal burner is completed by automatic rotating system by motor for every 2 minutes in $45^0$ rotating. Fume handling also installed into the kilns which operated when fume is going out from the hollow steel and discharged into the water in order to remove fumes by mixing into the water. Vertical and horizontal kilns is shown in Figure 12 and 13.

**Figure 12** Vertical kilns

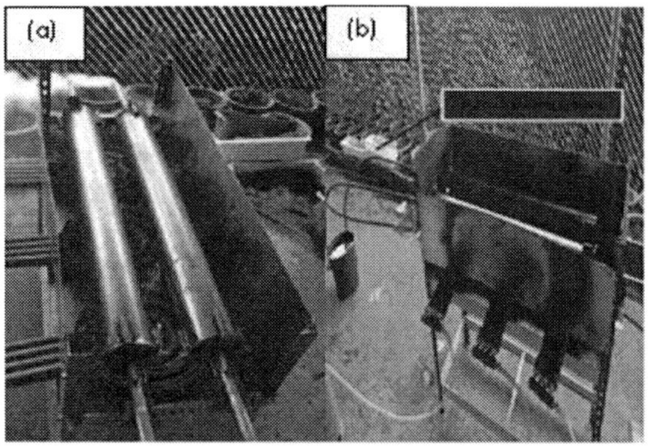


**Figure 13** Horizontal kilns (a) without fume handling system and (b) with fume handling system

## 4 Expected Result of Palm Shell Charcoal

Capacity charcoal production of this burner is 20 kg/cycle and 2 h/cycle for palm shell charcoal as well as 30 minutes/cycle for coconut shell charcoal. Temperature 300 – 600 $^{0}$C controlled by gas outlet rotating valve. Palm shell and coconut shell charcoal is shown in Figure 14.

**Figure 14** (a) Coconut shell charcoal and (b) Palm shell charcoal

Shorter time of the burning process as compared to existing kilns indicated that this kiln more efficiency regarding in time production. It also effective for handling fume to secure the environment in healthy level. Potential results of this charcoal refer to Rahman et al., (2012) [17] based on the surface area and pore diameter are shown in Figure 15 and Table 3. The charcoal is activated using phosphoric acid with the percentage 10, 20 and 30% in order to improve surface area and improve total pore volume. Reaction during carbonization can effect to the pore distribution and rough surface will effect to the lack pores development [17].

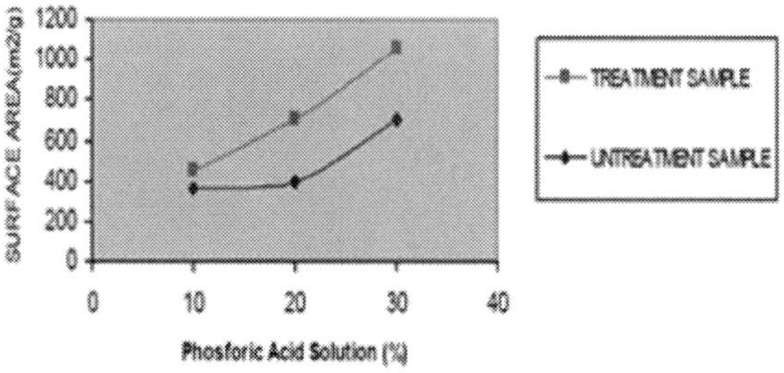


**Figure 15** The surface area of treated and untreated activated carbon [17]

**Table 3**. Effect of untreated, treated activated carbon on pore diameter and pore volume with different phosphoric concentration [17]

| Sample name | Average pore diameter($10^{-5}$ m) | Total pore volume (cc/g) |
|---|---|---|
| U10 | 2.137E+01 | 1.961E-01 |
| U20 | 2.126E+01 | 2.129E-01 |
| U30 | 2.063E+01 | 3.611E-01 |
| T10 | 2.147E+01 | 2.445E-01 |
| T20 | 2.143E+01 | 3.758E-01 |
| T30 | 2.064E+01 | 5.457E-01 |

Optimum surface area of untreated activated carbon is 700.1 $m^2g^{-1}$ while for treated activated carbon is 1058 $m^2g^{-1}$. It means that the treatment or activation process may increase the surface area and adsorption capacity of activated carbon.

## 5 Conclusion

Carbonization process for activated carbon production is conducted with new proposed design completed by auto-rotating and fume handling system. This kiln need shorter carbonization time of 30 minutes for coconut shell charcoal and 2 h for palm shell charcoal. Meanwhile existing burner/kilns need longer carbonization up to 24 h. Charcoal with shorter carbonization time indicated that it have homogeneity thermal distribution and it can effect to the increment of pore development and higher surface area which led to higher adsorption capacity of produced activated carbon

## Acknowledgement

The authors would like to thank the Ministry of Higher Education Malaysia and Universiti Tun Hussein Onn Malaysia (UTHM) through the funding supported ICI grant under No Vot. I002 and Centre for Graduate Studies – UTHM.

## References

1. W. Seifrits. Should we store carbon in charcoal. Int. J. Hydrogen Energy, 18, 5: 405-407 (1993).
2. B. Glaser, J. Lehmann and W. Wolfgang Zech. Ameliorating physical and chemical properties of highly weathered soils in the tropics with charcoal – a review. Biol Fertil Soils, 35, 219–230 (2002).
3. O.V. Adriana, S.L. Electo, R.Q. Quelbis, A.V. Ricardo and P.S.S. Thalis. A new technology for the combined production of charcoal and electricity through cogeneration. Biomass and bio energy, 69, 222–240 (2014).
4. S.E. Hosseini, and M. Abdul Wahid. Utilization of palm solid residue as a source of renewable and sustainable energy in Malaysia. Renewable and Sustainable Energy Reviews, 40, 621–632 (2014).
5. T.A. Metts and S.A. Batterman. Effect of VOC loading on the ozone removal efficiency of activated carbon filter. Chemosphere, 62, 34–44 (2006).
6. M.A. Sidheswaran, H. Destaillats, D.P. Sullivan, S. Cohn and W.J. Fisk. Energy efficient indoor VOC air cleaning with activated carbon fiber (ACF) filter. Building and Environment, 47, 357-367 (2012).
7. E. Gallego, F.J. Roca, J.F. Perales and X. Guardino. Experimental evaluation of VOC removal efficiency of a coconut shell activated carbon filter for indoor air quality enhancement. Building and Environment, 67, 14-25 (2013).
8. N.A. Fuadi, A.S. Ibrahim and K.N. Ismail. Review study for activated carbon from palm shell used for treatment of waste water, (2008).
9. S. Wu, B. Dong, T. Qiao and J. Zhang. Effect of a biological activated carbon filter on particle counts. J Zhejiang Univ Sci A. 9, 1,1576-1581 (2008).
10. I. Adedayo, A.O. Inegbenebor and H.I. Boyo. Comparison of the Adsorptive Capacity of Raw Materials in Making Activated Carbon Filter for Purification of Polluted Water for Drinking. ARPN Journal of Science and Technology, 2, 9, ISSN 2225-7217 (2012).
11. A.A. Ismaiel, M.K. Aroua and R. Yusoff. Palm shell activated carbon impregnated with task-specific ionic-liquids as a novel adsorbent for the removal of mercury from contaminated water. Chemical Engineering Journal, 225, 306–314 (2013).
12. M.A. Mujeebu, M.Z. Abdullah and A.A. Mohamad. Development of energy efficient porous medium burners on surface and submerged combustion modes. Energy, 36, 5132–5139 (2011)
13. O.A. Ortiz, S.E. Noriega, C.A. Mengual and N.D. Martínez. Steady State Simulation of a Rotary Kiln for Charcoal Activation. Latin American Applied Research, 33, 51-57 (2003)
14. E. Benanti, C. Freda, V. Lorefice, G. Braccio, and V.K. VinoSharma. Simulation of Olive Pits Pyrolysis in a Rotary Kiln Plant. Thermal Science, 15, 1, 145–158 (2011).
15. Y. Yang, R. Boom, B. Brijan Irion, D.J. Heerden, P. Kuiper and W. Hans. Recycling of composite materials. Chem. Eng. Process, (2011).
16. Alex English. Preliminary tests on charcoal making pyrolysis gas burners – high performance combustion with a simple technology. Energy for Sustainable Development, 6, 2. (2002).
17. M.M. Rahman, M. Awang, B.S. Mohosina, W.B. Kamaruzzaman, Wan Nik and C.M.C. Adnan. Waste Palm Shell Converted to High Efficient Activated Carbon by Chemical Activation Method and Its Adsorption Capacity Tested by Water Filtration. APCBEE Procedia, 1, 293 – 298 (2012).

 DOI: 10.1051/matecconf/20178702019

# Diffusion and Bonding Mechanism of Protective $\gamma$-$Al_2O_3$ on FeCrAl Foil for Metallic Three-Way Catalytic Converter

Dafit Feriyanto[1,a], A.M. Leman[1], Supaat zakaria[1], I. Baba[1], Fakhrurrazi Rahman[1] and F. Afiqah Jajuli[1]

[1]*Faculty of Engineering Technology, Universiti Tun Hussein Onn Malaysia (UTHM) 86400, Parit Raja, Batu Pahat, Johor, Malaysia*

**Abstract.** High pollutant level contributed by mobile sources/land transportation that become main problems for the human health. Improving exhaust emission system by improving catalytic converter properties is one of the most effective way to produce healthy air in our environment. It is conducted by two methods i.e. ultrasonic during electroplating (UBDEL) and electroplating process (EL) which are not fully investigated yet as catalytic converter coating process. UBDEL is conducted using sulphamate types electrolyte solution, Frequency of 35 kHz, current of 1.28A, Voltage of 12 V, and various time of 15, 30, 45, 60 and 75 minutes. Meanwhile El method is conducted using parameters of current of 1.28A, Voltage of 12 V, stirrer speed of 60 rpm and various time of 15, 30, 45, 60 and 75 minutes. Fully $\gamma$-$Al_2O_3$ bonding to the FeCrAl substrate is shown by UBDEL 75 minutes samples proved by SEM images and Ra and Rq are 4.01 μm and 5.64 μm, respectively. Ni present on the FeCrAl substrate as other protective layer generated by Ni electroplating process that will improve thermal stability of FeCrAl at high temperature of 1000 $^0$C. From the results, can summarized that UBDEL technique is promoted as an effective catalytic converter coating technique.

## 1 Introduction

The first strict regulation to control automobile emissions was introduced through a US Clean Air Act, 1970, in the United States of America. In addition, Mobile sources contribute for about 44% of outdoor toxic emissions, approximately 50% of cancer risk and at around 74% of non-cancer risk health problems. The technology for removing pollutants in the exhaust system is catalytic converter that first developed in 1965 [1, 2]. The location of the catalytic converter in the car is shown in Figure 1.

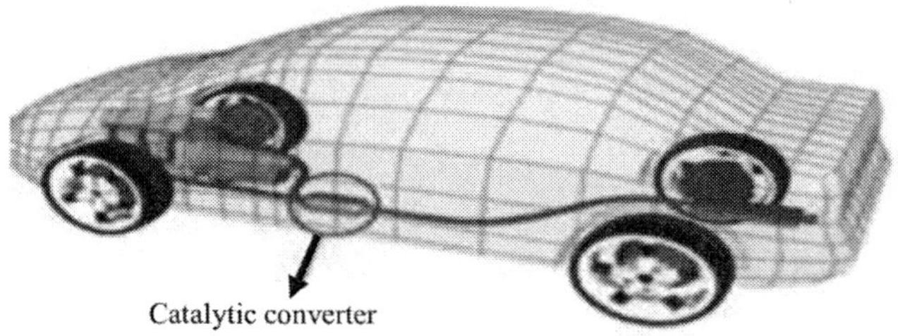


**Figure 1.** Location of catalytic converter in passenger car [3]

Nowadays, the metallic catalytic converter of FeCrAl was interesting component to explore due to their advantages in the high-temperature corrosion resistance, including strong adherence of oxide film on the surface of substrate [4, 5]. FeCrAl is consists of 5-8 wt% aluminium, 17-22 wt% chromium and reactive element to improve oxide adhesion and oxidation resistance [6]. Command washcoat material such as $Al_2O_3$, $SiO_2$, $TiO_2$, or $SiO_2$- $Al_2O_3$ [7]. However, $\gamma$-$Al_2O_3$ washcoat material not fully explored even though it promising several advantages such as low price, high surface area, good chemical stability, high micro-hardness, and wear resistance at high-temperature [8]. Therefore, in this research the FeCrAl is coated by $\gamma$-$Al_2O_3$ powder which can result higher surface area of about 100 $m^2/g$.

There are several method for coating FeCrAl substrate such as dip coatings, which combined with some pre-treatments, such as growing a number of textured alumina whiskers on the surface of the metal support and shortened the diffusion path before depositing the washcoat [9, 10]. The other methods are co-precipitation, sol-gel and spray-pyrolysis methods [11], electrophoretic deposition [12], solution combustion synthesis (SCS) [13], aluminizing technique [14], and hydrothermal method [15, 16]. That method presented good adherence to the substrate, decreasing surface roughness and closed

[a] Corresponding author: dafitferiyanto@yahoo.co.id

the surface porosity of the substrate [17]. However, the existing methods still have some limitations, especially due to the rather complicated method to applying the catalyst which is in the form of powder.

Material coated by $Al_2O_3$ promote high surface roughness which important to the filtration and catalyst application [17]. Roughness of the material indicates the surface topography which have several forms such as digs, pits, dust particle, scratches, machining marks on machine surface, polishing marks on optical surface, granularity, crystallite in film deposited in surface and electro-polishing [18]. According to Putrasari et al., (2010) [19] that surface roughness of FeCrAl coated by SiC and $Al_2O_3$ is successfully developed using ultrasonic technique for 10 min in frequency of 35 kHz and it improve high thermal stability of FeCrAl substrate and homogeneity of Ni electroplating layer. However, it still need improvement and special treatments. In other hand, diffusion of aluminium alloy (FeCrAl) is very challenging to investigate since some phenomena occurred. Less protective $\gamma$-$Al_2O_3$ powder is caused by slow diffusion of Al alloy to the metal oxide interface to compensate for the aluminum consumption. Oxidation behavior of the aluminium alloy is measure by diffusion coefficient [20]. The growth mechanism occurred by both cation and anion transport in the $Al_2O_3$ scales on Fe-Cr-Al. The alumina scale grows exclusively by oxygen diffusion along grain boundaries. Diffusion results were also obtained in alumina developed on $\gamma$-NiAl by cationic and anionic self diffusion tests and also by an electrochemical method [21].

Therefore, the diffusion phenomena of the coated FeCrAl and its surface roughness treated by electroplating approach is very challenging to investigated. It believe that electroplating approach to promote fully embedded of $\gamma$-$Al_2O_3$ powder to FeCrAl substrate that figured by SEM/EDS and surface roughness testing.

## 2 Methodology

FeCrAl foils was treated using various novel Nickel electroplating technique i.e. electroplating ultrasonic bath during electroplating technique. It coated by $\gamma$-$Al_2O_3$ powder with crystallite size of 121.68 μm which measured by X-Ray Diffraction (XRD) machine. In this research, electrolyte used in electroplating process is called by sulphamate types which consist of nickel (II) sulphamate 6-hydrate ($NiSO_4.6H_2O$), nickel (ii) chloride ($NiC_{l2}{\cdot}6H_2O$), boric acid ($H_3BO_3$), and sodium dodecyl sulfate ($C_{12}H_{25}OSO_3.Na$) which believed that it can optimize the electroplating process.

### 2.1 Nickel (Ni) electroplating process (EL sample)

Nickel electroplating was carried out using some equipments and components such as DC power supply, anode (Ni plate), cathode (FeCrAl foil), electrolyte solution, magnetic stirrer and hot plate for adjust temperature and stir speed as well as electronic microbalance. Ni plate and FeCrAl foil is cropped at the fix size of 50 mm x 10 mm and 40 mm x 20 mm respectively. Preparation of electrolyte solution with the composition is shown in Table 1.

**Table 1.** Composition of electrolyte solution

| Electrolyte solution | Weight (gr) | Composition (%) |
|---|---|---|
| $NiSO_4.6H_2O$ | 10.2 | 51.28 |
| $NiC_{l2}{\cdot}6H_2O$ | 0.2 | 0.85 |
| $H_3BO_3$ | 1.2 | 5.12 |
| $C_{12}H_{25}OSO_3.Na$ | 8.4 | 42.73 |

Electrolyte solution is prepared with distilled water at constant temperature of 60 $^0$C, distilled water volume of 0.25 L, holding time approximately of 10 minutes and pH value is adjusted of 2.5-5. Electroplating process was conducted using various times of 15, 30, 45, 60 and 75 minutes, current of 1.28 A, voltage of 12 V, distance between anode and cathode of 25 mm, mass of $\gamma$-$Al_2O_3$ powder of 5 g and total surface area of 1600 mm$^2$ in two sides. The schematic diagram of electroplating process is shown in Figure 2.

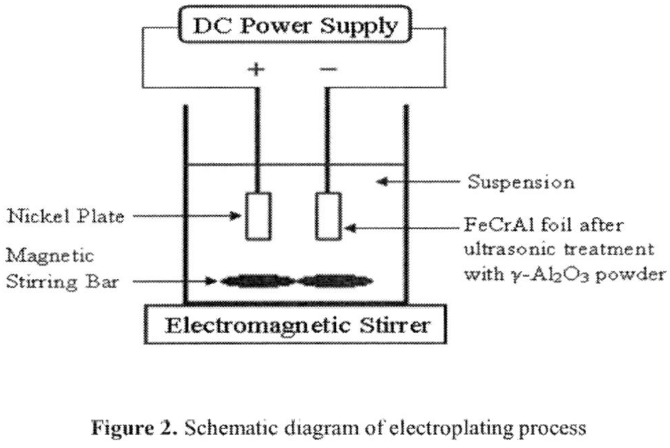


**Figure 2.** Schematic diagram of electroplating process

 DOI: 10.1051/matecconf/20178702019

### 2.2 Ultrasonic bath during electroplating technique (UBDEL)

Basic principle of UBDEL is similar with electroplating technique. However, in this technique have a unique methods where transducer as frequency sources is generated in electroplating technique as replacement of magnetic stirrer. Electrolyte solution is prepared with distilled water at constant temperature of 60 $^0$C, distilled water volume of 0.25 L, holding time approximately of 10 minutes and pH value is adjusted of 2.5-5. UBDEL technique was carried out using various times of 15, 30, 45, 60 and 75 minutes, frequency of 35 kHz, current of 1.28 A, voltage of 12 V, distance between anode and cathode of 25 mm, mass of $\gamma$- $Al_2O_3$ powder of 5 gr. The schematic diagram of UBDEL technique is shown in Figure 3. The specification of ultrasonic machine is listed in Table 2.

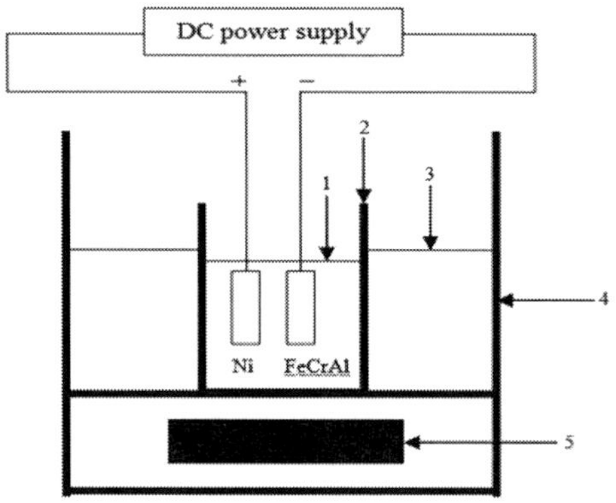


**Figure 3.** Schematic diagram of ultrasonic bath during electroplating where (1) Electrolyte solution; (2) Beaker; (3) Water; (4) ultrasonic tank; (5) Ultrasonic source

**Table 2.** Technical specification of ultrasonic bath machine

| Description | Nominal |
|---|---|
| Bath dimension | Ø 24.5 cm and 13 cm deep |
| Effective capacity | 5.6 litters |
| Insert tray | Ø 22.5, 11.5 cm deep with mesh 10 x 10 mm |
| Sound level | 2 x 240 Watt/ period or 35 kHz |
| Power input | 140 Watt |
| Weight | Net 5.5 kg, and gross 8 kg |
| Dimension w x d x h | 27 x 27 x 27 cm |

All the samples are characterized using Scanning Electron Microscopy (SEM) which is coupled by Energy-Dispersive Spectroscopy (EDS) and surface roughness machine. SEM is generated using magnification of 300 times while Surface roughness machine is arranged by $\lambda_c$ of 0.8 mm, $\lambda_s$ of 2.5 µm, eva-L of 4 mm and Roughness range of 800. Surface roughness and surface roughness mean squared could be measured manually using the Eq. 1 and Eq 2 respectively [22].

$$R_a = \frac{1}{n}\sum_{i=1}^{n}|y_i| \quad (1)$$

$$R_q = \sqrt{\frac{1}{n}\sum_{i=1}^{n} y_i^2} \quad (2)$$

Where;
Ra is surface roughness, Rq is surface roughness mean squared, n is equally spaced points along the trace, yi is is the vertical distance from the mean line to the $i^{th}$ data point.

## 3 RESULT AND DISCUSSION

### 3.1 Influence of electroplating approach to the coating surface and composition

Figure 4 to 14 show the surface morphology of coated and uncoated FeCrAl by $\gamma$-$Al_2O_3$ powder using ultrasonic during electroplating and electroplating technique. Figure 4 shows the uncoated sample with even surface that consist of C, O, Al, Cr and Fe. Figure 5 to Figure 9 show the samples that treated by ultrasonic during electroplating with the various times of 15, 30, 45, 60 and 75 minutes, respectively. There are many unrelated chemical is detected in those samples such as Sodium (Na) for 6-12 %mass and Sulfur (S) for 0-6%mass because the electrolyte used in this process is contain Na and S. Therefore, there is bonding activity between those chemical and coated FeCrAl. Nickel (Ni) present 11-32 % mass in the coated samples because Ni plate is performed during this process. From the UBDEL samples can be seen that the bonding activity for surface $\gamma$-$Al_2O_3$ powder and electrolyte chemical is successfully achieved. Fully embedded of $\gamma$-$Al_2O_3$ in line with the time increment. It approved by surface roughness analysis that the UBDEL 75 minute samples has highest surface roughness for 4.01 µm as compared to other UBDEL samples. The full bonding phenomena of UBDEL samples proof that the jet flush energy or high speed bubbles promote to the $\gamma$-$Al_2O_3$ powder to bond on the FeCrAl substrate [19]. Fully bonding activity is related to the high diffusion of $\gamma$-$Al_2O_3$ to the FeCrAl substrate since there is cation and anion transport which promote higher diffusion coefficient [20].

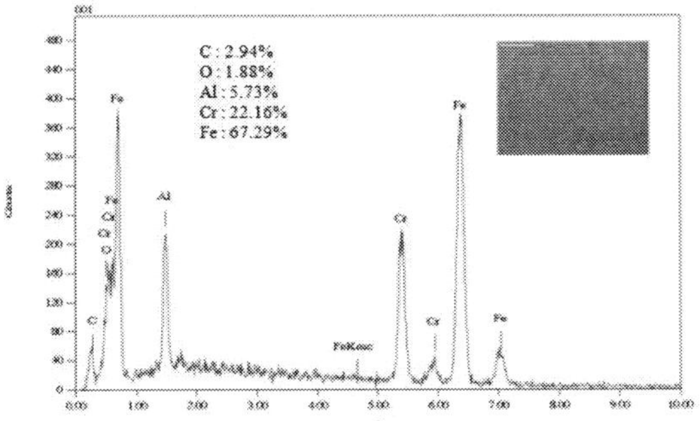


**Figure 4.** SEM & EDS image for Raw material

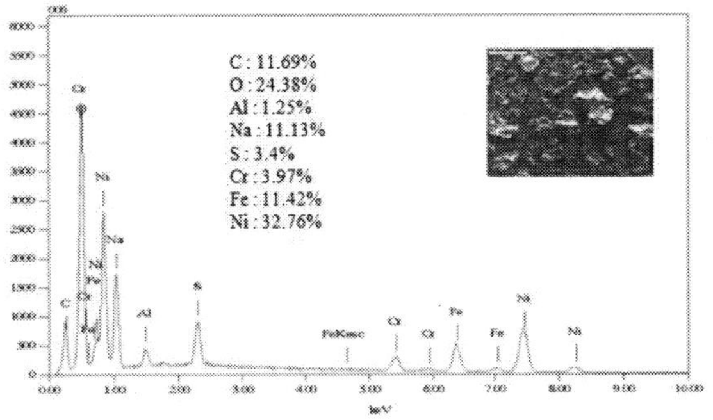


**Figure 5.** SEM & EDS for UBDEL 15 min

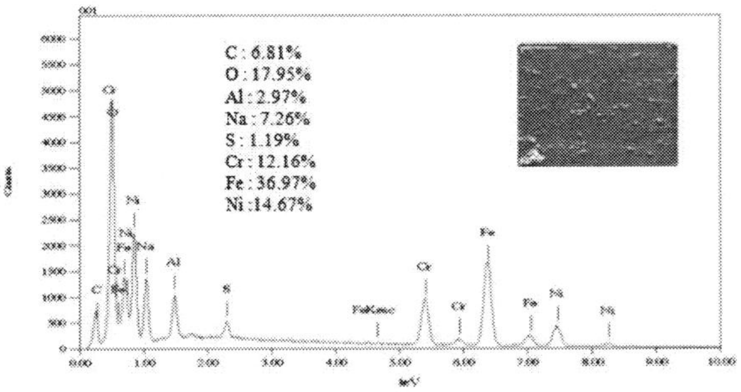


**Figure 6.** SEM & EDS for UBDEL 30 min

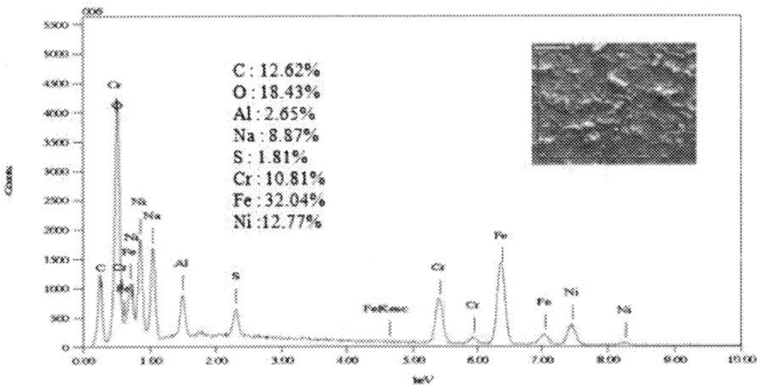


**Figure 7.** SEM & EDS for UBDEL 45 min

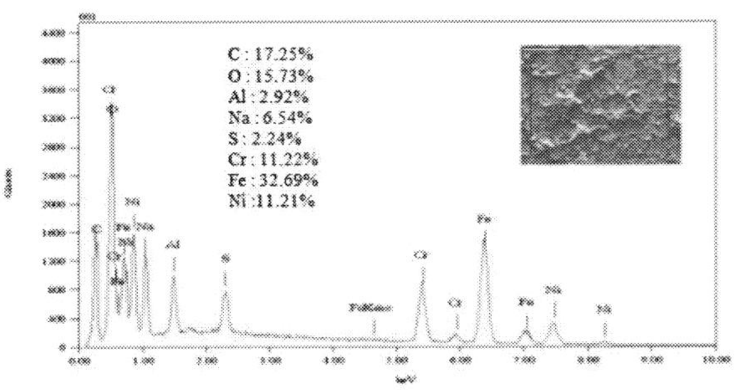


**Figure 8.** SEM & EDS for UBDEL 60 min

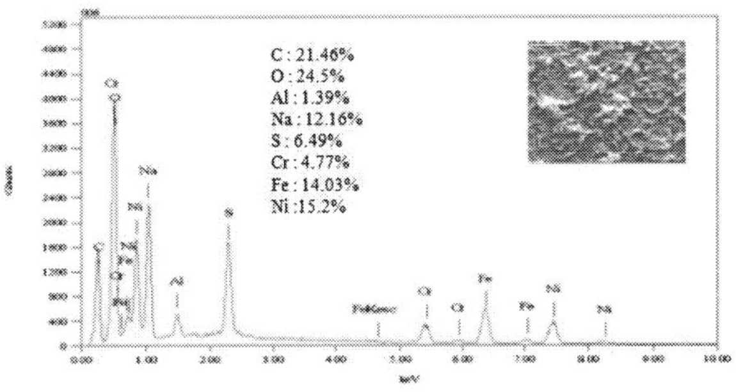


**Figure 9.** SEM & EDS for UBDEL 75 min

Figure 10 to Figure 14 shows the EL samples with the various electroplating times of 15, 30, 45, 60 and 75 minutes, respectively. The influence of electrolyte to electroplating process is shown in EDS analysis that the Na (2-13 %mass) and S (0-4 %mass) chemical present on coated FeCrAl surface. It will play the important role when the coated material is performed in high temperature of 1000 $^0$C. The $\gamma$-$Al_2O_3$ as main coated material is increased with time performed increased as well. It approved by surface roughness of 1.06 μm as the highest number as compared to other EL samples.

Besides that Ni has observed in the EDS analysis of coated FeCrAl for 6-25 %mass. It is promote the protection at high temperature [23] because appear intensive co-precipitated NiO-$Al_2O_3$ system [24]. Interaction between $\gamma$-$Al_2O_3$ and FeCrAl substrate is related to the diffusion activity where the alumina scale grows exclusively by oxygen diffusion along grain boundaries. Higher Al content and Oxygen content will promote higher diffusion phenomena of the samples [20].

DOI: 10.1051/matecconf/20178702019

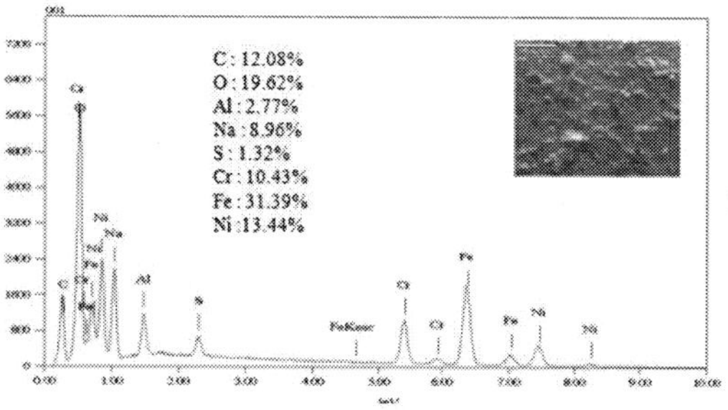


**Figure 10.** SEM & EDS for EL 15 min

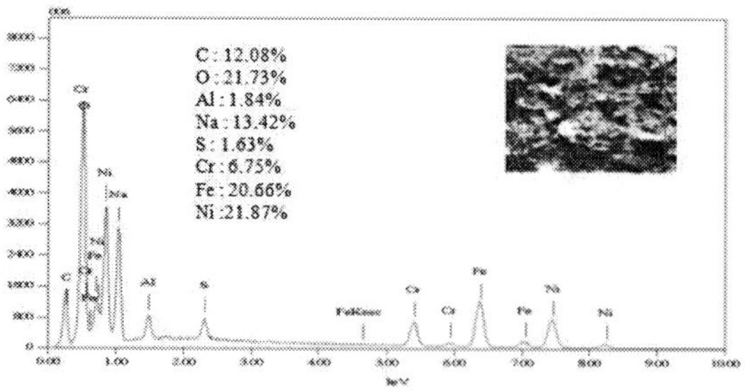


**Figure 11.** SEM & EDS for EL 30 min

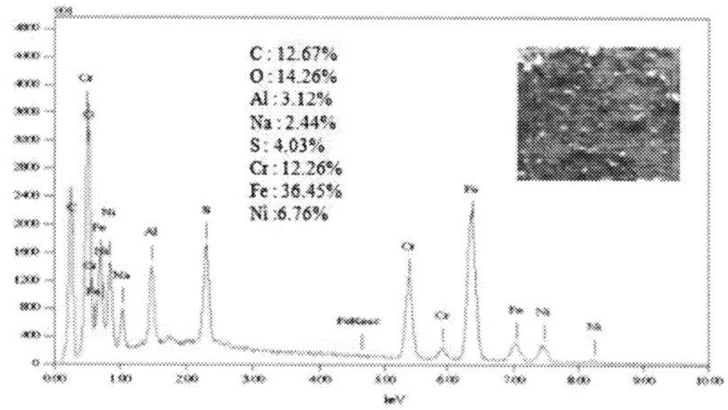


**Figure 12.** SEM & EDS for EL 45 min

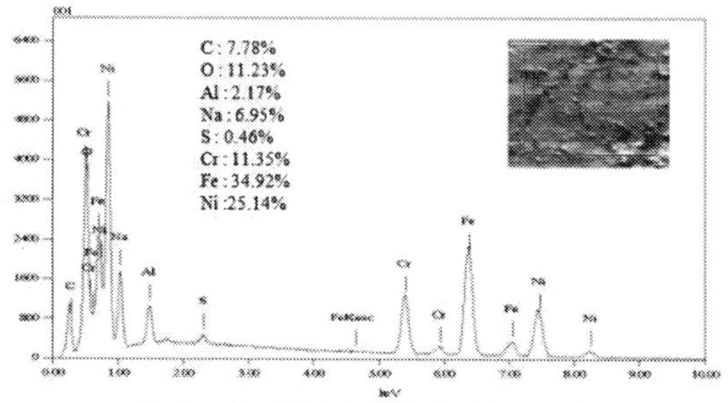


**Figure 13.** SEM & EDS for EL 60 min

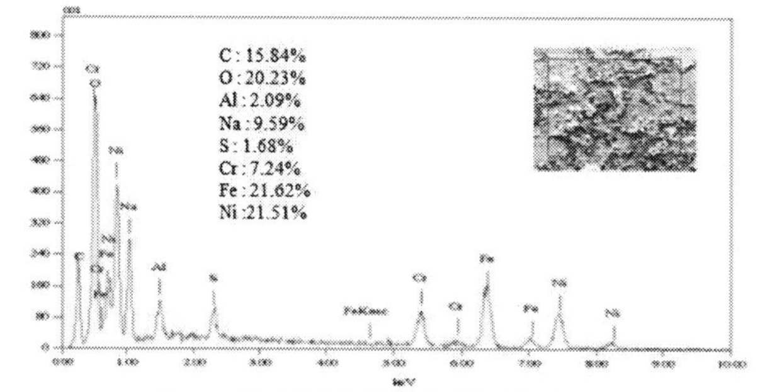


**Figure 14.** SEM & EDS for EL 75 min

## 3.2 Surface roughness (Ra) and surface roughness mean squared (Rq) analysis

Figure 15 shows the surface roughness (Ra) of coated and uncoated samples which called by raw material, UBDEL samples and EL samples. Those samples have different Ra influenced by different treatment times of 15, 30, 45, 60 and 75 minutes. surface roughness for sample raw material is 0.02 μm, for the UBDEL samples shown by Figure 15 b to f i.e. 3.35 μm, 1.22 μm, 2.04 μm, 1.59 μm and 4.01 μm, respectively. Surface roughness of El samples shown in Figure 15 g to k i.e. 1.04 μm, 0.7 μm, 0.92 μm, 1.05 μm and 1.06 μm respectively. The fluctuate Ra value by different treatment times is caused by many several factors such as electrolyte chemical bonding, unstable atmospheric environment and diffusion activity of $\gamma$-$Al_2O_3$ into FeCrAl during treatment.

a

b

c

d

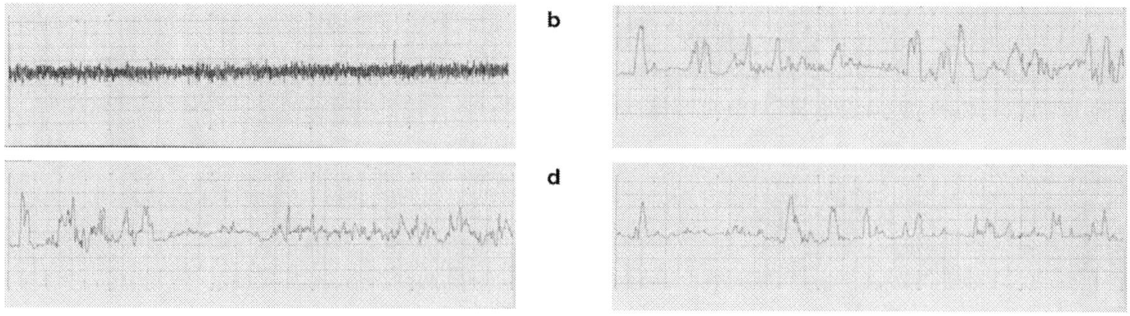

 DOI: 10.1051/matecconf/20178702019

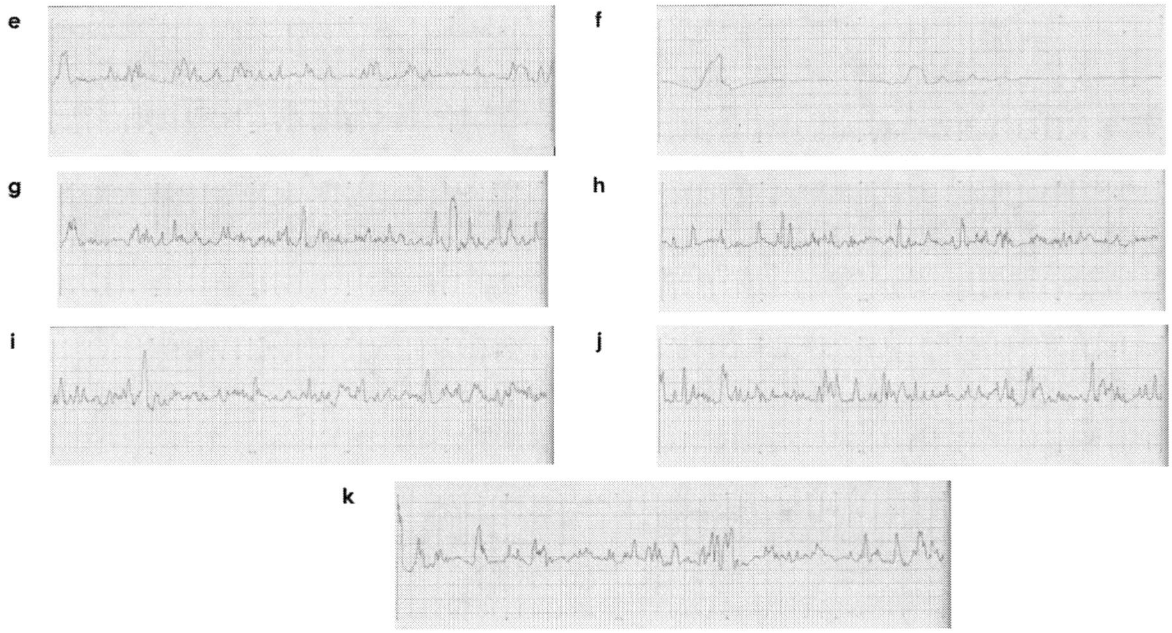


**Figure 15.** Surface roughness of a: Raw material, b:UBDEL 15 min, c: UBDEL 30 min, d: UBDEL 45 min, e: UBDEL 60 min, f: UBDEL 75 min, g: EL 15 min, h: EL 30 min, i: EL 45 min, j: EL 60 min and k: EL 75 min

The other parameter such as surface roughness mean squared (Rq) also presented by Figure 15 a to k i.e. 0.03 µm, 4.38 µm, 1.57µm, 2.87 µm, 2.05 µm, 5.64 µm, 1.46 µm, 0.94 µm, 1.25 µm, 1.42 µm and 1.46 µm, respectively. Lowest Ra and Rq are shown by raw material for 0.02 µm and 0.03 µm since there is no coating activity for the FeCrAl substrate. The highest Ra and Rq are shown by UBDEL 75 minutes for 4.01 µm and 5.64 µm. It shown the higher number as compared to other UBDEL samples and EL samples. It may caused by the combination between electrical activity sourced by DC power supply and ultrasonic frequency which led to effective coating activity. High speed bubbles combined by cathode-anode interaction may promote better bonding and diffusion process of $\gamma$-$Al_2O_3$ and FeCrAl substrate.

## 4 Conclusion

Coating process of FeCrAl substrate by $\gamma$-$Al_2O_3$ powder is successfully conducted using ultrasonic during electroplating and electroplating process. There are several phenomena tha has been observed such as:

1. Fully bonding of $\gamma$-$Al_2O_3$ into FeCrAl is shown by UBDEL 75 minutes sample where increasing holding time will increase bonding activity of washcoat material to the substrate material.
2. Surface roughness of the samples is defined by how many particle of powder and chemical agent which ebbed into the substrate. surface roughness of the EL samples is lower than UBDEL samples where highest Ra and Rq are 4.01 µm and 5.64 µm.
3. There are several chemical observed by EDS analysis such as O, C, Ni and Na which promote developing NiO- $\gamma$-$Al_2O_3$, C- $\gamma$-$Al_2O_3$ and NaO-$\gamma$-$Al_2O_3$ form. That form believe that it can be increased thermal stability of developed substrate.

## Acknowledgment

The authors would like to thank the Ministry of Higher Education Malaysia and Universiti Tun Hussein Onn Malaysia (UTHM) through the funding supported FRGS grant under No Vot. 1216 and Centre for Graduate Studies – UTHM.

## References

1. R.M. Heck, and R. J. Farrauto, Automobile Exhaust Catalyst. Applied Catalysis. A: General, **221**, 443-457 (2001).
2. S. Darwin, U. Puji, R. Hamimah Abd. and H. Shahrin. Conceptual Design of Catalytic Converter. International Advanced Technology Congress 2005, Kuala lumpur, Malaysia (2005).
3. J.M. Pardiwala, F. Patel and S. Patel. Review paper on Catalytic Converter for Automotive Exhaust Emission. International Conference On Current Trends In Technology, NUiCONE, 382-481 (2011).
4. D. Pilone. Ferritic Stainless Steels for High Temperature Applications in Oxidizing Environments. Recent Patents on Materials Science, **2**, 1, 27-31, ISSN 1874-4648, (2009).
5. T. Amano, Y. Takezawa, A. Shiino and T. Shishido. Surface Morphology of Scale on FeCrAl (Pd, Pt, Y) alloys. Journal of Alloys and Compounds, **452**, 1, 16-22 (2008)

6. J.R. Nicholls and W.J. Quadakkers. Materials Issues Relevant to the Development of Future Metal Foil Automotive Cataltic Converters. In: Materials Aspects in Automotive Catalytic Converters, Hans Bode (Ed.), 31-48 (2002).
7. R.M. Heck, R.J. Farrauto and S.T. Gulati. Catalytic Air Pollution Control Commercial Technology, 3rd ed, John Wiley & Sons, New Jersey, ISBN 978 -0-470-27503-0 (2002).
8. N.A. Badarulzaman, S. Purwadaria, A.A. Mohamad, & Z.A. Ahmad. The Production of Nickel–Alumina Composite Coating via Electroplating. Ionics, **15**, pp. 603-607 (2009).
9. S. Zhao, J. Zhang, D. Weng, and X. Wu. A Method to Form Well-Adhered γ- $Al_2O_3$Layers on FeCrAl Metallic Supports. Surface and Coating Technology, **167**, 1, 97-105 (2003).
10. L. Jia, M. Shen and J. Wang. Preparation and Characterization of Dip-Coated γ-Alumina Based Ceramic Materials on FeCrAl foils. Surface & Coatings Technology, **201**, 16-17, 7159–7165 (2007).
11. Z. Yanqing, X. Jieming, L. Cuiqing, X. Xin and L. Guohua. Influence of Preparation Method on Performance of a Metal Supported Perovskite Catalyst for Combustion of Methane. Journal of Rare Earths, **28**, 1, 54-58 (2010).
12. I. Corni, M.P. Ryan, and A.R. Boccaccini. Electrophoretic deposition: From Traditional Ceramics to Nanotechnology. Journal of the European Ceramic Society, **28,** 7, 1353–1367 (2008).
13. A. Specchia, A. Civera, and G. Saracco. In-Situ Combustion Synthesis of Perovskite Catalysts for Efficient and Clean Methane Premixed Metal Burner. Chemical Engineering Science, **59,** 22-23, 5091-5098 (2004).
14. X. Wu, D. Weng, S. Zhao and W. Chen. Influence of An Aluminized Intermediate Layer on The Adhesion of a Gamma-AL2O3 Washcoaton FeCrAl. Surface & Coatings Technology, **190,** 2-3, 434– 439 (2005).
15. J.M. Zamaro, M.A. Ulla, and E.E. Miro. ZSM5 Growth on a FeCrAl Steel Support. Coating Characteristics Upon the Catalytic Behavior in The NOx SCR. Microporous and Mesoporous Materials, **115**, 1-2, 113–122 (2008).
16. M.V. Sivaiah, S. Petit, M.F. Beaufort, D. Eyidi, J. Barrault, C. Batiot-Dupeyrat and S. Valange. Nickel Based Catalysts Derived from Hydrothermally Synthesized 1:1 and 2:1 Phyllosilicate as Precursors for Carbon Dioxide Reforming of Methane. Microporous and Mesoporous Materials, **140**, 1-3, 69-80 (2010).
17. T.Y. Liu, Lay-Gaik, Teoh, Chao-Kai Huang, Ying-Chieh Lee. A Study of Roughness Improvement of Al2O3 Substrates Using Sol-Gel method. Procedia Engineering **141**, 108–114 (2016).
18. L. Henke. An AFM determination of the effects on surface roughness caused by cleaning of fused silica and glass substrates in the process. Biosensors and Bioelectronics, **17**, 547-555 (2002).
19. Y. Putrasari, P. Untoro, S. Hasan, Huda, and D. Sebayang. Modification of Surface Roughness and Area of FeCrAl Substrate for Catalytic Converter using Ultrasonic Treatment. Journal of Mechatronics, Electrical Power, and Vehicular Technology, **01**, 2 (2010).
20. A. Heesemann, E. Schmidtke, F. Faupel, A. Kolb-Telieps, and J. Klower. Aluminum And Silicon Diffusion In Fe-Cr-Al Alloys. Scripta Materialia, **40**, 5, 517–522 (1999).
21. K. Messaoudi, A.M. Huntz and B. Lesage. Diffusion and growth mechanism of $Al_2O_3$ Fe-Cr-Al alloys scales on ferritic. Materials Science and Engineering, **A247**, 248–262 (1998).
22. E.P. Degarmo, J. Black and R.A. Kohser. Materials and Processes in Manufacturing **223**, ISBN: 0-471-65653-4 (2003).
23. L. Xiang, Y.L. Gong, Li, and Z.W. Wang. Influence of Hydrothermal Modification on the Properties of Ni/ $Al_2O_3$ Catalyst. Applied Surface Science, **239**, 94–100 (2004).
24. E. Kis, R. Marinkovic-Neducin, G. Lomic, G. Boskovic, D.Z. Obadovic, J. Kiurski, and P. Putanov. Structural and textural properties of the NiO- $Al_2O_3$ catalyst. Journal of Polvhedron, **17**, 1, 27-34 (1998).

DOI: 10.1051/matecconf/20178702020

# Advanced Catalytic Converter in Gasoline Enginer Emission Control: A Review

A.M.Leman [1,a], Afiqah Jajuli [1], Dafit Feriyanto [1], Fakhrurrazi Rahman [1] and Supaat Zakaria[1]

[1] *Faculty of Engineering Technology, Universiti Tun Hussein Onn Malaysia, 86400, Parit Raja, Batu Pahat, Johor, Malaysia*

**Abstract.** Exhaust emission from automobile source has become a major contributor to the air pollution and environmental problem. Catalytic converter is found to be one of the most effective tools to reduce the overwhelming exhaust pollutants in our environment. The development of sustainable catalytic converter still remains a critical issue due to the stringent exhaust emission regulations. Another issue such as price and availability of the precious metal were also forced the automotive industry to investigate the alternatives for producing a better replacement for the material used in catalytic converter. This paper aims at reviewing the present development and improvement on the catalytic converter used on the reduction of exhaust emission in order to meet the regulations and market demand. The use of new catalyst such as to replace the noble metal material of Platinum (Pt), Palladium (Pd) and Rhodium (Rh) has been reviewed. Material such as zeolite, nickel oxide and metal oxide has been found to effectively reduce the emission than the commercial converter. The preparation method of the catalyst has also evolved through the years as it is to ensure a good characteristic of a good monolith catalyst. Ultrasonic treatment with combination of electroplating technique, citrate method and Plasma Electrolytic Oxidation (PEO) has been found as the latest novel preparation method on producing an effective catalyst in reducing the exhaust emission.

## 1 Introduction

Air pollution in Malaysia is reaching a critical level as witnessed during haze and El Nino crisis. Automobile source has been found contributed nearly 82% to the air pollution [1]. Since in the earlier 1940 and 1950s the air quality problem that engulfing some developing cities has been caused by the increasing number of vehicles [2].According to the Annual Report of the Road Transport Department of Malaysia [3], the number of registered road vehicles had increased from approximately 6.8 million in 1995 to 12.2 million in March 2003. Figure 1 shows the number of registered vehicles in Malaysia as of 1987-2002.

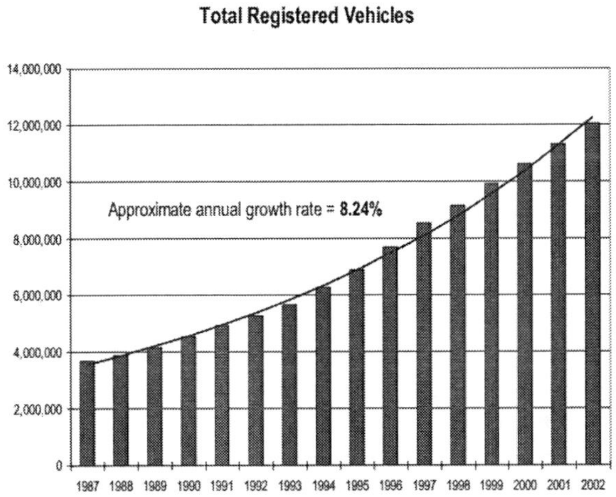


**Figure 1** : Analysis of total registered vehicles in 1987 -2002 [3]

Emission by gasoline has known as a major source of air pollution along urban traffic routes especially in developed countries. In Malaysia, it can be estimated that nearly 51% from 12 millions of registered vehicles using gasoline engine [4].Gasoline engines which are also known as internal combustion engine (ICE) where initiation of the combustion process of air–fuel mixture is ignited within the combustion chamber and the ignition is done either by spark ignition (SI) or compression ignition (CI) [5]. Emission from gasoline engine can be reduced with an improvement in engine design, combustion conditions and catalytic after treatment devices [6]. Since the introduction in the middle of 1970s, catalytic converter has been found to be the best option out of various existing technologies for the control of vehicle exhaust emission. These gases are eliminated by the basic reactions occur inside a catalytic converter through oxidation for Carbon Monoxide(CO), Hydrocarbon(HC) and reduction for Nitrogen Oxide ($NO_x$).

Over the recent years, automotive exhaust after-treatment processes a range of advanced technologies based on oxidation and three-way catalyst, adsorption, storage and filtration processes. This enables the reduction of CO, HC, $NO_x$ and particulate emissions (PM) from a gasoline or diesel engine in order to meet the demands of current and future exhaust emission regulations. In this paper, the advanced approach on the catalyst material and preparation of the catalyst were highlighted in order to identify the most suitable type of catalyst on to fulfill the requirements and market demand in automotive industry.

[a] Corresponding author: mutalib@uthm.edu.my

DOI: 10.1051/matecconf/20178702020

## 2 Three-Way Catalytic Converter

In the earlier 1980s, the automotive manufacturer has to include exhaust after treatment control system for four-stroke passenger car gasoline engines by three-way catalytic converter, which is capable of simultaneously reducing hydrocarbons, carbon monoxide and nitrogen oxides emissions [7]. This catalytic converter normally is placed after the exhaust manifold in-line with the exhaust system in order to be used to bring a desirable chemical reaction to take place in the exhaust flow [8].

The conventional Three-way catalyst (TWC) converters control CO, HC, and NOx effectively (80 to 90%) efficiency at or close to the stoichiometric air/fuel ratio which is normally used in the spark ignition engines [9] as shown in Figure 2. Three-way catalysts operate in a closed-loop system including a lambda or oxygen sensor to regulate the air: fuel ratio on gasoline engines [10].

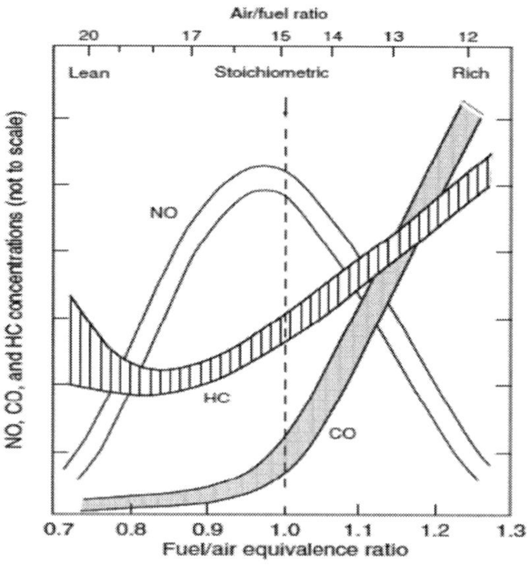


**Figure 2** .Air-fuel-ratio for exhaust SI gasoline emission

## 3 Advanced Catalytic Materials

In the last few years, the most dominant catalyst for gasoline vehicles has been the monolith or honeycomb structure catalyst. This catalyst consists of a cordierite skeleton coated with a highly porous washcoat of about 90% $\gamma$-$Al_2O_3$ and a mixture of alkaline-earth metals, oxides etc and last the noble metals (Platinum (Pt), Palladium(Pd) and Rhodium(Rh) which are fixed in the washcoat surface as shown in Figure 3.

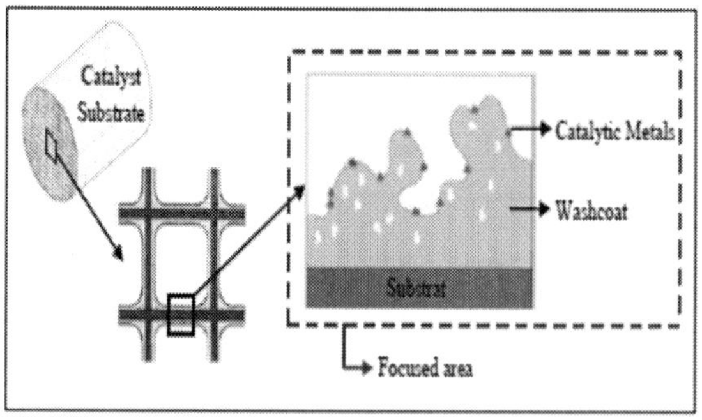


**Figure 3.** Washcoat and catalyst development focused area in catalytic converter [12].

The development in catalyst material was driven by the quest for the replacement of precious metal that limited in supply and higher cost. The noble metal catalytic converter has been found to cause several type of problems such as converter meltdown, carbon deposit, catalyst fracture and poisoning [13]. Amin and Rathod [14] have stated that the metal-base is more readily available and substantially cost effective than precious metal. Figure 4 shows the parameters that could affect the performance of the catalyst of precious metal [ 15].

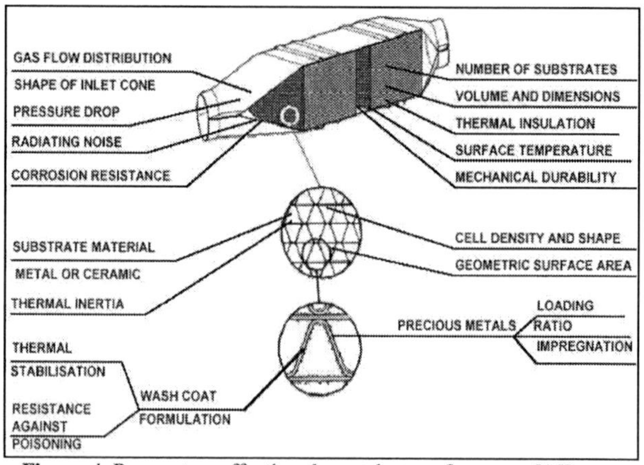


**Figure 4** .Parameters affecting the catalyst performance [15]

### 3.1. Zeolite

Zeolites in natural and synthetic form have been found on the removal of NOx pollutant. The combination of zeolites with vanadia/titania can produce a catalyst that is more resistant to deactivation with a wide DeNOx operating window [16]. Latest study by Karthikeyan [17] has presented the use of zeolite as a catalyst in catalytic reduction of S.I engine emission. Zeolites may be produced by the conversion of fly ash as shown in. This low cost fly ash based zeolites has gained new research interest among the automotive industry manufactures due to their unique properties in several applications such as ion-exchange, molecular sieves, adsorbents and catalyst. The result revealed that the CO and HC emissions were significantly reduced at all levels of load conditions. It is also observed that the trends for percent reduction in HC and CO are almost similar in speed test. This proves that the In-house made metal doped zeolite converter performs better than that of commercial converter. Figure 5 and 6 shows the variation of HC and CO conversion efficiency with brake power of the engine.

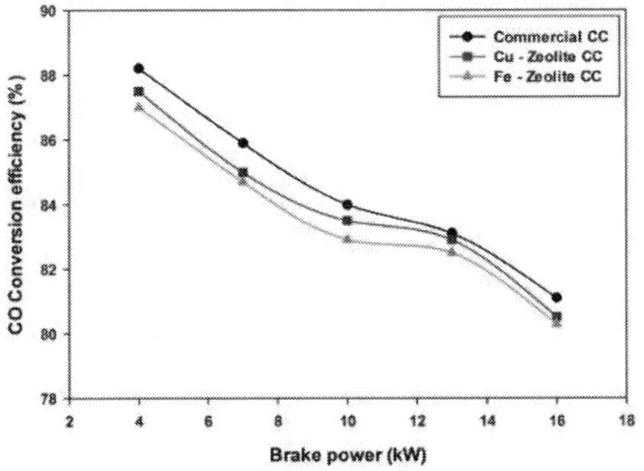


**Figure 5**. Hydrocarbon against Brake Power[17]

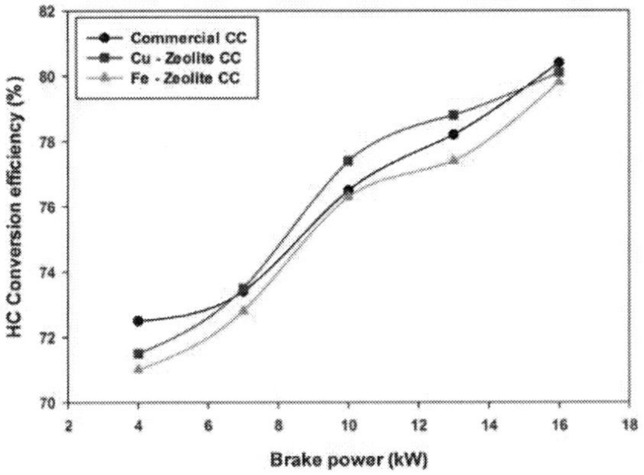


**Figure 6** . Carbon Monoxide against Brake Power[17]

### 3.2 Nickel Oxide

Sebayang et al.,[18] had studied the preparation of NiO catalyst on FeCrAl substrate. FeCrAl is known as metallic substrate due to their advantages of high temperature corrosion resistance. Nickel oxide (NiO) exhibited high activity and selectivity of methane due to the ability of NiO to undergo reduction process owing to the presence of defect sites of the surface. Despite of the fast catalyst deactivation and carbon deposition, NiO catalyst was favorable due to its high thermal stability and low price. Previous study by Kester et al.,[19] which investigates the structure, texture and reducibility of co-precipitated $NiO$-$Al_2O_3$ catalyst which is influenced by nickel oxide loading and temperature of heat treatment has found that interactions in co-precipitated $NiO$-$Al_2O_3$ system are intensive. Therefore, nickel oxide can be considered as the best catalyst material for catalytic converter. Nickel oxide application is also effectively for the NOx absorber catalyst in order to reduce the emissions of hydrogen sulfide ($H_2S$) during desulfation [20]

### 3.3 Titanium Dioxide

Ranganathan [21] in his study invstigate the performance of several type of converters which are the metal oxide such as titanium dioxide ($TiO_2$), cerium oxide ($CeO_2$) and copper nitrate ($Cu(NO_3)_2$) , Zirconium dioxide ($ZiO_2$) with wire mesh substrate. Both the catalyst material used in this are inexpensive in comparison with conventional catalyst such as noble metals like Platinum (Pt), Palladium (Pd) and Rhodium (Rh). The study show a positive results that is the new two catalyst catalytic converters are effective than the present conventional catalytic converter manufacture. The emission from exhaust is rectified up to 20% than the OEM fitted values. Figure 7 shows the titanium based catalytic converter.

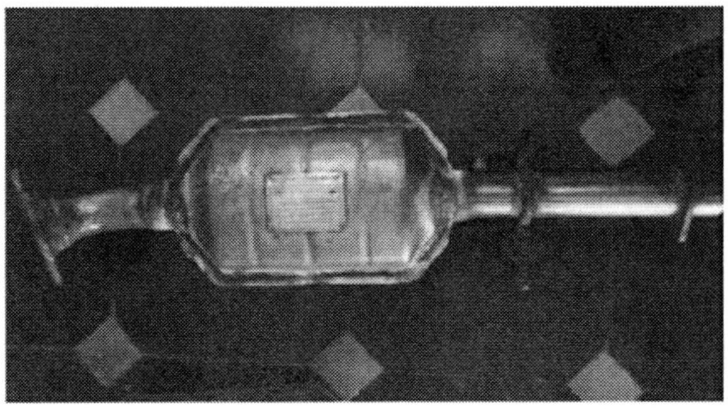

**Figure 7**. Titanium based catalytic converter [21]

## 4 Advances Preparation of Novel Catalyst

Preparation technologies for the manufacture a catalyst have evolved through the years. The characteristics of a good monolith catalyst should be met by managing preparation procedures.

### 4.1. Ultrasonic Treatment

Putrasari [22] investigates the new approach of coating technique using a combination of nickel electroplating, ultrasonic treatment and oxidation process. This research

DOI: 10.1051/matecconf/20178702020

discovered the preparation of NiO catalyst on FeCrAl substrate using various nickel electroplating process based on the weight gain during oxidation. The ultrasonic treatment is the power which has been converted to frequency by the transducer which generates a high cavitations bubble in a liquid system. This will result in the production of an intense shock out wave which flow through the liquid and strike out the material. The ultrasonic treatment prior to nickel electroplating was shown in Figure 8.

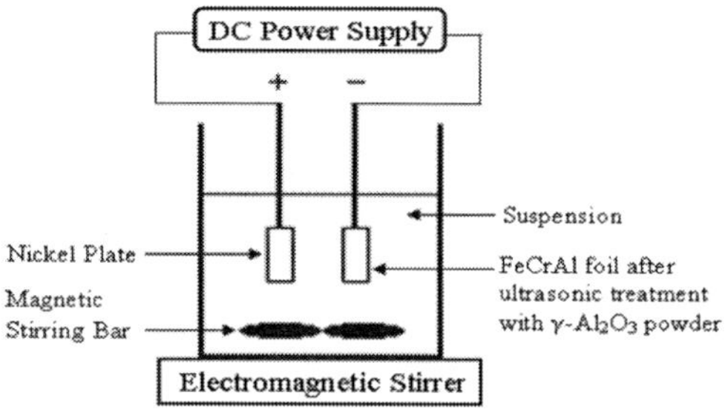


**Figure 8** . Schematic diagram of electroplating process [22].

This novel electroplating process was found to improved the high temperature oxidation resistance and the best method to adhere NiO on FeCrAl substrate.

### 4.2. Citrate Method

Benadda et al., [23] investigated the effect of the preparation method on the structural and catalytic properties of MnOx-CeO2 Manganese Cerium Mixed Oxides. The catalyst of MnOx-CeO2 was prepared by a-Carbonate route [MnCe-CPC] and b-Citrate route [MnCe- Cit]. The citrate method leads to the formation of a solid with smaller surface area, pore volume and average pore diameter as shown in Figure 9 and Figure 10.

**Figure 9**. SEM of the MnCe-CPC-500[23]

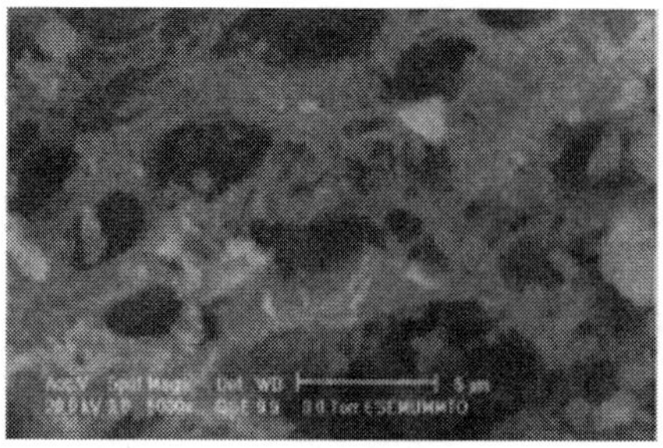

**Figure 10**. SEM of the MnCe-CIT-500[23]

The result shows that MnOx-$CeO_2$ mixed oxides prepared by the co-precipitation and citrate methods promotes to a good catalytic activity. This proves that the preparation method is strongly affect the textural properties and catalytic behavior of these oxides. The citrate method was found to form the solid with the best catalytic performance, a strong interaction between cerium and manganese and the small particle size seem to be at the origin of this performance.

### 4.3. Plasma Electrolytic Oxidation (PEO)

PEO is a novel method that being used in surface engineering technology. It has been considered as one of the methods that provide advantages in term of cost-effectiveness and environmentally friendly in which it can improve the corrosion and wear resistance of magnesium and magnesium alloys [24-26]. The coating layer that produced by PEO has a complex mechanism due to the combination of electro-, thermal-, and plasma-chemical reactions in the electrolyte [27].

Hussein et al. [28] conduct PEO process to form the ceramic coatings. During the process, the ceramic coating simultaneously grows inwards towards the alloy substrate and outwards towards the surface of the coating. The coating growth that been produced involved three processes including electrochemical reactions, the plasma chemical reactions and thermal diffusion. Figure 11 shows the schematic diagram of the coating process for PEO treatment.

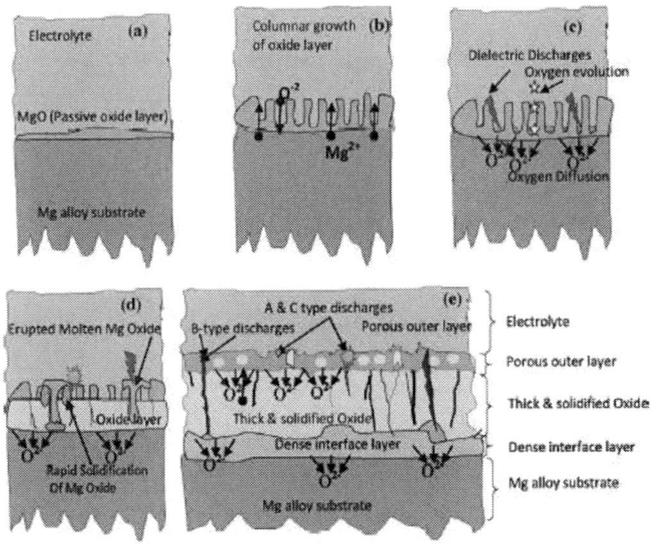


**Figure 11** .Schematic diagram of the coating process for PEO treatment [29]

Yang et al., [30] proposed the method of a process as shown in Figure 12 to form aluminium oxide on FeCrAl surface. This new approach of PEO use electric energy prior to oxidation time to formed the oxide particle .It was found that Al and O contents increased and Al oxide area was observed when applied energy increased because of enhancement of reactivity between insulating oxide and micro-charging in an electrolyte. The study presents a possibility of aluminium oxide formation by electro-chemical coating process without any pre-treatment of FeCrAl.

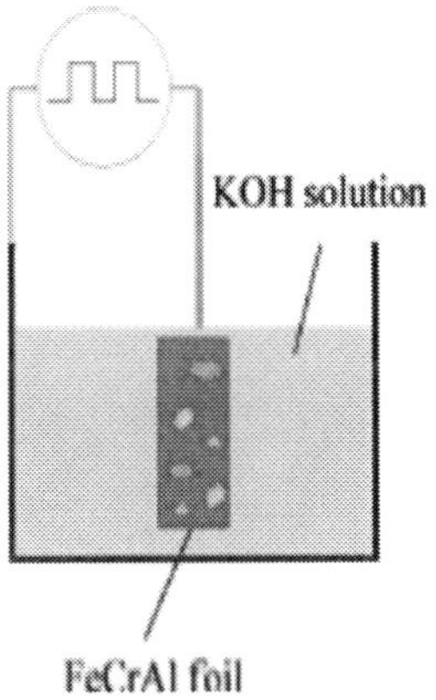


**Figure 12** . Schematic diagram of the plasma electrolytic oxidation process [30]

## 5 Conclusions

A list of new technologies was developed by the researchers to reduce exhaust emission and catalytic exhaust after treatment has been reviewed. As to meet the stringent exhaust emission regulations, research was centred around the material developed and the preparation method for the best replacement of the precious noble metal. Continuous improvement in materials modification technologies, as part of an integrated system comprising control of exhaust emission allows meeting more and more stringent combustion engines emissions legislations.

## Acknowlegement

The authors would like to thank the Ministry of Education and Universiti Tun Hussein Onn Malaysia (UTHM) through the funding supported FRGS grant under no Vot. 1216 and Centre for Graduate Studies-UTHM.

## References

1. Mohsin, R., Majid,Z.A., Shihnan,A.H., Nasri,N.S. and Sharer,Z. Effect of Biodiesel Blend on Exhaust Emission and Engine Performance of Diesel Dual Fuel Engine. Iranica Journal of Energy and Environment 6(3): 154-160 (2015).
2. Heagen Smit, A.J. Haagen-Smit, E.F. Darley, M. Zaitlin, H. Hull, W. Noble, Plant Physiol. 27 18 (1952).
3. Department of Environment. Malaysia Environmental Quality Report 2004.Department of Environment. Available: www.doe.gov.my (2004).
4. Road Transport Department Malaysia. Annual Report of the Road Transport Department of Malaysia. Available : Annual Report of the Road Transport Department of Malaysia (2003).
5. Kalam, M. and H. Masjuki. An experimental investigation of high performance natural gas engine with direct injection. Energy, 36(5): 3563-3571 (2011).
6. Bera,P.B. and Hedge,M.S. Recent advances in auto exhaust catalysis. Journal of the Indian Institute of Science. 90(2): 299-325 (2010).
7. Faiz,A.,Weaver,C.S., and Walsh,M.P. Air Pollution from Motor Vehicles. Standards and Technologies for Controlling Emissions. The World Bank Washington, D.C. (1996) .
8. Cooper,B.J. Challenges in Emission Control Catalysis for the Next Decade, Platinum Metal Rev. 38(l): 2-10 (1994).
9. Hasan,A.O. Influence of Prototype Three Way Catalytic Converter on Regulated and Unregulated Emissions from Gasoline HCCI/SI Engine. The University of Birmingham : Doctor Of Philosophy Thesis (2011) .

10. Shelef,M. and McCabe, R.W. Shelef. "Twenty-five years after introduction of automotive catalysts: What next?", Catalysis Today. 62 : 35-50(2000).
11. Favre,C., May, J., and Bosteels,D. Emissions Control Technologies To Meet Current And Future European Vehicle Emissions Legislation. Association for Emissions Control by Catalyst (AECC) (2003).
12. Twigg,M.V. Progress and future challenges in controlling automotive exhaust gas emissions. Appl Catal B-Environ. 70 : 2-15 (2007).
13. Firdianto,A . Ultrasonic Treatment with Nickel Electroplating Combined with Oxidation for Developing Gamma-Alumina Washcoat on Fe-Cr-Al Substrate. Universiti Tun Hussein Onn Malaysia, Malaysia: Master Thesis.(2012).
14. Amin,A. and Rathod,P . Catalytic Converter Based On Non-Noble Material. International Journal of Advanced Engineering Research and Studies. 1(2) : 118-120. (2012).
15. Kummer,J. "Oxidation of CO and C2H4 by base metal catalyst for honeycomb supports,"ACS Series 143, Catalyst for the Control of Automotive Pollutants, J.Mc-Evoy, ed.,American Chemical Society .2(3) :178-192.(1975).
16. Harkonen, M. Exhaust Gas Catalysts. Nanotechnology in Northern Europe Helsinki Fair Center. Ecocat Oy (2005).
17. Karthikeyan,D. Saravanan, C.G. and Jeyakumar,T. Catalytic Reduction of S.I. Engine Emissions Using Zeolite as Catalyst Synthesized From Coal Fly Ash. International Journal of Engineering and Technology. 6(2):62-68.(2016).
18. Sebayang, D., Putrasari, Y.,Sulaiman Hassan,&Untoro,P. Preparation NiO Catalyst on FeCrAl Substrate Using Various Technique at Higher Oxidation Process, Electroplating, ISBN -978-953-51-0471-1 : 1-25 (2012).
19. Keith B. Kester , Zagli,E., and Falconer,J.L. Methanation of carbon monoxide and carbon dioxide on Ni/Al2O3 catalysts: effects of nickel loading. Vol. 22(2). (1986).
20. Elwart,S. System and Method for Removing Hydrogen Sulfide from an Emissions Stream. U.S. Patent 7,104,045 B2. (2006).
21. Ranganatham,M. ., Remo, S.A.R., Kishore,U., Yuvaraj,S. and Arun,S . Development And Performance Analysis Of New Catalytic Converter, International Conference on "Advance Research and Innovation in Engineering, Science, Technology and Management" ICARSM' 15 . 1(3) (2015).
22. Yanuandri Putrasari. Substrate Using Various Techniques At Higher Oxidation Process. Universiti Tun Hussein Onn Malaysia : Master Thesis (2012).
23. Benadda,A.,Djadoun,A.,Guessis,H.,andBrama,A. Effect of the Preparation Method on the Structural and Catalytic Properties of MnOx-CeO2 Manganese Cerium Mixed Oxides. Proceedings Of The International Conference Nanomaterials: Applications And Properties.2(1):1-3 (2013).
24. Arrabal,R.,Matykina,E.,Hashimoto,T.,Skeldon,P. ,Thompson,G.E.Characterization of AC PEO coatings on magnesium alloys, Surf.Coating Technol.203(2009).
25. Hussein,R.O.,Zhang,P.,Xia,Y.,Nie,X.,Northwood,D .O.The effect of current mode and discharde type on the corrosion rsitance of plasma electrolytic oxidation (PEO) coated magenesium alloy AJ62, Surf.Coat.Technol,206(2011).
26. Ghasemi,A.,Raja,V.S.,Blawert,C.,Dietzel,W.,Kainr, K.U. Study of the structure and corrosion behaviour of PEO coatings on AM50 magnesium alloy by electrochemical impedance spectroscopy, Surf.Coat.Technol.202 (2008).
27. Yerokhin,A.L.,Shatrov,A.,Somsonov,V.,Shaskov,P. ,Pikington,A.,Leyland,A.,Mathews,A.Oxide ceramic coatings on aluminium alloys produced by a pulsed bipolar plasma electrolytic oxidation process.Durf.Coat.Technol.199(2005).
28. Hussein,R.O.,Nie,X. and Northwood,D.O. An investigation Of ceramic coatings gwoth mechanisms in plasma electrolytic oxidation (PEO) processing. Electrochimica Acta 112 (2013).
29. Hussein,R.O.,Nie,X. and Northwood,D.O.The application of plasma electrolytic oxidation (PEO) to the production of corrosion resistance coatings on magnesium alloy: a review,Corros.Mater.38(1)(2013).
30. Yang,H.S., Jang D.H and Lee K.J. Aluminum Oxide Formation On Fecral Catalyst Support By Electro-Chemical Coating. Archieves of Metallurgy and Materials. 60(2) DOI:10.1515/amm-2015-0161(2015).

# Emission Treatment towards Cold Start and Back Pressure in Internal Combustion Engine against Performance of Catalytic Converter: A Review

A.M. Leman[1,a], Fakhrurrazi Rahman[1], Afiqah Jajuli[1], Supaat Zakaria[1] and Dafit Feriyanto[1]

[1]*Faculty of Engineering Technology, Universiti Tun Hussein Onn Malaysia (UTHM) 86400, Parit raja, Batu Pahat, johor, Malaysia*

**Abstract.** Nowadays, regulation for the vehicles emissions has becoming more stringent in order to reduce the effect of pollutant gases that was being released by the vehicles exhaust. The development of catalytic converter is to resolve the pollution emission aspect . There are lots of improvements done by the researchers towards improving catalytic converter, yet there are still key issues which give negative impact to the environment. One of the problems that are being concern by among of researchers is the cold-start and back pressure problems that usually occur in the composition of catalytic converter. Presented here is a review of cold-starts and back pressure problems together with several alternatives taken by not affecting the performance of vehicles engine and fuel consumption. The review also includes alternative system development and selection of materials to resolve these problems.

## 1 Introduction

In 21$^{th}$ century, the number of vehicles grew more quickly which increases the total of air pollution from internal combustion. These pollutant gases influence to give bad effect to the human health and also surrounding environment [1]. Vehicles nowadays facing serious issues with the engines combustion gases that released high pollutant gases such as Unburned Hydrocarbon (HC), Caron Monoxide (CO) and Nitrogen oxide ($NO_x$) through the exhaust pipeline [2]. Most vehicles with gasoline fuelled were attached with three way catalyst in catalytic converter that gives an alternative way to reduce pollutant gaseous from the exhaust emission [3]. Mansha et al., (2013)[4] stated that there are two common methods that were usually being used in reducing the harmful gases that were adapted from the emission gases of automobiles exhaust by improving the technology of automobiles engine and composition of vehicles fuel. The catalytic converter was a special device that is originally designed to convert the pollution gaseous from being released into the atmosphere. The conversion still releasing amount of pollutant throughout the warm-up phases until the catalytic converter reaches out its light-off temperature. Thus, this kind of problem has becoming barriers to achieve future standard of emission [3]. Consequently, there are several ways of studies that involved the combustion of vehicles engine which also have a relation with the engine control and after treatment process that were been investigated to reduced cold-start emission and back pressure that usually occurs in the vehicles engine combustion [5][6]. According to Steven et al., (1995)[7] the cold start emission is closely related with the thermal control in catalytic converter. A better technique to treat the cold start emission is by using compact insulation which could continuously give a better impact towards thermal conductance and maintaining the heat between the combustion system by providing low conductivity and also high conductivity for the engines heat rejection from the converter during the engine operation. This paper aims to review the techniques by previous researchers to improve the performance of cold-start system, evaluating the findings and discuss the some strategies to overcome the cold start problems from the vehicles exhaust combustion based on the vehicles system performance and also materials optimization.

[a] Corresponding author: mutalib@uthm.edu.my

## 2 Catalytic Converter Design and Stability

Catalytic converter is a device that is used to reduce the pollutant gaseous that originally comes from the internal combustion of vehicles engine. Catalytic converter is a stainless steel container that is attached alongside the exhaust pipeline and inside the container is a porous ceramic or metallic structure through which the exhaust gas flows [8]. It is usually being placed between engine

DOI: 10.1051/matecconf/20178702021

manifold that connected with the tailpipe of vehicle exhaust system. The combustion occurs in the vehicles engine produced pollutant gases and will be pass through the tailpipe before undergo chemical reaction with the catalyst within the porous wash coat in the converter to produce a harmless gaseous which is readily to be released to the atmosphere. There are two types of catalytic converter which is classified as Two-Way or Three-Way catalytic converter. The Two-Way converter only works on two types of gases which are CO and unburned HC while the NOx is controlled by exhaust gas recirculation (EGR) system and by retarding the ignition timing [9]. Early manufactures used loose granular ceramic to manufactured ceramic converters with the gas passing through the packed spheres. Since it is difficult to keep the spheres in place, many converter developers opted for ceramic monolith which offers various advantages. It offers several advantages such as lower mass, smaller volumes and easily for packaging process [10]. The wash coat that is applied for the ceramic converters monolith walls composed of porous, high surface area inorganic oxides ZrO2 (Zirconia), $CeO_2$ (Ceria) and γ-Al2O3 (gamma alumina).

Noble metal catalysts, such as Platinum (Pt), Palladium (Pd) and Rhodium (Rh), are deposited on the surface and within the pores of the wash coat [11]. There are several models in predicting the pressure loss of channel substrate in the catalytic converter by approaching various types of cell shapes. Common model used by vehicles is the classic model discovered by Hagen-Poiseuille. This model purposed a laminar flow through a circular duct [12]. The evolution for design and optimization of the monolith in the catalytic converter is increasing from time to time. It is because the honeycomb material is available in various type of shapes and dimension which including the hexagon, square, triangle, circle and also sinusoidal. Figure 1 shows the single channel concept of monolith. The improved techniques in treating the honeycomb structure is by using the sub-grid scale modelling displays in Figure 2.

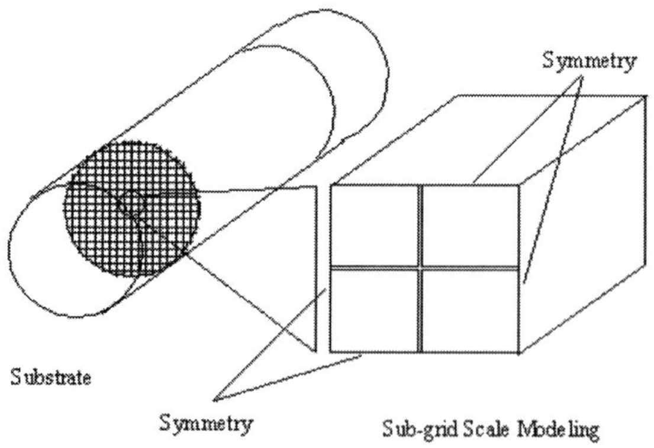


**Figure 1**. Single Channel Concept [13]

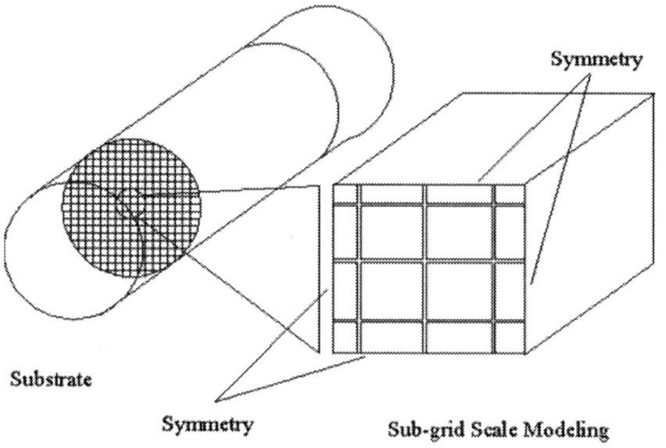


**Figure 2.** Sub-Grid Scale Channel Concept [14]

## 3 Cold Start Emission

The cold-start issues among the researchers in automotive engineering have become one of new topic that needs to be discuses and solve in order to develop new emissions treatment techniques. Until it reaches the light-off temperature, a catalytic converter remains essentially ineffective. The understanding about characteristics of catalytic converter is important especially during the cold-start period in order to improve the performances of cold-start. There are two common methods used in handling the cold-start which is passive and active methods. The passive method is referring on the optimization of the exhaust system design which including the modification made upon the catalytic converter or by changing the position of the converter relative to the vehicles engine. On the other hand, the active method is based on the additional energy used to increase the performance of converter especially for the exhaust system temperature during the cold start. This type of method requires heating the preheating process for the catalytic converter. Both methods usually do not need additional devices and also not increasing the cost to overcome the cold-start problems. The cold performance of vehicles is closely related to the fuel consumption of engine performance. According to Kunze et al., (2006) [15], when engine starts to warm-up from temperature between 25°C – 90°C, the fuel consumption during this period is decreasing in an average about 10%. Hence, for the vehicle that only involve in short distance trips, it needs to improve the fuel consumption especially during the warmup phase because the engine could not reach the optimum temperature for the catalytic converter to be activated which could causes cold start emission. (Roberts et al., 2014)[16] claims that the average of the cars travel in Europe is around 10 km while Burke et al. (2010) [17] state that in United State of America, the mean length of travelling distances of vehicles are less than 15km. The data presented in Figure 3 shows the energy thermal balance occurs in the vehicles cylinder during the warm-up stages.

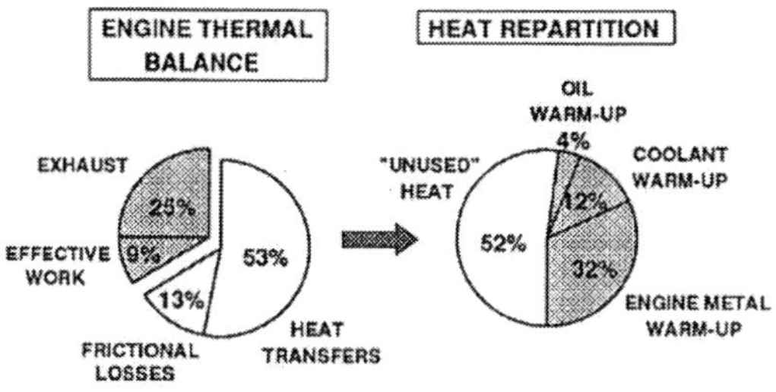


**Figure 3.** Energy balance that occurs in the vehicles engine during warm up stages and alternatives in solving cold start problems [18]

There is only 4% of heat that is remained in the lubricant during the energy transfer process occurs in the engine through combustion which resulted low warm-up rate. Temporarily, another 32% is reserved in the engine metal warm-up at the engine block that provides very little benefits except it will conserve to increase engines cylinder temperature (Roberts et al., 2014)[16]. Still there is 52% energy from the combustion is expelled from the engine system as waste heat. Clearly, the needs to improve cold start emission by utilizing the waste heat in the engines to ensure that the lubricant temperature may rise in quicker time. Brito et al., (2012)[19] stated that the released energy from the engine combustion will be then released to the exhaust gas about 15% up to 40%.

Cold start problems can be said that it is one of the difficult problems to be solved as the relation system in the engine combustion is complex. There is a need to apply a system that can fasten the warm-up temperature without making any additional of thermal inertia to the engine system as it can result in deterioration to the performance measure of engine setup. Hence, the pollutant emission gaseous can be significantly reduced by shorten the time taken for the catalytic converter to reach light-off temperature during a cold start. This is because most vehicles emit pollutant gaseous during the legislated driving cycles. (Moore et al., 1993)[20] stated that during Federal Test Procedure (FTP), the pollutant emissions released by the vehicles was about 50% to 80%. It is occurs before the catalytic converter reach light-off temperature. According to Heywood (1998) [21], it was found that the largest contribution of HC and CO emissions occur during the first minutes of engine operation. Figure 4 shows the efficiency of catalytic converter for HC and CO conversion against temperature of the engine combustion.

Leading to these issues, there were several approaches are taken to overcome the cold-start problems. The energy waste from the exhaust combustion can be one of the alternatives to heat up the catalytic converter temperature by having heat exchanger applications. Kumar et al., (2013)[22] investigate that the temperature of exhaust gases that flows from the internal combustion engine was from 300°C to 900°C which is depending on the engine load. The exhaust gas is one of the waste energy that is released and it would give benefits in rising up the catalytic temperature if the waste energy can be jointed to the catalytic converter which is also known as energy capture. It is not only achieved better accelerated temperature rising, the fuel consumption of the vehicles engine also can be reduced. It must be noted that the heat extraction from the exhaust temperature may rise up the light-off timing which also would increase the level of pollutant gaseous. The idea to conduct the energy capture must be carefully considered as the usual length between exhaust manifold and catalytic converter can be over than 0.1m length and Zhao and Winterbone (1993)[23] state that the temperature would reduce about 200°C between that distance. Another consideration also must be taken that any addition for energy capturing system will give some additional weight to the vehicles which would affect the fuel consumption to be increased.

Will and Boretti (2011)[24] had conducted an investigation on the heat exchanger that applied from the exhaust gas directly towards the lubricant device heat exchange. From their investigation, it showed an improvement for fuel consumption efficiency about 7% without installing heat exchanger system device from the exhaust combustion. Other than that, the lubricant heat exchange device also reduced 2% of HC emission together with reduction of 19% of $NO_x$ and 27% of CO emissions. In order to avoid emission distress, the location of heat exchanger is placed near to the catalytic converter. It is argued that by increasing the lubricant temperature may give other positives effects for the engine combustion by reducing friction in the engine piston.

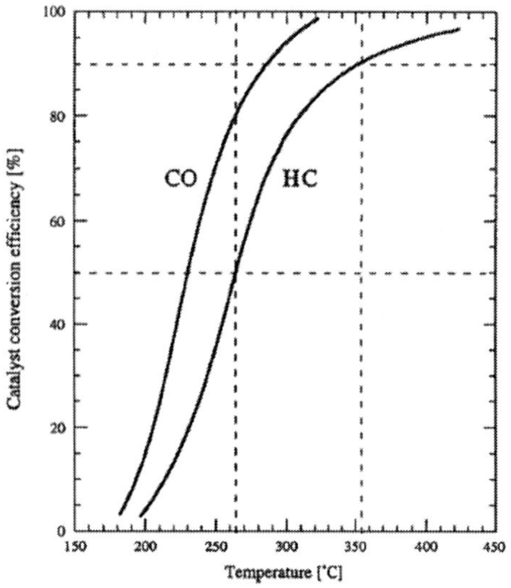


**Figure 4**. HC and CO emission against engine combustion temperature [24]

Another way to manage the cold-starts problems from the vehicles engine combustion is by using thermal electric applications. The application of thermal electricity is generated using the Seebeck Effect which can eventually operate thermal energy system [25]. It is caused by the existing of electric voltage that exists across the junction of two dissimilar materials which will result temperature difference such as shown in Figure 5. This alternative ways offers benefits on fuel consumption and produced better emissions which also related in reducing the engines load over engine alternator. Brito et al., (2012) [19] had managed the investigation for thermal electricity application which is also known as thermoelectric generator (TEG). The heat exchanger process occurs between the TEG and the exhaust gas is using the heat pipes which act as a medium. Conversion of the exhaust combustion gases to the electrical energy power output was still in limit order which just only 1%, showing that the conversion is still limited. Therefore, it can be conclude that the application of TEG to the engine combustion can brings new alternative ways for recapturing waste thermal energy from the exhaust combustion gases. Kumar et al., (2013)[22] proved that TEG may recover the energy in the range of 64% from the exhaust combustion gas the enters the device and about 58.7% from the inlet energy is being moved to the coolant system due to operating temperature limitations of the TEG. Thus, the electrical generation produced from the process can be only 3% from the total exhaust gas energy.

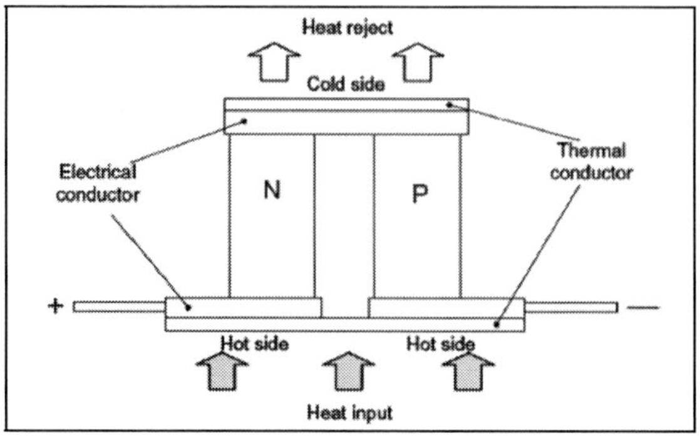


**Figure 5**. Thermoelectric junction of two dissimilar materials of p-type and n-type [16]

Kollman et al.,(1994)[26] overcome the cold-start problem by heating the catalytic converter with an external combustion Hydrogen gas that is added to the vacuum insulation which contains a small amount of metal hydride [27]. The function of the metal hydride is to electrically heat the pressure of the hydrogen. The temperature of the hydride is controlled which directly varies with the pressure of the hydrogen inside the catalytic converter. When the metal hydride is at low temperatures, the hydrogen gas is absorbed by the metal oxide and the heat conductance is only affected by the conductance of the insulation. When the metal oxide is heated, the hydrogen gas is released and its pressure begins to increase. Figure 6 shows the catalytic converter structure with the vacuum insulation.

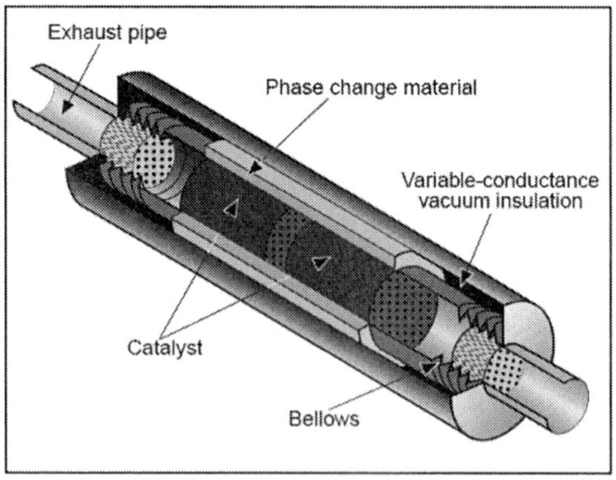


**Figure 6**. Vacuum insulation installed in catalytic converter [28]

## 4 Back Pressures

The installation of catalytic converter in the exhaust system has been approved by numerous researchers where the catalyst surface needs to have a sufficient area for treating the gases to meet the emission limits. However, this procedure increases the pressure drop, resulting in engine power losses and fuel wastage. Indeed, an increased pressure drop is challenges that need to overcome. Typically, an engine will lose about 300 w of power per 1000 pa of pressure loss [29]. The catalyst and filter materials placed inside the catalytic converter increase back pressure. This increase in back pressure causes more fuel consumption, and in most cases, engine stalling might happen. The filtration efficiency and back pressure are interrelated. If maximum filtration efficiency using very fine grid size wire meshes, is achieved, the back pressure will also be increased, which causes more fuel consumption. On the other hand, if larger grid size wire meshes are used, back pressure will be less, but the filtration efficiency will also be reduced, which does not help in meeting the present emission norms. With the help of computational fluids dynamics (CFD) analysis, it is attempted to find out the optimum solution to get maximum filtration efficiency with limited back pressure developed inside the catalytic converter [30]. According to Lashmikanti and Keck (2004)[31] there are two major components that are related to the pressure drop in the catalytic converter which are substrate and flow distribution devices. Figure 7 shows that the substrates commonly in pellet forms by the usage of $\gamma$-$Al_2O_3$ particles is used to replace the honeycomb monolith structure that gives positive impact in terms of ensuring a lower pressure drop by having a high open frontal area. Moreover, an increase in cell density is accompanied by a reduction in wall thickness to compensate for the increase in backpressure. The substrate length, cross-sectional area, and cell shape are also important parameters that have been investigated by few researchers. According to (Day, 1997) [32] and Miyairi et al., (2003) [33], they have identified the importance of cell shape in the overall performance of a catalytic converter. Pressure drop, heat, and mass transfer characteristics have been calculated in relation to different cell shapes.

Other investigation done by Rajadurai et al., (2006) [34] state that the Knitted wire mesh substrates with different geometry and channels gives positives impact towards the back pressure occurs in the catalytic converter. The basic requirements for better flow in the catalytic converter were lower in back pressure, system weight and also better flow of gaseous emissions.

**Figure 7.** Pellets that replaced substrate used as the conversion for pollutant gases in catalytic converters [32]

The combinations of these requirements will provide a better performance for the vehicles ventilation. Karuppusamy et al., (2013)[35] have conducted an experiment related to the flow in catalytic converter in order to minimize back pressure in the several types of converter based on different diameter, length and inlet cone angle of the monolith for the catalytic converter. It is observed that the back pressure for the converters that have larger diameter. The result from the research shows that increase in inlet cone length will reduces the backpressure and also reduces the recirculation zones in the catalytic converter. Figure 8a and 8b shows different inlet cone angle and different catalyst length of catalytic converter used to compare the back pressure in the catalytic converter.

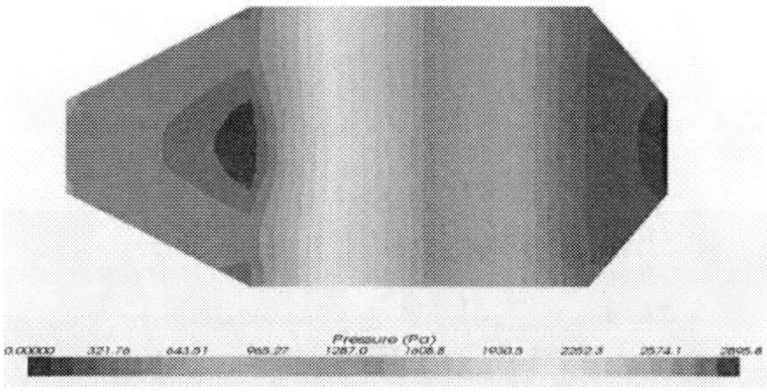


**Figure 8a.** Catalytic with higher cone length and angle [32]

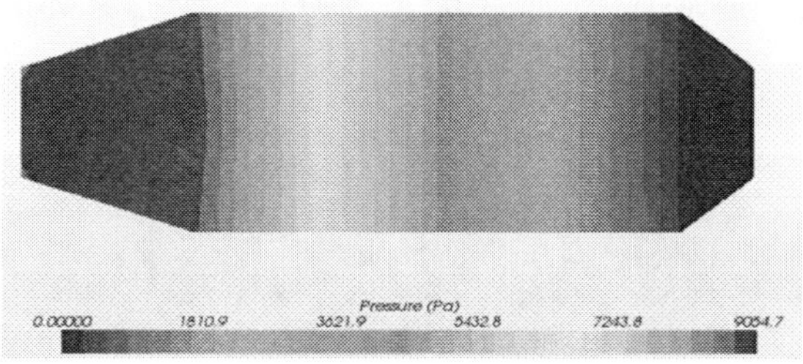


**Figure 8b.** Catalytic with less cone length and angle [32]

## 5 Conclusions

Through this review study, it is notice that cold-start is one of the important issues that need to be concern for the performance of vehicles engine combustion. There are a few strategies used by the researchers to improve the cold-start performance in a technical ways. The most important factors in solving cold start performance are to improve the light-off temperature during the warm-up stage. Thus, modification of catalytic converter is an important aspect that needs to be concern in order to reduce the pollutant gaseous from being exposed to the atmosphere especially during the cold-start period. Other than that, the flows of the emission gaseous in the catalytic converter also need to be discussed in order to make sure the flow of the emissions is treating in a better way to avoid engine power losses and fuel wastage.

## Acknowledgement

The authors would like to thank the Ministry of Higher Education Malaysia and Universiti Tun Hussein Onn Malaysia (UTHM) through the funding supported FRGS grant under No Vot. 1216 and Centre for Graduate Studies – UTHM.

## References

1. Seinfeld, J.H. Air Pollution: A Half Century Progress. Environmental and Energy Engineering, 50(6):1096-1108 (2004).
2. Kalam M. A. Masjuki H. H, Mohd. Redzuan, Fuad M. A., Mohibah M., Halim K.H., Mahlia T. M. I., Abdul Ishak. Development and test of a new catalytic converter for natural gas fuelled engine. Sadhana. Vol. 34, Part 3, pp 467-481 (2009).
3. Denis Igorevich Andrianov, Chris Manzie and Michael John Brear. A Methodology for Minimising Emissions Constrained Cold Start Fuel Consumption. SAE International (2012).
4. Mansha M., Qureshi A.H., Chaudry I.A. and Shahid E.M. Three Ways Catalytic Simulation of Engine-Out Exhaust Emission. Journal of Quality and Technology Management Volume IX, Issue I, pp 57 – 68 (2013).
5. Roberts C. E. and Stanglmaier R. H. Investigation of intake timing effects on the cold start behavior of a spark ignition engine, SAE International, Paper No.1999-01-3622 (1999).
6. H. Santoso and W. K. Cheng. Mixture preparation and hydrocarbon emissions behaviors in the first cycle of SI engine cranking, SAE International, Paper No. 2002-01-2805. (2002).
7. Steven D. Burch Thomas F. Potter Matthew A. Keyser. Reducing Cold-Start Emissions by Catalytic Converter Thermal Management. National Renewable Energy Laboratory. U.S. Department of Energy. (1995).
8. Ganessan, V. Internal Combustion Engine. Second Edition, McGraw Hill (2004).
9. Mansha M. Shahid. E.M., Qureshi A.H. Control of Combustion Generated Emissions from Spark Ignition Engines: A Review, Pakistan Journal of Engineering and Applied Sciences, Vol. 11, pp 114-128 (2012).
10. Heck, R.M. and Farrauto, R.J. Catalytic Air Pollution Control. New York: Van Nostrand Reinhold (1995).
11. Pontikakis, G.N. Modelling, Reaction Schemes and Kinetic Parameter Estimation in Automotive Catalytic Converters and Diesel Particulate Filters," Published Thesis (2003).
12. White F.M. Fluid Mechanics. Singapore:Mc-Graw Hill, Inc (2003).
13. Deutschmann O., Maier L. I., Riedel U., Stroemann A.H. and Dibble R.W. 2000, "Hydrogen assisted catalytic combustion of methane on platinum, " Catalysis Today. 59, pp. 141-150.
14. Miyairi Y., Aoki T., Hirose S., Yamamoto Y., Makino M., Miwa S., Abe and F.. 2003. Effect of Cell Shape on Mass Transfer and Pressure Loss. SAE International, 2003-01-0659 (2003).
15. Kunze K, Wolff S, Lade I, Tonhauser J. A systematic analysis of CO2-reduction by an optimized heat supply during vehicle warm-up. SAE technical paper 2006-01-1450; (2006).
16. Roberts A., Brooks R. and Shipway P. Internal combustion engine cold-start efficiency: A review of the problem, causes and potential solutions. Journal of energy conversion and management, Vol. 82, pp 327 – 350 (2014).
17. Burke RD, Brace CJ, Hawley JG, Pegg I. Review of the systems analysis of interactions between the thermal, lubricant, and combustion processes of diesel engines. Proc Inst Mech Eng Part D: J Automobile Engineering; 224(5):681–704 (2010).
18. Trapy JD, Damiral P. An investigation of lubricating system warm-up for the improvement of cold start efficiency and emissions of SI automotive engines. SAE technical paper 902089 (1990).
19. Brito FP, Martins J, Goncalves L, Sousa R. Temperature controlled exhaust heat thermoelectric generation. SAE technical paper 2012-01-1214 (2012).
20. Moore, Wayne R. and Mondt, J. Robert. "Predicted Cold Start Emission Reductions Resulting from Exhaust Thermal Energy Conservation to Quicken Catalytic Converter Lightoff," SAE International, paper No. 931087 (1993).
21. Heywood, J. B. Motor vehicle emissions control: past achievements, future prospects. In Handbook of Air Pollution from Internal Combustion Engines, Academic Press (1998).
22. Kumar S, Heister SD, Xu X, Salvador JR, Meisner GP. Thermoelectric generators for automotive waste heat recovery systems. Part I: numerical modelling and baseline model analysis. J Electron Mater; 42(4):665–74 (2013).

23. Zhao Y, Winterbone DE. A study of warm-up processes in SI engine exhaust systems. SAE technical paper 931094 (1993).
24. Will F, Boretti A. A new method to warm up lubricating oil to improve the fuel efficiency during cold start. SAE technical paper 2011-01-0318; (2011).
25. Yu C, Chau KT. Thermoelectric automotive waste heat energy recovery using maximum power point tracking. Energy Convers Manage 50(6):1506–12 (2009).
26. Kollman, K. Abthoff, J. and Zahn, W. Three-Way Catalysts for Ultra-Low-Emission Vehicles. Automotive Engineering, 17–22 (1994).
27. Cooper, B. The future of catalytic systems. Automotive Engineering, 9–12 (1992).
28. Burch S.S. and Keyser M.A. Applications and Benefits of Catalytic Converter Thermal Management. National Renewable Energy Laboratory (2003).
29. Pannone G.M and Mueller J.D. A comparison of conversion efficiency and flow restriction performance of ceramic and metallic catalyst substrate," SAE International, 2001-01-0926 (2001).
30. Balakrishna B.,Srinivasarao Mamidala. Design Optimization of Catalytic Converter to reduce Particulate Matter and Achieve Limited Back Pressure in Diesel Engine by CFD. International Journal of Current Engineering and Technology. E-ISSN 2277 – 4106, P-ISSN 2347 – 5161 (2014).
31. Lashmikantha M. and Keck M. Optimization of exhaust systems. SAE International, 2002-01-0059 (2002).
32. Day J.P. Substrate effects on light-off – part (ii) cell shape contributions, SAE International, 971024 (1997).
33. Miyairi Y., Aoki T., Hirose S., Yamamoto Y., Makino M., Miwa S., Abe and F.. 2003. Effect of Cell Shape on Mass Transfer and Pressure Loss. SAE International, 2003-01-0659 (2003).
34. Rajadurai S., Jacob S., Serrell C., Morin R. and Kircanski Z. Wiremesh Substrates for Oxidation.TWC and SCR Converters; GPC Advanced Propulsion and Emission (2006).
35. Karuppusamy P. Senthil R. Design, Analysis of Flow Characteristics of Catalytic Converter and Effects of Backpressure On Engine Performance. International Journal of Research in Engineering & Advanced Technology (IJERAT), Volume 1, Issue 1. ISSN: 2320 – 8791 (2013).

DOI: 10.1051/matecconf/20178702022

# Improving Road Safety of Tank Truck in Indonesia by Speed Limiter Installation

Hadi Pranoto[1,2a], A.M Leman[2],Ishak Baba[2],Dafit Feriyanto[2] and Gama Widya Putra[3]

[1]*Faculty of Mechanical Engineering, Mercubuana University,Jakarta,Indonesia*
[2]*Faculty of Engineering Technology, Universiti Tun Hussein Onn Malaysia 86400 Parit Raja, Batu Pahat, Johor, Malaysia*
[3]*Faculty of Public Health, Indonesia University, Jakarta Indonesia*

**Abstract.** Indonesia has one of the highest number of fatalities caused by traffic accident. It is become main concern since last decades. Approximately of 10% fatalities is caused by tank truck accident, it recorded by PT. Pertamina Persero, Indonesia in 2015 that 17% and 20% tank truck accident is caused by over speed and fatigue, respectively. Therefore, over speed has become main factor the occurrence of tank truck accident. Main objective of this research is to install speed limiter on the tank truck in order to improve safety engineering system, decrease accident and to maintain engine performance. This research is conducted in Indonesia especially in Java-Bali route travel. Speed limiter is installed to the tank truck engine which completed by fuel cut-off solenoid to reduce the speed automatically when it exceeding the maximum speed that has been determined. From the result shows that top speed which performed by driver up to 133 km/h when tank truck uninstalled by speed limiter. Meanwhile, when speed limiter is installed to the tank truck, top speed locked at 70 km/h even though the driver want to speed up. It means that fuel cut-off system is very effective to lock the speed at 70 km/h and it shown the improvement up to 65%. The monitoring activities observed that the decreasing number of fatalities caused by tank truck accident become 7% as compared to last year of 17%. It can be found that the speed limiter coupled by speed recorder was very efficient to improve safety engineering system of the tank truck.

## 1 Introduction

Traffic accident is one of the causes of the death in Indonesia because based on report of Indonesia Police Departement that close to 31.234 fatalites cause by trafic accident and it means that there are 3-4 fatalites/hour [1]. The rapid growth of vehicle and industrie will incerasse the signifiant proportion differences between vehicle and capacity of road. The average annual vehicle growth from 1996 to 2006 of 20% (Table 1) which may take a higher possibility in increasing road fatalities [2].

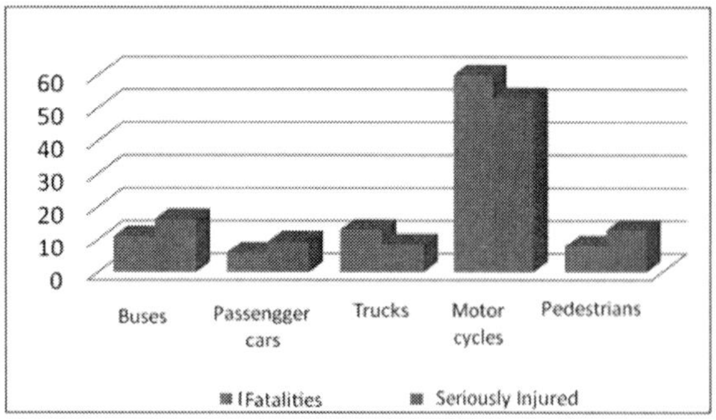


**Figure 1** Percentage number victims by vehicle types [2]

**Table 1** Number of vehicle growth in Indonesia [2]

| Year | Total registered vehicle | Annual growth of total vehicle (%) |
|---|---|---|
| 1996 | 14530095 | |
| 1997 | 16821076 | 16 |
| 1998 | 17644885 | 5 |
| 1999 | 18224149 | 3 |
| 2000 | 18975344 | 4 |
| 2001 | 21201272 | 12 |
| 2002 | 24671330 | 16 |
| 2003 | 32774929 | 33 |
| 2004 | 41986814 | 23 |
| 2005 | 47654826 | 13 |
| 2006 | 50102492 | 5 |

Figure 1 shows that fatalities caused by tank truck approximately of 10% and it higher than buses, passenger car and pedestrian, even though motorcycle is the highest number of fatalities in Indonesia. Therefore, vehicle safety of tank truck is quite needed in order to reduce number of fatalities [2].

There many causes of tank truck accident such as fatigue, engine problems, over speed, attitude and external causes as shown in Figure 2. Fatigue and over speed contributes 20% and 17% of total tank truck

[a] Corresponding author : hadi.pranoto@mercubuana.org

DOI: 10.1051/matecconf/20178702022

accident. Therefore, it very needed to solve that problems to improve safety driving and reduce tank truck accident [3].

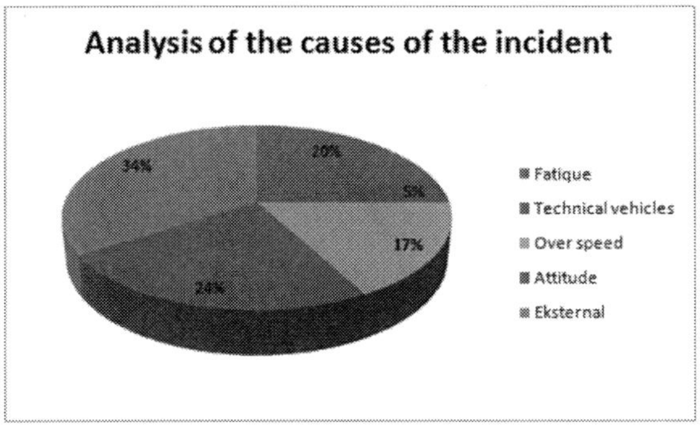


**Figure 2** Several factor of tank truck accident

According to KLH (1999) [4], Constitution No. 22 (2009) [5] and Ministry of Transportation, Indonesia (2015) [6] issued the regulation of speed limit based on area which consists of maximum speed of 100 km/h in highway, 80 km/h in urban area, 50 km/h in centre of urban area and 30 km/h in residential area. Moreover, PT. Pertamina Persero (2013) [7] has issued the regulation that the maximum speed of the tank truck in highway is 70 km/h. This regulation was issued to reduce the accident rate. However there are many driver not obey that regulation. Therefore, installing the additional part in the engine to limit the speed is quite needed, which may reduce the accident rate. In this research innitiate to develop new speed limiter which coupled by fuel cut-off solenoide to limit the speed by reducing supply fuel into the engine.

## 2 Methodology

### 2.1 Installation of speed limiter on the tank truck

Speed limiter is installed to the tank truck of PT. Pertamina Persero. It managed more than 6,000 tank truck scattered throughout Indonesia. This research focus on the Installation Jakarta Group (IJG) which located in Plumpang and they operates approximately 257 tank trucks with a various payload capacity from 16,000 to 40,000 liters.

### 2.2 Speed limiter installation

The modification of speed limiter installation has been done by improvement on electrical wire, which placed and connected with the engine. Speed limiter able to limit the speed at the certain speeds based on government regulation depending on the urban or highway speed. The speed limiter installation procedure is shown in Figure 3 and Figure 4 where it consists of buzzer, ECU, speed sensor, engine and speed limiter control box.

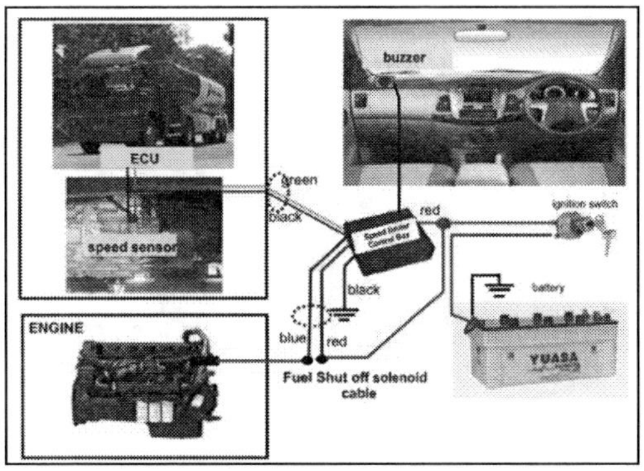


**Figure 3**. Diagram Installation engine truck

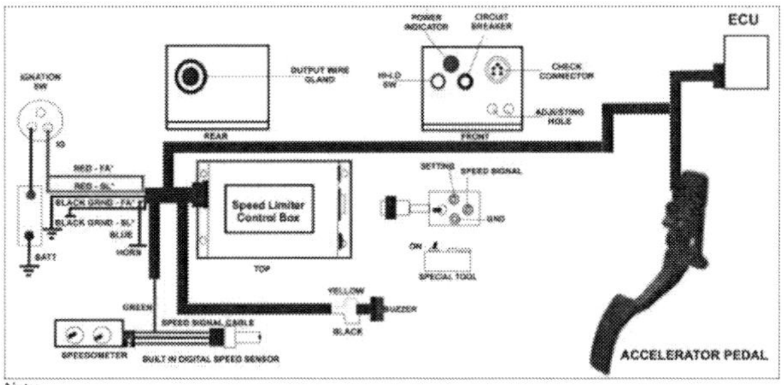


Note:
RED-FA= RED cable line-for fatigue alarm
RED-SL= RED cable line-for speed limiter
BLACK-GRAND FA= BLACK cable line-for fatigue alarm
BLACK-GRAND SL= BLACK cable line-for speed limiter

**Figure 4** Schematic diagram of speed limiter

The working principle of the speed limiter mounted on the tank truck machine. When the vehicle speed exceeds of a specified speed limit, speed sensor at the transmission output will send signal frequency to accelerator pedal. it will be transformed into a voltage signal at IC program and voltage signal will be transformed into the comparators components, forwarded to relay and followed by sound a warning alarm buzzer. When the driver still perform additions speed, fuel cut-off solenoid will reduce the fuel consumption into the engine and it will reduce the speed/rpm to the normal speed. Meanwhile, when the driver is slow down the speed, the buzzer alarm will be stop and back to the normal speed.

### 2.3 Speed reduction test

The speed limiter is installed to the engine and it controlled and adapted by government regulation. Basic principle of speed limiter is performed when the tank truck in over speed and it wills automatically cut-off the fuel. It make a tank truck speed will be operated under top speed regulation. Schematic diagram of speed limiter is shown in Figure 5

DOI: 10.1051/matecconf/20178702022

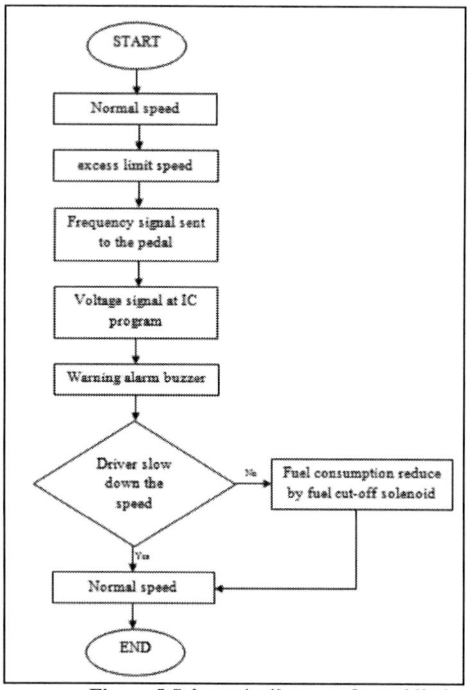


**Figure 5** Schematic diagram of speed limiter

# 3 Speed Limiter Development

## 3.1. Tank truck without speed limiter

The monitoring activity has been conducted for 1 month on selected tank truck before installed speed limiter. The data is summarized in Table 2 that the highest speed is shown by tank truck with ID B 9511 UFU for 133 km/h and the lowest speed shown by tank truck ID B 9211 SEH for 122 km/h. That speed is out of the regulation which mean that the attitude of the driver is not appropriate with safety driving guideline and government regulation. It causes many things such as fatigue, stress, angry that can make unstable driving condition.

**Table 2** Top speed of tank truck before installing speed limiter

| No. | Vehicle ID | Type of Vehicle | Top speed (km/h ) |
|---|---|---|---|
| 1 | B 9280 UU | Truck B3 | 125 |
| 2 | B 9281 UU | Truck B3 | 130 |
| 3 | B 9263 UU | Truck B3 | 130 |
| 4 | B 9264 UU | Truck B3 | 128 |
| 5 | B 9211 SEH | Truck B3 | 122 |
| 6 | B 9511 UFU | Truck B3 | 133 |

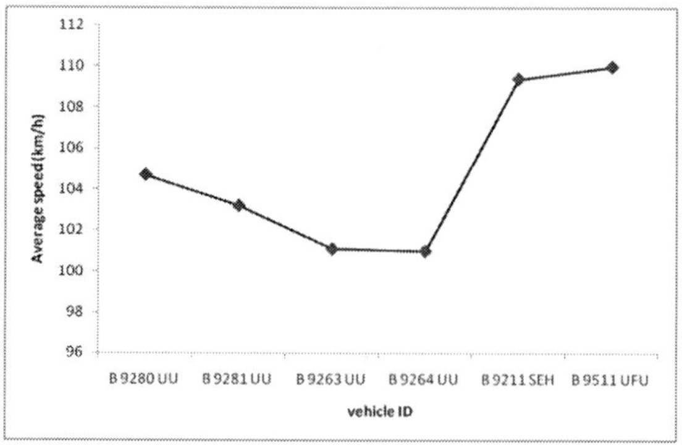


**Figure 6** Average speed of tank truck before installing speed limiter

From the Figure 6 shows that the average speed of tank truck before installed by speed limiter is in above 100 km/h. Meanwhile the regulation state that the maximum speed is 70 km/h. Tank truck which commonly perform highest speed is shown by B 9511 UFU with average speed of 110 km/h during 1 month monitoring activity and the lowest average speed performed by B 9264 UU for 101 km/h during 1 month. Those speed is very dangerous when performed in urban, centre of urban or residential area. It can enable for road capacity and it more possible to increase the fatalities which caused by road accident.

## 3.2 Tank truck after installed by speed limiter

Speed of tank truck monitored before installed by speed limiter has references point in speed limiter requirement. Therefore, this research investigates the influence of the speed limiter in limiting speed, improving safety riding and reducing possibility of accident. Speed limiter is installed to the tank truck, which coupled by fuel cut-off solenoid, which can automatically lock the speed in top speed regulation. The driver cannot speed up of 70 km/h because it controlled by speed limiter. Therefore, highest speed of the tank truck is 70 km/h. It is detail listed in Table 3 and the average of speed is shown in the Figure 7.

**Table 3** Top speed of the tank truck after installed by speed limiter

| No. | Vehicle ID | Type of Vehicle | Top speed (Km /h) |
|---|---|---|---|
| 1 | B 9280 UU | Truck B3 | 70 |
| 2 | B 9281 UU | Truck B3 | 70 |
| 3 | B 9263 UU | Truck B3 | 70 |
| 4 | B 9264 UU | Truck B3 | 70 |
| 5 | B 9211 SEH | Truck B3 | 70 |
| 6 | B 9511 UFU | Truck B3 | 70 |

 DOI: 10.1051/matecconf/20178702022

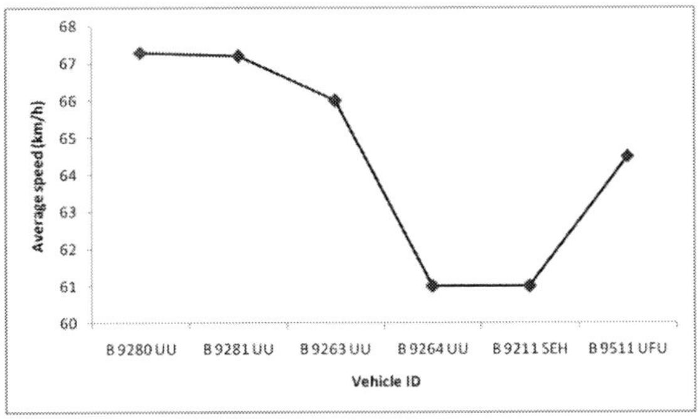


**Figure 7** Average speed of tank truck after installed by speed limiter

After speed limiter is installed in selected tank truck, top speed has been locked by 70 km/h and average speed also decreased because the driver cannot speed up. Lowest average speed is performed by B9264 UU and B 9211 SEH for 61 km/h and highest average speed is performed by B 9280 UU for 67.3 km/h. Therefore, the possibility of accident can be decreased because the driver in steady or relax condition. Top speed in adjusted by government regulation that the maximum speed for tank truck is 70 km/h in highway. Moreover, this system is purposed to the driver who able to full controlling speed of tank truck which led to safe driving could be performed by all drivers.

### 3.3 The improvement and analysis after install speed limiter

Speed limiter on the tank truck has monitored during 1 month and it can be used as references that the speed limiter is effective to control offer the speed and protect the performance of engine. Improvement is calculated from the comparison between average speeds of selected tank truck before and after installed by speed limiter shown in Figure 8. B 9263 UU shows the highest improvement for 65% and B 9211 SEH shows the lowest improvement for 56%. This data indicated that higher % improvement will increase percentage of safe driving has been performed by tank truck driver.

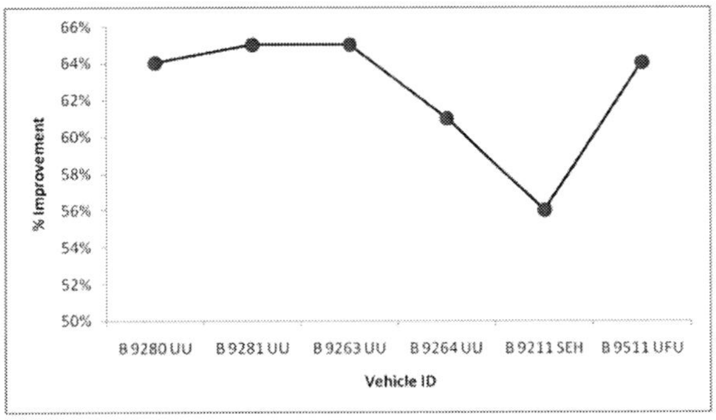


**Figure 8** Percentage of improvement in speed restriction

### 3.4 Report accident after installation speed limiter on the tank truck

Table 4 shows the number of incident which recorded by july 2016. There are several decreasing incident cases observed such as in over speed and environment for 7% and 0% as compared to last year (2015) report for 17% and 34%, respectively. Significant decreament number of incident is caused by speed limiter installation on the tank truck. Decreasing incident due to speed of the tank truck is limited by fuel cut off system. Therefore, the driver disable to speed up when the speed is on the top of regulation speed. When the buzzer is on indirectly influence to the concentration and alarm when the driver sleepy. Moreover, this product is very potential to applied in the tank truck in other country and it believed able to decrease the number of tank truck incident.

**Table 4** Percentage of tank truck incident for July 2016 [7]

| Incident Classification | Percentage | No. of cases |
|---|---|---|
| Over speed | 7 | 3 |
| Fatigue | 85 | 39 |
| Premature | 9 | 4 |
| Environment report | 0 | 0 |
| Total cases | 100 | 46 |

## 4 Conclusion

The effectiveness of the speed limiter in reducing the speed and control the driver is shown by speed decrement after speed limiter installed to tank truck. It performed by fuel cut-off solenoide which installed to the speed limiter system. Highest speed before installed speed limiter is 133 km/h and after installed by speed limiter, to speed is locked at 70 km/h. It has adapted by government regulation regarding to the transportation law. From the data, it can be concluded that the speed limiter is very suitable to install in the tank truck to improve safe driving, protect the engine performance, and reduce an accident.

## Acknowledgement

The authors would like to thank the Mercubuana University through the conference fees support and thanks to Pertamina Persero and Universiti Tun Hussein Onn Malaysia (UTHM).

## References

1. Rencana Umum Nasional Keselamatan (RUNK), Republic of Indonesia. Technical report 2011-2022 (2010)
2. T. Tjahjono. The effect of traffic and road condition to the fatality rates on rural road in eastern Indonesia. Journal of The Eastern Asia Society for Transportation Studies, **8** (2010).
3. B.S. Dhillon. Safety and Human Error In Engineering System, (2013).
4. Regulation Ministry of Environment (KLH) No.

74, (1999).
5. Constitution Number 22. Clause 203 Republic of Indonesia, (2009).
6. Regulation Manistry of transportation, Republic of Indonesia. PP No. 111, Procedures for determining vehicle speed limit, (2015).
7. PT. Pertamina Persero. Policy is expressed in an internal memo No. 154, (2013)
8. PT. Pertamina Persero Indonesia, Plumpang Site Report incident, (2016).

# Effect of Bottom Ash and Fly Ash as a Susceptor Material on The Properties of Aluminium Based Composites Prepared by Microwave Sintering

Wan Nur Azrina Binti Wan Muhammad[1 a], Muhammad Firdaus Bin Idris[1], Md. Nor Anuar Bin Mohamad[2], Saliza Azlina Binti Osman[3]

[1] *Department of Material Engineering & Design, Faculty of Mechanical Engineering & Manufacturing, Universiti Tun Hussein Onn Malaysia, Parit Raja, Johor, Malaysia*
[2] *Department of Thermofluid Engineering, Faculty of Mechanical Engineering & Manufacturing, Universiti Tun Hussein Onn Malaysia, Parit Raja, Johor, Malaysia*
[3] *Department of Manufacturing and Industrial Engineering, Faculty of Mechanical Engineering & Manufacturing, Universiti Tun Hussein Onn Malaysia, Parit Raja, Johor, Malaysia*

**Abstract.** The use of aluminium as a single material in automotive applications is not suitable without a mixture with reinforcement materials that can support the properties at high temperature. In this study, aluminium based composite were prepared with weight percentage of SiC reinforcement, varying from 5 to 20 wt%. Aluminium powder and reinforcement materials were mixed using ball milling machine with speed of 100 rpm for 2 hours. The powder mixture were then compressed at pressure 4 tonnes with 5 minutes holding time. The compact samples were sintered using microwave sintering technique. Microwave sintering techniques in this study using two different types of susceptor materials that are bottom ash and fly ash. Sintered aluminium based composites using bottom ash susceptor material involving the sintering temperature of 526 °C for 30 minutes whereas for the samples sintered using fly ash susceptor material, involving a temperature of 523 °C for 15 minutes. From the result, the sintered samples using fly ash susceptor material, showed higher density with a value of 2.1933 g/cm³ compared to bottom ash 2.0002 g/cm³ and having the higher hardness value 72.1315 HV compared to bottom ash 50.0511HV. The using of fly ash could affect the heating rate during the sintering process which could influence the properties of aluminium based composites. In conclusion, the type of susceptor could affect the physical and mechanical properties of aluminum-based composite reinforced with silicon carbide.

## 1 Introduction

Aluminium based composite is the material that is in category of metal matrix composites. The aluminium based composites material comprising aluminium as the matrix material and the reinforcing material that normally used silicon carbide SiC to change the properties of the original material to be stronger in terms of physical and mechanical properties. The automotive industry today has a lot of application which need the use of composite materials for replacing aluminium products such as vehicle components. Components such as disc brakes, pistons and engine block is a component that is usually manufactured using aluminium composites [1]. Silicon carbide (SiC) material, is a compound consisting of silicon and carbon. SiC is one of the types of reinforcement materials used in metal matrix composites. The purpose of SiC in the metal matrix composite as reinforcement to strengthen the existing microstructure and properties of the matrix material used. For example, the matrix material used in this study are aluminum. With the addition of reinforcement materials such as SiC, physical and mechanical properties of the material will be changed. With the use of applications that are often associated with friction such as disc brakes, the addition of SiC in metal matrix composite material is very suitable because it will increase the wear resistance properties of the material [2]. Generally, Powder metallurgy (PM) process is well known to be one of excellent metal forming techniques for producing high quality structural components with near-net shape and also can produce alloys and composites that cannot be fabricated by conventional melting and casting processes [3].

Starting with the process of mixing metal powder in which the matrix material and reinforcement material to be mixed by using ball milling machine. This mixing process is intended to provide a homogeneous mixture of matrix material and reinforcing material. In the usual PM production sequence, the powders are compressed into the desirable shape and then heated to cause bonding of the particles into a hard, rigid mass [3]. The heating process namely as sintering. Sintering is a heat treatment operation performed on the compact to bond its metallic particles, thereby increasing strength and hardness [4]. There are challenging demands from the PM industry for new and improved sintering process with finer microstructures and enhanced physical and mechanical properties. And thus, this is where the microwave technology is found to be advantages compared to the conventional sintering technique [5]. Microwave sintering involves a microwave energy, form an electromagnetic energy at frequencies between 300MHz rates up to 300 GHz [3]. In microwave heating, the material used for the sintering will work alongside the microwave absorbing electromagnetic energy which in turn is converted into heat. Theoretically, microwave heating is a method where heat is generated in the material first and then followed by heating the entire volume of material. In microwave sintering process, the susceptor material was turned the microwave energy into

[a] Corresponding author: wazrina@uthm.edu.my

 DOI: 10.1051/matecconf/20178702023

sensible infrared heat energy which was used to sinter the compacted sample. In addition, the susceptor material was helped in distributing microwave energy more evenly which reduced run away heating. According to previous studies by Oghbaei & Mirzaee [5] involving austenitic stainless steel (316L) and ferritic (434L) which had been sintered by both technique of sintering, microwave and conventional sintering, respectively. In their study, it found that the microwave sintering technique indeed has a rapid heating time compared to conventional sintering techniques. The sintering technique used also plays a role in which the microstructure formed by sintering technique depends on whether the techniques used a conventional sintering or microwave sintering technique. The microstructure formed for both of these techniques showed a different where microstructure of the sample that has been sintered by conventional sintering techniques looked coarser than the samples sintered by microwave sintering technique. This is because the microwave sintering restrict the microstructural coarsening. Upadhyaya et al. [5] compared a difference on the heating mode on microstructure and mechanical properties of materials 92.5W-6.4Ni-1.1Fe alloy. The rate of microwave heating is 20 ° C / min while the conventional furnace is heating by 5 ° C / min. The results obtained also showed a different microstructure for both sintering techniques. The microstructure of the sample sintered by microwave sintering technique was finer than the microstructure sintered by conventional sintering techniques. Besides that, from analysis phase in conventional sintering techniques, it indicate that the presence of brittle intermetallic phase but it does not seen at the samples that has been sintered by the microwave sintering [5]. The advantages of microwave sintering can be proved by the properties of a sintered material. This is due to the sintering method will influence the size of the microstructure and the pore size. In conventional sintering, the microstructure is coarser and the pore size is bigger whereas microwave sintering techniques produce smaller pores and lead to more robust power generated. The study has been proved by Dinesh Agrawal [6], where the test has been conducted on a sample hollow cylinder FC-208 proven sample using conventional techniques fail to load 320 lbs while the samples using microwave sintering technique failed at load 430 lbs. In this study, the advantages of the microwave sintering is used to prepared the aluminium based composites reinforced with silicon carbide by using two types of susceptor i.e. bottom ash and fly ash. The function of the susceptor in microwave sintering is important as it could heat quickly, retain and radiating the heat to the sample.

## 2 Experimental Procedure

In this study, the material that has been used was aluminium powder with various percentage of SiC from 5 wt%, 10 wt%, 15 wt% and 20 wt%. Both powders were mixed in ball milling at 100rpm for 2 hours. The mixture of the powder were compacted at 4 tonnes load and sintered in microwave oven.The microwave oven was modified by inserting three insulator board and used two type of susceptor i.e fly ash and bottom ash. Figure 1 shows the schematic diagram of microwave oven modification set up. Before sintering the materials, the temperature calibration has been done to determine the sintering temperature and heating rates. The compacted samples had been sintered at 526°C in 30 minutes for bottom ash susceptor and 523°C in 15 minutes for fly ash susceptor. The susceptor is used to ensure that green powder samples are being exposed sufficiently to the electric field at high temperature even though the penetration of susceptor decreases as the temperature risen. A susceptor is a component that has high dielectric loss while passing through the low loss material with little drop in energy instead of only absorbing energy. At lower temperature, susceptor material absorbs microwave and reaches higher temperature and transfer the heat to the sample [5]. Both sintering processes did not have a holding time because the process was too fast compared to conventional sintering technique. The sintered samples were 15 mm in diameter with 8 mm in thickness. The bulk density and apparent porosity of the sintered samples were measured according to Archimedes principle by immersion test and electronic balancer with the accuracy of + 0.0001g. Microstructure observation was conducted by using optical microscopy. The hardness of the sintered samples was measured by means of Vickers Hardness Tester with an indenting load of 30kgf and a holding time of 10s.

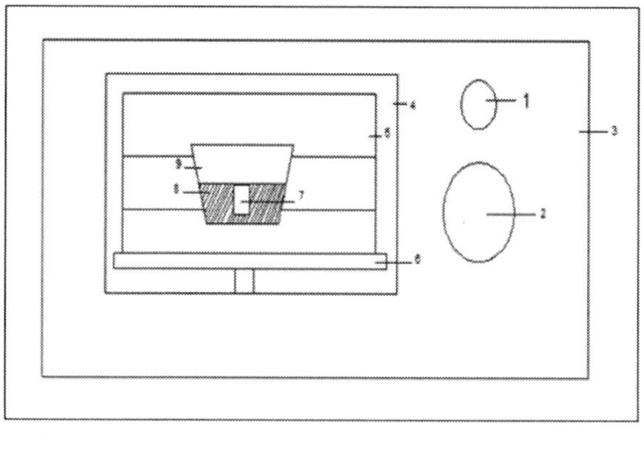

**Figure 1.** Temperature calibration profile of microwave sintering using bottom ash susceptor. 1) Temperature controller, (2) Time control button, (3) Microwave oven, (4) Controllled atmosphere chamber, (5) Insulator boards, (6) Turntable, (7) Sample, (8) susceptor, (9) crucible

## 3 Results and Discussion

### 3.1 Heating rate

Heating rate of the microwave sintering by using bottom ash and fly ash susceptors has been determined by performing the temperature calibration before conducting the sintering process. It is important to study the effect of type of susceptors to the heating rate during microwave sintering. During temperature calibration, temperature data has been taken in every 5 minutes as shown in Figure 2.

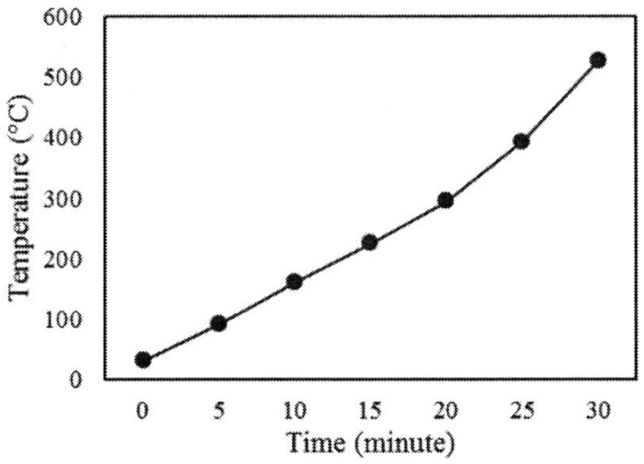


**Figure 2.** Temperature calibration profile of microwave sintering using bottom ash susceptor

In Figure 2, it shows the increase of temperature as time increased for microwave sintering using bottom ash as a susceptor material. It took 30 minutes to achieve the sintering temperature of 526°C.

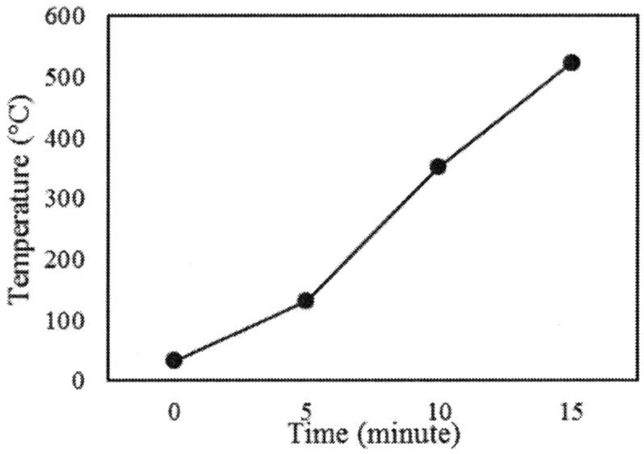


**Figure 3.** Temperature calibration profile of microwave sintering using fly ash susceptor

In Figure 3 shows the temperature calibration profile for sintering process using fly ash susceptor. It showed that the temperature increased up to 523°C only in 15 minutes. It means that the heating rate of the sintering process using fly ash susceptor is higher than bottom ash susceptor. The difference in heating rates has been attributed to the difference of composition of susceptor. Bottom ash and fly ash contain different percentage of silica. Fly ash has higher percentage of silica compared to bottom ash. The silica is an element that has high thermal conductivity. The higher heating rate has been attributed by the element silica in susceptor materials that has been used. Higher heating rate would improve the sintering process using microwave. This has allowed the microwaves to penetrate and then the susceptor will radiate the heat energy to the sample. The microwave simultaneously coupled with green body sample and penetrated heat internally due to penetrating feature of microwave in samples. This was led to volumetric heating of samples with short duration. [7]

## 3.2. Microstructure observation

Figure 4 and Figure 5 show the microstructure of aluminium based composites sintered by microwave sintering using bottom ash susceptor and fly ash susceptor, respectively. From the microstructure, white region refers to the aluminium matrix, grey region refers to SiC and black region refers to the presence of porosity. In the Figure 4, it was noted that the porosity of the sample sintered with bottom ash susceptor can be clearly seen compared to the samples that sintered with fly ash susceptor. In Figure 5, the microstructure is smooth and less pores. The presence of pores in the sample is related to microwave sintering technique which depend on the heating rate used in the microwave oven [7]. It is proved that the heating rate during microwave sintering is better when used fly ash as the susceptor.

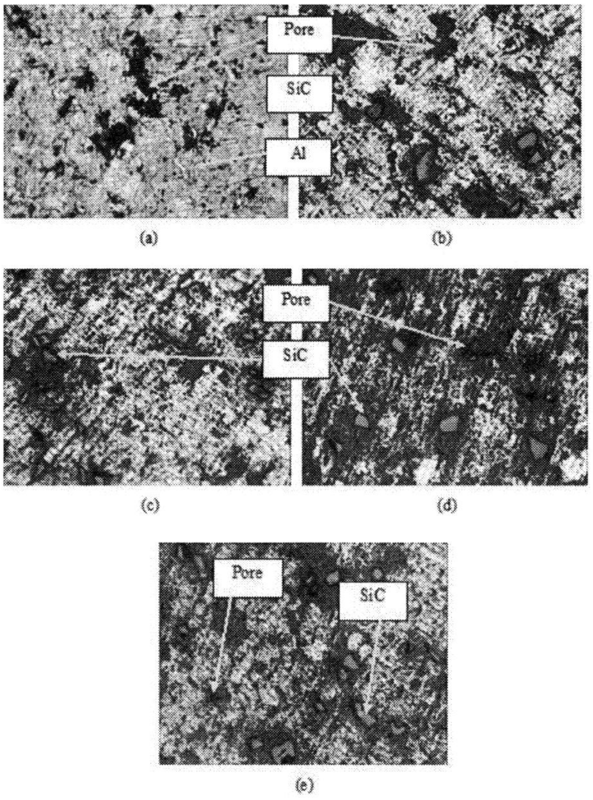


**Figure 4**: Microstructure behaviour at different composition by using bottom ash susceptor: (a) 0 wt%, (b) 5 wt%, (c) 10 wt%, (d) 15 wt%, (e) 20 wt%

DOI: 10.1051/matecconf/20178702023

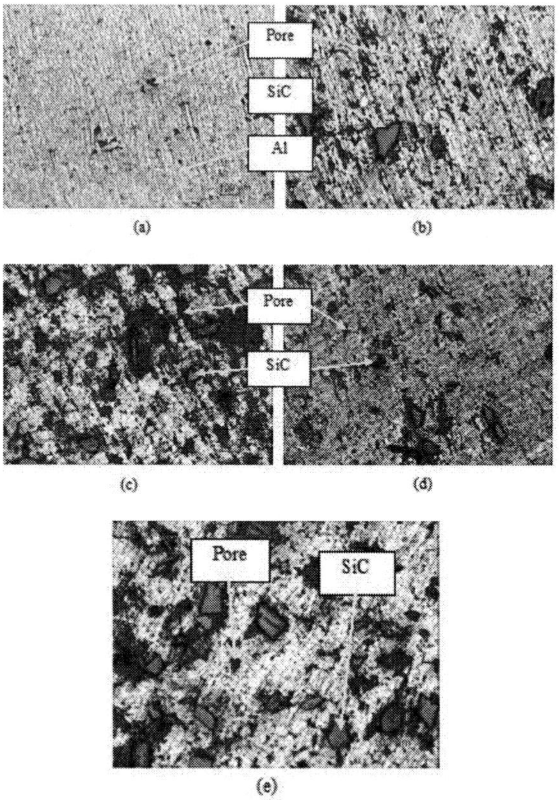


**Figure 5.** Microstructure of aluminium composites reinforced with SiC at different composition by using fly ash susceptor: (a) 0 wt%, (b) 5 wt%, (c) 10 wt%, (d) 15 wt%, (e) 20 wt%.

### 3.3 Density

In Figure 6 it shows the density of the aluminium based composites as function of percentage of SiC reinforcement. As it can be seen, the density of the sample for both type susceptor shows not much different but it shows the increment as SiC reinforcement increased. However, In previous studies, the density increased as the content of SiC reinforcement increased [9]. This is due to percentage of SiC reinforcement reduce the pores in aluminum matrix. Besides that, the increment of density showed the good densification in the sample by using microwave sintering and this could be related to the function of susceptor that controlled the heating rates during the sintering process. Rapid heating during microwave sintering generally improved the densification for certain sintered. The rapid heating allows atomic diffusion for sintering process. As a result, more atoms move to points of contact between powder particles to form sintering necks. When the necks grow, the pore sizes are reduced this causes an increase in the final density of the sintered products.

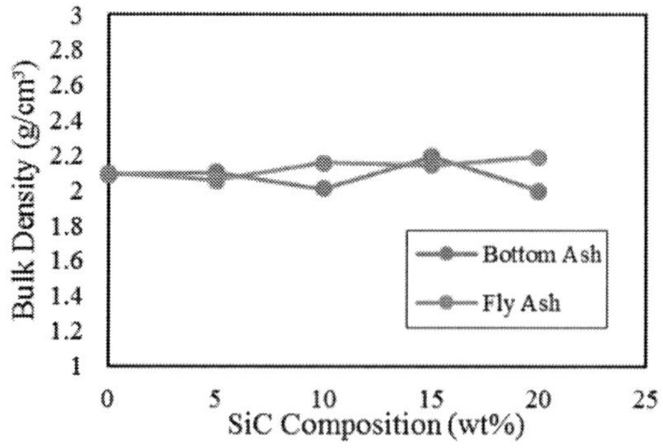


**Figure 6.** Bulk density of aluminium composites as function of SiC composition.

### 3.4 Porosity

Porosity test is a physical test to determine the percentage of porosity exist in the sample. Existing porosity of the material will affect the density of the sample. From Figure 7 it shows that aluminum composite samples sintered with fly ash has lower percentage of porosity compared to the aluminum composite material that sintered using bottom ash susceptor material. The value of apparent porosity is related to the value of bulk density. The presence of porosity reduced the densification in sintered aluminum composites. Microwave heating rate also plays a role which the high heating rates will contribute to a low porosity [8].

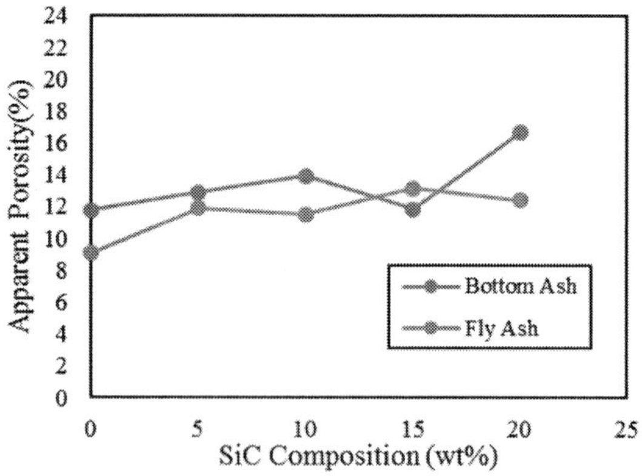


**Figure 7.** Porosity of aluminium composites as function of SiC composition.

### 3.5 Hardness

Figure 8 shows the relationship between hardness and SiC composition for aluminium based composites prepared by microwave sintering with two different susceptor used. The value of hardness was increased with increasing SiC composition for both type of susceptor. By using fly ash as susceptor, the value of hardness is higher than that of sintered by using bottom ash as susceptor material. It revealed that the use of fly ash could improve the properties of aluminium composites. The improvement of the hardness is related to the densification of the composites. High density and low porosity will give higher hardness value. In microwave sintering, the densification of the sample is related to the susceptor material that used during the process. The properties of susceptor material could affect the heating rate during sintering. Fly ash and bottom ash contain the silicon dioxide (silica) in its composition. The element of silicon dioxide in fly ash is higher (65.01 wt %) than bottom ash (51.18 wt %) [10]. Silicon dioxide has high thermal conductivity. Higher thermal conductivity contribute to the effectiveness of sintering process as the wave could be absorb more and transfer the heat in the sample [3]. As a consequence, the heating rate is increased and sintering of the composite samples were completed. The completion of sintering process could be measure by the improvement of the physical and mechanical properties of the composites.

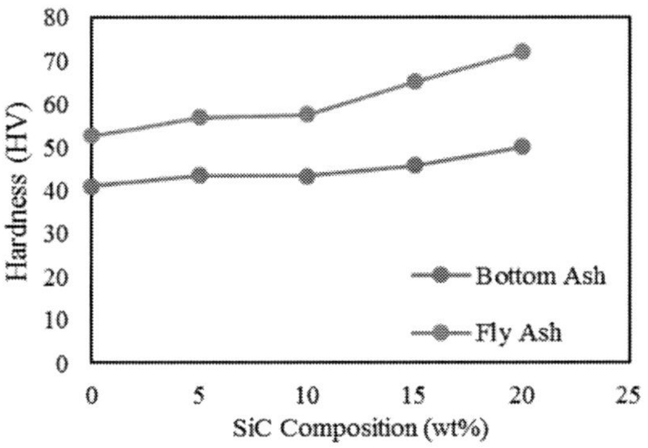


**Figure 8.** Hardness of aluminium composites as function of SiC composition.

## Conclusion

The main objectives of the study are to determine the effect of bottom ash and fly ash as susceptor during microwave sintering which could improve the physical and mechanical properties of sintered aluminium composites reinforced with silicon carbide. It was found that fly ash susceptor would increase the heating rates during microwave sintering of aluminum based composites samples. Fly ash susceptor has been achieved the higher density, lower porosity and highest hardness compared to the sample prepared by using bottom ash susceptor. The good quality of the sintered sample due to the composition of silica in fly ash is higher than bottom ash. The silica in susceptor material helps to improve the heating rates during sintering due to high thermal conductivity properties of silica. Overall, it has been concluded that sintered aluminium based composite reinforced with silicon carbide with fly ash susceptor during microwave sintering improved the mechanical and physical properties of the composites.

## Acknowledgement

The work was partly supported by University Tun Hussein Onn Malaysia under the Research Supporting Grant Scheme (RSGS) Vote U104.

## References

[1] M. Walczak, D. Pieniak, and M. Zwierzchowski, "The tribological characteristics of SiC particle reinforced aluminium composites," *Arch. Civ. Mech. Eng.*, vol. 15, no. 1, pp. 116–123, 2014.

[2] K. R. Padmavathi and R. Ramakrishnan, "Tribological behaviour of Aluminium Hybrid Metal Matrix Composite," *Procedia Eng.*, vol. 97, pp. 660–667, 2014.

[3] P.Perez, G.Garces., P.Adeva, Influence of texture on the mechanical properties of commercially pure magnesium prepared by powder metallurgy. Journal Material Science, 2007. 42: p. 39693976.

[4] Groover M.P (2002). "Fundamental of Modern Manufacturing."2nd .ed. John Wiley & Sons,Inc.187.

[5] Oghbaei, M., and O. Mirzaee. "Microwave versus conventional sintering: A revies of fundamentals, advantages and applications. " *Alloys and Compounds*, pp 175-189, 2010

[6] Dinesh Agrawal (The Pennsylvania State University), "Microwave Sintering of Ceramics, Composites and Metallic Materials, and Melting of Glasses," *Transactions of the Indian Ceramic Society*, vol. 65, no. 3. pp. 129–144, 2006.

[7] S. Leparoux. S. Vaucher, O. Beffort, Assessment of Microwave Heating for Sintering of Al/SiC and for in-situ Synthesis of TiC, *Adv. Eng Mater* 5, pp 449-453, 2003

[8] O. Ertugrul, H.-S. Park, K. Onel, and M. Willert-Porada, "Effect of particle size and heating rate in microwave sintering of 316L stainless steel," *Powder Technol.*, vol. 253, pp. 703–709, 2014.

[9] M. V. Krishna and A. M. Xavior, "An Investigation on the Mechanical Properties of Hybrid Metal Matrix Composites," *Procedia Eng.*, vol. 97, pp. 918–924, 2014.

[10] Y. Zarina, A.M.Mustafa Al Bakri, H. Kamarudin, I.K.Nizar and R. Rafiza, " Review on the various ash from palm oil waste as geopolymer material." *Rev. Adv. Mater. Sci.*, vol. 34. no.1, pp. 37-43, 2013.

MATEC Web of Conferences 87, 02024 (2017) DOI: 10.1051/matecconf/20178702024
*ENCON 2016*

# Similarities of lean manufacturing approaches implementation in SMEs towards the success: Case study in the automotive component industry

Rose, A.N.M[1], Ab Rashid, M.F.F[1], Nik Mohamed, N.M.Z[1]., Ahmad, H.[2]

[1]*Faculty of Mechanical Engineering, Universiti Malaysia Pahang, 26600 Pekan, Pahang, Malaysia*
[2]*TATI University College, 24000 Kemaman, Terengganu, Malaysia*

**Abstract;** Nowadays, manufacturing companies are striving for a better system like lean manufacturing (LM). The primary objective of LM is to identify and eliminate wastes. LM can be applied successfully in all industries providing a full understanding of lean ingredients i.e. concept, principles, and practices. There are a lot of practices which are necessary to be implemented in order to gain full benefits of LM. However, small and medium enterprises (SMEs) are lack of knowledge in LM and facing difficulties to adopt all of the LM principles. Therefore, it is necessary to the researchers to come out with a simple guideline for LM implementation. The objective of this paper is to explore the journey of LM implementation including preliminary, in process and post of LM. This research was conducted through multi-case study research. There were four SMEs and two large companies. The gathered information shows that the preliminary stage of LM implementation is similar to each other including large companies. The result shows SMEs still have a potential to success in LM. This finding might give an opportunity to SMEs to prepare the basis for LM implementation effectively. As a result, SMEs able compete in the competitive global marketplace and strive for world class performance through implementation of LM.

## 1 Introduction

Today, Small and Medium Enterprises (SMEs) can be considered as the backbone of an economy. SMEs are among the suppliers of goods and services to large companies. The challenging economy has forced SMEs to become more responsive and flexible in operational, tactical and strategic areas. In order to survive nationally and internationally, SMEs need to produce high-quality products at competitive prices. Hence, globalization has lead companies to find the right management system to improve current performance. According to [1]Rineheart *et al.*, Lean Manufacturing (LM) will be the standard management system of the 21$^{st}$ century. It will be acceptable and adaptable to all industries. The main advantage of LM is its ability to reduce production cost through elimination of wastes, and if well implemented, it can assist a company to become a world-class organization [2]. The elimination of wastes has given a substantial impact on manufacturing companies, resulting in higher performance enhancement and significant improvement in delivery, quality, flexibility and manufacturing cost [3]. Researchers and practitioners widely knew the benefits of LM after the book titled "The machine that changed the world" is published [4]. Later, LM became an established field of research. The majority of past studied had concentrated their investigation on lean performance, lean indicators and lean implementation [5],[6],[7],[8]. Farris *et al.* [9] have investigated the importance of critical success factors for effective and successful LM implementation. Up till now, the majority of LM researchers were focusing on large companies, and little has been researched on SMEs [8],[6],[10]. The objectives of this study are to explore the early preparation and ongoing development activities for LM implementation. Apart from that, this study also would like to explore the benefits and barriers of LM application. This was due to many implementations to date were based on trial and error. The study shows that most of the companies were applying same practices and follow similar steps towards the success.

## 2 Literature review

LM can be applied effectively with a strategic and systematic approach [11]. Therefore, the implementation and selection of the right LM practices must be based on knowledge and experience [12]. Currently, there are many of LM practices available and being practiced by the practitioner [13]. The company should prioritize and decide which practice to start first [14]. Large company do not have any difficulties and constraint in adopting LM practices and are likely to implement all of the LM practices, unlike SME [15]. Therefore, SMEs are encouraged to start with the feasible LM practices such as 5S, multifunction, continuous improvement and reduction in setup time [14][16][17]. These practices are considered simple, easy, least investment and able to be managed internally. Understanding and guidance on lean practices implementation are very crucial to SMEs. The top management and responsible person need to know how to get started, where to start and how to proceed [18]. Therefore, in this paper, the authors would like to identify the route to success for SMEs to implement LM such as early preparation for the appointment of steering committee, team formation and also the selection of production line which applies to them. A comprehensive literature review on LM enables the authors to categorize LM practices into three groups based on company size, piecemeal practices and number of years with established LM [15],[17],[16]. Piecemeal practices are those can be applied independently such as 5S, preventive maintenance and multifunction employees. Thus, the SMEs should take into consideration their ability before embarking on LM. Rose et al. [14] have proposed three categories of LM practices; basic, intermediate and advanced. This perhaps will assist SMEs to use it as a guide for LM implementation and will also overcome difficulty factors in LM implementation, such as lack of understanding and wrong methodology [19]. Various studies have been carried out on LM practices, such as based on lean bundles, company sizes, operation performance, infrastructure and location, but none of them discussed on feasible practices [20], [21],[7],[22],[17], [23]. The study of LM in Malaysia is still limited [24] and it is an excellent opportunity for researchers to explore on how LM can be implemented here. Apart from LM practices, the company also need to consider on early preparation, best implementation and evaluation [18]. Therefore in this study, the authors would like to explore the stage of preparation on LM implementation including focusing on feasible practices. In addition, the authors also try to highlight the significant activities need to be implemented in order to ensure the LM is succeeded.

## 3 Methodology

This research was conducted by using multi-case study. One of the advantages of case study research can examine how and why to the key personnel that involves in the research area [25]. The main difference of case study from other methods is that it is not sampling research [26]. In this research, six case study companies were investigated. The advantage of multi-case studies can provide robustness and compelling data as compared to a single case study [27]. Also, multi-case studies can generate a replication logic which could not be produced by a single case study. And it is good for comparison purposes [28]. Therefore, the authors confident that this methodology could provide good research to investigate and compare the distinguish characteristics of LM implementation among the multi-case studies.

The case study protocol was developed for interview purpose. The purpose of this protocol is to ensure the information is consistently asked without bias. In this protocol, the structured questions were design based on LM practices, preliminary preparation on LM, benefits and barriers. It also contains the overview of the case study, project including objectives, LM issues and relevant findings on LM implementation as prescribed by Yin [29]. All proposed LM practices were based on comprehensive literature. This protocol could be one of the methods to ensure the reliability is maintained [30]. On the validity of information, the first gathered information through first case study will be used as a reference for the next case study. This replication logic based on the first case study in the next case study could be considered as part of external validity [30].

The six companies were selected based on the list of the respondent that show interest during last survey questionnaire which was conducted earlier. All companies were informed within two weeks before the visits and also emailed a case study protocol for early preparation. These will give enough time for preparing and reserving their time for the visits. The interviews were conducted with LM coordinator and production managers. All transcribed information were checked immediately by the interviewee, as to ensure all recorded data are correct and genuine. Besides the interviews, observations were also carried out on documents, notice board, poster, and company environment. The details of company name are treated confidentially. In this paper, the companies are known as Company A, B, C, D, E and F.

## 4 Company background

Table 1 shows the summary of case study companies. There were four SMEs and two large companies. All companies are automotive component background. Most of the case study companies implemented LM in the second trial since the first trial was failed due to poor commitment from top management and employees. All companies were established more than 20 years but are considered new in LM implementation with 4 to 5 years experiences especially SMEs. However, large companies have implemented LM to more than ten years.

**Table1.** Companies background

| | Company A | Company B | Company C | Company D | Company E | Company F |
|---|---|---|---|---|---|---|
| Trial | 2 | 1 | 1 | 2 | 2 | 2 |
| Type of product | Interior | Metal | Mechanical | Metal | Mechanical | Electronic/ Mechanical |
| Age (year) | 39 | 22 | 49 | 27 | 35 | 35 |
| Lean establishment (year) | 4 | 5 | 5 | 5 | 12 | 17 |
| Ownership | Local | Local | Local | Local | Local | Foreign |
| Size | SME | SME | SME | SME | Large | Large |
| MAJAICO | Member (2011) | Member (2009) | Member (2009) | Member (2009) | Model (2006) | Model (2007) |

All case study companies started LM from Malaysia Japan Automotive Industries Cooperation (MAJAICO). This cooperation is under the Malaysia-Japan Economic Partnership Agreement, which was signed in July 2006. The objective of MAJAICO is to inculcate, develop and improve Malaysian automotive companies to become more competitive as manufacturer players. One of the MAJAICO's objectives is to promote LM in Malaysian automotive operation. Two companies, E and F were recognised as LM model companies. The new company can learn and get advice from these two companies on any LM matters. All case study companies are local except one company is considered foreign which headquarter is in Japan. They are certified ISO9001 and TS16949. In brief, Company A is SME that manufactures the interior parts of a car. This is the second trial on LM implementation. The objective of Company A in LM application is to maximize company's profit. Company B is also under SME category and it is the first time in LM implementation. This company produces metal parts. Company B is very confident in LM ability towards the elimination of wastes. Company C is also considered as SME. This company manufactures mechanical parts. The interviews show this company is very committed to LM implementation. Company D is also SME. This company produces metal parts. The main reason for implementing LM is to maximize profits through reduction of operation cost included the elimination of wastes. This is a second time for LM implementation. The first time implementation failed due to poor employees' commitment and lack of knowledge on LM. There are two large companies in this case study for comparison purpose, Company E and F. Company E produces mechanical parts. This company has implemented LM twice. Again the same problems occurred on the failure of LM implementation due to lack of employees' commitment and lack of knowledge. Company F is a large company and is considered as foreign ownership. This company produces electronic and mechanical parts. According to the person in charge in LM, the first trial implementation was failed due to the management not serious in LM application. Despite of economy quite slow in the year of 1998, the top management insisted LM implementation as part of strategies to eliminate production wastes.

## 5 Discussion

Multiple case studies were designed to explore the similarity approaches which were carried out by the case study companies in LM implementation. During these investigations, the researchers discovered the similarities on the initial implementation, LM activities, relevant issues, benefits and challenges in LM implementation. Table 2 shows the summary of activities that were carried out by the case study companies. All case study companies were applying similar activities starting from basic activities such as the formation of steering committee and formation of LM unit as highlighted in Table 2.

**Table 2.** Lean Manufacturing implementation in case study companies

| Particulars | Company A | Company B | Company C | Company D | Company E | Company F |
|---|---|---|---|---|---|---|
| **Basic** | | | | | | |
| Formation steering committee | √ | √ | √ | √ | √ | √ |
| Formation unit of LM | *√ | √ | √ | √ | √ | √ |
| Incentive scheme | √ | √ | √ | √ | √ | √ |
| Formation of LM team | √ | √ | √ | √ | √ | √ |
| Production line selection | √ | √ | √ | √ | √ | √ |

LM could not run successfully without a steering committee. In this study, all case study companies have started with the formation of steering committee. This committee comprised of top management including production manager, quality manager and sales manager led by the senior manager. The function of this committee is to develop policy, budget, resources and company direction based on LM. Even, they should always support the LM team and employees on obstacle matters. To support LM implementation, all case study companies agreed that the formation of LM unit is a must for better coordination purpose. Most of the case study companies appointed LM coordinator for coordinating LM implementation including training arrangement. Company A, D, E and F have experienced on the failure of LM due to no dedicated unit is assigned for LM management. Most of their employees were discouraged due to lack of morale support and motivation from top management on LM implementation. Therefore, the company who want to success in LM implementation should always motivate their employees through providing an incentive scheme which able to boost the employee initiative in LM activities. All case study companies confirmed that they are providing incentive scheme to their employees who involved in LM in term of financial, recognition, free parking and present such as hamper. All companies agreed that the company should select a production line as a model for LM implementation. In this study, all case study companies started with a production line which has a significant problem on wastes. Company A, C, and D had chosen the production line which has high work in progress. However, Company B, E, and F decided the line which has high lead time. Then, the successful approaches in this production line will be applied to another production line. This approach could increase the employee's spirit and able to motivate them to run LM effectively.

On the LM implementation, most of the case study companies were implemented same practices such as 5S, continuous improvement, reduce set up time, visual display, value stream mapping, cell layout and improvement team as tabulated in Table 3.

**Table 3.** Preliminary implementation of lean manufacturing

| Company | A | B | C | D | E | F |
|---|---|---|---|---|---|---|
| 5S | √ | √ | √ | √ | √ | √ |
| Continuous improvement | √ | √ | √ | √ | √ | √ |
| Reduce set up time | √ | √ | √ | √ | √ | √ |
| Visual display | √ | √ | | √ | √ | √ |
| Value stream mapping | √ | √ | √ | √ | √ | √ |
| Improvement team | √ | √ | √ | √ | √ | √ |
| Cell layout | | √ | √ | | √ | √ |

5S is considered the key to success in LM implementation [31]. It is a good starting point for 5S implementation in a selected production line. All interviewees highlighted that 5S implementation is easy and feasible to be applied, provided all employees are very serious and committed. Through 5S, all problems are visible compared to before 5S implementation. Like Company C, their workers are happy with the new environment which is clean and very tidy. Apart from that, all case study companies managed to reduce searching time on their tools when needed. Next practice which was highlighted very important is continuous improvement. The company should encourage their employees to implement continuous improvement through suggestion scheme. Continuous improvement can be in any activities as long it could provide benefits to the company and employees. Good and feasible suggestion will be rewarded as approved in the incentive scheme. Company A and Company B similarly provides an incentive scheme to a good suggestion by giving a certain amount of cash per suggestion. Other than financial scheme, Company E has provided a special parking lot for employees who give an excellent suggestion.

Table 4 shows the benefits gained by the case study companies. Most of these companies can identify non-value added activities through VSM. Like company A and B could reduce waiting time by 10% after reallocating inspection area. In average the case study companies able to reduce lead time by 10% to 15% through set-up time reduction. All activities and achievement were displayed in notice board. Indirectly, this information could motivate the employees to implement LM especially when there is some improvement on the action taken. Similarly, sharing information through notice board is also practiced in the ceramic company for displaying their quality achievement [32].

**Table 4.** The benefits of lean manufacturing

| Particulars | Company A | Company B | Company C | Company D | Company E | Company F |
|---|---|---|---|---|---|---|
| Reduced inventory | | √ | √ | √ | √ | √ |
| Reduced floor space utilization | √ | | | | √ | √ |
| Reduced lead time | √ | √ | √ | √ | √ | √ |
| Delivery on time | √ | √ | √ | | √ | √ |
| Increased productivity | √ | √ | √ | √ | √ | √ |
| Effective cash flow | | √ | | | √ | √ |
| Increased quality | | √ | √ | | √ | √ |

All companies agreed that the highest achievement benefits from LM implementation are reduced inventory, lead time, increase productivity and deliver on time. The improved percentage of inventory was discovered about 20% to 40%. Apart from that, the lead time also has improved from 40% to 50%. As a result, their productivity has increased about 20% to 30%. However, some companies did not gain the benefits of floor space utilization due to still under transition period. They were focusing on inventory management and lead time reduction. Frequent delivery arrangement to the customer. Like Company B has planned two trips morning and afternoon for customer delivery. If missed delivery, the production manager has to check and find out what's wrong with the production. By practicing good practice, the company able to monitor and control the production plan effectively. Thus, their cash flow improved when the inventory is maintained at the minimum level compared to previously.

**Table 5.** The barriers of lean manufacturing implementation

| Particulars | Company A | Company B | Company C | Company D | Company E | Company F |
|---|---|---|---|---|---|---|
| Frequent quit the job | √ | √ | √ | √ | √ | √ |
| Lack of employee involvement | √ | √ | √ | √ | √ | √ |
| Lack of knowledge in LM | √ | √ | √ | √ | √ | √ |
| Organisation culture | √ | √ | √ | √ | √ | √ |
| Lack of employee commitment | | √ | √ | √ | √ | √ |

All case study companies claimed that five barrier factors that caused the failure of LM implementation as highlighted in Table 5. These factors are high employee turnover, lack of employee involvement, lack of knowledge in LM, organization culture and lack of employee commitment. One of the reasons employees keep changing the job is due to better offers from other company. Despite this, the companies have to retrain new employees on LM implementation. This factor has caused the companies have to spend money and time twice on the same activities. The companies agreed that the training is very importance to be provided to their employees including top management. Most of them revealed that their employees were unmotivated on LM due to lack of understanding on the LM concept and practices.

Additionally, this factor causes poor employees commitment to the LM implementation, especially when the current system has to change to a better system such as using pull system. However, they were still practicing the old system. The challenge of employee commitment is they do not understand the LM system. The case studies have shown that the training on LM is considered compulsory to employees and top management. Through systematic training, the employees able to foreseen the potential improvement included the contrasting between LM and the current system.

Apart from that, the observation and interviews also discovered the importance of regular meeting and follow up on LM implementation. These activities could overcome any facing problem in the premise due to LM. All responsible persons have to report on LM achievement and problems in the meeting. Normally all problems engagement are solved in the meeting. One of the interviewees highlighted the meeting could motivate their employees and also overcome the problem which affects the process of LM implementation. Therefore, it is good to a new company to consider the LM practices, early preparation and LM activities as were carried out by the case study companies in order to ensure LM could be effectively implemented.

## 6 Conclusion

The multi-case studies have explored the successful of LM implementation in SMEs. Apart from that, this method also highlighted the benefits and barriers along the journey of LM implementation. It is challenging part for SMEs to overcome the barriers, especially on employees' loyalty. Initially, the planning and executing on LM preparation are almost the same. However, some companies have implemented and developed based on their environment as to suit with the production requirement. The LM implementation should not copy exactly as the successful company did but they need to adopt and adapt based on their needs.

The comparison between SMEs and large companies on LM approaches looks similar, but the way of implementation depends on their uniqueness and creativity. All case study companies agreed that the LM could reduce their inventory to the minimum level by ensuring all wastes are eliminated accordingly. The case study companies still not satisfied with the current achievement. It is due to most of them are considered new and still at early stage of LM, compared to established company like Toyota Motor. Therefore, they are eagerly striving for LM excellence through empowering people and select the best LM practice. The case study companies also inculcate LM thinking in their activities which reflected to the LM principles. Due the budget constraint, the case study companies have groomed their LM coordinator to be knowledgeable in LM and able to train their employees internally. They can be independent in LM management without engaging on a consultant. All companies agreed that the journey of LM implementation is still a long way to be accomplished. This paper has discussed the results and key findings of the case studies. It is found that the primary objective of implementing LM is to eliminate wastes through educating people and improving current system based on to the best practice as being practiced by the successful company. Though the companies have practiced different approaches and implementation, it is understood that LM has changed the way of thinking towards the elimination of wastes in every aspect. The research finding in this paper will be used for developing a framework for SME reference. This framework will hopefully be applicable within the context of Malaysian SME automotive component industry environment.

The authors wish to Automotive Excellence Centre (AEC), Universiti Malaysia Pahang (UMP), 26600 Pekan, Pahang, Malaysia for providing the financial support (Project No. RDU150339).

## References

[1] J, Rineheart, C. Huxley and D. Robertson, *Just another Car Factory? Lean Production and its Contents* (Cornell University Press, Ithaca, NY 1997)

[2] T.C. Papadopoulu & M. Ozbayrak, Leanness: experiences from the journey to date, Journal of Manufacturing Tech. Mgt **16**(7), 784-806(2005).

[3] R.R. Fullerton.& C.S. McWatters, The production performance benefits from JIT implementation, J. of Operations Management, **19**: 81–96(2001).

[4] J.Womack, D.T. Jones & D. Roos, *The machine that changed the world,* (Rawson Associates, New York, NY 1990).

[5] S. Bhasin & P. Burcher, Lean viewed as a philosophy, J. of Manufacturing Technology Management, **17**(1): 56-72 (2006).

[6] A.M. Sanchez, & P. Perez, Lean indicators and manufacturing strategies, Int. J. of Operations & Production Management, **21**(11): 1433-1452(2001).

[7] R. Shah, P.T. Ward 2007. Defining and developing measures of lean production. *Journal of Operations Management,* 25: 785–805.

[8] P. Achanga, E. Shehab, R. Roy, & G. Nelder, Critical success factors for lean implementation within SMEs, J. of Manufacturing Technology Management, **17**(4): 460-471(2006).

[9] J. A. Farris, E.M.V. Aken, T.L. Doolen, & J. Worley, Critical success factors for human resources outcomes in Kaizen event: An empirical study. International Journal of Production Economics, **117**(1): 42-65(2009)

[10] P.Y. Yang, & Y. Yu, The Barriers to SMEs' Implementation of Lean Production and Countermeasures - Based on SMS in Wenzhou. Int. J. of Innovation, Management and Technology, **1**(2): 220-225(2010).

[11] N. Nordin, Md. B. Deros, & D.A. Wahab, A Survey on Lean Manufacturing Implementation in Malaysian Automotive. Industry, Int. J. of Innovation, Management and Technology, **1**(4): 374-380(2010).

[12] A.N.M. Rose, Md. B. Deros, Abd. M.N. Rahman, Abd. M.N., Lean Manufacturing Practices Implementation in Malaysian's SME Automotive Component Industry. Applied Mechanics and Materials, 315: 686-690(2013).

[13] S.J. Pavnaskar, J.K. Gershenson, & A.B. Jambekar, Classification scheme for lean manufacturing tools. Int. J. of Production Research **41**(13): 3075-3090(2003).

[14] A.M.N. Rose, Md. B. Deros, Abd. M.N.Rahman, Lean manufacturing best practices in SMEs. *Proceedings of the International Conference on Industrial Engineering and Operations Management (2011.*

[15] R. Shah, & P.T. Ward, Lean manufacturing context, practice bundles and performance. J. of Operations Management **21**(12): 129-149(2003).

[16] A. Gunasekaran, L. Forker, & B. Kobu, Improving operations performance in a small company: a case study. Int. J. of Operations & Production Management **20**(3): 316-335(2000).

[17] M. Kumar, & A. Antony, Comparing the quality management practices in UK SMEs, Industrial Management & Data Systems **108**(9): 1153-1166(2008).

[18] A. Anvari, N. Zulkifli, R.M. Yusuff, S.M.H. Hojjati & Y. Ismail, A proposed dynamic model for a lean roadmap, African Journal of Business Management **5**(16): 6727-6737(2011).

[19] S. J. Pavnaskar, J. K. Gershenson, & A.B. Jambekar, Classification scheme for lean manufacturing tools, Int. J. Production Research **41**(13), 3075–3090(2003).

[20] F. Ferdousi & A. Ahmed, An investigation of manufacturing performance improvement through lean production: A study on Bangladeshi garment firms, Int. J. of Business and Management **4**(9), 106-114(2009).

[21] J.E. Matson & J.O. Matson, Just-in-time implementation issues among automotive suppliers in the southern USA. Supply Chain Management: An International Journal **12**(6): 432–443(2007).

[22] R.E. White, J.N. Pearson & J.R. Wilson, JIT Manufacturing: A Survey of implementations in small and large U.S. manufacturers, Management Science **45**(1): 1-14(1999).

[23] S. Sakakibara, B.B. Flynn, R.C. Schroeder, & W.T. Morris,, The Impact of Just-In-Time Manufacturing and Its Infrastructure on Manufacturing Performance, Management Science **43**(9): 1246-1257(1997).

[24] Y.C. Wong, K.Y. Wong, & A. Ali, A study on lean manufacturing implementation in the Malaysia electrical and electronics industry. European Journal of Scientific Research **38**(4), 521-535(2009).

[25] R.K. Yin, *Case study research, (*Sage Publication, 2nd edition 1994).

[26] R.E. Stake, *The art of case study research,* (Thousand Oaks, CA: SAGE Publications 1995).

[27] U. Flick, *An Introduction to Qualitative Research (SAGE Publications 2002)*

[28] A. Bryman, *Social Research Methods, Oxford Publication 3rd Edition 2008).*

[29] R.K. Yin, *Case Study Research: Design and Methods* (Sage Publication, 3rd edition, 2003).

[30] R.K. Yin, *Case study research* (Sage Publication, 1st edition, 1989).

[31] P. Hines, M. Holweg, & N. Rich, Learning to evolve - A review of contemporary lean thinking. Int. J. of Operations & Production Management **24** (10), 994-1011(2004).

[32] T. Bonavia & J.A.Marin, An empirical study of lean production in the ceramic tile industry in Spain, Int. J. of Operations & Production Management, **26**(5), 505-531(2006).

DOI: 10.1051/matecconf/20178702025

# Environmentally Benign Syntheses and Characterization of 4-Aryldihydropyrimidin-2(1H)-ones

Syed Salman Shafqat[1], Amir Azam Khan[1], Misbahul Ain Khan[2], Nicholas Kuan Hoo Tien[1], Mohd Syahmi Jamaludin[1], Pang Suh Cem[3]

[1]*Department of Mechanical and Manufacturing Engineering, Faculty of Engineering, Universiti Malaysia Sarawak (UNIMAS), 94300 Kota Samarahan, Sarawak, Malaysia.*
[2]*Department of Chemistry, The Islamia University of Bahawalpur, Pakistan.*
[3]*Department of chemistry, Faculty of Resource Science & Technology Universiti Malaysia Sarawak (UNIMAS), 94300 Kota Samarahan, Sarawak, Malaysia.*

**Abstract.** A general, simple, efficient, cost-effective and environmental benign procedure for the syntheses of 4-aryl-3,4-dihydropyrimidin-2(1H)-ones (DHPMs) has been developed. The reaction is efficiently catalysed by using various amino acids viz., glycine, L-proline, L-cysteine, leucine and tryptophan. Reactions give excellent yields under reflux conditions. The chemical structures of these compounds were identified by Fourier Transform-Infra Red Spectroscopy (FT-IR), Gas Chromatography-Mass Spectroscopy (GC-MS) and Nuclear Magnetic Resonance Spectroscopy ($^1$H NMR). Physical parameters such as m.p., mixed m.p. as well as TLC also helped to characterize the expected products. The progress of the reaction time and purity was checked by TLC and the products were purified by recrystallization using ethyl acetate as a solvent. The synthesized molecule has numerous biological applications as it is anti-microbial, anti-inflammatory and anti-hypertensive in character.

**Key Words**: Green Approach, Amino Acids, 4-Aryldihydropyrimidinone.

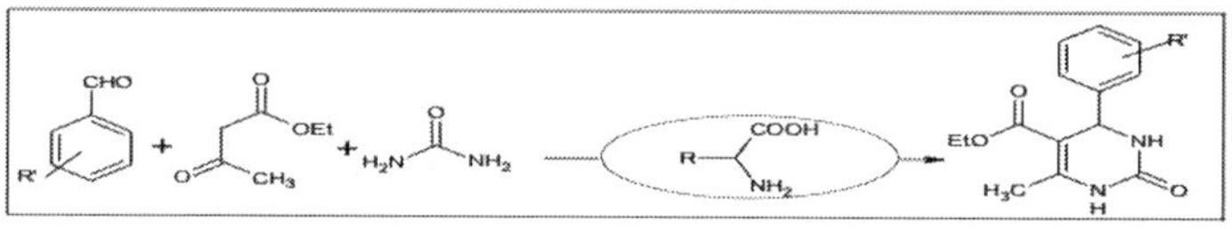


## 1 Introduction

The synthesis of dihydropyrimidinones (DHPMs) has gained importance in recent years due to structural similarities between dihydropyrimidinones and pyrimidine [1]. Pyrimidines are one of biologically important families. Pyrimidine bases such as cytosine (C), thymine (T), are basic constituents in nucleic acids [2]. This class of heterocyclic compounds has also very important pharmaceutical activities such as antimicrobial [3], anti-inflammatory [4], antitumor [3]antihypertensive agents [5], chemical modulators, calcium channel modulators as well as melanin concentrating hormone receptor, antagonists[6], antifungal[7], as inhibitors to replicate hepatitis B, and inhibitors of specific organic acid transporters [8], [9].

In1893, Pietro Biginelli reported synthesis of 4-dihydropyrimidin-2(1H)-one by cyclocondensation of ethyl acetoacetate, benzaldehyde, and urea by simply heating a mixture of these three components in the presence of acid as a catalysts under reflux conditions [10].

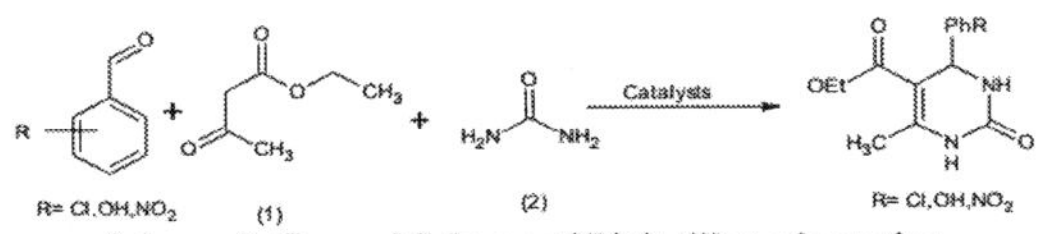


**Scheme I: General Scheme of Biginelli condensation**

Now Biginelli reaction is an important multicomponent reaction for the preparation of biologically active dihydropyrimidinone derivatives. Usually the synthesis of 4-aryldihydropyrimidin-2*(1H)*-ones (DHPMs) is carried out by the condensation of an aromatic aldehyde, ethyl acetoacetate, urea/thiourea in the presence of metal salts [11], [12], Bronsted and Lewis acids [13], ionic liquids [14]microwave irradiations [15] or refluxing acetic acid for few hours[16] .

Many of these classical methods, however, suffer from drastic reaction conditions, lower yields, difficult workup process, excessive use of chemicals as well as catalyst and environmental impact such as carcinogenic solvents and non-biodegradable catalysts [17]. Although a variety of methods are reported for the synthesis of DHPMs, only few examples were developed with the aim of replacing conventional toxic and polluting Bronsted and Lewis acid catalysts with efficient and green catalysts such as recyclable catalyst under solvent-free conditions[18] , amino acid catalyzed [19], [20], 21], Mesoporous SBA-15 nanoparticles[22] and Quaternary Ammonium-treated clay in water catalyzed[23]synthesis of DHPMs. Drawbacks of these classical methods and catalytic efficacy of environmental benign catalyst gave us a direction to introduce new efficient catalysts for prepration of 4-aryl-3,4-dihydropyrimidin-2*(1H)*-ones with cost effectiveness and mild conditions with high yields.

We had earlier communicated our preliminary finding that amino acids can be efficiently used for these reactions [19]. Dihydropyrimidinone (DHPM) synthesis was carried out by using benzaldehyde and urea/thiourea.

Amino acids are also being reported as efficient catalysts in the Knoevenagel condensations, reduction, oxidation, electrophilic α-fluorination and amination, transamidation, as well as promoter of some of the reactions [24], [25].

We are extending this Biginelli, amino acid catalyzed reaction, using some other aryl aldehydes which are now being communicated.

## 2 Experimental Work

Most of the chemicals used during the development of this environmentally benign synthesis were of laboratory grade and were used as it as. Progress of the reactions were monitored by using Silica TLC plates and UV light was used as a spot locating agent. IR were recorded on Perkins Elmer FTIR (spectrum RXI) spectrometer. Mass spectra were recorded GC-MS Schimadzu QP2010.$^{1}$H NMR spectra were recorded on and (NMR; Jeol JNM–ECA500).

### 2.1 Synthesis of 4-Aryldihydropyrimidin-2(1H)-ones:

**General Method:** A reactions mixture containing 40 mmol of aromatic aldehyde, 40 mmol of ethyl acetoacetate, 45 mmol of urea, 0.01 mmol of an amino acid were dissolved in 10 ml of ethanol. The reaction mixture was heated under reflux for two hours. The reaction mixture was work up by allowing to cool down to room temperature and was diluted with ice cold water and the precipitated product was isolated. The product was recrystallized from ethyl acetate o purify.

Four aromatic aldehydes were used by using this general procedure to syntheses four different 4-Aryl-dihydropyrimidin-2(1H)-ones by employing five different amino acids as catalysts.

**2.1.1 By using *p*-chlrobenzyldehyde (4):** 4.5g of *p*-chlrobenzyldehyde **(4)** was mixed with 5.2 g of ethyl acetoacetate **(1)**, 4g of urea **(2)** and .01mmol of glycine in 10 ml of ethanol. The reaction mixture was heated in reflux conditions for 2 hours. The reaction was allowed to stand to room temperature and then crushed ice was added to it, pale yellow crystals of 5-Ethoxycarbonyl-6-methyl-4-*p*-chlorophenyldihydropyrimidin-2-one **(4)** were collected and recrystallized to purify the product. The product was characterized through its m.p., IR, Mass spectrum and $^{1}$H-NMR. After this the reaction was carried out by employing few other amino acids. These products were identical in their physical and chemical characteristics to the product obtained by using glycine. The proposed reaction is presented in scheme **(scheme II)** and results are described in **(Table 1)**.

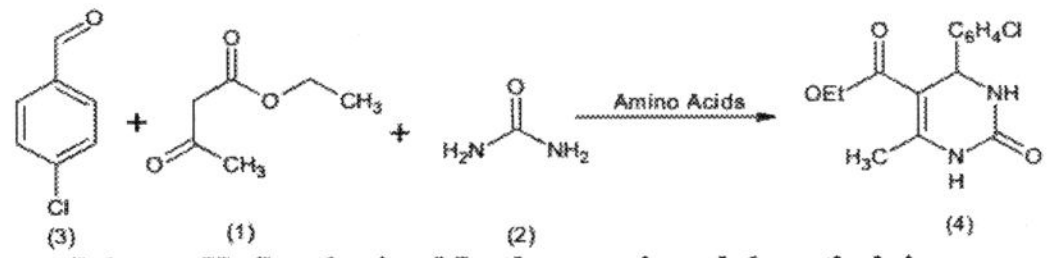


**Scheme II: Synthesis of 5-ethoxycarbonyl-6-methyl-4-*p*-chlorophenyldihydropyrimidin-2-one**

**2.1.2: By using *p*-nitrobenzaldehyde (5):** Condensation of *p*-nitrobenzaldehyde **(5)** with ethyl acetoacetate **(1)** and urea **(2)** was conducted by using Glycine as a catalyst as a test reaction. This reaction gave good yield of 5-Ethoxycarbonyl-6-methyl-4-*p*-nitrophenyldihydropyrimidin-2-one **(6)**. So this reaction was carried out by employing other available amino acids. The products were characterized through m.p., IR, Mass spectrum and $^{1}$H-NMR. The proposed reaction is presented in scheme **(scheme III)** and results are summarized in **(Table 2)**.

 DOI: 10.1051/matecconf/20178702025

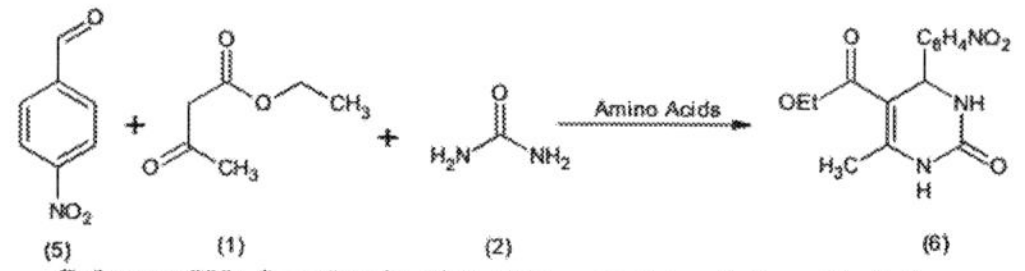

**Scheme III: Synthesis of 5-ethoxycarbonyl-6-methyl-4-*p*-nitrophenyldihydropyrimidin-2-one (6)**

### 2.1.3 By using 2-hydroxybenzaldehyde (7):

Condensation of 2-hydroxybenzaldehyde **(7)** with ethyl acetoacetate **(1)** and urea **(2)** was carried out in the presence of Glycine as catalyst using ethanol as a solvent. This reaction gave a very good yield of 5-Ethoxycarbonyl-6-methyl-4-*o*-hydroxyphenyldihydropyrimidin-2-one **(8)**. The same reaction was also carried out in the presence of other available amino acids. The product **(8)** was characterized through its m.p., IR, Mass spectrum and $^1$H-NMR and verified its structure. The proposed reaction is presented in scheme **(scheme IV)** and results are summarized in **(Table3)**.

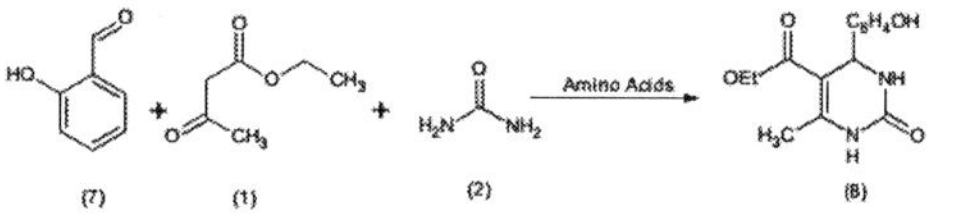

**Scheme IV: Synthesis of 5-ethoxycarbonyl-6-methyl-4-*o*-hydroxyphenyldihydropyrimidin-2-one (8)**

### 2.1.4 By using 4-hydroxybenzaldehyde (9):

Condensation of 4-hydroxybenzaldehyde **(9)** with ethyl acetoacetate **(1)** and urea **(2)** in the presence of Glycine gave 5-Ethoxycarbonyl-6-methyl-4-*p*-hydroxyphenyldihydropyrimidin-2-one **(10).** Some other amino acids were also used to prepare product **(10)**. The product **(10)** was characterized through its TLC, m.p., IR, Mass spectrum and $^1$H-NMR. All the products were identical in the characteristics. The proposed reaction is presented in scheme **(scheme V)** and results are presented in **(Table 4)**.

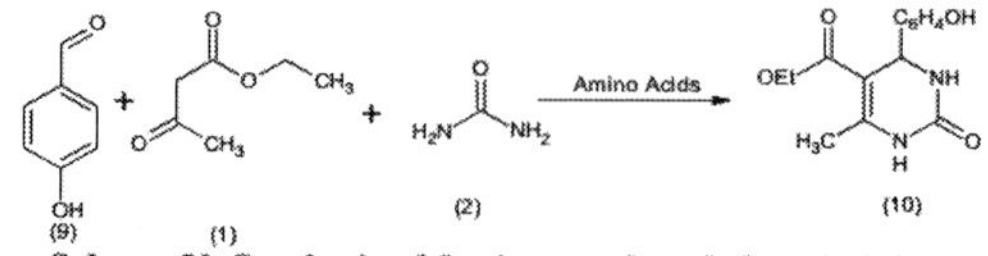

**Scheme V: Synthesis of 5-ethoxycarbonyl-6-methyl-4-*p*-hydroxyphenyldihydropyrimidin-2-one (10)**

**Table 1: Characterization of 5-Ethoxycarbonyl-6-methyl-*p*-chlorophenyldihydropyrimidin-2-one (4)**

| Entry | Catalyst | Yield % | m.p. °C<br>Lit [26]<br>m.p210 | IR $cm^{-1}$ | MS(m/z) |
|---|---|---|---|---|---|
| 1 | glycine | 92 | 207 | 3236 (N-H )<br>3113 (aro C-H)<br>2985(ali C-H)<br>1710 (C=O)<br>1646 (C=C)<br>1461 ( C-H)<br>1281(C-O)<br>776 (Cl) | 294(M+)<br>265(30%)<br>221(25%)<br>183(100%) |
| 2 | L-cysteine | 94 | 206 | | |
| 3 | L-proline | 95 | 208 | | |
| 4 | *B*-alanine | 93 | 206 | | |
| 5 | tryptophan | 90 | 205 | | |

**Table 2: Synthesis and characterization of 5-Ethoxycarbonyl-6-methyl-4-nitrophenyldihydropyrimidin-2-one (6)**

| Entry | Catalyst | Yield % | m.p°C<br>Lit 208[26] | IR cm$^{-1}$ | MS (m/z) | $^{1}$H NMR |
|---|---|---|---|---|---|---|
| 1 | glycine | 95 | 207 | 3283 (N-H)<br>3102(aro C-H)<br>2985(ali C-H)<br>1717(C=O)<br>168 C=O)<br>1642(C=C)<br>1502 and<br>1323(NO2)<br>1232 (C-O ) | 305(M+)<br>288(20%)<br>276(15%)<br>183(100%) | NMR (500MHz, CDCl3)1H<br>8.18(2H) J=9.17 (d)<br>2H 7.50 (2H) J=9.17 (d)<br>3H5.552(1H) J=3.06 (d)<br>4H 5.74(1H) (s)<br>5H 2.37(3H) (s)<br>6H 4.10(2H) (q)<br>7H 1.17(3H) 6.88<br>8H 7.4159 (1H) incomplete splitting |
| 2 | L-cysteine | 96 | 208 | | | |
| 3 | L-proline | 92 | 208 | | | |
| 4 | B-alanine | 93 | 207 | | | |
| 5 | tryptophan | 94 | 206 | | | |

**Table 3: Synthesis and characterization of 5-Ethoxycarbonyl-6-methyl-2-hydroxyphenyldihydropyrimidin-2-one (8)**

| Entry | Catalyst | Yield % | m.p°C<br>Lit 265[26] | IR cm$^{-1}$ | MS (m/z) | $^{1}$H NMR |
|---|---|---|---|---|---|---|
| 1 | glycine | 92 | 260 | 3232 (O-H)<br>3220 (N-H)<br>3101 (aro C-H)<br>2978 (ali C-H)<br>1727 (C=O)<br>1682 (C=O)<br>1653 (C=C),<br>1513 (C-H)<br>1212 (C-O). | 276 (M+)<br>247(20%)<br>229(30%)<br>183(50%)<br>57(100%) | (NMR500 MHz,DMSO-d6, 9.35 (1H) (s)<br>9.12 (1H) (s)<br>7.63(1H)(s) incomplete splitting<br>7.02(2H)(d) J=8.41<br>6.68(2H)(d) J=8.41<br>5.03 (1H) (d)<br>3.97 (2H) (qss)<br>2.22 (3H) (s)<br>1.09 (3H) (t) |
| 2 | L-cysteine | 90 | 261 | | | |
| 3 | L-proline | 93 | 259 | | | |
| 4 | B-alanine | 90 | 261 | | | |
| 5 | tryptophan | 94 | 259 | | | |

**Table 4: Synthesis and characterization of 5-Ethoxycarbonyl-6-methyl-4-hydroxyphenyldihydropyrimidin-2-one (10)**

| Entry | Catalyst | Yield % | m.p°C<br>Lit 227[27] | IR cm$^{-1}$ | MS (m/z) | $^{1}$H NMR |
|---|---|---|---|---|---|---|
| 1 | glycine | 92 | 220 | 3232 (O-H)<br>3220 (N-H)<br>3101 (aro C-H)<br>2978 (ali C-H)<br>1727 (C=O)<br>1682 (C=O)<br>1653 (C=C),<br>1513 (C-H)<br>1212 (C-O). | 276 (M+)<br>247(20%)<br>229(30%)<br>183(50%)<br>57(100%) | (NMR500 MHz,DMSO-d6, 9.35 (1H) (s)<br>9.12 (1H) (s)<br>7.63(1H)(s)<br>7.02(2H)(d) J=8.41<br>6.68(2H)(d) J=8.41<br>5.03 (1H) (d)<br>3.97 (2H) (qss)<br>2.22 (3H) (s)<br>1.09 (3H) (t) |
| 2 | L-cysteine | 90 | 220 | | | |
| 3 | L-proline | 93 | 219 | | | |
| 4 | B-alanine | 95 | 220 | | | |
| 5 | tryptophan | 91 | 222 | | | |

## 3 Results and Discussion

In this study as part of our continued effort to develop environmental benign and new catalysis systems with a reduced environmental impact, the utility of five amino acids *viz*. glycine, L-proline, L-cysteine, lucien and tryptophan is explored as catalysts for the multi component synthesis of 4-aryldihydropyrimidin-2*(1H)*-ones. All the amino acids gave promising results with high yields under reflux conditions. Initially, a model reaction was conducted by heating a mixture of *p*-chlrobenzyldehyde, ethyl acetoacetate and urea using glycine as a catalyst and excellent results were obtained. The product 5-ethoxycarbonyl-6-methyl-4-*p*-chlorophenyldihydropyrimidin-2-one (**4**) was obtained in almost quantitative yield (92%). Keeping in view of this successful result various other amino acids were employed for the reaction with equal success. All the product were characterized through its m.p., IR, mass spectrum and $^1$H NMR **(Table-1)**.The efficacy of amino acid catalysts was further tested in the synthesis of diverse 4-aryldihydropyrimidin-2(1H)-ones with comparable results **(Table1-4)**.Plausible mechanism for the glycine -catalyzed synthesis of 4-aryldihydropyrimidin-2(1H)-ones has been proposed **(Scheme VI)**. The reaction may proceed through the protonation of aromatic aldehyde (A) by the amino acid which provides this proton to catalyze the reaction. Further reaction proceeds via intermediates B and C as depicted in the scheme.

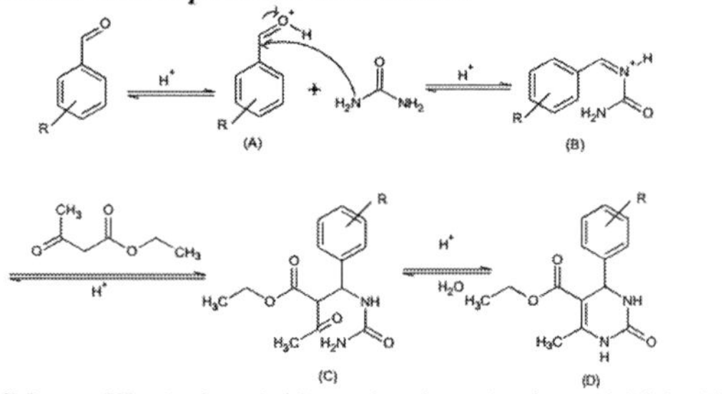

**Scheme VI: Amino Acid catalyzed mechanism of Biginelli Condensation**

## 4 Conclusion

In summary, amino acids were successfully employed as catalyst for the synthesis of various 4-aryldihydropyrimidin-2(1H)-ones. Simple and mild reaction conditions, low cost, readily available and biodegradable catalyst are the important features of this protocol. For the present work, only a few representative aromatic aldehydes and urea were used to check the feasibility of amino acids as a catalyst in these reactions. The yields in all these reactions were comparable and the use of these catalysts can be extended to study the utilization of other aldehydes in Biginelli reactions.

## Acknowledgments

We gratefully acknowledge Universiti Malaysia Sarawak (UNIMAS) to support this work under the grant No. DPI 02 (DPI08)/824/2011(08).

## References

[1] I. Phucho, A. Nongpiur, S. Tumtin, R. Nongrum and R. Nongkhlaw, ChemInform 2010, 41, i.

[2] J. D. Watson and F. H. Crick, Nature 1953, 171, 737-738.

[3] M. Jalali, M. Mahdavi, H. Memarian, M. Ranjbar, M. Soleymani, A. Fassihi and D. Abedi, Research in pharmaceutical sciences 2012, 7, 243.

[4] R. H. Tale, A. H. Rodge, G. D. Hatnapure, A. P. Keche, K. M. Patil and R. P. Pawar, Medicinal Chemistry Research 2012, 21, 4252-4260.

[5] K. S. Atwal, B. N. Swanson, S. E. Unger, D. M. Floyd, S. Moreland, A. Hedberg and B. C. O'Reilly, Journal of medicinal chemistry 1991, 34, 806-811.

[6] C. O. Kappe, W. M. Fabian and M. A. Semones, Tetrahedron 1997, 53, 2803-2816.

[7] D. Duguay, M. Zamora, J. Blacquiere, F. Appoh, C. Vogels, S. Wheaton, F. Baerlocher, A. Decken and S. Westcott, Open Chemistry 2008, 6, 562-568.

[8] T. Kadre, S. R. Jetti, A. Bhatewara, P. Paliwal and S. Jain, Archives of Applied Science Research 2012, 4, 988-993.

[9] R. F. Canto, A. Bernardi, A. M. O. Battastini, D. Russowsky and V. L. Eifler-Lima, Journal of the Brazilian Chemical Society 2011, 22, 1379-1388.

[10] P. Biginelli, Gazz. Chim. Ital 1893, 23, 360-413.

[11] S. R. Narahari, B. R. Reguri, O. Gudaparthi and K. Mukkanti, Tetrahedron letters 2012, 53, 1543-1545.

[12] C. Navya, D. Rajeswari, E. U. Rajitha, I. Jyothsna, K. Manasa, K. S. Kumar and A. L. Rao, International Journal of Applied Pharmaceutical Sciences 2016, 3, 301-305.

[13] H. Kefayati, K. Rad-Moghadam, M. Zamani and S. Hosseyni, Letters in Organic Chemistry 2010, 7, 277-282.

[14] Z. Liu, R. Ma, D. Cao and C. Liu, Molecules 2016, 21, 462.

[15] T. Raj, B. Kaur Billing, N. Kaur and N. Singh, RSC Advances 2015, 5, 46654-46661.

[16] Faryal Chaudhry, Syed Salman Shafqat, Amir Azam Khan, and M. A. M. M. A. Khan, Synthetic Communications 2016, 46.

[17] A. R. Bhat, A. H. Shalla and R. S. Dongre, Journal of Saudi Chemical Society 2014.

[18] K. Aswin, S. S. Mansoor, K. Logaiya, P. N. Sudhan and R. N. Ahmed, Journal of Taibah University for Science 2014, 8, 236-247.

[19] Ansa Madeeha Zafar, Muhammad Naeem Khan, Muhammad Azad, Munawar Ali Munawar and Misbahul Ain Khan., Asian Journal of Chemistry 2013, 25, 3244.

[20] M. Bararjanian, S. Balalaie, B. Movassag and A. Amani, Journal of the Iranian Chemical Society 2009, 6, 436-442.

[21] S. P. Bahekar, P. B. Sarode, M. P. Wadekar and H. S. Chandak, Journal of Saudi Chemical Society 2015.

[22] Y. Sarrafi, F. Pazokie, S. Azizi, K. Alimohammadi, E. Mehrasbi and E. Chiani, Current Chemistry Letters 2014, 3, 97-102.

[23] S. Sayyahi, S. Jahanbakhshi and Z. Dehghani, Journal of Chemistry 2013, 2013.

[24] S. S. Shafqat, M. A. Khan, A. Zulkharnain, S. Hamdan, H. Rigit, and A. A. khan, Asian Journal of Chemistry 2014, 26.

[25] B. List, Tetrahedron 2002, 58, 5573-5590.

[26] S. D. Sharma, P. Gogoi and D. Konwar, Green Chemistry 2007, 9, 153-157.

[27] T. S. Jin, H. X. Wang, C. Y. Xing, X. L. Li and T. S. Li, Synthetic communications 2004, 34, 3009-3016.

DOI: 10.1051/matecconf/20178702026

# Microstructural effects of banana leaf on adhesion and self cleaning behaviour

Hasrawati Abu Hassan[1], Mariyam Jameelah Ghazali[1,a], Che Husna Azhari[1] and Cevdet Meric[2]

[1] *Department of Mechanical and Materials Engineering, Faculty of Engineering and Built Environment, Universiti Kebangsaan Malaysia, 43600 UKM Bangi, Selangor, Malaysia*
[2] *Faculty of Engineering, Fatih University, 34500 Buyukcekmece/Istanbul/Turkey*

**Abstract.** A recent discovery on banana leaf that has high potential as the lotus effect has intrigued further investigation particularly on the mechanism of its self-cleaning behaviour. In this paper, the adhesive and the contact angle effects of the banana leaf, named *Musa Parasidiaca L.* (*berangan*) surface in determining the self-cleaning effect is discussed especially at high temperatures. Samples were slid against a heated metal plate at two different temperatures; 100°C and 200°C, under different loads; 0.5 kg and 1 kg. Correlation between the leaf morphologies, adhesion force and contact angles were discussed. Measurements were carefully carried out by using the various pressure scanning electron microscopy (VP-SEM), an atomic force microscope (AFM) and a drop shape analysis (DSA). The adhesion force and contact angle were decreased at the highest temperature and load. The adhesion force was found to be reduced from 5.6 nN to 1.9 nN and the value of contact angles was decreased from 108.9° to 99.4°. In short, it is hope that the new discovery of this study can shed light on its importance to mankind particularly in surface modifications for applications at higher temperature.

## 1 Introduction

Over recent years, there are several interests in nature that include many plants and animals exhibit superhydrophobic property with water contact angle (WCA) more than 150° and sliding angle (SA) lest than 10°, such as lotus leaves, rice leaves, red rose petals, gecko feet and desert beetle [1]. *Lotus effect* occurs when falling water drops onto the leaves can bead up and roll off, washing dirt from the lotus leaves so that they are self-cleaning [2].

Lotus leaf is a typical natural super-hydrophobic surface with WCA and SA on the surface of lotus leaf is about 161.0 ± 2.7° and 2°, respectively because of the hierarchy surface structures [3]. Figure 1 shows a scanning electron micrograph (SEM) image of the lotus leaf surface where papillas with branch-like nanostructures were found on this surface in a random distribution efficiently increases the roughness.

Hydrophobicity of leaf is due to the surface that contained a layer of hydrophobic wax and also surface topography [4]. Wax is one of the materials that can construct hydrophobic (water-repellent) surfaces [5]. The existence of wax will increase the surface roughness and contact angle.

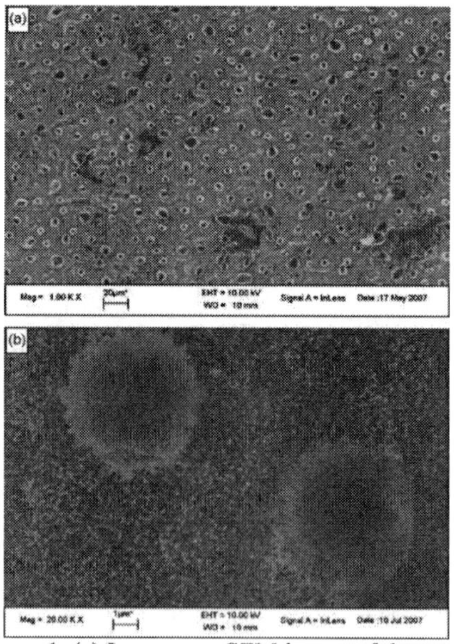

**Figure 1.** (a) Large-area SEM image of the surface of lotus leaf. (b) Enlarged view of lotus leaf from (a)[3].

The wettability of plant surfaces may change with age and maturity with a little variation in contact angle [6]. In this study, the leaf is taken from the plants age between 7 to 9 months and the layer number 3 down the profile [7].

[a] Corresponding author: mariyam.j.ghazali@gmail.com

Meaning that the leaf is selected from the third layer of the banana leaves counted from upper part of the plant. The size and distribution of wax differed among the types of banana. The irregular shape of wax deposits form walls that may create the air pocket that contribute to the surface roughness and wettability.

Even though lotus and rose petal have the superhydrophobic behaviour, but both show different adhesive force [8]. The reason why they have different adhesive forces is that the pitch values of microstructures and the density of nanostructures of their surfaces are varies, thus giving rise to different states of water droplet on the two surfaces.

Recently, the effect of high temperature and loading condition on the plant surface is not yet discover. This paper is comparing the results of two different temperatures (100°C and 200°C) and two different applied loads (0.5kg and 1.0kg). The purpose of this paper is to stimulate the further investigation on the banana leaf as similar to lotus and other plants surface that have been established.

## 2 Methology

### 2.1 Starting Materials

A dessert type of banana, *Musa Parasidiaca L.* (*Berangan*) was collected from Taman Pertanian Universiti (TPU), Universiti Pertanian Malaysia to obtain the standard age and leave taken form the banana cultivar. Leaves were sectioned to 20cm x 20cm and ready for the next steps.

### 2.2 Sliding Test

Metal plates with two different loads (0.5kg and 1.0kg) were heated in a furnace at two different temperatures (100°C and 200°C) as a medium that acted similar as the typical home iron. The parameters were selected as a continuity from the study that have been done using typical home iron [9]. Sliding were conducted constantly for 20s on the adaxial (top) of the banana leave. Adaxial side is selected because it is exposed to the environment and during the ironing work; the adaxial side is attached to the iron.

### 2.3 Microstructure and Surface Examination

Banana leaf samples were prepared for detailed microstructural observations under a scanning electron microscope (LEO 1450VP). Prior to observations, samples were rinsed in 0.1 M sodium cacodylate buffer, dehydrated in a graded series of ethanol, and then dried in hexametyldisilazano. The dried samples were mounted on stubs and stored in dry cabinet before being coated with gold and examined.

### 2.4 Adhesion Force and Surface Roughness

A commercial Atomic Force Microscope (Park System) with contact cantilever (PPP-CONTSCR 10M) is used to obtain the surface roughness and adhesion force of the treated banana leave. The surface roughness is obtained from the amount of bumps and the adhesion force was determined from the pull-off force of the force distance curve measured with the AFM [10].

### 2.5 Contact Angle Measurement

Samples are placed horizontally on glass slides. Water contact angle is determined as per standard ASTM D7334 using Drop Shape Analysis.

## 3 Result and Discussion

### 3.1 Microstructure and surface examination

Figure 2 shows the SEM results of the banana leaf surface.

| Temperature (°C) | Load (kg) | |
|---|---|---|
| | 0.5 | 1.0 |
| 100 | 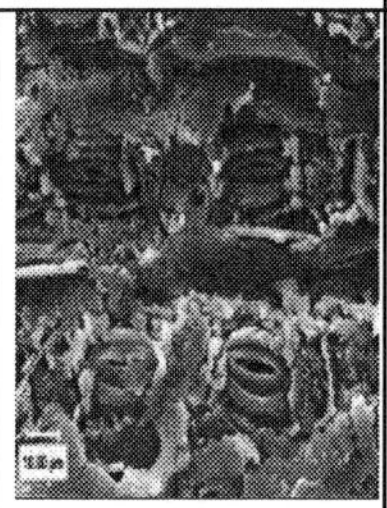 | 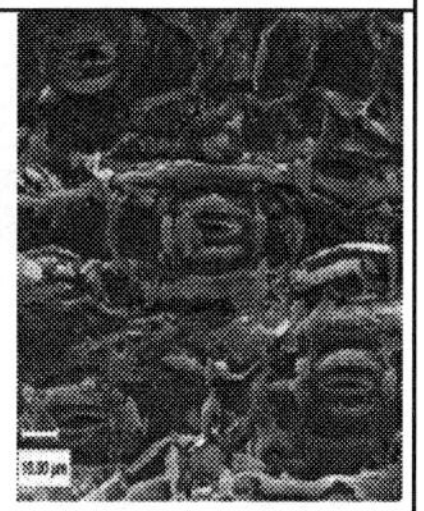 |
| 200 |  | 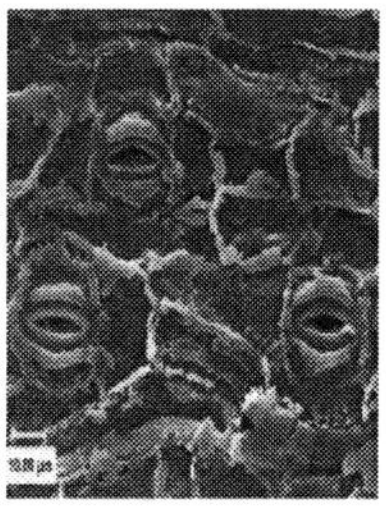 |

Figure 2. SEM images of the banana leaf surface (*Berangan*) at different load and temperature.

The melting point of wax that is below 100°C did show in the micrograph in Figure 2. Even the wax is melted, the form of air pocket from the wax wall still maintain till 200°C. The formation of air pocket contributes to the hydrophobic and self-cleaning behavior of banana leaf.

### 3.2 Adhesion Force and Surface Roughness

Figure 3 shows that the adhesion force is decrease with the increasing of temperature, same as the load applied. The applied loads have compressed the wall of the air pocket and lower the adhesion force of the surface. The higher temperature also lowers the force as well. Melted wax caused the adhesion to decrease. Hydrophobicity surface includes low adhesion, high contact angle and high surface roughness [5].

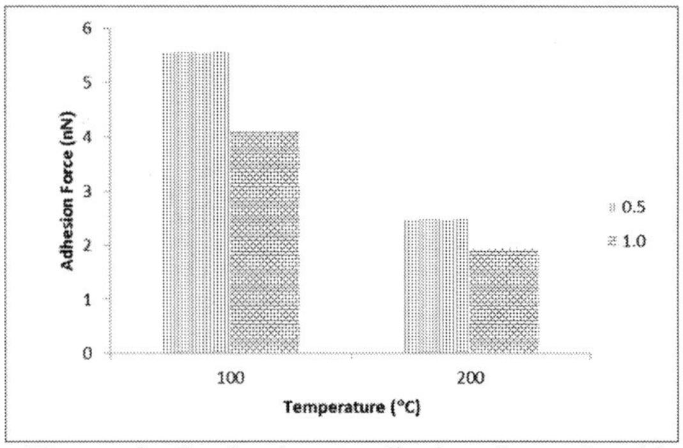


**Figure 3** Graph shows adhesion force for *Musa Parasidiaca L.* (*Berangan*) at different load and temperature.

Surface roughness is shown in Figure 4. Melting of wax caused the roughness of the surface to increase for the lower load and decrease in the higher one.

In order to determine the adhesion force, contact tip is use. As the principle to get adhesion force is from the pull-off force of the curve measured, at higher temperature and load, the wax is melted and compressed. When the tip comes into contact with the leaf area, the sample is not easily deformed as the moisture is gone. This will cause the leaf to dry and give a rougher surface when the temperature is increase. For the higher load, the decrease in the surface roughness may cause by higher load that smoothing the surface.

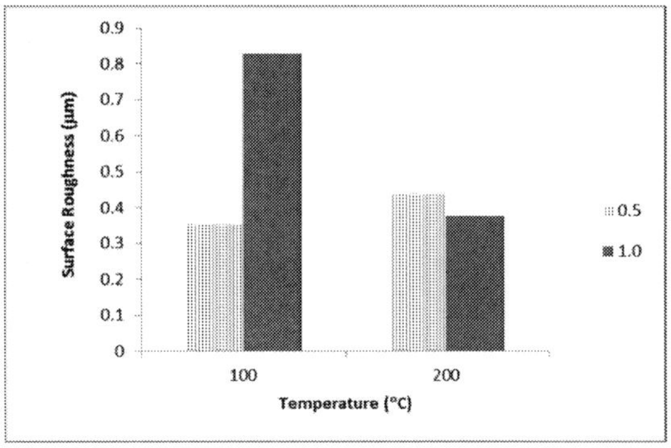


**Figure 4** Graph shows surface roughness for *Musa Parasidiaca L.* (*Berangan*) at different load and temperature

### 3.3 Contact Angle

The continuity from the results of the micrograph and adhesion force showed that temperature and load also affects the wettability of the surface. Value of contact angle also exhibit reduction as the increment of temperature and load. Decrease in contact angle due to the melting of wax and minimized the rough surface that contribute to the wetting behaviour. The value of contact angle is shown in Figure 5.

| Temperature (°C) | Load (kg) | |
|---|---|---|
| | 0.5 | 1.0 |
| **100** |  108.9 ± 0.31 |  107.1 ± 0.29 |
| **200** |  101.6 ± 0.17 |  99.4 ± 0.95 |

Figure 5. Images shows contact angle for *Musa Parasidiaca L.* (*Berangan*) at different load and temperature.

## 4 Conclusions

Considering all of this evidence, it seems that at 100°C and 200°C with applied loads 0.5kg and 1.0kg did not affect much in micrograph and contact angle as the form of stomata is still obvious in the macrograph. The adhesion force and surface roughness is donated by the deformation of the wax after applying certain load and temperature. Even though the melting point is lower than 100°C, the waxes are still there giving the hydrophobic behavior as the contact angle still give the reading more than 90°. Further research should focus on another applied temperatures and loads to determine the feasibility of banana leaves as a bio-adhesion lubricant at high temperature.

## Acknowledgement

This work was financially supported by FRGS/2/2013/TK04/UKM/02/2 and the authors are thankful to Prof Dr Yakub Bakis, Coordinator of BioNano Technology Centre and Dr. Isa Seker from Fatih University, Istanbul for all their full supports.

## References

1. J. Liu, X. Xiao, Y. Shi, C. Wan. Liu, J., Xiao X., Shi, Y., C. Wan, Appl. Surf. Sc., **297** (2014)
2. M. Zhang, S. Feng, L. Wang and Y. Zheng, Biotribology, **5** (2016)
3. W. Xi, Z. Qiao, C. Zhu, A. Jia, A and M. Li, Applied Surface Science, **255** (2009)
4. J. Wang, H. Chen, T. Sui, A. Li and D. Chen, Plant Science, **176** (2009)
5. Z. Burton, and B. Bhushan, Ultramicroscopy, **106** (2006)
6. R.E. Gaskin, K.D. Steele and W.A. Forster,(2005). New Zealand Plant Protection **58** (2005)

DOI: 10.1051/matecconf/20178702026

7. W. Zakaria and A.R. Razak, J. Trop. Agric. and Fd. Sc. **27**, 2 (1999)
8. G. Wang, Z. Guo and W. Liu, Journal of Bionic Engineering, **11** (2014)
9. F. Azura, *Final Year Theses* (Universiti Kebangsaan Malaysia, 2014)
10. K. Meine, K. Kloß, T. Schneider, T. and D. Spaltmann, D. (2004) Surf. Interface Anal. **36** (2004)

# Microstructure and mechanical properties of plasma sprayed $Al_2O_3$ – 13%$TiO_2$ Ceramic Coating

Juyana A Wahab[1,2], Mariyam Jameelah Ghazali[1,a] and Ahmad Firdaus Shamsul Baharin[1]

[1]*Department of Mechanical & Materials Engineering, Faculty of Engineering and Built Environment, Universiti Kebangsaan Malaysia, 43600 UKM Bangi, Selangor, Malaysia*
[2]*School of Materials Engineering, Universiti Malaysia Perlis, Kompleks Pusat Pengajian Jejawi 2, Taman Muhibah, 02600 Jejawi, Arau, Perlis, Malaysia.*

**Abstract.** This paper focused on the effect of deposition conditions on the microstructural and mechanical properties of the ceramic coating. In this study, $Al_2O_3$ – 13%$TiO_2$ coated mild steel were prepared by using atmospheric plasma spray technology with different plasma power ranging from 25 kW to 40 kW. The as-sprayed coatings consist of γ-Al2O3 phase as the major phase and small amount of the titania phase existed in the coating structure. High degree of fully melted region was observed in the surface morphology for the coating sprayed with high plasma power, which lead to the high hardness and low percentage of porosity. In this study, nanoindentation test was carried out to investigate mechanical properties of the coating and the results showed that the coatings possess high elastic behaviour, which beneficial in engineering practice.

## 1 Introduction

The research and manufacturing of ceramic coatings had gained great attention due to their excellent characteristics and potential applications. Most of ceramic material presenting useful features such as low density, excellent wear resistance, high inertness, high stability and high temperature strength [1, 2]. These features resulted by high atomic strength of ceramic that led to its great hardness, strength and stiffness. Ceramics also possess high resistance to oxidation as well as corrosion in extreme condition and thus make them having a wide range of applications especially in structural and engineering fields [3].

Development of new materials has affected ceramic film and coating technology. The relatively high hardness and inertness of ceramic coating led to its excellent performance as a protective layer on substrate surface in order to mitigate corrosion, oxidation and wear problem especially in marine environment. Oil and gas industry, shipping industry and the marine structures suffer from corrosion due to the exposure of aggressive ions in seawater. Combinations of wear, oxidation and corrosion problems contribute to failure factors that would eventually cause total shutdown, economic loss and hazardous effects [4]. For these reasons, applying coating protection method to material surface is very important. Various types of ceramic coating gained attention to overcome these problems such as alumina ($Al_2O_3$), titanium carbide (TiC), silicon nitride ($Si_3N_4$) [5], titanium nitride ($TiNiO_3$) [6], alumina-titania ($Al_2O_3$-$TiO_2$) [7-10], alumina-titania-zirconia ($Al_2O_3$-$TiO_2$-$ZrO_2$) [11], chromium oxide ($Cr_2O_3$-Ag) [12] and titanium diborate ($TiB_2$). Such ceramic coatings can be produced using thermal spraying technology, CVD, PVD and sputtering. Previous studies stated that among these techniques, atmospheric plasma spraying technique is an effective technique for deposition of thick coating on the substrates [13]. It is the most commonly used thermal spraying technology due to its high flexibility, high deposition consistency and high deposition rates. It is known that the condition and properties of plasma sprayed coating is strictly influenced by plasma spraying process parameters, which are stand-off distance, plasma power and gas pressure. These parameters directly affect the heat and mass transfer between powder particles and plasma and thus influence the degree of powder particles melting and the structure of droplets on the coating surface [14].

The mechanical properties and the surface morphology of the coating are affected by the spraying conditions that are plasma spray power, stand-off distance and powder feeding rate. These parameters affect the thermal and kinetic energy of the powder particles. If the powder particles gained an extra thermal energy, the vapourisation of the powder particles take place in the plasma jet before impact onto the substrate surface in the fully-melted condition [15]. However, if the powder particles gained small amount of thermal energy, they will reach onto the substrate surface in a partially-melted condition. Both of these two condition will affect the characteristics of the coating. Venkataraman et al. [16] investigated the effect of porosity, pore size and pores distribution on the hardness

[a] Corresponding author: mariyam@ukm.edu.my

of plasma sprayed ceramic coatings. They claimed that among several microstructural features, porosity generated by partially melted particles has a major effect on the hardness of the coating.

In this work, nanoindentation is utilised in order to measure mechanical properties of materials. It has recently begun to be utilized widely in thermally sprayed materials but the results are always complicated to analyse. This technique measures the coating hardness as well as elastic modulus by considering the degree of the material deformation. The deformation mechanisms upon mechanical loading during the test always considered as properties that depends on the testing conditions. Therefore, in this study, the nanoindentation technology was utilized to determine the effect of indentation load and depth of the measured mechanical properties of thermally sprayed $Al_2O_3$ - 13%$TiO_2$ coating.

## 2 Materials and method

### 2.1 Plasma spraying

Mild steel (ASTM A36) was used in this study as substrate material with dimension of 25 x 10 x 3 mm. The commercial feedstock $Al_2O_3$ – 13%$TiO_2$ (Metco 130SF) was mechanically clad having a fine particle distribution (–45+5 µm) was deposited vertically onto the substrates by atmospheric plasma spraying (APS) using a plasma torch (Praxair SG-100) mounted on a programmable robot. The surface of the substrates was grit-blasted with $Al_2O_3$ before coating to increase the roughness and the mechanical interlocking between coating and substrate. The thickness of coating was fixed to be within 200–300 µm. The deposition parameters set for the spraying process are shown in Table 1.

**Table 1.** Plasma spraying parameter.

| Parameter | Value |
|---|---|
| Plasma power (kW) | 25 – 40 |
| Primary gas pressure (Psi) | 80 |
| Secondary gas pressure (Psi) | 60 |
| Carrier gas pressure (Psi) | 40 |
| Powder feed rate (Rpm) | 3 |
| Coating speed (mm/s) | 250 |
| Stand-off distance (mm) | 100 |

### 2.2 Characterisation and testing

For the characterisation techniques, all metallography samples were ground and polished to obtain mirror-like surface finish. The morphology of feedstock and as-sprayed coating was observed under a LEO 1430 VP scanning electron microscope (VPSEM). The phase composition existed in the powder feedstocks and the coatings was examined using an X-Ray powder diffractometer (XRD) model Bruker, D8- Advance 2009, Germany with CuKα (λ=0.1541 nm) radiation source. A software of image analysis was used to calculate the coating porosity. The quote values are an average of 10 areas of coating cross-section images observed at 1000× magnification. The porosity of coating was calculated based on the standard of ASTM E2109-01. A HMV-2 T Vickers microhardness tester was used to measure the microhardness of the coating. 25 readings were taken at different dense locations with the 300 g applied load and the dwell time was set to 15 s. For measurement of nanomechanical properties, the indentations were performed on polished cross section of coating samples using Micro Materials Nanotest$^{TM}$ indenter equipped with a Berkovich diamond tip. Ten indentations were made on each sample. The indentation load was set at 50mN and the maximum load was held constant for 5s. The loading rate was applied at 1mN/s.

## 3 Results and discussion

### 3.1 Phase composition and microstructure

Five levels of plasma spray power were applied to examine the characteristic of microstructure and nanoindentation behaviour of the coating. As shown in Figure 1, alumina-titania powder is consisted of α-$Al_2O_3$ phase, while the as-sprayed coating is consisted of γ-$Al_2O_3$ phase (major) and α-$Al_2O_3$ (minor). This result showed that the powders have experienced a phase transformation from α phase to γ phase during the deposition process. This transformation was due to a rapid cooling effect. Some amount of titania phases in the form of aluminium titanate ($Al_3Ti_5O_2$) were also identified. However, the amount of $Al_2TiO$ phase was very low due to the reaction of alumina as titania was dissolved into γ-$Al_2O_3$ during the deposition process [17].

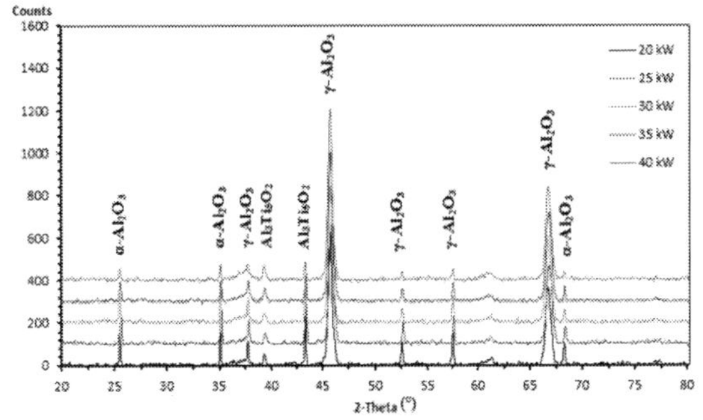


**Figure 1.** XRD pattern of as-sprayed $Al_2O_3$ -13%$TiO_2$ coating

Figure 2 shows the SEM micrographs of as-sprayed coating at four different plasma powers. It can be seen that high amount of unmelted particles with larger size was retained within the surface at low plasma power coating. The deposition exhibited uniform and large sized droplets with less porosity at high plasma power coating. At high plasma power, it was clearly noted that there was larger fully melted region of powder particles resulting thinner and better contact between splats of the coating surface.

The cross sectional of the coatings is shown in the Figure 3. It was revealed that higher plasma power resulted in less porosity within the coating. High plasma power generated higher temperature thus resulting an increase in the fully melted region on the coating structure [18]. This factor contributed to the denser coating structure with low percentage of porosity. Figure 4 illustrates the average percentage of porosity formed in the coating. The average percentage of porosity in the coating gradually decreased as the plasma power increased.

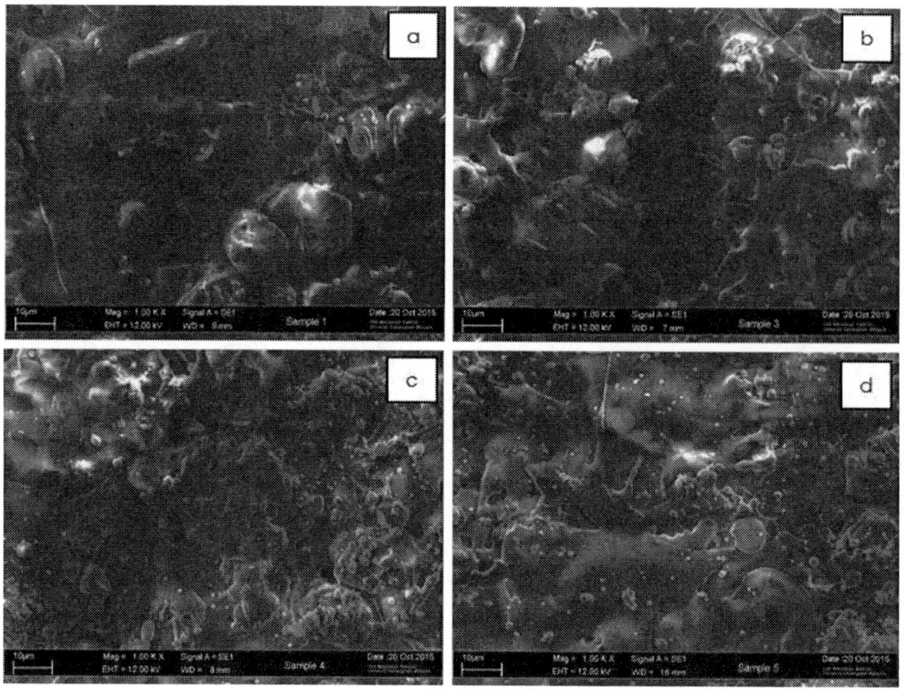


**Figure 2.** Surface morphology of $Al_2O_3$ -13%$TiO_2$ coating a) 25 kW, b) 30 kW, c) 35 kW and d) 40 kW of plasma power.

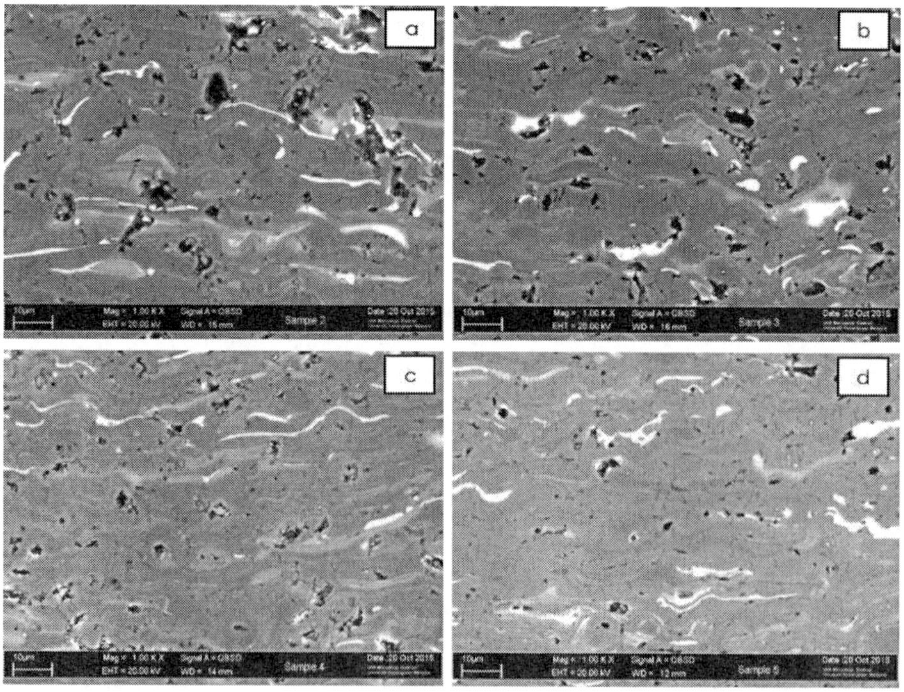


**Figure 3.** Cross sectional of $Al_2O_3$ -13%$TiO_2$ coating a) 25 kW, b) 30 kW, c) 35 kW and d) 40 kW of plasma power.

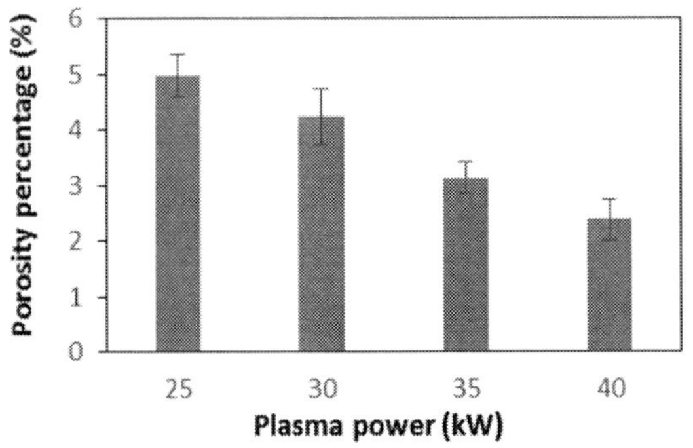


**Figure 4.** Percentage of porosity of plasma sprayed $Al_2O_3$ -13%$TiO_2$ coating

### 3.2. Nanoindentation

As stated in earlier sections, plasma power is one of the key factors that affect the microstructure of plasma sprayed ceramic coating. Generally, high plasma power induced high current that led to the high deposition temperature. This would increase the melting fraction of powder particles and reduced the porosity. According to Yusoff, et al. [19], porosity in the coating affected the mechanical properties of the coating particularly, the hardness properties. Figure 5 presents the hardness of the coating. It was noted that a slight increase had occurred as the plasma power was raised up to 40 kW. Thus, this indicates that high percentage of coating porosity contributed to a lower hardness of the coating.

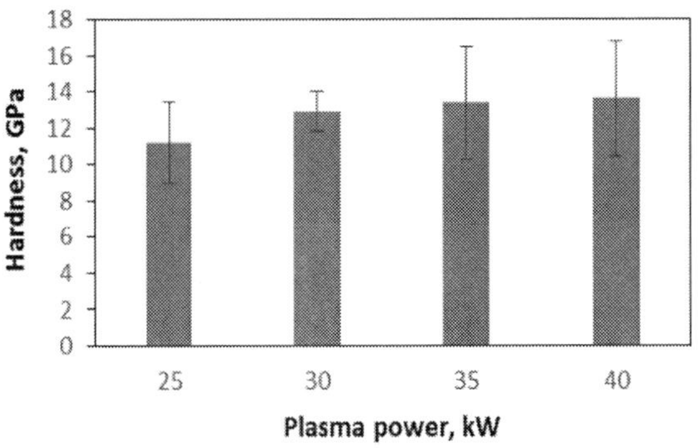


**Figure 5.** Hardness of plasma sprayed $Al_2O_3$ -13%$TiO_2$ coating

Figure 6 shows the loads as a function of indentation depths of the coating. Differences in hardness of the coating were indicated by the depth attained at the maximum load. It was found that at 50 mN load, the penetration depth for 25 kW plasma power coating was far greater than 500 nm, compared to the higher plasma power coating which were only around 450 nm. Other than that, the ceramic coatings show high degree of elastic recovery during unloading which contributed to the low plastic deformation during the indentation process. Large elastic recovery in coatings is beneficial in reducing the damage by impact and abrasive wear.

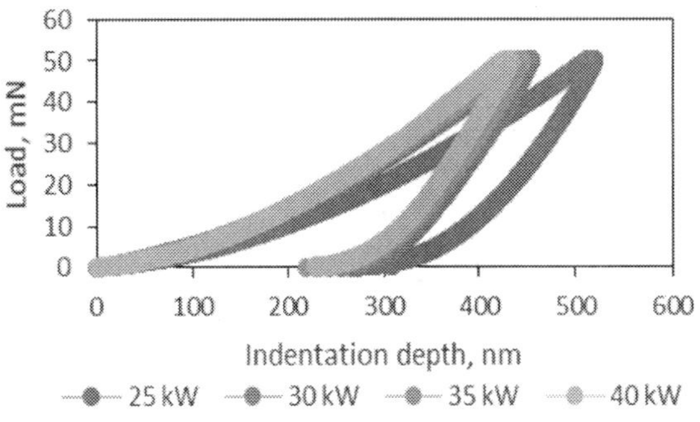


**Figure 6.** Load-displacement curve of plasma sprayed $Al_2O_3$ -13%$TiO_2$ coating

In the previous section, it was clearly noted that the spraying power directly affect the heat and mass transfer between powder particles and plasma and then influence the degree of powder particles melting and the structure of droplets on the coating surface. High plasma power resulted in denser coating with low porosity. It occurs when high current induced by high plasma power generates high energy to facilitate more ionization of the gases flowing between the electrodes to form plasma plume. As the temperature of the plasma plume increase, a tremendous level of thermal energy is released so that the feedstock material has sufficient energy to melt completely when injected into hot plasma plume and then propelled towards the target substrate to form the denser coating.

## 4 Conclusion

$Al_2O_3$ -13%$TiO_2$ coatings were deposited onto mild steel using atmospheric plasma spray technology by varying the plasma power from 25 kW to 40 kW. The microstructure and mechanical properties of the coatings varied with the applied plasma power. It can be concluded that high temperatures resulted a high melting degree of the powder particles deposition that led to a denser coating structure with lower percentage of porosity. From the nanoindentation test, the high plasma power at 40 kW resulted the highest hardness of coating; 13.63 GPa, whereas at 25 kW, little hardness of coating about 11.21 GPa, was recorded. In short, all coatings showed good elastic properties, which are beneficial in most of engineering applications.

## Acknowledgement

This work was financially supported by the Universiti Kebangsaan Malaysia grant number (DIP-2014-003). The author also acknowledges the Universiti Malaysia Perlis, UniMAP and Ministry of Education Malaysia (KPM) and for their meaningful supports and co-operations.

## References

1. Y. Xing, J. Deng, X. Feng, and S. Yu, Materials & Design, **52**, 234-245 (2013)
2. A. N. Samant and N. B. Dahotre, Journal of the European Ceramic Society, **29**, 969-993 (2009)
3. J. B. Wachtman and R. A. Haber, *Ceramic Films and Coatings* (United States, Noyes Publications, 1993)
4. M. A. Zavareh, A. A. D. M. Sarhan, B. B. A. Razak, and W. J. Basirun, Ceramics International, **40**, 14267-14277 (2014)
5. Y. Xing, J. Deng, Z. Wu, and H. Cheng, Applied Surface Science, **265**, 823-832 (2013)
6. Y. Xing, J. Deng, Y. Zhou, and S. Li, Surface and Coatings Technology, **258**, 699-710 (2014)
7. Y. Yang, Y. Wang, W. Tian, D.-r. Yan, J.-x. Zhang, and L. Wang, Materials & Design, **65**, 814-822 (2015)
8. E. Bannier, M. Vicent, E. Rayón, R. Benavente, M. D. Salvador, and E. Sánchez, Applied Surface Science, 316, 141-146 (2014)
9. W. Tian, Y. Wang, and Y. Yang, Tribology International, **43**, 876-881, (2010)
10. E. P. Song, J. Ahn, S. Lee, and N. J. Kim, Surface and Coatings Technology, **202**, 3625-3632 (2008)
11. Y. Yang, D. Yan, Y. Dong, X. Chen, L. Wang, Z. Chu, Vacuum, **96**, 39-45 (2013)
12. E. Sadri and F. Ashrafizadeh, Surface and Coatings Technology, **236**, 91-101 (2013)
13. G. Di Girolamo, A. Brentari, C. Blasi, and E. Serra, Ceramics International, **40**, 12861-12867 (2014)
14. D. Thirumalaikumarasamy, K. S. Kamalamoorthy, and V. B. Visvalingam, Journal of Magnesium and Alloys, **3**, 237-246 (2015)
15. Wang, H.-d. Ma, J.-l. Li, G.-l. Kang, J.-j. and. Xu, B.-s , Applied Surface Science, **314**, 468-475 (2014)
16. R. Venkataraman, G. Das, S. R. Singh, L. C. Pathak, R. N. Ghosh, B. Venkataraman, Materials Science and Engineering: A, **445–446**, 269-274 (2007)
17. D. Goberman, Y. H. Sohn, Shaw, Jordan, and M. Gell, Acta Materialia, **50,** 1141–1152 (2000)
18. M. J. Ghazali, S. M. Forghani, N. Hassanuddin, Muchtar, and A. R. Daud, Tribology International, **93,** 681-686 (2016)
19. N. H. N. Yusoff, M. J. Ghazali, M. C. Isa, A. R. Daud, A. Muchtar, and S. M. Forghani, Materials & Design. 39: 504-508 (2012)

MATEC Web of Conferences **87**, 03001 (2017) DOI: 10.1051/matecconf/20178703001
*ENCON 2016*

# Effects of Swirl Bubble Injection on Mass Transfer and Hydrodynamics for Bubbly Flow Reactors: A Concept Paper

Ahmad Salam Farooqi[1a], Ariny Demong[1], Khairuddin Sanaullah[1], Shaharin A. Sulaiman[3], Andrew Ragai Rigit[2], Shanti Faridah Saleh[1], Shah Jehan Gillani[4] and Afrasyab Khan[1]

[1]*Department of Chemical and Energy Sustainability, Faculty of Engineering, Universiti Malaysia Sarawak (UNIMAS), Malaysia*
[2]*Department of Mechanical and Manufacturing Engineering, Faculty of Engineering, Universiti Malaysia Sarawak (UNIMAS), Malaysia*
[3]*Department of Mechanical Engineering, Universiti Teknologi Petronas, 32610 Bandar Seri Iskandar, Perak, Malaysia*
[4]*Department of Chemical Engineering, Pakistan Institute of Engineering and Applied Sciences (PIEAS), 45650 Islamabad, Pakistan*

**Abstract.** Bubble flow reactors (BFR) are commonly used for various industrial processes in the field of oil and gas production, pharmaceutical industries, biochemical and environmental engineering etc. The operation and performance of these reactors rely heavily on a range of hydrodynamic parameters; prominent among them are geometric configurations including gas injection geometry, operating conditions, mass transfer etc. A huge body of literature is available to describe the optimum design and performance of bubbly flow reactors with conventional bubble injection. Attempts were made to modify gas injection for improved efficiency of BFR's. However, here instead of modifying the geometry of the gas injection, an attempt has been made to generate swirl bubbles for gaining larger mass transfer between gas and liquid. Here an exceptionally well thought strategies have been used in our numerical simulations towards the design of swirl injection mechanism, whose paramount aspect is to inhibit the rotary liquid motion but facilitates the swirl movement for bubbles in nearly stationary liquid. Our comprehension here is that the swirl motion can strongly affect the performance of bubbly reactor by identifying the changes in hydrodynamic parameters as compared to the conventional bubbly flows. In order to achieve this bubbly flow, an experimental setup has been designed as well as computational fluid dynamic (CFD) code was used with to highlight a provision of swirl bubble injection by rotating the sparger plate.

## 1 Introduction

Bubble column reactors are a class of multiphase reactors that are commonly used in different industrial applications such as chemical, petrochemical, metallurgical, pharmaceutical, food environmental and wastewater treatment [1][2]. The multiphase reactors are generally categorized into three types namely, the trickle bed reactor (fixed or packed beds), fluidized bed reactors and the bubble column reactors [1]. In general, bubble column offers many advantages over other multiphase reactors, which include the simplicity of their construction and maintenance, low energy consumption, minimal space requirements, good mass transfer properties, and high thermal stability [3]. Interfacial mass transfer coefficient is an important parameter that affects design and operation of the columns, and this is greatly influenced by air flow rate, fluid dynamics (gas void fraction, liquid/bubble velocity, etc.), physical properties (density, viscosity, surface tension, etc.) as well as apparatus geometry [4].

The characteristics and hydrodynamics of bubble column are strongly depend on bubble size distribution, superficial gas velocity and gas distributor configuration are a few factors which govern the performance of a bubbling system. Among all these, bubble size is the most important parameter because it not only affects the bubble rise velocity but also has a direct influence on the gas void fraction and interfacial area; and also an important criterion for evaluating the efficiency of a bubble column reactor [5]. Gas void fraction is defined as the fraction of gas occupied in the total volume of the gas and liquid mixture in bubble column [6]. Due to the complex nature of interfacial processes between gas and liquid, it is an uphill task to optimize the operating parameters (e.g. gas and liquid rates, sparger geometry, bubble size and shape, liquid & bubble turbulence etc.) [7].

In generalized gas–liquid two-phase flows, bubbles are observed in different sizes and shapes, behave differently in terms of relative motion and interaction mechanisms [8]. Bubbles are categorized into various groups with its

[a] Corresponding author: 15020293@siswa.unimas.my

own transport phenomena. For a special case of bubbly flows, all of the bubbles are in spherical or distorted shape. The injection of air bubbles can increase the wall shear stress which bubbles travelling close to the wall create a periodic perturbation. The small bubbles will tend to move to the wall, hence more bubble will move to the wall. As more air bubbles travel in the wall region, the mean shear stress increases [19].

To the best of our knowledge many investigations studied the effect of conventional bubble injection to predict the hydrodynamics parameters but there are few open literature to study hydrodynamics in bubble columns with swirl gas injection. So this study especially focuses on the effect of swirl injection on the hydrodynamics parameters as well as the mass transfer coefficient for both experimentally and numerical investigations. The findings of this work are derived from a novel bubble column design and fabrication and the CFD study using Ansys Fluent Code.

## 2 Literature review

In the bubble column, swirl flow causes an increase in mass transfer by modifying the hydrodynamic characteristics. However, the slip velocity decreases in such conditions [9]. Pressure drop across the height of the column also has been affected by the swirling flow as well as the centrifugal effects due to the swirl motion. Sreevisanan and Raghavan (2002) have found that the pressure drop in swirling regime is not constant, but it can increase with increase in gas flow rate. The dominant process occurring inside the column is the mass transfer between bubble and liquid [10]. Introducing swirl flow in the bubble column can promote axial and radial flows, which in turn can increase the gas void fraction as the small bubbles coalesce into large bubbles [11].

Mixing can be accomplished in a vessel by gas injection with no agitation. Mixing time also can be decreased with increasing the gas velocity [12]. The gas bubbles tend to concentrate more in the central core of the column, which causes an increase in liquid circulation [13]. Bubble column with perforated plate distributors causes a swarm of bubbles to rise through the liquid and this gives an agitation effect to the liquid phase. Therefore, swirl flows have wide range of applications in various engineering areas such as mechanical and chemical mixing and separation devices, combustion chambers, turbo machinery, pollution control devices, etc. Better utilization of swirl flows may lead to the heat and mass transfer enhancements [14]. Swirling flow can enhance the mixing of reactants, this may lead to the increase in the mixing efficiency as well as mass transfer rate of the reactants and hence the production yield. Such swirling flow is used in the chemical process industry to enhance heat and mass transfer in pipe flow and can improve the mixing and the hydrodynamic characteristics [15].

Gas void fraction is also an important parameter to characterize the hydrodynamics of gas – liquid flow in a bubble column. From the literature, gas void fraction is linear with the gas flow rate and the superficial gas velocity. As the gas velocity increases, the amount of gas introduced per unit time increases, this leads to an increase in gas void fraction [16][17][13]. Interfacial area between gas and liquid also increases, with increase in gas void fraction [11].

Increase in the bubble buoyancy contributes to the water velocity profile that gives rise to an increased wall shear stress. The pressure and shear stress at the wall are related to the flow in the boundary layer at the wall [18]. The wall shear stress is locally determined by the velocity gradient adjacent to the wall and thus, the wall pressure is a weighted integral of the effects of the velocity field. Generally, the magnitude of the wall shear stress increases as the rotational rate increases. [19]. The injection of macroscopic air bubbles can enhance the wall shear stress because small bubbles tend to move in the wall region, hence more bubbles travel close to the wall [20]. Wall shear stress depends on the size of the bubbles injected. Small bubbles cause a higher wall shear stress than large bubbles. Larger gas fractions shows the shear stress fluctuations that seem to be independent of local gas fraction, $\varepsilon_g$, which has been observed in the experiment of Magaud et.al (2001) and Moursali et. al. (1995) [24]. Murai et.al. (2007) have shown that the instantaneous evolution of the wall shear stress in air-water flows is directly related to the passage of air bubbles. As more bubbles travel in the wall region, more spikes occur, and the mean shear stress thus increases [14].

Apart from that, pressure drop in bubble column also has been affected by the superficial gas velocity and the wall shear stress [21][22]. In the presence of bubbles, the pressure drop increases as the phase velocity is higher than the superficial velocity and this causes the turbulence further higher [22].

A wide range of CFD studies on bubble column reactors have been conducted in recent years. Mainly two mathematical approaches namely, Euler-Euler and Euler-Lagrange have been employed to calculate bubbly flow characteristics. In the *Euler–Euler (E-E) approach* bubbles are not tracked individually, but the dynamics of the dispersed phase are ensemble averaged, to obtain a set of Eulerian equations, similar to the equations for the continuous phase The *Euler – Lagragian (E-L)* approach gives a direct physical interpretation of the fluid particle interaction, but it is computationally intensive, and hence needs large computing time for simulating bubbly flows, and systems having high dispersed phase volume fraction.

*Pfleger, (2001)* investigated the hydrodynamic simulation of the bubbly flow in a cylindrical bubble column operated in the homogenous flow regime by the use of a commercial CFD package (CFX 4.3). They applied Eulerian-Eulerian scheme with a k–ε turbulence model. For turbulence closure, computed results mainly focused on the local-liquid-phase velocities and gas void fractions. Whereas, *Ranade and Tayalia, (2001)* investigated the influence of sparger design on the hydrodynamics of a bubble column using the Euler-Euler approach with the single phase k–ε turbulence model. However, *Tiefeng WANG, (2011)* used population balance model (PBM). It is an effective approach to predict the bubble size distribution, and great efforts have been made in recent years to couple the PBM to CFD simulations. The CFD-PBM coupled model with the bubble break up and coalescence schemes and the interphase force formulations in his work has the ability of predicting the complex hydrodynamics in different flow regimes and thus provide a unified description of both the homogenous and heterogeneous regimes.

The literature on the swirl gas injection in bubble columns is scarce, so this study is an attempt to introduce a more efficient gas injection mechanism which can serve to increase the hydrodynamic characteristics as well as the efficiency of the bubble column reactor. Here in the present ongoing investigation, will contain both computations and experiments to identify the effect of swirl bubble on the performance of bubbly reactor by capturing the changes in hydrodynamic parameters as compared to the conventional bubbly flows. A bubbly flow rig has been designed with a provision of swirl bubble injection by rotating the sparger. Bubble properties (i.e. void fraction, bubble velocity etc.), mass transfer coefficient and flow parameters (e.g. shear stress, liquid velocity etc.) will be measured by means of videography (e.g. Charge couple diode (CCD), particle image velocimetry (PIV), Laser Doppler anemometry (LDA) etc.) and pressure sensors.

# 3 EQUIPMENT AND METHODS

## *3.1 Computational Schemes and Swirl Bubble Column Design*

CFD (Computational Fluid Dynamics) uses numerical methods and algorithms to solve out the problems involving single and multiphase flows. We have started with a computational part of the project by the use of workbench to draw geometry of a bubble column, which is shown in figure 1. Then the preliminary simulations, using CFD commercial code ANSYS FLUENT 14 have been processed. The dimensions of the column selected for simulations are; 45.72 cm diameter and 121.92 cm height same as the fabricated bubble column design, shown in figure 3. Keeping in view, these dimensions are chosen based on the physical size of pressure sensor and the mechanism of traversing them across the column. And these dimensions have been consistent with industrial and academic importance [23][22][24]. A porous glass sparger having diameter of 12.7cm is stationed at the center of the bottom of vessel. The next step has been to attempt rotation of bubbles in a vessel. This has been achieved by making use of awareness of possible effects of different types of fins mounted on the collar that surrounds the air sparger. The dimensions of the collar chamber are; dia 12cm and 20cm high. It is obvious that with the rotation of sparger the water around it will also revolve and this will reduce the interaction of the air bubbles with the surrounding water. We have planned to test different fin configurations such as converging-diverging, fins with tubular openings and fins with angled tubular openings which are shown in figure 2 to optimize swirl bubble injection. The fundamental purpose of all these is to dissipate the effects of the revolving sparger on to the water in its surroundings in order to ensure enhanced interaction of bubbles with the water.

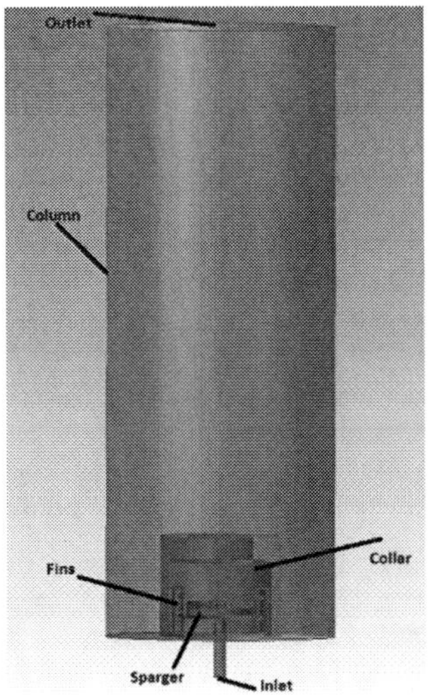


**Figure 1:** Geometry of the bubble column reactor with collar and fins

**Figure 2:** Different types of fins

## *3.2 Computations*

The important hydrodynamic parameters that characterize the bubbly flow are gas void fraction, bubble size and velocity, and mean shear stress and its distribution across the column. Gas void fraction is single most important operating parameter because it not only represents phase fraction and governs gas-phase residence time but is also crucial for mass transfer between liquid and gas. Gas void fraction depends chiefly on gas flow rate, but also to a great extent on the vessel geometry and configuration. The void fraction at any spatial location is due to the probability of appearance of gas compared to the liquid.

DOI: 10.1051/matecconf/20178703001

$$\epsilon_g = \frac{t_g}{t_T} \quad (1)$$

Where; $t_g$ is time when gas appears and $t_T$ is total time. The relation of mean gas void fraction to superficial gas velocity which is defined as;

$$\overline{\epsilon_g} = \frac{u_{gs}}{u_g} \quad (2)$$

Where $u_{gs}$ is the gas superficial velocity and $u_g$ is the gas average velocity when the air and water flow as mixture in the column.

Thus parameters such as shear stress and bubble properties (e.g. void fraction, bubble size and velocity etc.) computed by the use of ANSYS FLUENT 14.0 will be validated with similar measurements conducted by the experiments.

### *3.3 Experimental Setup*

Figure 3 shows schematic diagram of the bubbly flow reactor design. The bubbly flow column is equipped with a sparger and collar with fins. The sparger is connected to the rotating rod, where the rod is driven by the use of variable speed motor to produce the swirl gas injection. When the sparger rotates, the water around the sparger moves along with sparger rotation. So the purpose of collar and fins is to dissipate the momentum and energy associated with the water movement inside the collar, and thus this serves to eliminate the water from rotating as the sparger rotates. The efficient scenario here is to enable bubbles to drift radially as they move axially up the column in a body of water, which is nearly stationary. Some preliminary results are shown in figure 4. In the current simulation, only two fins are used to dissipate the liquid momentum inside the collar chamber.

The experiments are carried out in a bubble column made of transparent Perspex of 45.72 cm diameter and 121.92 cm height, equipped with a porous gas distributor. Tap water at room temperature is used as liquid phase and gas is taken from air compressor. Volumetric gas flowrate is measured by using a rotameter, and the rotation of the sparger is controlled by means of a variable speed motor. The pressure sensor is flushed along axial bubble movement. The location of the pressure sensors both axially and radially can be controlled by the specially designed mechanical setup. A high speed digital imaging system will be used to acquire bubble size and bubble properties. The schematic of the measurement set-up is shown in Figure 5.

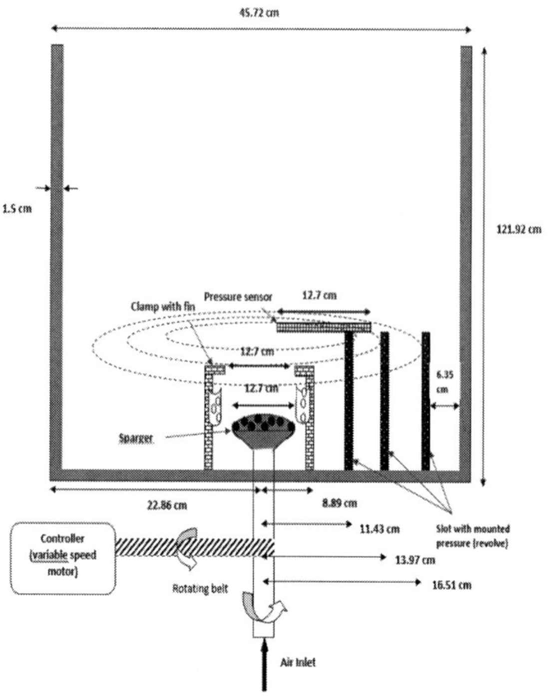


**Figure 3:** Schematic diagram of air water bubble column

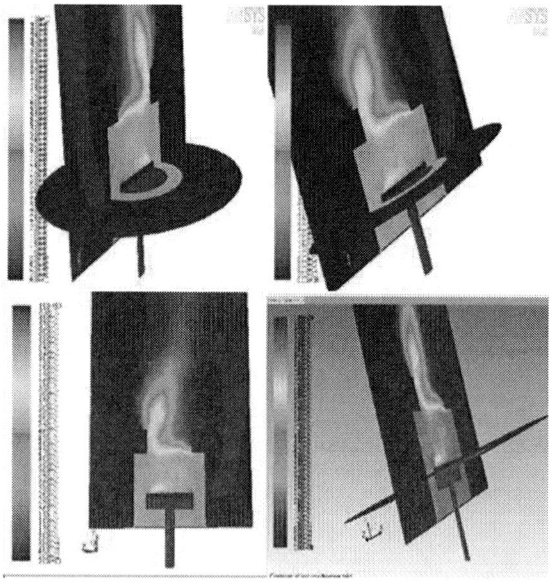

**Figure 4:** Different views of contour of volume fraction of air

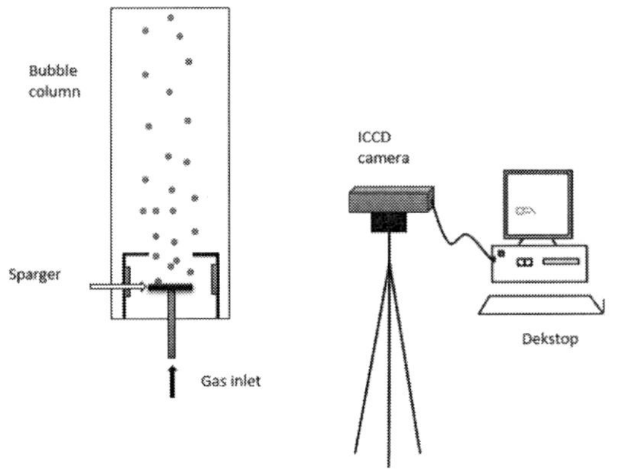


**Figure 5:** Experimental set-up for characterization gas-liquid in a bubble column

### *3.4 Proposed Measurements*

Specially configured cold experiments by the utilization of liquid-gas phases have been designed to study the hydrodynamic behaviour in bubble columns. The complex hydrodynamics due to the interaction of the two phases along with additional complexity due to local bubble entrainment and turbulence can be studied by the use of non-intrusive optical instrumentation (i.e. CCD videography, LDA, PIV etc.) as well as Electrical Resistance Tomography (ERT). The former is used to provide information for bubble (i.e. size, shape, velocity etc.) and mean shear characteristics including the turbulence intensity. Whereas ERT techniques are useful to measure volume fraction of the phases (i.e. gas and liquid) along with interaction between phases.

From these experiments, parametric effects due to gas and liquid velocity, gas volume fractions as well as geometry of gas distributor will be studied. Conclusive impacts of these parameters on the hydrodynamics will help to optimize as well as scale up of the reactor. Sampling ports will be provided along axial and radial locations of the bubble reactor for mass transfer measurements.

## 4 Conclusion

In this paper, both numerical and experimental methods have been outlined for proposed investigations involving the effects of swirl gas injection on the hydrodynamic parameters and mass transfer for BFRs. The commercial code ANSYS FLUENT 14 has been used for the simulations to infer information towards the design of swirl gas injection mechanism. The proposed methods based on optical as well as pressure sensors and Resistivity based tomography have also been briefed here and these will be used to find the hydrodynamic characteristics (bubble velocity, gas void fraction etc.). Measurements and computations will both be undertaken for conventional and swirl bubble injection.

## References

[1] N. Kantarci, F. Borak, and K. O. Ulgen, "Bubble column reactors," *Process Biochem.*, vol. 40, no. 7, pp. 2263–2283, 2005.

[2] G. L. Salierno, M. Maestri, S. Piovano, M. Cassanello, M. A. Cardona, D. Hojman, and H. Somacal, "Discrete axial motion of a radioactive tracer reconstructed from the response of axially aligned detectors: Application to the analysis of a bubble column dynamics," *Chem. Eng. Sci.*, vol. 100, pp. 402–412, 2013.

[3] H. Jin, Y. Lian, L. Qin, S. Yang, G. He, and Z. Guo, "Parameters measurement of hydrodynamics and CFD simulation in multi-stage bubble columns," *Can. J. Chem. Eng.*, vol. 92, no. 8, pp. 1444–1454, 2014.

[4] Z. Zhou, Q. Yuan, X. Jia, W. Feng, and J. Wen, "Experimental Study and CFD Simulation of Mass Transfer Characteristics of a Gas-Induced Pulsating Flow Bubble Column," vol. 27, no. 2, pp. 167–175, 2013.

[5] F. Lehr, M. Millies, and D. Mewes, "Bubble-size distributions and flow fields in bubble columns," *AIChE J.*, vol. 48, no. 11, pp. 2426–2443, 2002.

[6] I. Ismail, A. Shafquet, and M. N. Karsiti, "Void Fraction Estimation by Using Electrical Capacitance Tomography and Differential Pressure in an Air-Water Co-Current Bubble Column," *Aust. J. Basic Appl. Sci.*, vol. 5, no. 11, pp. 1533–1541, 2011.

[7] M. Irani and M. A. Khodagholi, "Investigation of Bubble Column Hydrodynamics Using Cfd Simulation ( 2D and 3D ) and Experimental," *Pet. Coal*, vol. 53, no. 2, pp. 146–158, 2011.

[8] X. Wang and X. Sun, "Three-dimensional simulations of air-water bubbly flows," *Int. J. Multiph. Flow*, vol. 36, no. 11–12, pp. 882–890, 2010.

[9] P. Kaushik, S. Pati, S. K. Som, and S. Chakraborty, "International Journal of Heat and Mass Transfer Hydrodynamic and thermal transport characteristics of swirling flows through microchannels with interfacial slip," vol. 55, pp. 4359–4365, 2012.

[10] M. Sivaiah, R. Parmar, S. K. Majumder, and A. E. Setup, "Hydrodynamics of Bubbly Flow in a Modified Reactor," vol. 6, no. 9, pp. 82–85, 2012.

[11] K. Shanmugam, K. Saravanan, V. Ramamoorthy, and R. Balasubramani, "Hydrodynamic studies in stirred bubble column," pp. 113–118, 2008.

[12] W. D. Deckwer and A. Schumpe, "Improved tools for bubble column reactor design and scale-up," *Chem. Eng. Sci.*, vol. 48, no. 5, pp. 889–911, 1993.

[13] S. M. Wagh, M. E. A. Ansari, and P. T. Kene, "Axial and Radial Gas Holdup in Bubble Column Reactor," vol. 35, no. 6, pp. 1703–1705, 2014.

[14] S. Martemianov and V. L. Okulov, "On heat transfer enhancement in swirl pipe flows," *Int. J.*

*Heat Mass Transf.*, vol. 47, no. 10–11, pp. 2379–2393, 2004.

[15] T. F. Ayinde, "A generalized relationship for swirl decay in laminar pipe flow," *Sadhana - Acad. Proc. Eng. Sci.*, vol. 35, no. 2, pp. 129–137, 2010.

[16] K. Koide, K. Shibata, H. Ito, S. Y. Kim, and K. Ohtaguciii, "Gas holdup and volumetric liquid-phase mass transfer coefficient in a gel-particle suspended bubble column with draught tube," *J. Chem. Eng. Japan*, vol. 25, no. 1, pp. 11–16, 1992.

[17] S. C. Cardona, F. López, A. Abad, and J. Navarro-Laboulais, "On bubble column reactor design for the determination of kinetic rate constants in gas-liquid systems," *Can. J. Chem. Eng.*, vol. 88, no. 4, pp. 491–502, 2010.

[18] H. G. Nepomuceno and R. M. Lueptow, "Pressure and shear stress measurements at the wall in a turbulent boundary layer on a cylinder," vol. 9, no. February 1996, 1997.

[19] M. M. E. S. Alek, P. O. S. Attari, and R. O. J. M. Artinuzzi, "Analysis of Fluid Flow and Wall Shear Stress Patterns Inside Partially Filled Agitated Culture Well Plates," 2011.

[20] M. N. Descamps, R. V. A. Oliemans, G. Ooms, and R. F. Mudde, "Air-water flow in a vertical pipe: Experimental study of air bubbles in the vicinity of the wall," *Exp. Fluids*, vol. 45, no. 2, pp. 357–370, 2008.

[21] A. A. Al-dughaither, A. A. Ibrahim, and W. A. Al-masry, "Investigating pressure drop across wire mesh mist eliminators in bubble column," pp. 1–9, 2011.

[22] V. S. Vitankar, M. T. Dhotre, and J. B. Joshi, "A low Reynolds number k–ε model for the prediction of flow pattern and pressure drop in bubble column reactors," *Chem. Eng. Sci.*, vol. 57, no. 16, pp. 3235–3250, 2002.

[23] N. Kantarci, F. Borak, and K. O. Ulgen, "Bubble column reactors," vol. 40, pp. 2263–2283, 2005.

[24] A. Shaikh, "I NTERNATIONAL J OURNAL OF C HEMICAL A Review on Flow Regime Transition in Bubble Columns," vol. 5, 2007.

# Measurement of Streaming Potential in Downhole Application: An Insight for Enhanced Oil Recovery Monitoring

Tengku Amran Tengku Mohd[1,2], Mohd Zaidi Jaafar[1,a], Azad Anugerah Ali Rasol[1], Mohd Fauzi Hamid[1]

[1]*Faculty of Chemical and Energy Engineering, Universiti Teknologi Malaysia, Skudai, MALAYSIA*
[2]*Faculty of Chemical Engineering, Universiti Teknologi MARA, Shah Alam, MALAYSIA*

**Abstract.** Downhole monitoring using streaming potential measurement has been developing in order to respond to actual reservoir condition. Most studies have emphasized on monitoring water flooding at various reservoir condition and improving the approaches of measurement. Enhanced Oil Recovery (EOR) could significantly improve oil recovery and the efficiency of the process should be well-monitored. Alkaline-surfactant-polymer (ASP) flooding is the most promising chemical EOR method due to its synergy of alkaline, surfactant and polymer, which could enhance the extraction of residual oil. However, limited studies have been focused on the application of streaming potential in EOR processes, particularly ASP. Thus, this paper aims to review the streaming potential measurement in downhole monitoring with an insight for EOR application and propose the potential measurement in monitoring ASP flooding. It is important for a preliminary study to investigate the synergy in ASP and the effects on oil recovery. The behaviour of streaming potential should be investigated when the environment of porous media changes with respect to ASP flooding. Numerical model can be generated from the experimental data to forecast the measured streaming potential signal during production associated with ASP flooding. Based on the streaming potential behaviour on foam assisted water alternate gas (FAWAG) and water alternate gas (WAG) processes, it is expected that the streaming potential could change significantly when ASP flooding alters the environment and surface properties of porous media. The findings could provide new prospect and knowledge in the relationship between streaming potential and ASP mechanisms, which could be a potential approach in monitoring the efficiency of the process.

## 1 Introduction

Real-time permanent monitoring and control is important in maximizing the hydrocarbon production from oil and gas wells. Permanent downhole sensors have been previously investigated to optimize reservoir management [1,2,3]. According to Brinsden (2005), an effective development and optimization of oil and gas reservoirs requires long-term reservoir monitoring and control, which is important in both new and mature fields [4]. Therefore, many downhole monitoring techniques have been proposed and developed in order to resolve any difficulties, cost and maintenance associated with the installation of the tools.

Streaming potential measurement applying electrodes permanently installed downhole has been a promising reservoir-monitoring technology [5,6]. Jaafar et al. (2009) have measured the streaming potential coupling coefficient in sandstones saturated with higher brine salinity and reported the values of streaming potential coupling coefficient changed with respect to rock texture or mineralogy, injected and formation brine salinity [7]. Jackson et al. (2011) found that the streaming potential at the well changed when the encroached water on a production well was several tens to hundreds of meters away [8]. The contrast with most other downhole monitoring techniques provides great advantages in monitoring the efficiency of secondary or tertiary recovery processes. Jaafar et al. (2013) found the point of zero charge (pzc) of two types of core samples (sandstone and carbonate), where the voltage difference was measured at different pH of solution, but there were no further study on the effect of the degree of alkalinity to the streaming potential [9]. In 2014, Walker et al. applied new pressure transient approach as a faster approach than the conventional measurement methods, where the measured streaming potential coupling coefficients and zeta potential were aligned with the findings by Vinogradov et al. (2010) at high salinity [10,11]. Results obtained by Esmaeili et al. (2016) agreed with the published experimental data, and much lower scattering in the data points was achieved for different saline injecting water [12]. Streaming potential changes accordingly with the wetting state of the rock surface. When the wettability of rock partially changes to the oil-wet, the overall zeta potential and coupling coefficient approach to their values of completely oil-wet sand particles [13]. Surfactant has the ability to reduce interfacial tension, as well as changing the wetting state

[a] Corresponding author: mzaidi@utm.my

of the rock surface, which require further studies on the influences to the streaming potential.

It is evident that original rock surface properties and its modification during the EOR process determine the efficiency of oil recovery from reservoirs. The minerals and surface properties of the reservoir are very sensitive to any alteration of the fluids properties. ASP changes the environment in porous media, including IFT reduction, wettability alteration, ionic adsorption at the interface and degree of alkalinity [14], which could improve sweep efficiency. ASP provides synergy of alkaline, surfactant and polymer [15]. ASP mechanisms should be well understood, since they have significant effects on the behaviour of streaming potential. Zeta potential is one of the critical parameters to characterize the surface properties, where its average measurement of reservoir rock could be potentially estimated using streaming potential measurement as a new promising tool. [16]. There were no further studies from past researchers on how streaming potential behaves and its significant uncertainties when ASP changes the properties of porous media. Thus, the purpose of this paper is to review the streaming potential measurement in downhole monitoring with an insight on EOR application and propose the potential of streaming potential application in ASP flooding. This could provide new prospect and knowledge in the relationship between streaming potential and ASP mechanisms.

## 2 Streaming Potential

Electrokinetic is a group of physicochemical phenomena, which involves the transport of charge, action of charged particle, effect of applied electrical potential and fluid transport in porous media to achieve a desired fluid flow or migration. The phenomena included are streaming potential, electroosmosis, electrophoresis and reduction potential. Streaming potential is widely used in detection of contamination in the subsurface water table and currently study extended to enhanced oil recovery application. Streaming potential is defined as a result of electric current generated whenever an electrolyte move with respect to stationary solid that is in contact with in a case water flow through earth and rock [17]. This was first observed in capillary tubes by Quincke in 1859 [18], and a theoretical model for the effect in capillaries was later developed by Helmholtz in 1879 [19]. That model is still being applied currently and proved to be valid for flow though porous media.

The mechanism of streaming potential is fluid flow in porous media cause the electric conduction current which flow in reverse direction through both liquids and solid matrix [20]. The conduction current acts to cancel the convection current in a steady state. Solid matrix that electrically conductive in any degree will reduce the magnitude of the streaming potential. When a fluid flows with respect to a solid that it is in contact with, the solid surfaces are induced to be electrically charged. The fluid carries with it charged particles that are attracted to the charged surface of the solid, forming a diffuse layer containing excess of countercharge [4]. Extra double layers could be produced at fluid/fluid interfaces with the presence of more than one phase [21].

### 2.1 Streaming Potential Measurement

The electrical double layer formed at solid/fluid interface is the origin of streaming potentials in porous media [21]. If an external pressure gradient induces the fluid to flow, the flow will carry some excess charge within the diffuse layer, which could increase the streaming current. Density divergence of the streaming current produces an electrical potential, known as streaming potential [11]. The coupling coefficient (C) provides the relationship between the fluid ($\nabla P$) and electrical ($\nabla V$) potential gradients at zero total current density (j), as expressed in Eq. 1 [22].

$$C = \left.\frac{\partial V}{\partial P}\right|_{j=0} \qquad (1)$$

Thus, the magnitude of the streaming potential could be predicted if the fluid potential is known. The coupling coefficient is a function of the electrical conductivity of the brine ($\sigma_w$) and brine-saturated rock ($\sigma_{rw}$), permittivity ($\varepsilon_w$) and viscosity ($\mu_w$) of the brine, and the zeta potential ($\zeta$), which is the microscopic electrical potential subjected to the excess charge in the double layer. It can be shown as in Eq. 2 [21].

$$C = \frac{\varepsilon_w \zeta}{\mu_w \sigma_{rw} F} \qquad (2)$$

F is the formation factor ($\sigma_w/\sigma_{rw}$) measured when surface conductivity is negligible (associated with very saline brine).

### 2.2 Numerical Simulation of Streaming Potential

Numerical simulation of streaming potentials is important in understanding the phenomenon and for self-potential anomalies interpretation. Many simple models involving charged bodies of various shapes have been constructed, but not realistic since they did not consider the nature of all sources that generated the potentials. As mentioned previously, streaming potentials result from electric currents created by the flow of a liquid with respect to a solid. The current flow, defined by Ohm's law and the fluid flow, defined by Darcy's law, can be correlated by the theory of coupled flows. With reference to the earlier work of Marshall and Madden (1959) [23], Nourbehecht (1963) has produced the first model applying this approach, where solutions were formulated in terms of a total- or pseudo-potential [24]. However, the pseudo-potential approach misrepresented the actual mechanisms that generate streaming potentials, and only applicable to hydraulically driven fluid flow. Its application to more complex geometries can be difficult

A new approach for modelling streaming potentials was developed by Sill (1983), who also applied the coupled flow theory and formulated explicit expressions for the involved current sources [22]. His approach is straight-forward even when application on complex geometries, and can be applied as well to non-hydraulically driven fluid flow such as that generated by temperature gradients. Sill's method is more analogous to the actual physical process involved and forms the basis for most modern modelling of the streaming potential phenomenon.

## 3 Electrical Double Layer (EDL) and Zeta Potential

Electrical double layers (EDL) is very important factor in electrokinetic phenomena. A fundamental of understanding the physical properties of EDL is zeta potential which can be defined as electrical potential in electrolyte. A simple model describing the potential distribution in EDL is illustrated in Fig. 1. The slipping (shear) plane is defined as the point which the liquids begin to flow in a small distance from the solid surface. By definition, zeta potential is the electrical potential as this shear plane. Zeta potential is directly proportional to the surface charge that developed on the solid surface when in it is in contact with electrolyte [25]. However if the solid material already possesses a surface charge due to an imbalanced crystal structure, the two main chemical processes that act to develop a surface charge are the adsorption of ions and the hydrolysis of surface hydroxyl groups. Both of these processes typically occur simultaneously and depending on the chemical compositions of both the electrolyte and the solid. The chemistry at the interface can be treated by general chemical equilibrium equations and the equilibrium constants for many materials can be found in chemistry texts. The pH value of the electrolyte influences the surface charge and hence zeta potential in a solid and liquids chemical composition. The pH has a large effect on the hydrolysis of the surface hydroxyl group and lesser in ion adsorption [26]. By chemical equilibrium equation, the surface charge density is decreased with increasing pH. As a result, the zeta potential also decrease with increasing pH.

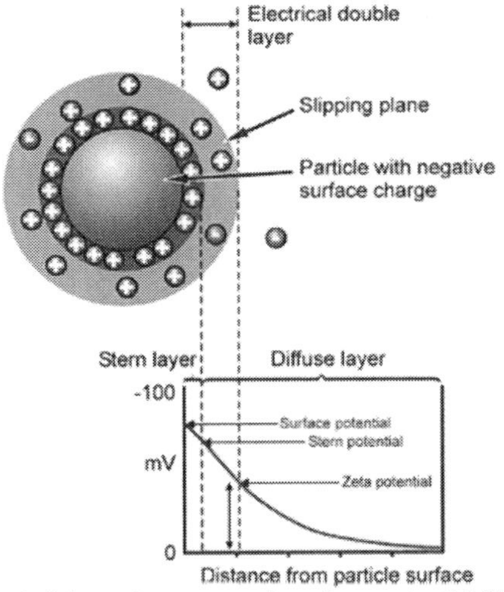


**Figure 1.** Schematic representation of zeta potential (ζ) [27].

## 4 Enhanced Oil Recovery

Enhanced Oil Recovery (EOR) or tertiary recovery is has been applied to achieve higher oil recovery from a reservoir that could not be produced by primary and secondary recoveries [28]. According to Collins (1977), EOR is the incremental production of oil with the introduction of artificial energy into the reservoir and can be divided into three major categories that are thermal EOR, chemical EOR and miscible EOR [29]. The performance of EOR processes is assessed based on the reservoir rocks characteristics and properties of displacing and displaced fluids. The reservoir rocks characteristics include rock porosity, wettability, as well as oil and water permeability of the porous media while the properties of displacing and displaced fluids are such as flow velocities and interfacial tension at fluid interface [30]. Hill et al. (2010) suggested modification of the reservoir matrix and fluid properties such as the viscosities of liquid phases, interfacial tension between two immiscible phases, rock wettability, and the capillary forces within the matrix [31]. This could increase the oil recovery, decrease water cut and reduce the residual oil saturation. ASP flooding is one of the promising chemical EOR methods, which provides the synergy of alkaline, surfactant and polymer in enhancing oil recovery.

### 4.1 Alkaline-Surfactant-Polymer Flooding

ASP flooding is a chemical EOR process which involves the injection of surfactant, alkaline and polymer into an oil reservoir targeted to produce the remaining residual oil, which could not be recovered by conventional water flood [32]. In addition, most of the oil fields have been under water flooding, which favour the implementation of chemical EOR method. Therefore, the facilities required to add chemicals in the injection water could be reduced [15]. ASP flooding is subjected to several processes in enhancing oil recovery, including interfacial tension (IFT) reduction between trapped oil phase and ASP slug [32,33], improving mobility control and sweep efficiency [34] as well as wettability alteration by surfactant to more favourable state [35].

ASP flooding ensures more efficient displacement through IFT reduction between oil and water with the addition of surfactant, while addition of polymer could improve the oil and water mobility [36]. Addition of alkali could reduce adsorption of the surfactant onto the rock surface besides controlling the salinity for minimum IFT and rock wettability alteration [36]. Thus, oil recovery could be significantly improved by ASP flooding [37].

#### *4.1.1 Mechanisms of Polymer Flooding*

Polymer flooding results in reducing the mobility with addition of polymer to the water during water flooding. Viscosity of the aqueous phase is increased by the polymer and water permeability is reduced due to mechanical entrapment, resulting in more favorable mobility ratio. With a more viscous phase, a more stable displacement can be achieved, which ensures the easier movement of the oil through the reservoir and into the producing well, improving the sweep efficiency [38,39]. It has been reported that a successful polymer application can achieve an incremental recovery factor of higher than 5% OOIP (original oil in place) [40,41]. Due to polymer viscoelastic properties, a larger pull force is exerted on oil droplets or oil films, which reduces residual oil saturation [42].

### 4.1.2 Mechanisms of Surfactant Flooding

The surfactant flooding aims to produce the residual oil trapped by the capillary action, which could not be recovered by a conventional water flooding. Its key mechanism is the reduction in the IFT between oil and water, which aid in mobilizing the residual oil. Since the higher the capillary number, the lower residual oil saturation [43], the three ways to increase the capillary number are by increasing the velocity of injection fluid, enhancing the viscosity of displacing fluid as well as ensuring the IFT reduction. IFT reduction is the most preferable method due to its practicality, where IFT reduction between oil and a surfactant solution can be achieved in the range of 20-30 to $10^{-3}$ mN/m. Instead, the injection fluid velocity is limited by pump capacity or formation injectivity leading to formation fracture, while increasing injection fluid viscosity using too high polymer concentration is associated with solubility issue [15]. Crude oil contains organic acid and salts, alcohols and other natural surface-active agents. When the crude oil is in contact with brine or water, an adsorbed film is formed by the accumulation of natural surfactants at the interface, which reduces the IFT of the crude oil/water interface [44].

### 4.1.3 Mechanisms of Alkaline Flooding

The injected alkaline or caustic solutions will react with the natural acids (naphthenic acids) contain in the crude oils to form in-situ surfactants (sodium naphthenate). The produced surfactants act similarly with the synthetic surfactants injected into the reservoir to ensure the IFT reduction between oil/water [44]. This generated surfactant is referred as soap in order to distinguish it from the synthetic surfactant. The reaction equation is as follows [15]:

$$HA + OH^- \rightarrow A^- + H_2O \qquad (3)$$

where HA is known as pseudo-acid component and $A^-$ is the soap component. In ASP process, alkali has significant role in reducing the amount of surfactant adsorbed on the grain surfaces through the formation and sequestering of divalent ions. Therefore, less surfactant injection is required, which ensures the surfactant to work more efficiently. Total Acid Number (TAN) of the oil affects how the alkali works in ASP flooding. Nelson et al. (1984) found the reduction of oil-water IFT during alkaline injection for oil with high TAN due to the combination effects of the generated soap with the injected surfactant [45]. However, alkali can still significantly reduce surfactant adsorption for oil with low TAN [46]. Liu et al. (2008) has proved that soap-to-surfactant ratio is important, which has significant contribution in order to understand the ASP process [47]. Besides, formation wettability can be altered with the presence of alkali to become more oil-wet or water-wet. Other mechanisms involved are such as emulsification, oil entrainment and bubble entrapment [48].

### 4.1.4 Mechanisms and Synergy in ASP

An emulsion is formed from the alkali-surfactant mixture with the oil, where a displacement takes place using a polymer drive to be recovered from the reservoir. Addition of surfactant in ASP flooding ensures the IFT reduction between oil and water, where the mobility ratio is further improved by the addition of polymer, resulting in improving displacement [36]. Addition of alkali to the water could minimize the amount of adsorbed surfactant onto the rock surface, which could ensure minimum IFT and rock wettability alteration [36]. An important mechanism in ASP is the synergy provided between synthetic surfactant and the generated soap. Generally, the synthetic surfactant has high optimum salinity, whereas the soap has unrealistically low optimum salinity. However, an increased range of salinity in which IFT reaches its low values could be achieved when they are functioning together [45]. Synergistic IFT reduction is produced by the surfactant and alkali while polymer improves sweep efficiency by emulsions [14]. The summary of mechanisms and synergies in ASP is shown as follows [15]:

1. Amount of adsorbed surfactant is reduced by alkaline injection.
2. In-situ soap is generated from the reaction of alkali and crude oil, which functioning together with synthetic surfactant to provide an increase of salinity range in which IFT reaches its low values.
3. Improvement of sweep efficiency by the emulsions, since IFT is reduced by soap and surfactant while emulsions are stabilized by high viscous polymer to retard coalescence.
4. Surfactant adsorption is reduced by addition of polymer, or vice versa, as there is an adsorption competition between polymer and surfactant [49].
5. Addition of polymer improves the sweep efficiency of alkaline and surfactants.
6. The decrease of liquid production was not only due to increase of the displacing fluid viscosity, but also related to emulsification and scaling after injection of ASP slug [50].

## 5 Measurement of Streaming Potential in Monitoring EOR

Measurement of streaming potential in downhole application using electrodes has recently been introduced, but there is still a significant degree of uncertainty in measuring and interpreting the streaming potential results. Two field tests were successfully carried out by Chen et al. (2006) for the first time. They have measured the streaming potential generated from pressure transients by electrodes located in the borehole [51] and found two responses of the recorded streaming potential on the pressure changes. These resulted from water flooding into high permeability fracture and rock matrix. Therefore, this indicated that streaming potential measurement has high potential application in petroleum industry for reservoir monitoring as well as reservoir characterization. Streaming potential measurement has been a promising reservoir monitoring technique, but many studies have been focused on the water flooding brine-saturated porous media and the recent analysis of streaming potential at varying brine salinity [7,12,13]. Limited studies have been conducted to measure streaming potential in monitoring EOR processes, such as water alternate gas (WAG) and foam assisted water alternate gas (FAWAG) by Anuar et al. and Omar et al. [52,53,54,55].

Basically WAG is a process where intermittent slugs of gas and water are injected into the reservoir. Three-phase (gas, oil and water) flow is better than two-phase flow in recovering residual oil in the pore space [56]. Water can control the mobility for more stable displacement which could improve the gas injection sweep efficiency. WAG could control gas fingering thus enhancing the vertical sweep efficiency. While FAWAG could improve the sweep efficiency during gas injection by reducing the Gas Oil Ratio (GOR) and maximizing hydrocarbon production rate in the production tubing [57]. In addition, foam improves gas mobility control by delaying the early gas breakthroughs.

Omar et al. (2013) has measured the streaming potential resulted from the flowing fluid in FAWAG process to be correlated with the foam stability [53]. The principle is similar with the water flooding to monitor the process, but it provides indirect measurement of foam stability in order to monitor the efficiency of the EOR process. Since foamability and stability are the key factors for enhancing the recovery [58], measuring streaming potential could aid in monitoring the foam stability from the surface. Using the same principle from Kostoglou et al. (2010) [59], their research was based on theory that the FAWAG process flowing in the porous medium could produce an electric current, where electrokinetic signal pattern varied with the changes of foam stability [53]. In 2014, Omar et al. found that the increase in surfactant concentration, sodium dodecyl sulphate (SDS) has decreased the voltage (electrokinetic potential) and decreased as shown in Fig. 2 [54]. Thus, electrokinetic methods have been applied to understand the conductive properties of foam as well as for investigation of the initial expansion and subsequent stability of various foams. Fig. 3 shows the relationship between electrokinectic signal and foam rupture time during the FAWAG process in the porous media at various SDS concentration. It was found that the voltage increased when increase in foam stability, thus electrokinectic potential has significant effects in monitoring foam stability during FAWAG process.

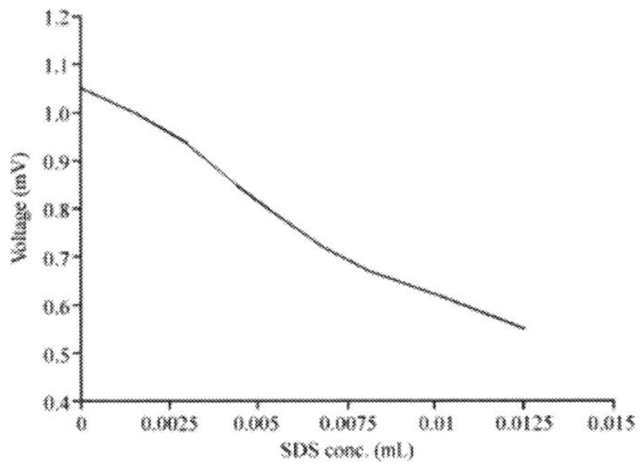


**Figure 2**. Relationship between voltage and SDS concentration [54].

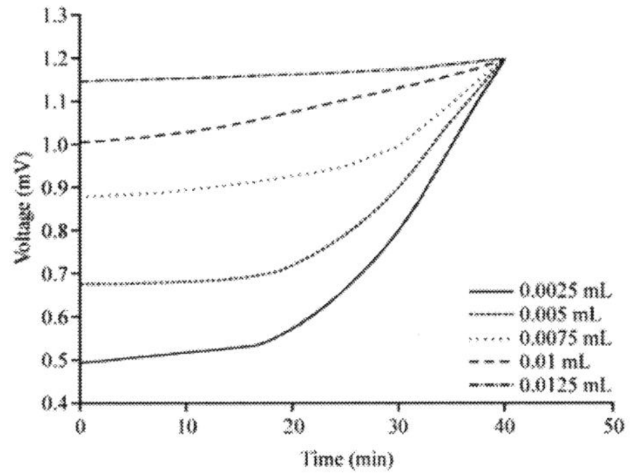


**Figure 3**. Relationship between electrokinetic signal and foam stability with time [54].

Anuar et al. (2013) proposed the application of streaming potential in monitoring WAG process as it could provide real-time monitoring [52], overcoming the longer process taken by tracer system [60]. The research was conducted based on the principle that water is conductive, while gas is non-polar and resistive, thus different effects could be observed on the voltage measurement (streaming potential). Fig. 4 illustrates the streaming potential profile during WAG process, indicating high streaming potential signals when water is flowing, while low signals when the present of gas. Streaming potential signals at different electrodes are shown in Fig. 5, where similar profile was observed but different responses between each other due to different location of electrodes. It can be concluded that streaming potential has significant effects on WAG process, which could effectively monitor the progress of water and gas in enhancing oil recovery [55].

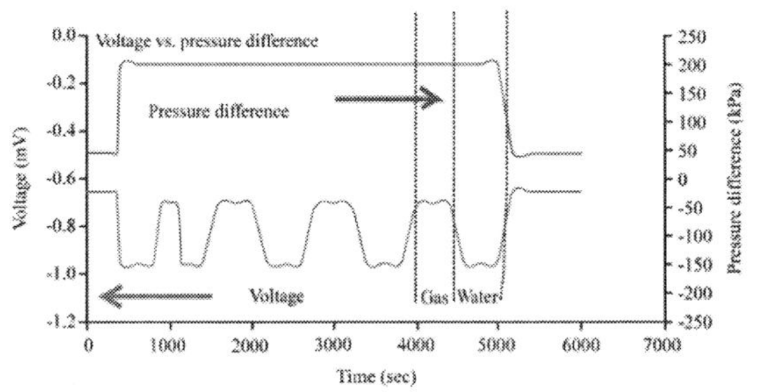


**Figure 4.** Streaming potential profile during WAG process [55].

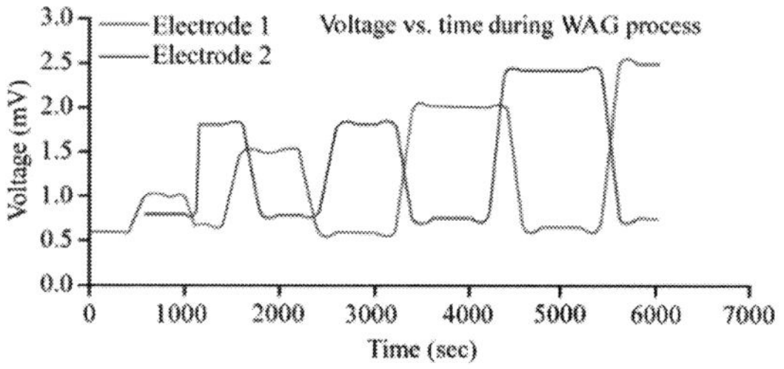


**Figure 5.** Streaming potential signals at different electrodes [55].

Many studies have been conducted to measure streaming potential related to groundwater movement and hydrocarbon production. However, there is a significant degree of uncertainty in the behaviour of streaming potential when the porous media changes, particularly with respect to alkalinity, ionic adsorption onto the rock surface and the interfacial tension (IFT) between the fluids residing in the pore space during ASP flooding. From the mechanisms of alkaline, surfactant and polymer flooding as well as the synergy in ASP, the interaction between the chemicals, the reservoir fluids and the rock surface could enhance the oil recovery [32]. The original rock surface properties and its modification during the EOR process will affect the efficiency of oil recovery from reservoirs. The minerals and surface properties of the reservoir are very sensitive to any alteration of the fluids properties.

Thus, there is a need for future studies to investigate streaming potential in monitoring ASP flooding, since ASP could provide the synergy, resulting in further enhancing oil recovery. Understanding the mechanisms involved in ASP flooding is vital in predicting the behaviour of streaming potential with respect to the interactions. This EOR process should be monitored effectively to ensure that the process is at optimum condition. In addition, the experimental data acquired can be applied in a numerical model to predict the streaming potential signal that would be measured during production associated with ASP flooding. From the review of the streaming potential theory and concept, as well as the ASP and its individual mechanisms, streaming potential measurement is believed to have a great potential in monitoring ASP flooding.

## 6 Conclusion

Approaches of streaming potential measurement in water flooding application have been improved. There is a need for further studies in EOR monitoring, since streaming potential behaviour has changed significantly during FAWAG and WAG processes. The synergy of alkaline, surfactant and polymer has revealed the ASP flooding to be an attractive and the most promising chemical EOR method. Thus, the process should be monitored effectively to ensure optimum recovery of residual oil. There are significant uncertainties in the streaming potential behaviour when the porous media changes, particularly with respect to alkalinity, ionic adsorption onto the rock surface and the interfacial tension (IFT) between the fluids residing in the pore space, potentially produced by ASP flooding. It is critical to understand the streaming potential behaviour and its relation with ASP mechanisms for efficient monitoring of ASP flooding. The streaming potential signals measured during production could be potentially predicted using numerical modelling. Application and measurement of streaming potential to monitor ASP flooding in the real field could benefit the oil industry in term of making the EOR process more efficient and economic.

## Acknowledgment

The authors would like to express appreciation to the Faculty of Chemical and Energy Engineering, UTM Skudai for the facilities provided and constant encouragement. This work was financially supported by Fundamental Research Grant Scheme (FRGS) Vote: 4F561.

## References

1. Chalaturnyk, R. and Moffatt, T. (1995). Permanent instrumentation for production optimization and reservoir management. Paper SPE 30274-MS proceedings of the SPE International Heavy Oil Symposium, Calgary, Alberta, 19-21 June.
2. Wu, Z., Alpak, F. O. and Torres-Verdin, C. (2003). A quantitative study to assess the value of pressure data acquired with in-situ permanent sensors in complex 3D reservoir models subject to two-phase fluid flow. Paper SPE 84375-MS proceedings of the SPE Annual Technical Conference and Exhibition, Denver, Colorado, 5-8 October.
3. Silva, M. I. and Kato, E. T. (2004). Reservoir management optimization using permanent downhole gauge data. Paper SPE 90973-MS proceedings of the SPE Annual Technical Conference and Exhibition, Houston, Texas, 26-29 September.
4. Brinsden, M. S. (2005). A New Wireless Solution to Real Time Reservoir Surveillance. Presented at the SPE Middle East Oil and Gas Show and Conference, Manama, Bahrain, 12-15 March. SPE-93512-MS.
5. Jackson, M. D., Saunders, J. H., and Addiego-Guevara, E. A. (2005). Development and Application of New Downhole Technology to Detect Water Encroachment toward Intelligent Wells. Paper SPE 97063 presented at the SPE Annual Technical Conference and Exhibition, Dallas, 9–12 October.
6. Saunders, J. H., Jackson, M. D. and Pain, C. C. (2008). Fluid flow monitoring in oil fields using downhole measurements of electrokinetic potential, Geophysics, Vol. 73(5).
7. Jaafar, M. Z., Vinogradov, J., and Jackson, M. D. (2009). Measurement of streaming potential coupling coefficient in sandstones saturated with high salinity NaCl brine. Geophysical Research Letters 36: L21306.
8. Jackson, M. D., Vinogradov, J., Saunders, J. H. and Jaafar, M. Z. (2011). Laboratory Measurements and

Numerical Modeling of Streaming Potential for Downhole Monitoring in Intelligent Wells, SPE 120460, September 2011 SPE Journal.
9. Jaafar, M., Nasir, A. and Hamid, M. (2013). Point of Zero Charge for Sandstone and Carbonate Rocks by Streaming Potential. International Journal of Petroleum & Geoscience Engineering (IJPGE), 1(2): 82-90.
10. Walker, E., Glover, P. and Ruel, J. (2014). A transient method for measuring the DC streaming potential coefficient of porous and fractured rocks. Journal of Geophysical Research: Solid Earth, 119(2): 957-970.
11. Vinogradov, J., Jaafar, M. and Jackson, M. (2010). Measurement of streaming potential coupling coefficient in sandstones saturated with natural and artificial brines at high salinity. Journal of Geophysical Research: Solid Earth (1978–2012), 115(B12).
12. Esmaeili, S., Rahbar, M., Pahlavanzadeh, H. and Ayatollahi, S. (2016). Investigation of streaming potential coupling coefficients and zeta potential at low and high salinity conditions: Experimental and modeling approaches. Journal of Petroleum Science and Engineering, 145, 137–147.
13. Sadeqi-Moqadam, M., Riahi, S. and Bahramian, A. (2016). Monitoring wettability alteration of porous media by streaming potential measurements: Experimental and modeling investigation. Colloids and Surfaces A: Physicochemical and Engineering Aspects. Elsevier B.V.
14. Sheng J. J. (2011). Modern chemical enhanced oil recovery: theory and practice. Elsevier; 2011.
15. Sheng, J. J. (2013). SPE 165358, A Comprehensive Review of Alkaline-Surfactant-Polymer (ASP) Flooding, Paper SPE 165358 proceedings of the SPE Western Regional & AAPG Pacific Section Meeting, Monterey, California, USA, 19-25 April 2013.
16. Chen, L., Zhang, G., Wang, L., Wu, W. and Ge, J. (2014). Zeta potential of limestone in a large range of salinity. Colloids and Surfaces A: Physicochemical and Engineering Aspects, 450: 1-8.
17. Wurmstich, B., Morgan, E. D., Texas, A., Merkler, G., Lytton, R. L., Texas, A., and G, E. (1994). Engineering and Groundwater 3 : Case Histories II Potentials Due to Seepage : Study of a Dam, 3–5.
18. Quincke, G. (1859). Concerning a new type of electrical current. Annalen der Physics and Chemie (Poggendorff's Annal., Ser. 2) 107, 1-47.
19. Helmholtz, H. (1879). Studien uber electrische grenzschichten. *Ann. Phys. Chem.* 7(ser.3):337-382.
20. Jackson, M., Gulamali, M., Leinov, E., Saunders, J. and Vinogradov, J. (2012). Spontaneous Potentials in Hydrocarbon Reservoirs During Waterflooding: Application to Water-Front Monitoring. SPE Journal, 17(1), 53–69.
21. Hunter, R. J. (1981). Zeta Potential in Colloid Science, Academic, New York.
22. Sill, W. R. (1983). Self-potential modeling from primary flows, Geophysics, 48, 76–86.
23. Marshall, D. J. and Madden, T. R. (1959). Induced polarization, a study of its causes, Geophysics, 24(4): 790-816.
24. Nourbehecht, B. (1963) Irreversible thermodynamics effect in inhomogeneous media and their applications in certain geoelectric problems, Ph.D. thesis, Mass. Inst. Of Technol., Cambridge.
25. Luong, D. T., and Sprik, R. (2014). Examination of a Theoretical Model of Streaming Potential Coupling Coefficient, 2014.
26. Zhang, W., Yao, J., & Sun, H. (2015). Electrokinetic coupling in single phase flow in periodically changed capillary with a very small throat size. International Journal of Heat and Mass Transfer, 84, 722–728.
27. Chilingar, G. V. Haroun, M., Shojaei, H. and Shin, S. (2014). Introduction to Electrokinectics, *Electrokinetics for Petroleum and Environmental Engineers*. Scrivener Publishing.
28. Stosur, G. J., Hite, J. R., Carnahan, N. F., and Miller, K. (2003). The Alphabet Soup of IOR, EOR and AOR: Effective Communication Requires a Definition of Terms. SPE International Improved Oil Recovery Conference in Asia Pacific, Kuala Lumpur, Malaysia, October 20-21, 2003.
29. Collins, A. (1977). Enhanced-Oil-Recovery Water Injections. In SPE-AIME International Symposium on Oilfield and Geothermal Chemistry. La Jolla, California.
30. Zhang, T., Davidson. D., Bryant S. L. and Huh. C. (2010). Nanoparticle-Stabilized Emulsions for Applications in Enhanced Oil Recovery, SPE/DOE Paper 129885, to be presented at SPE Improved Oil Recovery Symp., Tulsa, OK., Apr. 26-28
31. Hill, D. G., Wittle J. K., Fricker, D. J., and Chilingar, G. V. (2010). Moving Goo: Direct Electric Current Oil Recovery (EEOR)—A New Approach to Enhancing Oil Production, Sacramento Petroleum Association, November 17.
32. Austine, J., Van Batenburg, D. W., Southwick, J. G., Zarubinska, M. A., Paramanathan, S., Bouwmeester, R. C. M., Kechut, N. I., Viig, S. O., Haugen, O. B., and Brandvoll, Ø. (2015). Laboratory Evaluation of Inter-Well Partitioning Tracers for the determination of remaining oil saturation after ASP flooding. Paper SPE-174610-MS proceedings of the SPE Enhanced Oil Recovery Conference, Kuala Lumpur, Malaysia, 11-13 August 2015.
33. Flaaten, A. K., Nguyen, Q. P., Pope, G .A and Zhang, J. (2009). A Systematic Laboratory Approach to Low-Cost, High-Performance Chemical Flooding. SPE Reservoir Evaluation & Engineering. Volume 12, Number 5. October 2009. pp. 713-723. SPE paper 113469-PA.
34. Shen, P., Wang, J., Yuan, S., Zhong, T. and Jia, X. (2009). Study of Enhanced-Oil-Recovery Mechanism of Alkali/Surfactant/Polymer Flooding in Porous Media from Experiments. SPE Journal. Volume 14, Number 2. June 2009. pp. 237- 244. SPE paper 126128-PA.
35. Spinler, E. A. and Baldwin, B. A. (2000). Surfactant Flooding in Enhanced Oil Recovery. In Schramm, L. L. (Ed.) Surfactants Fundamentals and Applications

in the Petroleum Industry. 1ed. Cambridge, Cambridge University Press.
36. Hirasaki, G. J., Miller, C. A., Puerto, M. (2011). Recent advances in surfactant EOR. Soc Pet Eng J 2011;16:889-907.
37. Stoll, W. M., Al Shureqi, H., Finol, J., Al-Harthy, S. A. A., Oyemade, S. and De Kruijf, A. (2011). Alkaline/surfactant/polymer flood: from the laboratory to the field. SPE Reserv Eval Eng 2011;4(6):704-712.
38. Sorbie, K. S. (1991). Polymer-Improved Oil Recovery, CRC Press, Inc., Boca Raton, Florida.
39. Gao, S., Li, H., Li, H. (1995). Laboratory investigation of combination of alkali/ surfactant/polymer technology for Daqing EOR. SPE Resrv Engg 10:194–197
40. Rai, K., Mccomb, T., Rodriguez, E., Withers, R., Company, C. E. T. (2012). Development of a tool to predict technical Success of polymer flooding applications. Society of Petroleum Engineers, SPE; 2012. SPE 153878.
41. Singhal A. (2011). Preliminary review of IETP projects using polymers. 2011.
42. Wang, D. M., Cheng, J. C., Yang, Q. Y., Gong, W. C., Li, Q., and Chen, F. M. (2000). Viscous-elastic polymer can increase microscale displacement efficiency in cores, paper SPE 63227 presented at the SPE Annual Technical Conference and Exhibition, Dallas, Texas, 1–4 October.
43. Stegemeier, G. L. (1977). Mechanisms of entrapment and mobilization of oil in porous media, In Improved Oil Recovery by Surfactant and Polymer Flooding, (D.O. Shah and R.S. Schechter, eds.), pp. 55–91, Academic Press, New York.
44. Olajire, A. A. (2014). Review of ASP EOR (alkaline surfactant polymer enhanced oil recovery) technology in the petroleum industry: Prospects and challenges. Energy, 77, 963–982.
45. Nelson, R. C., Lawson, J. B., Thigpen, D. R., and Stegemeier, G. L. (1984). Cosurfactant-enhanced alkaline flooding. Paper SPE 12672 presented at the SPE/DOE Fourth Symposium on Enhanced Oil Recovery held in Tulsa, Oklahoma, 15–18 April.
46. Southwick, J. G., van den Pol, E., van Rijn, C. H. T., van Batenburg, D. W., Boersma, D. M., Svec, Y., Mastan, A. A., Shahin, G. T., and Raney, K. (2014). Ammonia as alkali for ASP floods – Comparison to sodium carbonate. In SPE Improved Oil Recovery Symposium, Tulsa, Oklahoma, USA. SPE 169057.
47. Liu, S., Zhang, D. L., Yan, W., Puerto, M., Hirasaki, G. J. and Miller, C. A. (2008). Favorable attributes of alkaline-surfactant-polymer flooding. Soc Pet Eng J 2008;13(1):5-16.
48. Johnson Jr., C. E. (1976). Status of caustic and emulsion methods. JPT (January), 85–92.
49. Wang, J., Han, M., Fuseni, A. B., & Cao, D. (2015). Surfactant Adsorption in Surfactant-Polymer Flooding for Carbonate Reservoirs. SPE-172700-MS: SPE Middle East Oil & Gas Show and Conference, Manama, Bahrain, 8–11 March 2015.
50. Zhu, Y. Y., Hou, Q. F., Liu, W. D., Ma, D. S. and Liao, G. Z. (2012). Recent progress and effects analysis of ASP flooding field tests, paper SPE 151285 presented at the SPE Improved Oil Recovery Symposium, 14-18 April, Tulsa, Oklahoma.
51. Chen, M., Raghuraman, B., Bryant, I. and Supp, M., (2006). Streaming potential applications in oil fields, SPE Annual Technical Conference and Exhibition, an Antonio, Texas, USA.
52. Anuar, S. M. M., Jaafar, M. Z. and Sulaiman, W. R. W. (2013). Monitoring Water Alternate Gas (WAG) Process Using Streaming Potential Measurement, Paper SPE 165300 proceedings of the SPE Enhanced Oil Recovery Conference held in Kuala Lumpur, Malaysia, 2-4 July 2013.
53. Omar, S., Jaafar, M. Z., Ismail, A. R., & Sulaiman, W. R. W. (2013). Monitoring Foam Stability in Foam Assisted Water Alternate Gas (FAWAG) Processes Using Electrokinetic Signals, Paper SPE 165312 proceedings of the SPE Enhanced Oil Recovery Conference held in Kuala Lumpur, Malaysia, 2-4 July 2013.
54. Omar, S., Jaafar, M. Z., Ismail, A. R. and Sulaiman, W. R. W. (2014). Relationship between Foam Stability and The Generated Electrokinetic Signals during FAWAG (Foam Assisted Water Alternate Gas) Process. Journal of Applied Sciences, 14(11): 1123-1130.
55. Anuar, S. M. M., Jaafar, M. Z., Sulaiman, W. R. W. and Ismail, A. R. (2014). Correlation Study between Streaming Potential and Waterfront Progression during Water Alternate Gas (WAG) Injection. Journal of Applied Sciences, 14(17): 1959-1965.
56. Tewari, R. D., Riyadi, S., Kittrell, C., Kadir, F. A., Bakar, M. A., Othman, T. R. T. and Banu, N. (2010). Maximizing Oil Recovery through Immiscible Water Alternating Gas (IWAG) in Mature Offshore Field, SPE-133345-MS. Presented at the SPE Asia Pacific Oil & Gas Conference and Exhibition held in Brisbane, Queensland, Australia, 18–20 October.
57. Tunio, S. Q. and T. A. Chandio (2012). Recovery enhancement with application of FAWAG for a Malaysian field. J. Applied Sci. Eng. Technol., 4:8-10
58. Qingfeng, H., Youyi, Z., Yousong, L., and Rui, W. (2012). Studies on Foam Flooding EOR Technique for Daqing Reservoirs After Polymer Flooding. Paper SPE 151955 presented at the 18th SPE Improved Oil Recovery Symposium held in Tulsa, Oklahoma, USA, 14-18 April.
59. Kostoglou, M., Georgiou, M., and Karapantsios, T. D. (2010). A New Device for Accessing Film Stability in Foams: Experiment and Theory. Colloids and Surfaces A: Physicochemical and Engineering Aspects, 382(2011), 64-73.
60. Abdullah, Z. Z., Zain, Z. M., Anua, N. A., and Singhal, A. (2011). Application of Radiactive and Chemical Tracers for Offshore WAG Pilot Project. Paper SPE 143391 presented at the SPE Enhanced Oil Recovery Conference held in Kuala Lumpur, Malaysia, 19-21 July.

# Characterizing local Garden Croton ('pokok puding') leaf powder as low cost biomaterial for Chromium(VI) uptake

Sannasi Palsan[1,a], Wan Nurul Aisyah Wan Jusoh[2] and Hemagowsalya Subramaniam[2]

[1]*Faculty of Agro Based Industry, Universiti Malaysia Kelantan, Jeli Campus, Locked Bag No. 100, 17600 Jeli, Kelantan, Malaysia*
[2]*Fostem, Inti University, Putra Nilai, 71800 Nilai, N. Sembilan, Malaysia*

**Abstract.** This paper reports the first ever characterization of the use of raw Garden Croton (locally known as 'pokok puding') leaf powder as low cost biomaterial for Cr(VI) removal. Biosorption experiments were carried out at ambient temperature to study the effects of initial pH (1-7), contact time (15, 30, 45, 60 min), leaf powder's particle size (≤ 150, 151-300, 301-500 μm), and initial Cr(VI) concentrations (10, 20, 40, 60, 80 and 100 mg/L) towards Cr(VI) uptake. Optimum Cr(VI) uptake was observed at pH 1 (2.41 mg/g) and 2 (2.19 mg/g) ($p > 0.05$). The leaf powder in the 151-300 μm size displayed better Cr(VI) uptake up to 2.77 mg/g (77.78%; $C_o$ = 100 mg/L) at pH 2, after 30 min ($p < 0.05$). Equilibrium metal uptake fitted best ($R^2 = 0.927$) to the Freundlich isotherm model ($K_f = 0.21$ mg/g, $n$ = 1.23) indicating biosorption onto a heterogeneous surface of active sites. Morphological and elemental analyses by SEM/EDX and FTIR confirmed Cr retention, and presence of active chemical moieties. Interestingly, XRD revealed that Cr was immobilized within the biomass leaf powder in both forms, i.e. hexavalent, Cr(VI) as chromatite ($CaCrO_4$) and trivalent Cr(III), as chromium(III) oxide ($Cr_2O_3$). Overall findings support the feasibility of the leaf powder as potential biomaterial for water pollution abatement.

## 1 Introduction

The contamination of water bodies from heavy metal ions have become a major concern because of their tendency to persist and accumulate in the ecosystem and exerting toxicity to human and nature. Among the heavy metal ions, hexavalent chromium or Cr(VI), which predominantly occurs as anionic hydrogen chromate ($HCrO_4^-$), chromate ($CrO_4^{2-}$), and dichromate ($Cr_2O_7^{2-}$) in natural waters have been in greater focus as it is classified as one of the top sixteen toxic pollutants [1, 2]. It is an oxidizing agent, highly soluble and known to be carcinogenic, mutagenic and teratogenic to living organisms [3]. By regulation, the concentration of Cr(VI) in industrial effluents should not exceed 0.1 mg/L for releases into inland waters and 0.05 mg/L for potable waters [4]. Discharge of toxic chromate and dichromate ions originates from industries involved with textile manufacturing, chrome electroplating, tanning, petrochemicals manufacturing, wood preservation, and various metal-based activities [5]. The treatment of heavy metal-laden waste can generally be divided into physico-chemical and biological approach. The former includes chemical precipitation, adsorption, ion exchange, electrolysis, reverse osmosis, membrane technology, and such [2, 6]. The physico-chemical approach requires high investment and maintenance with limited operating conditions especially at very low metal concentrations [7] and generates toxic chemical sludge which necessitate further remedial actions. Biosorption is an environmental-friendly, alternate technology for treating heavy metal polluted water and industrial wastewater. Many researchers have described the use of active and inactive biomass of microbial origins i.e. bacteria, fungi, and yeast [8, 9]. The use of active microbial cells as was the norm before have proven to be more tedious and cost demanding in the long run. Living cells need to be supplied with nutrient to ensure sustained growth. Even the use of non-living microbial cells necessitates the cultivation of the cells prior to their use. Nowadays, the trend has shifted to use low cost biowaste or biomass materials as biosorbents for heavy metal biosorption and sequestration. Some of the work involving agricultural or plant waste include papaya wood, maize, peanut hull, grape stalk, sago, neem, rice husks, rice bran, wheat bran, paddy straw, palm oil fibre, sawdust, alfa grass, and moss peat [1, 2, 7, 10, 11]. However, not many studies have explored using garden refuse as potential biosorbents material. This is encouraging because naturally occurring garden refuse can be utilized effectively and economically. Low cost treatment options will be better received for implementation by the industry operators. Finding a useful purpose for the garden refuse can eliminate the burning of leaves and other yard refuse which contributes to environmental pollution and deteriorating aesthetic value of the location.

Garden Croton (*Codiaeum variegatum;* Croton, CV) or locally known as 'pokok puding' belongs to the

[a] Corresponding author: palsan.abdullah@umk.edu.my

DOI: 10.1051/matecconf/20178703003

Euphorbiaceae family. It is characterised by the 'batik' patterned mixture of red, yellow, and green leaf colour. In this study, unmodified withered fallen dead leaves were used. So far, literature has yet to capture the use of 'puding' or Garden Croton leaves as potential biosorbent material highlighting the study novelty. Leafs of different plants are highly versatile in terms of their natural organic and inorganic compounds. Several inter-dependent parameters governing Cr(VI) uptake and removal, i.e. pH, particle size of biosorbent, and contact time for equilibrium were studied. This work is aimed to explore the potential of the local variety of Garden Croton ('pokok puding') plant's leaf powder to uptake Cr(VI) from aqueous solutions.

## 2 Methods

### 2.1 Chemicals and reagents

Reagents were of analytical grade and prepared using distilled water. The standard, working, and test solutions of varying Cr(VI) concentrations were made by diluting potassium dichromate ($K_2Cr_2O_7$) stock solution (1000 mg/L). The initial solution pH was adjusted by addition of either 1.0 M sulfuric acid ($H_2SO_4$) or 1.0 M sodium hydroxide (NaOH) as required.

### 2.2 Leaf sourcing and biomaterial preparation

The withered fallen dead leafs of Garden Croton ('pokok puding') plant were sourced from the shrubs growing at the university's garden, backyard, and from the local kindergarten compound. Leafs were torn into smaller pieces (the petiole and stalks were detached), washed repetitively with distilled water to remove dirt, and other adherents. Then left to air-dry overnight before placed in the convection oven for further drying for two days at 60° C. The dried leaf pieces were ground with a high speed grinder (26 000 rpm, 2-3 min) (Shanghai Jingke Scientific Instrument Co., Ltd., China) to powder form. The leaf powder was sieved through a series of sieves with mesh numbers of 35, 48, and 100, corresponding to particle size ranges of 301-500, 151-300, and ≤ 150 µm, respectively. The sieved leaf powder was kept in air-tight containers.

### 2.3 pH, $pH_{pzc}$, and moisture content determination

About 0.5 g of the Croton leaf powder was mixed with 100 mL distilled water in a covered beaker, agitated for 10 minutes and left to stand for 1 h. After the hour, the clear aqueous solution was separated and the pH measured.

A simplified mass potentiometric titration method was used to determine the point of zero charge pH ($pH_{pzc}$), the pH where the net total charge will be zero. Two identical 0.1 M NaOH (100 mL) solutions were first prepared. One will be the blank solution, and the other to contain the leaf powder (test solution). For the blank solution, 0.5 mL of 1 M NaOH was added and pH was recorded. For the test solution, 0.05 g of leaf powder was added, followed by 0.5 mL of 1 M NaOH, and the initial pH was noted. Both solutions were then titrated with 1 M HCl (0.2 mL additions each time) and the pH were recorded. A graph of pH against volume of 1 M HCl was plotted. The $pH_{pzc}$ of the Croton leaf powder was estimated to be at the point where both titration curves crossed, modified from [12].

For moisture content, the initial weight of 0.5 g of Croton leaf powder was recorded as wet weight (A, g) of sample. The leaf powder was dried in a micro oven (at 200 watts) for two minutes until the sample attained constant weight and left to cool. Then it was re-weighed and the weight recorded as dry weight (B, g) of sample. The moisture content (% M) of the leaf powder was calculated as: (A-B)/B × 100%.

### 2.4 Chromium(VI) assay

The standard assay using the Cr(VI)-specific reagent 1,5-diphenylcarbazide (DPC) was followed, as modified from [6]. At equilibrium time, the content of each flask was filtered through Whatman No.1 (125 mm) filter paper. The residual Cr(VI) in the filtrate was measured spectrophotometrically at 540 nm (Biochrom Libra S12, USA) and compared to a standard curve ($R^2$ = 0.999). The % Cr(VI) removal is given as = $C_o - C_e/C_o \times 100$, and Cr(VI) uptake ($q_e$, mg/g) = $C_o - C_e/w \times$ volume of solution (L); where $C_o$ (mg/L) is the initial concentration of Cr(VI), $C_e$ (mg/L) is the residual concentration of Cr(VI) at equilibrium, and $w$ is the dry weight of the leaf powder (g) [6, 13].

### 2.5 Experimental design

The experimental design is presented in Table 1.

**Table 1.** Experimental design and test parameters.

| Test parameters | Initial pH | Particle size range (µm) | Contact time (min) | Initial [Cr(VI)] (mg/L) |
|---|---|---|---|---|
| Effect of pH and contact time | 1-7 | 151-300 | 15-60 | 10 |
| Effect of particle size | 2 | ≤150-500 | 30 | 10 |
| Effect of initial [Cr(VI)] | 2 | 151-300 | 30 | 10-100 |

The test parameters were evaluated in batch mode. Biosorbent amount was fixed at 0.1 g and final volume of flask was 25 mL. The flask content was agitated at 200 rpm (orbital shaker; Smith, China) in ambient temperature (26 ± 2°C). Appropriate controls (without the leaf powder) were assigned to each run.

### 2.6 Batch biosorption isotherm study

The effect of initial concentrations of Cr(VI) and isotherm studies were conducted at $C_o$ = 10, 20, 40, 60, 80 and 100 mg/L; pH 2, with 0.1 g leaf powder (151-300

DOI: 10.1051/matecconf/20178703003

μm) for 30 min and agitated at 200 rpm in ambient temperature. Experimental data were fitted to the Langmuir and Freundlich isotherm models. The linearized Langmuir model, Eq. (1) assumes a monolayer adsorption onto a surface with a fixed number of identical sites which are equivalently distributed over the biosorbent surface without interaction between the adsorbed ions [5]. The isotherm parameters ($b$ and $Q_{max}$) can be determined from the intercept and slope of the linearized Langmuir plot.

Linearized form of Langmuir equation:

$$\frac{C_e}{q_e} = \frac{1}{Q_{max}} C_e + \frac{1}{Q_{max}\, b} \quad (1)$$

Where $q_e$ (mg/g) is the amount of Cr(VI) uptake, $Q_{max}$ (mg/g) is Langmuir monolayer adsorption capacity, $b$ (L/mg) is Langmuir constant-affinity of binding sites for metal ions and $C_e$ (mg/L) is the residual concentration of Cr(VI) at equilibrium.

Freundlich isotherm assumes that the uptake of metal ions occurs on a heterogeneous biosorbent surface with interactions between the adsorbed ions [13]. The experimental data were tested with the linearized form of Freundlich isotherm, Eq. (2). The isotherm parameters ($K_f$ and $n$) can be ascertained from the intercept and slope of the linearized Freundlich plot.

Linearized form of Freundlich equation:

$$\ln q_e = \ln K_f + 1/n \ \ln C_e \quad (2)$$

where $q_e$ (mg/g) is the amount of Cr(VI) uptake, $K_f$ (mg/g) is a constant related to Cr(VI) uptake at lower equilibrium concentration of $C_e$ = 1 mg/L (log $C_e$ = 0), and $n$ is a constant for adsorption intensity [8].

### 2.7 Characterization of leaf powder

The morphological features of the raw leaf powder (151-300 μm) and the leaf powder after 30 min exposure to 10 mg/L Cr(VI) were examined with a high resolution field emission scanning electron microscope (SEM; Hitachi S-3400N, Japan), and the Cr(VI)-exposed leaf powder was subjected to energy-dispersive X-Ray (EDX; Thermo Scientific, USA). Functional group scan was performed with attenuated total reflectance-Fourier transform infrared (ATR-FTIR; Spectrum 100, Perkin Elmer, USA) in the range of 650 – 4000 $cm^{-1}$.

### 2.8 Cr(VI) retention study

This study was to test Cr(VI) retention capability of the leaf powder. This capability is important when dealing with biosorbent systems in aqueous solutions. The test sample, i.e. exposed to Cr(VI) was stored in a plastic specimen bottle for more than a year at room temperature (uncontrolled). The sample was then subjected to SEM/EDX (Hitachi TM3030Plus, Japan), X-ray diffraction (XRD; Bruker D2 Phaser, Germany with DIFFRAC.SUITE software) and x-ray fluorescence (XRF; Bruker S2 Ranger, Germany) examinations to verify and to determine the form in which Cr was immobilized by the leaf powder. Samples were prepared in pellet form as described in the instrument's operations manual.

### 2.9 Statistical analysis

All tests described above were carried out in triplicates and the mean is reported. The experimental data were subjected to t-tests, one-way ANOVA, and post hoc tests (IBM SPSS version 21). Significance level was set at 95% (α = 0.05).

## 3 Results and discussion

### 3.1 pH, moisture content and $pH_{pzc}$

The pH of the leaf powder was 6.93 ± 0.03. The moisture content was 9.14 ± 0.49 %. According to Pradhan et al., [14], the leaf of a plant from the genus *Saraca asoca* has a pH of 5 to 7 and a moisture content of around 5%. The higher moisture content due to the presence of water can affect adsorption process and decrease the adsorption capacity. Figure 1 indicates $pH_{pzc}$ as 5.11 ± 0.17. This notation can be useful in explaining the mechanism of biosorption in aqueous solutions. Biosorption of cations is more preferred when the solution pH > $pH_{pzc}$ while biosorption of anions are better when solution pH < $pH_{pzc}$. Chromate is an oxyanion which means it is negatively charged, hence for the adsorption of this oxyanion, the pH of the solution should be lower than 5.11. When pH < $pH_{pzc}$ protonation of specific functional groups will occur and the biosorbent will resemble a positively charged affinity resin which attracts negatively charged ions in the solution [12].

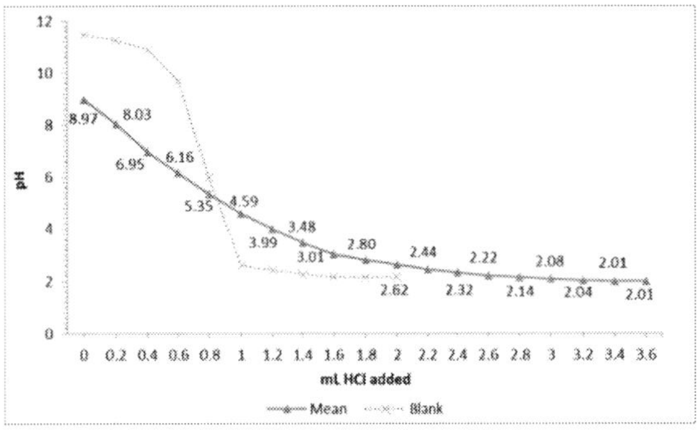


**Figure 1.** Titration result of *Croton* leaf powder for the determination of $pH_{pzc}$

### 3.2 Effect of pH, contact time and particle size

The effect of initial pH and time towards Cr(VI) uptake by the leaf powder can be seen in Figure 2. Three distinct groups were noticeable based on Cr(VI) uptake ability ($p$ < 0.05), i.e. pH 1 and 2; pH 3; and pH 4-7.

DOI: 10.1051/matecconf/20178703003

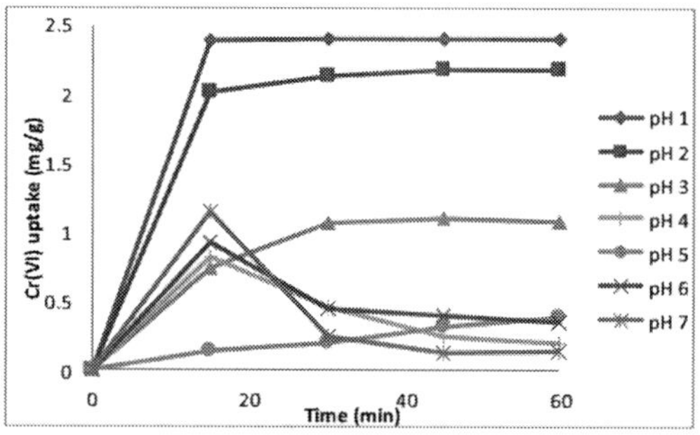


**Figure 2.** The effect of initial pH and time towards Cr(VI) uptake (mg/g) at $C_o$ = 10 mg/L.

The highest Cr(VI) uptake of 2.41 mg/g was read at pH 1 after 30 min, then at pH 2 (2.19 mg/g) after 45 min. However the difference in Cr(VI) uptake between pH 1 and pH 2 was insignificant ($p > 0.05$). In very acidic solutions, the $HCrO_4^{2-}$ ions with a smaller ionic size will be higher in concentration thus can easily diffuse and adsorb onto the biosorbent surface [11]. The excess $H^+$ ions cause the protonation of the chemical groups on the sorbent surface. At pH 2, Cr(VI) exist oxyanions; predominated by $HCrO_4^-$ , and followed by $CrO_4^{2-}$, and $Cr_2O_7^{2-}$. The preference for lower pH for Cr(VI) uptake is advantageous for application as electroplating baths, Cr containing solutions, and waste streams are generally acidic (pH < 3) in nature [13]. The solution's pH is one of the most significant aspect that affects the capacity of biosorbents in aqueous solution. The weakening electrostatic force between the metal ions and the surface of the leaf powder as pH increases results in decrement in metal uptake. In terms of reaction time, saturation was rapid within 15 min for all tested pHs; but slightly higher uptake was apparent after 30 min but there was no significant Cr(VI) uptake difference between 30 and 45 min at pH 1-3 ($p > 0.05$). The extent of adsorption was rapid earlier on as all the active sites are vacant with high amount of available adsorbates. As time passes, the sites become fully occupied and as the biosorbent quantity is constant, no further adsorption can occur and equilibrium is achieved. On the other hand, Cr(VI) uptake decreased drastically at higher pHs with time. At pH above 3, the uptake is higher in the first 15 minutes before dropping and leveling off thereafter. This could be due to the instability and dissociation of the Cr complexes due to chemical speciation at higher pH as well as changes with the functional groups (i.e. carboxylate, phosphate and amine) present on the biomass. This may be explained due to the higher deprotonating effect of the leaf powder surface over time hence repelling the negatively charged chromates. Initial pH of solution is anticipated as an important factor governing the removal of Cr(VI). It influences Cr(VI) and heavy metals ionic speciation, surface charge of the leaf powder due to dissociation of active functional sites, and possible reduction of Cr(VI) to Cr(III) [2]. As pH increase, the competitive environment created by the presence of $OH^-$ and $CrO_4^{2-}$ leads to reduction in overall biosorption capacity [1]. There have been reports of simultaneous occurrence of adsorption and reduction [2]. The Cr chemistry may have also played a role, as the initially adsorbed Cr(VI) can be reduced to Cr(III) by interaction with the leaf powder surface. The predominant anionic $Cr_2O_7^{2-}$ and $HCrO_4^-$ have the tendency to oxidize the leaf powder surface upon contact and hence become reduced [7]. The reduced Cr(III) will then be released and precipitated into the solution at higher pHs. X-ray analyses by XRD and XRF gave a better insight on this aspect (section 3.4). Based on this outcome, subsequent studies were carried out at pH 2 for 30 min. Similarly, Vinodhini and Das [13] have observed optimal removal of Cr(VI) at pH 2 with neem sawdust, mango sawdust, orange peel, wheat husk, and sugarcane bagasse. Recently, Gupta and Mote [7] had reported that the optimal pH for Cr(VI) uptake by sugarcane bagasse was pH 1, saw dust at pH 1-2, orange peels at pH 1, and rice husk ash at pH 1-2. However, Rezaei [3] when using *Spirulina* sp. observed optimal Cr(VI) uptake at pH 5. As mentioned above, the $pH_{pzc}$ of Croton leaf powder is 5.11 and this explains the better Cr(VI) uptake capacity at below this pH. In conditions when pH > $pH_{pzc,}$ deprotonation of the functional groups on the biosorbent will turn the biosorbent to become more negatively charged hence more preferable to bind cations. The pH adjustment will be useful for desorption of the anions for metal ion recovery and biomass regeneration process.

As for particle size, as seen in Table 2, the highest Cr(VI) removal and uptake was 97.34% and 1.62 mg/g, respectively. This was observed with the mixture containing the leaf powder in the 151-300 µm size. There was no significant difference between the 301-500 µm and 151-300 µm particle size ranges. The lowest removal was seen with the smallest powder particles, ≤ 150 µm ($p < 0.05$). Although, the leaf powder with the smallest particle size is expected to have the largest surface area, it did not favour more Cr(VI) removal or uptake. The leaf powder in the particle size range of 151-300 µm was chosen for the following adsorption isotherm modelling study.

**Table 2.** Effect of particle size towards Cr(VI) uptake at $C_o$ = 10 mg/L after 30 min.

| Leaf powder particle size (µm) | Cr(VI) uptake (mg/g) |
|---|---|
| 301-500 | 1.61 ± 0.006 |
| 151-300 | 1.62 ± 0.006 |
| ≤ 150 | 1.56 ± 0.027 |

### 3.3 Effect of initial Cr(VI) concentration and adsorption isotherm studies

This investigation was carried out at an initial Cr(VI) concentrations of 10-100 mg/L with the optimal conditions ascertained earlier, i.e. at pH 2, with particle size of 151-300 µm for 30 min. The other conditions were as before. As observed from Figure 3, the % removal of Cr(VI) gradually decreased from 77.78% at 10 mg/L to 62.67% at 100 mg/L as the initial Cr(VI) was increased. This is due to the limited and constant number of available active binding site in a batch system as opposed to the increasing concentration of Cr(VI). At equilibrium, saturation is expected to occur, with the leaf powder maximally filled with Cr ions per unit of surface area [11]. This results in the 10 fold increase of Cr(VI) uptake from 0.27 mg/g at $C_o$ = 10 mg/L to 2.77 mg/g at $C_o$ = 100 mg/L.

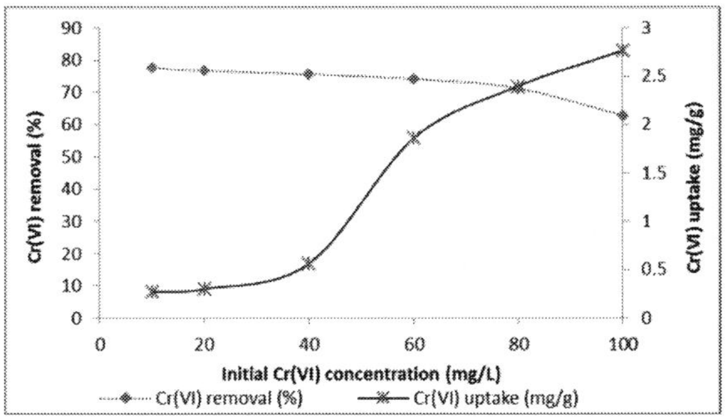


**Figure 3.** Effect of initial concentration of Cr(VI) towards Cr(VI) removal and uptake by 'puding' leaf powder.

The Cr(VI) uptake data fitted best to the Freundlich model ($K_f$ = 0.21 mg/g, $n$ = 1.23; $R^2$ = 0.927) as shown in Figure 4 as opposed to the Langmuir model which resulted in a very poor fit ($R^2$ = -1.198; plot not shown).

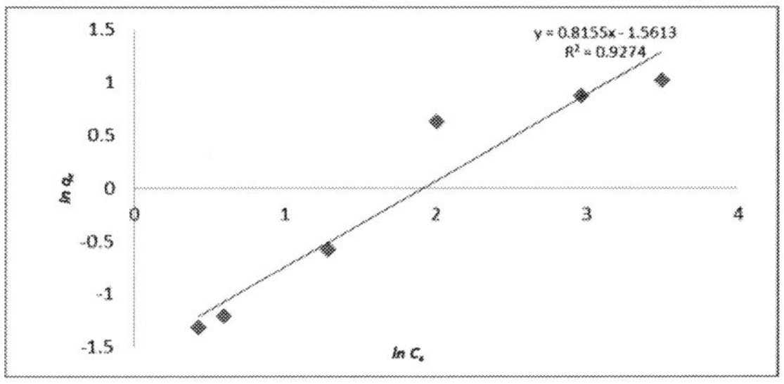


**Figure 4.** The Freundlich isotherm plot. Experimental conditions were as follows: biomass dose: 0.1 g; particle size: 151-300 μm; initial Cr(VI) concentration: 10, 20, 40, 60, 80, 100 mg/L; pH: 2; ambient temperature: 26 ± 2°C; agitation speed: 200 rpm; contact time: 30 min.

This indicated that Cr(VI) was adsorbed onto a heterogeneous surface of the leaf powder and biosorption was favourable as $1 < n < 10$. As the value of $n$ was below unity, chemisorption is implied for the uptake of Cr(VI). The Freundlich maximum uptake capacity, $q_m$ was calculated to be 8.876 mg/g at $C_o$ = 100 mg/L by the following formula; $K_f = q_m/C_o^{1/n}$ [15]. Overall, although sorption was favourable ($1 < n < 10$), but uptake and removal capacity were comparatively lower with other reported biomaterials, i.e. sugarcane bagasse ($K_f$ = 1.38 mg/g), orange peels ($K_f$ = 1.97 mg/g), sawdust ($K_f$ = 2.12 mg/g), dried rice husk ($K_f$ = 3.22 mg/g), and a bacterial consortium culture ($K_f$ = 2.86-3.85 mg/g), and wheat husk ($K_f$ = 4.57 mg/g) [7, 8, 13]. The higher the values of $K_f$ and $n$; the higher is the affinity of the biomass. To address this shortcoming, activation of the biosorbent either by physical means or chemical means or in combination, can be carried out to enhance uniformity of the biosorbent, create micro pores, and increase active surface area that can increase adsorption capacity. Chemical modification has been reported to improve adsorption capacity of adsorbents [10]. On the other hand, the low Cr(VI) uptake may also be due to the possibility that the reduced Cr(III) as suggested earlier is released and remains in the solution. In terms of mechanism governing biosorption, Rezaei [3] have outlined that it may be one or a combination of ion exchange, complexation, coordination, adsorption, electrostatic interaction, chelation and micro precipitation.

### 3.4 Characterization of the Leaf Powder

#### *3.4.1 SEM/EDX analysis*

The SEM micrographs of the Croton leaf powder before Cr(VI) exposure (a) and after Cr(VI) exposure (b) are shown in Figure 5. Generally, the raw leaf powder's surface was shown to be uneven, full of nooks and pockets which can support Cr(VI) adsorption and binding. Full formed pores or micro pores were not clearly visible. Nevertheless the pocket folding observed added with the natural chemical functional groups would provide the necessary binding sites.

**(a)** **(b)**

**Figure 5.** Scanning electron microscopy (SEM) image of the 'puding' leaf powder, (a) before Cr(VI) exposure; (b) after Cr(VI) exposure.

The leaf powder surface imaging after Cr(VI) exposure was shown to be more uneven and irregular. The high contrast area as observed in the image was assumed to be Cr deposits. Energy-dispersive X-ray (EDX) spectra and the quantitative tabulation are presented in Figure 6 and Table 3. The quantitative chemical analysis of the Cr(VI) exposed 'puding' leaf powder by EDX revealed the following: carbon (56.66%), oxygen (37.82% w/w), calcium (4.23 % w/w), chlorine (0.11 % w/w), Zn (1.00 % w/w), and chromium (0.18 % w/w). The abundance of calcium, chlorine, and zinc present on the leaf powder biomass may favour both ionic exchange and covalent bonding of the Cr ions. The detected Ca can be traced to the CaO compound as one of the natural constituents of leaf material. This is an apparent advantage in using natural biomaterial. The loaded Cr detected confirmed the capability of garden croton leaf powder as a potential biosorbent. Further extensive work are in progress in the laboratory to elaborate upon the mechanisms of biosorption within the leaf powder.

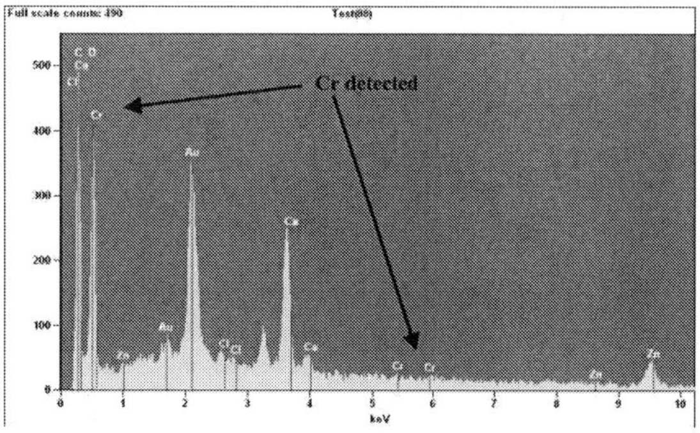


**Figure 6.** The SEM/EDX spectrum of test 'puding' leaf powder [after Cr(VI) exposure]. The elements map spectra was generated from samples shown in Fig. 5(b).

**Table 3.** Quantitative chemical analysis of test 'puding' leaf powder by EDX [after Cr(VI) exposure].

| Element | C | O | Ca | Cl | Zn | Cr | T % |
|---|---|---|---|---|---|---|---|
| Weight % | 56.66 | 37.82 | 4.23 | 0.11 | 1.00 | 0.18 | 100 |
| Atom % | 51.95 | 46.18 | 1.55 | 0.05 | 0.23 | 0.05 | 100 |

### *3.4.2 Cr(VI) retention*

The SEM/EDX imaging of (a) control and (b) test leaf powder samples stored over a year is presented in Figure 7.

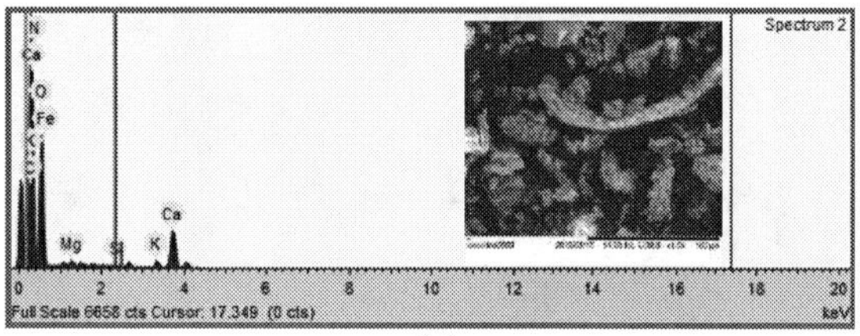


**(a)**

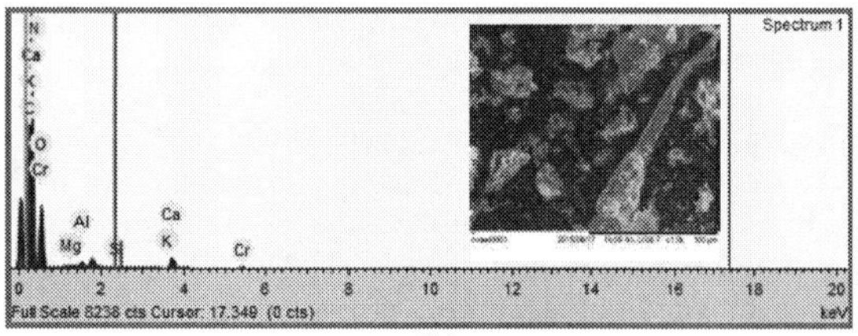


**(b)**

**Figure 7.** The SEM/EDX images and spectrum of the 'puding' leaf powder, (a) control sample (after more than 1 year); (b) test sample (after more than 1 year).

The EDX spectrum of the test sample clearly marked the loaded and retained Cr. Chemical analysis of both the control and test stored samples by EDX provided quantitative indication of element content fluctuations over the year. This also verified the retention of Cr (0.41%) quantitatively in the test sample. The other major elements measured (in comparison with control and test) were as follows: C increased (42% to 56.90%), O decreased (34.14% to 45.87%), N decreased (7.47% to 6.49%), and Ca decreased (3.77% to 1.56%). The ability of a biosorbent system to retain the adsorbate over time is to be given due attention. The question whether a biosorbent especially those intended to work in an aqueous solution might release its load will have to be investigated, and not many studies have actually reported on this matter. The leaching of the adsorbed toxic material into the environment is hazardous to the well-being of all living entity. For environmental remedial application, this is one factor that needs to be considered strongly. These findings will strengthen any suggested future use of the Croton leaf powder as potential biosorbent material. Furthermore, this revealed that over time, the adsorbed hazardous heavy metal, i.e. chromium will not be leached into the environment after the biosorption process. Thus, the safety of humankind and animals are better assured.

Through XRD observation (Figure 8), it was confirmed that Cr was immobilized within the biomass leaf powder. Interestingly, the adsorbed Cr was in both hexavalent, Cr(VI) and trivalent Cr(III) forms.

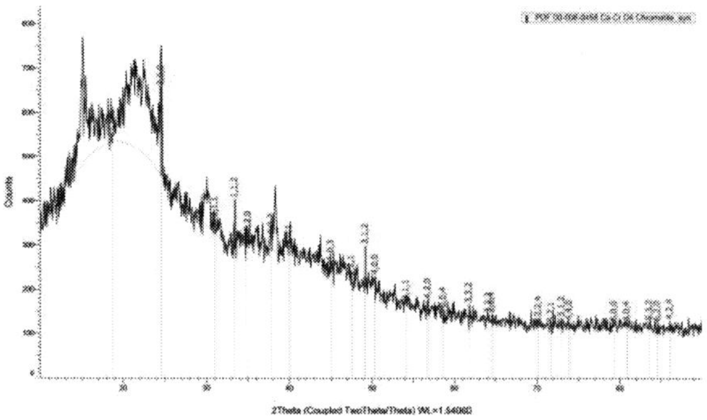

**(a)**

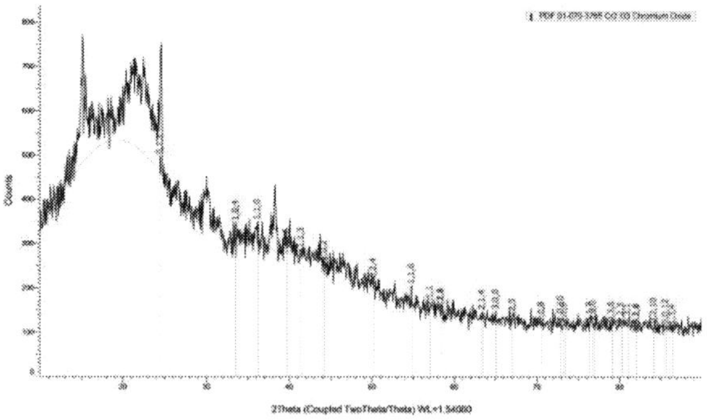

**(b)**

**Figure 8.** XRD spectrum of test 'puding' leaf powder after biosorption indicating (a) the complexation of Cr(VI) to Ca to form chromatite ($CaCrO_4$) and (b) the product of non-enzymatic reduction of Cr(VI) to Cr(III) in the form of chromium oxide ($Cr_2O_3$).

The hexavalent form existed as a complex with Ca, i.e. chromatite ($CaCrO_4$). Whereas the reduced trivalent form was seen as chromium(III) oxide ($Cr_2O_3$). This would be made possible by the presence of oxygen (Figure 7) as determined earlier acting as the electron donor. This suggests the non-enzymatic reduction of toxic hexavalent chromium to the much less toxic, and less soluble Cr(III). As to date, to the best of our knowledge, this is the first ever report and finding of possible non-enzymatic reduction of Cr(VI to Cr(III) with garden croton leaf powder. However, comparison between the amount of Cr adsorbed and the amount of Cr reduced prior to adsorption is not reported in this work. Further to this, in order to quantitatively elucidate the chemical form in which the elements were found, quantitative analysis of the compound types present in the biomass was also performed by XRF test (plot not shown). The data outputs clearly showed that the two highest concentration of oxides in the leaf powder were CaO (57.5%) and $Cr_2O_3$ (10.8%). The other oxide compounds read were: $SO_3$ (9.83%), $SiO_2$ (6.12%), $P_2O_3$ (4.96%), Cl (3.7%), $K_2O$ (2.25%), $Al_2O_3$ (1.67%), $Fe_2O_3$ (1.67%), BaO (0.22%). The abundance of Ca in the leaf powder correlated well with its ability to form complexes with Cr as a way to immobilize the toxic metal ion. This verified Cr(VI) uptake from aqueous solution by the garden croton leaf powder. The cation ($Ca^{2+}$) accepted the electrons from the anionic chromate ($CrO_4^{2-}$) to form the chromatite complex. The complex was proven to be stable even after more than a year. This characteristic is important for toxic ion removal as to restrict any untoward leakage back to the environment, and this will prove to be beneficial as well for time-constrained storage before possible metal ion recovery can be opted by the industry.

### *3.4.3 FTIR analysis*

The FTIR peak attributes are presented in Table 4. The FTIR scan spectra for both (a) control and (b) test leaf powders are depicted in Figure 9. In general, the 'puding' leaf powder displayed many peaks that can be assigned to major –OH stretch of alcohols, phenols and carboxylic acids, C=O stretch of carboxylic acids and aldehydes, and -N-H stretch of both aliphatic and aromatic amines, and the common alkanes (-CH), and acyl (-C-O) moieties, among others, characteristics of natural biological compound origin. Comparisons can be made between the control [not exposed to Cr(VI)] and the test [after Cr(VI) exposure] spectra (Figure 9); the shifts observed in the wavelength of the dominant peaks i.e. peak no. 1, 2, 4, 5, 6, and 7 denoted there was Cr binding process taking place at the surface of the leaf powder. Interestingly, two new peaks (peak no. 3 and 11) were detected after Cr(VI) exposure. The peaks can be assigned to –C-N stretch of aliphatic amines or –C-O stretch; two common groups significant for the binding of heavy metal ions and complexation. This and the shifts in the group's chemical group stretch indicate formation of new chemical bonding or association with the Cr ion. The vibration stretch observed with OH, NH and C, O groups reflect their involvement in aiding to Cr biosorption [3]. As the solution is acidic, the excess $H^+$ ions cause the protonation of the chemical groups on the biomass surface. At pH 2, Cr(VI) exist in anionic form; predominated by $HCrO_4^-$ , and followed by $CrO_4^{2-}$, and $Cr_2O_7^{2-}$. The more positively charged surface and the protonated carboxyls, carbonyls, amines, and amides promote electrostatic attractions with the anionic Cr(VI) ions in the solution [5]. It has been postulated that Cr(VI) reduction to Cr(III) is possible with non-living biomass as well. Chromium(VI) carry a very high positive redox potential (+1.33 V) in acidic conditions thus behaves as strong oxidizing agents, and unstable in the presence of electron donors [2] which are the reactive –OH, –NH, and –C-O or –C=O groups found on the leaf powder. These reactive groups will then reduce Cr(VI) to Cr(III) complex for a possible electrostatic adsorption coupled reduction complexation mechanism. The occurrence of non-enzymatic reduction of Cr(VI) to Cr(III) was established by the observation of reduced Cr(III) in the form of $Cr_2O_3$ within the leaf powder as reflected in the XRD and XRF outputs. However, the final pH of the test solution after 30 min increased up to pH 4-4.49 (data not shown) at 1-100 mg/L of Cr(VI). This points to probable precipitation of $Cr(OH)_3$ and/or exchange of $OH^-$ ions with $HCrO_4^-$ apart from or in addition to the reduction of Cr(VI) to the oxide form. This might explain the lower Cr(VI) uptake and loading ($K_f$ = 0.21 mg/g) with this leaf powder as the adsorbed Cr(VI) ions are precipitated back into the solution after undergoing reduction. In addition, anionic exchange can also be anticipated with the presence of Cl on the leaf powder. The C-Cl stretch (peak no. 13) matched the Cl detected in the EDX and XRD analyses (Figure 6; Table 3). A biosorbent prepared from biowaste materials can sequester metal ions through a variety of mechanisms mainly adsorption, complexation, and ion exchange, and can actually be dubbed as a biophysico-chemical technique. The findings strongly supports the suitability of the leaf powder as a low cost biosorbent material due to the abundance in the chemical groups present.

**Table 4.** Classification of functional groups on CV control [leaf powder not exposed to Cr(VI)] and CV test [leaf powder exposed to Cr(VI)] determined by FTIR scanning.

| Peak no. | CV (control) ($cm^{-1}$) | CV (test) ($cm^{-1}$) | Probable functional groups [2] |
|---|---|---|---|
| 1 | 3402 | 3408 | -OH (alcohols, phenols), -NH, $-NH_2$ |
| 2 | 2969, 2904, 2834 | 2962, 2915, 2844 | -CH, -OH (carboxylic acids) |
| 3* | | 2321 | -PH, $R_2HP$=S, -OH, -NH |
| 4 | 2013 | 2007 | R-N=C=S |
| 5 | 1768, 1717 | 1765, 1719 | -C=O (aldehydes, carboxylic acids) |
| 6 | 1620 | 1634 | -NH (1° amines) |
| 7 | 1433 | 1442 | -C-C- (aromatics) |
| 8 | 1376 | 1377 | -CH |
| 9 | 1319 | 1319 | -NO, C=O, C-N (amide III) |
| 10 | 1223 | 1223 | -C-N (aliphatic amines) |
| 11* | | 1164 | -C-N (aliphatic amines), -CH, C-O |
| 12 | 1036 | 1037 | -C-N (aromatic amines), S=O |
| 13 | 780, 682 | 781, 680 | -C-Cl |

* New peaks detected after exposure to Cr(VI)

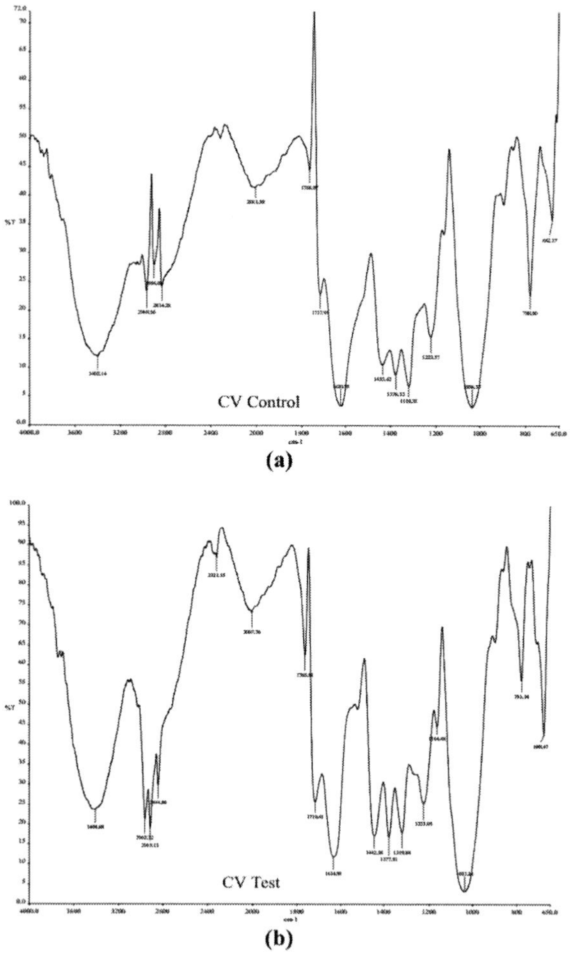


**Figure 9.** FTIR spectra of the 'puding' leaf powder; (a) control [not exposed to Cr(VI)] and (b) test [after exposure to Cr(VI)].

## 4 Conclusion

Although many studies have been carried out to study biosorption following similar schemes, it is evident that different organisms or materials do behave differently. The chemical makeup of each biosorbent material and the interacting immediate environment dictates the outcome of the overall process. Based on the findings of this study, the following can be concluded of the local garden croton leaf powder: (1) Cr(VI) uptake and removal is governed by initial solution pH, with pH 1 and 2 being the better options; (2) The $pH_{pzc}$ is 5.11 thus anionic biosorption is best below this pH, and cationic biosorption best above this pH; (3) The optimal time for the biosorption process is 30 min; (4) The leaf powder biomass in the particle size range of 150-300 µm displayed higher Cr(VI) uptake capacity; (5) Biosorption followed the Freundlich model of heterogeneous surface binding with a $K_f$ of 0.21 mg/g and $n$ = 1.23; (6) The raw leaf powder displayed many nooks and pockets but void of any visible pores; (7) The leaf powder is rich in chemical functional groups that can facilitate Cr(VI) binding and adsorption; (8) The potential of the leaf powder to adsorb both Cr(VI) and Cr(III) and retain the adsorbed Cr had been proven; (9) An electrostatic adsorption (complexation) coupled non-enzymatic reduction of Cr(VI) to Cr(III), with anionic exchange seems to be the apparent governing mechanisms for Cr(VI) removal; and (10) Cr was immobilized within the biomass leaf powder in both forms, i.e. hexavalent, Cr(VI) as chromatite ($CaCrO_4$) and trivalent Cr(III), as chromium(III) oxide ($Cr_2O_3$). Although biosorption was favourable ($1 < n < 10$), but uptake capacity was comparatively lower with other reported biomaterials; future studies will look into activation options to augment the Cr(VI) uptake capability of the leaf powder. This will be seconded with a column study to understand process kinetics and possible upscaling. Nevertheless, the local garden Croton ('puding') leaf litter or garden waste is proven to have bioadsorbent properties that is worthy to be developed further as low cost biomaterial for pollution abatement.

**Acknowledgment**

This work is supported by Universiti Malaysia Kelantan [R/SGPB/A07.00/01397A/001/2015/000305] and Inti.

## References

1. M. Nameni, M. Moghadam, M. Arami, Intl. J. Environ. Sci. Tech. **5**, 161 (2008)
2. B. Catalin, D. Bilba, M. Macoveanu, Cellulose Chem. Technol. **43**, 99 (2009)
3. H. Rezaei, Arab. J. Chem. Retrieved from http://www.sciencedirect.com/science/article/pii/S18 7853521300378X (2013)
4. M. J. Amiri, E. Fadaei, A. Baghvand, Z. Ezadkhasty, Int. J. Environ. Res. **8**, 2 (2014)
5. V. Vinodhini, V. Anabarasu, N. Das, Ind. J. Nat. Prod. Res. **1**, 174 (2010)
6. P. Sannasi, C. S. Fern, S. Bernadine, L. F. Shiang, J. Trop. Res. Sustain. Sci. **3**, 35 (2015)
7. A. Gupta, S. R. Mote, Chem. Process Eng. Res. **19**, 47 (2014)
8. P. Sannasi, J. Kader, B.S. Ismail, S. Salmijah, Bioresour. Technol. **97**, 740 (2006)
9. K. Selvam, K. Arungandhi, B. Vishnupriya, T. Shamuga, M. Yamuna, Biosci. Discov. **72** (2013)
10. W. S. Wan Ngah, M. A. K. M. Hanafiah, Bioresour. Technol. **99**, 3935 (2008)
11. N. Tazrouti, M. Amrani, BioResour. **4**, 740 (2009)
12. A. M. C. Pena, J. G. Ibanez, R.V. Medrano, Retrieved from http://www.electrochemsci.org/papers/vol7/707614 2.pdf (2012)
13. V. Vinodhini, N. Das, ARPN J. Agric. Biol. Sci. **4**, 19 (2009)
14. P. Pradhan, L. Joseph, V. Gupta, R. Chulet, H. Arya, R. Verma, A. Bajpai, J. Chem. Pharma. Res. **1**, 62 (2009)
15. O. Hamdaoui, E. Naffrechoux, J. Hazard Mat. **147**, 381 (2007)

# HC-SCR: $NO_x$ Reduction using Mn and Cu Catalysts Impregnated in Coconut and Palm Kernel Shell Activated Carbon

Sherra Bellina Barrabas[1], Rubiyah Baini[1,a], Norsuzailina Mohamed Sutan[2], Ibrahim Yakub[1], Khairul Anwar[1], Nur Syuhada Ahmad Zauzi[1], Noraziah Abdul Wahab[1], and Hapiszul Sulaiman[1]

[1] *Department of Chemical Engineering and Energy Sustainability, Universiti Malaysia Sarawak, Kuching, Sarawak.*
[2] *Department of Civil Engineering, Universiti Malaysia Sarawak, Kuching, Sarawak*

**Abstract.** The characteristics of catalysts impregnated in coconut shell (CS) and palm kernel shell (PKS) activated carbon were determined as potential precursors of catalysts used in a flue gas denitrification system at low temperature. In this study, Manganese (Mn) and Copper (Cu) with metal loading of 8% were impregnated in the activated carbon (AC) before undergoing low temperature calcination process. The morphological properties of samples was analysed using Scanning Electron Microscopy (SEM) and Brunauer, Emmett and Teller (BET) was used to determine the surface area and pore size of samples. The exhaust gas from a diesel engine at a constant flow rate of 4L/min was passed through in a fixed-bed catalytic reactor containing the catalyst, and the concentration of $NO_x$ was measured for temperatures ranging from 150°C to 250°C. It was found that the CS catalysts (CS-Mn and CS-Cu) and PKS catalysts (PKS-Mn and PKS-Cu) have the potential to reduce $NO_x$ concentration, and results showed that the metal loading of 8% resulted $NO_x$ reduction ranging from ~48% to 64%.

## 1 Introduction

Nitrogen oxides ($NO_x$) are among the severe contaminants released by automobiles and other vehicles, and by manufacturing industries. Diesel-powered vehicles are becoming more sought after due to their excellent fuel efficiency and less emission of $CO_2$ compared to their stoichiometric spark triggered gasoline counterpart [1]. Three ways catalysts (TWCs), one of the strategies for $NO_x$ conversion, is reported to produce excess of oxygen contributing to the incapability of catalyst in reducing $NO_x$ from lean burn engine exhaust [2]. Hence, a major challenge today is to minimize $NO_x$ emissions from lean-burn diesel engine exhausts.

Selective Catalytic Reduction (SCR) has been experimented and applied as one of the efficient ways to reduce $NO_x$ emission into the environment [3]. The most extensively used in the industries are $NH_3$-SCR, which uses ammonia gas as reductant, and HC-SCR that is using hydrocarbon as the alternative for reducing agent. In this case, HC-SCR is the most promising method to control the emission of $NO_x$ from diesel engine vehicles due to the simplicity of system as it does not require an extra gas tank installed because hydrocarbon (HC) itself can act as reducing agent and onboard fuel as reductant [4]. This means, the system, which operated under lean conditions, exploits the unburned hydrocarbon that are already present in the gas stream of the diesel engine exhaust [5].

Catalytic reduction systems usually use precious metal-based catalysts, for instance, Rhodium(Rh), Platinum(Pt), and Palladium(Pd) supported-oxide. Owing to their abilities that is water vapor resistant and $SO_x$ poisoning, these metals are capable of performing efficient $NO_x$ abatement in low operation temperature (lower than 300°C). A study claimed that Pt based catalysts, however, exhibit two downsides namely low $N_2$ selectivity and narrow temperature window system for reduction of $NO_x$ [6]. On the other hand, Rh based catalysts are mostly used in three ways catalysts (TWCs) operation as Rh metal promotes the best $NO_x$ reduction under low activity temperature; however, it is also said that the catalysis activity involved in synthesis gas conversion process [7].

Other metals that have also been studied for SCR catalysis process are such as Manganese (Mn), Titanium (Ti), Nickel (Ni) and Copper (Cu). These metals are believed to have the capability to remove $NO_x$ gas under low operation temperature especially from stationary sources [8]. However, the potential of these metals have yet to be broadly looked into hydrocarbon SCR systems as they are mostly investigated in urea-SCR. In this study, Mn and Cu are selected as metallic catalysts for the removal of $NO_x$ under low catalytic activity temperature owing to their chemical properties such as alkali metal resistance of catalysts for diesel exhaust treatment [9].

[a] Corresponding author: ruby@unimas.my

DOI: 10.1051/matecconf/20178703004

The mechanism of HC-SCR is complex. As concluded by Lamacz et al. (2013) in a proposed model, when hydrocarbons are unable to undergo oxidation process to $CO_2$ and $H_2O$ at the dissociation temperature of NO, which contributing to the formation of $N_2$, it is obligatory to make use of three-function catalysts. Oxidation of NO to $NO_2$ is the first function of this catalyst. The second function denoted by the hydrocarbons undergoing mild oxidation by $NO_2$ to oxygenate species which is presumed to be vital for regeneration process of catalyst. In the third function, $N_2$ is being released in the dissociation of $NO_x$ (via a dinitrosyl-adsorbed intermediate) prior to the $N_2$ formation and generation of adsorbed oxygen species. The overall activated reductant ($C_xH_yO_z$) oxidation process resulting in the removal of adsorbed oxygen species. As stated by this research group, function 2 and 3 are said to happen simultaneously, however, both of these functions are not in the same cycle of catalysis at molecular level [10].

A novel establishment of an alternatively sourced carbon-supported catalysts, for instance, activated carbon, bio-char and virgin biomass waste residue, has been developed which is believed to contribute to a more environmental friendly and cost effectiveness [11] for managing the emission of NO in SCR de-$NO_x$ process under low operation temperature [12]. In this study, a laboratory scale reactor was carried out for the purpose of determining the efficiency of $NO_x$ reduction under HC-SCR low temperature using activated carbon derived from coconut and palm kernel shell as catalysts support for Mn and Cu metals. The relation of the physical properties of the catalysts support with the reduction of $NO_x$ was also analysed.

## 2 Materials and Method

### 2.1 Preparation of catalyst

Prior to the impregnation process the coconut shell activated carbon (CSAC) and the palm kernel shell activated carbon (PKSAC) was sieved to give uniform particle size of 1 mm. Both CSAC and PKSAC were impregnated with metal nitrate of an appropriate concentration to obtain 8% of metal content per gram of activated carbon. In this study, manganese (II) acetate tetrahydrate ($Mn(CH_3COO)_2{\cdot}4H_2O$) from ACROS ORGANIC, and copper (II) nitrate-trihydrate ($Cu(NO_3)_2.3H_2O$) from SYSTERM ChemAR were used as the metal precursors.

During the impregnation, the solution of metal nitrate was continuously mixed with activated carbon for 5 hours. Then, the samples were heated to 70 °C while being constantly stirred until all the liquid evaporated. After that, the samples were dried in an oven at 110 °C for a period of 12 hours. Finally, the prepared samples were heat-treated at 300 °C for 4 hours. The catalysts that impregnated with 8% (w/v) of metallic catalysts denoted as CS-Mn8, CS-Cu8 and PKS-Mn8, PKS-Cu8.

### 2.2 Catalyst performance test

Experiments were carried out in a fixed-bed reactor that was filled with the catalyst. The schematic diagram of the experimental setup is shown in Figure 1. The prepared catalysts of CS and PKS were each weighed 10 g and placed on borosilicate glass wool (0.5 g) in the center of the sample holder. The sample holder containing the catalyst was placed inside the fixed-bed reactor after the set temperature was reached. Temperature used here was between 150 – 200 °C.

A stream gas from diesel engine was passed through the prepared catalyst, which act as sorbent and also as SCR catalyst . The feed flow through the sample holder was controlled at 4L/min. The concentration of NOx in the inlet and outlet of the reactor was measured using Flue Gas Analyzer for a duration of 5 minutes. The reduction efficiency was calculated using the relationship denoted by the following equation:

$$\%\ \text{of}\ NO_x\ \text{reduction} = \frac{NO_{x,in} NO_{x,out}}{NO_{x,in}} x100\% \quad (1)$$

$NO_{x,in}$ denotes the inlet $NO_x$ concentration (vol %) in the stream gas introduced into the fixed-bed reactor and $NO_{x,out}$ denotes the outlet $NO_x$ concentration (vol %) in the stream gas out of the reactor.

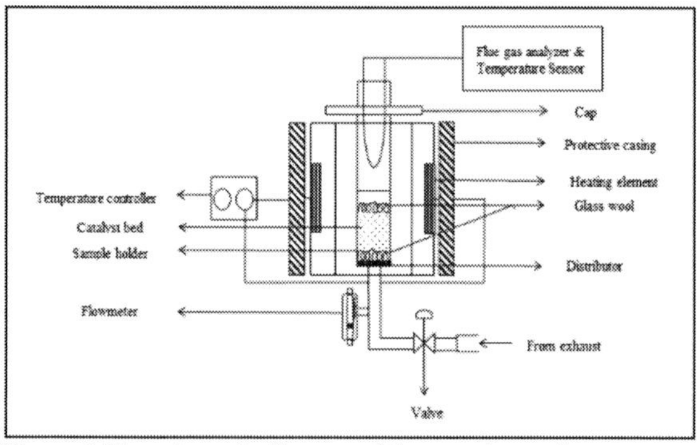


**Figure 1** The schematic diagram of the experimental apparatus.

### 2.3 Catalysts characterizations

The properties of the activated carbon that was impregnated with 8% of Mn and 8% of Cu were analysed based on their surface area, pore size and the images showing the morphological properties that obtained from the experimental work done using Brunauer, Emmett and Teller (BET) and Scanning Electron Microscopy (SEM) apparatus. The metallic catalysts derived from palm kernel shells and coconut shells activated carbon denoted as PKS-Mn8, PKS-Cu8, CS-Mn8 and CS-Cu8.

## 3 Results and Discussions

### 3.1 BET analysis

DOI: 10.1051/matecconf/20178703004

**Table 1** Physical properties of raw CSAC, PKSAC and CS- and PKS- based catalysts.

| Sample | SA BET ($m^2/g$) | Pore volume ($cm^3/g$) |
|---|---|---|
| Raw CSAC | 1244 | 0.416 |
| CS-Mn8 | 997 | 0.365 |
| CS-Cu8 | 1003 | 0.373 |
| Raw PKSAC | 1231 | 0.399 |
| PKS-Mn8 | 976 | 0.338 |
| PKS-Cu8 | 992 | 0.350 |

Physical properties of CS and PKS-based catalysts were investigated in BET analysis as shown in **Table 1**. From the analysis, it can be seen that the specific surface areas and the volume of micropore of both types of activated carbon samples decreased after the process of impregnation of metal. It is presumed that the partial pores were blocked by the particles of metals. This has been proven by several studies saying that metal embedded onto the support filled the micropores but will not affect the performance of the reduction activities [13 - 15]. Impregnation method has proved that the metal particles are successfully located in the pores especially in the internal part, hence, plugging the tapered microporosity. As a result, this fact contributes to the first proof of dispersion of metal on the surface of both activated carbons in high degree.

The difference in the surface area and pores in BET analysis was supremely low among the impregnated of both CSAC and PKSAC stipulating the percentage of metal loadings impregnated onto the surface of activated carbons were almost equal.

### 3.2 SEM analysis

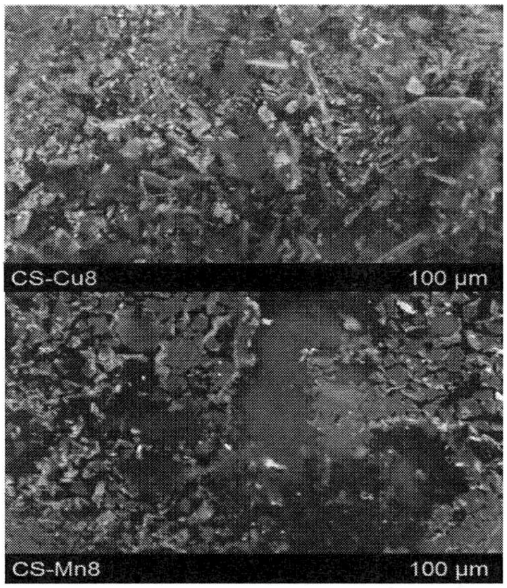


**Figure 2** SEM result for PKS-Cu8, PKS-Mn8, CS-Cu8 and CS-Mn8.

The catalysts' surfaces and morphologies which undergone the process of impregnation of metals are shown in **Figure 2**. Comparing between the two types of activated carbon support, smoother and better consistency dispersion of metal was observed for the coconut shells activated carbon support than those obtained for palm kernel shells activated carbon. This can be deducted that the nature of the CSAC reacted well with both Mn and Cu metals. Consequently, better dispersion of metals can be observed over the surface found for CSAC, which is also predicted to promote better catalytic efficiency. Tang et al. (2015) had proved that the impregnation method to prepare catalyst provided a better dispersion and formation of metal solution which will encourage the catalytic activity [13].

### 3.3 $NO_x$ reduction

In **Figure 3**, comparing between the performance of raw CSAC and PKSAC, it is shown that CSAC adsorbed NOx better than PKSAC with percentage adsorption of NOx estimated to be 28% and 22%, respectively. However, in this case, only adsorption process can be observed because there is no reaction between metal catalysts with hydrocarbon (HC). This is because the obvious difference in the surface area and size of pores of the AC to facilitate the adsorption of $NO_x$ molecules in **Table 1**. In addition to that, owing to favorable surface area and pores size, the performance of metal catalysts impregnated on CSAC shows better efficiency if compared to that in PKSAC. This trend is aligned with findings by Guera et al. (2016) where limited surface area and low porosity of activated carbon small amount of adsorbed $N_2$ [14].

According to SCR mechanisms by M. Adamowska-Tessier (2015), there are 3 functions of SCR mechanisms, which is respectively the oxidation of NO to $NO_2$ in the first function, followed by mild oxidation of hydrocarbon (HC) in the second and, finally, the conversion of $NO_2$ to $N_2$ happens in the third function. **Figures 4 (a)** and **(b)**, show the percentage of $NO_x$ reduction versus time. The reading recorded at zero minute is the reading taken for the $NO_x$ concentration before the exhaust gas entered the reactor that was filled with catalyst, which gives zero $NO_x$ reduction. The concentration of $NO_x$ in the exhaust gas that came out from the reactor was measured for 5 minutes, and the % of $NO_x$ reduction was found to be higher for the activated carbon impregnated with Cu compared to that impregnated with Mn. The addition of metal onto the activated carbon resulted in the increase in the % of $NO_x$ reduction as illustrated by results shown in Figures 3 and 4; The % of NOx reduction for raw CS was found to be about 28%, as shown in Figure 3 while that for CS with Cu and Mn was increased to above than 50%. Here, the increase in the % of $NO_x$ reduction indicates the effect of SCR reaction, which can be observed in the first 30 seconds as the actual conversion of $NO_x$ into $N_2$, as in the third function of the mechanisms, happened during that time frame. The % of $NO_x$ reduction was found to decrease gradually, which may be due to the trapped $NO_x$ that filled the pores of the activated carbon.

Based on studies conducted by Surjit Singh et al. (2013), he found that the optimum reduction of $NO_x$ was only 25% under temperature of 200 °C. The flow rate of the flue gas used in his study was at 1L/min, which may contribute to the low $NO_x$ reduction. However, the percentage of $NO_x$ reduction in the study increased to 67% as the temperature was elevated to 280 °C, which may indicate higher $NO_x$ conversion [12]. The % of $NO_x$ reduction for different samples of catalyst analysed in this work was compared as shown in **Figure 5**.

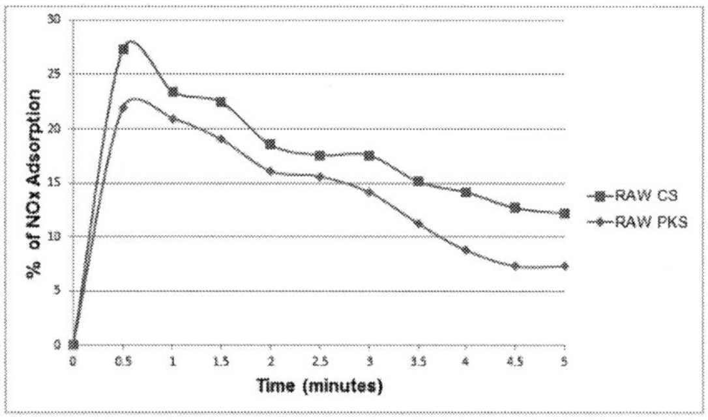


**Figure 3** $NO_x$ adsorption percentage as the function of time for raw CSAC and PKSAC at temperature of 200 °C and flow rate of 4L/min.

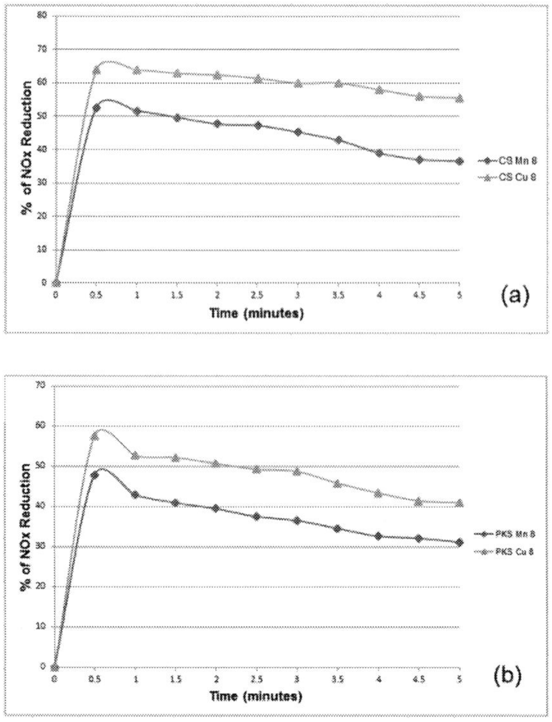


**Figure 4** The % of $NO_x$ reduction as a function of time for (a) CS-Mn8 and CS-Cu8, (b) PKS-Mn8 and PKS-Cu8 under temperature of 200 °C, flow rate of 4L/min.

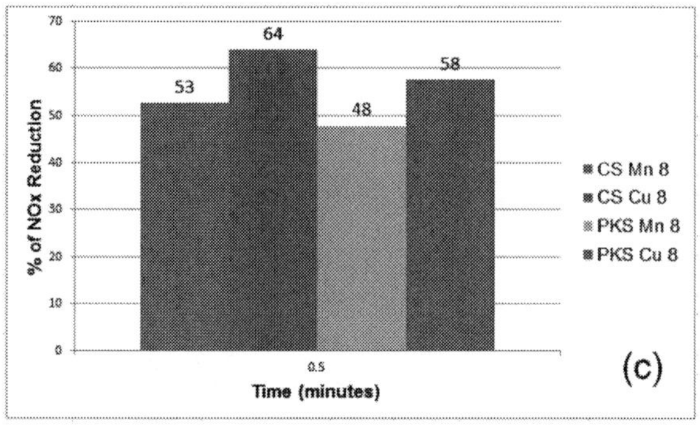


**Figure 5** The overall result of $NO_x$ reduction as a function of time for CS-Mn8 (53%), CS-Cu8 (64%), PKS-Mn8 (48%) and PKS-Cu8 (58%) under temperature of 200 °C, flow rate of 4L/min.

### 3.4 Effect of metal type

**Figure 4 (a)** and **(b)** showed the results of $NO_x$ reduction using the catalyst impregnated onto the activated carbon derived from coconut shells, as shown in Figure 4 (a) and from the palm kernel shells, as shown in Figure 4 (b) with 8% of Mn and Cu. Comparing between the potential of metal Cu and Mn in catalytic activities, it was found that both curves in (a) and (b) showed Cu gives better reduction of $NO_x$; The % of $NO_x$ reduction was estimated to be 64% and 58% for CS-Cu8 and PKS-Cu8, respectively.

Results obtained here may suggest that Cu that has higher electronegativity has a greater tendency to attract electron toward its atom resulting in higher $NO_x$ reduction. Consequently, owing to this nature, Cu makes a better catalytic metal than Mn.

## 4 Conclusions

This study emphasized on the effect of manganese and copper impregnated onto biomass waste derived activated carbon combined as Selective Catalytic Reduction (SCR) by using hydrocarbon (HC) as the reduce agent. Based from the observations and results obtained, it can be concluded that coconut shells activated carbon based metal catalysts was found to give better reduction of $NO_x$ compared to palm kernel shells activated carbon based catalysts.

CSAC based catalysts shows a catalytic effect towards the removal of $NO_x$ as well as active, selective towards $NO_x$ and was observed stable in laboratory testing condition. From the results, Cu is a better catalyst for the SCR reaction than Mn as it has been proven that the combination of Cu and coconut shell activated carbon showed better $NO_x$ removal.

## Acknowledgements

The authors would like to acknowledge Ministry of Higher Education for the fund aid FRGS/TK04(02)/1138/2014(05) as well as Universiti Malaysia Sarawak for supporting this project through research facilities.

## References

[1] C. Zhu, S. Liu, H. Liu, J. Yang, X. Liu, and G. Xu, "NOx emission characteristics of fluidized bed combustion in atmospheres rich in oxygen and water vapor for high-nitrogen fuel," *Fuel*, vol. 139, pp. 346–355, 2015.

[2] P. M. More, N. Jagtap, A. B. Kulal, M. K. Dongare, and S. B. Umbarkar, "Applied Catalysis B : Environmental Magnesia doped Ag / Al 2 O 3 – Sulfur tolerant catalyst for low temperature HC-SCR of NO x," *"Applied Catal. B, Environ.*, vol. 144, pp. 408–415, 2014.

[3] B. Sawatmongkhon, a. Tsolakis, S. Sitshebo, J. Rodríguez-Fernández, M. Ahmadinejad, J. Collier, and R. R. Rajaram, "Understanding the Ag/Al2O3 hydrocarbon-SCR catalyst deactivation through TG/DT analyses of different configurations," *Appl. Catal. B Environ.*, vol. 97, no. 3–4, pp. 373–380, 2010.

[4] H. Gu, K. M. Chun, and S. Song, "The effects of hydrogen on the efficiency of NOx reduction via hydrocarbon-selective catalytic reduction (HC-SCR) at low temperature using various reductants," *Int. J. Hydrogen Energy*, vol. 40, no. x, pp. 1–9, 2015.

[5] Y. J. Kim, H. J. Kwon, I. Heo, I. S. Nam, B. K. Cho, J. W. Choung, M. S. Cha, and G. K. Yeo, "Mn-Fe/ZSM5 as a low-temperature SCR catalyst to remove NOx from diesel engine exhaust," *Appl. Catal. B Environ.*, vol. 126, pp. 9–21, 2012.

[6] P. M. More, D. L. Nguyen, P. Granger, C. Dujardin, M. K. Dongare, and S. B. Umbarkar, "Activation by pretreatment of Ag–Au/Al2O3 bimetallic catalyst to improve low temperature HC-SCR of NOx for lean burn engine exhaust," *Appl. Catal. B Environ.*, vol. 174–175, no. x, pp. 145–156, 2015.

[7] M. Adamowska-Teyssier, A. Krztoń, P. Da Costa, and G. Djéga-Mariadassou, "SCR NOx mechanistic study with a mixture of hydrocarbons representative of the exhaust gas from coal combustion over Rh/Ce0.62Zr0.38O2 catalyst," *Fuel*, vol. 150, no. x, pp. 21–28, 2015.

[8] M. Kang, E. D. Park, J. M. Kim, and J. E. Yie, "Manganese oxide catalysts for NOx reduction with NH3 at low temperatures," *Appl. Catal. A Gen.*, vol. 327, no. 2, pp. 261–269, 2007.

[9] H. Taghavifar, H. Taghavifar, A. Mardani, and A. Mohebbi, "Modeling the impact of in-cylinder combustion parameters of di engines on soot and NOx emissions at rated EGR levels using ANN approach," *Energy Convers. Manag.*, vol. 87, pp. 1–9, 2014.

[10] A. Łamacz, A. Krztoń, and G. Djéga-Mariadassou, "Study on the selective catalytic reduction of NO with toluene over CuO/CeZrO2. A confirmation for the three-function model of HC-SCR using the temperature programmed methods and in situ DRIFTS," *Appl. Catal. B Environ.*, vol. 142–143, no. 2, pp. 268–277, 2013.

[11] C. K. Seo, H. Kim, B. Choi, M. T. Lim, C. H. Lee, and C. B. Lee, "De-NOx characteristics of a combined system of LNT and SCR catalysts according to hydrothermal aging and sulfur poisoning," *Catal. Today*, vol. 164, no. 1, pp. 507–514, 2011.

[12] S. Singh, M. A. Nahil, X. Sun, C. Wu, J. Chen, B. Shen, and P. T. Williams, "Novel application of cotton stalk as a waste derived catalyst in the low temperature SCR-deNOx process," *Fuel*, vol. 105, pp. 585–594, 2013.

[13] W. Tang, X. Wu, S. Li, X. Shan, G. Liu, and Y. Chen, "Co-nanocasting synthesis of mesoporous Cu-Mn composite oxides and their promoted catalytic activities for gaseous benzene removal," *Appl. Catal. B Environ.*, vol. 162, pp. 110–121,

2015.

[14] N. Gutiérrez-Guerra, L. Moreno-López, J. C. Serrano-Ruiz, J. L. Valverde, and A. de Lucas-Consuegra, "Gas phase electrocatalytic conversion of CO2 to syn-fuels on Cu based catalysts-electrodes," *Appl. Catal. B Environ.*, vol. 188, pp. 272–282, 2016.

[15] S. Sumathi, S. Bhatia, K. T. Lee, and a. R. Mohamed, "Selection of best impregnated palm shell activated carbon (PSAC) for simultaneous removal of SO2 and NOx," *J. Hazard. Mater.*, vol. 176, no. 1–3, pp. 1093–1096, 2010.

# Physical, Mechanical, Thermal and Morphology Properties of Biodegradable Polymer Nanocomposites and Its Comparison

Md. Rezaur Rahman[1,a], Josephine Lai Chang Hui[1] and Sinin Hamdan[2]

[1]*Department of Chemical Engineering and Energy Sustainability, Faculty of Engineering, Universiti Malaysia Sarawak, 94300, Kota Samarahan, Sarawak, Malaysia.*
[2]*Department of Mechanical and Manufacturing Engineering, Faculty of Engineering, Universiti Malaysia Sarawak, 94300, Kota Samarahan, Sarawak, Malaysia.*

**Abstract.** Polyvinyl alcohol (PVA) and Polylactic acid (PLA) were fabricated with the addition of nanofiller such as nanoclay and fumed silica through suitable technique namely solution intercalation film casting. These nanocomposites undergo Fourier transform infrared spectroscopy, scanning electron microscopy, tensile test and thermogravimetric analysis. FT-IR results showed that both nanocomposites were well intercalated with fumed silica and clay with the reduction of hydroxyl groups. From SEM results, it proved that clay 1.28E is more suitable to be intercalated with PLA matrix while clay 1.30E is more suitable to be introduced into PVA matrix. The addition of both nanofillers improved the tensile properties of the nanocomposites. TGA results showed that polyvinyl alcohol/fumed silica/clay (PVA/fsi/clay) had better thermal stability compared to polylactic acid/fumed silica/clay (PLA/fsi/clay) nanocomposites. Both nanocomposites are applicable in the biomedical field.

## 1 Introduction

The great improvement of the thermal and mechanical properties on the polymer nanocomposites leads them to be widely applied in many fields such as paper coating, articular cartilage, thickener materials in paints, detergent and adhesives due to its with excellent biocompatibility and attractive properties [1-2].

Biodegradable polymers such as polyvinyl alcohol (PVA) and polylactic acid (PLA) were greatly applied to form nanocomposites. PVA is prepared using the polymerization of vinyl acetate followed by alcoholysis while PLA is linear aliphatic polyester [3]. However, pure PVA and PLA had poor physical and mechanical properties [4]. Therefore, both polymer matrices should be deeply improved by introducing various types of filler.

Nowadays, nanofiller such as fumed silica and nanoclay were introduced into polymer matrix to enhance the properties of the nanocomposites. Fumed silica particles usually exist in a form of fine powder or colloid suspension. It was embedded in a polymer matrix for its large surface area and smooth nonporous surface [5]. For nanoclay, it is defined as a class of materials made up of layered silicates. Montmorillonite (MMT) was one of the common nanoclay that was introduced into polymer matrix to improve the physical and mechanical properties of nanocomposites [6].

The Na ion-exchanged clay was easily dispersed into PVA matrix which significantly enhanced the thermal and mechanical properties [7]. Besides, the solvent casted PVA/clay nanocomposites showed better surface morphology and stronger polymer-clay bonding that reduced water absorption [8]. Addition of fumed silica into PVA matrix improved the thermal degradation of the PVA/fumed silica nanocomposites [9]. Besides, incorporation of clay into PLA matrix enhanced the thermal stability of the nanocomposites by increasing the glass transition temperature consistently [10]. Sol-gel PLA/fumed silica nanocomposites were fabricated which improved the tensile properties as well as the thermal degradation with the introduction of fumed silica [11]. In addition, the fumed silica was incorporated into PLA matrix to improve the tensile strength and thermal stability of the nanocomposites [12].

In this present work, solution-intercalation and solution-intercalation film-casting technique were used to produce PVA/fsi/clay and PLA/fsi/clay nanocomposites respectively using different two types of clays (1.28E and 1.30E). Physical, mechanical, thermal and morphological properties of nanocomposites were investigated. The compatibility of the clays towards the polymer matrices was also investigated.

## 2 Methodology

### 2.1 Materials

The silica powders used in this study was obtained in a mesh size of 8 microns which is white to off-white in colour. The chemicals used polyvinyl alcohol (PVA), polylactic acid (PLA), Nanoclay, Nanomer 1.28E and

[a] Corresponding author: rmrezaur@feng.unimas.my

DOI: 10.1051/matecconf/20178703005

1.30E were supplied by Sigma-Aldrich. Nanomer 1.28E is montmorillonite clay surface was modified with 25-30 wt.% trimethyl stearyl ammonium. Nanomer 1.30E is montmorillonite clay was surface modified with 25-30 wt.% octadecylamine. The bulk density of both clays was 200 to 500 kg/m3 and average particle size was around 20 microns.

### 2.2 Methods

#### *2.2.1 Preparation of PVA nanocomposites by solution intercalation*

The PVA/fsi/clay nanocomposites were prepared as shown in Table 1. Silica was dried at 200 for 48 hours in a vacuum oven. PVA and clay were dried at 50 and 60 respectively for 24 hours in a vacuum oven. The samples were prepared in a deionized water/methanol (3:1) solvent mixture. The mixtures were first stirred at 70 for 24 hours. The homogeneous solutions were then ultrasonicated in an ultrasonic bath at room temperature for 1 hour. The solid films were produced by casting a sample onto a glass mold by slowly evaporating the solvent in air for 3 days. The films with a thickness of ~150 m were put in a vacuum oven at 50 for 24 hours in order to get rid of the extra solvent. The dried films were stored in desiccators prior to characterization [13].

#### *2.2.2 Preparation of PLA nanocomposites by solution intercalation film-casting*

The PLA/fsi/clay nanocomposites were prepared as shown in Table 1. PLA was dissolved in dichloromethane. Dichloromethane was added into four types of clay in four separated beakers to form clay dispersions. Both the PLA solution and clay suspension were ultrasonicated separately for 30 min with an ultrasonicator at room temperature and subsequently mixed. Fumed silica was added before the final mixture was further sonicated for 30 min. The mixture was then casted on a glass surface and kept in vacuum oven for controlled evaporation of the solvent for 2 days. The clear nanocomposite films with thickness ranging from 500 to 700 m were obtained and subsequently dried at 80 under vacuum for 2 days. The nanocomposites were kept in desiccator for further characterizations and analysis [14].

**Table 1.** Preparation of PVA/fsi/clay and PLA/fsi/clay nanocomposites

| Amount of PVA/PLA (g) | Amount of fumed silica (g) | Amount (g) and type of clay |
|---|---|---|
| 27.00 (PVA) | 2.50 | 0.50 (1.28E) |
| 27.00 (PVA) | 2.50 | 0.50 (1.30E) |
| 24.45 (PLA) | 0.05 | 0.50 (1.28E) |
| 24.45 (PLA) | 0.05 | 0.50 (1.30E) |

### 2.3 Microstructural Analysis

#### 2.3.1 Fourier Transform Infrared Spectroscopy (FT-IR)

The infrared spectra of the monomer systems were recorded on a Shimadzu IRAffinity-1. The transmittance range of the scan was 700 to 4000cm-1 [13-14].

#### *2.3.2 Scanning Electron Microscopy*

The interfacial bonding between the PVA or PLA matrices, clay and fumed silica were examined using a scanning electron microscope (SEM) (JSM-6710F) supplied by JEOL Company Limited, Japan. The specimens were first fixed with Karnovsky's fixative and then take through a graded alcohol dehydration series. Once dehydrated, the specimen was coated with a thin layer of gold before being viewed microscopically. The micrographs were taken at magnifications ranging from 250 to 1500 [13-14].

#### *2.3.3 Tensile Testing*

Mold shape thin films were cut with a rectangular die and tested in a AG-X Plus Series Precision Universal Testers (300kN Floor Model) supplied by Shimadzu Corporation at room temperature. The gauge length, width and thickness of the samples were 15, 2 and 0.15 mm respectively. The cross head speed used was 1mm/min. The quoted results were averaged over at least five specimens [13-14].

#### *2.3.4 Thermogravimetric Analysis (TGA)*

Thermogravimetric analysis (TGA) measurements were carried out on 5-10mg of both PVA/fsi/clay and PLA/fsi/clay nanocomposites at a heating rate of 10 /min in a nitrogen atmosphere using a Thermogravimetric Analyzer (TA Instrument SDT Q600). Thermal decomposition of each sample occurred in a programmed temperature range of 0 to 700 . The continuous weight loss and temperature were recorded and analysed [13-14].

## 3 Results and Discussion

### 3.1 Fourier Transform Infrared Spectroscopy (FT-IR) Analysis

The IR spectra of PVA/fsi/clay and PLA/fsi/clay nanocomposites were shown in Figure 1. The characteristic peaks between 1100-1000cm-1 was attributed to v(Si-O-Si) respectively [15]. The characteristic band 1650cm-1 was attributed to v(C=C) [16]. Besides, the peaks between 3500 and 3000cm-1 are mainly related to hydroxyl groups. From Figure 1(a-b), the IR spectrum of PVA/fsi/clay (1.28E) nanocomposite was broader compared to PVA/fsi/clay (1.30E) nanocomposite. This was due to the side chained-silica particles hampered the hydrogen bonds linkage in the

PVA chains. In addition, the amplitudes of Si-O-Si are increased compared to the original silica nanoparticles due to the intercalation between the PVA matrix with the clay and fumed silica.

Apart from that, Figure 1(c-d) presented the FT-IR spectra of PLA/fsi/clay (1.28E) and PLA/fsi/clay (1.30E) nanocomposites. PLA/fsi/clay (1.28E) nanocomposite showed broader peaks at 1080cm-1, 1450cm-1, 1750cm-1 and 2946cm-1 compared to PLA/fsi/clay (1.30E) nanocomposite. The IR peak at 1080cm-1 was associated with silanol groups which overlapped by the bands at 1042cm-1 of C-O-C and C-CH3 of PLA. This was attributed to the C-O stretching vibrations that impeded in this region [17]. Another strong absorption band at 1750cm-1 was corresponded to the C=O stretching vibration as detected in Figure 1(d) [18]. Moreover, the peak at 2946cm-1 was also visibly detected in Figure 1(d) assigned to the –CH groups in PLA [19].

From this result, it was successfully proved that both PVA and PLA matrices were compatible with fumed silica and clay. Besides, this result could concluded that clay (1.28E) was more compatible to be introduced into PLA/fsi matrix compared to clay (1.30E) while clay (1.30E) was more compatible to be introduced into PVA/fsi matrix compared to clay (1.28E).

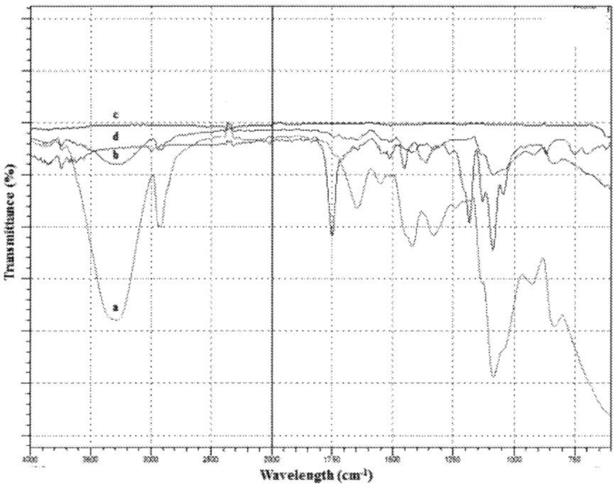


**Figure 1.** FT-IR spectra of (a) PVA/fsi/clay (1.28E) nanocomposite, (b) PVA/fsi/clay (1.30E) nanocomposite, (c) PLA/fsi/clay (1.28E) nanocomposite and (d) PLA/fsi/clay (1.30E) nanocomposite.

## 3.2 Surface Morphology

The SEM micrographs of PVA/fsi/clay and PLA/fsi/clay nanocomposites were shown in Figure 2(a-d). The uniform surface of PVA-si-clay (1.28E) nanocomposites in Figure 2(a) indicated better compatibility among the PVA/fsi with clay. Figure 2(b) showed that PVA/fsi/clay (1.30E) nanocomposite was dispersed unevenly with the presence of agglomeration due to high surface energy and poor adhesion between clay and PVA/fsi matrix [20]. Besides, it was found that PLA/fsi/clay (1.28E) nanocomposite appeared to have uniform and homogeneous dispersion without aggregation of fumed silica and clay [21]. This proved that PLA/fsi matrix had strong covalent bond with clay (1.28E). Figure 2(d) presented that PLA/fsi/clay (1.30E) nanocomposite showed uneven surface due to the non-homogeneous particle distribution.

From Figure 2, it was clearly showed that clay (1.28E) had great compatibility with PLA/fsi matrix while clay (1.30E) showed better compatibility with PVA/fsi matrix as there was no aggregation formed at the surface of PVA/fsi/clay (1.30E) and PLA/fsi/clay (1.28E) nanocomposites. The strong interfacial bonding between PVA/fsi/clay (1.30E) and PLA/fsi/clay (1.28E) nanocomposites was reflected in the higher mechanical and thermal properties.

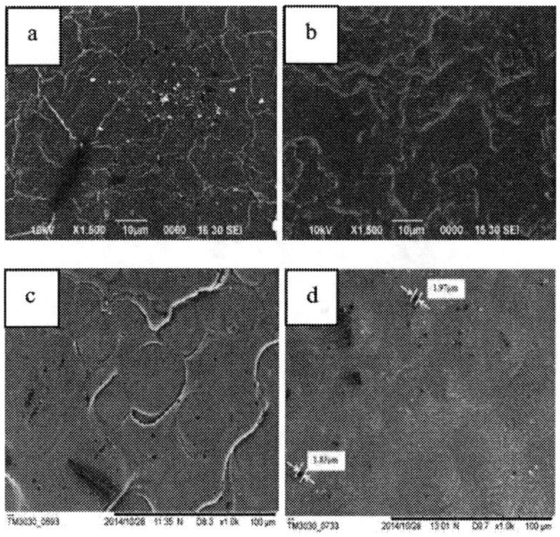


**Figure 2.** SEM micrographs of (a) PVA/fsi/clay (1.28E) nanocomposite, (b) PVA/fsi/clay (1.30E) nanocomposite, (c) PLA/fsi/clay (1.28E) nanocomposite and (d) PLA/fsi/clay (1.30E) nanocomposite.

## 3.3 Tensile Properties

The tensile strength and tensile modulus of PVA/fsi/clay and PLA/fsi/clay nanocomposites were shown in Figure 3 and 4 respectively. From Figure 3, PVA/fsi/clay (1.30E) and PLA/fsi/clay (1.28E) nanocomposites had higher tensile strength comparatively. This was due to the addition of clay towards the PVA and PLA matrices that reduced the inter-particle distance which enhanced the tensile strength of the nanocomposites [22]. When the suitable clay was introduced, the agglomeration could be reduced and less pores to be formed that improved the tensile strength of the nanocomposites.

According to Figure 4, PVA/fsi/clay (1.30E) nanocomposite had higher tensile modulus compared to PVA/fsi/clay (1.28E) nanocomposite. This was because the clay (1.28E) contained trimethyl stearyl ammonium, which imparted anti-wetting property and cause the surface of PVA-si-clay (1.28E) nanocomposites to become hydrophobic [23].

The improvements in tensile strength and modulus were ascribed to the presence of clay nanoparticles. Clay nanoparticles provided good compatibility and interfacial bonding strength toward PVA/fsi and PLA/fsi matrices. In addition, a stable hindrance layer was formed between clay particles which inhibited the agglomeration and thus improved the dispersion of the clay within PVA/fsi and PLA/fsi matrices [24].

Overall, tensile properties proved that clay (1.28E) was well introduced into PLA/fsi matrix while clay (1.30E) was well intercalated with PVA/fsi matrix. The outcome of the better compatibility of the nanocomposites was reflected in the thermal stability as well as surface morphology.

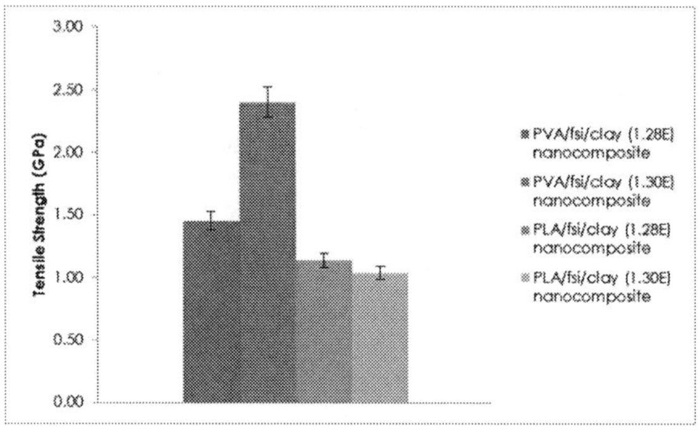


**Figure 3.** Tensile strength of PVA/fsi/clay and PLA/fsi/clay nanocomposites.

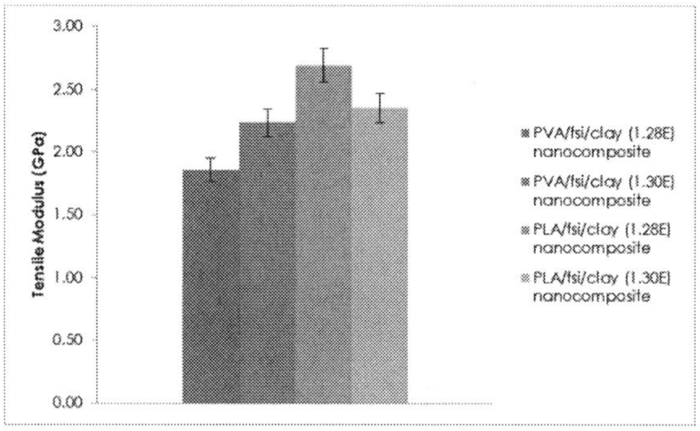


**Figure 4.** Tensile modulus of PVA/fsi/clay and PLA/fsi/clay nanocomposites.

## 3.4 Thermogravimetric analysis (TGA)

The thermal stability of PVA/fsi/clay and PLA/fsi/clay nanocomposites was shown in Figure 5. Different types of clay led to different weight loss as there was a restriction in polymer chain mobility and suppression of decomposition. The clay and silica are well intercalated with polymer matrix by covalent bonds that prevented complete decomposition of the nanocomposites [25].

According to the TGA thermographs, the weight percent loss for the first step degradation was about 5% for PVA/fsi/clay (1.28E) and PVA/fsi/clay (1.30E) nanocomposites, respectively due to the removal of moisture. The second step degradation occurred with the weight loss of 7% for PVA/fsi/clay nanocomposites. The final degradation weight loss was 1.5% and 1.3% for PVA/fsi/clay (1.28E) and PVA/fsi/clay (1.30E) nanocomposites, respectively. The bonding compatibility of PVA/fsi and clay (1.28E) was higher than clay (1.30E) as proven by tensile strength and modulus.

From Figure 5, the first step degradation of PLA/fsi/clay nanocomposites occurred at 200 while the second step degradation occurred at 350 . PLA/fsi/clay (1.28E) nanocomposite exhibited a small weight percent loss of 3.8% below 300 as the absorbed water on the surface of the nanocomposites was removed. At second stage, PLA/fsi/clay (1.28E) nanocomposite showed higher weight percent loss (12.05%) compared to PLA/fsi/clay (1.30E) nanocomposites. The improved thermal stability of PLA/fsi/clay (1.28E) nanocomposite was attributed to the clay acting as crosslinking agent that retarded the motion of polymer chain [26].

The activation energy can be helpful in reaching conclusions about the thermal stability of PVA-si-clay nanocomposites. The Arrhenius equation was used to determine the activation energy [27], which was summarized in Table 2. It was found that the activation energy of PVA/fsi/clay (1.30E) and PLA/fsi/clay (1.28E) nanocomposites wre significantly higher than PVA/fsi/clay (1.28E) and PLA/fsi/clay (1.30E) nanocomposites. The higher activation energy implied better thermal stability. Overall, the incorporation of clays especially clay (1.28E) and clay (1.30E) into PLA/fsi and PVA/fsi matrices successfully improved thermal stability of PVA/fsi/clay and PLA/fsi/clay nanocomposites that were reflected in other results.

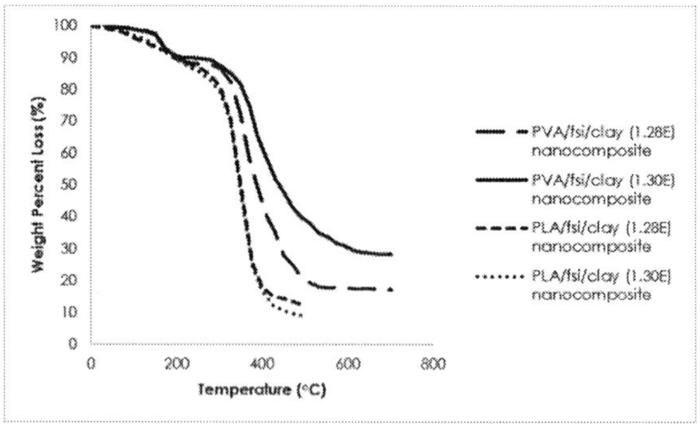


**Figure 5.** TGA curves of PVA/fsi/clay and PLA/fsi/clay nanocomposites.

**Table 2.** Activation energy of PVA/fsi/clay and PLA/fsi/clay nanocomposites.

| Samples | PVA/fsi /clay (1.28E) | PVA/fsi /clay (1.30E) | PLA/fsi /clay (1.28E) | PLA/fsi /clay (1.30E) |
|---|---|---|---|---|
| $T_i$(°C)$^a$ | 280.0 | 303.0 | 108.9 | 103.2 |

| | | | | |
|---|---|---|---|---|
| $T_m(°C)^b$ | 347.0 | 431.0 | 316.6 | 304.0 |
| $T_f(°C)^c$ | 410.0 | 511.0 | 421.1 | 418.1 |
| $W_{Ti}(°C)^d$ | 89.1 | 86.0 | 96.2 | 97.9 |
| $W_{Tm}(°C)^e$ | 66.0 | 52.0 | 80.4 | 79.4 |
| $W_{Tf}(°C)^f$ | 43.5 | 38.0 | 15.2 | 12.2 |
| $E_a$ | 25.574 | 31.660 | 6670.5 | 5523.5 |

[a] Temperature corresponding to the maximum rate of mass loss
[b] Temperature corresponding to the beginning of decomposition
[c] Temperature corresponding to the end of decomposition
[d] Mass loss at temperature corresponding to the maximum rate of mass loss
[e] Mass loss at temperature corresponding to the beginning of decomposition
[f] Mass loss at temperature corresponding to the end of decomposition
$E_a$ Activation Energy (kJ/mol)

## 4 CONCLUSION

PVA/fsi/clay and PLA/fsi/clay nanocomposites were prepared via solution intercalation and solution-intercalation film-casting. The resultant nanocomposites were characterized by FT-IR, SEM, tensile test and TGA. The reduction of hydroxyl group and the dispersion of clay for PVA/fsi/clay nanocomposites were confirmed by FT-IR spectroscopy. Besides, FT-IR spectrum indicated that PLA/fsi/clay (1.28E) nanocomposite showed broader peaks (especially the vibration of C-O-C group confined by silica network) than PLA/fsi/clay (1.30E) nanocomposite. The uniform surface of PVA/fsi/clay (1.30E) and PLA/fsi/clay (1.28E) nanocomposites indicated better compatibility among the matrices with clay. Higher mechanical properties were found in PVA/fsi/clay (1.30E) and PLA/fsi/clay (1.28E) nanocomposites compared to others nanocomposites. TGA result revealed that the thermal stability with higher activation energy of PVA/fsi/clay (1.30E) and PLA/fsi/clay (1.28E) nanocomposites were significantly improved.

## Acknowledgement

The authors would like to acknowledge the financial support from COERE Fund with Grant no. COERE/GRANT/2013/06

## References

1. R. Solaro, A. Corti, E. Chiellini, Polym. Adv. Technol. **11**, 873-878 (2010)
2. E. Chiellini, A. Corti, S. D'Antone, R. Solaro, Prog. Polym. Sci. **28**, 963-1014 (2003)
3. R.E. Drumright, P.R. Gruber, D.E. Henton, Adv. Mater. **12**, 1841-1846 (2000)
4. N.N. Chaloupi, *Development of Polylactide-clay nanocomposites for food packaging applications* (University of Montreal, Canada, 2011)
5. M. Conradi, Mater. Technol. **47**, 285-293 (2013)
6. R.T. Yang, N. Tharappiwattananon, R.Q. Long, J. Appl. Catalyst B. Environ. **19**, 289-304 (1998)
7. J.H. Chang, T.G. Jang, K.J. Ihn, W.K. Lee, G.S. Sur, J. Appl. Polym. Sci. **90**, 3208-3214 (2003)
8. A.A. Sapalidis, F.K. Katsaros, Th.A. Steriotis, N.K. Kanellopoulos, J. Appl. Polym. Sci. **123**, 1812-1821 (2012)
9. Z. Peng, L.X. Kong, Polym. Degrad. Stabil. **92**, 1061-1071 (2007)
10. J. Zhang, J. Lou, S. Ilias, P. Krishnamachari, Polym. **49**, 2381-2386 (2008)
11. A.R. McLauchlin, N.L. Thomas, Polym. Degrad. Stabil. **94**, 868-872 (2009)
12. S.M. Lai, Y.T. Hsieh, Journal of Macromolecular Science, Part B: Physics. **55**, 211-228 (2016)
13. J.C.H. Lai, M.R. Rahman, S. Hamdan, F.K. Liew, M.M. Rahman, M.F. Hossen, J. Appl. Polym. Sci. **132** (2015)
14. J.C.H. Lai, M.R. Rahman, S. Hamdan, Int. J. Polym. Sci. **2015** (2015)
15. S. Mansur Herman, L. Orefice Rodrigo, A.P. Mansur Alexandra, Polym. **45**, 7193-7202 (2004)
16. A.P.V. Perrira, W.L. Vasconcelis, L. Orefice Rodrigo, J. Non-Crystalline Solids. **273**, 180-185 (2000)
17. M. Liu, Y. Zhang, C. Zhou, Appl. Clay Sci. **75-76**, 52-59 (2013)
18. S. Bocchini, K. Fukushima, A. Di Blasio, A. Fina, A. Frache, F. Geobaldo, Biomacromol. **11**, 2919–2926 (2010)
19. E. Kontou, M. Niaounakis, P. Georgiopoulus, J. Appl. Polym. Sci. **122**, 1519-1529 (2011)
20. Y. Luo, Y. Zhao, Y. Cai, S. Du, Mater. Des. **33**, 405-412 (2012)
21. I.R. Mustapa, R.A. Shanks, I. Kong, Int. J. Adv. Sci. Eng. Technol. **3**, 192-199 (2013)
22. K. Nakane, T. Yamashita, K. Iwakura, F. Suzuki, J. Appl. Polym. Sci. **74**, 133-138 (1999)
23. K.M. Seeni Meera, R. M. Sankar, A. Murali, S.N. Janiskar, A.B. Mandal, Colloid. Surfaces B Biointerf. **90**, 204-210 (2012)
24. Y. Zheng, Y. Zheng, R. Ning, Mater. Lett. **57**, 2940-2944 (2003)
25. X. Jia, Y. Li, Q. Cheng, S. Zhang, B. Zhang, Eur. Polym. J. **43**, 1123-1131 (2007)
26. R. Benlikaya, M. Alkan, I. Kaya, Polym. Compos. **30**, 1585–1594 (2009)
27. C.V. Chanmal, J.P. Jog, eXPRESS Polym. Lett. **2**, 294-301 (2008)
28. Luigi T.De Luca, *Propulsion physics* (EDP Sciences, Les Ulis, 2009)
29. F. De Lillo, F. Cecconi, G. Lacorata, A. Vulpiani, EPL, **84** (2008)

# Investigation of Nanoparticles Dispersion in Sodium Hydroxide (NaOH) Solvent

Sia Chee Wee, Nur Asyraf Md Akhir[a] and Arif Luqman Rozman
*Petroleum Engineering Department, Universiti Teknologi PETRONAS, Perak, Malaysia*

**Abstract.** Recently, the study on nanoparticles application in enhanced oil recovery (EOR) starts to growth. Nanoparticles have given better indication for EOR development such as in foam stability as its nano size particles can be feasibly dispersed in aqueous solution and easily flow through porous media. Aggregation of nanoparticles are said to be a major contributor for paralyzing nanoparticles dispersion deep into the formation. Hence, in this research sodium hydroxide (NaOH) is used as stabilizing solvents or carrier fluids in enhancing nanoparticles properties to prevent coagulation of nanoparticles when mixed to create a nanofluid. The dispersion of various concentration of silica oxide ($SiO_2$) and aluminium oxide ($Al_2O_3$) are examined by using turbidity test. Results from this research show that the silicon dioxide nanoparticles are at best to be mixed in NaOH solvent to retain longer retention time.

## 1 Introduction

Nanoparticles are being used in industries in a considerable amount of time. It has been proven to give significant results that could be projected into the oil and gas industry specifically in tertiary recovery or enhanced oil recovery (EOR). With its nano-sized, these particles are theoretically able to disperse further into the formation covering a substantial amount of drainage area. It could also act as a stability agent for some EOR applications such as in foam flooding.

Nanotechnology was first popularized in the 1980's by a physicist named K. Eric Drexler, it was then talking about building machines on a molecular scale and even whole computers in cell sizes. Over the years, as nanotechnology became an accepted concept, the meaning of the word shifted to incorporate the simpler kinds of nanometer-scale technology for example nanoparticles. In the subject of nano-particles stream in permeable media has turned into another go for the headway in petroleum study. The development of nanotechnology application is truly sudden as of late. The disclosures of the nano-innovation potential to wind up arrangements towards a few issues in petroleum industry has ended up empowering according late research. EOR is the most theorized region for the potential improvement by this nanoparticle application. The ascent of vitality request in worldwide scale which anticipated that would happen in the oil and gas industry has made EOR the most essential technique as to meet this desire and demand from the market. The salient point of nanoparticles in EOR is its capacity to adjust certain component in the arrangement and liquid properties.

However, the dispersion of nanoparticles in porous media are still in questions. This is due to the aggregation of nanoparticles after a certain amount of time. Hence, the aim of this research is to evaluate the dispersion of various concentrations of hydrophilic and metal oxide nanoparticles for different concentrations of stabilizing fluids in brine. The scope of study with regards to this project would comprise the factors contributing to the dispersion of nanoparticles into the reservoir formation by using various concentrations of stabilizing fluids/agents.

## 2 Literature Review

### 2.1 Nanofluids

The nano-fluids are shaped by the addition of nanoparticles to fluids for amplification and development of some properties at low volume concentrations of the dispersing medium [1]. Nanofluids can be delegated as a strong nanofluid and weak nanofluid as to the quantity of nanoparticles added into the liquid. Advanced nanofluids are the liquids with more than one nano-sized added substance in the interim of basic nanofluid (the liquid with one nano-sized added substances). Nanoparticles can remain with multifunctional or single use of sort. Developing applications of nanotechnology in the industry involve new types of these nanofluids for various applications, particularly for EOR purposes [2].

[a] Corresponding author: asyraf.akhir@petronas.com.my

The test was conducted on Newtonian oil displacement showed a significant recovery utilizing nanofluids in oil recovery in a high pressure column filled with quartz sand of permeability 1 Darcy and porosity of 26% [1]. Concentration of nanofluids derive an essential key factor in optimizing EOR of low permeability water wet Berea sandstone by reducing contact angle as nanofluid concentration increases resulting in a desired nanofluid concentration of 0.01, 0.03 and 0.05 wt. % [3].

## 2.2 Nanoparticles Selection

In this study, silica nanoparticles for hydrophilic nanoparticles and aluminum oxide as metal oxide nanoparticles are chosen to be utilized as it has shown a great potential to be used in EOR (Table 1). Silicon dioxide, otherwise called silica, is an oxide compound that of silicon with the chemical formulae $SiO_2$. Silica nanoparticles possess a noticeable position in systematic research, in view of their simple readiness and their wide range of uses in different applications, for example, catalysis, electrics and electronics, pharmaceutical and various sensors [9]. Silica nanoparticles are also ventured for in enhanced oil recovery applications. These are due to the main component of sandstone comprises primarily of silica which directly makes silica nanoparticles an environmental friendly substance. Silica nanoparticle dispersion is said to have good stability due to its properties in the ability to counterbalance the gravity force effect [10].

**Table 1.** Nanoparticles Selection

| Nanoparticles | Findings |
|---|---|
| Silicon Oxide | Dispersed in ethanol improved recovery through change in rock wettability [4]. |
| | Homogeneous and heterogeneous water free-oil recovery increased better with surfactant dispersing agent [1]. |
| | Reduction of interfacial tension and wettability alteration using polymer coated $SiO_2$ [5]. |
| | Adjusting the surface charge density of the nanoparticles, stable fluids can be employed to more effectively displace oil from flow impaired locations [6]. |
| | Lower interfacial tension (IFT) was observed with increasing nanofluid concentration from 0.01 to 0.05 wt. % and reduces oil saturation by 13% [3]. |
| Aluminum Oxide | Reduce viscosity of the oil [7]. |
| | Results demonstrate that the wettability alteration plays a more dominant role in the oil displacement mechanism using nano-EOR using Polyvinylpyrrolidone (PVP) as stabilizing agent [8]. |
| | Produced oil lighter than injected oil in terms of viscosity reduction [4]. |

In a research conducted in 2013, the trend showed that an increased in hydrophilic silica nanofluid concentration will increase water-wetness in a low-permeability Berea sandstone core plug due to the electrostatic repulsion force between particles will be higher with an increased amount of nanoparticles [3]. The nanofluids will spread along the solid surface and adsorption may be occurred, decreases contact angle and displace most of the trapped oil that remains after secondary flooding with brine [11]. Torsaeter also concluded that the concentration of the nanoparticles comprised in the nanofluid affects the effectiveness in displacing oil as a tertiary recovery method. Although nanoparticles have a tendency to block pore network at higher concentration (e.g. > 0.06 wt%) in low-permeability Berea cores.

## 2.3 Retention and Aggregation

The major retention mechanism is the irreversible attraction to the rock grain surfaces [12]. The paper predicted a rock grain surface retention with the colloids, a constant first order rate coefficient is preferably used. On the other hand, it was also specified that when the colloids and rock grain have repulsive force, the coefficient has been found to be underestimated.

Retention and aggregation of nanoparticles are the major factors governing the dispersion of nanoparticles through the reservoir formation, hence, affecting drastically the recuperation of oil. Retention time indicates the time taken for any suspended particles to settle down or sediment in the bottom of the container. Theoretically, as the retention time is longer, more particles are left suspended within a certain fluid which indicates a more stable nanofluid. Transport in reservoir rock has two major components which comprises 1) the nanoparticle retention which quantifies the fraction of injected nanoparticles that reach the target zone and 2) the mobility of the nanoparticle dispersion through the porous media under operating conditions to bring the injected nanoparticles through the desired path and time to the target location [13].

### 2.4 Nanoparticles in Enhanced Oil Recovery (EOR)

Although the use of nanoparticles in EOR is very recently ventured, there are extensive researches done on the ability of nanoparticles to aid in EOR applications. Foam flooding is an advancement in EOR. Foams have been proposed for utilization as mobility control and to enhance oil recovery in several secondary recovery and EOR, for example, steam, $CO_2$ and nitrogen flooding. Distinctive surfactants are obliged to create foams that are tolerant to oil, electrolyte and steady at downhole demanding conditions. Sodium dodecyl sulfate (SDS) foam was used in a specific experiment where the foam stability was increased with the aid of $SiO_2$ nanoparticles. It also shows a better tolerance to downhole temperature whereby the foam bubbles are able to maintain a spherical shape with time through the permeable media [14]. Findings indicated how dispersed nanoparticles in a fluid mixture can alter the interfacial properties of the fluid/fluid frameworks [3, 15]. The study indicated that the surface of the whole formation can be altered in a sense of ionic changeability or nanofluid coating.

In a research done, it was observed that the interfacial tension (IFT) between the crude oil and nanofluids decreases as the nanofluid concentration increases [16]. On top of that, aluminum nanoparticles have a tendency to displace oil through their capability in reducing oil viscosity when used with fresh water and brine as dispersing fluids [4, 16]. In a research conducted using a spinning drop to observe a decrease in the interfacial tension between synthetic oil and a brine/nanofluid, a 0.01 wt % silica nanoparticle solution in brine will decrease the interfacial tension from 14.7 mN/m to 9.3 mN/m. Furthermore, when increasing the weight percent to 0.05% resulted further reduction of the interfacial tension to 5.2 mN/m. A reduction in IFT is one of the potential applications by which nanoparticles may aid in enhanced oil recovery due to less energy required to mobilize and remove oil trapped in the formation [15, 17].

## 3 Methodology

### 3.1. Nanofluid preparation

The nanoparticles used are silica oxide ($SiO_2$) and aluminium oxide ($Al_2O_3$). These nanoparticles are prepared by mixing them in carrier fluids or 3 wt% brine (NaCl) to investigate the effect of dispersion due to different concentrations of nanoparticles and stabilizing agent (NaOH). Silica and aluminium nanoparticles are prepared by mixing different concentrations of each nanoparticles in various concentrations of stabilizing agents (NaOH). These arrangements are then blended with a magnetic stirrer bar at various velocities for 4 hours [18]. Preparations for the nanofluid are shown in Table 2.

**Table 2.** Nanofluid preparation.

| Nanoparticles | Nanoparticles (wt. %) | NaOH (wt. %) |
|---|---|---|
| SiO2 + 3 wt.% Brine (NaCl) | 0.01 | 1.00 |
| | | 5.00 |
| | 0.05 | 1.00 |
| | | 5.00 |
| $Al_2O_3$ + 3 wt.% Brine (NaCl) | 0.01 | 1.00 |
| | | 5.00 |
| | 0.05 | 1.00 |
| | | 5.00 |

### 3.2. Nanofluid stability

A turbidity meter is used to confirm the aggregation effects for each samples. Prior to inserting the nanofluids into the turbidity meter, the nanofluids are stirred with a magnetic stirrer for another 4 hours in total of 8 hours. The turbidity meter reading was recorded every 10 minutes for one hour period in order to analyse the stability for each nanofluid mixtures.

Once prepared, the nanofluids stability were then tested via Nephelometric Turbidity Unit (NTU) readings on a turbidity meter. Theoretically, a higher NTU reading shows more particles suspended within a liquid hence showing a more stable nanofluid. Although, the high readings must retain its value throughout a certain time period, in this case one hour, to show that the suspended particles stay suspended instead of just settling down.

## 4 Results and Discussion

As shown in Figure 1 to Figure 4, a much stable nanofluid will represent a small or no deflection on the graph of NTU against time. Although aluminum oxide nanoparticles at 0.05 wt% have a much higher initial value of NTU, the nanoparticles was not able to retain that value within the one hour period instead most of the particles settled down.

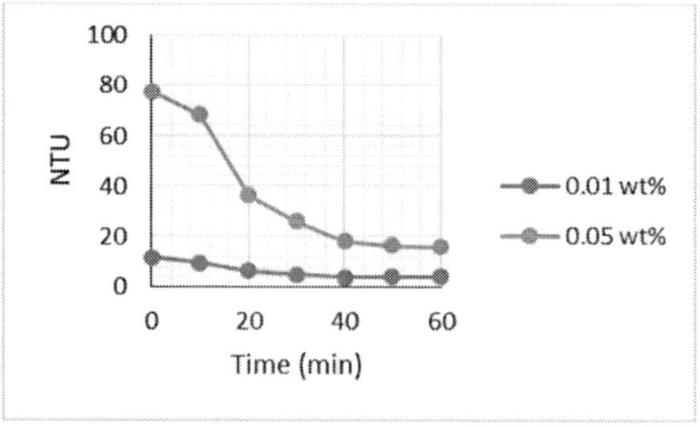


**Figure 1.** Aluminum oxide nanoparticles stability in 1 wt% sodium hydroxide (NaOH).

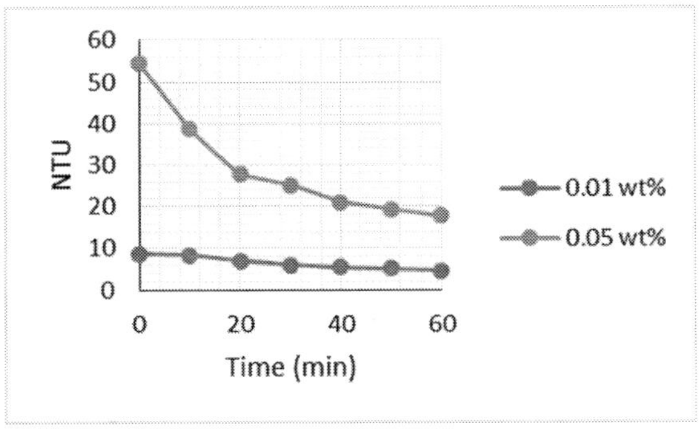


**Figure 2.** Aluminum oxide nanoparticles stability in 5 wt% sodium hydroxide (NaOH).

Figure 1 and 2 show the stability of aluminum oxide nanofluid in different concentrations of solvents and nanoparticles. As discussed, a higher NTU value results in more suspended particles within the fluid. Aluminum oxide nanofluid of 0.05 wt% mixed in 1 wt% NaOH resulted in the highest NTU value. The value was not stable and declined rapidly after 20 minutes and fluctuated 30 to 15 NTU. These values proves that the nanofluids prepared has more suspended aluminum oxide nanoparticles and stabilizes after 20 minutes.

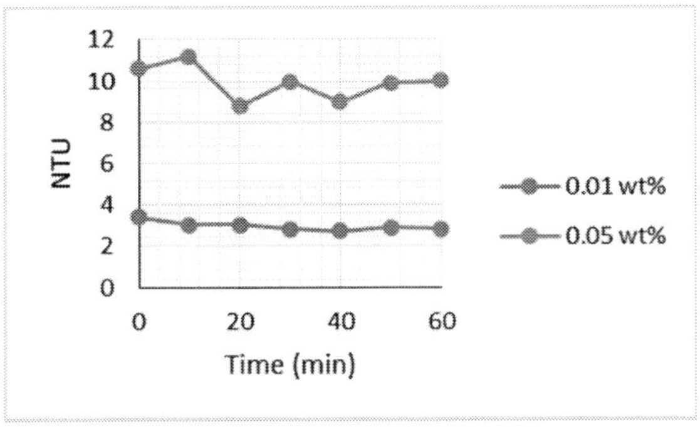


**Figure 3.** Silicon oxide nanoparticles stability in 1 wt% sodium hydroxide (NaOH).

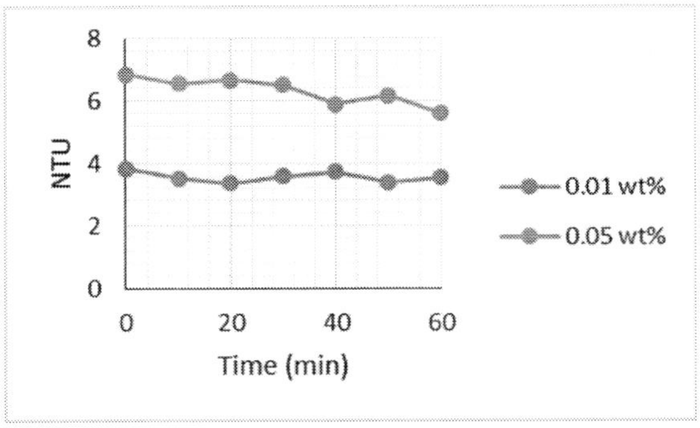


**Figure 4.** Silicon oxide nanoparticles stability in 5 wt% sodium hydroxide (NaOH).

Figure 3 and 4 illustrate silicon dioxide nanofluids in different concentrations of nanoparticles and NaOH solvent. The NTU values for silicon dioxide nanofluids are low resulting in a much lower amount of suspended particles within the fluid. Although displaying low values ranging at highest 13 NTU, these nanofluids are stable from the start until the end of one hour period. These results indicate that silicon dioxide nanoparticles are at best to be mixed in NaOH solvent to retain a longer retention time.

## 5 Conclusion

Fundamentally in concluding this project, majority of the literature reviews resulted in a positive reaction to the renowned properties of silica nanoparticles in aiding the effectiveness in EOR applications. Silica oxide nanoparticles was chosen as a representative for hydrophilic oxides. While aluminum oxide nanoparticles represents metal oxides. This way, the difference in effectiveness from both groups could be determined. Although aggregation is a major factor in the dispersion criteria, nanoparticles with the aid of different concentrations of stabilizing agents in this sense, sodium hydroxide (NaOH) will opt for a better result to overcome the retention effect.

## References

[1] Suleimanov, B., Ismailov, F., & Veliyev, E. Nanofluid for enhanced oil recovery. *Journal of Petroleum Science and Engineering,* **78(2)**, 431-437 (2011).

[2] Kong, X., & Ohadi, M. *Applications of Micro and Nano Technologies in the Oil and Gas Industry - Overview of the Recent Progress.* In *Abu Dhabi international petroleum exhibition and conference.* Society of Petroleum Engineers (2010).

[3] Torsaeter, O., Li, S., & Hendraningrat, L. *Enhancing Oil Recovery of Low-Permeability Berea Sandstone through Optimised Nanofluids Concentration.* In *SPE Enhanced Oil Recovery Conference*. Society of Petroleum Engineers (2013).

[4] Ogolo, N. A., Olafuyi, O. A., & Onyekonwu, M. O. *Enhanced Oil Recovery Using Nanoparticles.* In *SPE Saudi Arabia Section Technical Symposium and Exhibition*. Society of Petroleum Engineers (2012).

[5] Ragab, A. M. *Investigating the Potential of Nanomaterials for Enhanced Oil Recovery: State of Art.* Journal of Science and Technology, **6(1)** (2014).

[6] McElfresh, P. M., Wood, M., & Ector, D. *Stabilizing Nano Particle Dispersions in High Salinity, High Temperature Downhole Environments.* In *SPE International Oilfield Nanotechnology Conference and Exhibition*. Society of Petroleum Engineers. (2012).

DOI: 10.1051/matecconf/20178703006

[7] Bennetzen, M. V., & Mogensen, K. *Novel Applications of Nanoparticles for Future Enhanced Oil Recovery*. In *International Petroleum Technology Conference*. International Petroleum Technology Conference (2014).

[8] Hendraningrat, L., & Torsaeter, O. *Unlocking the Potential of Metal Oxides Nanoparticles to Enhance the Oil Recovery*. In *Offshore Technology Conference-Asia*. Offshore Technology Conference (2014).

[9] Rao, K. S., El-Hami, K., Kodaki, T., Matsushige, K., & Makino, K. A novel method for synthesis of silica nanoparticles. *Journal of Colloid and Interface Science,* **289(1)**, 125-131 (2005).

[10] Li, S., & Torsaeter, O. *The Impact of Nanoparticles Adsorption and Transport on Wettability Alteration of Intermediate Wet Berea Sandstone* In *SPE Middle East Unconventional Resources Conference and Exhibition*. Society of Petroleum Engineers (2015).

[11] Torsater, O., Li, S., & Hendraningrat, L. *A Coreflood Investigation of Nanofluid Enhanced Oil Recovery in Low-Medium Permeability Berea Sandstone*. In *SPE International Symposium on Oilfield Chemistry*. Society of Petroleum Engineers (2013).

[12] Caldelas, F. M., Murphy, M., Huh, C., & Bryant, S. L.. *Factors Governing Distance of Nanoparticle Propagation in Porous Media*. In *SPE Production and Operations Symposium*. Society of Petroleum Engineers (2011).

[13] Rodriguez Pin, E., Roberts, M., Yu, H., Huh, C., & Bryant, S. L. *Enhanced Migration of Surface-Treated Nanoparticles in Sedimentary Rocks*. In SPE Annual Technical Conference and Exhibition. Society of Petroleum Engineers (2009).

[14] Sun, Q., Li, Z., Li, S., Jiang, L., Wang, J., & Wang, P. Utilization of Surfactant-Stabilized Foam for Enhanced Oil Recovery by Adding Nanoparticles. *Energy & Fuels,* **28(4)**, 2384-2394 (2014).

[15] Hendraningrat, L., & Shidong, L. *A Glass Micromodel Experimental Study of Hydrophilic Nanoparticles Retention for EOR Project*. In SPE Russian Oil and Gas Exploration and Production Technical Conference and Exhibition. Society of Petroleum Engineers (2012).

[16] Alomair, O. A., Matar, K. M., & Alsaeed, Y. H. *Nanofluids Application for Heavy Oil Recovery* In *SPE Asia Pacific Oil & Gas Conference and Exhibition*. Society of Petroleum Engineers (2014).

[17] Bramer, V., & Christopher, W. *Nanoparticle dispersion flow for enhanced oil recovery using micromodels*. Doctoral dissertation (2014).

[18] Li, D., & Kaner, R. B., Shape and Aggregation Control of Nanoparticles: Not Shaken, Not Stirred. *Journal of the American Chemical Society,* **128(3)**, 968-975 (2005).

DOI: 10.1051/matecconf/20178703007

# Comparisons of the physicochemical and functional properties of commercially and traditionally processed sago starch

M.Mustafa Kamal[1], R.Baini[1,a], S.Mohamaddan[2], A.R.Henry Ragai[2], L. Soh Fong[1], N. Abdul Rahman[1], N. Mili[1], S.N.L.Taib[3], A.K.Othman[4], M.O. Abdullah[1]

[1]Department of Chemical Engineering and Energy Sustainability, Universiti Malaysia Sarawak, 94300 Kota Samarahan, Sarawak, Malaysia
[2] Department of Mechanical and Manufacturing Engineering, Universiti Malaysia Sarawak, 94300 Kota Samarahan, Sarawak, Malaysia
[3] Department of Civil Engineering, Universiti Malaysia Sarawak, 94300 Kota Samarahan, Sarawak, Malaysia
[4] Department of Electrical and Electronic Engineering, Universiti Malaysia Sarawak, 94300 Kota Samarahan, Sarawak, Malaysia

**Abstract.** Commercial sago starch was obtained from nearby supermarket while traditional sago starch (lemantak) was obtained from local wet-market. In this study, the physicochemical (moisture content, colour, particle size and microstructure) and functional properties (FT-IR spectrometer) of both types of starch were compared. Traditional sago starch (lemantak) exhibits moisture content two times higher than the commercial sago starch. Providing that commercial sago flour has been bleached in the factory, its LAB value also shows higher value than lemantak. Moreover, the particle size of sago flour is also smaller than lemantak, as the effect of multiple-refining in the factory. Meanwhile, the microstructural analysis shows significant difference that is cluster-form-granule in lemantak and disassociated-form-granule in commercial sago flour, which proof that commercial sago flour has been in an extensive process. Finally, the FT-IR results indicated that both sago flour and lemantak has similar spectrometer patterns but differ in the peak intensities, which also indicates the changes in functional properties of sago flour. Results obtained here show that the modern processing equipment as well as the chemical and drying treatment in factory, had led to the loss in some of the purity and functional characteristic in sago starch.

## 1 Introduction

Sago starch, derives from the stem of sago palm (*Metroxylon sagu*) or commonly called as "rumbia"; a tropical plant with huge trunk that may reach height up to 25 m and a diameter of 40cm; with pinnate-leaves up to 9 m long; though can tolerate wet-swampy lands such as peat soil (flood-immuned); however, it reaches commercial maturity after 9 to 12 years of planting (Karim, Tie, Manan, & Zaidul, 2008). According to Singhal et al. (2008), sago palm has been known as the highest in productivity among the starchy crops in the world (250kg per palm) and also considered as the 'starch crop of the 21st century' by many scientists, as it is economically acceptable, environmentally friendly, and it promotes a socially stable agroforestry system. Sago starch contains mainly carbohydrate, which is higher than rice and wheat flour; therefore, it has becomes the staple food since ancient times, It is an important raw materials in food industry, especially in Asian region.

In the food industry level, it is predicted that demand of sago in the future will increase as it has becomes an important raw food material (Articles & Issues, 2016). Malaysia is one of the world's leading sago producer, while Sarawak is the largest sago-growing areas in Malaysia. Until year 2000, there are 21 modern sago processing plants all around Sarawak (Palm, 2016). In addition to sago industry in Sarawak, it is the only state in Malaysia known to produce sago starch both commercially with modern techniques and traditionally for local needs (Shin & Collins, 2015). However, these modern sago processing plants had slowly wiped-out the production of traditional sago starch called "lemantak", a wet-type sago flour (Palm, 2016). This may be due to the low quality standards of 'lemantak' produced by local people. The raw material of both materials is the same but 'Lemantak' that is mixed with warm water will produce certain viscosity, which cannot be obtained using the commercial sago flour. This characteristic of 'lemantak' making it becomes favourable than the commercial sago flour among users for certain application. Sago flour is not only used as food, but also used in the processing of glue cosmetics. Hence, understanding the characteristics of sago flour, which may be influenced by the processing steps, can be important that will determine its appropriate applications.

The principles and methods in commercial sago processing are quite similar with the methods used in traditional level, but differing in the scale of operation and drying stage. The commercial sago flour manufacturer needs to meet the Malaysian standard for food product, therefore, some technique's modification as well as new technologies had been adapted in sago processing (Karim et al., 2008). In brief, the processing of sago started by feeding a segment of sago log into a slicer to separate

[a] Corresponding author: ruby@feng.unimas.my

between pith and bark. Then, the debarked section is fed into a mechanical rasper to produce finer pieces which are then fed in hammer mill via conveyor belt. The resulting starch slurry which still containing coarse fibre, is made to pass through a series of centrifugal sieves. Further purification is still needed by separation in a nozzle separator through sieve bends, as well as a series of cyclone separators, to obtain very pure starch. Most importantly, dewatering of starch is carried out using a rotary vacuum drum dryer followed by hot air drying (Karim et al., 2008). In addition, for wider acceptance, additional processing efforts have been developed, such as the production of noodles, edible film, sago pearls, bread and biscuits (Bantacut, 2014).

On the other hand, in traditional method of starch extraction, 'skinning' of the pith is made manually before chopping the pith and grating it into material resembling sawdust. The pith are then kneaded with water and filtered through sieves to extract the relatively large (20–60 lm) starch granules. After been subjected to several washings and strainings, and the starch milk is allowed to run into troughs. The starch settles in the troughs, which is subsequently sun- or kiln-dried (Singhal et al., 2008). The experiences of local people from Sarawak in consuming sago have proven that sago, when preparing from the traditional process has wide utilization and **taste even greater** than the commercially processed sago flour. For instances, sago starch has been consumed by mixing hot water and stirring to form dough called 'linut', which is then directly eaten with hot and spicy-protein based side dishes. Further instances of this is, lemantak can be used to make sago pancake or so called as 'tumpik', a baked starch cake, mix with shredded coconut, and top with protein-based food such as fish and prawn.

Interestingly, the compositions of sago is highly beneficial as it has low calorie and fat content which is good for diet; low protein content, therefore, are casein and gluten-free, which is good for gluten-allergic people; low glycemix index, which means suitable for diabetic people as it does not raise blood glucose levels immediately; and also as a resistant starch, sago can prevent the risk of constipation and colon cancer (Articles & Issues, 2016). For this reason, development of sago processing is necessary by upgrading the productivity and quality of lemantak, which consequently not only can give opportunity to home industry to grow, but also indirectly can give economic impact to the farmers. Furthermore, lemantak can be considered as the best choice to have the fully benefit of sago starch as it has not goes through any heavy-treatment during processing that is suspected to change some of the physicochemical and functional properties of sago starch. This work presents the comparison of the physicochemical and functional properties of commercially and traditionally processed sago starch. The way it is processed affect its characteristics, and in the future, it may become the interest of industry to upgrade the traditional process for commercial production of 'lemantak'. Sarawak is the largest producer of sago flour in Malaysia, meaning there is opportunity for the local people who produced 'lemantak' to generate income.

## 2 Materials and Methods

### 2.1 Sago starch sample preparation

Commercial sago flour, a product from S.G. Trading, Kuching, Sarawak; branded as 'Tepung Sagu Cap Bunga Daffodil'; was bought from local supermarket in Samarahan, Sarawak.

Meanwhile, traditional sago flour, 'lemantak' was bought from local wet-market in Samarahan, Sarawak; however, was processed from Sebuyau; a rural area which is located about 2 hours drive from Samarahan.

### 2.2 Physicochemical properties

#### *2.2.1 Moisture content*

The determination of moisture content in sago flour and lemantak were analysed by drying 3g of sample in gravity oven at 105°C until constant weight according to the AOAC, 2000 standard method. The moisture content was calculated in drying basis formula;

$$mc = \frac{M_i - M_t}{M_i} x100\%$$

Where $M_i$ and $M_t$ are the initial mass of sample (g) and the mass of sample (g) at a time (t).

#### *2.2.2 Colour analysis*

An analysis of colour was carried out manually using Matlab software which can measure the pixel data of samples image in RGB value. The RGB values were then converted into CIE-L*ab value by using a RGB-to-Lab converter program available from the internet at http://colormine.org/convert/rgb-to-lab. In addition, ''L'' represents lightness, ''a'' represents redness (+) or greenness (-), and ''b'' represents yellowness (+) or blueness (-). Whiteness index were calculated according to [6].

$$WI = 100 - \sqrt{(100 - L)^2 + a^2 + b^2}$$

#### *2.2.3 Scanning Electron Microscopy*

Granules morphology was examined using a scanning electron microscope (SEM). A thin layer of starch granule was mounted on a specimen holder; and the specimen holder was loaded in a sputter coater. Once metalized, the sample was viewed under a scanning electron microscopy (JSM-6701F) supplied by the JEOL Company Limited, Japan. The captured images were printed and saved. All optical measurements were performed at room temperature under ambient conditions.

#### *2.2.4 Particle size distribution*

The mean size distribution of sago flour and lemantak were measured by using CILAS 1090 Laser Particle Size Analyzer at room temperature. Deionized water was used as the medium for wet size measurement of flours. Sonis measurements were performed to avoid the aggregatation of flours.

### 2.3 Functional properties

#### *2.3.1 FT-IR*

Fourier transform infrared (FT-IR) spectroscopy was used to evaluate structural differences in the samples. The infrared spectra of the commercial and traditional sago starch were recorded on a Shimadzu Fourier Transform Infrared Spectroscopy (FTIR) 81001 Spectrophotometer. An appropriate amount of starch powder was pressed into a crystal window and the samples were analysed with a resolution of $4cm^{-1}$ and average scanning time of 1 min. Furthermore, the spectral resolution range of scan was 4000 to 400 $cm^{-1}$ (mid-infrared region) with ATR mode being utilized. The obtained spectra were transferred into a data analysis package. All optical measurements were performed at room temperature under ambient conditions.

## 3 Results and Discussion

### 3.1. Moisture Content

Results obtained from the moisture content experiment show that commercial sago flour exhibit 11.46% of moisture content per dry basis. The value meet the Malaysian Standard (MS470 1992) requirement for edible sago starch, which stated that the maximum moisture content is 13%. In addition, edible sago starch was defined as fine powder-starch derived from the trunk of sago palm through the process of extraction and purification. However, it should be noted that the starch produced by small factories is usually vary in term of quality including the moisture content; in fact, some often being spoilt through microbial activity [5]. On the other hand, the content of moisture in lemantak is 38.8%, which is significantly three times higher than sago flour. Karim et al. (2008) stated that lemantak is a low-quality starch; produced as the effect of inefficient extraction and processing method; far from international market requirement [1]. However, in local market, it has been understood that high moisture content is related to the local demand especially to make the 'linut'-a meal cooked by mixing lemantak and hot boiling water, which need certain level of consistency.

### 3.2 Colour

The L*, a* and b* characteristic of commercial sago flour and traditional lemantak are shown in Table 1. The colour characteristics of lemantak samples are all in the same agreement of values: L* (below 80), a* (1.5-4.5), and b* (9-16.5). The result showed that lemantak were not white, provided that it has low L* value, less red, provided that it has low a* value; but it were yellow, provided that it has significantly high b* value. In fact the average value of whiteness was 60.21. According to Aboubakar et al., (2008), the variation in b* value among samples may be attributes to the amount of carbohydrate and proteins content in the starch [7]. Both carbohydrate and proteins play an important role in the development of non-enzymatic browning in starch. However, there are also another factors that may affected color development of starch such as the presence of polyphenol compound and the cleanliness during washing step.

**Table 1.** The colour of commercial sago flour and traditional lemantak in triplicates as plotting randomly in Matlab software.

| Sample | L* | a* | b* | Whiteness |
|---|---|---|---|---|
| Commercial sago flour | 84.56 | 0.004 | -0.009 | 84.82 |
| | 87.05 | 0.005 | -0.009 | 87.05 |
| | 79.91 | -0.01 | 1.11 | 79.88 |
| Traditional lemantak | 49.69 | 4.004 | 16.19 | 46.99 |
| | 60.00 | 2.86 | 15.26 | 57.09 |
| | 78.72 | 1.88 | 9.69 | 76.54 |

In contrast, the colour characteristics of commercial sago flour showed that the flour were white, as it showed high L* value; neither red nor yellow, as it showed very small a*and b*value. In fact the average value for the whiteness were 83.92. It should be noted that commercial sago starches have the lowest whiteness among industrial starches, which accounts for the lower price. Therefore, in order to increase the whiteness property of the starch, several chemicals were added during the manufacturing process of the starch. According to Ranjith & Balakrishnan (2015), the residue of the chemicals present in the starch will cause several side effects in human beings [8].

### 3.3 Microstructure

Scanning electron micrograph showed that lemantak has an oval shape of diameter 50 μm, in agreement with most of the previous study which stated that native starch has oval, spherical and irregularly shaped granules with 6–60 μm [9]. Figures 1(a) showed that traditionally processed sago starch, lemantak, are associated as cluster or compound. Furthermore, in Figure 1 (b) and (c), lemantak show the uniformity in granular shape as the amylose and amylopectin inside the granules are tightly packed. The aggregation is a natural property for native starch, hence shows that lemantak's property is much closed to native starch.

Scanning electron micrographs commercially processed sago flour are shown in Fig. 2. Processing effected

DOI: 10.1051/matecconf/20178703007

granule's association due to extraction that has been done in an alkaline solution has known to disassociate sago starch granules, as shown in Figure 2(a). Sago flour is refined, purified or slightly bleached by oxidation with small quantities of hypochlorites (0.25-6% chlorine on dry basis). Ranjith & Balakrishnan (2015) had warned that the chemicals added to bleach sago flour, may cause side effects to human being [9]. Figure 2(b) and (c) clearly showed granules that formed into an aggregated mass comprising of several small granules; in fact, some of the granulars appearance totally disappears.

(a)

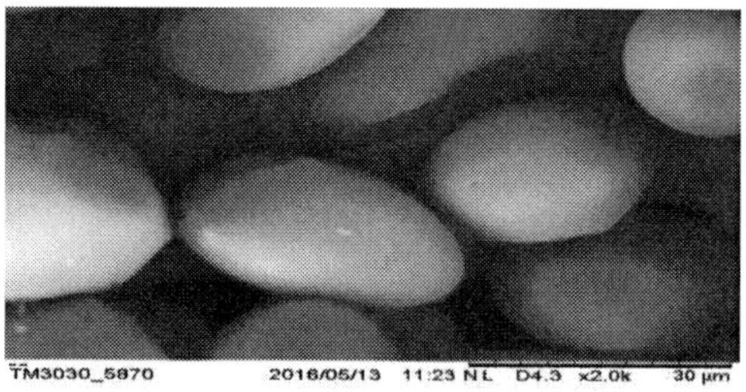


(b)

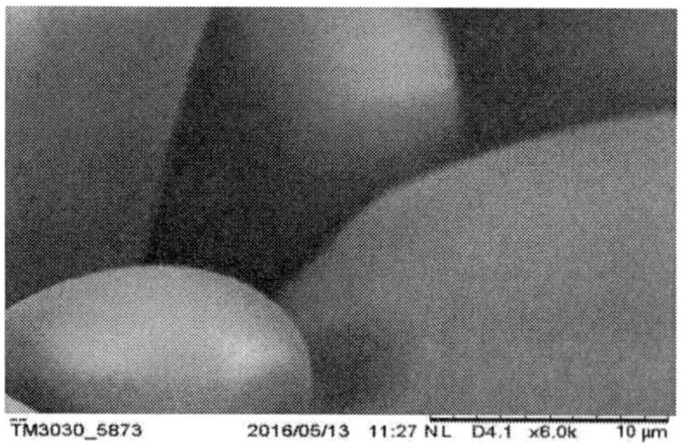


(c)

**Figure 1.** Scanning electron microscopy (SEM) of starch granules from lemantak. (a) x500, (b) x2000 and (c) x6000

According to Yadav et. al., (2006), the disruption is the result of agglomeration during drying process [10]. In flour industry, disaggregation aimed to improve the flour into free flowing, substantially dustless, more readily "wet-able", packaging, handling and sufficient bulk density to permit measuring without sifting [11]. Apart from this, there were few hollowness observed in sago flour samples indicated the occurrence of gelatinisation during drying processes [12]. Gelatinisation occurs when starch, with the presence of water, is heated at or more than its gelatinization temperature. In addition, moisture content, the variable of interest in this future work, can be one of the factors that affect the aggregation and disaggregation of granular in starch.

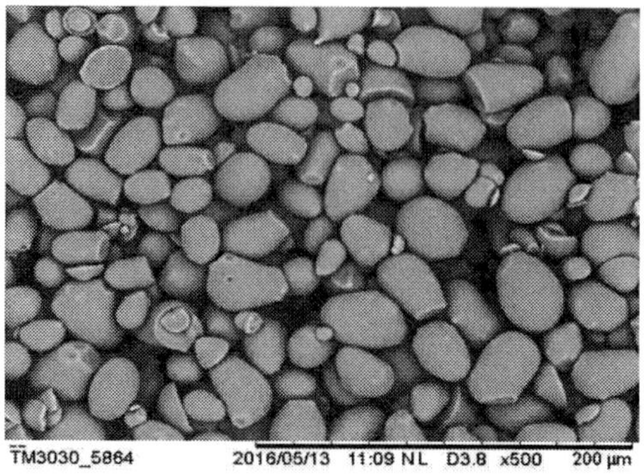


(a)

(b)

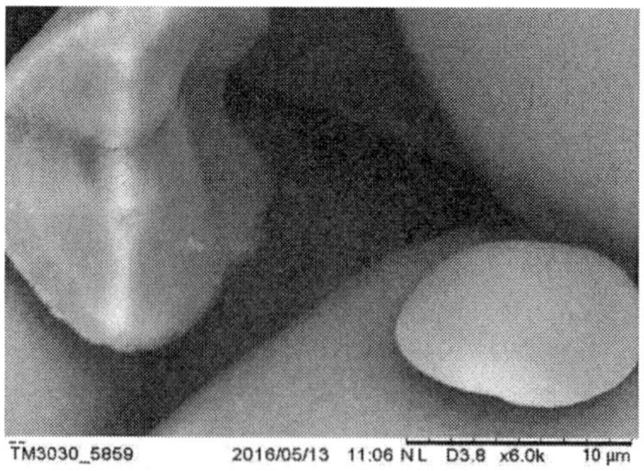


(c)

**Figure 2.** Scanning electron microscopy (SEM) of starch granules from commercial sago flour. (a) x500, (b) x2000 and (c) x6000

### 3.4 Particle size distribution

DOI: 10.1051/matecconf/20178703007

The granule size distribution of commercial sago flour is shown in Figure 3, while the granule size distribution of traditional sago flour is shown in Figure 4. The size of commercial sago flour is in similar manner with traditional sago flour which is in ranged from 1- 100 μm. The size of commercial sago flour granules were in the range of 15 – 40 μm and the samples displayed a bimodal size distribution with the first mode under 7 μm and the second mode between 7 and 100μm.

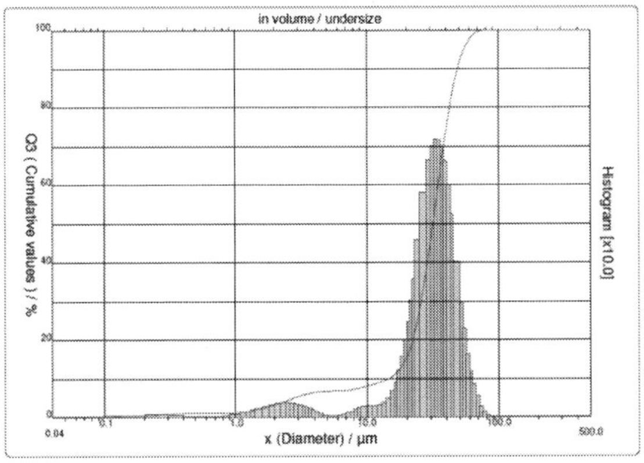


**Figure 3.** Starch granule size distribution of commercially processed sago starch

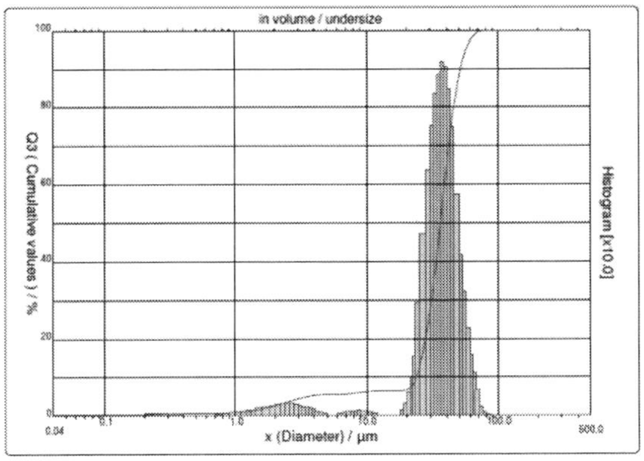


**Figure 4.** Starch granule size distribution of traditionally processed sago starch (lemantak)

On the contrary, the size of traditional sago flour granules were in the range of and 23 – 50 μm and the samples displayed a trimodal size distribution with the first mode under 7 μm, the second mode between 7 and 12 μm, and the third mode between 19 and 100μm. Moreover, the diameter at 10%, 50% and 90% for commercial sago flour are 14.79, 31.64 and 48.90 μm, respectively while the mean diameter was 31.74 μm. In contrast, the diameter at 10%, 50% and 90% for traditional sago flour are 23.61, 35.98 and 51.36 μm, respectively while the mean diameter was 35.91 μm.

Sago starch granules were oval with diameter of 50μm. Therefore, it is suggested that the presence of granules with diameter up to 98μm are the result of agglomeration during drying process. According to Njintang et al. (2007), starch with lower the particle size exhibit more hardness, adhesiveness and chewiness, as in food cooked with commercial sago starch [11]. The particle size of lemantak is more than sago flour, hence explain the grainy texture of delicacy made from it.

### 3.5 FT-IR

In order to get information on the possible difference in starch granules structures between commercial and traditional sago starch, their FT-IR were recorded. The spectra obtained from both samples were similar based on the form, but differed in the intensity of the major peaks as shown in Figure 5 and 6.

The changes in functional properties between commercial sago flour and lemantak are attributed to the agglomeration of starch granule [12]. The spectra show high absorption at the wave numbers 854, 929, 997, 1076, 1149, 1338, 2910 and 3277 $cm^{-1,}$ which confirming the carbohydrate nature of the samples. According to Aboubakar et al. (2008), the bands at 1047 and 1022 $cm^{-1}$ described the crystalline and amorphous properties of starch, respectively while their ratios indicate the degree of starch order, which were appeared in this study [7].

## 4 Conclusion

The results obtained showed some differences on the structure and properties between commercially and traditionally processed sago starch. The study revealed that sago flours are mainly made of carbohydrate and particularly starch. The granule colour, structure and size varied significantly between each other as the effect of processing method which will be discussed in further work.

## Acknowledgements

This work was supported by FRGS/TK02(04)/1304/2015(210) from Ministry of Higher Education.

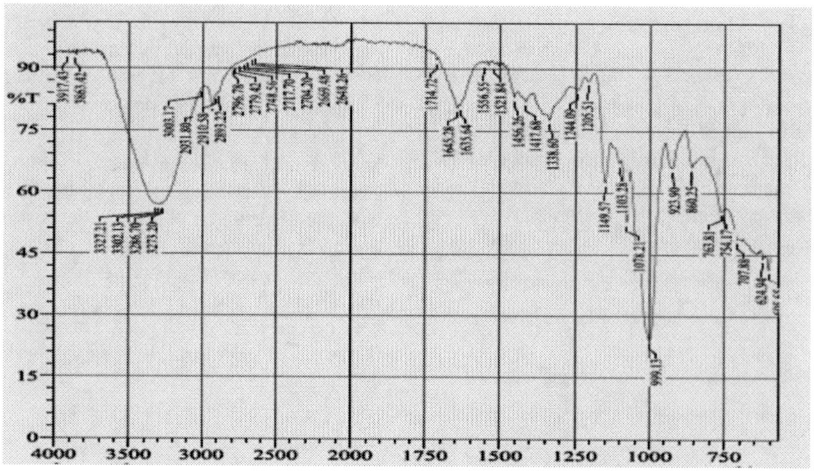


**Figure 5**. Infrared spectroscopy of commercially processed sago starch

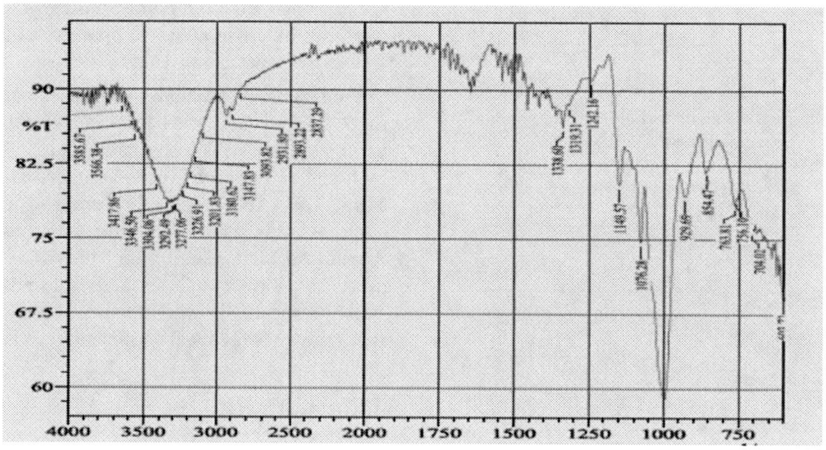


**Figure 6**. Infrared spectroscopy of traditionally processed sago starch (lemantak)

## References

1. A.A.Karim, A. Pei-Lang Tie, D.M.A. Manan, I.S.M. Zaidul, *Comprehensive Reviews in Food Science and Food Safety,* (Institute of Food Technologist) *7*(2008).
2. Rekha S. Singhal, John F. Kennedy, Sajilata M. Gopalakrishnan, Agnieszka Kaczmarek, Charles J. Knill, Putri Faridatul Akmar, Carbohydrate Polymer *72*, 1–20 (2008)
3. Rolf G. Gidlow, North St. Paul, Robert L. Teders, *United States Patent Office* (1966)
4. Chong Shin, James T. Collins, Mediterranean J. Social Sci. **6**(6), 136–144 (2015)
5. Bantacut, T. J. Envi. Earth Sci. **4**(21), 202–216 (2014)
6. Chancelle B. Ndangui, Jeremy Petit, Claire Gaiani, Jean-Mathurin Nzikou, Joel Scher, *Food Bioprocess Technol*, 7, 3618–3628 (2014)
7. Aboubakar, Y. N. Njintang, J. Scher, C. M. F. Mbofung, *J. Food Eng.* ***86***(2), 294–305 (2008)
8. D. Ranjith,P.A.Balakrishnan, *National Conferences on Research Advances in Communication, Computation, Electrical Sciences and Structures 2015* (2015)
9. Pei-Lang Tie, *Physicochemical properties os sago starch.* (2008)
10. Ramesh Yadav, Manisha Guha, R.N. Tharanathan, R.S. Ramteke, Eur Food Res Technol 223, 553-560 (2006)
11. N.Y. Njintang, M.L. Parker, G.K. Moates, C.B. Faulds, A.C. Smith, K.W. Waldron, C.M.F. Mbofung, J. Scher, J. Food Eng. 87, 172-180 (2008)

MATEC Web of Conferences **87**, 03008 (2017) DOI: 10.1051/matecconf/20178703008
*ENCON 2016*

# Effect of activated carbon in PSF-PEI-Ag symmetric membrane

Khairul Anwar Mohamad Said[1], Genevieve Gadung George[1], Nurul Ain Mohamed Alipah[1], Nor Zakirah Ismail[1], Ramizah Liyana Jama'in[1]

[1]*Department of Chemical Engineering and Energy Sustainability, UNIMAS, 94300, Kota Samarahan, Sarawak, Malaysia*

**Abstract.** Polysulfone (Psf) composite membrane consist of activated carbon, polyethyleneimine and silver nitrate was prepared by phase inversion. The activated carbon (AC) act as adsorbent to adsorb heavy metal present in synthetic waste water while polysulfone membrane act as support. Phase inversion was carried out on different composition of activated carbon from 0 to 0.9% while other component are remain constant. The surface morphology of composite membrane was characterized by scanning electron microscopy (SEM) while heavy metal absorption was quantified by atomic absorption spectrometer (AAS). The SEM image show symmetric membrane matrix with sponge structure. The composite membrane with 0.9wt% AC has the highest water flux as well as removal of heavy metal (chromium, lead, silver and cadmium) compare to composite membrane with 0.3wt% AC and 0.5wt% AC. The percentage of heavy metal reduction by composite membrane 0.9wt% AC was 35% cadmium, 19% chromium, 16% silver and 2% lead. The result indicated that the introduction of 0.9wt% AC indeed plays an important role towards enhancing the adsorption of heavy metal in water.

## 1 Introduction

Membrane comes in different meanings where it is defined as an intervening phase separating two phases acting as a barrier to transport matter between phases. Membrane separates substances through its pore using momentum such as pressure difference, temperature difference, concentration and electric potential difference [1].

Membranes have gained an important place in chemical technology and have been utilized on broad range of applications. Among the application including manufacturing, medical, water treatment and fuel cells [2]. For example, in wastewater treatment, membrane successfully treat wastewater up to 98% rejection of heavy metals [3].

Malaysia get 2,000 to 3,000mm rainfall a year, recorded as the seventh highest in the world. But, most of it is lost to surface run-off, evaporation and groundwater recharge that is often used for drinking water. In Malaysia, water resources can be obtained from rainfall, surface runoff, groundwater recharge, dams, aquifers and evapotranspiration. On top of that, streams and river also contribute 98 % of total water used in Malaysia.

Problem arises when heavy metals from industrial and consumer enter water into the streams, lakes, rivers and groundwater. In the end, water are polluted with heavy metal residue and possess danger to aquatic organism. Heavy metals are categorized as metallic elements having atomic weights between 63.5 and 200.6, and a specific gravity greater than 5.0. Unlike other organic contaminants, heavy metals are not biodegradable.

A major concern involved heavy metal with carcinogenic effect such as chromium, cadmium, lead, zinc, and copper [4]. In addition, these heavy metal are usually found in industrial wastewater and at large amount of discharge would lead to significant effect to aquatic life either acute or chronic toxicity [5]. Therefore, it is vital to treat and process the effluent before discharge

There are several proposed methods available for treatment of heavy metal in wastewater including microbial system, electrochemical process, chemical precipitation, coagulation, adsorption, membrane filtration and etc [6]. All the proposed treatment are divided into three categories mainly chemical, physical and biological treatment where the latter are widely known for its low economic implication and environmental friendly [7]. Despite the advantages, biological treatment would require large land area to fully utilize its performance which brings physical treatment in main picture as the substitute.

Membrane technology is one of the promising physical treatments that would solve the wastewater problem. Apart from membrane, activated carbon are being widely used by various researchers for the removal of heavy metal. Activated carbon is a carbonaceous material with a highly developed porous structure and large surface functional group which makes it capable of distributing toxic elements on its surface [8,9].

The adsorption process usually associated with activated carbon happened when molecule of gases or liquids (adsorbent) attached to the surface of activated carbon (adsorbate). In theory, there are three step associated with adsorption process start with macro transport which refer to movement of particle with size of more than 50nm followed by micro transport for particles size of less than 2nm and finally sorption process. Sorption involved a physical attachment of organic material onto adsorbate surface. Example of industrial application utilizing adsorption process were water purification by using synthetic resin and activated charcoal as adsorbate [10].

Hence, activated carbon has been known widely as a good adsorbate and the idea of incorporating activated carbon into membrane matrix has not been fully explored. The properties of activated carbon as a highly amorphous solid with micro crystallite and graphic lattice would increase adsorption performance especially in this context the heavy metal present in wastewater [10]. Therefore, this study proposed a composite membrane consist of polysulfone, polyethyleneimine, n-methyl-2-pyrrolidone, silver nitrate and activated carbon to reduce heavy metals present in the wastewater.

# 2 Methodology

## 2.1 Materials and Chemicals

Polysulfone (PSf, MW~ 35,000 gmol-1), Polyethyleneimine (PEI, MW~25,000 gmol-1) in powder form, and N-Methyl-2-pyrrolidone (NMP, 99.5%) are obtained from Sigma-Aldrich. Silver nitrate (AgNO3, MW~169.87 gmol-1) and granular activated carbon (AC) synthesized from palm kernel shell.

## 2.2 Preparation of Composite Membrane

Silver nitrate, polyethyleneimine and activated carbon are slowly added in PSf/NMP solution and continued stirred until completely dissolved. These solutions were further agitated for another 24 hours to reach homogeneous mixing. Then, the membrane solutions were cast on a clean, dry, level glass plate. Casting was carried out at ambient atmosphere, 30°C. 5 ml of casting solution was poured onto a glass plate for gap setting of 100 µm and left for evaporation process. After 1 minute evaporating time, membranes were immersed in deionized water for 1 hour at room temperature to allow membrane formation. Membrane were washed in deionized water to remove remaining solvent and finally leaved to dry at room temperature for 1 day. There were five membranes to be fabricated. Table 1 shows the membrane composition of all composite membrane (CM).

**Table 1.** Composite membrane composition

| Type | Psf (wt. %) | NMP (wt. %) | PEI (wt. %) | $AgNO_3$ (wt. %) | AC (wt. %) |
|---|---|---|---|---|---|
| CM 1 | 15.0 | 85.0 | - | - | - |
| CM 2 | 15.0 | 84.2 | 0.3 | 0.5 | 0.0 |
| CM 3 | 15.0 | 83.9 | 0.3 | 0.5 | 0.3 |
| CM 4 | 15.0 | 83.7 | 0.3 | 0.5 | 0.5 |
| CM 5 | 15.0 | 83.3 | 0.3 | 0.5 | 0.9 |

## 2.3 Membrane Characterization

Morphologies of composite membranes with different weight of activated carbon were observed by scanning electron microscopy (SEM, JEOL JSM-6390LA). SEM was used to analyse the surface and cross section of the membrane. It was operated at magnification of 100,000x and membrane were coated with gold sputter under vacuum condition prior to SEM. For analysing the cross-section, membrane was prepared in liquid nitrogen and placed vertically onto SEM stubs covered with carbon tape. Meanwhile for a surface analysis, membranes were placed horizontally on SEM stubs covered with carbon tape.

For water flux test, each flat sheet membrane was cut into circular form and immersed in deionized water for 5 minutes prior to dead-end filtration test. Filtration test was conducted by pouring 30ml of water into cell with membrane at the end as filter under pressure of 2 bar while running for 10 minutes.

The flow of filtered water was directly perpendicular to the membrane surface. Equation 1 shows the formula of flux, JW, is

$$J_w = \frac{V}{A\Delta t} \quad (1)$$

where V is volume of permeated water (L), A is the effective area of membrane ($m^2$) and Δt is the time taken (hour) of permeated to filter.

## 2.4 Adsorption Test

Four different wastewater were prepared for adsorption test; chromium (Cr), lead (Pb), silver (Ag) and cadmium (Cd). The synthetic wastewater containing Cr, Pb and Cd was prepared by dissolving 0.05 g of $K_2Cr_2O_7$, $Pb(NO_3)_2$, $Ag(NO_3)$ and $Cd(NO_3)_2.4H_2O$ in 500 ml deionized water. The initial concentration of elements in synthetic wastewater solution was determined by Atomic Absorption Spectrometer (AAS). Prepared synthetic waste water was stored in tight bottle to prevent changes of concentration through evaporation/condensation.

Heavy metal adsorption test was conducted with dead end filtration system where 50 ml of synthetic wastewater solution was poured into the cell. The composite membranes were cut into circular with a diameter of 44.5 mm. The test was conducted under pressure of 3 bar for a duration of 30 minutes. Final concentration of permeated solution was determined by AAS.

AAS is used to measure the concentration of chemical elements usually for environmental related samples. The removal efficiency is obtained from equation 2:

$$R(\%) = \frac{C_0 - C_t}{C_0} \times 100 \quad (2)$$

where $C_0$ and $C_t$ are the initial and final concentrations of heavy metals present in the wastewater before and after adsorption.

 DOI: 10.1051/matecconf/20178703008

The silver leached out from the membranes was also investigated by using AAS. Membranes were cut into 1 cm × 1 cm area and subsequently immersed in 10 ml of sodium chloride solution. After 1 to 2 days of being in contact with sodium chloride solution, the membranes were removed and the aqueous solutions analyzed.

## 3 Result and Discussion

### 3.1 Membrane Morphology

According to the literature, membrane structure has two forms either a symmetric or an asymmetric structure. Symmetric membrane is a membrane with a uniform structure while asymmetric membrane has a gradient structure[1]. Besides that, symmetric membrane have cylindrical, sponge and dense film structure. For asymmetric, it can be categorized into integral-asymmetric and thin film composite[11].

Figure 1 presented the surface and cross-section of SEM images of membranes with different composition. CM1 was a membrane with PSf/NMP only. CM 2, CM 3, CM 4 and CM 5 were a composite membrane with 0%, 0.3%, 0.5% and 0.9% weight fraction of activated carbon, respectively.

The cross section areas of membranes induced by phase inversion showed a symmetric structure with a lot of micro void appear across membrane matrix. According to Pinnau and Freeman (2000), cross-section of a typical polysulfone membrane prepared with phase inversion are usually has symmetric structure along with formation of micro void within matrix. On top of that, increased AC content to 0.9 wt. % (CM5) were a clear evident for formation of sponge structure as shown Figure 1 (i) and (j). which aligned with stipulation stated the increase amount of activated carbon, cause pore to broaden/enlarge, thus increasing membrane structure porosity [12].

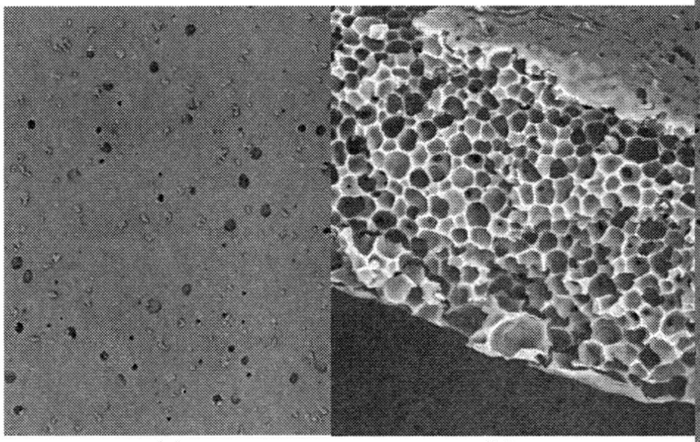

(a) (b)

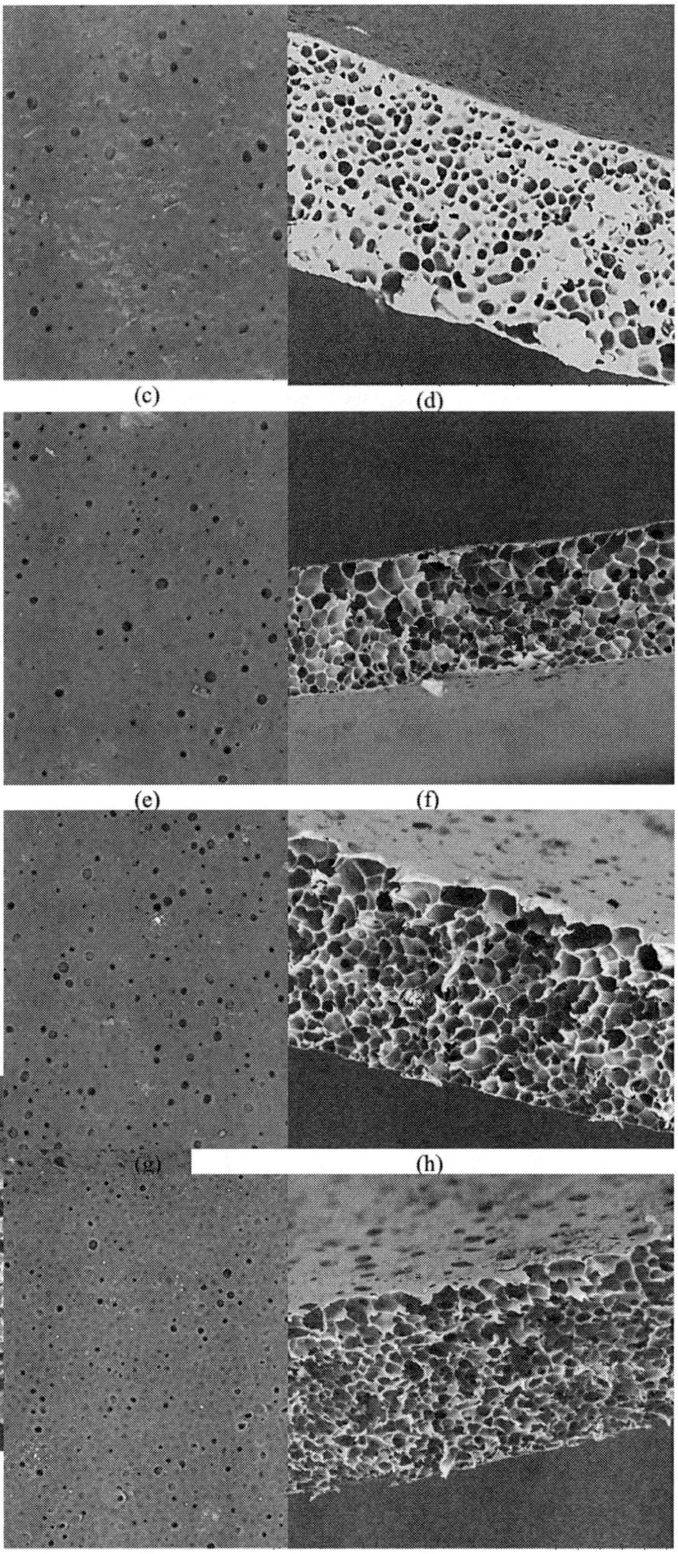

(c) (d) (e) (f) (g) (h) (i) (j)

**Figure 1.** SEM images of top surface and cross section of composite membrane CM1(a,b), CM2(c,d), CM3(e,f), CM4(g,h) and CM5(i,j)

Therefore, by adding different composition of activated carbon will result with increased in surface areas, pore volume, and sizes of pores. Comparing the surface and cross sectional morphologies of CM 1 to CM 5, it was clearly shows that the membrane structures were changed as AC content increased.

### 3.2 Water flux test

Water flux and rejection are two most important parameters for membrane applications [13]. The water flux testing provided an indication of the membrane hydraulic performance. To recap, membrane were subjected to filtration test at pressure of 2 bar for 10 minute continuously. The result for water flux is as shown in Table 2.

**Table 2.** Water flux of various composite membrane

| Membrane | Activated carbon (wt%) | Water flux (L/m$^2$.hr) |
|---|---|---|
| CM1 (Control) | - | 5.54 |
| CM2 | 0 | 9.94 |
| CM3 | 0.3 | 11.15 |
| CM4 | 0.5 | 22.25 |
| CM5 | 0.9 | 49.33 |

The water flux increased when the AC was incorporated into the membranes. This result is attributed to the increase of membrane hydrophilicity and possibly by the increase in porosity [14]. Hydrophilicity refer to the affinity of the membrane material for water or ability of the membrane to become wetted with water [15].

CM 1 shows the lowest water flux than CM 5 with a difference of 43.79 L/m2 .hr which is a significant increase of 88% water flux performance. Based on Figure 1, the large difference of water flux was expected because CM1 has a dense structure with a distinct honeycomb pattern unlike CM5 with porous-like structure. Also, the addition of silver nitrate has contribute to increase in water flux for CM2 from 5.54 L/m2 .hr (CM1) to 9.94 L/m2 .hr (CM2).

Andrade et al. (2015) studied water flux performance of polysulfone membranes modified with silver nanoparticles prepared by phase inversion method. In the event of silver nitrate addition, a significant increase of water flux was observed with increase composition of silver in membrane.

On the other hand, studies developed by Hwang et al. (2013), water flux of composite membrane increased as the AC particle content increased. Authors proposed that the addition of AC particles may affect the pore size, pore distribution, and filtration flux. All this findings are aligned with result observed in this study which shows an increase of water flux with addition of activated carbon and silver nitrate.

Figure 2 clearly shows that the water flux was increasing with increasing activated carbon content. Control membrane alone has the lowest water flux with 5.54 L/m2.hr and CM 5 with 0.9 wt. % of AC content show the highest water flux with 49.33 L/m2.hr. The incline in flux may be attributed to the increasing of membrane structure porosity.

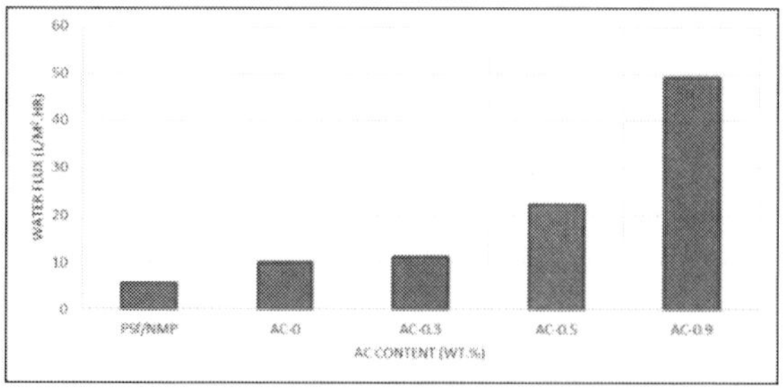


**Figure 2.** Water flux of different composite membranes

### 3.3 Removal of heavy metal

The adsorption studies were determined by atomic absorption spectrometry (AAS). In this studies, AAS was used to measure the initial and final concentration of heavy metal presence in synthetic wastewater after membrane filtration. To recap, filtration of synthetic wastewater was operated at 3 bar pressure for 30 minutes. Table 3 show the result of heavy metal (chromium, cadmium, lead and silver) rejection for all composite membrane.

**Table 3.** Rejection percentage of various composite membrane towards chromium, cadmium, lead and silver.

| Membrane | Initial (mg/L) | Final (mg/L) | Rejection (%) |
|---|---|---|---|
| Chromium | | | |
| CM 1 | 159.74 | 3.76 | 97.64 |
| CM 2 | | 137.83 | 13.72 |
| CM 3 | | 135.22 | 15.35 |
| CM 4 | | 131.79 | 17.50 |
| CM 5 | | 128.82 | 19.36 |
| Cadmium | | | |
| CM 1 | 86.78 | 1.23 | 98.5840 |
| CM 2 | | 14.23 | 83.5982 |
| CM 3 | | 55.88 | 35.6111 |
| CM 4 | | 55.49 | 36.0554 |
| CM 5 | | 56.03 | 35.4371 |
| Lead | | | |
| CM 1 | 285.17 | 3.35 | 98.83 |
| CM 2 | | 9.82 | 96.56 |
| CM 3 | | 243.73 | 14.53 |
| CM 4 | | 272.72 | 4.36 |
| CM 5 | | 277.74 | 2.61 |
| Silver | | | |
| CM 1 | 164.47 | 1.75 | 98.93 |
| CM 2 | | 42.29 | 74.14 |
| CM 3 | | 137.77 | 16.24 |
| CM 4 | | 137.66 | 16.30 |
| CM 5 | | 137.43 | 16.44 |

Rejection of heavy metal are consistent for CM1 between 97 to 98% most likely caused by small pore size and dense structure shown in figure 1 therefore increase ability of CM1 to entrapped or filtered heavy metal. However the trade-off having high rejection resulting

CM1 to have a low water flux at 5.54 L/m$^2$.hr. In general, the addition of activated carbon has reduce the ability of membrane to filtered heavy metal. CM5 having 0.9wt% activated carbon which has the highest water flux of 49.33 L/m$^2$ .hr can only manage to remove 19% of chromium although the rejection value was the highest among membrane with activated carbon but it still lack 78% from 97% rejection of control membrane (CM1). Result of cadmium and silver also show CM5 capable of rejecting heavy metal at high water flux but result of lead removal show that at high water flux almost all lead has passed through CM5 as filtrate.

Figure 3 shows the result in graph to observe the trends of heavy metals rejection. In general the trends of rejection show a sudden drop from 75 to 80% reduction after addition of activated carbon except for chromium rejection.

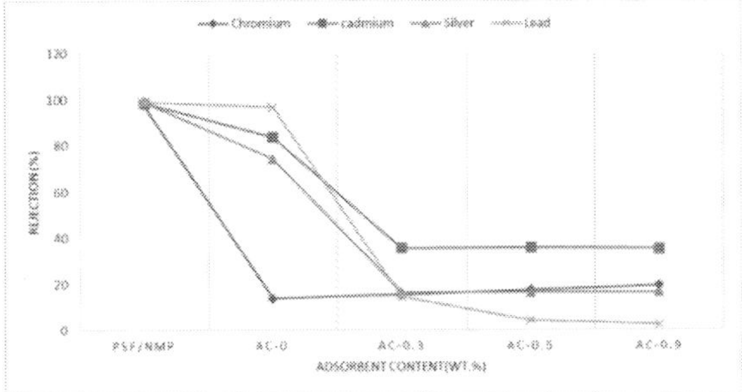


**Figure 3.** Heavy metal rejection for different membrane composition

Chromium rejection show an increase from approximately 13% (CM2) to 19% (CM5) although the gradient was not apparent but the increase of rejection might as well contributed by activated carbon inside the membrane matrix that absorb heavy metal while membrane filtered the remaining. All heavy metal result show approximately 20% rejection for CM3 while cadmium has more than average rejection at 35%. The result might as well explained by ionic radii of metal ions. The ionic radii follows the following sequence: $Cr^{3+}$ (130 pm) > $Ag^{+}$ (126 pm) > $Pb^{2+}$ (120 pm) > $Cd^{2+}$ (97 pm). The rejection of cadmium was the highest among other metal for CM3 might cause by small ionic radii that give advantage for activated carbon to adsorb by allowing cadmium to enter its pore. Activated carbon has surface area in range of 1000 to 3000 m$^2$/g depend on type of activation where the higher surface area refer to small pore size and dense quantity of pore [16–18]. Henceforth, small ionic radii of 97pm cadmium ion can be adsorb more into pore of activated carbon. After addition of activated carbon, the rejection of cadmium, silver and chromium show a consistent trends within 15 to 19 % for chromium and silver while cadmium show steady rejection at approximately 35% for CM3 to CM5. Contrasting, the increase of activated carbon composition has reduce the ability of membrane to remove lead where the reduce performance can be observed from figure 3 starting from 14% (CM3) rejection and eventually drop to 2% (CM5). Therefore, addition of 0.3wt% activated carbon was the optimum composition for heavy metal removal where further increase of composition would either reduce or provide insignificant improvement to heavy metal rejection.

Figure 4 shows the effect of water flux to heavy metal rejection. In general, the increase of flux has significant effect to heavy metal rejection with the lowest flux contribute to highest rejection up to 98% and sudden reduction was observed for chromium which show steady rejection for the rest of the flux averaging 16%. The high rejection at low flux might contribute to small pore that restrict flow rate which eventually increase rate of separation which are usually based on molecular size. The walls of the pore would change during compaction/operation resulting with denser and closer matrix would be another reason for the increase of rejection while addition of activated carbon has reduce the effect of compaction [19].
In the range of 11 to 49 L/m$^2$.hr, rejection are constant or steady with value below average from 2% to 35% rejection. Consequently, the reduce effect of compaction following addition of activated carbon has allow membrane to swell and increase water flux. Membrane swelling has reduce its ability to filter metal ions while high water flux has reduce the adsorption capacity of activated carbon [20]. It is believed a low water flux would increase the chances of metal ions/heavy metal to adsorb onto the surface of impregnated activated carbon.

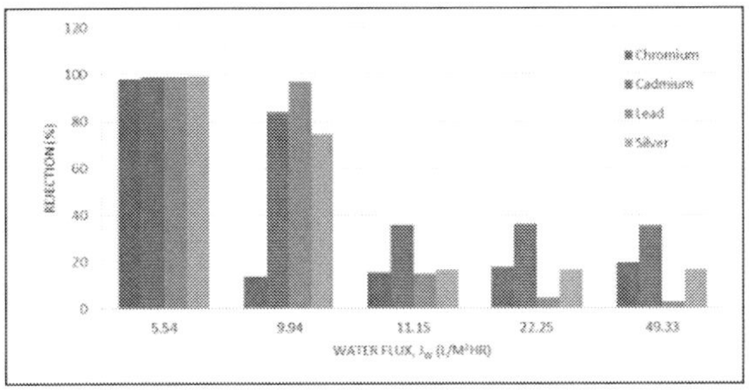


**Figure 4.** Effect of water flux to heavy metal rejection

Table 4 shows the silver leaching test for CM2 to CM5. The main purpose of conducting silver leaching test was to quantify amount of silver adsorb or filtered by membrane by identifying the rate of silver leach beforehand. Silver leaching refer to detachment of silver from membrane surface or membrane matrix and leach out as filtrate into product stream or to upstream process.

**Table 4.** Silver leaching test

| Membrane sample | Silver leaching (mg/L) | | Total |
|---|---|---|---|
| | Day 1 | Day 2 | |
| CM 2 | 6.31 | 0.69 | 7.01 |
| CM 3 | 2.32 | 0.33 | 2.65 |
| CM 4 | 1.93 | 0.69 | 2.62 |
| CM 5 | 1.85 | 0.56 | 2.42 |

Silver leach reach 7mg/L for CM2 while constant for the rest of composite membrane. CM2 show 5mg/L more silver leach, an increase of 71% which explain the addition of activated carbon has increase the attachment of silver onto membrane matrix and eventually reduce the amount of silver leach. Result of heavy metal rejection (Table 3) has taken into consideration of the amount of silver leach from membrane and subtract it from result presented in Table 4.

## 4. CONCLUSION

With the addition of activated carbon, the membrane morphology was modified with increase of water flux. The dense structure of polysulfone membrane (CM1) were changed to a more porous with a lot of macrovoid formation with the addition of activated carbon especially 0.9 wt% (CM5). The improving porosity of CM3 to CM5 were proved by increasing water flux. Water flux value of CM3, CM4 and CM5 were 55%, 77% and 88% higher while the cadmium rejection of CM3, CM4, CM5 were 63%, 62% and 63% lower that of CM1, respectively. This is because composite membrane (CM) with added activated carbon experience higher porosity therefore increase the possibility of heavy metal ions to pass through during filtration whereas the high flux has reduce the time for heavy metal to be adsorbed onto activated carbon impregnated in the membrane matrix. Based on the obtained result, it can be concluded that an optimization study to fine-tune the membrane porosity against activated carbon content/composition is desirable to ensure the developed composite membrane (CM) can operate efficiently by mean of filtration application with adsorption capability.

## Acknowledgement

This research is fully supported by FRGS grant, FRGS/TK04(03)/1140/2014(07). The authors fully acknowledged Ministry of Higher Education (MOHE) and Universiti Malaysia Sarawak for the approved fund which makes this important research viable and effective.

## References

1. H. Strathmann, L. Giorno, E. Drioli, An Introduction to Membrane Science and Technology, 2006.
2. C.M. Lewandowski, A Brief Mindfulness Intervention on Acute Pain Experience: An Examination of Individual Difference., 2015.
3. Y. Yurekli, Removal of heavy metals in wastewater by using zeolite nano-particles impregnated polysulfone membranes, J. Hazard. Mater. 309 (2016) 53–64. doi:10.1016/j.jhazmat.2016.01.064.
4. F. Fu, Q. Wang, Removal of heavy metal ions from wastewaters: A review, J. Environ. Manage. 92 (2011) 407–418. doi:10.1016/j.jenvman.2010.11.011.
5. E. Bernard, a Jimoh, J.O. Odigure, Heavy Metals Removal from Industrial Wastewater by Activated Carbon Prepared from Coconut Shell, Res. J. Chem. Sci. 3 (2013) 3–9.
6. M.J.K. Ahmed, M. Ahmaruzzaman, A review on potential usage of industrial waste materials for binding heavy metal ions from aqueous solutions, J. Water Process Eng. 10 (2016) 39–47. doi:10.1016/j.jwpe.2016.01.014.
7. S. Fazal, B. Zhang, Z. Zhong, L. Gao, X. Chen, Industrial Wastewater Treatment by Using MBR ( Membrane Bioreactor ) Review Study, J. Environ. Prot. (Irvine,. Calif). 6 (2015) 584–598.
8. H.A. Hegazi, Removal of heavy metals from wastewater using agricultural and industrial wastes as adsorbents, HBRC J. 9 (2013) 276–282. doi:10.1016/j.hbrcj.2013.08.004.
9. M. Karnib, A. Kabbani, H. Holail, Z. Olama, Heavy metals removal using activated carbon, silica and silica activated carbon composite, in: Energy Procedia, 2014: pp. 113–120. doi:10.1016/j.egypro.2014.06.014.
10. D. Lakherwal, Adsorption of Heavy Metals: A Review, Int. J. Environ. Res. Dev. 4 (2014) 2249–3131. doi:10.1007/s11270-007-9401-5.
11. A.F. Ismail, A.R. Hassan, Effect of additive contents on the performances and structural properties of asymmetric polyethersulfone ( PES ) nanofiltration membranes, 55 (2007) 98–109. doi:10.1016/j.seppur.2006.11.002.
12. L.L. Hwang, J.C. Chen, M.Y. Wey, The properties and filtration efficiency of activated carbon polymer composite membranes for the removal of humic acid, Desalination. 313 (2013) 166–175. doi:10.1016/j.desal.2012.12.019.
13. X. Lin, K. Wang, Y. Feng, J.Z. Liu, X. Fang, T. Xu, H. Wang, Composite ultrafiltration membranes from polymer and its quaternary phosphonium-functionalized derivative with enhanced water flux, J. Memb. Sci. 482 (2015) 67–75. doi:10.1016/j.memsci.2015.02.017.
14. P.F. Andrade, A.F. de Faria, S.R. Oliveira, M.A.Z. Arruda, M. do C. Gonçalves, Improved antibacterial activity of nanofiltration polysulfone membranes modified with silver nanoparticles, Water Res. 81 (2015) 333–342. doi:10.1016/j.watres.2015.05.006.
15. H.L. Richards, P.G.L. Baker, E. Iwuoha, Metal Nanoparticle Modified Polysulfone Membranes for Use in Wastewater Treatment: A Critical Review, J. Surf. Eng. Mater. Adv. Technol. 02 (2012) 183–193. doi:10.4236/jsemat.2012.223029.
16. L. Li, S. Liu, J. Liu, Surface modification of coconut shell based activated carbon for the improvement of hydrophobic VOC removal, J. Hazard. Mater. 192 (2011) 683–690. doi:10.1016/j.jhazmat.2011.05.069.
17. X. Yao, J. Liu, G. Gong, Y. Jiang, Q. Xie, Preparation and modification of activated carbon for benzene adsorption by steam activation in the presence of KOH, Int. J. Min. Sci. Technol. 23 (2013) 395–401. doi:10.1016/j.ijmst.2013.05.015.
18. J. Alcañiz-Monge, M.J. Illán-Gómez, Insight into hydroxides-activated coals: Chemical or physical activation?, J. Colloid Interface Sci. 318 (2008) 35–41. doi:10.1016/j.jcis.2007.10.017.
19. E. Eren, A. Sarihan, B. Eren, H. Gumus, F.O. Kocak, Preparation, characterization and performance enhancement of polysulfone ultrafiltration membrane using PBI as hydrophilic modifier, J. Memb. Sci. 475 (2015) 1–8. doi:10.1016/j.memsci.2014.10.010.
20. P.S.S. Goh, A.F.F. Ismail, S.M.M. Sanip, B.C.C. Ng,

 DOI: 10.1051/matecconf/20178703008

M. Aziz, Recent advances of inorganic fillers in mixed matrix membrane for gas separation, Sep. Purif. Technol. 81 (2011) 243–264. doi:10.1016/j.seppur.2011.07.042.

# Palm oil mill effluent treatment using coconut shell – based activated carbon: Adsorption equilibrium and isotherm

Sherlynna Parveen Deshon Kaman[1], Ivy Ai Wei Tan[1,a] and Leonard Lik Pueh Lim[2]

[1]*Department of Chemical Engineering and Energy Sustainability, Faculty of Engineering, Universiti Malaysia Sarawak, 94300 Kota Samarahan, Sarawak, Malaysia.*
[2]*Department of Civil Engineering, Faculty of Engineering, Universiti Malaysia Sarawak, 94300 Kota Samarahan, Sarawak, Malaysia.*

**Abstract.** The current ponding system applied for palm oil mill effluent (POME) treatment often struggle to comply with the POME discharge limit, thus it has become a major environmental concern. Batch adsorption study was conducted for reducing the Chemical Oxygen Demand (COD), Total Suspended Solids (TSS) and Color of pre-treated POME using coconut shell-based activated carbon (CS-AC). The CS-AC showed BET surface area of 744.118 m²/g, with pore volume of 04359cm³/g. The adsorption uptake was studied at various contact time and POME initial concentration. The CS-AC exhibited good ability with average percentage removal of 70% for COD, TSS and Color. The adsorption uptake increased over time and attained equilibrium in 30 hours. The equilibrium data were analyzed using the Langmuir, Freundlich, Temkin and Dubinin–Radushkevich isotherm models. Based on the coefficient regression and sum of squared errors, the Langmuir isotherm described the adsorption of COD satisfactorily, while best described the TSS and Color adsorption; giving the highest adsorption capacity of 10.215 mg/g, 1.435 mg/g, and 63.291 PtCo/g respectively. The CS-AC was shown to be a promising adsorbent for treating POME and was able to comply with the Environmental Quality Act (EQA) discharge limit. The outcome of treated effluent using CS-AC was shown to be cleaner than the industrial biologically treated effluent, achieved within shorter treatment time.

## 1 Introduction

The vast development of palm oil industry has led Malaysia to become the largest palm oil exporter to more than 100 countries [1]. Consequently, this increases the waste discharged, i.e. partly treated palm oil mill effluent (POME) into waterways. According to Malaysian Palm Oil Board (MPOB), as of December 2012, there were 429 palm oil mills nationwide; generating an average of 0.65 $m^3$ POME from every processed ton of oil palm fresh fruit bunches (FFB). Based on annual production of 9,288,000 tons of FFB in Sarawak; this results in an estimated annual effluent generation of 6,037,200 $m^3$.

Fresh POME is a hot (80-90°C), acidic (pH 4-5) slury with unpleasant odor, and high colloidal suspension; which consists of mainly water, oil and total solids including suspended solids from debris of palm fruit residues [2]. POME is the result from processing of FFB mainly from physical processes such as sterilization, clarification and separation process of palm oil. Some researchers suggested that the dark, brownish color of POME could be attributed to the breakdown of lignocellulosic from the raw effluent into lignin and tannin [3, 4]. POME also contains various types of organic compounds which are mainly pigments extracted from sterilization of FFB, such as anthocyanin and carotene pigment [5].

The discharge of POME without appropriate treatment into waterways has been an environmental concern, as it affects the water quality. The high quantity of organic substance in POME has translated into high detection of Chemical Oxygen Demand (COD) and Biological Oxygen Demand (BOD) which has been a challenge for the treatment of this effluent. The degradation of POME in waterways can produce by-products which can be toxic to aquatic life and therefore, there is an urgency to treat POME properly before being discharged.

The typical characteristics of raw POME, biologically treated POME and the corresponding standard discharge limit of Environmental Quality Act (EQA) 1974, Department of Environment (DOE), Malaysia are shown in Table 1. The biologically treated POME characteristics showed non-compliance with the standard discharge limits. In this study, the collected POME was at pH 8, however the effluent was yet to comply with other parameters of EQA discharge limit.

In Malaysia, most of the operating palm oil mills are using the conventional ponding system for POME treatment in view of advantages such as low capital and operating cost. However, it requires large area due to the long hydraulic retention time (HRT), typically about 1–2 months. In addition, the current system has low efficiency in treating POME and sometimes could not meet the final

[a] Corresponding author: awitan@unimas.my

discharge limit. Therefore, there is a need to upgrade the existing POME treatment system to ensure regular compliance with the stringent limit imposed by DOE Malaysia.

Various techniques have been innovated and applied by researchers for POME treatment, especially at polishing stage prior to discharge. Adsorption using activated carbon (AC) is a popular technique due to its high adsorptive capacity. Hence, broad application of AC as adsorbent can be often seen in catalysis or separation processes [6]. However, the usage of AC is not cost effective due to the utilization of costly raw material such as coal which is non-renewable. Lately, rising attention is put on the the utilization of dumped biomass waste as precursor for AC such as oil palm shell [7], banana peel [4], palm kernel shell and empty fruit bunch [5]. The use of agricultural waste-based AC is much more economical as its raw material is low cost and can be easily obtained. Besides, it gives an added value to the generated waste, as this practice leads to a reduction of agricultural waste disposal, and adds functionality of the by-product.

This research explores the adsorptive potential of coconut shell-based AC (CS-AC) for reduction of COD, Total Suspended Solids (TSS) and Color from POME of different initial concentration and contact time. The equilibrium data of the adsorption process were then evaluated by fitting to four isotherm models to determine the adsorption mechanism of the pollutant molecules onto the CS-AC.

## 2 Experimental

### 2.1 Materials

#### *2.1.1 Preparation of POME*

The POME sample was supplied by a palm oil mill in Kota Samarahan, Sarawak, Malaysia from their anaerobic pond, at the end of anaerobic treatment. The samples were then cooled at a temperature of 4°C, and let to sediment for total solid removal. The supernatant was withdrawn for tests and analysis.

#### *2.1.2 Preparation of Activated Carbon*

The source of adsorbent is a commercial AC derived from coconut shell, produced via physical activation using steam, obtained from a local manufacturer. The CS-AC was sieved to sizes between 1 mm to 2 mm, and then washed 5-6 times to neutralize the pH and remove residual ash. Finally, the cleaned CS-AC was oven-dried at 110°C over night and stored in air-tight bottles. Distilled water was utilized for AC washing and preparation of reagents and solutions.

### 2.2 Method

#### *2.2.1 Batch Adsorption Study*

The effects of two variables were studied; contact time for adsorption and initial POME concentration. Six different POME initial concentrations were examined at a range of contact time from 0.5 to 48 hours. POME was diluted to five different ratios (50%, 60%, 70%, 80% and 90%) using distilled water. The original concentration was presented as 100%. 5.0 g of CS-AC was added to each flask containing 100 mL anaerobic treated POME (at natural pH 8) and were agitated (120 rpm at 30 ± 1 °C). The percentage removal and the adsorbed amount of COD, TSS and Color at equilibrium, $q_e$ (mg/g) were determined using Eq. (1) and (2), respectively.

$$\text{Removal (\%)} = \frac{(C_o - C_e)}{C_o} \times 100 \quad (1)$$

$$q_e = \frac{(C_o - C_e) \times V}{W} \quad (2)$$

where $C_o$ and $C_e$ (mg/L) is POME concentration at initial and equilibrium, respectively. $V$ (L) is the solution volume, and $W$ (g) is the mass of CS-AC used.

**Table 1.** Typical POME Characteristics [8, 9, 4]

| Parameter | Unit | Raw | Biologically Treated | Standard Discharge Limits |
|---|---|---|---|---|
| Chemical Oxygen Demand | mg/L | 51000 | 4700 | |
| Total Suspended Solid | mg/L | 18000 | 1800 | 400 |
| Color | PtCo/L | | 9900 | |
| pH | | 4.2 | 8.4 | 5 - 9 |
| Biological Oxygen Demand ($BOD_3$ at 30°C) | mg/L | 25000 | 1350 | 100 |

#### *2.2.2 Adsorption Isotherm Model*

Optimization of an adsorption system design of can be obtained through a suitable correlation for their equilibrium curves. In this study, CS-AC was used as adsorbent for anaerobic treated POME with different initial concentrations of COD, TSS and Color. The data acquired from adsorption equilibrium experiments were interpreted using four models; namely Langmuir, Freundlich, Temkin and Dubinin–Radushkevich (DR) equations utilizing correlation method.

#### *2.2.3 Validity of Isotherm Model*

The adsorption isotherm of POME onto the CS-AC was verified at different initial concentrations. Each model validity was measured by the sum of squared errors (SSE, %) (Eq. 3).

$$\mathrm{SSE} = \sqrt{\frac{\sum (q_{e,\exp} - q_{e,cal})^2}{N}} \qquad (3)$$

where N represents the number of data points. Lower SSE value indicated a better fit.

### 2.3 Analytical Procedure

#### *2.3.1 Activated Carbon Characterization*

The Brunauer–Emmett–Teller (BET) surface area and pore volume of the CS-AC were measured by multi-point BET from $N_2$ adsorption isotherms, using Automated Gas Sorption System (Quantachrome Instruments). The morphology of the CS-AC surface was determined by utilizing Scanning Electron Microscope (SEM). SEM micrographs were taken using Hitachi TM3030 with 500x magnification.

#### *2.3.2 Analysis*

Standard methods were applied for the detection of COD, TSS and Color. Method No. 8006 was used to determine COD concentration of the sample carried in COD vials by APHA potassium dichromate method (HR 20–1500 mg/L) using DR2800 HACH spectrophotometer and DRB 200 reactor. TSS was determined in conformity with the standard method No. 8000 using DR2800 HACH spectrophotometer.This Spectrophotometer was also used to detect Color according to the procedure 8025 Platinum–Cobalt standard method [1].

## 3 Results and Discussion

### 3.1 Activated Carbon Textural Characterization

From the CS-AC textural characterization results, the BET surface area was 744.118 $m^2/g$ and total pore volume was 0.436 $cm^3/g$. These characteristics were comparable to commercial AC produced industrially which reported BET surface area between 900 to 1000 $m^2/g$, as well as total pore volume of 0.50 and 0.60 $cm^3/g$, respectively [10].

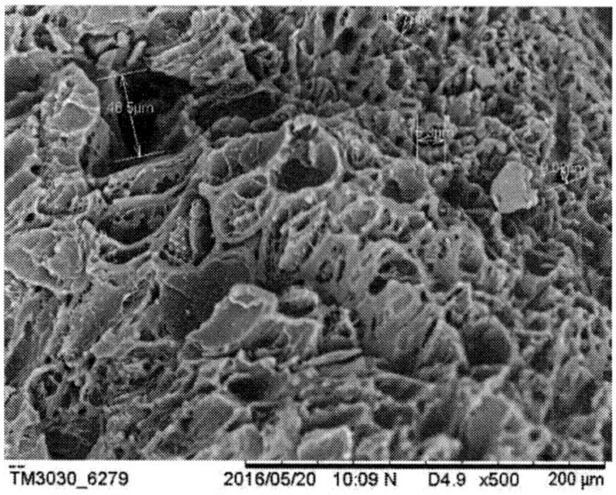


**Figure 1.** SEM image of CS-AC (x500)

The surface morphology of the CS-AC is shown by the SEM image in Figure 1. It was noticed that the CS-AC surface morphology reflected the common AC characteristic with the presence of various sizes pores, as reported in literature [11].

### 3.2 Adsorption Study

From the batch adsorption studies, about 70% removal of COD, TSS and Color was achieved from POME using CS-AC. This result was comparable with a study using palm kernel shell AC (PKS-AC) for POME pre-treatment which showed similar percentage removal about 70%. However, PKS-AC took shorter equilibrium time possibly due to the lower initial concentration of POME used [12].

#### *3.2.1 Initial POME concentration and contact time effects on adsorption equilibrium*

The adsorption uptake at time t, $q_t$ (mg/g) at various POME concentrations is presented in Figure 2. COD, TSS and Color at all initial concentrations showed rising adsorption uptake with contact time. At early stage, the uptake increased rapidly, indicating the availability of plenty active sites. It then eventually showed a saturation curve upon reaching equilibrium, possibly due to monolayer coverage of pollutants onto the surface of the adsorbent [5]. The adsorption uptake, or the amount of pollutants adsorbed by CS-AC is utmost at equilibrium time of 30 hours (Figure 2), under the studied operating conditions [13].

The $q_e$ values increased from 5.21 to 9.47 mg/g, 1.01 to 1.32 mg/g, and 20.80 to 39.10 PtCo/g for COD, TSS and Color, respectively when POME concentration was increased from 50% to 100%. However, the percentage of COD, TSS and Color removal reduced with increasing POME concentration. This phenomenon indicated reduction in adsorption of pollutants due to the blocked active sites on the CS-AC for treating the concentrated POME [14], besides due to the limited capacity of CS-AC to adsorb high amount of pollutants.

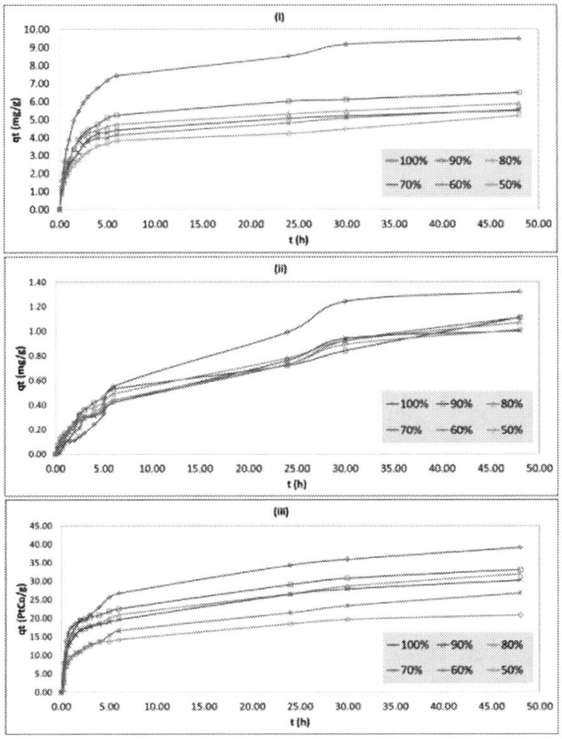


**Figure 2.** Adsorption uptake of: (i) COD, (ii) TSS, and (iii) Color over contact time at various POME concentration

## 3.3 Adsorption Isotherm

The adsorption isotherm indicates effluent distribution between the solid and solution phases at a particular temperature upon reaching the equilibrium. Isotherm data can be correlated via empirical or theoretical equations following the Langmuir, Freundlich, Temkin and DR isotherm models. The CS-AC characteristics such as adsorption capacity and surface properties can be made known from constants values obtained from each isotherm model.

### *3.3.1 Langmuir Isotherm*

Langmuir isotherm and Freundlich isotherm [15] have been broadly applied in vast sorption processes [16]. The Langmuir isotherm equation is as shown in Eq. (4):

$$\frac{C_e}{q_e} = \frac{1}{Q_o b} + \frac{1}{Q_o} C_e \quad (4)$$

where $C_e$ referes to POME concentration at equilibrium (mg/L; except for Color, denoted as PtCo) and $q_e$ is the amount of adsorbed pollutants (mg/g). $Q_o$ (mg/g) is the adsorption capacity related to complete monolayer coverage, and Langmuir constant is denoted as $b$ (L/g). Monolayer coverage means that each active site on the AC holds only one molecule of pollutant, which is a basic assumption for Langmuir isotherm. A linear plot between ($C_e/q_e$) and $C_e$ provides the values of $b$ and $Q_o$. The separation factor, $R_L$ describes essential characteristis of Langmuir isotherm as in Eq. (5) [17]:

$$R_L = \frac{1}{(1 + bC_o)} \quad (5)$$

where $C_o$ is the POME initial concentration (g/L). The value of $R_L$ between 0 and 1 indicates that the isotherm is favorable. The isotherm is unfavourable if $R_L > 1$, linear if $R_L = 1$, and irreversible if $R_L = 0$ [17]. The values of $R_L$ obtained in this study were between 0 to 1 (Table 2), which proved that the Langmuir isotherm was a favorable model to describe the adsorption of POME in terms of COD, TSS and Color, onto the CS-AC under the conditions used in this study.

**Table 2.** Isotherm constants and correlations for adsorption of COD, TSS and Color onto CS-AC

| Isotherms | Parameter | Constants | | | | |
|---|---|---|---|---|---|---|
| Langmuir | | | | | | |
| | | **Qo (mg/g)** | **b (L/g)** | **R2** | **SSE (%)** | **$R_L$** |
| | COD | 10.215 | 0.011 | 0.686 | 1.103 | 0.11 |
| | TSS | 1.435 | 0.051 | 0.935 | 0.085 | 0.10 |
| | | **Qo (PtCo/g)** | **b (L/PtCo)** | | | |
| | Colour | 63.291 | 0.001 | 0.930 | 1.363 | 0.21 |
| Freundlich | | **KF (mg/g (L/mg)^n)** | **1/n** | **R2** | **SSE (%)** | |
| | COD | 1.280 | 0.317 | 0.524 | 1.037 | |
| | TSS | 1.839 | 0.168 | 0.500 | 2.624 | |
| | | **KF (PtCo/mg (L/PtCo)^1/n)** | | | | |
| | Colour | 1.149 | 0.532 | 0.937 | 9.962 | |
| Temkin | | | | | | |
| | | **A (L/g)** | **B** | **R2** | **SSE (%)** | |
| | COD | 7.794 | 2.163 | 0.467 | 8.945 | |
| | TSS | 4.519 | 0.194 | 0.497 | 0.075 | |
| | | **A (PtCo/g)** | | | | |
| | Colour | 106.449 | 15.352 | 0.944 | 143.320 | |
| Dubinin-Radushkevich | | **qs (mg/g)** | **E (J/mol)** | **R2** | **SSE (%)** | |
| | COD | 6.944 | 40.825 | 0.318 | 1.243 | |
| | TSS | 1.181 | 111.803 | 0.316 | 0.088 | |
| | | **qs (PtCo/g)** | | | | |
| | Colour | 38.652 | 5.103 | 0.936 | 1.724 | |

This result was consistent with the work of Mohammed & Chong [4], using banana peel AC (BP-AC) for POME treatment, which reported that Langmuir isotherm was more favourable for COD, TSS and Color removal. However, the adsorption capacity, $Q_o$ of CS-AC (10.215 mg/g, 1.435 mg/g and 63.291 PtCo/g for COD,

TSS and Color respectively) was much lower than that of BP-AC (200 mg/g, 63 mg/g and 135 PtCo/g for COD, TSS and Color respectively). This could be due to different type of raw material and carbon activation processes used to synthesize the AC. As reported by Mohammed & Chong [4], the BP-AC was a result of two-step activation, which was carbonization (physical) and chemical activation. This led to higher surface area and pore development of the AC, thus having higher capacity to adsorb pollutants. However, the usage of chemical is costly, especially when it deals with massive amount of AC preparation. Therefore, AC from single-step physical activation was implemented in this study.

The experimental data sufficiently fitted the Langmuir isotherm for both TSS and Color in accordance with the correlation coefficients $R^2 > 0.9$, except for COD. The greater values of $R^2$ indicated the relevence of the adsorption model for pollutants removal in POME.

Referring to Table 2, Langmuir isotherm model described the adsorption of COD satisfactorily. Yet, this model was the best in describing the adsorption equilibrium of TSS and Color, since the SSE values reported were the lowest with high correlation of $R^2 > 0.9$. These findings confirmed that the active sites were homogeneously distributed on the CS-AC surface, since the application of Langmuir equation assumed homogeneous surface of adsorbent [5].

### *3.3.2 Freundlich Isotherm*

Freundlich isotherm model (Eq. 6) considers non-uniform heat distribution on the absorbent surface and hence it is called as a heterogeneous adsorption [18]. Freundlich constants representing adsorption capacity and intensity are indicated as $n$ and $K_F$ [mg/g (L/mg) $^{1/n}$] respectively.

$$\log q_e = \log K_F + \left(\frac{1}{n}\right)\log C_e \qquad (6)$$

The extent of correlation between solution concentration and adsorption depends on the adsorption intensity, $n$. The adsorption can be linear ($n = 1$), chemical ($n < 1$), and favaroble physical process ($n > 1$) [19]. A normal Freundlish isotherm is indicated if $1/n < 1$, and indicates a cooperative adsorption if vise versa. [20]. A straight line obtained from log $q_e$ vs log $C_e$ plot gives the slope ($1/n$) values of 0.17-0.53 for COD, TSS and Color, which indicated their favorable adsorption onto the CS-AC at the experimental conditions studied.

Comparable adsorption intensity ($K_F$) between COD, TSS and Color was also observed. Although the adsorption capacity and intensity according to Frendluich isotherm showed similar results between CS-AC and BP-AC [4], however this model was found to be less favourable as a result of lower $R^2$ and greater SSE values.

### *3.3.3. Temkin Isotherm*

Temkin isotherm concernes the effects of indirect adsorbent-adsorbate interactions, in which it causes the molecules heat of adsorption to reduce with increased coverage in the layer [21]. This model is expressed by Eq. (7):

$$q_e = \left(\frac{RT}{b}\right)\ln(AC_e) \qquad (7)$$

where $RT/b$ signifies the constant B. A linear plot between $q_e$ and ln $C_e$ determines the values of both constants A and B. This model resulted in lower $R^2$ values (except for Color), thus it was not suitable to describe the overall POME adsorption onto CS-AC.

### *3.3.4 Dubinin–Radushkevich (DR) Isotherm*

The DR equation (Eq. 8) can also be applied for the analysis of isotherm [22].

$$q_e = q_s \exp(-B\varepsilon^2) \qquad (8)$$

where ε can be correlated using Eq. (9)

$$\varepsilon = RT\ln\left(1 + \frac{1}{C_e}\right) \qquad (9)$$

where $R$ represents gas constant (8.31 J/mol K) and $T$, the absolute temperature. The constants $q_s$ and $E$ can be obtained from the ln $q_e$ versus $\varepsilon^2$ plot. Constant $B$ conveys the mean free energy, $E$ of COD, TSS and Color molecule sorption when it is transferred to the surface of the CS-AC from infinity in the effluent. The value of E can be calculated using Eq. (10). However, similar to Temkin isotherm, this model was found to be less favourable due to lower correlation for adsorption, especially for COD and TSS.

$$E = \left(\frac{1}{\sqrt{2B}}\right) \qquad (10)$$

## 4 Conclusion

The present investigation showed CS-AC was a promising agricultural waste-based adsorbent for the reduction of COD, TSS and Color from POME over a range of 50% to 100% POME concentrations, with average percentage removal of 70% for COD, TSS and Colour after 48 hours. The BET surface area of 744.118 m²/g and pore volume of 0.436 cm³/g showed that the CS-AC was porous with well-developed pores. This was confirmed by the SEM image of the CS-AC. The adsorption uptake for the removal of COD, TSS and Color onto the CS-AC were 10.215mg/g, 1.435mg/g and 63.291 PtCo/g, respectively. Langmuir isotherm was proven to satisfactoryly describe the overall adsorption of COD, TSS and Color in POME.

## Acknowledgement

The authors acknowledge the research grant awarded by Universiti Malaysia Sarawak under Special Grant Scheme F02/SpGS/1406/16/7 and financial assistance supported by Ministry of Higher Education Malaysia under MyBrain15.

## References

1. R.R. Mohammed, M.R. Ketabachi, G. McKay, Combined magnetic field and adsorption process for treatment of biologically treated palm oil mill effluent (POME). *Chem. Eng. J.* **243**, 31–42 (2014)
2. A.L. Ahmad, C.Y. Chan, Sustainability of Palm Oil Industries: An Innovative Treatment via Membrane Technology. *J. Appl. Sci.* **9(17)**, 3074-3079 (2009)
3. R. Chanida and P. Poonsuk, Decolorization and Phenol Removal of Anaerobic Palm Oil Mill Effluent by Phanerochaete chrysosporium ATCC 24725. TIChE International Conference 2011. Thailand (2011)
4. R.R. Mohammed, M.F. Chong, Treatment and decolorization of biologically treated Palm Oil Mill Effluent (POME) using banana peel as novel biosorbent. *J. Environ. Manage.* **132**, 237–249 (2014)
5. R.R. Mohammed, Decolorisation of Biologically Treated Palm Oil Mill Effluent ( POME ) Using Adsorption Technique. IRJES. **2(10)**, 1–11 (2013)
6. C. Saka, BET, TG-DTG, FTIR, SEM, iodine number analysis and preparation of activated carbon from acorn shell by chemical activation with ZnCl2. *J. Anal. Appl. Pyrol.* **95**, 21-24 (2012)
7. S.G. Herawan, M.S. Hadi, Md. R. Ayob, A. Putra, Characterization of Activated Carbons from Oil-Palm Shell by $CO_2$ Activation with No Holding Carbonization Temperature. *Scientific World J.* (2013)
8. MPOB, Monthly Production of Oil Palm Products Summary for the Month of May 2014. Retrieved 03 May, 2016. http://bepi.mpob.gov.my/index.php/statistics/production/125-production-2014/660-production-of-oil-palm-products-2014.html (2014)
9. N. Saifuddin, S. Dinara, Pretreatment of palm oil mill effluent (POME) using magnetic chitosan, *E-J. Chem.* **8**, 67–78 (2011)
10. S. Wang, Z.H. Zhu, A. Coomes, F. Haghseresht, G.Q. Lu, The physical and surface chemical characteristics of activated carbons and the adsorption of methylene blue from wastewater, *J. Colloid Interface Sci.* **284**, 440–446 (2005)
11. R. Sitthikhankaew, D. Chadwick, S. Assabumrungrat and N. Laosiripojana, Effects of humidity, $O_2$ and $CO_2$ on $H_2S$ adsorption onto upgraded and KOH impregnated activated carbons. *Fuel Process. Technol.* **124**, 249-257 (2014)
12. N.S. Azmi, K.F.M. Yunos, Wastewater Treatment of Palm Oil Mill Effluent (POME) by Ultrafiltration Membrane Separation Technique Coupled with Adsorption Treatment as Pre-treatment. *Agric. Agric. Sci. Procedia.* **2**, 257–264 (2014)
13. I.A.W. Tan, B.H. Hameed, A.L. Ahmad, Equilibrium and kinetic studies on basic dye adsorption by oil palm fibre activated carbon. *Chem. Eng. J.* **127(1-3)**, 111–119 (2007)
14. M. Jibril, J. Noraini, L.S. Poh, A.M. Evuti, Removal of colour from waste water using coconut shell activated carbon (CSAC) and commercial activated carbon (CAC). *Jurnal Teknologi (Sciences and Engineering)*, **60**, 15–19 (2013)
15. L.S. Frank, Adsorption Technology: A Step-by-Step Approach to Process Evaluation and Application. Marcel Dekker, NewYork. (1985)
16. A.L. Ahmad, S. Sumathi, B.H. Hameed, Adsorption of residue oil from palm oil mill effluent using powder and flake chitosan: Equilibrium and kinetic studies. *Water Res.* ***39*(12)**, 2483–2494 (2005)
17. K.R. Hall, L.C. Eagleton, A. Acrivos, T. Vermeulen, Pore and solid diffusion kinetics in fixed bed adsorption under constant pattern conditions. *Ind. Eng. Chem. Fundam.* **5**, 212–218 (1966)
18. H. Freundlich, W. Heller, The adsorption of cis- and trans- azobenzene. *J. Am. Chem. Soc.* **61**, 2228–2230 (1939)
19. P.S. Kumar, S. Ramalingam, C. Senthamarai, M. Niranjanaa, P. Vijayalakshmi, S. Sivanesan, Adsorption of dye from aqueous solutionby cashewnut shell: studies on equilibrium isotherm, kinetics and thermodynamics of interactions. *Desalination,* **261**, 52–60 (2010)
20. K. Fytianos, E. Voudrias, E. Kokkalis, Sorption-desorption behavior of 2,4-dichlorophenol by marine sediments, *Chemosphere* **40**, 3–6 (2000)
21. M. Hosseini, S.F.L. Mertens, M. Ghorbani, M.R. Arshadi, Asymmetrical Schiff bases as inhibitors of mild steel corrosion in sulphuric acid media, Mater. *Chem. Phys.* **78,** 800 (2003)
22. S. Rengaraj, Y. Kim, C.K. Joo, K. Choi, J. Yi, Batch adsorptive removal of copper ions in aqueous solutions by ion exchange resins: 1200H and IRN97H, *Korean J. Chem. Eng.* **21 (1)**, 187–194 (2004)

# Adsorption of colour, TSS and COD from palm oil mill effluent (POME) using acid-washed coconut shell activated carbon: Kinetic and mechanism studies

Yong Yin Sia, Ivy Ai Wei Tan[a] and Mohammad Omar Abdullah

*Department of Chemical Engineering and Energy Sustainability, Faculty of Engineering, Universiti Malaysia Sarawak, 94300 Kota Samarahan, Malaysia*

**Abstract.** The disposal of palm oil mill effluent (POME) without proper treatment before being discharged into natural water sources has become undesirable because of high concentration of suspended solid (SS), oil and grease (O&G), chemical oxygen demand (COD) and biological oxygen demand (BOD). This study investigated the feasibility of removing colour, total suspended solid (TSS) and COD using acid-washed coconut shell based activated carbon (CSAC) through the evaluation of the adsorption uptake as well as the adsorption kinetics and mechanism. The percentage removal of colour, TSS and COD from POME onto CSAC were 61%, 39%, 66%, respectively achieved within 48 hours of contact time. The kinetic models studied were pseudo-first-order (PFO), pseudo-second-order (PSO), and Elovich models. The intra-particle diffusion (IPD) model was studied to interpret the adsorption diffusion mechanism. The adsorption of colour, TSS and COD onto CSAC were best interpreted by the PSO model, and well fitted by the Elovich model. The IPD and Boyd plots indicated that IPD and film diffusion controlled the adsorption of colour, TSS and COD onto the CSAC.

## 1 Introduction

Palm oil industry is a main economic contributor in Malaysia. Palm oil mill effluent (POME) is the largest waste contributor in the palm oil mills. Oil palm waste is also generated from the refined fresh fruit bunch (FFB). FFB consists of 30% fibre, 6% shell, 3% decanter cake and 28.5 % empty fruit bunch (EFB). In Malaysia, approximately 30 million tonnes of POME were released by 381 palm oil production mills in 2004 [1]. Malaysia produced at least 44 million tonnes of POME in 2008 [2]. An efficient and sustainable management system is therefore required for treatment of POME to protect the environment and monitor the degradation of river water in order to prevent environmental pollution [1].

In Malaysia, ponding system has been applied in most palm oil production mills to treat POME [3]. This is mainly because Malaysia has large areas of land for the ponding system. The ponding system is more economically feasible and can stand a larger limit of organic loading rate (OLR) [4]. However, large amounts of methane ($CH_4$) and carbon dioxide ($CO_2$) are released in an uncontrollable manner from open ponds into the atmosphere. This might worsen the global warming effect as 6.54 kg $CH_4$/tonne FFB would be emitted from the ponds [5]. The equivalent amount of $CO_2$ released into the air is 137.4 kg. In 2020, POME treatment ponds are estimated to release 0.36 million tonnes of $CH_4$ [6]. $CH_4$ has approximately 20 times more greenhouse effect than $CO_2$ with a possible 100 year global warming effect [7]. Therefore, by the year of 2020, the ponding systems in Malaysia would cause greenhouse effect corresponding to 7.2 million tonnes $CO_2$.

Activated carbon (AC) has been proven to be efficient in treating various wastewaters. Nevertheless, its constraint is the high cost of commercial AC. Therefore, derivation of AC from agriculture wastes has been emphasized in recent years since the interest in low-cost AC derived from renewable raw materials is growing. A past study reported that the removal efficiency of suspended solids and turbidity was 89.1% and 91.4%, which was respectively achieved by utilizing powdered activated carbon (PAC) obtained from EFB, with the following specific conditions: 3.5 g PAC/100 mL bio-treated POME, contact time of 45 minutes and mixing speed of 150 rpm [8]. Another study on column adsorption of dye wastewater was conducted with a bottom layer of granular activated carbon (GAC) and an upper layer of zeolite in a column of 6.35 cm diameter at 2.84 mL/cm$^2$.min surface loading rate and 10 cm bed height [9]. The reduction of chemical oxygen demand (COD), ammonia and colour were 59.46%, 60.82% and 58.4%, respectively. A study on colour adsorption from biologically treated POME by utilizing AC derived from palm kernel shell which was produced by following chemical and microwave method (PKSCM) reported colour removal of 100% at the optimum condition of 5 g AC/100 mL POME, pH 2, and equilibrium time of 12

[a] Corresponding author: awitan@unimas.my

hours [10]. Therefore, the aim of this study is to investigate the feasibility of acid-washed CSAC to remove colour, TSS and COD from POME through evaluation of the adsorption uptake, kinetics and mechanism.

## 2 Materials and methods

### 2.1 Materials

The POME sample was obtained from the second anaerobic pond of a palm oil mill located at Kota Samarahan, Sarawak, Malaysia. The POME sample was kept in a refrigerator at 4°C. CSAC with mesh size 8x30 and iodine number 1000 was used as the adsorbent to treat POME. Distilled water was utilized for the solutions and reagents preparation.

### 2.2 Batch adsorption studies

In the batch adsorption study, 5 g CSAC was added to 100 mL POME solution in 250 mL closed flask. The sample was stirred at 150 rpm utilizing a rotary orbital shaker over a time duration ranging from 15 minutes to 48 hours at room temperature. For kinetic studies, samples were withdrawn at certain regular times using a glass pipette. The adsorbed quantity of colour, TSS and COD, $q_t$ (mg/g) at time $t$ and $q_e$ (mg/g) at equilibrium stage were determined through Eq. (1) and Eq. (2), respectively:

$$q_t = \frac{(C_0 - C_t)V}{M} \quad (1)$$

$$q_e = \frac{(C_0 - C_e)V}{M} \quad (2)$$

where $C_0$, $C_t$ and $C_e$ (mg/L) are the POME concentrations at initial, certain time interval and equilibrium stage, respectively. $V$ is the POME volume (L) and $M$ is CSAC mass (g).

### 2.3 Adsorption kinetic studies

#### *2.3.1 Kinetic model: Pseudo-first-order*

The pseudo-first-order (PFO) kinetic model is defined according to Eq. (3) [11], which can be shown as a linear expression by Eq. (4) through integration over boundary conditions such as the colour, TSS and COD adsorbed quantity $q = 0$ at $t = 0$ and $q = q_e$ at $t = t$.

$$\frac{dq}{dt} = k_1(q_e - q) \quad (3)$$

$$\ln(q_e - q_t) = \ln q_e - k_1 t \quad (4)$$

where $q_t$ and $q_e$ (mg/g) are the adsorbed quantities of adsorbate at any time, $t$ (h) and equilibrium phase. $k_1$ ($h^{-1}$) is the constant of adsorption rate. $k_1$ and $\ln q_e$ are obtained from the slope and intercept of the graph of ln $(q_e\text{-}q_t)$ versus $t$.

#### *2.3.2 Kinetic model: Pseudo-second-order*

According to Ho et al. [12], the pseudo-second-order (PSO) kinetic model is expressed as Eq. (5). The Eq. (5) can be expressed as Eq. (6) by separating the variables. Further integrating Eq. (6) over boundaries $q = 0$ at $t = 0$ and $q = q_e$ at $t = t$, Eq. (6) turns into Eq. (7) in which $k_2$ (g/mg h) is the second-order adsorption rate constant.

$$\frac{dq}{dt} = k_2(q_e - q)^2 \quad (5)$$

$$\frac{dq}{(q_e - q)^2} = k_2 dt \quad (6)$$

$$\frac{t}{q_t} = \frac{1}{k_2 q^2_e} + \frac{1}{q_e} t \quad (7)$$

The linear graph between $t/q_t$ and $t$ provides the slope $1/q_e$ and corresponding intercept $1/k_2 q^2_e$.

#### *2.3.3 Elovich model*

As expressed by Eq. (8), the Elovich model is defined according to Aharoni et al. [13] in linear form, where $x$ and $y$ are the constants of the Elovich model, which can be obtained from $q_t$ versus ln $t$ graph.

$$q_t = \frac{1}{y}\ln(xy) + \frac{1}{y}\ln t \quad (8)$$

### 2.4 Adsorption mechanism

Weber and Morries [14] defined the intra-particle diffusion (IPD) model as in Eq. (9).

$$q_t = k_{pi} t^{1/2} + C_i \quad (9)$$

where $k_{pi}$ (mg/g $h^{1/2}$) is the diffusion rate constant of stage $i$, obtained from the graph of $q_t$ versus $t^{1/2}$. $C_i$ is the thickness of boundary layer. If the intercept value is higher, the boundary layer effect will be greater [15]. The IPD occurs as a result of linear graph. The rate of adsorption is governed by the IPD if the straight line intersects with the origin point. On the contrary, IPD occurs together with some other mechanisms [16].

The actual slow step involved can be estimated by employing the Boyd model as expressed in Eq. (10) and Eq. (11).

$$B = 1 - (6/\pi^2)\exp(-F_t) \quad (10)$$

$$B = \frac{q_t}{q_0} \quad (11)$$

where $q_0$ (mg/g) is the adsorbed adsorbate quantity at infinite time and $q_t$ (mg/g) is the adsorbed adsorbate quantity at any time $t$ (h), $B$ represents adsorbed solute fraction at any time $t$, and $F_t$ is a mathematical function of $B$. Eq. (10) is simplified by substituting Eq. (11) in Eq. (10) and becomes Eq. (12) for $B$ values that are greater than 0.85. While, for $B$ values that are smaller than 0.85, Eq. (13) is used to calculate $F_t$.

$$F_t = -0.4977 - \ln(1-B) \quad (12)$$

$$F_t = \left(\sqrt{\pi} - \sqrt{\pi - \left(\frac{\pi^2 B}{3}\right)}\right)^2 \quad (13)$$

Referring to Boyd's theory, adsorption process has three steps in sequence, namely film diffusion, particle diffusion, and adsorption on the adsorbent inner surface [17, 18]. The third step cannot be considered as a step that governs the adsorption rate because it is a very fast step. Particle diffusion controls the rate if exterior transport is greater than the inner transport. Film diffusion controls the adsorption rate if exterior transport is lesser than inner transport and the moving of adsorbate molecules to the boundary at a significant rate may not occur if exterior transport is equivalent to that of inner transport.

### 2.5 Error analysis

An error function is required to be determined for evaluating the model suitability to fit the experimental results. Sum of squared errors (SSE, %) as shown in Eq. (14) is used for analysis of error calculation.

$$SSE = \sqrt{\frac{\sum (q_{e,\exp} - q_{e,cal})^2}{N}} \quad (14)$$

where $N$ is the number of collected data, $q_{e,exp}$ and $q_{e,cal}$ (mg/g) are equilibrium adsorption capacity values according to the experimental and theoretical results, respectively. The lower the SSE, the better a fit is.

### 2.6 Analytical methods

The POME colour was analysed according to platinum-cobalt standard method 8025 at 455 nm wavelength by utilizing a spectrophotometer (HACH, DR900) [19]. A 10 mL POME sample was placed in a square quartz sample cell. TSS was determined spectroscopically using the American Public Health Association (APHA) Standard method 8006 through HACH spectrophotometer (DR900) by placing 10 mL POME sample in the same sample cell used for colour analysis [19]. COD was determined by APHA Standard method 8000 using spectroscopic method. A COD vial (HR 20-1500 mg/L) was used for adding 2 mL POME sample. Then, the sample was digested for 2 hours by using a HACH digester (DRB200). The COD concentration was then measured using a HACH spectrophotometer [19].

## 3 Results and discussion

### 3.1 Effect of POME initial concentration and contact time on colour, TSS and COD adsorption

Figures 1(a)-(c) show the adsorption uptake of colour, TSS, and COD onto CSAC at various POME initial concentrations and adsorption times. The plots showed that the adsorption of colour, TSS and COD were generally increased over time with reference to POME initial concentrations ranging 50% to 100%. At the initial stage, the adsorption of colour, TSS, and COD was fast and became slower while approaching equilibrium. The fast adsorption phenomenon could be correlated with the availability of significant number of vacant active surface sites on the CSAC at the beginning process. Over a certain time period, the leftover marginally available vacant surface sites of CSAC had become difficult to be filled up due to significant forces of repulsion present between the pollutants on the CSAC as well as in the POME solution. The amount of colour, TSS and COD adsorbed onto CSAC increased with POME initial concentration. This was because the rising POME initial concentration increased the mass transfer driving force.

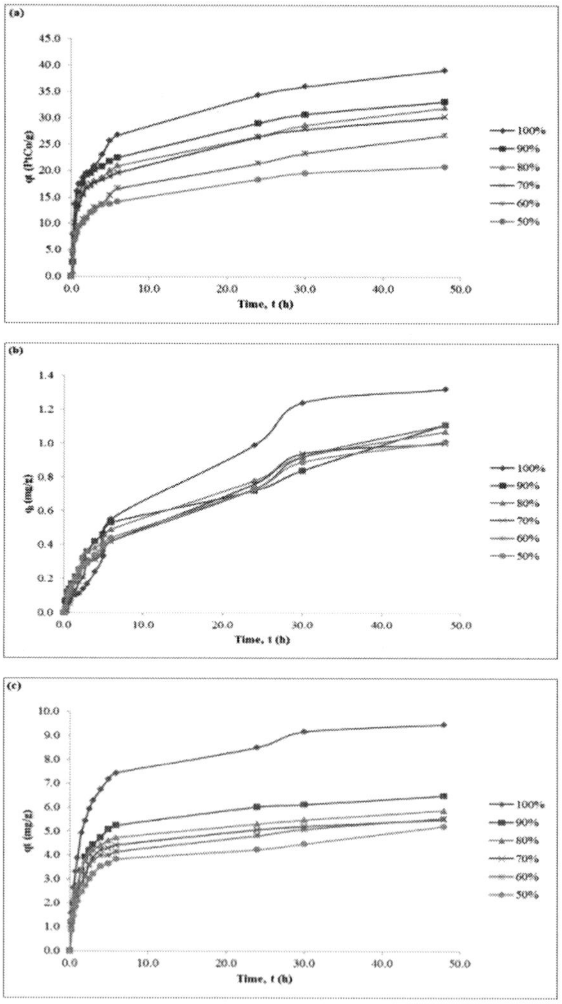


**Figure 1.** Adsoprtion uptake of (a) colour, (b) TSS, and (c) COD onto CSAC.

The percentage reductions of colour, TSS and COD from POME onto CSAC achieved within 48 hours were 61%, 39% and 66%, respectively as indicated in Figure 2.

The operating conditions were pH 8 and 5g CSAC/100 mL POME. In a past study on treatment of biologically treated POME using banana peel activated carbon (BPAC), the reduction efficiency of colour (95.96%), TSS (100%), COD (100%), BOD (97.41%), and tannin and lignin (76.74%) were reported. These results were obtained at optimized process condition of pH 2, contact time of 30 hours and 30 g BPAC/100 mL POME [20]. Another study on treatment of biologically treated POME using commercial AC reported removal efficiency of 28.35%, 29.02%, and 65.9% for colour, TSS and COD, respectively which were achieved using an AC dose of 5 g/100 mL POME and an equilibrium time of 5 hours [19].

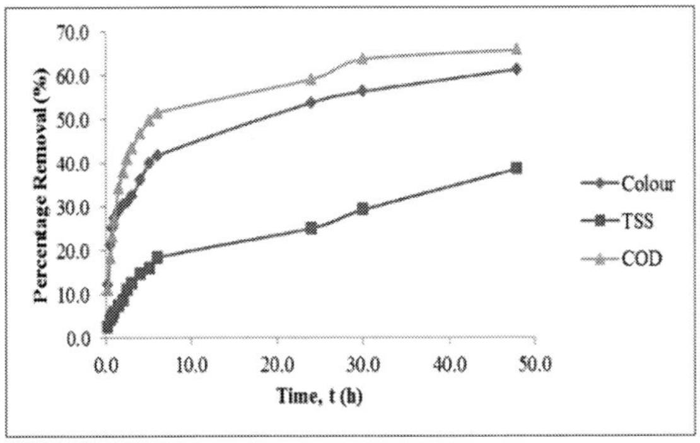


**Figure 2.** Percentage removal of colour, TSS, and COD from POME onto CSAC.

### 3.2 Adsorption kinetics study

The kinetics study of colour, TSS and COD adsorption onto CSAC as well as experimental data correlation analysis were carried by employing PFO, PSO, and Elovich kinetic models. Table 1 shows the kinetic parameters, correlation coefficients ($R^2$) and SSE obtained by using the three models. The $R^2$ and SSE were used to express the agreement between the results as obtained from model analysis and experiments. High $R^2$ and low SSE signified that the kinetic models fitted the adsorption process. For removal of colour and COD, the $R^2$ values of PSO model were higher than the $R^2$ of the other two models. Although the $R^2$ value of the PFO model obtained for adsorption of TSS was higher than the $R^2$ of the other two models, the SSE value of the PFO model was also relatively higher. Therefore, the kinetics of colour, TSS and COD adsorption onto CSAC were better interpreted by the PSO and Elovich models due to the comparable $R^2$ values and SSE values obtained for both models.

Figure 3 shows the comparison of different kinetic model analysis with the experimental results for the adsorption of colour, TSS and COD of POME onto CSAC. The $q_e$ values calculated by using the PSO and Elovich kinetic models were in accordance with $q_e$ of experimental results. This indicated a good correlation between the PSO and Elovich kinetic models with respect to adsorption of colour, TSS and COD of POME onto CSAC. The suitability of the PSO kinetic model indicated that the step that controls the adsorption rate was mainly chemical sorption. The PSO kinetic model also indicated that the colour, TSS and COD adsorption rates were affected more by the number of available adsorption sites on the CSAC surface than the effect of POME solution concentration [21]. The adsorption rate was governed by a second-order rate law. The $k_2$, PSO rate constant, for TSS adsorption was higher than that of colour and COD. The Elovich model also indicated chemisorption. The calculated $q_e$ values from the Elovich model fitted the experimental $q_e$ values well (Figure 3) which indicated that the colour, TSS and COD adsorption onto CSAC was highly heterogeneous system [22].

In a study on the use of improved EFB-based PAC for the reduction of suspended solids and turbidity from bio-treated POME, the adsorption kinetics were also discovered to be better interpreted using the PSO model with $R^2$ value of 0.9998. This meant the sorption trend was due to the chemisorption [8]. Another study on the reduction of colour, TSS, BOD, COD, and lignin and tannin from biologically treated POME using natural banana peel, methylated banana peel and BPAC also revealed that the adsorption kinetics were governed by the PSO model with $R^2$ values of more than 0.99 and very low relative average errors [20]. Besides, a study on adsorption of colour, TSS and COD from biologically treated POME onto AC also expressed that the adsorption kinetics fitted the PSO model well with high $R^2$ and low $\chi^2$ values. The kinetics of colour, TSS and COD adsorption onto the AC also fitted the Elovich model with $R^2$ values of 0.9407-0.9968 and $\chi^2$ of 0.054931-1.919106 [19].

**Table 1.** Kinetic parameter analysis results using PFO, PSO and Elovich models

| Variables | | Colour | TSS | COD |
|---|---|---|---|---|
| $q_{e,exp}$ (mg/g) | | 39.1000 | 1.1100 | 9.4700 |
| PFO model | $q_{e,cal}$ (mg/g) | 29.0291 | 1.0542 | 7.4902 |
| | $k_1$ ($h^{-1}$) | 0.1545 | 0.1026 | 0.2450 |
| | $R^2$ | 0.8600 | 0.9852 | 0.9374 |
| | SSE | 10.4137 | 0.0592 | 2.0270 |
| PSO model | $q_{e,cal}$ (mg/g) | 28.6533 | 0.7775 | 8.8968 |
| | $k_2$ (g/mg h) | 0.0453 | 0.3799 | 0.0910 |
| | $R^2$ | 0.9845 | 0.9306 | 0.9994 |
| | SSE | 1.4269 | 0.0189 | 0.0571 |
| Elovich model | $q_{e,cal}$ (mg/g) | 36.6072 | 0.7649 | 11.5074 |
| | x | 119.3950 | 0.5940 | 15.8458 |
| | y | 0.1912 | 6.9061 | 0.5196 |
| | $R^2$ | 0.9639 | 0.9234 | 0.9962 |
| | SSE | 0.9298 | 0.0383 | 0.1092 |

DOI: 10.1051/matecconf/20178703010

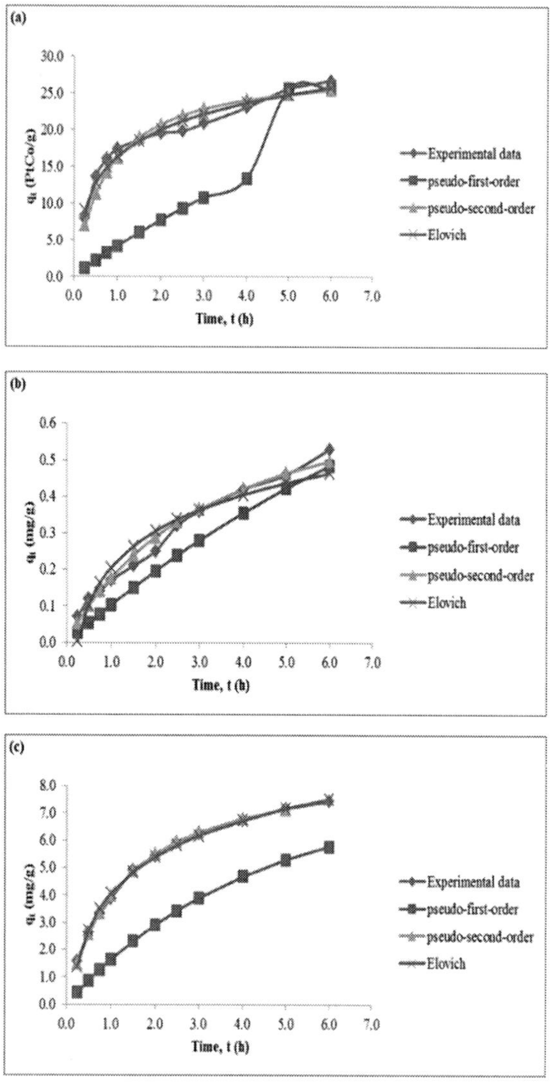


**Figure 3.** Comparison of different kinetic models with experimental data of (a) colour, (b) TSS, and (c) COD adsorption onto CSAC.

## 3.3 Adsorption mechanism

As shown in Figure 4, outer surface adsorption occurred in the first sharper section [23-25]. The second step described the gradual adsorption for the removal of colour, TSS and COD. The rate controlling of the second section was IPD. The linear lines of the second section as shown in Figure 4(a)-(c) did not intersect the origin due to the difference in the initial and final mass transfer rate of adsorption [23-25]. This implied that the rate of adsorption might be controlled by both IPD and film diffusion [24,26-27]. The actual mechanisms that controlled the adsorption rate were further evaluated through Boyd plots as shown in Figure 5. The linear lines for adsorption of colour, TSS and COD onto CSAC did not intersect with the origin point. This confirmed that the colour, TSS and COD adsorption onto CSAC were controlled by both IPD and film diffusion [24].

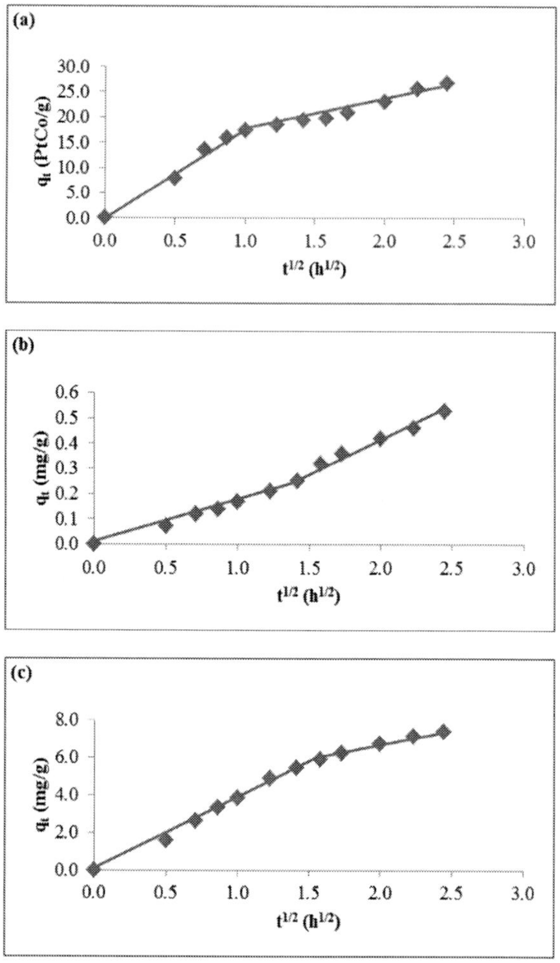


**Figure 4.** IPD plots of (a) colour, (b) TSS and (c) COD adsorption onto CSAC.

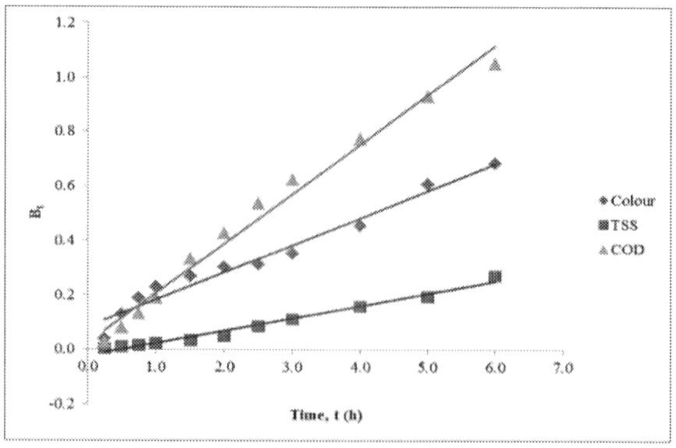


**Figure 5.** Boyd plots of colour, TSS and COD adsorption onto CSAC.

## 4 Conclusion

Acid-washed CSAC was successfully used to remove colour, TSS and COD from POME, with percentage removal of 61%, 39% and 66%, respectively. The adsorption kinetics of colour, TSS and COD onto CSAC were best interpreted by the PSO and Elovich models. The fit of the IPD and Boyd models explained that IPD and film diffusion controlled the adsorption of colour, TSS and COD onto CSAC.

## Acknowledgements

The authors acknowledge Universiti Malaysia Sarawak (UNIMAS) for providing research grant F02/SpGS/1406/16/7 under the Special Grant Scheme and the Ministry of Higher Education Malaysia (MOHE) for providing financial support under MyBrain15.

## References

1. J.D. Bala, J. Lalung, N. Ismail, *Palm oil mill effluent (POME) treatment "microbial communities in an anaerobic digester": A review*, IJSRP **4** (6) (2014) 1-24.
2. T.Y. Wu, A.W. Mohammad, J.Md. Jahim, N. Anuar, *Pollution control technologies for the treatment of palm oil mill effluent (POME) through end-of-pipe processes*, J. Environ. Manage. **91** (7) (2010) 1467-1490.
3. A.N. Ma, A.S.H. Ong, *Pollution control in palm oil mills in Malaysia*, J. Am. Oil Chem. Soc. **62** (2) (1985) 261-266.
4. P.E. Poh, M.F. Chong, *Development of anaerobic digestion methods for palm oil mill effluent (POME) treatment*, Bioresour. Technol. **100** (1) (2009) 1-9.
5. F. Schuchardt, K. Wulfert, Darnoko, T. Herawan, *Effect of new palm oil mill processes on the EFB and POME utilization*, Journal of Oil Palm Research Special Issue (2008) 115-126.
6. Y. Shirai, M. Wakisaka, S. Yacob, M.A. Hassan, S. Suzuki, *Reduction of methane released from palm oil mill lagoon in Malaysia and its countermeasures,* Mitig. Adapt. Strategies Glob. Chang. **8** (3) (2003) 237-252.
7. L. Milich, *The role of methane in global warming: Where might mitigation strategies be focused*, Glob. Environ. Chang. **9** (3) (1999) 179-201.
8. M.K. Amosa, M.S. Jami, M.F.R. Alkhatib, T. Tajari, D.N. Jimat, R. U. Owolabi, *Turbidity and suspended solids removal from high-strength wastewater using high surface area adsorbent: Mechanistic pathway and statistical analysis*, Cogent Eng. **3** (1) (2016) 1-18.
9. S. Syafalni, A. Ismail, I. Dahlan, K.W. Chan, *Treatment of dye wastewater using granular activated carbon and zeolite filter*, Modern Applied Science **6** (2) (2012) 37-51.
10. R.R. Mohammed, *Decolorisation of biologically treated palm oil mill effluent (POME) using adsorption technique*, IRJES **2** (10) (2013) 1-11.
11. S. Langergren, B.K. Svenska, *Zur theorie der sogenannten adsorption geloester stoffe*, Veternskapsakad Handlingar **24** (4) (1898) 1–39.
12. Y.S. Ho, G. McKay, *The kinetics of sorption of basic dyes from aqueous solutions by sphagnum moss peat*, Can. J. Chem. Eng. **76** (1998) 822–826.
13. C. Aharoni, F.C. Tompkins, in: D.D. Eley, H. Pines, P.B. Weisz (Eds.), *Advances in Catalysis and Related Subjects vol. 21* (Academic Press, New York, 1970).
14. W.J. Weber, J.C. Morris, *Proc. Int. Conf. Water pollution symposium vol.2*, pp.231–266 (Pergamon, Oxford, 1962).
15. D. Kavitha, C. Namasivayam, *Experimental and kinetic studies on methylene blue adsorption by coir pith carbon*, Bioresour. Technol. **98** (2007) 14–21.
16. H.K. Boparai, M. Joseph, D.M. O'Carroll, *Kinetics and thermodynamics of cadmium ion removal by adsorption onto nanozerovalent iron particles*, J. Hazard. Mater. **186** (2011) 458–465.
17. G.E. Boyd, A.W. Adamson, L.S. Meyers, *The exchange adsorption of ions from aqueous solution by organic zeolites. II. Kinetics*, J. Am. Chem. Soc. **69** (1947) 2836–2848.
18. D. Reichenberg, *Properties of ion exchange resins in relation to their structure. III. Kinetics of exchange*, J. Am. Chem. Soc. **75** (1953) 589–597.
19. R.R. Mohammed, M.R. Ketabchi, G. McKay, *Combined magnetic field and adsorption process for treatment of biologically treated palm oil mill effluent (POME)*, Chem. Eng. J. **243** (2014) 31-42.
20. R.R. Mohammed, M.F. Chong, *Treatment and decolorization of biologically treated Palm Oil Mill Effluent (POME) using banana peel as novel biosorbent*, J. Environ. Manage. **132** (2014) 237-249.
21. J. Gao, D. Kong, Y. Wang, J. Wu, S. Sun, P. Xu, *Production of mesoporous activated carbon from tea fruit peel residues and its evaluation of methylene blue removal from aqueous solutions*, BioResour. **8** (2013) 2145–2160.
22. Y.S. Ho, G. McKay, *Application of kinetic models to the sorption of copper (II) on to peat*, Adsorpt. Sci. Technol. **20** (2002) 797–815.
23. R. M. Ali, H. A. Hamad, M. M. Hussein, G. F. Malash, *Potential of using green adsorbent of heavy metal removal from aqueous solutions: Adsorption kinetics, isotherm, thermodynamic, mechanism and economic analysis,* Ecol. Eng. **91** (2016) 317-332.
24. H. Wang, X. Yuan, Y. Wu, G. Zeng, H. Dong, X. Chen, L. Leng. Z. Wu, L. Peng, *In situ synthesis of $In_2S_3$@MIL-125(Ti) core-shell microparticle for the removal of tetracycline from wastewater by integrated adsorption and visible-light-driven photocatalysis,* Appl. Catal., B: Environ. **186** (2016) 19-29.
25. T. Anitha, P. S. Kumar, K. S. Kumar, B. Ramkumar, S. Ramaligam, *Adsorptive removal of Pb(II) ions from polluted water by newly synthesized chitosan-polyacrylonitrile blend: Equilibrium, kinetic, mechanism and*

*thermodynamic approach,* Process Saf. Environ. Prot. **98** (2015) 187-197.

26. S. Duan, R. Tang, Z. Xue, X. Zhang, Y. Zhao, W. Zhang, J. Zhang, B. Wang, S. Zeng, D. Sun, *Effective removal of Pb(II) using magnetic* $Co_{0.6}Fe_{2.4}O_4$ *micro-particles as the adsorbent: Synthesis and study on the kinetic and thermodynamic behaviors for its adsorption,* Colloids Surf., A: Physicochem Eng. Asp. **469** (2015) 211-223.
27. S. Nethaji, A. Sivasamy, A. B. Mandal, *Adsorption isotherms, kinetics and mechanism for the adsorption of cationic and anionic dyes onto carbonaceous particles prepared from Juglans regia shell biomass,* Int. J. Environ. Sci. Technol. **10** (2013) 231-242.

 DOI: 10.1051/matecconf/20178704001

# A Comparison Study of Students' Performance in Pre and Post Result of A Mathematics Competency Test

Najiyah Safwa Khashi'ie[1,a], Rahaini Mohd Said[1], Nurul Amira Zainal[1] and Nor Hamizah Miswan[1]

[1] *Faculty of Engineering Technology, Universiti Teknikal Malaysia Melaka, Hang Tuah Jaya, 76100 Durian Tunggal, Melaka, Malaysia*

**Abstract.** The objective of this study was to compare the performance of Engineering Technology (ET) students' in a Pre and Post Mathematics Competency Test during their first year of study at the Faculty of Engineering Technology (FTK), Universiti Teknikal Malaysia Melaka (UTeM). This article also aims to examine the student's understanding in the area of Mathematics, particularly in Algebra, Trigonometry and Functions. A total of 176 first year students from the Department of Mechanical Engineering Technology (JTKM) were chosen to answer the Mathematics Competency Test. The pre-test was held in the first week of the semester, while the post-test was held during the second semester. The results showed that the students' performance in post-test was better compared to that pre-test. However, statistical analysis on students' performance by each question showed that most of the students did not understand the basic concepts in Algebra, Trigonometry and Functions.

## 1 Introduction

These days, it is important to increase the number and competency of students in science, technology, engineering and mathematics (STEM) for future career development. The Faculty of Engineering Technology (FTK), Universiti Teknikal Malaysia Melaka (UTeM) is one of the universities in the Malaysian Technical University Network (MTUN) which provide Engineering Technology Programs (ETP) [1]. This program is oriented towards applications whereby more practical and hands-on activities are assessed, and introductory mathematics courses, science courses and qualitative introductory to engineering fundamentals are provided. One of the important subjects that support and relates to ETP is mathematics [1, 2]. The aims of this paper are to compare the performance of ET students while learning Mathematics in their first year of study at FTK, UTeM and examine their understanding in the area of Mathematics, particularly in the topics of Algebra, Trigonometry and Functions.

Students frequently encounter problems in learning mathematics involving calculations and understanding the concept and method used [2]. According to [3], factors affecting difficulties in learning and understanding Mathematics are psychological incidences like fear and lack of confidence, the curriculum of mathematics itself which fails to show much relevance in real life applications and the way academicians or mathematicians attract students to learn mathematics. It is believed that a good mathematical background is a prerequisite in understanding engineering concepts, especially for engineering and engineering technology students. The lack of understanding in basic mathematical principles can result in the inability to solve numerous subjects which require mathematics [4]. Furthermore, Mathematics has always been perceived as a difficult and demanding subject [5]. Difficulties in mathematics are often caused by lacking of mathematics skills [6]. The understanding of the knowledge, skills and commitment of teachers are keys to succeed in mathematics [7]. The preconceived notion that Mathematics is difficult might have affected the students' performance in mathematics. Coupled with this is the fact that mathematics anxiety is brought about by low self-esteem and the fear for failure. Perhaps these negative preoccupations have adversely affected the students' performance.

Academic performance is the ability to study and remember facts and the capability to communicate acquired knowledge verbally or on paper, both of which are commonly measured by means of grades [8]. Algebra, Trigonometry and Function are few examples of broad parts in mathematics. Although these topics are basics and fundamental, but they are important especially for ET students who will apply in other subjects such as Physics, Chemistry, Thermodynamics and Fluid Mechanics. According to a previous study by [8], students who took the Algebra and Trigonometry subjects for two academic years were not able to achieve the required criteria. Furthermore, studying technique was a common factor affecting the performance of the students in mathematics. A paper by [9] compared students' performance in a mathematics test and the circuit analysis course. The mathematics test consisted of several prerequisite topics in circuit analysis such as basic algebra, complex numbers, integral and differential calculus, and was held in the beginning and end of the semester. Results based on the mathematics and electric circuit tests were

[a] Corresponding author: najiyah@utem.edu.my

DOI: 10.1051/matecconf/20178704001

compared and showed that there were correlations between both subjects. In this study, Algebra Trigonometry and Function topics were chosen as the basis for research evaluation and assessment benchmark since these parts of mathematics are useful and relevant for engineers and technologists.

## 2 Methodology

This study was conducted to compare the performance of ET students while learning mathematics during the first year of study at FTK, UTeM. The pre-test for the Mathematics Competency test was held in the first week of the semester, while the post-test was held in the second semester. The Mathematics Competency Test consisted of 40 questions; 20 questions (No 1-20) focused on the topics of Algebra, Trigonometry, Solving Equation and Function, while 20 questions (21-40) focused on the topics of Derivation and Integration.

Data were collected from a total of 176 first year students from the Department of Mechanical Engineering Technology (JTKM) as samples for the entire population who took the competency test in the first year. A summary of students' demographics is shown in Table 1; 76 students were studying for Bachelor's Degree in Mechanical Engineering Technology (Automotive Technology), 38 students were studying for Bachelor's Degree in Mechanical Engineering Technology (Refrigeration and Air-Conditioning Systems) and 62 students were studying for Bachelor's Degree in Mechanical Engineering Technology (Maintenance Technology). Table 2 exhibits the contents description in the Mathematics Competency Test (Questions 1-20). There are 30% questions on Algebra and solving equation and 20% questions on Trigonometric and Function.

**Table 1.** Total of First Year Students in Department of Mechanical Engineering Technology by Gender and Programs.

| Department | Number of Students | | |
|---|---|---|---|
| | **Male** | **Female** | **Total** |
| Bachelor's Degree in Mechanical Engineering Technology (Automotive Technology) | 62 | 14 | 76 |
| Bachelor's Degree in Mechanical Engineering Technology (Refrigeration and Air-Conditioning Systems) | 30 | 8 | 38 |
| Bachelor's Degree in Mechanical Engineering Technology (Maintenance Technology) | 52 | 10 | 62 |

**Table 2.** Description of content in Mathematics Competency Test (Questions 1-20).

| Content | Question No | No. of Item | Weightage |
|---|---|---|---|
| Algebraic Expression | 1,2,3,4,5,11 | 6 | 30 |
| Trigonometry | 13,14,15,16 | 4 | 20 |
| Solving Equation | 6,7,8,9,10,12 | 6 | 30 |
| Function and Domain | 17,18,19,20 | 4 | 20 |

## 3 Results And Discussion

### 3.1 Overall Result

Figure 1 shows a comparison of students' performance in the Pre and Post Competency Test (Question 1-20) by program in JTKM. Result show that there was only a slight increment between the pre- and post-tests whereby less than 50% of students could understand and answer the questions correctly.

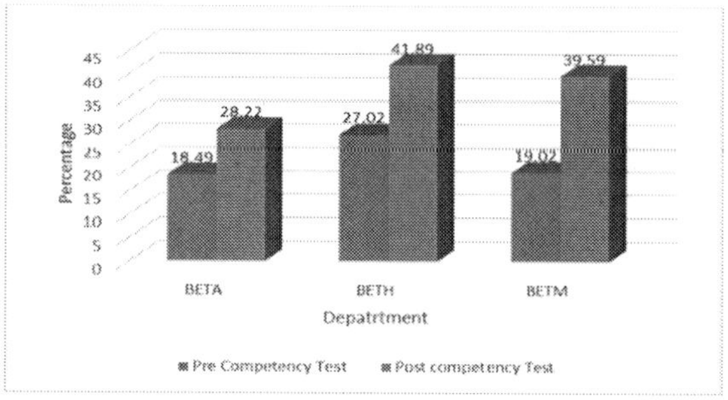


**Figure 1.** Students' performance in the Pre and Post Competency Test (Question 1-20) by program in JTKM

Figure 2 illustrates the percentage of JTKM students' performance for pre and post competency test by question and content. The graph shows changes in the percentage of students' obtaining correct answers between the pre- and post-tests. The graph also shows that more than 40% of students understood and correctly answered Question 1 (algebraic expression), Question 7 (solving equation), Question 10 (solving equation) and Question 14 (trigonometry), while less than 10% correctly answered Question 6 (solving equation), Question 19 (function and domain) and Question 20 (function and domain). Based on this graph, it can be seen that students had lack of understanding and knowledge in the topics of algebraic expression, trigonometry, solving equation and function since the percentage of students who were able to answer the question correctly was low. There were many factors affecting these results, such as the different backgrounds of study before entering FTK and psychological incidences like mathematics anxiety.

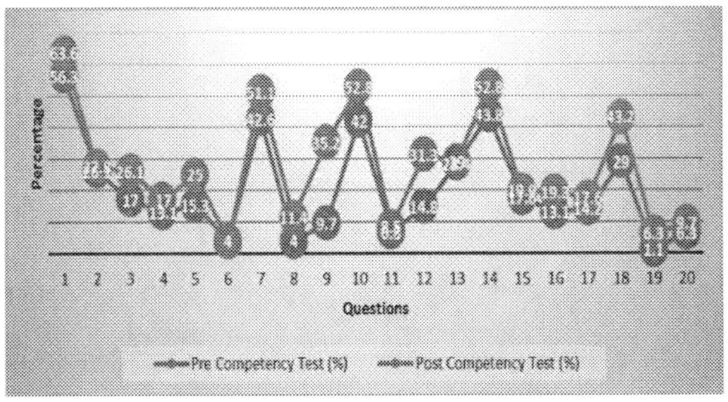


**Figure 2.** Percentage of JTKM students' performance for pre- and post- competency test by question and content

**Table 3.** Paired Samples Statistics : Means.

| | Mean | Number of samples | Standard Deviation |
|---|---|---|---|
| Pre-Test | 4.05 | 176 | 3.243 |
| Post-Test | 6.94 | 176 | 3.607 |

**Table 4.** Paired Samples Statistics : Correlation.

| | r | Sig. Value |
|---|---|---|
| Pre-Test – Post-Test | 0.641 | 0.000 |

From the results shown in Table 3, the mean for post-competency test score was approximately higher by 7 than that of pre competency test, while the Pearson correlation between pre- and post-tests (Table 4) shows moderate and positive correlation with the value of r = 0.641, with a significant value of $\rho < 0.05$. This indicates that the students' performance in post-test was better than pre-test after taking the Mathematics course in FTK for one semester.

### 3.2 Analysis of Variance, ANOVA

ANOVA is a statistical method used to test the significant difference among means. ANOVA is also a hypothesis test that consists of a null hypothesis ($H_0$) and an alternative hypothesis ($H_1$). If the $\rho$ value is greater than the critical value, which is 0.05, then the null hypothesis ($H_0$) is rejected and the alternative hypothesis ($H_1$) is accepted. In this article, ANOVA is conducted at 95% confidence levels to test the significant difference in students' performance after taking the pre- and post-competency test. ANOVA is employed to test the following hypothesis:

$H_0$: There are no significant difference between pre competency test and post competency test.
$H_1$: There are significant difference between pre competency test and post competency test.

Table 5 below shows the critical value, $\rho$ less than 0.05 which was 0.000. Thus the null hypothesis ($H_0$) is rejected, which also implies that there were significant differences between pre competency test and post competency test.

**Table 5.** Paired T-Test for Pre and Post Test

| | Mean | Standard Deviation | P value |
|---|---|---|---|
| Pre Test-Post Test | 0.14 | 0.1460 | 0.000 |

## 4 Conclusion

The focus of this study was to determine the relative relationship in student's achievement in Mathematics Competency Test between 'before' (early of semester) and 'after' (in second semester). There were four components in the test, which were Algebraic Expression, Solving Equation, Trigonometric Function and lastly Function and Domain. The results showsed there were slightly significant difference between 'pre' and 'post' tests. However still, the percentage of students getting the right answers was low.

For algebraic expression component, knowledge on factorization, indices and inequalities are really important to solve a problem. Based on the findings, we realized that students failed to express and simplify the given question in the simplest way. Students mostly repeated the same error in pre- and post-tests probably because they misunderstood algebraical concepts as mentioned above. The second component involved solving equations. The questions wanted the student to solve giving equation and to find the values (unknown) that fulfil a condition stated depending on the problem. The percentage of students getting right answer was still below than 50%. Ultimately, the result was lower than the first touch component. However the percentage of getting the right answers was higher when comparing pre- and post- results for each question. This indicates that students were able to gain more conceptual understanding of mathematics by the end of the semester. This also shows that students are improving their skills and abilities in solving mathematics problems.

As for trigonometric questions, the results showed no significance different for each question, even though the percentage of the post- result was higher. To simplify this, there are two main reasons as to why this could happened. Firstly, students completely failed to understand the questions, and secondly, students did not use the trigonometric function very often in daily life. Therefore, lack of practise definitely could have led to poor skills or bad performance as shown in the findings. The last component was Function. Findings showed that more than half of the samples answered the questions incorrectly for this topic, both in pre- and post-tests. There are several reasons that could have triggered this event. For example, the inability to understand the relation between domain and range in a graph, as well as misunderstanding the basic concepts in Mathematics.

In a nutshell, engagement of preconceptions in Mathematics is important in order to understand and solve Mathematical problems. If this does not happened, students will not achieve the full comprehension of the new material related to the previous concepts. Worst, students may reject all new information or struggle to implement the wrong concepts, resulting in a misunderstanding of the contents. In conclusion, eliminating students' misconceptions is a crucial goal in this new era.

The authors would like to acknowledge Faculty of Engineering Technology and Universiti Teknikal Malaysia Melaka (UTeM) for financial support under the UTeM short term grant PJP/2015/FTK(15B)/S01427.

## References

1. N. S. Khashi'ie, K. Hamzah, I. Waini, F. Salim, M. F. Mukhtar and N. A. Zainal. International Journal of Science and Research (IJSR) **3**(10), 1017-1019 (2014)
2. K. Hamzah, N. S. Khashi'ie, N. H. Hussin, M. F. Wahid, C. W. M. Che Wan Mohd Zalani and S. K. Sayed Nordin, International Journal of Humanities and Social Science (IJHSSI) Volume **3** Issue 3, 25-28 (2014)
3. A. Nordin, Prosiding Simposium Kebangsaan Sains dan Matematik ke V, Universiti Teknologi Malaysia, 20-22 (1992)
4. M. Bursal and L. Paznokas, School Science and Mathematics **106**(4), 173-180 (2006)
5. M. H. Ashcraft and K. S. Ridley, Am. Psychol. Soc., 181-185 (2002)
6. S. F. Mokhtar, Z. M. Yusof and M. Misiran, Journal of Applied Sciences Research **8**(8), 4133-4137 (2012)
7. V. Suthar, R. A. Tarmizi, H. Midi and M. B. Adam, Proceedings of the 4th International Conference on Research and Education in Mathematics, 171-178 (2009)
8. B. L. Dinglasan and A. D. C. Patena, Alberta School Of Business, Research Paper No. 2013-**1308** (2013)
9. J. Jalkio, ASEE Annu. Conf. Expo, 14.15.1-14.15.7 (2009)

# FTK Students' Performance in Mathematics: Comparison between SPM and First Year Exam

Fadzilah Salim[1,a], Aminah Ahmad[1], Iskandar Waini[1] and Nor Hamizah Miswan[1]

[1]*FTK, Universiti Teknikal Malaysia Melaka, Hang Tuah Jaya, 76100 Durian Tunggal, Melaka, Malaysia*

**Abstract.** In the last 20 years, mathematics teaching and learning encounters quite a big problem, especially at the tertiary level. The main concern always surrounds the students' achievement in the subject matter. Students' performance in mathematics at first year is reflected by the students' mathematical background prior to the admittance into the university. The study examined 165 first year students in the Faculty of Engineering Technology (FTK) who took the Mathematics Competency test upon entering the university at the beginning of their first semester. A test consisted of 40 fundamental mathematical questions which students have learned them before. From the result, 84% of these students failed this test. However, looking at their Sijil Pelajaran Malaysia (SPM) mathematics result during Form Five ($12^{th}$ grade) in school, the majority did quite well in that exam. These students also took a first year mathematics course which is Technical Mathematics at the same semester. At the end of the semester, the result of their Technical Mathematics course seemed to be quite good. The performance of these three mathematics results was being compared and studied.

## 1 Introduction

Mathematics is a very important subject matter. It is used in our daily lives as well as it is taught at all academic institutions in the world; kindergartens, primary schools, secondary schools, pre-university colleges as well as at higher learning institutions. Even at home, among the first thing that is taught to a baby or a toddler is a counting number. However, since the last two decades, Mathematics teaching and learning seemed to face a lot of big challenges. Nowadays, the same scenario is happening to a lot of institutions of higher learning, be it in Malaysia or outside this country, the students' performance in this subject matter is being raised and questioned. This really the biggest concern, especially during the first year at the university [1].

There has been much research being studied on factors that affect students' performance in Mathematics at the tertiary level [2,3,4]. However, mathematical background of students entering the university, perhaps, is one of the key problems faced by the educators [1]. In a report released about the Trends in International Mathematics and Science Study (TIMSS) in 2011 among Grade 8 students in 42 participating countries, Malaysia's ranking in Mathematics fell from 16th (1999) to 10th (2003), 20th (2007) and 26th (2011). Similar patterns can also be seen in some other countries. This is very alarming

Studies made in the United States and Australia, both pointing to the instructors, classrooms and education environment for being among the factors affecting students' achievements in mathematics [5] beside other factors. However, it is undeniable that students' prior knowledge and academic achievement which result from students' prior cognitive mathematical ability influence very strongly in their mathematics achievements [6]. Students' secondary school mathematics background, not their university entrance score, has a spectacular consequence on pass rates [1]. Research shows that students hold remarkable perceptions regarding the knowledge of mathematics and science from the secondary school, that can influence considerably on the achievement of the mathematics courses at the university [7].

Mathematics can also be somewhat difficult to learn at times. It involves in applying formulae, using measurements, writing out phases of calculations, writing numbers, and spatial perception problems. A lot of understanding of the basic mathematical concepts and knowledge is required at the early stage of mathematical education. Students may easily get bored and dislike mathematics if they could not solve mathematical problems. However, the study made by Thien and Ong [8] found that Malaysian students loved studying Mathematics and they believed that mathematics is necessary for their future careers despite of their mathematics competency.

For this reason, mathematics educators play a very important role in ensuring students acquired a good grasp of the basic knowledge. What constitutes good mathematics teaching are having a very experienced educator who possesses comprehensive knowledge of

[a] Corresponding author: fadzilah@utem.edu.my

 DOI: 10.1051/matecconf/20178704002

mathematics, fosters mathematical comprehension, connects and inspires students, and entails successful management competence in mathematics [9].

## 2 Methodology

This study started with the collection of relevant information to investigate students' mathematics performance in SPM (Sijil Pelajaran Malaysia) examination, first year Mathematics Competency test and first year Technical Mathematics exam. The population was targeted to the first year university students in a Technical University in Malaysia, specifically at the Faculty of Engineering Technology or *Fakulti Teknologi Kejuruteraan* (FTK). In this study, the stratified sampling was used to obtain the sample. The sample consisted of 165 first year technical students enrolled in semester 1, 2015/2016 academic year.

The data were collected from both secondary school and university sources. Data on students' secondary school were obtained from the student's past record. The students were asked to key in their results in google form. Meanwhile, the Mathematics Competency test results were obtained from the respective lecturers who had compiled the results previously, and first year Technical Mathematics exam results were extracted from the university database system.

The results for SPM mathematics were taken from Modern Mathematics exam, which is a compulsory subject taken by all SPM candidates. The topics learnt in this subject includes basic Statistics, Trigonometry and Algebra. The students' mathematics grades collected were based on a grade scale, ranging from A to F.

The Mathematics Competency test is a test taken by all first year university students to gauge their mathematics competency based on their prior knowledge. The main objective is to ensure that the students can continue their engineering or engineering technology studies smoothly because mathematics is the main foundation of this field. The test was executed at the beginning of the first semester first year, upon entering into the university. The test consisted of 40 questions to be answered in 90 minutes. There were 2 sections in the question paper. 50% of the questions were a combination of topics like basic functions, algebra and trigonometry, while another 50% of the questions were on calculus, such as basic derivatives and integrations. Students were not allowed to use any calculators during the test.

Technical Mathematics course, on the other hand, is the first out of four compulsory mathematics courses offered in the engineering technology faculty. The main course contents cover topics like algebra and trigonometry. The course is taught for 14 weeks of instructions with a breakdown of 28 hours lectures and 22 hours tutorials. The course work consists of 3 assignments worth 30%, 1 mid-semester test worths 30% and 1 final exam worths 40%, in which altogether make up to a total of 100% of the course marks. The result used in this study was the final exam results, which is the 100% total, obtained from the university database system.

## 3 Results and Discussion

The overall results of mathematics achievements are shown in Table 1 below.

**Table 1.** Percentage of Mathematics result performance.

| Grade | SPM (%) | Mathematics Competency Test (%) | Technical Mathematics course (%) |
|---|---|---|---|
| A+ | 43.75 | 0.00 | 0.00 |
| A | 40.63 | 0.00 | 34.38 |
| A- | 3.13 | 0.00 | 9.38 |
| B+ | 9.38 | 0.00 | 9.38 |
| B | 0.00 | 0.00 | 15.63 |
| B- | 0.00 | 0.00 | 3.13 |
| C+ | 0.00 | 15.63 | 3.13 |
| C | 3.13 | 0.00 | 12.50 |
| C- | 0.00 | 0.00 | 3.13 |
| D+ | 0.00 | 0.00 | 3.13 |
| D | 0.00 | 0.00 | 6.25 |
| E | 0.00 | 84.38 | 0 |

As depicted in Table 1 above, the highest percentage of mathematics performances for three different categories are an A+ grade for the SPM Modern Mathematics exam (43.75%), an E grade for the first year Mathematics Competency test (84.38%), and an A grade for the first year Technical Mathematics exam (34.38%). In addition, the grading range of SPM result was from A+ to B+ and only a few students obtained a C grade. Based on the SPM result, we can conclude that students prior knowledge of mathematics during secondary school are very good. Surprisingly, Mathematics Competency test results showed otherwise. In fact, there were only two types of grades obtained by these students, that is, a C+ grade and an E grade. It shows that the majority of these students did very badly upon coming to the university. Approximately 84% of the students failed this test (E grade). However, at the end of the semester, all of them passed their first mathematics course which is Technical Mathematics. Their results were considered quite good.

One of the reasons why these results turned out the way they were, perhaps, the nature of the SPM exam is that the two years duration of mathematics learnt during Form Four and Form Five ($12^{th}$ grade) will be tested at the end of the two years of study. During this period, students had enough time to fully understand the subject matter. In other word, they would be able to make a good preparation for the SPM examination. Usually, students will be asked to do a lot of questions, exercises, and related problems which can also be the mathematics past years' questions, two or three months prior to this examination. Therefore, students were expected to do well in the SPM examination.

Basically, after the SPM results came out, students would be enrolled in either a pre-university or matriculation programme or Diploma programme. Most of the students entered their first year of a bachelor's degree programme in the university two or three years after the SPM examination. The Mathematics Competency test was carried out upon their admission to the university, that is, first year of a bachelor's degree programme, to see how well they could answer the fundamental questions of mathematics that they have learnt before. In other words, the university anticipates that the students should possess a certain level of mathematical knowledge prior to pursue their degree programmes. This is to prepare them for the mathematics and engineering technology courses at the degree level, so that, should there be any problems arise, some measures can be taken to help them to proceed smoothly and successfully in their studies. Unexpectedly, the majority of them failed this test, that is, 84% obtained E grades. Only a few of them, about 16% obtained C+ grades. The feedback received was the majority of the students had forgotten what they had learnt before, even though the majority obtained good results in the SPM mathematics examination.

As for the Technical Mathematics results, this is quite a normal result obtained for any courses studied in a particular semester. In fact, this result can be considered quite a good achievement. This course runs every week for 14 weeks with a few assignments and one midterm test to help the students master the subject matter prior to the final examination. The breakdown assessments for the course is 60% course work, while final exam only carries 40% of the total course. By the time the students sat for the final examination, the knowledge of what they have been taught in that semester were still not forgotten. Thus, like any other courses, the final results normally would have a range of A to D. In this case, no one failed this course.

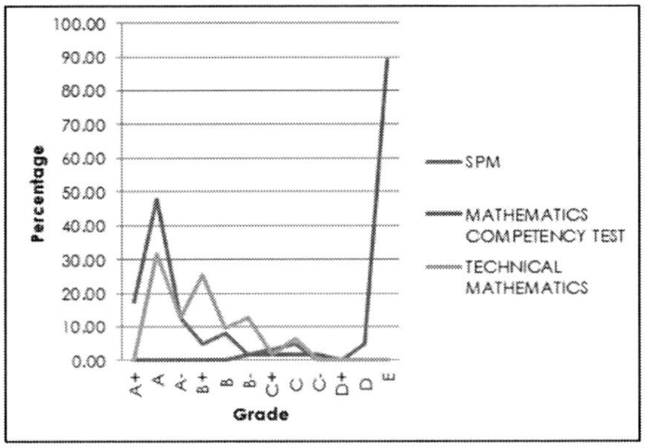


**Figure 1.** Line graph for mathematics performance comparison between SPM, Competency test and Technical Mathematics.

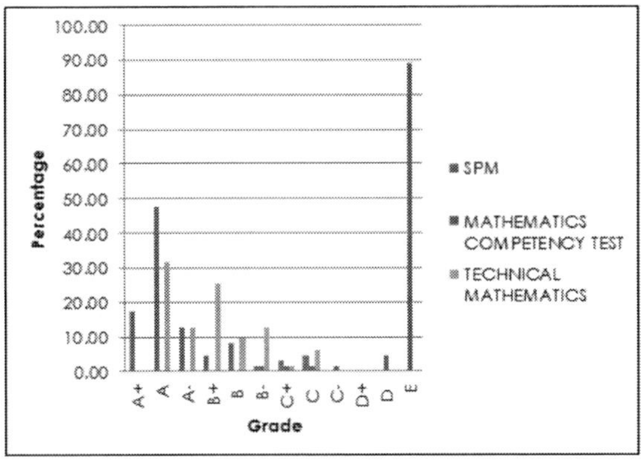


**Figure 2.** Histogram for mathematics performance comparison between SPM, Competency test and Technical Mathematics.

Figure 1 and Figure 2 shows the line graph and histogram of the mathematics performance in three categories respectively. The patterns show a very significant abnormalities in the first year Mathematics Competency test where the graph is skewed to the left with the line or bar to the further right shoot up very high. However, an approximate normal pattern was found for Technical Mathematics subject where the exam results were distributed between A to D grades.

The results also show a very significant difference between the Mathematics Competency test and SPM results. Most of the first year students obtained an E grade in their Mathematics Competency test and it was contradict to their SPM results. In SPM, most of them obtained A+ grade. One of the reasons was that the students were trained to answer a lot of past years' SPM questions for a few months, right before the SPM examination. Their minds and memory were not being interfered by other stuff which might cause the recollection of memories of what recently taught would be gone [10], right before the examination. In contrast, the Mathematics Competency test was carried out upon entering the university which is after students had quite a long break in between. After the SPM examination, some students might enter a pre-university courses such as Matriculation program or STPM (Sijil Tinggi Persekolahan Malaysia) for two years, or Diploma program for three years before entering the university, still there are a few months gap before they enter the university. That is the main reason why the majority of the students failed the Mathematics Competency test which is due to the lost of short term memory of the mathematical knowledge learnt before. In other words, they have forgotten what they have been taught before.

**Table 2.** Mean and standard deviation of mathematics performance.

| | SPM | Mathematics Competency Test | Technical Mathematics course |
|---|---|---|---|
| Mean | 3.62 | 0.23 | 3.04 |

DOI: 10.1051/matecconf/20178704002

| | | | |
|---|---|---|---|
| Standard Deviation | 5.71 | 11.25 | 4.26 |

Table 2 represents the mean and standard deviation of mathematics achievements based on SPM, first year Mathematics Competency test and first year Technical Mathematics examination results. These mean and standard deviation were calculated based on the grade point score for every grade in the grading system as shown in Table 3 below. From Table 2, the mean for SPM, first year Mathematics Competency test and first year Technical Mathematics exam results were 3.62, 0.23 and 3.04, respectively. For SPM result, the mean score suggests that the majority of the students obtained a range of A- to B+ grade, which is a very good achievement.

Meanwhile, the mean score of the Mathematics Competency test result suggests that the majority of the students failed this test because the value of the mean score would reflect to the range of D to E grade, which is a very bad result. However, the mean score for the Technical Mathematics result suggests that the majority was in the range of B to B+, which shows a significant improvement after they have started to learn again the mathematics course, doing all the assignments and other course work, and sitting for the final examination at the end of the semester.

**Table 3.** A Standard grading system used in university in Malaysia.

| Grade | Point |
|---|---|
| A+ | 4.00 |
| A | 4.00 |
| A- | 3.67 |
| B+ | 3.33 |
| B | 3.00 |
| B- | 2.67 |
| C+ | 2.33 |
| C | 2.00 |
| C- | 1.67 |
| D+ | 1.33 |
| D | 1.00 |
| E | 0.00 |

## 4 Conclusion

Although having a sound mathematical knowledge at the secondary school is very crucial, other factors would also contribute to the successful achievement of students in mathematics and other mathematics related courses at the tertiary level. The main conclusion of this study is that based on the result of the Mathematics Competency test which was carried out upon admittance to the university, the majority of the first year students did not acquire sufficient knowledge in their basic mathematics during admission to the university. Therefore, the first impression to the mathematics educators at the tertiary level is that the basic mathematical knowledge of the majority of the students when they first enter the university are very poor. Thus, it is anticipated that the students may face some degree of difficulties in their future study in the university.

However, after investigating their SPM results, the perception of math educators towards students' background knowledge of mathematics contradicts with the Mathematics Competency test results. The majority of the first year students who entered the faculty of engineering technology possessed a very good result in Mathematics during their SPM examination. Thus, most of mathematics educators at the tertiary level have always questioned the validity and the reliability of the SPM results. Not to mention, some might wonder the way the mathematical subjects were being taught at schools. Then again, the results for Technical Mathematics course taken in the same semester shows that students' knowledge in mathematics has improved quite drastically.

The contradiction might happen because the students did not feel the essence of learning. Essence-based learning is unique in the way that it simultaneously addresses and triggers many important aspects of the learner's essential self: thinking, memory, emotions, ethics, aesthetics, the physical, and the senses. These factors personally and holistically engage the learner. All the factors must come together to produce great learner [11].

According to one of the five aspirations in Malaysian Education Blueprint (Higher Education) 2015-2025 [12], the government is very serious to improve the quality of education. Malaysia aimed to become the top 3$^{rd}$ of countries in international assessment such as TIMSS and PISA (Program for International Student Assessment) in 15 years of time. The STEM (Science, Technology, Engineering and Mathematics) education has been emphasized. The national examination and assessments have been revised. The focus is now more on the level of higher-order thinking skills in STEM related subjects.

As a suggestion to the future study, there are still many areas that can be explored or researched further. One of them is the bridging knowledge that the students should acquire after SPM and before coming to the university. As mentioned before, most of the students will undergo a matriculation or pre-university programme before entering into the university. Thus, it will be interesting to compare their pre-university mathematics results prior to their entry into the university. Another interesting aspect that is worth to investigate is about the psychological effects like why many students nowadays can easily forget what they were being taught yesterday. There might be a few reasons which lead to the forgetfulness or short-term memory among students since they are being busy with many gadgets nowadays.

As part of the improvement strategy in teaching mathematics, researchers like Stigler and James [13] have looked into using the video survey to enhance the mathematics teaching and learning. Using technology in the teaching and learning Mathematics at the university may also help to better understand this subject matter [14]. However, Prins emphasized on the importance to consider the classroom size and students' preparedness before implementing any new style of teaching and learning [15]. Above all, studying the characteristics of students, teachers and school, educational aids and resources, as well as students' attitudes and motivation towards the learning of Mathematics would be essential in order to improve students' understanding and performance in Mathematics [16].

## Acknowledgement

We wish to thank Universiti Teknikal Malaysia Melaka (UTeM), in particular, the Faculty of Engineering Technology for the help and support in carrying out the Mathematics Competency test for all the first year students last semester, i.e., semester 1, 2015-2016.

## References

1. L. J. Rylands and C. Coady, "Performance of students with weak mathematics in first-year mathematics and science," *Int. J. Math. Educ. Sci. Technol.*, vol. 40, no. 6, pp. 741–753, (2009).
2. H. Othman, I. Asshaari, N. M. Tawil, N. A. Ismail, Z. M. Nopiah, and A. Zaharim, "Analysis on Mathematics Fundamental Knowledge for Mathematics Engineering Courses based on a Comparative Study of Students' Entry Performance," *Procedia - Soc. Behav. Sci.*, vol. 60, pp. 365–371, (2012).
3. P. Sastre-Vazquez, R. D. Andrea, Y. Villacampa, and F. J. Navarro-Gonzalez, "Do First-year University Students Understand the Language of Mathematics?," *Procedia - Soc. Behav. Sci.*, vol. 93, pp. 1658–1662, (2013).
4. G. Anthony, "Factors influencing first-year students' success in mathematics," *Int. J. Math. Educ. Sci. Technol.*, vol. 31, no. 1, pp. 3–14, (2000).
5. S. Lamb and S. Fullarton, "Classroom and school factors affecting mathematics achievement: A comparative study of Australia and the United States using TIMSS," *Aust. J. Educ.*, vol. 46, no. 2, pp. 154–171, (2002).
6. J. Murray, "The Factors that Influence Mathematics Achievement at the Berbice Campus," Int. *J. Bus. Soc. Sci.*, vol. 4, no. 10, pp. 150–164, (2013).
7. K. E. Matthews, P. Adams, and M. Goos, "Putting it into perspective: Mathematics in the undergraduate science curriculum," *Int. J. Math. Educ. Sci. Technol.*, vol. 40, no. 7, pp. 891–902, (2009).
8. L. M. Thien and M. Y. Ong, "Malaysian and Singaporean students' affective characteristics and mathematics performance: evidence from PISA 2012," *Springerplus*, vol. 4, no. 1, p. 563, (2015).
9. P. S. Wilson, T. J. Cooney, and D. W. Stinson, "What constitutes good mathematics teaching and how it develops: Nine high school teachers' perspectives," *J. Math. Teach. Educ.*, vol. 8, no. 2, pp. 83–111, (2005).
10. J. T. Wixted, "A theory about why we forget what we once knew," *Curr. Dir. Psychol. Sci.*, vol. 14, no. 1, pp. 6–9, (2005).
11. A. Bokas and R, Rock. (2015, Feb 24). Edutopia [Online]. Available: http://www.edutopia.org/discussion/what-essence-learning
12. E. Summary, "Malaysia Education Blueprint 2013-2025," *Vasa*, vol. 2025, (2012).
13. J. W. Stigler and J. Hiebert, "Improving Mathematics Teaching," *Educ. Leadersh.*, vol. 61, no. 5, pp. 12–17, (2004).
14. Z. Lavicza, "The examination of technology use in university-level mathematics teaching," Proc. Symp. Occas., (2008).
15. S. C. Bates Prins, "Student-Centered Instruction in a Theoretical Statistics Course," *J. Stat. Educ.*, vol. 17, no. 3, pp. 1–13, (2009).
16. N. A. Ismail and H. Awang, "Mathematics Achievement among Malaysian Students : What Can They Learn from Singapore ?," *Int. Educ. Stud.*, vol. 2, no. 1, pp. 8–17, (2009).

# The Dependency of Engineering Technology Student's towards the Usage of Calculator in Mathematics

Nor Hafizah Hussin[1,a], Siti Haryanti Hairol Anuar[1], Khairum Hamzah[1], Mohd Fariduddin Mukhtar[1] and Irianto[1]

[1]*Faculty of Engineering Technology, Universiti Teknikal Malaysia Melaka, Hang Tuah Jaya, 76100 Durian Tunggal, Melaka, Malaysia*

**Abstract.** Calculators are one of the important technology used to solve mathematical computations. It also can be the tool for learning mathematics if it is used appropriately. However, too much depends on calculator can be harmful to students ability to solve simple mathematical problem. The purpose of this study is to examine the dependency of students in Faculty of Engineering Technology (FTK), Universiti Teknikal Malaysia Melaka, on the usage of calculator to solve the mathematical problems. A sample of 383 first year Engineering Technology (ET) students' taking mathematics subject are selected from five different course. Students were examined based on the results of Mathematic Competency Test and the survey from a questionnaire that covers questions regarding the students' enjoyment on the usage of calculator and the usefulness of calculator in mathematic activities. The investigation yield a result showing that the students has a high dependency on using calculator to solve mathematical problem.

## 1 Introduction

Mathematics is a subject that involves calculation and the calculating device have existed for thousands of years ago. Calculator is one of a calculating device used for so many years and has evolved dramatically for every new generation. It is known to be the best tools to aid mathematics learning because it allows the students to access the mathematical concepts and experiences from which they were previously limited with only paper and pencil [1]. Calculator is also proven to be the tools to help increase the effectiveness of learning mathematics. Few researchers have examined the attitudes of students towards calculator and the effect of calculator towards the students' achievement in mathematics. It is proven that the use of calculator resulted in better students' attitude thus enhances students' mathematic self-concept and maintain their motivation to learn [2].

Calculator helps to solve simple and complicated mathematics computation. To solve mathematical problem, students' need to apply their mathematics knowledge before using calculator to do the calculation. Despite all the capabalities and benefits, calculators can never replace the human capabilities in solving mathematics. It does not helps in understanding the problem, choosing the equation to solve the problem and finding the correct answer. Calculator will only be helpful when the student feed the information to it. However, students' nowaday are still using calculator even if it is just a simple mathematical problem. This is because, according to findings in [3], majority of the students' enjoy using calculator and considered mathematics to be easier when a calculator is used to solve problems. This shows that the students' has a high dependency towards calculator. The study in [4] indicates that it is vital that students' gain proficiency in basic skills before being allowed access to technology which can short-cut their understanding. Calculator helps students to access the lesson very quickly, however, it also can hinder students' ability to focus on problem solving methods.

The objective of this study is to examine the dependency of engineering technology (ET) students' in Faculty of Engineering Technology (FTK), Universiti Teknikal Malaysia Melaka, towards the usage of calculator for solving mathematical problem. ET is a field of applied physics that focus on applying knowledge in physics to solve technical problems. On the other hand, mathematics is the language of physics and it is use widely in this subject. Hence, in order for the students to pursue their studies in ET undergraduate course, they are required to be proficient in mathematics [6]. Mathematics is an important subject for the ET students because it has a high relationship with ET field. Students who took ET course will use a lot of mathematical concept in their study. Hence, it is important for the students to enhance their knowledge of concept in algebra, trigonometry, calculus and statistics.

According to [7], ET program is one of the new education field in Malaysia and was first offered by a public university in June 2011, which is Universiti Teknikal Malaysia Melaka (UTeM). Students that entered this program came from three type of programs from post-secondary education institutions which are Matriculation, Malaysia Higher School Certificate (STPM), and Diploma. Mathematics is a core subject for

[a] Corresponding author: norhafizah.hussin@utem.edu.my

these three programs and all the students are required to pass the subject for completion [8],[9]. Through out the sessions, students are able to use the calculator to aid the learning process as well as during the examination. The student has been exposed to the calculator since the secondary school level and have been using it in higher level institutions. This will some how effect their skills on solving basic mathematics question without using a calculator. The objective of this study is to examine the dependency of the first year ET students' towards calculator for solving mathematics by comparing the results of their mathematics competency test with an analysis from a survey.

## 2 Methodology

Three hundred-eighty three first year ET students' in FTK, UTeM from five different course were involved in this study. The Mathematic Competency test is a test for all new first year students in University Teknikal Malaysia Melaka (UTeM). The main purpose of this test is to measure the students background in mathematics such as the knowledge as well as the competency of the students when entering UTeM. Besides that, it helps to acknowledge the lecturers about the students level of understanding so that the subject can be carried out smoothly. This test consist of 40 fundamental questions that divided into two subtopics which is algebra and calculus. During the test, students are not allowed to use a calculator. This test is conducted in semester one for the 2015/2016 session and the students need to answer all the questions without using a calculator.

A set of questionnaire obtained from [10] were given to the student to assess their attitudes towards the usage of calculator focusing on the dependency of students on using calculator to solve mathematics problem. The questionnaire consist of two parts which is the enjoyment when using calculator and the usefulness of calculator in learning mathematics. According to [10], the term enjoyment is referred to as whether students liked using calculators during mathematics. The questions are as follows:

i. I enjoy using a calculator in mathematics activities
ii. A calculator is interesting, fascinating and easy to use
iii. enjoy investigating mathematics problems using a calculator
iv. A calculator is very interesting and challenging to use

The term usefulness in [10] is referred to as whether students viewed calculator as helpful tools in learning mathematics. The questions are as follows:

i. I do not have any use for calculators on a day to day basis
ii. I do not think a calculator will be useful to me in my future job
iii. I do not see how a calculator can help me to learn some new skills.
iv. Anything that a calculator can be used for, I can do just as well some other way.

All 383 students answer the test and questionnaire successfully. These results were analyzed using Microsoft Excel to obtain the statistical results for each of the questions.

## 3 Result and Discussion

The purpose of this study is to examine the dependency of first year ET students' in FTK on the usage of calculator to solve mathematical problem. The author compares the result from the mathematics competency test and result from the questionnaire to examine the dependency. Table 1 below shows the result for mathematic competency test for the selected first year ET students' from electric, electronic, mechanical and manufacturing students under the Bachelor of Engineering Technology program in FTK, UTeM.

**Table 1.** Students' results for Mathematic Competency Test.

| No. of student | Average Marks | Standard Deviation | No. of students | |
|---|---|---|---|---|
| | | | Less than 20 marks | More than 20 marks |
| 383 | 7.11 | 1.19 | 365 | 18 |

The results listed in Table 1 above shows that the students did not perform during the test yielding an average marks of only 7.11 over 40. Among the 383 students selected, only 18 students get a mark higher that 20 showing that only 4.7% of the students perform during the test. The standard deviation measures the dispersion of the data. A small standard deviation indicates that the data points tend to be close to the mean while a large standard deviation indicates that the data points are spread out over a wider range of values. Based on the results in Table 1, a small value of the standard deviation which is 1.19 indicates that the data points tend to be close to the average marks. This shows that among the 383 students selected, most of them get a marks around the average marks which is 7.11 out of 40, proving that the students did not perform well during the test.

Based on this results, the lecturer in FTK come out with a hypothesis that the students cannot answer the question due to the restriction of using calculator during the test because students nowadays prefer to use calculator to solve mathematics problem. To prove that the lecturer's claim is true, a set of questionnaire was designed to assess student's perception towards calculator for solving mathematical problem. Generally, most of the respondents agree that they prefer to use calculator to solve mathematics problem, as shown in Table 2.

The majority of the students, which is, 88.8% agree that they enjoy using calculator in mathematics activities. 334 out of 383 respondents, which is 87.2% agree that learning and solving mathematics will be more enjoyable with the use of calculator and 86.9% of the students enjoy investigating mathematics problem using calculator.

81.3% of the students agree that calculator is very interesting to use especially when it comes to solving a mathematical problems. These statements shows that the students enjoy using calculator in learning and solving mathematics. Hence, we can see the dependency among the students towards the usage of calculator in solving mathematic problems.

**Table 2.** Students' Enjoyment of Using Calculator.

| Statement | Percentage (%) | |
|---|---|---|
| | Agree | Disagree |
| I enjoy using a calculator in mathematics activities | 88.8 | 11.2 |
| A calculator is interesting, fascinating and easy to use | 87.2 | 12.8 |
| I enjoy investigating mathematics problems using a calculator | 86.9 | 13.1 |
| A calculator is very interesting and challenging to use | 81.3 | 18.7 |

**Table 3.** Students' Perception on the Usefulness of Calculator.

| STATEMENT | PERCENTAGE (%) | |
|---|---|---|
| | Agree | Disagree |
| I do not have any use for calculators on a day to day basis | 20.8 | 79.2 |
| I do not think a calculator will be useful to me in my future job | 18.2 | 81.8 |
| I do not see how a calculator can help me to learn some new skills. | 25.1 | 74.9 |
| Anything that a calculator can be used for, I can do just as well some other way | 58.1 | 41.9 |

Result from Table 3 above shows that most of the students agree that the calculator is useful in mathematics as well as their daily basis. Among 383 students, only 20.8% agree that calulator do not have any use in daily basis. It shows that most of the students think that besides using calculator to solve mathematic problems, calculator also helps them in other field. 81.8% of the students disagree with the statement that they do not think a calculator will be useful for their future job. This shows that the calculator will be useful to them not only in learning mathematics, but also in the future undertaking. 74.9% disagree with the statement that calculator in not helping in learning some new skills shows that the students think that they can also learn new skills using calculator because some of the mathematics problem can solve straight away from the calculator.

Based on the result obtained in Table 1, Table 2 and Table 3, we can say that the ET students' in FTK really depends on calculator when solving mathematics problem and the fact that they need to answer the mathematics competency test without using calculator makes them feel uncomfortable and not confident to anwer the question. This factors then affect the students marks for Mathematic Competency test as shown in Table 1.

## 4 Conclusion

The finding of this investigation shows that the calculator plays an important role to the students approach in learning and solving mathematics. The main focus of this study is to investigate the dependency of first year ET students' in solving mathematics using calculator. Hence, the result of this study supported the authors claims proving that the first year ET students' in FTK, UTeM have a high dependency on using calculator to solve mathematical problems. The calculator can be a tool to motivate the students to enjoy learning mathemathics especially in solving complicated problems. However, a high dependency on the calculator will limit the students ability to solve simple and basic mathematical problems. An inferential statistics may also be perform for future study, which is by making a hypothesis testing about the dependency of the students towards the calculator by comparing the data from the new first year ET students' in FTK with the existing data. Making a hypothesis testing about this will helps to prove whether or not the first year ET students' in FTK really depends on calculator when it comes to solving a mathematical problems. The results of this study must be interpreted with caution as it is only consist of first year ET students. For future study, a larger sample size should be considered so that a stronger conclusion can be made about the dependency of the first year ET students' in FTK towards calculator in solving mathematics.

**Acknowledgement**

All the authors acknowledge support from the Faculty of Engineering Technology (FTK), Universiti Teknikal Malaysia Melaka (UTeM) and those who are in help with the ideas and distribution of mathematics competency test and mini survey.

## References

1. H. Pomeratz. Prepared for the Urban Systemic Initiative/Comprehensive Partnership for Mathematics and Science Achievement (USI/CPMSA). Superintendents Forum. Dallas, Texas. (1997)
2. A. J. Ellington, Journal for Research in Mathematics Education. **34**(5), 433-463 (2003)
3. L. Avernon and J. Bobis. University of Sydney. (1995)
4. Z. K. Mbugua, M. W. Muthomi, and M. O. Okere. Int. Journal of Humanities and Social Science. **1**(13), 131-136 (2011)
5. J. Clark. Research Paper. SUNY Oswego (2011)
6. E. Goold. . *Paper Presented at 41st SEFI Conference.* Leuven, Belgium. (2013)

7. M. S. Yahaya, A. Noordin, S. Ahmad, and M. R. Sapiee. Proc. International Conference on Teaching and Learning in Higher Education, Johor, Malaysia. (2012)
8. N. Misran, S. N. S. Sahuri, S. Arsad, H. Hussain, W. M. D. W. Zaki, and N. A. Aziz. Asian Social Science. **8**(16), 222-231 (2012)
9. M. Alias. (Doctoral Dissertation). Retrieved from https://core.ac.uk/download/files/434/12007833.pdf (2012)
10. L. M. Kaino and E. B. Salani. In *Proceedings of the 28$^{th}$ Conference of International.* **3**, 113-120. (2004)

# Assessing Urban and Rural Teachers' Competencies in STEM Integrated Education in Malaysia

Ahmad Zamri Khairani[1,a]

[1]*School of Educational Studies, 11800 Universiti Sains Malaysia, Malaysia*

**Abstract.** In order to fulfil the need of sizeable skill workers, Malaysia will introduce STEM integration education in mainstream schools throughout the country. However, like any educational reform, one important issue that needs to be taken into account is the teachers' readiness especially in terms of their skills and competency in implementing the reform. As such, the purpose of this study is to assess differences between teachers' competency for STEM integration education between urban and rural teachers. A total of 244 teachers (urban = 129, rural = 115) are employed as sample in this cross-sectional quantitative study. Responses from an 18-item questionnaire were analysed using Rasch Model analysis to determine characteristics of item that measure competency between urban and rural teachers. The DIF analysis shows that items related to competency in (1) ICT integration, and (2) organizing co-curricular activities showed a significant difference in their measures between both sets of teachers. The result from this study would certainly provide useful information to relevant stakeholders, especially with regards to providing training for the teachers in the designated areas.

## 1 Introduction

In the quest to become a developed nation in 2020, Malaysia requires a sizeable number of skilled workers in science and technology. To meet the requirement, the government has turned its attention to encourage students to enrol in Science, Technology, Engineering, and Mathematics (STEM) subjects through the integration of STEM education in teaching and learning in the year 2017. STEM integrated education is defined as an approach to explore teaching and learning between any two or more STEM components or between one STEM components with another branch of knowledge [1]. The integration can occur in a particular unit of learning as well as through other related activities [2]. This is an effort to produce students with STEM literacy – defined as the ability to identify and integrate concepts from science, technology, engineering, and mathematics to understand complex problems and to make innovations to solve the problem. In order to achieve this, students need to have good experience with STEM subjects at school levels.

The roadmap for STEM integrated education in Malaysia is well documented in the Malaysian Education Blueprint 2013-2015 [3]. The blueprint underlines three steps to strengthened STEM education in Malaysia. Firstly, is to increase students' interest through new learning approaches as well as improved curriculum. Strategies that can be implemented range from the inclusion of higher order thinking skills to making subject content relevant to everyday life. Secondly, by improving teachers' skills and competencies through continuous training. Thirdly, the Ministry of Education has also put efforts to increase both students and parents awareness about the importance and opportunities in STEM fields.

According to reference [4], integrated curriculum such as STEM integrated education helps to provide more relevant and open experience for the students. They are required to apply problem-solving skills as well as their knowledge about STEM content to relate their learning experience with real life activities [5]. Reference [6] concludes that the main purpose of STEM education is to enhance one's ability to address STEM-related personal, social, as well as global issues. To do this, STEM education should emphasize on problem-based learning [7]. Apart from problem-solving, reference [8] mentions that STEM education is also necessary to produce students with the following knowledge and skills: innovation, creative, independent, logic, and IT-savvy.

Nevertheless, the proposed STEM integration education is expected to face some difficulties. Firstly, Malaysia is experiencing a decline in enrollment of science students at secondary school and university levels. In 2012, only about 33-40% students were learning science at school level [9]. As such, the aim to produce 500,00 scientists by 2020 might not be achieved. Secondly, Malaysian students' performance in science and mathematics at international level is far from encouraging. For example, in the Program for International Student Assessment (PISA), nearly 60% of the Malaysian respondents fail to achieve minimum benchmark in mathematics while 43% get a similar result

[a] Corresponding author: ahmadzamri@usm.my

in science [3]. Thirdly, apart from difficulties related to students' enrollment in STEM fields, Malaysia is also facing problems regarding the teachers' readiness in educational reform. For example, one of the factors for discontinuation of national-wide educational reform named *Teaching Mathematics and Science in English* (PPSMI) is attributed to the fact that teachers are not ready to teach those subjects in English. Rather, they are more comfortable to teach in their mother tongue [10]. While the Ministry of Education has taken many initiatives to overcome the two recent challenges, less attention, however, is focussed on the issue of teachers' readiness especially in assessing their competencies for STEM integration education.

## 2 Teachers' Competency for STEM Integration Education

Teacher is considered as an important factor in influencing successfulness in many educational reform such as integration of STEM education [11]. Reference [12] meanwhile, quotes that for STEM educators to function effectively, the teachers are required not only to understand their content taught but also have the ability to explain concepts and procedures. In addition, teachers themselves also need to immerse in technology so that they are able to explain how fun STEM integrated education is. This is important since there is a common fear that STEM is difficult both to learn and to teach. This misconception needs to be addressed to encourage more students to enrol in STEM-related areas. More importantly, there is a need to investigate teachers' readiness in the integration of STEM education, especially regarding their knowledge and skills so that this high-stake national-level educational reform can achieve its objectives.

One possible problem that may arise in implementing STEM integration education is the gap between urban and rural areas especially regarding facilities and other resources. Like many other countries, urban schools in Malaysia are said to enjoy better facilities and resources compared to their rural compatriot. The gap then leads to differences in students' achievement, provision of infrastructures as well as dropout among students learning science and technology. This will eventually result in socioeconomic imbalance between urban and rural areas. Also, it is also well known that urban schools enjoy better quality teachers. Despite this imbalance, the proposed STEM integration education is expected to produce similar results between the two set of schools. The existing gaps, therefore, is hypothesized as an important variable that determines the successfulness of STEM integration education. As such, the present study is conducted to assess differences teachers' competencies for integrating STEM education between urban and rural teachers. The findings provide information that may be useful to relevant stakeholders to be taken into consideration in implementing STEM integration education.

## 3 Method

### 3.1 Sample

Sample of the present study consists of 129 (52.9%) urban school and 115 (47.1%) rural school teachers. It comprises of 206 (84.4%) female teachers and 38 (15.65%) male teachers. The mean of their teaching experience is 13.6 years (SD = 7.3 years). The teachers are from the states of Penang, Kedah, Perak, and Selangor. Information on the main subject taught by the teachers is given in the following Table 1.

**Table 1.** Main Subject taught by the Sample.

| | Subject | N | % |
|---|---|---|---|
| 1 | Mathematics | 64 | 26.2 |
| 2 | Science | 63 | 25.8 |
| 3 | Physics | 14 | 5.7 |
| 4 | Biology | 15 | 6.1 |
| 5 | Chemistry | 25 | 10.2 |
| 6 | Others | 63 | 25.8 |
| | | 244 | 100.00 |

### 3.2 Instrument

This study gauges responses of the sample using an 18-items questionnaire developed specifically for this study. The instrument covers various aspects of teachers' competencies such as academic, knowledge, as well as skills in integrating STEM education into their teaching and learning. Response on the item is measured using 6-point Likert scale continuum, where '1' indicates Strongly Disagree, while '6' denotes Strongly Agree responses.

### 3.3 Data Analysis

The data was analyzed using Rasch Model measurement framework. In a nutshell, this modern test theory provides significant advantages for a better precision in measurement of a unobservable construct such as teachers' competency. More specifically, the model transforms raw scores into an equal-interval unit of measurement such as in a ruler. This transformation is done in a procedure called calibration where both item difficulty and teachers' competency parameters are estimated. Scores from the parameters estimation process are called 'measures' and are defined in logits unit. The equal-interval property is important in this study since it does not only provide identification of the differences in

DOI: 10.1051/matecconf/20178704004

teachers' competency but also able to demonstrate how much the differences is.

Even though it provides a huge opportunity for precision in measurement, Rasch Model comes with strong assumptions. Two important assumptions must be met before the data can be treated as equal-interval property. Firstly, the data collected must fit the model's expectation and secondly, the construct must pose a unidimensionality property [13]. Model – data fit is an important assumption for many estimation procedures like Rasch Model analysis. It helps to identify discrepancies between the model's expectation and the data collected to ensure that the measurement procedure can measure the intended construct adequately and at the same time keeping the unintended constructs in a reasonable proportion. In this study, model-fit issues are addressed by investigating the infit and outfit, mean-squares (MNSQ) by adopting the criteria of 0.6 – 1.4 logits set by reference [14]. Unidimensionality, meanwhile, assumes that items in a test measure a single construct [15]. In this study, a principal component analysis (PCA) of residuals procedure helps to identify the second construct, apart from the intended one that may become a threat to unidimensionality assumptions. This study applies guideline suggested by reference [13], where the unexplained variance from the first factor in the PCA of residuals procedure should be less than the strength of 5 items.

Apart from the assumption of model – data fit and unidimensionality, this study also run differential item functioning (DIF) analysis of every item in the instrument. The procedure is aimed at identifying how much each item is different between the urban and rural teachers. DIF contrast statistic from Rasch Model analysis offers the straightforward procedure to detect item that shows evidence of DIF. The values of > ±.5 logits indicate that the particular item function differently between urban and teachers [14].

## 4 Results and Discussion

The Rasch Model analysis from Winsteps 3.63 identifies two items (Item 10 and Item 11) which do not fit the model's expectation, and thus removed from further analysis. Table 2 shows final items' statistics from the Rasch Model analysis. All items demonstrate acceptable infit and outfit MNSQ values of between 0.6 – 1.4 logits. Meanwhile, based on Figure 1, the unexplained variance from the first factor is 2.8 which is about the strength of 3 items. Since the data fulfilled both assumptions of model-data fit and unidimensionality, the items' measure from the analysis demonstrates the property of equal-interval.

Table 3 demonstrates DIF analysis of all 18 items based on urban (examinee class 1) and rural teachers (examinee class 2). For Q1 (*I have the relevant academic qualifications to integrate STEM in teaching and learning*), DIF measure for urban teachers is –1.56 logits while rural teachers recorded a slightly lower measure of –1.84 logits. Even though there is a different of .28 logits [–1.56 – (–1.84 logits)] between both set of scores, the difference is not significant based on DIF Contrast statistic of .27 logits which is less than .5 logits [14]. Thus, it can be concurred that both urban and rural teachers endorse that they have the competency in terms of relevant academic qualification to integrate STEM education into teaching and learning. Overall, the present study recorded a satisfactory result where 14 out of 16 items (87.5%) function similarly across both urban and rural teachers, with only two items (Q13 and Q14), show otherwise.

With regards to Item Q13 (*I am able to incorporate ICT in integrating STEM in teaching and learning*), urban teachers show significantly more competence (measure= –1.15 logits) compared to rural teachers (measure= –.34 logits). The result is rather not unexpected since it is well known that urban areas enjoy

**Table 2.** Item Statistics

| ENTRY NUMBER | RAW SCORE | COUNT | MEASURE | MODEL S.E. | INFIT MNSQ | INFIT ZSTD | OUTFIT MNSQ | OUTFIT ZSTD | PTMEA CORR. | item |
|---|---|---|---|---|---|---|---|---|---|---|
| 1 | 958 | 232 | -1.68 | .12 | 1.31 | 3.0 | 1.18 | 1.7 | .70 | Q1 |
| 2 | 839 | 244 | .61 | .11 | .73 | -3.2 | .72 | -3.1 | .86 | Q2 |
| 3 | 800 | 244 | 1.07 | .11 | .86 | -1.5 | .82 | -1.9 | .84 | Q3 |
| 4 | 876 | 244 | .16 | .11 | .65 | -4.2 | .67 | -3.6 | .87 | Q4 |
| 5 | 830 | 244 | .72 | .11 | .77 | -2.7 | .74 | -2.8 | .87 | Q5 |
| 6 | 821 | 244 | .83 | .11 | 1.10 | 1.0 | 1.05 | .5 | .83 | Q6 |
| 7 | 886 | 244 | .03 | .11 | .72 | -3.3 | .74 | -2.7 | .84 | Q7 |
| 8 | 887 | 244 | .02 | .11 | .70 | -3.5 | .68 | -3.5 | .84 | Q8 |
| 9 | 975 | 227 | -2.21 | .13 | 1.19 | 1.9 | 1.12 | 1.2 | .71 | Q9 |
| 10 | DELETED | | | | | | | | | Q10 |
| 11 | DELETED | | | | | | | | | Q11 |
| 12 | 877 | 244 | .15 | .11 | .91 | -.9 | .89 | -1.1 | .83 | Q12 |
| 13 | 948 | 244 | -.79 | .12 | 1.27 | 2.6 | 1.31 | 2.8 | .76 | Q13 |
| 14 | 882 | 244 | .08 | .11 | 1.38 | 3.6 | 1.33 | 2.9 | .73 | Q14 |
| 15 | 885 | 244 | .05 | .11 | .82 | -2.0 | .83 | -1.8 | .83 | Q15 |
| 16 | 788 | 244 | 1.21 | .11 | 1.26 | 2.7 | 1.26 | 2.5 | .80 | Q16 |
| 17 | 896 | 244 | -.10 | .11 | .93 | -.7 | .83 | -1.7 | .81 | Q17 |
| 18 | 900 | 244 | -.15 | .11 | 1.27 | 2.6 | 1.13 | 1.2 | .77 | Q18 |
| MEAN | 878.0 | 242.2 | .00 | .11 | .99 | -.3 | .96 | -.6 | | |
| S.D. | 51.8 | 4.9 | .89 | .01 | .25 | 2.6 | .23 | 2.3 | | |

| | | Empirical | | Modeled |
|---|---|---|---|---|
| Total variance in observations | = | 75.8 | 100.0% | 100.0% |
| Variance explained by measures | = | 59.8 | 78.9% | 77.9% |
| Unexplained variance (total) | = | 16.0 | 21.1% | 22.1% |
| Unexpl var explained by 1st factor | = | 2.8 | 3.7% | |

**Figure 1.** Principal Component Analysis of the Residuals

**Table 3.** DIF Analysis

| EXAMINEE CLASS | DIF MEASURE | DIF S.E. | EXAMINEE CLASS | DIF MEASURE | DIF S.E. | DIF CONTRAST | JOINT S.E. | t | d.f. | Prob. | item Number | Name |
|---|---|---|---|---|---|---|---|---|---|---|---|---|
| 1 | -1.56 | .17 | 2 | -1.84 | .19 | .27 | .25 | 1.09 | 230 | .2768 | 1 | Q1 |
| 1 | .77 | .15 | 2 | .42 | .17 | .35 | .22 | 1.58 | 242 | .1165 | 2 | Q2 |
| 1 | 1.16 | .14 | 2 | .96 | .16 | .21 | .22 | .95 | 242 | .3432 | 3 | Q3 |
| 1 | .22 | .15 | 2 | .08 | .17 | .15 | .23 | .65 | 242 | .5141 | 4 | Q4 |
| 1 | .68 | .15 | 2 | .77 | .16 | -.09 | .22 | -.42 | 242 | .6773 | 5 | Q5 |
| 1 | 1.00 | .14 | 2 | .61 | .17 | .39 | .22 | 1.77 | 242 | .0787 | 6 | Q6 |
| 1 | .11 | .15 | 2 | -.07 | .17 | .18 | .23 | .80 | 242 | .4250 | 7 | Q7 |
| 1 | .02 | .15 | 2 | .02 | .17 | .00 | .23 | .02 | 242 | .9858 | 8 | Q8 |
| 1 | -2.05 | .17 | 2 | -2.41 | .19 | .36 | .26 | 1.39 | 225 | .1666 | 9 | Q9 |
| 1 | .16 | .15 | 2 | .13 | .17 | .02 | .23 | .10 | 242 | .9181 | 12 | Q12 |
| 1 | -1.15 | .16 | 2 | -.34 | .17 | -.82 | .24 | -3.47 | 242 | .0006 | 13 | Q13 |
| 1 | -.23 | .15 | 2 | .47 | .17 | -.70 | .23 | -3.11 | 242 | .0021 | 14 | Q14 |
| 1 | -.02 | .15 | 2 | .13 | .17 | -.16 | .23 | -.69 | 242 | .4896 | 15 | Q15 |
| 1 | 1.06 | .14 | 2 | 1.39 | .16 | -.33 | .21 | -1.56 | 242 | .1202 | 16 | Q16 |
| 1 | -.02 | .15 | 2 | -.19 | .17 | .16 | .23 | .72 | 242 | .4732 | 17 | Q17 |
| 1 | -.11 | .15 | 2 | -.19 | .17 | .07 | .23 | .32 | 242 | .7500 | 18 | Q18 |

better ICT facilities and internet connections, which in turns encourage them to integrate ICT in teaching and learning [16]. As rightly observes by reference [17], digital disparities between urban and rural school can be seen in terms of (1) funding, and (2) ready access to technology. Meanwhile, reference [18] argues that teachers at rural schools are also lacking regarding skill and experience as well as technical training. However, one of the most important issues regarding ICT integration is the (poor) internet connection at rural schools [19]. Therefore, one might speculate that STEM integration in rural school may face difficulty to implement the STEM integration education. In order to address this issue, researcher such as reference [20] have suggested that school leaders to adopt technology leadership style – where they took the initiative to lead technology integration in various aspect in school organisation such as in the (1) vision and leadership, (2) teaching and learning, (3) professional practices and productivity, (4) infrastructure, and (5) support, management and control of ICT. These comprehensive aspects of leadership are said to be a major factor that influences technology integration in schools.

Item Q14 (*I can organize co-curricular activities related to STEM integration for students*), urban teachers (measure = □–.23 logits) also show significant more competency compared to rural teachers (measure = .47 logits). This is surprising since co-curricular activities is an essential component of the school curriculum and comprehensive guidelines have been provided to schools and teachers. For example, teachers can choose to rather extracurricular activities conduct activities for 45 type of clubs and society, 43 sports and games as well as 19 uniform units with time allocation between 60 – 120 minutes each week [21]. The result, however, certainly needs to be investigated further since co-curricular activities are identified as one important avenue to attract students' interest in science and technology especially when it is aligned with what been taught in the classroom.

## 5 Conclusion

Teacher is an important factor in determining quality as well as successfulness of STEM integration program. Therefore, it is important for them to acquire knowledge and skills especially the STEM-specific methods of teaching such as project-based learning. This study reports that in most cases, there is no significant differences between in teachers' competencies for both urban and their rural counterparts. However, the researcher is also able to identify two areas where gaps between urban and rural teachers regarding their competency in integrating STEM education exist. It was found that the rural teachers lack competency in incorporating ICT in teaching and learning. In contrast, the rural teachers are more competent in organizing co-curricular activities related to STEM integration compared to the urban teachers. The finding is rather important since if left unattended, it is highly unlikely that rural schools are able to get the benefits promised by STEM integration education. Ultimately, it will also affect the aspiration for 50% reduction in achievement gaps between urban and rural students come the year 2020.

## References

1. K. Becker, K. Park, Effects of integrative approaches among science, technology, engineering, and mathematics (STEM) subjects on students' learning: A preliminary meta-analysis." J STEM Educ, **12**, (2001)
2. C. Merrill, J. Daugherty, STEM education and leadership: A Mathematics and Science partner approach. J Technol Educ **21**, 2 (2010)
3. Ministry of Education, *Malaysian Education Blueprint 2013-2025.* (Kementerian Pendidikan Malaysia, Putrajaya, 2013).
4. J. Furner, D. Kumar, The mathematics and science integration argument: a stand for teacher education. Eurasia J Maths, Sci Technol, **3**, 3 (2007)
5. N. Tspuros, R, Kohler, J. Hallinen, *STEM education: A project to identify the missing components.* (STEM Education and Leonard Gelfand Centre for Service Learning and Outreach, Carnegie Mellon University, Pennsylvania, 2009)
6. R. W. Baybee, Technol Eng Teach **70**, 1 (2010)
7. J. M. Breiner, S. S. Harkness, C. C. Johnson, C. M. Koehler, 'What is STEM?' A discussion about conceptions of STEM in education and partnership. School Sci and Math, **112,** 1 (2012)
8. J. Morrison, *STEM education monograph series: Attributes of STEM education.* (Teaching Institute for Essential Science. Baltimore, 2006)
9. Ministry of Science, Technology, and Innovation, *Science and human capital technology roadmap towards 2020* (Ministry of Science, Technology, and Innovation, Putrajaya, 2012
10. S. Asri, E. Ahmad, S. S. Sholeh, A. Asliaty, J. Educ, **16**, (2011).
11. M. Stohlman, T. J. Moore, G. H. Roehig, Considerations in teaching integrated STEM education. J Pre-College Eng Educ Res, **2** 2 (2012)
12. H. Lilia, S. A. S. I. Shahrina, M. M. Subahan, J of Sci Educ Technol, **23**, 2 (2014)
13. J. M. Linacre, *A user's guide to WINSTEPS: Rasch Model computer programs.* (MESA Press, Chicago, 2005)
14. T. G. Bond, C. M. Fox, *Applying the Rasch Model: Fundamental measurement in the human sciences* (3rd Ed.) (Routledge, New York, 2007)
15. B. D. Wright, G. N. Masters, Rating scale analysis: Rasch measurement (Chicago, MESA Press. 1982).
16. Gomathi, G. (2003). Penggunaan internet di dalam pengajaran dan pembelajaran Sains: Sikap, pengetahuan dan kemahiran Sains guru-guru luar bandar. Projek Sarjana Pendidikan, Fakulti Pendidikan, Universiti Kebangsaan Malaysia, Bangi.
17. I. Zulkefli, S. Alnin, T. M. Faziharudean, Bringing the internet to the rural area: A Case Study of the 'Kedaikom' project in Malaysia in J. Steyn, J. P. van Belle, E. V. Mansilia (Eds), *ICTs for global development and sustainability: Practice and applications.* (IGI Global, 2010).
18. M. E. Herselman, *Proceedings of Informing Science,* (South Africa, 2003)
19. A. Noraini, K. Hamidon, M. H. I Mohd Izham, Amalan kepimpinan teknologi pengetua dalam pengintegrasian ICT sekolah menengah kebangsaan di Malaysia. *Proceeding of the 3rd Clobal Summit on Education.* (Kuala Lumpur, 2015)
20. H. D. Mehlinger, S. M. Power, *Technology and teacher education: A guide to policymakers.* (Houghton Mifflin Company, 2002)
21. Ministry of Education, *Buku Panduan Pengurusan Kokurikulum,* (Kuala Lumpur, A. G. Grafik Sdn Bhd, 2009)

# Academics and Learners' Perceptions on Blended Learning as a Strategic Initiative to Improve Student Learning Experience

Adeline Ng Ling Ying[1a], Irene Yang[1]

[1]*Swinburne University of Technology Sarawak, Malaysia*

**Abstract.** The increasingly tighter shift of socio-economic constraints on higher education sectors in the recent years has called for greater flexibilities in student learning experience both locally and abroad. To this end, we have recently implemented a Blended Learning Initiative in an attempt to provide better learning support and greater flexibility to our students. This initiative is also in line with the University's aim of having 50% of our learning and teaching delivered on-line by 2020. In this report, we present our findings on academics and learners' perceptions on the approach which were obtained through surveys. Results showed that blended learning approach was new to the academics and the factors for successful blended learning implementation were identified. Results also showed that learners appreciated the approach as it made learning more accessible and flexible. Furthermore, they also enjoyed the interesting online activities incorporated into their units. In addition, learners were also able to review and pace their own learning. They also perceived that they have the access to the resources and technical ability to cope with online learning materials and activities. Nonetheless, the survey also revealed that learners still prefer to have academics delivering information to them directly rather than a flipped classroom model. In conclusion, findings from this study provide insights that blended learning could be effective to supplement courses offered by the faculty.

## 1 Introduction

With the emergence of technology, universities are currently embracing transformation in learning delivery. In fact, incorporating technology into education is essential to address diverse needs of learners. For instance, learners may differ in terms of education background, abilities, culture and learning preferences. As learners' needs were being identified, there is a need to seek an effective approach to facilitate quality learning.

Blended learning is an innovative instructional approach which combines several delivery strategies of online and face-to-face components [1]. Interactive capabilities of information and communication technologies (ICTs) in blended learning offer the opportunities to increase flexibility of access to learning resources and enhance self-directed learning at any time and pace. When thoughtfully integrating ICTs in blended learning, this learning approach as indicated in various researches has significant education implications due to its effectiveness for course delivery in higher education. Specifically, there is evidence that overall students enrolled in blended classes achieve higher grades than their counterparts in fully online or face to face courses [2]. Likewise, student satisfaction also tends to be higher in blended courses when compared to traditional lecture courses [3].

Due to the praises blended learning has received, many higher learning institutions are now shifting towards implementing blended learning as a strategic initiative to improve student learning experience. There is compelling evidence of significant growth of blended learning in the last decade with the increase of theoretical studies in journal articles published as early as 1996 [2, 4]. Recognizing the benefits of blended learning in education, it was emphasized as one of the key ideas in the Malaysian Education Blueprint 2015 – 2025 [5] to enhance the quality of learning and teaching through the use of Open Education Resources (OER), Open Course Ware (OCW) and Massive Open Online Courses (MOOC) from renowned higher education institutes.

The university uses ICTs in learning and teaching ever since it was established 16 years ago. The learning management system (LMS) adopted was Blackboard. It is a learning platform that allows academics to communicate with learners, monitor learners' participation while assessing their performance. The university requires all instructional materials to be uploaded on Blackboard for learners to access online. Besides that, Blackboard is also used to manage grades for a course by using the online grade center tool. This tool efficiently reduces double handling work and students could access their results and feedback easily. In addition, the management also uses Blackboard as the formal platform for making announcement. Some academics have also taken a step further to create online assessments and student forums on Blackboard. These approaches have the potential to motivate students to

[a] ang@swinburne.edu.my

engage in learning while monitoring their progress in an ongoing manner.

Although blended learning has the potential to increase quality of learning, integration of technology in learning requires clear understanding of pedagogical goals. In other words, educators need to select suitable instructional technology to assist students in achieving the learning objectives while providing them with the best possible learning and teaching experiences. The attempt to use technology tools as a delivery vehicle without educational goal is bound to fail. Besides that, there must also be an alignment of goals at all levels of an institution from senior administration to students for successful implementation of blended learning [6,7].

As most studies on blended learning reported were from the United States and Australia [2-4, 6-7], it is crucial to study and understand the perceptions and factors contributing to its success locally. Consequently, a working group was established to start a Blended Learning Project spearheaded by the Faculty of Engineering, Computing and Science to determine academic staff and student perceptions on blended learning and also to identify the factors for successful implementation in the local environment.

## 2 Blended learning project

The transition to adopt and implement blended learning in learning and teaching in the faculty started in year 2014 as a strategic initiative to align course delivery with its main campus in Melbourne. The primary aim was to improve student learning experience and to provide students with more flexible and interactive modes of delivery.

As a start, two units were selected to trial blended learning delivery. These units were first year common units with a large number of students. The components identified as beneficial to be delivered online were ongoing assignments and tutorials. Instead of the traditional physical assignments, online practice quizzes were created. The quizzes were rolled out after each topic were delivered and the questions were taken from a pool of online multiple choice questions (MCQ) which were created and randomized in Blackboard. Each online assignment was opened to students for a week and students were allowed multiple attempts. The rationale behind this approach was to encourage learners to have more practice and to familiarize themselves with the concepts applied. Since MCQ assignments were used in Blackboard, marking for the assignments was automated and students could receive immediate feedback on their understanding of the topics. Furthermore, automated marking also avoided ongoing workload for academics.

In addition to online assessments, tutorial videos were also recorded using the Camtasia recording software and posted on Blackboard for students to review at their convenience. This enabled students to pace their own learning and was especially useful in large scale classes of students with diverse learning abilities and background. The recorded videos also facilitated to reduce the workload of academics who only need to present only once.

In 2015, three other units were identified to implement blended learning in the unit delivery. Besides online assignments and recorded videos, online discussion boards and student forums were created on Blackboard for information sharing and peer review. This environment motivated learners to have an ongoing and active engagement through collaborative learning as learners communicate to discuss their work using the Blackboard platform. Students were also introduced and provided the link to relevant videos and reading materials available on the OER. The purpose was to foster students' learning about the topics discussed. Additionally, technology tools such as iPad and Clickers were also applied in classes to attract and engage students. All the efforts mentioned above were used as supplement and did not replace any contact hours in class.

## 3 Research objective and methodology

This study aims to evaluate both academics and learners' perceptions on blended learning in the university. This investigation will provide better understanding of factors and issues to be considered when adopting blended learning.

Academics' perceptions survey on blended learning was conducted in the beginning of the implementation of blended learning. The survey was rolled out to a focus group. The aim of the survey was to investigate academics' understanding and to identify the types of support they require in the implementation of blended learning. The survey was divided into two sections; qualitative and quantitative. The quantitative section consisted of four Likert items on a scale rated from 1 to 5 ("Strongly Disagree" to "Strongly Agree").

On the other hand, students' perceptions survey was conducted to investigate their learning experiences in blended learning activities. The survey was conducted at the end of the semester. The students' satisfactions were determined from different perspectives such as motivation, technical support, and flexibility and preparation time. Similar to the academics' perceptions survey, the students' perceptions survey was also divided into two sections; qualitative and quantitative, measured with the same Likert scale rated from 1 to 5 ("Strongly Disagree" to "Strongly Agree").

## 4 Results and discussions

### 4.1. Academics' perceptions on blended learning

Although advances in technology provide opportunities for academics to design and deliver their courses in ways that support and enhance a teacher's role, the integration of technology in blended learning has also changed the role expectations of teachers and students. According to Norberg, Dziuban & Moskal [8], teachers embraced new role as facilitators and students had access to seemingly limitless source of informational assets through the internet. Conversely, leaders of educational institution are challenged to meet the demands to facilitate meaningful learning outcomes. Hence, Garrison & Kanuka [1] emphasized that a clear institutional direction is vital to successfully adopt blended learning initiative.

According to the feedback from the academics' perceptions survey, most academics responded that they were new to blended learning and had never implemented blended learning initiative in their units. The implementation of blended learning is certainly a new challenge for most educators as they have to change their delivery method to accommodate the use of technology. Hence, it was not surprising that some academics had expressed drawbacks to the implementation of blended learning.

In the survey, some academics had also expressed concern that technology will gradually substitute teachers in learning institution. Academics lacked understanding that even though technology seems to be a powerful and engagement tool, if it is not tied with pedagogical principles, learning will remain futile. Besides that, another concern raised was that blended learning approach would increase academics' workload as learning, planning and implementation stage would require significant amount of effort and time. In addition, academics' also perceived that courses which require more technical or practical skills would be harder to teach online, whereas less technical courses would be well-suited to the blended environment.

Moreover, it is important to take all other stakeholders' expectation into consideration. These include the requirements of professional bodies and the expectation of parents in education. Academics were concerned that the implementation of blended learning will affect learners' feedback with regards to teaching and learning at the end of the semester. As a result, they were hesitant to introduce and employ blended learning in their units. Nonetheless, they agreed that blended learning made learning more accessible and flexible to students. It could also increase student engagement in learning and in the long run, blended learning is time and cost effective compare to traditional face-to-face learning environment.

Apart from academics' understanding of blended learning, other factors which determine the successful implementation of blended learning were also identified through the survey. Most academics perceived themselves as beginners in the use of technology tools in teaching and learning activities. Some lacked technology skills and thought that it would be a challenge to apply the approach in their unit delivery. Thus, there is a need for academic support in the use of technology in order to implement blended learning.

The types of support that academics expect in the implementation of blended learning are shown in Table 1. Specifically, they require a dedicated personal to assist them to gain experience with technology. Apart from this, they would also like to have professional development training and workshops in designing blended activities and the use of technology. In addition to the support, it is also essential for academics to have adequate access to infrastructures and technology tools in the faculty for implementation.

**Table 1**. Descriptive statistics for types of support in the implementation of blended learning.

| Items | Mean | Standard deviation |
|---|---|---|
| 1) Educational support group | 4.00 | 0.93 |
| 2) Library and media specialists | 4.13 | 0.64 |
| 3) Professional development and training/workshop in designing blended module and the use of technology | 4.13 | 0.99 |
| 4) Demo/sharing/learning session | 4.13 | 0.99 |

In conclusion, although the adoption of blended learning has the potential to enhance quality of delivery in learning, academics' perceptions survey revealed some concerns with regards to blended learning. These concerns were similar to the major barriers in technology integration such as: 1) resources, 2) educator's knowledge and skills and 3) educator's attitudes and beliefs [9]. In fact, the most notable concern was related to the availability of essential resources which include (a) access to technological tools; (b) technical and learning instruction support (c) professional development training and preparation time [10].

## 4.2 Learners' perceptions on blended learning

Learners showed positive responses to the blended learning initiative applied in the two trial units identified in 2014. The components identified as beneficial to be delivered online were ongoing assignments and tutorials. Due to these blended learning activities, learners were able to pace their own learning as they had flexible access to the learning resources they required. Besides that, they were also able to review and test their understanding of a topic whenever they needed. As automated marking on Blackboard was used, learners were also able to get feedback and grades on their assignments immediately.

In 2015, the blended learning approach was introduced to the whole faculty and three additional units were selected to participate in the delivery method. The participants were undergraduate degree students who enrolled in the respective units in the faculty. The type of blended learning activities implemented in the units include online quizzes, videos, online forum or discussions, online peer feedback and use of technological tools such as iPad and Clickers.

The descriptive statistics for items (mean and standard deviation) with regards to the blended learning activities in the three units were shown in Table 2. It is essential to obtain learners' feedback on their learning experience from the aspect of resources, content, design of the blended learning activities to identify areas for future improvements.

**Table 2.** Descriptive statistics for blended learning activities.

| Items | Mean | Standard deviation |
|---|---|---|
| 1) Accessibility and flexibility | 4.00 | 0.84 |
| 2) Self-directed learning | 3.88 | 0.87 |
| 3) Motivation | 3.42 | 0.91 |
| 4) Content (interesting) | 3.75 | 0.90 |
| 5) Resources (technological tools and access to study) | 4.30 | 0.72 |
| 6) Technical ability | 3.83 | 0.85 |
| 7) Preparation time | 3.60 | 0.89 |
| 8) Preferences over face-to-face instruction | 3.07 | 0.95 |
| Overall Satisfaction of Blended Learning Activities | 3.93 | 0.76 |

According to the survey, most learners reacted positively to the blended learning initiatives applied in the units as shown in Table 2 with overall satisfaction mean score of 3.93. Blended learning activities allowed learners to access learning resources and receive online feedback easily. This is especially helpful as the classes here were dominated by Asian students. The students rarely volunteer to ask or answer a teacher's question in class.

According to Lee [11], the concept of "face" which is associated with pride and self- esteem is related to this phenomenon. The learners' lack of confidence to speak in class may be due to one of these three conditions: (1) students are afraid of losing face if they provide a wrong answer or ask an obvious question, (2) students are afraid of being considered as showing-offs if they respond to too much in class and (3) students think that their opinions are not worth sharing in class and are only comfortable to be shared with close friends. Hence, blended learning approach could reduce the fear and anxiety of asking questions in public.

With the presence of online discussion boards, learners were able to share freely in Blackboard. They may be evaluated by their peers who provide valuable feedback on their work without feeling challenged. In this way, learners could identify the areas for improvements whereas academics could monitor learners' progress by making use of peer review. This was shown by the number of posts and views of the site. Besides that, learners also agreed that the blended learning initiatives had also improved their accessibility and flexibility in their learning (item 1) such as reviewing the recorded videos whenever they need to. In addition, online assessments had also encouraged learners to practice more as they were allowed to perform multiple attempts on the assessment items and were able to receive prompt feedback. This had encouraged learners to learn independently (item 2) and helped to improve their understanding on the topics.

The findings also suggest that generally learners were neutral to the extent blended learning activities could motivate (item 3) them in learning. Many researches had shown that students' attention varies during lectures and thus, teachers need to develop ways to maintain student interest in the classroom [12]. With the use of iPad and Clickers, academics were able to capture learners' attention and enhance their engagement in class. The instruments were used to promote active learning through questions. Instead of verbal answers, learners provided their response anonymously in class. The use of iPad also enabled academics to move freely around the class and be approached by learners who may want to have a quick clarification on the topic but were reluctant to speak up. While academics intend to increase learners' motivation in learning, this requires learners to have discipline, maturity and continuous commitment in learning. As indicated by Marino, Eager & Draxler [13], some students experienced difficulty adjusting to the structure of online courses, managing their time in such environments and maintaining self-motivation.

Moreover, the survey revealed that the activities designed for blended learning were interesting (item 4). Besides that, learners also indicated that they had the resources and technical ability to cope with online materials and activities (items 5 & 6). Although not all new generations were tech savvy, the results obtained were expected as most learners considered themselves from a technology savvy generation. While most learners had an overall good experience with blended learning, some were concerned that this approach will increase their preparation time for class (Item 7). Generally, they still prefer to have academics delivering information to them directly rather than a flipped classroom model. This is evidenced by the lowest mean score 3.07 obtained in the survey (item 8). Lastly, some learners had expressed concerned on the fairness in grading of group assignments conducted online.

In conclusion, the learners' perceptions survey revealed that the factors of the acceptance of blended learning approach mainly depend on system functionality, technology self-efficacy, content feature, effective interaction and motivation as reported by Wu, Tennyson & Hsia [14].

### 4.3. University's blended learning initiatives

It is shown in the study that to facilitate blended learning in the university, academics and learners ought to have the shared vision, understanding and expectation in the adoption of the approach. Specifically there is a need to have a clear definition of blended learning. There are many different models of blended learning found [4,7] but the model adopted by the university was developed by Alammary, Sheard, & Carbone [4]. In their model, three distinct design approaches were identified:

(1) Low-impact blend: adding extra activities to an existing course

(2) Medium-impact blend: replacing activities in an existing course

(3) High-impact blend: building the blended course from scratch.

Once the blended learning vision has been identified and recognized, it is essential to develop and reinforce plans for university to embark this new blended learning initiative. An academic practice team was established to lead this initiative and strategically address issues and challenges that may emerge during blended learning implementation. A blended learning site was also created in Blackboard to share information and resources in the faculty. Furthermore, the university also appointed faculty-based digital learning technologists to support technical development in the faculty. Recording rooms completed with necessary recording facilities were set up and recording software and relevant broadcasting system were identified and installed.

In order to facilitate blended learning, a minimum standard of Blackboard usage must be achieved. It is agreed that the LMS must be used to manage content effectively. The environment should be attractive and user friendly so that students can easily access the required information. Besides, information about assessments must be available and grade center should be utilized to provide timely feedback to class. Discussion board on Blackboard could also be used to create opportunities for interactions between learners-learners and instructor-learners. The activities created should also supplement and not overload the students as it has been reported that E-learning often suffers from making large amounts of information available for students to absorb independently [15].

## 5 Conclusions

The present study has several implications in education. First, it indicates that blended learning has the potential to meet learners' educational needs. The mixture of face-to-face and online components in blended learning provides learners with the opportunities to learn from multiple delivery method. The overall learners' satisfaction survey also showed that blended learning increased student learning flexibility while maintaining social connections in the traditional classroom.

Secondly, the study indicates that academics were challenged to meet with the growing expectations required in blended learning. It is undeniable that an educator is a significant key to determine the success in the implementation of blended learning. Designing a blended delivery course requires time, effort, technical skills and clear pedagogical goal with the integration of technology. Consequently, there is a need to provide professional development program and support for academics.

In general, the findings also suggested that when implementing blended learning, various factors need to be considered. The factors to be considered by academics and learners are summarized in Table 3.

This study determines the readiness of academics and learners to embrace blended learning and the factors for successful implementation. Moving on, future studies would focus to evaluate the effectiveness and quality of blended learning approach in the university. Blended learning elements could be evaluated from the relationship between various factors such as learning outcomes, resources, content, assessments, delivery methods, pedagogy and learner's achievements. In addition, the context of the studies could be expanded to address the perceptions of internal (leaders and administration in education) and external (educational community) stakeholders. In conclusion, blended learning is a promising approach to facilitate learning experience with the condition that academics are willing to embrace change and commit in delivery strategy. Ultimately the goal is for academics to focus on pedagogy-based learning instruction to ensure success in blended learning approach.

**Table 3.** Factors to be considered in implementing blended learning.

| | Academic Staff | Learners |
|---|---|---|
| 1. | Definition & Expectation | Definition & Expectation |
| 2. | Facilities | Facilities |
| 3. | Time management | Time management |
| 4. | Technical development and support | Technical ability and support |
| 5. | Online materials/resources management | Free riders/fairness in grading |
| 6. | Stakeholders' expectation (Professional bodies) | Stakeholders' expectation (Parents) |

## References

1. D.R. Garrison, H. Kanuka. IHE, **7**(2), 95–105 (2004)
2. B. Means, Y. Toyama, R. Murphy, M. Bakia, K. Jones. *Evaluation of evidence based practices in online learning: A meta-analysis and review of online learning studies.* Washington, DC: U.S. Department of Education, Office of Planning, Evaluation, and Policy Development (2009)
3. E. Martinez-Caro, F. Campuzano-Bolarin. EJEE, **36**(5), 473–483 (2011)
4. A. Alammary, J. Sheard, A. Carbone. AJET, **30**(4), 440-454 (2014)
5. M. E. Blueprint. *Blueprint 2015–2025*. Ministry of Education Malaysia (2015)
6. R. Owston. IHE., **18**, 1–3 (2013).
7. P. Moskal, C. Dziubna, J. Hartman. IHE, **18**, 15-23 (2013).
8. A. Norberg, C. D. Dziuban, P.D. Moskal. A time-based blended learning model. On the Horizon, **19**, 207-216 (2011)
9. K. F. Hew, T. Brush. ETRD, **55**(3), 223-252 (2007)
10. J. Groff, C. Mouza. AACE J., **16**(1), 21-46 (2008)
11. J. Y. Lee. IES, **4**(2), 75-81 (2011)
12. K. Wilson, J.H. Korn. TOP, **34**(2), 85-89 (2007).
13. T. A. Marino, M. Eager, T. Draxler. NTLF **9**(4), 4-6 (2000)
14. J. H. Wu, R. D. Tennyson, T. L. Hsia. Comput Educ, **55**(1), 155-164 (2010)
15. G. L. Waddoups, S. L. Howell. IRRODL, **2**(2) 52/108 (2002).

# AUTHOR INDEX

# AUTHOR INDEX

# AUTHOR INDEX